THE ELEMENTS*

P9-AOY-958

	Symbol	Atomic No.	Atomic Mass		Symbol	Atomic No.	Atomic Mass
Actinium	Ac	89	[227]†	Mercury	Hg	80	200.59
Aluminum	Al	13	26.98154	Molybdenum	Mo	42	95.94
Americium	Am	95	[243]†	Neodymium	Nd	60	144.24
Antimony	Sb	51	121.75	Neon	Ne	10	20.179
Argon	Ar	18	39.948	Neptunium	Np	93	237.0482‡
Arsenic	As	33	74.9216	Nickel	Ni	28	58.71
Astatine	At	85	[210]†	Niobium	Nb	41	92.9064
Barium	Ba	56	137.34	Nitrogen	N	7	14.0067
Berkelium	Bk	97	[249]†	Nobelium	No	102	[254]†
Beryllium	Be	4	9.0128	Osmium	Os	76	190.2
Bismuth	Bi	83	208.9804	Oxygen	O	8	15.9994
Boron	B	5	10.81	Palladium	Pd	46	106.4
Bromine	Br	35	79.904	Phosphorus	P	15	30.97376
Cadmium	Cd	48	112.40	Platinum	Pt	78	195.09
Calcium	Ca	20	40.08	Plutonium	Pu	94	[242]†
Californium	Cf	98	[251]†	Polonium	Po	84	[210]†
Carbon	C	6	12.011	Potassium	K	19	39.098
Cerium	Ce	58	140.12	Praseodymium	Pr	59	140.9077
Cesium	Cs	55	132.9054	Promethium	Pm	61	[147]†
Chlorine	Cl	17	35.453	Protactinium	Pa	91	231.0359‡
Chromium	Cr	24	51.996	Radium	Ra	88	226.0254‡
Cobalt	Co	27	58.9332	Radon	Rn	86	[222]†
Copper	Cu	29	63.546	Rhenium	Re	75	186.2
Curium	Cm	96	[247]†	Rhodium	Rh	45	102.9055
Dysprosium	Dy	66	162.50	Rubidium	Rb	37	85.4678
Einsteinium	Es	99	[254]†	Ruthenium	Ru	44	101.07
Erbium	Er	68	167.26	Samarium	Sm	62	150.4
Europium	Eu	63	151.96	Scandium	Sc	21	44.9559
Fermium	Fm	100	[253]†	Selenium	Se	34	78.96
Fluorine	F	9	18.99840	Silicon	Si	14	28.086
Francium	Fr	87	[223]	Silver	Ag	47	107.868
Gadolinium	Gd	64	157.25	Sodium	Na	11	22.98977
Gallium	Ga	31	69.72	Strontium	Sr	38	87.62
Germanium	Ge	32	72.59	Sulfur	S	16	32.06
Gold	Au	79	196.9665	Tantalum	Ta	73	180.9479
Hafnium	Hf	72	178.49	Technetium	Tc	43	98.906‡
Helium	He	2	4.00260	Tellurium	Te	52	127.60
Holmium	Ho	67	164.9304	Terbium	Tb	65	158.9254
Hydrogen	H	1	1.0079	Thallium	Tl	81	204.37
Indium	In	49	114.82	Thorium	Th	90	232.0381‡
Iodine	I	53	126.9045	Thulium	Tm	69	168.9342
Iridium	Ir	77	192.22	Tin	Sn	50	118.69
Iron	Fe	26	55.847	Titanium	Ti	22	47.90
Krypton	Kr	36	83.80	Tungsten	W	74	183.85
Lanthanum	La	57	138.9055	Uranium	U	92	238.029
Lawrencium	Lr	103	[257]†	Vanadium	V	23	50.9414
Lead	Pb	82	207.2	Xenon	Xe	54	131.30
Lithium	Li	3	6.941	Ytterbium	Yb	70	173.04
Lutetium	Lu	71	174.97	Yttrium	Y	39	88.9059
Magnesium	Mg	12	24.305	Zinc	Zn	30	65.38
Manganese	Mn	25	54.9380	Zirconium	Zr	40	91.22
Mendelevium	Md	101	[256]†				

* Only 103 elements are listed, as there is no international agreement for the names of elements 103–109.

† Mass number of most stable or best known isotope.

‡ Mass of most commonly available, long-lived isotope.

Chemistry and Society

FIFTH EDITION

Chemistry and Society

FIFTH EDITION

MARK M. JONES
Department of Chemistry
Vanderbilt University
Nashville, Tennessee

DAVID O. JOHNSTON
Department of Chemistry
David Lipscomb College
Nashville, Tennessee

JOHN T. NETTERVILLE
Williamson County Schools
Franklin, Tennessee

JAMES L. WOOD
Consultant
Resource Consultants, Inc.
Brentwood, Tennessee

MELVIN D. JOESTEN
Department of Chemistry
Vanderbilt University
Nashville, Tennessee

Saunders Golden Sunburst Series

SAUNDERS COLLEGE PUBLISHING

Philadelphia   New York   Chicago
San Francisco   Montreal   Toronto
London   Sydney   Tokyo   Mexico City
Rio de Janeiro   Madrid

Address orders to:
383 Madison Avenue
New York, NY 10017

Address editorial correspondence to:
210 West Washington Square
Philadelphia, PA 19105

Text Typeface: Century Schoolbook
Compositor: Progressive Typographers
Acquisitions Editor: John Vondeling
Project Editor: Martha Hicks-Courant
Copyeditor: Nanette Bendyna
Art Director: Carol C. Bleistine
Text Designer: Edward A. Butler
Cover Designer: Lawrence R. Didona
New Text Artwork: J & R Technical Services
Production Manager: Tim Frelick
Assistant Production Manager: JoAnn Melody

Cover Credit: Solar Energy by JANEART LTD./© 1987 THE IMAGE BANK

**Library of Congress
Cataloging-in-Publication Data**

Chemistry and society.
    (Saunders golden sunburst series)
    Rev. ed. of: Chemistry, man, and society.
4th ed. 1983.

    Includes index.

    1. Chemistry. I. Jones, Mark Martin, 1928–
II. Title: Chemistry, man, and society.
QD31.2.C418 1986   540   86-20214

ISBN 0-03-008139-4

CHEMISTRY AND SOCIETY   0-03-008139-4

7890   032   987654321

CBS COLLEGE PUBLISHING
Saunders College Publishing
Holt, Rinehart and Winston
The Dryden Press

Preface

The fifth edition of *Chemistry and Society* has been prepared for students wishing a one- or two-semester course in college chemistry. The fundamental approach of the previous editions has been preserved—physical and chemical discoveries are presented along with the impact of these discoveries on our way of life. These human investigations of nature have produced the modern chemical world. Throughout the text the student is confronted with applications of chemical knowledge that dramatically affect the quality of human life.

No previous knowledge of chemistry is assumed or required in this presentation. However, the approach is sufficiently different to challenge and interest the student with a background in high school chemistry.

To the beginning student there may be a mystery in chemistry, but to leave the workings of the chemist a mystery argues that the liberally educated person must be dependent on the chemist for those chemical decisions that affect society as a whole. *Chemistry and Society* is based on the belief that the liberal arts student can see and appreciate the chain of events leading from chemical fact to chemical theory and the ingenious manipulation of materials based on chemical theories. Thoughtful students will then see that the intellectual struggles in chemistry are closely akin to their own personal intellectual pursuits and will feel that each educated individual should and can have a say in how the applications of chemical knowledge are to affect the human experience.

The topics covered in this book have been selected on the basis of what we have observed to be student interests. As a team of authors, and as individual teachers of chemistry at the college level for many years, we have found that our students are interested in the following:

1. Understanding the causes of natural phenomena
2. Understanding scientific bases for making important personal choices about the use of chemicals and chemical products
3. Participating on a rational basis in societal choices that affect the quality of human life
4. Helping preserve and restore the quality of the environment
5. Developing an insight into the perplexing problem of chemical dependency
6. Sensing the balance involved in population control, the chemical control of disease, and the ability of the world to produce food
7. Choosing personal habits in exercise programs and in nutritional selections that are compatible with healthful living

8. Taking thoughtful advantage of the latest developments in technical applications of material in information recording and management
9. Going places in vehicles that reflect the best uses of the materials used in transportation
10. Using present energy reserves at a sensible rate as new energy sources are developed for the long term

These paramount interests, as well as many of lesser note, are featured in this chemistry text on material substances and their uses.

The fifth edition of *Chemistry and Society* is a major revision. Without sacrificing our successful approach, which emphasizes that structure causes function, chemical periodicity, and consequent material properties, we have carefully selected that thread of chemical history that shows chemistry to be a human endeavor. Much of the text has been rewritten in line with the essential chemical story to be told and with the extensive critical reviews we have received through the years. The philosophical setting for the presentation is established in Chapter 1, allowing the text and teacher to whet the appetites of unsuspecting students for an understanding of what they may have previously thought belonged only to the scientific elite. Major new topics have demanded attention they did not receive in earlier editions: there are new chapters on the nutritional basis of healthy living, the chemical recording and manipulation of information and its display in images and symbols, the agricultural production of food for a hungry world population, and the chemistry of transportation.

A concerted effort has been made to inform the reader about certain vital matters that have bases in chemistry. For example, the vexing problem of designer drugs comes into focus with an understanding of the principles concerning molecular structural properties and minimum structural change in modifying complex molecules. A chemical understanding of the general viral structure and the defenses of the body against such structures leads to an understanding of viral diseases and what possible chemical cures there might be for diseases ranging from the common cold to AIDS. A rational relationship between diet and the major killer diseases, such as heart disease, unfolds as a chemical story. We hope readers will find many of these items of compelling interest.

The teaching aids have been further refined to help the book communicate. Boldface for new terms and concepts, along with marginal notes, adds emphasis to focus reader attention. Numerous interesting features have been added throughout the text. Self-tests, which have proved extremely helpful in measuring retention and comprehension, have been revised and expanded. Questions at the ends of the chapters provide for additional study and opportunities for extended research. Matching sets are included at the ends of chapters to help students keep the necessary vocabulary in mind during chapter review. Numerous illustrations, a logical extension of the text, often communicate better than words.

We are deeply grateful to all who have contributed to the improvement of each edition of this book. For this fifth edition, we especially acknowledge the reviews and suggestions of Dennis A. Kohl, Norman Fogel, Keith J. Harper, Ralph A. Zingaro, Paul G. Seybold, Charles Atwood, Jerry L. Mills, Mabel-Ruth Stephanic and Eugene G. Rochow. The creative ideas provided by Professor Rochow's special insights into chemical education have enriched our work.

Thanks to Robert E. Haltman, with the Food and Drug Administration, and Quincy N. Styke III, with the Air Pollution Control Division of the Tennessee

Department of Health and Environment, for updating material. Also, thanks to Villa M. Mitchell for sharing with us her expertise in nutrition.

To the entire staff at Saunders College Publishing we give our thanks. Special consideration should be given to two of them. Kate Pachuta has facilitated the flow of information and ideas, prompted the necessary decisions, pushed for deadlines in a very nice manner, and worked for excellence in every facet of editorial control. Martha Hicks-Courant, with her superb ability to provoke harmonious and creative work by the various professionals involved, has been outstanding in bringing it all together to turn the authors' material into a book.

To John Vondeling, Associate Publisher at Saunders, we give our biggest bundle of thanks for the confidence he has expressed in us over the years. He can "read the market" as well as any and has wisely combined this marketing ability with a healthy respect for intellectual independence in academic matters.

Much help has come our way, but of course the responsibility for the contents of the text rests entirely on us.

As in all of our previous works, we dedicate this effort to our spouses and gratefully acknowledge their support and understanding during the preparation of this manuscript.

MARK M. JONES
DAVID O. JOHNSTON
JOHN T. NETTERVILLE
JAMES L. WOOD
MELVIN D. JOESTEN

Contents
Overview

Contents

4 Elements in Useful Order — The Periodic Table   74

5 Chemical Bonds — The Ultimate Glue   92

6 Some Principles of Chemical Reactivity   120

7 Nuclear Reactions — Electrons, Protons, Neutrons, and More!   136

8 Energy and Our Society   165

9 Acids and Bases — Chemical Opposites   204

10 Oxidation-Reduction: Electron Transfer Chemistry   224

15 Biochemistry — Chemistry of Living Systems   352

16 Toxic Substances — Interferences in Biochemical Systems   395

17 Water — Plenty of It, But of What Quality?   429

18 Clean Air — Should It Be Taken for Granted?   462

19 Agricultural Chemistry   490

20 Nutrition — The Basis of Healthy Living   522

21 Medicines and Drugs — Saving Lives with Calculated Risks   576

22 Beauty and Cleansing Agents — Chemical Formulations for Looks, Health, and Comfort   607

23 Transportation Chemistry — From Here to There and Quickly About It!   635

Chapter 1

IMPACT OF SCIENCE AND TECHNOLOGY ON SOCIETY

WHAT IS SCIENCE?

Science can be defined in a number of ways. Perhaps the place to start is with the word *science,* which is derived from the Latin *scientia,* meaning "knowledge." Science is a human activity involved in the accumulation of knowledge about the universe around us. Pursuit of knowledge is common to all scholarly endeavors in the humanities, social sciences, and natural sciences. Historically, the natural sciences have been closely associated with our observations about nature, our physical and biological environment. Knowledge in this context is more than a collection of facts; it involves comprehension, correlation, and an ability to explain established facts, usually in terms of a physical cause for an observed effect.

Figure 1–1 is a classification for the natural sciences. There is no sharp distinction between the two groups of sciences or among members within a group. The scheme in Figure 1–1 should not be viewed as rigid, since new disciplines emerge that bridge areas at different levels. In the physical sciences, for example, there are biophysicists, geochemists, bioinorganic chemists, and chemical physicists. Some of these names define broad interdisciplinary fields, while others refer to more specialized subfields. The dynamic character of science is illustrated by the emergence of new disciplines such as bioinorganic chemistry, the study of the function of metal ions in biological systems.

WHAT IS CHEMISTRY?

There are many different kinds of science because there are many different ways to focus on the world around us. In this book we shall study the science of chemistry, which is one of the physical sciences. **Chemistry** is concerned with the study of matter and its changes. Since matter is the material of the universe, every object we see or use is part of the chemical story. Our body is a sophisticated chemical factory with hundreds of chemical reactions occurring even as you read this page. Because chemistry is so intimately involved in every aspect of our contact with the material world, chemistry can be regarded as the central science, an integral part of our culture.

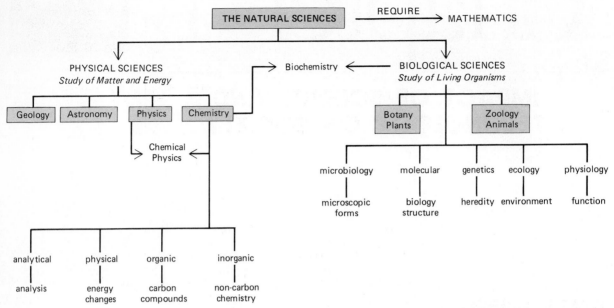

FIGURE 1–1   **Organizational chart for the natural sciences with detailed emphasis on chemistry.**

WHAT IS TECHNOLOGY?

A deeper understanding comes when we distinguish among basic science, applied science, and technology. The difference between basic and applied science is determined by the motivation for doing the work. **Basic science** is the pursuit of knowledge about the universe with no short-term practical objectives for application. An example of **basic research** is to seek the answer to the question "What is penicillin?" by determining the molecular structure of penicillin. **Applied science** has well-defined, short-term goals related to solving a specific problem. For example, after the antibacterial action of penicillin was discovered, scientists conducted **applied research** on the effectiveness of penicillin against different types of bacterial infections. Thus both basic and applied scientific research produces new knowledge.

Technology is the use of scientific knowledge to manipulate nature for advantage. This may involve production of (1) new drugs, (2) better plastics, (3) safer automobiles, (4) nuclear weapons, and (5) chemical warfare agents. Using the penicillin example, engineers developed economical methods for large-scale production of penicillin from a knowledge of the results of basic and applied research on the chemistry and biology of penicillin.

The important point to realize is that technology, like science, is a human activity. Decisions about technological applications and priorities for their development are made by men and women; whether scientific knowledge is used to promote good or bad technological applications depends on those persons in industry and government who have the authority to make such decisions. **In a democratic society the voters may influence these decisions. Therefore, it is important to have an informed citizenry who can critically evaluate societal issues that are consequences of technology.**

The image most people have of science is strongly influenced by their familiarity with technological advances and, in most instances, is their only view of scientific progress. It is important to get beyond the everyday image of technology in order to recognize the symbiotic relationship of science and technology. Modern science depends on technological advances, especially in the development of more sophisticated instrumentation, to examine in greater depth the unanswered questions about the universe. Examples of this interrelationship are given throughout this book.

> Symbiosis is the close association of two dissimilar things in a mutually beneficial relationship.

Perhaps more easily seen are the advances in technology that have occurred whenever new scientific discoveries are made. Regardless of the type of scientific discovery, there is a delay between a discovery and its technological application. The incubation period depends on (1) rate of information transmittal, (2) recognition of the applicability of the discovery, (3) development of the invention for application of the new science to a technological use, and (4) actual large-scale manufacture of the new invention. The incubation times for several ideas or scientific discoveries are given in Table 1–1. Although there are exceptions, innovations based on applied research tend to have a shorter time frame than those developed from basic research.

HOW DO SCIENTISTS COMMUNICATE?

Scientific knowledge is cumulative, and progress in science and technology depends on access to this body of knowledge. Since the earliest beginnings of science, this knowledge has been transmitted primarily by the written word. The invention

TABLE 1–1   How Long It Has Taken Some Fruitful Ideas to Reach Technological Realization

INNOVATION	CONCEPTION	REALIZATION	INCUBATION INTERVAL (YEARS)
Antibiotics	1910	1940	30
Automatic transmission	1930	1946	16
Ballpoint pen	1938	1945	7
Cellophane	1900	1912	12
Cisplatin, anticancer drug	1964	1972	8
Dry soup mixes	1943	1962	19
Filter cigarettes	1953	1955	2
Heart pacemaker	1928	1960	32
Hybrid corn	1908	1933	25
Instant camera	1945	1947	2
Instant coffee	1934	1956	22
Liquid shampoo	1950	1958	8
Long-playing records	1945	1948	3
Nuclear energy	1919	1945	26
Nylon	1927	1939	12
Photography	1782	1838	56
Radar	1907	1939	32
Recombinant DNA drug synthesis	1972	1982	10
Roll-on deodorant	1948	1955	7
Self-winding wristwatch	1923	1939	16
Video tape recorder	1950	1956	6
Xerox copying	1935	1950	15
X rays in medicine	Dec, 1895	Jan, 1896	0.08
Zipper	1883	1913	30

Most scientific journals
are publications in which
the results of scientific
research are reported or
reviewed.

Abstracting journals
publish summaries of sci-
entific publications.

of the printing press led to the development of scientific journals and other publications collectively known as the **scientific literature.** The explosive expansion of the scientific literature since the 1940s makes information management an essential part of modern science and technology. The scientific journals have been the fastest growing segment of the scientific literature (Fig. 1–2).

The chemical literature explosion can be illustrated more dramatically by examining the growth rates of *Chemical Abstracts* (CA), which provides comprehensive coverage of the chemical literature worldwide. Figure 1–3 shows the growth rate of papers abstracted per year. Note the sharp increase since 1947. The number of journals monitored by CA has grown from 400 in 1907 to over 12,000 in 1985. In addition, CA monitors patent documents issued by 26 nations and two international bodies, conference and symposium proceedings, dissertations, government reports, and books. The publications abstracted by CA include leading physical, biological, medical, and other technical journals as well as chemistry-oriented journals. As a result, CA can be used for comprehensive searches of subject matter that crosses interdisciplinary lines.

An example of a CA abstract is shown in Figure 1–4. At present, CA publishes approximately 450,000 abstracts each year, and 16% of these are patent abstracts. This is one measure of the ratio of scientific knowledge to technological application. CA indexes make this mountain of information accessible. Over 80

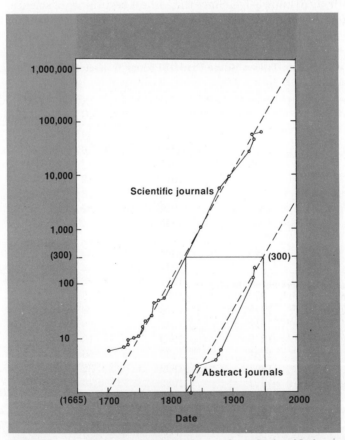

FIGURE 1–2   Rate of increase in the number of scientific journals since 1665. (Reprinted with permission of Yale University Press.)

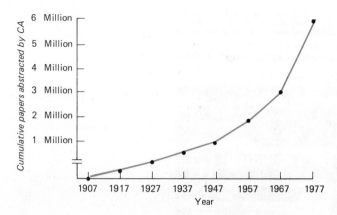

FIGURE 1–3 Growth rate of papers abstracted by *Chemical Abstracts*. By 1984 the figure was up to 10 million.

volumes make up the tenth Collective Index of Chemical Abstracts for the period 1977 to 1981, and this collective index covers over 2.5 million documents.

An example of how technology has aided science is in the retrieval of information from the massive volume of published works. The advent of the computer has resulted in the development of computer information retrieval as an efficient and accurate way to gain access to the literature. The information explosion makes the use of computer information retrieval essential to all persons who need to maintain an awareness of the literature in their particular area of interest.

Two companies that provide databases for computer information retrieval are DIALOG Information Retrieval Service and Chemical Abstracts Service (CAS) ONLINE. DIALOG has more than 200 databases (Fig. 1–5) that cover natural sciences, social sciences, humanities, technology, engineering, medicine, business, government, law, economics, and current events.

Chemical Abstracts Service (CAS) ONLINE is the online equivalent of the printed *Chemical Abstracts*. This is the world's largest file of information on substances and is updated weekly. Each registered substance is assigned a unique CAS Registry Number that permits retrieval of information about formula, structure, and name, any one of which can be used to retrieve information about a particular substance. For example, a chemist can search by structure or common name through more than 6.5 million substances cited in the chemical literature in the past 20 years and do so comprehensively and quickly; the search can take as little as 5 minutes, and most searches can be completed within 15 minutes.

The CAS Registry Number is the "Social Security Number" of the unique substance. Table salt (NaCl) is 14762-51-7. In 1984 there were 1074 abstracts that referred to table salt.

Inhibition of cell division in Escherichia coli by electrolysis products from a platinum electrode. Barnett Rosenberg, Loretta Van Camp, and Thomas Krigas (Michigan State Univ., Lansing).   *Nature* 205(4972), 698–9(1965)(Eng).   *E. coli* cells grown in a continuous culture app. with an esp. designed chamber contg. 2 Pt electrodes were subjected to an elec. field in an O atm. With an applied voltage of 500 to 6000 cycles/sec., cell division ceased and filamentous growth occurred.   A similar effect was shown with a soln. of 10 ppm. $(NH_4)_2PtCl_6$.   Pt(IV) was found in the electrolyzed medium in concns. of 8 ppm.   Other transition metal ions, including Rh, in concns. of 1–10 ppm. inhibited cell division in *E. coli* while not interfering with growth.

Gloria H. Cartan

FIGURE 1–4   Example of abstract found in *Chemical Abstracts*. This one was published in Volume 62, abstract 13543a, 1965. (This CA citation is copyrighted by The American Chemical Society and is reprinted by permission. No further copying is allowed.)

FIGURE 1–5 "Sea of disk drives" for DIALOG computers. The DIALOG disk storage contains over 100 million records of information. (Reprinted courtesy of DIALOG® Information Services, Inc.)

The items you need to search database systems are shown in Figure 1–6. They include (1) a telephone line, (2) a modem, (3) a computer terminal (microcomputers can be used), and (4) an account with the information system.

Regardless of your chosen profession, you should become familiar with computer information retrieval. It is essential for efficient information management. Leaders in any field will have to manage information with technical devices, since the human brain cannot process the information fast enough.

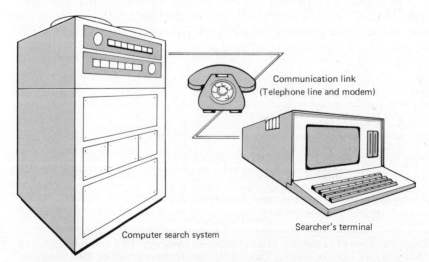

Communication link
(Telephone line and modem)

Searcher's terminal

Computer search system

FIGURE 1–6   Equipment needed to perform a computer search of a database.

Obviously, the growth of scientific information and the consequent technological application are facilitated by communication; and new knowledge and its application are growing exponentially.

HOW IS SCIENCE DONE?
Scientific Method

The methodology of science is often summarized by the term *scientific method*. The scientific method is a logical approach to solving scientific problems that includes (1) **observation,** facts gathered by experiment, (2) **inductive reasoning** to interpret and classify facts by a general statement **(law),** (3) **hypothesis,** speculation or idea about how to explain facts or observations, (4) **deductive reasoning** to test a hypothesis with carefully designed experiments, and (5) **theory,** a tested hypothesis or model to explain laws. Although variations of this approach are often presented as the ideal method for scientific research, the imagination, creativity, and mental attitude of the scientist are more important than the procedure, and there is no unique method used by all scientists.

Inductive reasoning: from specific facts to generalization.

Deductive reasoning: from generalization to specific facts.

Another outline of the scientific method: observe, generalize, theorize, test, and retest.

The attitude requires the strictest intellectual honesty in the collection of observable facts and in the arrangement of these facts into a pattern to see if the pattern reveals the underlying basis for the *observed* behavior. The data normally must be collected under conditions that can be reproduced anywhere in the world. Then new data can be obtained to confirm or to refute the correctness of the suggested pattern (geologists and astronomers may sometimes be excused from these rigid requirements). The results represent a unique type of objective truth because they are independent of differences in language, culture, religion, or economic status of the various observers. Such established truth is appropriately referred to as **scientific fact.**

More on Facts, Laws, Theories

A **scientific fact** is an observation about nature that usually can be reproduced at will. For example, carbon in all forms will readily burn in the presence of air at a sufficiently high temperature. If you have any doubt about this fact it is easy enough to set up an experiment that will readily demonstrate the fact anew. You would only need some carbon, air, and a source of heat. The repeatability of a scientific fact distinguishes it from a historical fact, which obviously cannot be reproduced. Of course, some scientific facts are also historical facts — such as the movement of heavenly bodies — and are not repeatable at will.

A scientific fact can be verified independently of any particular observer.

Often a large number of related scientific facts can be summarized into broad, sweeping statements called natural **laws.** The law of gravity is a classic example of a natural law. This law, that all bodies in the universe have an attraction for all other bodies that is directly proportional to the product of their masses and inversely related to the square of their separation distance, summarizes in one sweeping statement an enormous number of facts. It implies that any object lifted a short distance from the surface of the earth will fall back if released. Such a natural law can only be established in our minds by inductive reasoning; that is, you conclude that the law applies to all possible cases, since it applies in all of the cases studied or observed. A well-established law allows us to predict future events.

A scientific law summarizes a large number of related facts.

A scientific law predicts what *will* happen. A governmental law describes what people *should* do.

When convinced of the generality of a scientific law, we may reason deductively, based on our belief that if the law holds for all related situations, it will surely hold for the events in question.

The same procedure is used in the establishment of chemical laws, as can be seen from the following example. Suppose an experimenter carried out hundreds of different chemical changes in closed, leakproof containers, and suppose further that he weighed the containers and their contents before and after each of the chemical changes. Also, suppose that in every case he found that the container and its contents weighed exactly the same before and after the chemical change occurred. Finally, suppose that he repeated the same experiments over and over again, obtaining the same results each time, until he was absolutely sure that he was dealing in reproducible facts. It can be understood then that the experimenter would reasonably conclude: **"All chemical changes occur without any detectable loss or gain in weight."** This is indeed a basic chemical law and serves as one of the foundations of modern scientific theory.

After a natural law has been established, its explanation must be sought. Chemists are not satisfied until they have explained chemical laws logically in terms of the submicroscopic structure of matter. This is indeed a difficult process, and until recently its progress has been painfully slow because of our lack of direct access into the submicroscopic structure of matter with our physical senses. All we can do is collect information in the macroscopic world in which we live, and then try, by circumstantial reasoning, to visualize what the submicroscopic world must be like in order to explain our macroscopic world. Such a visualization of the submicroscopic world is called a **theoretical model.** If the theoretical model is successful in explaining a number of chemical laws, a major scientific theory is built around it. The atomic theory and the electron theory of chemical bonding are two such major theories, and both will be discussed in relation to chemical laws in later chapters.

Theories are ideas or models used to explain facts and laws.

Consider again the chemical law concerning the conservation of weight in chemical changes. What is a possible theoretical model that could explain this law? If we assume that matter is made up of atoms, which are grouped in a particular way in a given pure substance, we can reason that a chemical change is simply the rearrangement of these atoms into new groupings without loss or destruction of atoms, and, consequently, rearrangement into new substances. If the same atoms are still there, they should have the same individual characteristic weight, and hence the law of conservation of weight is explained.

Chemical theories use the concepts of atoms to explain chemical observations.

For a scientific theory to have much value, it must not only explain the pertinent facts and laws at hand, but it must also be able to explain or accommodate new facts and laws that are obviously related. If the theory cannot consistently perform in this manner, it must be revised until it is consistent, or, if this is not possible, it must be completely discarded. You must not allow yourself to think that this process of trying to understand nature's secrets is nearing completion. The process is a continuing one.

The word **theory** is often used in a different sense from the one discussed previously. If a student is absent from the chemistry class, his neighbor may say, "I do not know why he is absent, but my theory is that he is sick and unable to come to class." The speculative guess of the student about his absent friend is what scientists call a **hypothesis** and is vastly different from the broad theoretical picture used to explain a number of laws. The reader should be alert for the considerable amount of confusion that has resulted from the different meanings associated with this word. In this book the word *hypothesis* is used when speaking of a speculation about a particular event or set of data, and the word *theory* is reserved for the broad imaginative concepts that have gained wide acceptance through test and proof.

Experimental Methods

The methods of collecting observed, objective data can be grouped generally into about four classes: trial and error, planned research, "let's do some experiments and see what happens," and accident or serendipity.*

Planned research is based on fact and theory and expects predicted results.

Discovery by trial and error begins by having a problem to solve and doing various experiments in the hope that something desirable will emerge. The next set of experiments then depends on the results obtained. The discovery of the Edison battery by Thomas Edison's group is an example of discovery by trial and error. His group performed more than 2000 experiments, each guided by the last, before settling on the composition of Edison's battery.

Discovery by planned research comes from plotting out specific experiments to test a well-defined hypothesis. The carcinogenic nature of some compounds is determined by progressing through a set pattern of experimental tests.

Carcinogens are substances that cause cancer.

The method of "let's try some experiments and see what happens" lacks the formality of the other methods. Experimentation evolves with vague hypotheses or no specific hypotheses because no well-defined pathway is obvious. Joseph Priestley, who discovered oxygen in 1774, utilized this approach on occasion.

Discovery by accident may be a misnomer. The investigator is usually actively involved in investigating nature through experimentation, but "accidentally" finds some phenomenon not originally imagined or conceived. Thus the "accident" has an element of serendipity and is not seen unless the investigator is a trained observer. Pasteur said, "Chance favors the prepared mind."

The discovery of one of the leading anticancer drugs, cisplatin, is an example of an "accidental" discovery. In 1964 Barnett Rosenberg and his co-workers at Michigan State University were studying the effects of an electric current on bacterial growth. They were using an electrical apparatus with platinum electrodes to pass a small alternating current through a live culture of *Escherichia coli (E. coli)* bacteria. After an hour, they examined the bacterial culture under a microscope and observed that cell division was no longer taking place. After thorough analysis of the culture medium and additional experimentation, they determined that traces of several different platinum compounds were produced during electrolysis from the reaction of the platinum electrodes with chemicals in the culture medium.

Figure 1–4 is an abstract of the first published report of this discovery.

Careful observation was essential since platinum electrodes are commonly regarded as inert or unreactive, and only a few parts of platinum compounds per million parts of culture medium were present. Additional testing indicated a compound known as cisplatin was responsible for inhibiting cell division in *E. coli*. Approximately two years after its initial discovery, the Rosenberg group had the answer to the question "What caused inhibition of cell division in *E. coli* bacteria?" At this point they had the idea that cisplatin might inhibit cell division in rapidly growing cancer cells. The compound was tested as an anticancer drug, and in 1979 the Food Drug Administration approved its use as such. The drug has now been

* A. B. Garrett: The discovery process and the creative mind. *Journal of Chemical Education*, Vol. 41, pp. 479–482, 1964.

Chemists represent the
cisplatin molecule as

$$Cl \diagdown \diagup NH_3$$
$$Pt$$
$$Cl \diagup \diagdown NH_3$$

proved to be effective alone or in combination with other drugs for the treatment of a variety of cancers.

Another interesting aspect of the story is that cisplatin, a compound that contains two chloride ions, two ammonia molecules, and platinum, was first prepared in 1845. Although its chemistry had been thoroughly studied since then, the biological effects of cisplatin or inhibition of cell division were not discovered until "the accident" 120 years later.

WHAT ARE YOUR ATTITUDES?

Before beginning the study of chemistry and its relationship to our culture, each of us needs to examine our prejudices (if any) and attitudes about chemistry, science, and technology. Many nonscientists regard science and its various branches as a mystery and have the attitude that they cannot possibly comprehend the basic concepts and consequent societal issues. Many also have a fear of unleashed chemicals and a feeling of hopelessness about the environment. Many of these attitudes are the result of reading about harmful effects of technology. Some of these harmful effects are indeed tragic. However, what is needed is a full realization of both the benefits and the harmful effects that can be attributed to science and technology. In the analysis of these pluses and minuses, we need to determine why the harmful effects occurred and whether the risk can be reduced for future generations as we seek advantages offered by the human understanding of nature.

TECHNOLOGY AND THE INDUSTRIAL REVOLUTION

Over the last 200 years, accumulated scientific and technical knowledge has been put to use on an extensive scale in Europe and in those areas of the world that have had the means and the will to follow the examples set by England, where the first "industrial revolution" occurred. The result has been the development of a society largely dependent upon and supported by a constantly changing technology. The first consequence of this technology has been to increase the *rate* at which things can be produced. This, in turn, has continually changed the occupational patterns of millions of human beings and has brought forcefully to mind the persistence of *change* in our pattern of life.

These changes have influenced profoundly the way people think about their material wants and the ways they can be satisfied. For example, there seems to be little argument with the statement, "If the number of human beings on the Earth could be stabilized, a much higher standard of living could prevail over most of the Earth." A statement such as this would have been greeted with widespread derision 500 years ago. While depending on technology, people today are beginning to doubt its ability to solve both personal and social problems on a long-range basis. It is obvious that confusion exists on this point since the cries about the curses of technology come from people who are highly dependent on it and who are even asking for more from technology.

Almost as soon as the industrial revolution began in England, the public realized that technological progress brought with it a series of problems. The first to be noticed was the necessity for progress to be accompanied by changing patterns of employment.

It is obvious that if a machine makes as much thread as 100 workers can make, the workers are released to do other work. The 100 workers, however, do not look on this as an advantage, especially if they are settled in their place of employment with their families. The new opportunities that result from such a machine are rarely of benefit directly to the workers displaced. The wealth of their country is increased since there are now 100 workers able to do other work. However, the initial reaction of the workers in 18th-century England was to riot and break up the machinery.

The increased use of fuels of all sorts, especially the introduction of coal and coke into metallurgical plants and then the use of coal to fuel engines, led to widespread problems with air pollution that were recognized and discussed over 200 years ago (Fig. 1–7).

The most important point of these results is the realization that technological progress is always obtained at some cost, and the cost may not be obvious at the outset.

Technological unemployment results when people are displaced from their jobs by machines.

Modern application: robots for routine assembly work.

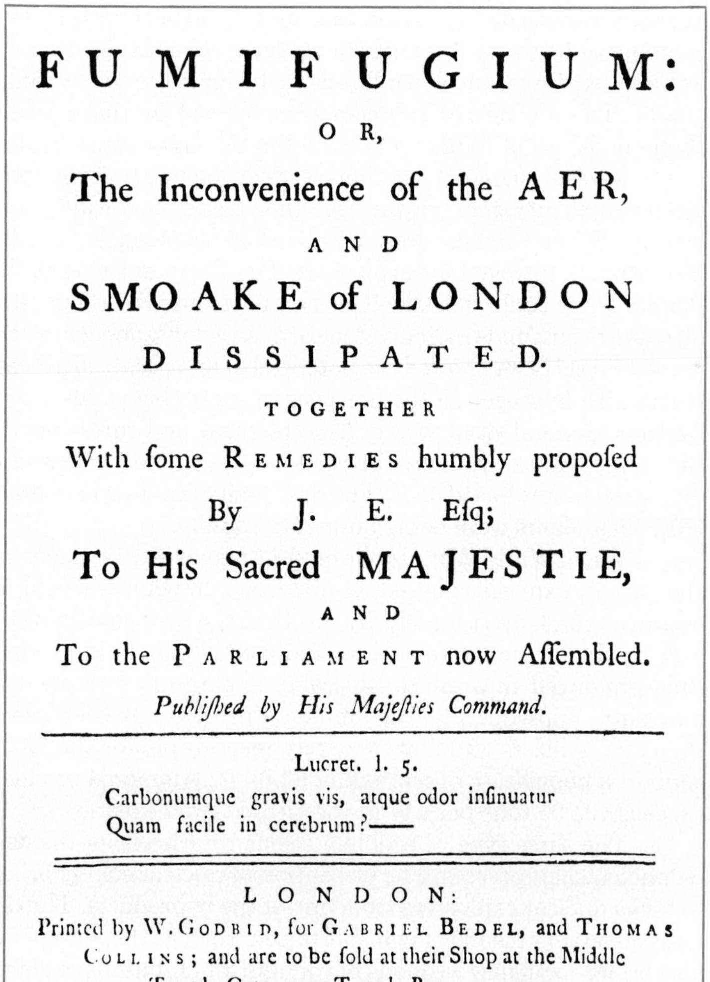

FUMIFUGIUM:

OR,

The Inconvenience of the AER,

AND

SMOAKE of LONDON

DISSIPATED.

TOGETHER

With fome REMEDIES humbly propofed

By J. E. Efq;

To His Sacred MAJESTIE,

AND

To the PARLIAMENT now Affembled.

Publifhed by His Majefties Command.

Lucret. l. 5.

Carbonumque gravis vis, atque odor infinuatur
Quam facile in cerebrum?——

LONDON:
Printed by W. GODBID, for GABRIEL BEDEL, and THOMAS COLLINS; and are to be fold at their Shop at the Middle Temple Gate, neer Temple Bar. M.DC.LXI.
Re-printed for B. WHITE, at Horace's Head, in Fleet-ftreet.
MDCCLXXII.

FIGURE 1–7 Title page from J. Evelyn, F.R.S., *The Smoake of London.* The Latin quotation is from the Roman poet Lucretius (97–53 B.C.). It may be translated, "How easily the heavy potency of carbons and odors sneaks into the brain!" (Courtesy A. E. Gunther and the University Press, Oxford.)

The amount of food that can be grown is related to the nitrogen content of the soil.

An important technological development was recognized as necessary in 1890 by Sir William Crookes, who addressed the British Association for the Advancement of Science on the problem of the fixed nitrogen supply (that is, nitrogen in a chemical form that plants can use). At the time, scientists recognized that the world's future food supply would be determined by the amount of nitrogen compounds made available for fertilizers. The source of these nitrogen supplies was then limited to rapidly depleting supplies of guano (bird droppings) in Peru and to sodium nitrate in Chile. It was realized that when these were exhausted, widespread famine would result unless an alternative supply could be developed. This problem was recognized first by English scientists as a potentially acute one, because by the 1890s England had become very dependent on imported food supplies. The Industrial Revolution allowed the population to grow rapidly, so the number of hungry people soon outstripped the domestic food supply.

The N_2 molecule is rather unreactive because of its strong triple bond ($N \equiv N$).

Widespread interest in this problem led to research on a number of chemical reactions to obtain nitrogen from the relatively inexhaustible supply present in air. Air is 21% oxygen and 79% nitrogen. The nitrogen in the air is present as the rather unreactive molecule N_2, and in this form it can be used as a source of other nitrogen compounds by only a few kinds of bacteria. Some is also transformed into nitrogen oxide by lightning, and when this is washed into the soil by rain, it can be utilized by plants. The amount of nitrogen transformed by these processes into chemical compounds useful to plants is quite limited and cannot be increased easily.

The chemical transformation of air nitrogen into useful nitrogen compounds for plant consumption is *nitrogen fixation.*

The Haber process transforms the nitrogen of the air into ammonia.

Several chemical reactions were developed to form useful compounds from atmospheric nitrogen, but the best known and most widely used one has an ironic history. While England was interested in nitrogen for fertilizers, Germany was interested in nitrogen for explosives. The German General Staff realized that the British Navy could blockade German ports and cut them off from the sources of nitrogen compounds in South America. As a consequence, when a German chemist named Fritz Haber showed the potential of an industrial process in which nitrogen reacts with hydrogen in the presence of a suitable catalyst to form ammonia, the German General Staff was quite interested and furnished support through the German chemical industry for the study of the reaction and the development of industrial plants based on it. The first such plant was in operation by 1911, and by 1914 such plants were being built very rapidly.

$N_2(g) + 3H_2(g) \rightleftharpoons 2NH_3(g)$

When World War I broke out in August of 1914, many people thought that a shortage of explosives based on nitrogen compounds would force the war to end within a year. Unfortunately, by this time the nitrogen-fixing industry in Germany was capable of supplying the needed compounds in large amounts. This process thus prolonged the war considerably and resulted in an enormous increase in mortality. Subsequently, the ammonia process has been used on a huge scale to prepare fertilizers and now is largely responsible for the fact that the Earth can support a population of more than 5 billion. Ammonia production by this process exceeds 40,000 tons per day in the United States alone.

Ammonia is the number three commercial chemical.

The same type of problem seems to arise from the development of many technological processes. The utilization of nuclear energy brings with it the ability to make nuclear explosives from one of the byproducts. The development of rapid and convenient means of transportation, such as the automobile and the airplane, also brings forth new weapons of war and air pollution problems. We must learn to control our technology in such a manner as to maximize its benefits and minimize its disadvantages. These problems arise with *all* technological developments, even the most primitive. The discovery of the techniques necessary to the manufacture of iron led first to the development of weapons (swords) by their discoverers, the Hittites, who then proceeded to conquer their neighbors and lead the first successful invasion of Egypt (ca. 1550 B.C.).

Now let's turn our attention to the present age and see where chemistry and technology are taking us.

Understanding nature does not necessarily lead to a better world.

THE CHIP AND THE SPLICE

Two major technological revolutions currently under way are the microelectronic revolution and the biotechnology revolution.

The Microelectronic Revolution

The chip, nickname for the integrated circuit, is a small slice of silicon that contains an intricate pattern of electronic switches (transistors) joined by "wires" etched from thin films of metal. Some are information storers called memory chips. Others combine memory with logic function to produce computer or microprocessor chips. These two applications make the chip, like the mind, capable of essentially infinite application. A microprocessor chip, for example, can provide a machine with decision-making ability, memory for instructions, and self-adjusting controls.

In everyday life we see many examples of the influence of the chip: the digital watches, the microwave oven controls, the new cars with their carefully metered fuel-air mixtures, the hand calculators, the cash registers that total bills, post sales, and update inventories, the computers of a variety of sizes and capacity—all of these make use of the chip. By looking at Figure 1–8, one can appreciate the technological advancement represented by the chip. A typical microprocessor chip holds 30,000 transistors but is small enough to be carried by a large ant.

The small chip stores and processes vast amounts of information at amazing speeds.

The story of the chip starts with the invention of the transistor in 1947. The transistor is a semiconductor device that acts either as an amplifier or as a current switch. Although transistorized circuits were a tremendous improvement over vacuum tubes, large computer circuits using 50,000 or more transistors and similar numbers of diodes, capacitors, and resistors were difficult to build because com-

FIGURE 1–8 Photograph (17× magnification) of a typical microprocessor chip small enough to be carried by a large ant. (Courtesy of North American Philips Corporation.)

puters had to be wired together in a continuous loop, and a circuit with 100,000 components could easily require a million soldered connections. The cost of labor for soldering and the chance for defects were high. In the late 1950s, the Navy's newest destroyers required 350,000 electronic components and millions of hand-soldered connections. It was clear that the limit to supercircuits using transistors was the number of individual connections that were required.

In 1958 Jack Kilby at Texas Instruments and Robert Noyce at Fairchild Semiconductor, working independently, came up with the solution. Make the semiconductor (silicon or germanium) in the transistor serve as its own circuit board. If all the transistors, capacitors, and resistors could be integrated on a single slice of silicon, connections could be made internally within the semiconductor and no wiring or soldering would be necessary. Figure 1–9 shows Kilby's hand-made chip and a modern chip with 200,000 transistors on one piece of silicon.

The Biotechnology Revolution — Designer Genes

Recombinant: genetic recombination.

The biotechnology revolution began after the first successful gene splicing and gene cloning experiments produced recombinant DNA in the early 1970s. In Chapter 15 we will discuss the biochemistry of DNA and the genetic code. The present discussion will focus on the potential of recombinant DNA technology for solving three of the world's greatest problems: hunger, sickness, and energy shortages.

The process for forming and cloning recombinant DNA molecules is outlined in Figure 1–10. The basic idea is to use the rapidly dividing property of common bacteria, such as *E. coli,* as a microbe factory for producing recombinant DNA molecules that contain the genetic information for the desired product. Rings of DNA called plasmids are isolated from the *E. coli* cell. The ring is cut open with a cutting enzyme, which also cuts the appropriate gene segment from the desired

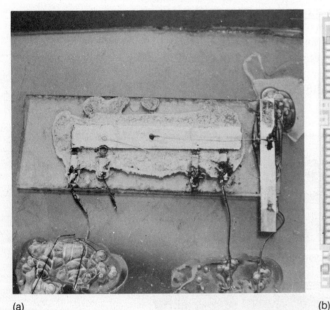

(a)

(b)

FIGURE 1–9 *(a)* Original integrated circuit built by Jack Kilby (Courtesy of Texas Instruments, Inc.) *(b)* Photomicrograph of a microprocessor chip with 200,000 transistors on one piece of silicon. (Courtesy of Motorola, Inc.)

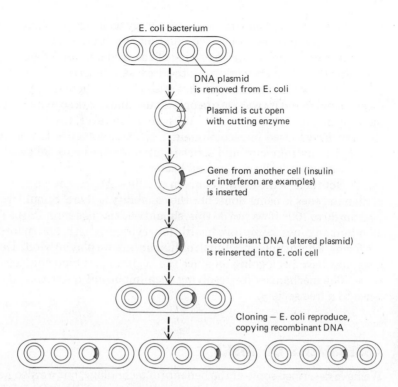

E. coli bacterium

DNA plasmid
is removed from E. coli

Plasmid is cut open
with cutting enzyme

Gene from another cell (insulin
or interferon are examples)
is inserted

Recombinant DNA (altered plasmid)
is reinserted into E. coli cell

Cloning — E. coli reproduce,
copying recombinant DNA

**FIGURE 1–10
Synthesis and cloning
of recombinant DNA
molecules. DNA
plasmids are small,
circular, duplex DNA
molecules that carry a
few genes.**

human, animal, or viral DNA. The new gene segment is spliced into the cut ring by a splicing enzyme. The altered DNA ring (recombinant DNA) is reinserted into the host *E. coli* cell. Each plasmid is copied many times in a cell. When the *E. coli* cells divide, they pass on to their offspring the same genetic information contained in the parent cell.

The control of life, not just the environment, is at hand!

The commercial potential of recombinant DNA technology was recognized very early, and several biotechnology companies were started by the scientists who did key experiments in gene splicing and gene cloning. There are more than 200 biotechnology companies, but most of them are still doing development work and have no products to sell. The risk is high, but the payoff of successful ventures will also be high. Some of the early applications illustrate how biotechnology can help to solve some of the world's problems in combatting disease, hunger, and energy shortages.

One of the earliest benefits of recombinant DNA was the biosynthesis of human insulin in 1978. Millions of diabetics depend on the availability of insulin, but many are allergic to the animal insulin, which was the only previous source of insulin. The biosynthesized human insulin is now being marketed by a firm called Genentech. Biotechnology firms are also producing human growth hormone, which is used in treating youth dwarfism, and interferon, which is a potential anticancer agent.

In 1985, Japanese scientists demonstrated that silkworm larvae can be used to biosynthesize human interferon, a potential anticancer drug.

Genetic engineers are trying to modify crops so they will make more nutritious protein, resist disease and herbicides, and even provide their own fertilizer. Another example is the use of recombinant DNA techniques to produce vaccines against diseases that attack livestock.

Strains of bacteria are being developed that will convert garbage, plant waste material such as cornstalks, and industrial wastes into useful chemicals and fuels. This not only helps to solve the energy shortage but also provides a way to recycle wastes and lower the accumulation of solid wastes.

A decade ago, at the beginning of the recombinant DNA era, many people, including the scientists working in the area, saw danger in biotechnology. Since *E. coli* is an intestinal bacterium, what if some of the genetically engineered *E. coli* escaped and found its way into people's intestines? These fears led to an 18-month moratorium on recombinant DNA research. However, the evidence to date shows that the *E. coli* used in recombinant DNA technology is too delicate to survive outside its environment, and strict regulations are being followed in the experiments with genetically engineered bacteria.

But there is the deeper concern of ethics. At present, no genetic engineering of human genes is being done, but the capability is there. Should we raise a child's IQ from 80 to 100? If we can do this, should we raise IQs from 120 to 160? Should we alter human genes to improve health, longevity, strength, and so forth? By altering life forms, whether plant, animal, or human, are we playing God? Debates on these questions have been going on since the beginning of biotechnology and will continue. The mechanism for resolving such important questions offers a new challenge to a free society.

ASSESSMENT OF RISKS

We have described some of the benefits of technology, but we also need to examine the risks of technology. The public is well aware of the dangers of chemicals in the environment. Many persons have developed **chemophobia** because of careless industrial practices such as improper disposal of hazardous wastes; environmental pollution of air, water, and earth; catastrophic industrial accidents such as the tragic one in Bhopal, India, in 1984, which claimed over 2000 lives. We will take a brief look at some of these, and a more thorough treatment will be given later, after you have learned some basic chemical concepts.

Careless disposal of hazardous wastes has been the cause of many human health problems. Love Canal in Niagara Falls, New York, the Times Beach community in Missouri, and the Minamata Bay and Jinzu River in Japan are just a few locations where serious health problems have resulted from improper disposal of hazardous wastes.

Love Canal, the neighborhood that in 1977 discovered it was built on a toxic chemical dump, was the first publicized example of the problems of chemical waste dumps. In the mid-1970s heavy rains and snows seeped into the dump and pushed an oily black liquid to the surface. The liquid contained at least 82 chemicals, 12 of which were suspected carcinogens.

The entire community of Times Beach, Missouri, was bought by the U.S. Environmental Protection Agency (EPA) in 1983, and the 2200 residents were relocated because dioxins, a group of toxic chemicals produced in small amounts during the synthesis of a herbicide, were found in the soil at concentrations as high as 1100 times the acceptable level.

In the 1950s, tons of waste mercury were dumped into the bay at Minamata, Japan. In the next few years thousands of persons in the Minamata area suffered paralysis and mental disorders, and over 200 people died. Several years passed before it was determined that these people suffered from poisoning by methyl mercury compounds. Anaerobic bacteria in the sea bottom converted mercury to methyl mercury compounds, which were eaten by plankton. The methyl mercury

The living must accept risk; the question is: How much?

compounds were carried up the food chain and eventually accumulated in the fatty tissue of fish. Since fish are a major part of the Japanese diet, intake of methyl mercury compounds reached levels that caused the sickness now known as Minamata disease.

Itai-Itai disease makes bones brittle and easily broken. It is caused by cadmium poisoning and was also first observed on a major scale in Japan. *Itai* means "it hurts" in Japanese and graphically illustrates the pain associated with this disease. Many cases were observed downstream from a zinc-refining plant on the Jinzu River in Japan. Cadmium is a byproduct of the zinc-refining industry and is used in various alloys and in nickel-cadmium rechargeable batteries.

The worst industrial accident in history occurred in the early morning hours of December 3, 1984, at a Union Carbide insecticide plant in Bhopal, India. Over 2000 people were killed and tens of thousands were injured when methyl isocyanate, a deadly gas used in the preparation of pesticides, escaped from a storage tank. Several violations of recommended safety procedures contributed to the disastrous leak, including an inoperative refrigeration system to keep the methyl isocyanate cool, an inoperative scrubbing tank to neutralize any escaping gas, and an inoperative flare tower designed to burn any gas that escaped from the storage tank.

The important point is that our present environmental problems often stem from decades of neglect. The industrial revolution brought prosperity, and little thought was given to the possible harmful effects of the technology that was providing so many visible benefits. However, as shown by the Bhopal incident, serious accidents still occur despite the development of safety procedures that, if enforced, would prevent such accidents.

Those in responsible positions in business and government now have a greater awareness of the need to solve the environmental problems associated with technological production and to assess the risks of technology. The chemical industry must take the lead in demonstrating its willingness to help solve the problems caused by its predecessors' lack of foresight. This should be done through cooperation and not confrontation. We need a science policy that is based on input from responsive leaders, from both industry and government, who provide a forum to examine the facts and then reach responsible decisions that lead to prompt action.

The control of chemical hazards is essential for everyone's well-being.

Is this possible? You may doubt it, but you have a responsibility to future generations to do your part in seeing that responsible action is taken. We cannot and should not "turn off" science and technology. Those who long for the "good old days" should remember what that means — diseases such as malaria, smallpox, and polio, which took many lives; no antibiotics for infections; none of the modern fertilizers to increase crop yields needed to feed the world's population. You could add to this list many things that are of a humanitarian nature before you even start listing the technical advances that have raised the comfort level of our lives.

Risk management requires value judgments that integrate social, economic, and political issues with the scientific assessment of the risk. The determination of the acceptability of the risk is a societal issue, not a scientific one. It is up to all of us to weigh the benefits against the risks in an intelligent and competent manner. The assumption of this text is that the wit to deal with environmental problems caused by uncontrolled technology is to be found in the educated public at large, not in the select group that stands to make a short-term profit that is either financial or political. Always keep in mind that, except in the case of some radioactive wastes, the knowledge is available to "clean up" after any industrial operation; it is just a matter of cost, energy, and values.

Society must determine how much risk it will accept.

It is apparent that we need citizens to take responsibility for being informed about the technological issues that affect society. Albert Gore, Jr., U.S. senator from Tennessee, said in support of better science education,

> Science and technology are integral parts of today's world. Technology, which grows out of scientific discovery, has changed and will continue to change our society. Utilization of science in the solution of practical problems has resulted in complex social issues that must be intelligently addressed by all citizens. Students must be prepared to understand technological innovation, the productivity of technology, the impact of the products of technology on the quality of life, and the need for critical evaluation of societal matters involving the consequences of technology.

SUMMARY

How can we summarize this chapter as it concerns you, the student?

1. We need an informed citizenry to use and to evaluate scientific and technological advances.
2. To be informed in chemical problems requires basic knowledge in what matter is really like and what matter does.
3. More sophisticated chemical problems require a deeper understanding of the workings (facts and theories) of chemistry.
4. You should be involved. As an educated person, you have a responsibility and a privilege.

This book will give you the basics in chemistry, which we hope will give you a more satisfying life through understanding and a richer, healthier life through wise decisions about problems that concern our world.

SELF-TEST 1 – A*

1. The ultimate test of a scientific theory is its agreement with _____.
2. Different workers, in different countries, who carry out a particular laboratory experiment in exactly the same way should get _____ result.
3. The chip is the nickname for _____.
4. The common bacterium used as the microbe factory in recombinant DNA technology is _____.
5. Arrange from most abstract to general to specific: laws, facts, theories.
   a. _____ b. _____ c. _____

SUBJECTIVE QUESTIONS

Rate the following on a scale: good, bad, indifferent. Be prepared to state your reasons.

1. Computer information retrieval _____

* Use these self-tests as a measure of how well you understand the material. Take a test only after careful reading of the material preceding the test. Do not return to the text during the self-test, but reread entire sections carefully if you do poorly on the self-test on those sections. The answers to the self-tests are at the end of the text.

2. Nuclear energy _____
3. Coal as a fuel _____
4. Petroleum as a fuel _____
5. Birth control pills _____
6. Fertilizers _____
7. Plastic containers _____
8. Synthetic foods _____
9. Solar energy _____
10. Recombinant DNA technology _____
11. Government control of scientific research _____
12. Government control of applied research _____

MATCHING SET

_____ 1. Scientific theory
_____ 2. Transmission of scientific knowledge
_____ 3. Allowed World War I to be prolonged
_____ 4. Source of the element nitrogen
_____ 5. Chemistry
_____ 6. Unsatisfactory way to establish scientific truth
_____ 7. Computer information retrieval company
_____ 8. Increases rate of change in life
_____ 9. Used in explosives
_____ 10. Needed to grow crops
_____ 11. Scientific methods
_____ 12. Basis of scientific truth
_____ 13. Cisplatin

a. Appeal to authority
b. Study of matter and its changes
c. Technology
d. Nitrogen compounds
e. Used to study nature
f. Printed journals
g. Haber process
h. Air
i. Observation
j. DIALOG
k. Anticancer agent
l. Used to explain facts and laws

QUESTIONS

1. Distinguish between theory and law in chemistry.
2. Give an example of a chemical fact.
3. Give an example of a chemical law.
4. Give an example of an assumption made to explain an observed fact.
5. How many times do you think a given experiment should give the same result before a scientific fact is established?
6. Suppose a mother and her children discover the family car missing on returning from an afternoon ball game, even though the father rode to work with a neighbor that morning. The mother says to the children, "Don't worry, Dad must have had an unexpected need for the car and got it after lunch." Would you call the statement made by the mother a theory or a hypothesis? Why?
7. Persons often confuse science with scientism. Look up the definition of *scientism* in a dictionary and discuss why it is important to society that science not be confused with scientism.

Chapter 2

THE CHEMICAL VIEW OF MATTER — ORDER IN DISORDER

Matter occupies space and has weight.

Do you enjoy the material things around you? Sometimes yes and sometimes no, right? Think beyond your immediate setting: how many materials and things are out there in this universe of ours? Too many to count? These things — all of them — that we can see, touch, and weigh are made of **matter.** Although an uncountable number of materials and objects exist, are there any order and simplicity in the makeup of matter? Can we reasonably hope to control matter to make life more pleasant for the human race? The science of **chemistry** addresses these fundamental questions.

Most of the things we use in our daily life are very different from the materials that are an obvious part of our natural surroundings. Practically everything we use has been changed from a natural state of little or no utility to one of very different appearance and much greater utility. The processes by which the materials found in nature can be changed and a detailed description of such changes are highly intriguing. This is a basic dimension of the science of chemistry: the **changes** in matter.

Chemical change alters the kind of matter without changing the amount of matter.

Examples:
a. Chemical change — burning, rusting, souring
b. Physical change — melting, boiling, cutting
c. Nuclear change — production of nuclear energy and radioisotopes

Our attention will be focused on a particular kind of change — **chemical change.** In any chemical change, the starting material is changed into a different kind of matter. Matter may also undergo other kinds of changes: **physical changes,** which result in new forms of the same material, and **nuclear changes,** which often change matter into energy while producing new materials. Do not worry about definitive definitions of these types of changes at this point; they will follow. Also, as you study further about physical and chemical changes, you will come to realize that the categorizing of material change is not as clear-cut as it first appears.

What causes changes to occur in matter? It is **energy!** Examples of energy are heat, light, sound, and electricity. Energy and matter are not the same even though they are closely related. Energy has the ability to move matter (engines, eardrums, and motors, for example). Matter is converted into energy in nuclear reactors and nuclear bombs. Energy can infiltrate matter and manifest itself through the actions of the matter. A sample of hot water contains more energy than the same sample when it is cold. Some forms of energy can exist apart from matter; examples are light and radiant heat. It appears that all forms of energy are generated by changes in matter, and that matter, in turn, can absorb energy to produce other physical and chemical changes. Indeed, energy by definition is that which can produce change in matter. It follows, then, that a study of chemistry involves still another dimension: the *energy* associated with chemical changes.

It is difficult, in a few words, to establish the exact bounds of chemistry. Even so, it will be helpful to think of **chemistry** as **the study of the kinds of matter and the changes of one kind of matter into another with the associated energy changes.**

Since the feature used to recognize a chemical change is the production of a different kind of matter, recognition of a chemical change requires a recognition of different kinds of matter. In a natural state the kinds of matter are usually mixed together, and the separation of such mixtures has to precede their systematic classification. After an examination of the methods of separating such mixtures into their components, we can appreciate some of the problems involved in an accurate definition of the terms *kinds of matter* and *chemical change.*

MIXTURES AND PURE SUBSTANCES

Most natural samples of matter are **mixtures.** Often, it is easy to see the various ingredients in a mixture (Fig. 2–1). Some mixtures are obviously heterogeneous, as the uneven texture of the material is clearly visible. Some mixtures appear to be homogeneous when actually they are not. For example, the air in your room appears homogeneous until a beam of light is scattered, revealing floating dust particles. Milk is another example. It appears smooth in texture to the eye, while magnification reveals an uneven distribution of materials. **Colloids** are mixtures that appear to be homogeneous in normal lighting but actually are heterogeneous (Fig. 2–2). Homogeneous mixtures do exist; such mixtures are **solutions.** No amount of optical magnification will reveal a solution to be heterogeneous, for heterogeneity in solutions exists only at atomic and molecular levels, where the individual particles are too small to be seen with ordinary light. Examples of solutions are clean air (nitrogen and oxygen), sugar-water, and brass, a homogeneous mixture of copper and zinc.

When a mixture is separated into its components, the components are said to be *purified.* However, most efforts at separations are incomplete in a single

Homogeneous—of smooth texture, the same throughout.

Heterogeneous—of nonuniform texture, not the same at every observed point.

FIGURE 2–1 *(a)* This NASA photograph of a moon rock and many similar ones show that lunar materials, like the solid formations in the crust of the earth, tend to be mixtures of more basic substances. It is likely that this is characteristic of crust materials in the universe. *(b)* The isolation of pure carbon in nature, the chemical makeup of diamond, is the exception, as there are relatively few pure substances isolated in natural formations. (*a*: NASA—JCS photo. *b*: Courtesy of the Gemological Institute of America.)

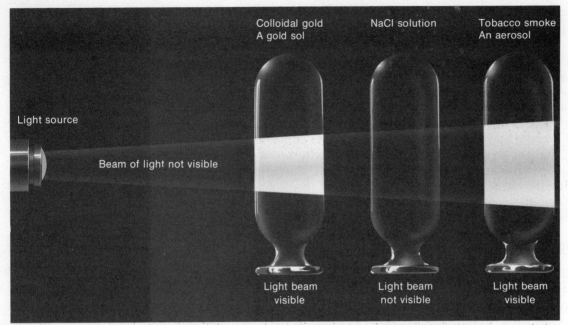

FIGURE 2–2 A colloid in a clear liquid or gas will scatter light because of the relatively large size of the dispersed particles. In contrast, a solution, which is a dispersion at the molecular level, will pass the light with no scatter.

Bits of iron and sulfur mixed

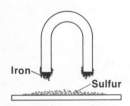

Purification separates the kinds of matter.

About 6 million pure substances have been identified.

Most materials in nature are mixtures; a few are relatively pure substances.

operation or step, and repetition of the purification process results in a better separation. Ultimately in such a procedure the experimenter may arrive at pure substances, samples of matter that cannot be purified further. For example, if sulfur and iron powder are ground together to form a mixture, the iron can be separated from the sulfur by repeated stirrings of the mixture with a magnet. When the mixture is stirred the first time and the magnet removed, much of the iron is removed with it, leaving the sulfur in a higher state of purity. However, after just one stirring the sulfur may still have a dirty appearance due to a small amount of iron that remains. Repeated stirring with the magnet, or perhaps the use of a very strong magnet, will finally leave a bright yellow sample of sulfur that apparently cannot be purified further by this technique. In this purification process a property of the mixture, its color, is a measure of the extent of purification. After the bright yellow color is obtained, it could be assumed that the sulfur has been purified.

Drawing a conclusion based on one property of the mixture may be dangerous because other methods of purification might change some other properties of the sample. It is safe to call the sulfur a pure substance only when all possible methods of purification fail to change its properties. This assumes that all pure substances have a set of properties by which they can be recognized, just as a person can be recognized by a set of characteristics. **A pure substance, then, is a kind of matter with properties that cannot be changed by further purification.**

There are some naturally occurring pure substances. Rain is very nearly pure water, except for small amounts of dust, dissolved air, and various pollutants. Gold, diamond, and sulfur are also found in very pure form. These substances are special cases. The human, a complex assemblage of mixtures, lives in a world of mixtures—eating them, wearing them, living in houses made of them, and using tools made of them.

Although naturally occurring pure substances are not common, it is possible to produce many pure substances from natural mixtures. Relatively pure substances are now very common as a consequence of the development of modern purification techniques. Common examples are refined sugar, table salt (sodium chloride), copper, sodium bicarbonate, nitrogen, dextrose, ammonia, uranium, and carbon dioxide — to mention just a few. In all, about 6 million pure substances have been identified and catalogued.

STATES OF MATTER

As we examine mixtures and pure substances, it is easy to recognize three of the four states of matter — **solids, liquids,** and **gases** — and to note that both mixtures and pure substances can exist in the different states. Solids have definite shapes and volumes, liquids have definite volumes but indefinite shapes, and gases will take any shape or volume imposed by the containing vessel. **Plasmas** constitute a fourth state of matter, which is not as common in our everyday experience. However, it is probably the most prevalent state of matter in the universe (Fig. 2–3). A plasma is much like a gas except it is composed of charged particles called ions, which dramatically respond to electric and magnetic forces. Natural materials in the plasma state include flames, the outer portion of the earth's atmosphere, the atmosphere of the stars, much of the material in nebular space, and part of a comet's tail. The aurora borealis offers a dazzling display of matter in the plasma state streaming through a magnetic field.

An ion is a charged atom or group of atoms.

FIGURE 2–3 The Orion Nebula. Light from such nebula is produced by highly energized atoms, molecules, and ions.

DEFINITIONS: OPERATIONAL AND THEORETICAL

Chemistry begins with observation and experiments.

A pure substance is identified with an **operational** definition, or a definition using specific experiments or operations. When further purification efforts are unsuccessful in changing the properties of a substance, it is said to be a pure substance. It is evident, then, that operational definitions result from performing operations or tests on matter and summarizing the results in a statement. For example, iron is a magnetic metal that melts at 1535°C, boils at 3000°C, and is 7.86 times denser than water. These properties come from the operations of applying heat to the pure substance and measuring temperatures, weights, and volumes. When all of the properties of pure iron have been listed, we find that the pure substance has been characterized in a way that distinguishes it from all other pure substances. Although there are millions of pure substances, no two of them have exactly the same set of properties.

A pure substance also can be defined in theoretical terms, that is, in terms of the molecules, atoms, and subatomic particles that compose it. Both types of definitions are important in the study of chemistry, and both are used in this text. The **theoretical** definitions follow the development of the theory on which they are based.

SELF-TEST 2–A

1. Four common materials that cannot be pure substances are:
   a. _____   c. _____
   b. _____   d. _____
2. In the human experience, which usually comes first, the operational definition or the theoretical definition? _____
3. Four common materials that are very nearly pure substances are:
   a. _____   c. _____
   b. _____   d. _____
4. The properties of two different pure substances could all be identical. True ( ) or False ( )
5. Three types of changes that are fundamental to nature are:
   a. _____   b. _____   c. _____
6. A homogeneous mixture is a _____.
7. Solutions may exist in solid, liquid, or gaseous states. True ( ) or False ( )

MATCHING SET

_____ 1. Definite shape and volume
_____ 2. Indefinite shape and volume
_____ 3. Composed of charged particles
_____ 4. Indefinite shape, definite volume

a. gas
b. liquid
c. solid
d. plasma

SEPARATION OF MIXTURES INTO PURE SUBSTANCES

The separation of mixtures is usually more difficult than the magnetic separation of iron and sulfur described previously. Most beginning chemistry students would

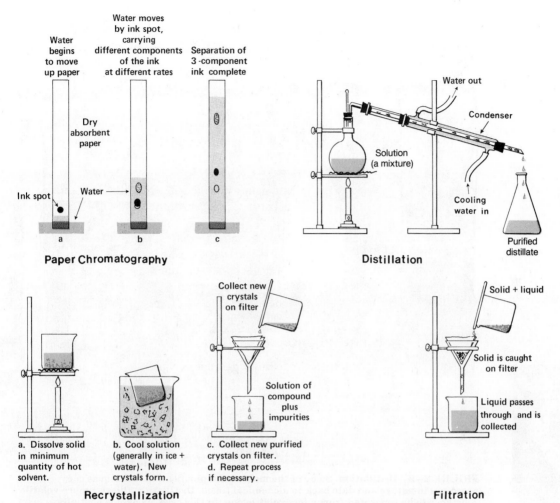

Paper Chromatography

Water begins to move up paper

Water moves by ink spot, carrying different components of the ink at different rates

Separation of 3-component ink complete

Dry absorbent paper

Ink spot   Water

a   b   c

Distillation

Water out

Condenser

Solution (a mixture)

Cooling water in

Purified distillate

Recrystallization

a. Dissolve solid in minimum quantity of hot solvent.

b. Cool solution (generally in ice + water). New crystals form.

Collect new crystals on filter

Solution of compound plus impurities

c. Collect new purified crystals on filter.
d. Repeat process if necessary.

Filtration

Solid + liquid

Solid is caught on filter

Liquid passes through and is collected

FIGURE 2–4 **Four methods of purifying mixtures of elements and compounds. Paper chromatography. Owing to the absorbent character of paper, water moves against gravity and carries the ink dyes along its path. If the ink dyes move at different rates because of differing attraction to the paper, they will be separated in the developed chromatogram. Distillation. Sodium chloride dissolves in water to form a clear solution. When heated above the boiling point (indicated by thermometer), water will vaporize and pass into the condenser. Cool water injected into the glass jacket of the condenser circulates over the inner tube, causing the steam to liquefy and collect in the flask. In this simple example pure water collects in the receiving flask, while the salt remains in the boiling flask. Recrystallization. This can be used to separate some solid mixtures. Filtration. The separation of a solid from a liquid by filtration.**

find it bewildering to separate a piece of granite into pure substances. Indeed, a trained chemist would find this difficult. Since each of the pure substances in the granite has a set of properties unlike those of any other pure substance, it should be possible to use these properties to separate the pure substances, just as the attraction of iron to a magnet is used to separate it from sulfur.

Many different methods have been devised to separate the pure substances in a mixture. In each case, differing properties of the pure substances are exploited to effect the separation. Figure 2–4 illustrates four commonly used methods: chromatography, distillation, recrystallization, and filtration. Figure 2–5 illustrates a separation by distillation that has been of interest to many. Table 2–1 shows that

FIGURE 2–5 Distillation. Some of the most useful purification techniques copy processes in nature and date back to alchemical times. Distillation allows a more volatile substance to be separated from a less volatile one. In this case, alcohol is partially separated by evaporation from water and other ingredients. One distillation can produce a mixture that is 40% alcohol from one that is only 12%. Further distillations would produce an even better separation.

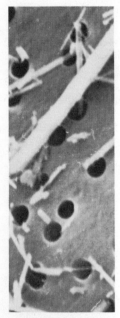

Scanning electron micrograph of particles of asbestos filtered from a sample of air by a small-pore filter.

These are operational definitions of "element" and "compound."

the cost for "pure" elements is a function of the purity desired as well as of the availability of the substance.

ELEMENTS AND COMPOUNDS

Experimentally, pure substances can be classified into two categories: those that can be broken down by chemical change into simpler pure substances and those that cannot. Table sugar (sucrose), a pure substance, will decompose when heated in the oven, leaving carbon, another pure substance, and evolving water. No chemical operation has ever been devised that will decompose carbon into simpler pure substances. Obviously sucrose and carbon belong to two different categories of pure substances. Only 89 substances found in nature cannot be reduced chemically to simpler substances; 20 others are available artificially. These 109 substances are called **elements.** Pure substances that can be decomposed into two or more differ-

TABLE 2–1 Costs for Selected Elements*

ELEMENT	FORM	HIGHEST PURITY (%)	SMALLEST PACKAGE (g)	PACKAGE COST ($)	GRAM COST ($)
Aluminum	Wire	99.999	10.5	9.00	0.86
Arsenic	Mesh	99.9999	10	30.00	3.00
Barium	Stick	99	10	27.50	2.75
Beryllium	Flake	99.99	10	60.00	6.00
Boron	Powder	99.9995	0.500	67.00	134.00
Calcium	Turnings	99.5	100	18.40	0.18
Cesium	Ingot	99.95	1	27.50	27.50
Copper	Rod	99.9998	56	58.00	0.97
Gallium	Splatter	99.99999	5	27.00	5.40
Gold	Powder	99.999	0.250	27.50	110.00
Iodine	Flake	99.9999	20	20.00	1.00
Iridium	Powder	99.9	0.500	39.50	79.00
Iron	Powder	99.999	10	18.50	1.85
Lead	Foil	99.9995	28	16.30	0.58
Mercury	Liquid	99.99999	50	22.00	0.44
Phosphorus	Lump	99.99995	5	30.00	6.00
Platinum	Powder	99.999	1	59.00	59.00
Potassium	Ingot	99.95	1	26.00	26.00
Rhodium	Powder	99.999	1	180.00	180.00
Selenium	Pellets	99.9999	20	12.00	0.60
Silicon	Powder	99.999	5	10.75	2.15
Silver	Shot	99.999	5	18.00	3.60
Sodium	Ingot	99.95	5	38.00	7.60
Sulfur	Powder	99.9999	50	16.00	0.32
Zinc	Shot	99.9999	25	27.50	1.10

* The costs of pure samples of the elements are a function of the availability of the elemental source and the degree of difficulty in the isolation of the element in the purity desired. Prices published by Aldrich Chemical Co., Milwaukee, Wisconsin, in 1984. On the world's bullion markets, gold sells for about $12.30 per g; silver sells for about $0.32 per g.

ent pure substances are referred to as **compounds.** Even though there are presently only 109 known elements, there appears to be no practical limit to the number of compounds that can be made from the 109 elements.

Elements are the basic building blocks of the universe and the world in which we live. Table 2–2 lists the properties of some of the common elements; a complete list of the elements is found inside the back cover of this text. Several elements are found as the elementary substance in nature; examples include gold, silver, oxygen, nitrogen, carbon (graphite and diamond), copper, platinum, sulfur, and the noble gases (helium, neon, argon, krypton, xenon, and radon). Many more elements, however, are found chemically combined with other elements in the form of compounds.

Elements in compounds no longer show all of their original, characteristic properties, such as color, hardness, and melting point. Consider ordinary sugar, which is properly called sucrose, as an example. It is made up of three elements: carbon (which is usually a black powder), hydrogen (the lightest gas known), and oxygen (a gas necessary for respiration). The compound sucrose is completely

Figure 2–6 lists the most common elements in the universe and on Earth.

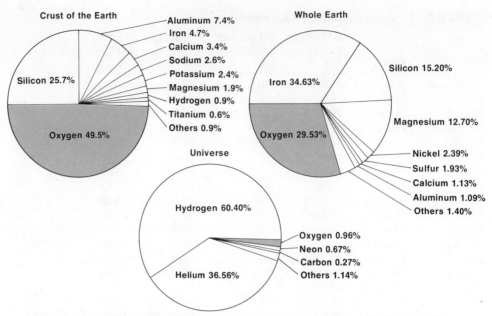

FIGURE 2-6   Relative abundance (by mass) of the most common elements in the Earth's crust, the whole Earth, and the universe. Note that the Earth's crust differs significantly from the cosmic array of elements.

unlike any of the three elements; it is a white crystalline powder that, unlike carbon, is readily soluble in water.

A careful distinction should be made between a *compound* of two or more elements and a *mixture* of the same elements. The two gases hydrogen and oxygen can be mixed in all proportions. However, these two elements can and do react chemically to form the compound water. Not only does water exhibit properties peculiar to itself and different from those of hydrogen and oxygen, but it also has a definite percentage composition by weight (88.8% oxygen and 11.2% hydrogen). In addition to the distinctly different properties between compounds and their parent elements, there is this second distinct difference between compounds and mixtures: **Compounds have a definite percentage composition by weight of the combining elements.**

> Compounds have a fixed composition of the elements they contain.

WHY STUDY PURE SUBSTANCES AND THEIR CHANGES? COMFORT, PROFIT, AND CURIOSITY!

Perhaps by now you are wondering why we should be interested in elements and compounds and their chemical properties. There are two basic reasons. The first is the belief that the knowledge of chemical substances and chemical changes will allow us to bring about desired changes in the nature of everyday life. Two hundred years ago most of the materials surrounding a normal person could be changed only by physical means. Only a few useful materials, such as iron and pottery, were the product of control over chemical change. By contrast, today's synthetic fibers, plastics, drugs, latex paints, detergents, new and better fuels, photographic films, and audio and video tapes are but a few of the materials produced by controlled

TABLE 2–2   Some Common Elements*

NAME	SYMBOL	PROPERTIES OF PURE ELEMENTS
		METALS

(A metal is a good conductor of electricity, can have a shiny or lustrous surface, and in the solid form usually can be deformed without breaking.)

NAME	SYMBOL	PROPERTIES OF PURE ELEMENTS
Iron Latin, *ferrum*	Fe	Strong, malleable, corrosive
Copper Latin, *cuprum*	Cu	Soft, reddish-colored, ductile
Sodium Latin, *natrium*	Na	Soft, light metal, very reactive, low melting point
Silver Latin, *argentum*	Ag	Shiny, white metal, relatively unreactive, good conductor of electricity and heat
Gold Latin, *aurum*	Au	Heavy, yellow metal, very unreactive, ductile, good conductor
Chromium	Cr	Resistant to corrosion, hard, bluish-gray, brittle

NONMETALS

(A nonmetal is often a poor conductor of electricity, normally lacks a shiny surface, and is brittle in crystal-solid form.)

NAME	SYMBOL	PROPERTIES OF PURE ELEMENTS
Hydrogen	H	Colorless, odorless, occurs as a very light gas (H_2), burns in air
Oxygen	O	Colorless, odorless gas (O_2), reactive, constituent of air
Sulfur	S	Odorless, yellow solid (S_8), low melting point, burns in air
Nitrogen	N	Colorless, odorless gas (N_2), rather unreactive
Chlorine	Cl	Greenish-yellow gas (Cl_2), very sharp choking odor, poisonous
Iodine	I	Dark purple solid (I_2), sublimes easily

* Chemists usually use the symbol rather than the name of the element. In addition to denoting the element, the chemical symbol has a very specialized meaning, which is described later in this chapter. A complete list of the elements with their symbols can be found inside the back cover of this book.

chemical change. (We shall return to examine the chemistry of many of these later.) You will find it difficult to find more than a few objects in your home that have not been altered by a desirable chemical change. Not only is it important to bring about desirable changes, but also in the areas of toxicity and pollution it is important to avoid undesirable changes. **Applied research,** directing itself to the profitable control of chemicals, is conducted by those interested in the business and commerce of human needs; such research is product centered.

The second reason for chemical studies is even more important — curiosity. Chemicals and chemical change are a part of nature that is open to investigation, and, like the mountain climber, we will find this task both interesting and challenging simply because it is there. If we hope to understand matter, the first steps are to discover the simplest forms of matter and to study their interactions. Many chemists are drawn toward **basic research,** activities calculated to uncover basic information about nature without much, if any, regard for products and the profit motive. Basic research is carried on by individuals, universities, governments, philanthropic groups, and industries that realize the long-term profit in expanding the scope of human knowledge.

THE STRUCTURE OF MATTER EXPLAINS CHEMICAL AND PHYSICAL PROPERTIES

For reasons that are partly theoretical and partly practical, we are deeply interested in the **structure of matter**—that is, the minute parts of matter and how these parts are fitted together to make larger units. Why does an element or compound have the properties it has? Why does one element or compound undergo a change that another element or compound will not undergo? Inanimate matter is the way it is because of the nature of its parts. A watch is what it is because of the nature of its individual parts. So is a car, a refrigerator, and the salt in your salt shaker. The individual parts (smaller than the whole) are what determine the nature (actions and properties) of the whole. The most basic parts of matter, as we shall see, are very small. If we even hope to understand the nature of matter, it is absolutely necessary that we have some understanding of these minute parts and how they are related to each other.

The causes of the properties of matter lie in the structure and composition of its parts.

A very large portion of today's research in chemistry is aimed at sorting out and elucidating the structure of matter. Indeed, the basic theme of this text is the relationship between the structure of matter and its properties. This theme of structure and related properties is of great interest because if we know exactly how and with what strength the minute parts of matter are put together, we can discover exact relationships between structure and properties. Armed with this understanding, we can make changes that result in new substances and predict the properties of these substances. Such knowledge can save many months of trial and error that otherwise may be required to prepare a product with the desired qualities. While this day of predicting chemical changes based on structural characteristics has not arrived completely, such significant advances have been made that the practice of modern chemistry would not be possible without such knowledge.

Submicroscopic structures help to explain chemistry.

Samples of matter large enough to be seen and felt and handled, and thus large enough for ordinary laboratory experiments, are called **macroscopic** samples, in contrast to **microscopic** samples, which are so small that they have to be viewed with the aid of a microscope. The structure of matter that really interests us, however, is at the **submicroscopic** level. Our senses have no direct access into this small world of structure, and any conclusions about it will have to be based on circumstantial evidence gathered in the macroscopic and microscopic worlds (Fig. 2–7).

Scope of chemistry.

We can now extend our concept of the science of chemistry. It is that science that investigates the properties and changes of pure substances. Chemistry is also deeply concerned with structure, both **macrostructure** and **submicrostructure,** in an effort to give plausible reasons for properties and change, with emphasis on chemical change.

CHEMICAL, PHYSICAL, AND NUCLEAR CHANGES — A CLOSER LOOK

Chemical and physical changes were originally defined in an operational sense while nuclear changes were first conceived in theoretical terms. We are now prepared to consider these important transformations in matter from both points of view, even though your theoretical understandings will develop further as you study the following chapters.

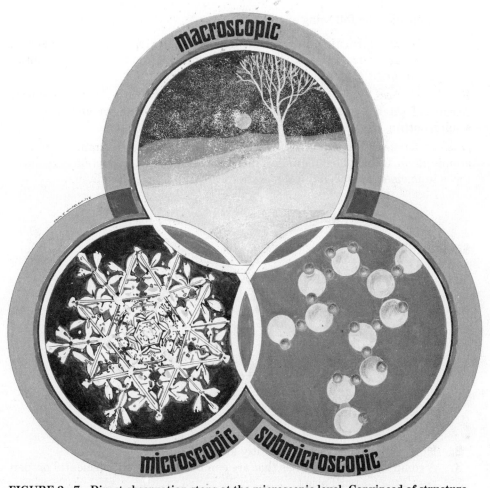

FIGURE 2–7 Direct observation stops at the microscopic level. Convinced of structure beyond the microscopic level, the chemist employs circumstantial evidence to construct the world of molecules, atoms, and subatomic parts in the mind's eye.

A chemical change involves the disappearance of one or more pure substances and the appearance of one or more other pure substances. In theoretical terms, a chemical change produces a new arrangement of the atoms involved without a loss or gain in the number of atoms. In a physical change, the pure substance is preserved from an operational point of view even though it may have changed its physical state or the gross size and shape of its pieces. In theory, the physical change may break the gross arrangement on the overall structure of the theoretical particles (cutting a diamond) or even separate the particles (melting and vaporizing a metal), but the fundamental particles remain and their tendency to interact with each other is retained. A nuclear change occurs when atoms of one type are changed into atoms of another type. Physically, the nuclear change produces new pure substances, as does the chemical change; however, nuclear changes are generally associated with conversion of elements into other elements, radioactive emissions, and very large energy transformations relative to those involved in chemical and physical changes. Also, matter is destroyed in many nuclear changes in that matter is converted into energy.

Consider the following examples:

Chemical changes. Rusting of iron, burning of gasoline in an automobile engine, preparation of caramel by heating sugar, preparation of iron from its ores, solution of copper in nitric acid, ripening and souring of fruit.

Physical changes. Evaporation of a liquid, melting of iron, drawing of metal wire, freezing of water, crystallization of salt from sea water, grinding or pulverization of a solid, cutting and shaping of wood, bending, breaking, molding.

Nuclear changes. Production of the elements in stars, splitting of uranium atoms in atomic bomb, fusion of light atoms in thermonuclear bomb, production of radioisotopes in nuclear reactor, natural radiation from uranium and radium.

SELF-TEST 2−B

1. The two large categories into which elements can be divided are:

   a. _____ and b. _____

   A half-dozen examples of each are:

   a. _____   a. _____

   b. _____   b. _____

   c. _____   c. _____

   d. _____   d. _____

   e. _____   e. _____

   f. _____   f. _____

2. How many elements are presently known? _____

3. A compound has properties that are combinations of the elemental properties. True (   ) or False (   )

4. Four chemical changes not listed in the chapter are:

   a. _____

   b. _____

   c. _____

   d. _____

5. Four physical changes not listed in the chapter are:

   a. _____

   b. _____

   c. _____

   d. _____

6. A chemical change always produces a new _____ .

7. _____ structures explain chemical properties.

8. Put in order of decreasing size: 1. microscopic, 2. molecular, and 3. macroscopic.

   a. _____   b. _____   c. _____

9. In what type of natural change are atoms altered in their internal structure?

   _____

10. Which is more closely associated with the development of products for market? Applied research (   ) or Basic research (   )

11. Which element is most abundant:

    a. In the crust of the Earth? _____

    b. In the bulk of the Earth? _____

    c. In the universe?   _____

12. Name four separation techniques (or purification methods).

    a. _____

    b. _____

    c. _____

    d. _____

13. List as many elements as you can recall without looking at a reference.

_____

THE LANGUAGE OF CHEMISTRY

Symbols, formulas, and equations are used in chemistry to convey ideas quickly and concisely. These shorthand notations are merely a convenience and contain no mysterious concepts that cannot be expressed in words. Certain characters are used often, and a general familiarity with them will help in reading chemical text.

    A **chemical symbol** for an element is a one- or two-letter term, the first letter a capital and the second a lowercase letter. The symbol represents three concepts. First, the symbol stands for the element in general. H, O, N, Cl, Fe, Pt are shorthand notations for the elements hydrogen, oxygen, nitrogen, chlorine, iron, and platinum, respectively, and it is useful and timesaving to substitute these symbols for the words themselves in describing chemical changes. Some symbols originate from Latin words (such as Fe, from *ferrum,* the Latin word for iron); others came from English, French, and German names. Second, the chemical symbol stands for a single atom of the element. The **atom** is the smallest particle of the element that can enter into chemical combinations. Third, the elemental symbol stands for a mole of the atoms of the element. The **mole** is a term in chemical usage (it is derived from the Latin for "a pile of" or "a quantity of") that means the quantity of substance that contains 602 sextillion identical particles. Just as a dozen apples would be 12 apples, a mole of atoms would be 602,200,000,000,000,000,000,000 atoms or 6.022×10^{23} atoms. How big is this number? A mole of textbooks like this one would cover the *entire surface of the continental states* to a height of 190 miles! To match the population density on Earth, a mole of people would require 150 trillion planets. It takes 134,000 years for a mole of water drops to flow over Niagara Falls at a flow rate of 112,500,000 gallons per minute. A mole is a very large number. Yet, it turns out that a mole of very small atoms is usually a convenient amount for laboratory work. Thus, the symbol Ca can stand for the element calcium, or a single calcium atom, or a mole of calcium atoms. It will be evident from the context which of these meanings is implied.

    Atoms can unite (bond together) to form molecules. A **molecule** is the smallest particle of an element or a compound that can have a stable existence in the close presence of like molecules. One or more of the same kind of atom can make

"chemistry"

Chemical symbols are abbreviations for the different elements.

A mole contains 6.02×10^{23} particles.

A mole of water molecules in the liquid state occupies only about four teaspoonfuls. Are molecules small . . . or are they small?

a molecule of an element. For example, two atoms of hydrogen will bond together to form a molecule of ordinary hydrogen, and eight sulfur atoms will form a single molecule. Subscripts in a chemical formula show the number of atoms involved: H_2 means a hydrogen molecule is composed of two atoms, and S_8 means a sulfur molecule is composed of eight atoms. The noble gases, such as helium (He), have monatomic molecules (monatomic = one atom).

When unlike atoms combine, as in the case of water (H_2O) or sulfuric acid (H_2SO_4), the formulas tell what atoms and how many of each are present in a molecule of the compound. For example, H_2SO_4 molecules are composed of two hydrogen atoms, one sulfur atom, and four oxygen atoms. A **formula** can stand not only for the substance itself but also for one molecule of the substance or for a mole of such molecules, depending on the context.

When elements or compounds undergo a chemical change, the formulas, arranged in the form of a **chemical equation,** can present the information in a very concise fashion. For example, carbon can react with oxygen to form carbon monoxide. Like most solid elements, carbon is written as though it had one atom per molecule; oxygen exists as diatomic (two-atom) molecules, and carbon monoxide molecules contain two atoms, one each of carbon and oxygen. Furthermore, one oxygen molecule will combine with two carbon atoms to form two carbon monoxide molecules. All of this information is contained in the equation

$$2\,C + O_2 \rightarrow 2\,CO$$

The arrow is often read "yields"; the equation then states the following information:

1. Carbon plus oxygen yields carbon monoxide.
2. Two atoms of carbon plus one diatomic molecule of oxygen yield two molecules of carbon monoxide.
3. Two moles of carbon atoms plus one mole of diatomic oxygen molecules yield two moles of carbon monoxide.

The number written before a formula, the **coefficient,** gives the amount of the substance involved, while the **subscript** is a part of the definition of the pure substance itself. Changing the coefficient only changes the amount of the element or compound involved, whereas changing the subscript would necessarily involve changing from one substance to another. For example, 2 CO means either two molecules of carbon monoxide or two moles of these molecules, whereas CO_2 would mean a molecule or a mole of carbon dioxide, a very different substance.

MEASUREMENT

Establishing natural facts and laws is obviously dependent on accurate observations and measurements. While one can be as accurate in one language as in another or in one system of units as in another, there has been an effort since the French Revolution to have all scientists embrace one simple system of measure. The hope was and is to facilitate communication in science. The metric system, which was born of this effort, has two advantages. First, it is easy to convert from one unit to another since subunits and multiple units differ only by factors of ten. Consequently, to change millimeters to meters, one has only to shift the decimal three places to the left. Compare the difficulty of the decimal shift to the problem of

H_2SO_4

4 atoms of oxygen

1 atom of sulfur

2 atoms of hydrogen

Chemical equations summarize information on chemical reactions in a concise fashion.

changing inches to miles. The second advantage, and one that has not been fully achieved, is that standards for measurements are defined by phenomena of nature rather than by the length of the king's foot or some other such changeable standard. For example, length is now defined in terms of the length of a particular wavelength of light, a number we believe to be invariant.

The International System of Units, abbreviated SI, was adopted by the International Bureau of Weights and Measures in 1960. It is an extension of the metric system, retaining its ease of unit conversion but doing a better job in the definition of units based on physical phenomena.

Seven fundamental units are required to describe what is now known about the universe. The concept to be measured, the name of the unit, and its symbol are:

1. Length   Meter   m
2. Mass   Kilogram   kg
3. Time   Seconds   s
4. Temperature   Degree Kelvin   K
5. Luminous intensity Candle   cd
6. Electric charge   Coulomb   c
7. Molecular quantity Mole   mol

Other units that are needed are derived from these seven. For example, volume is defined in terms of cubic length.

In this book, we shall frequently employ five units and a sixth one to a lesser degree. Along with a suitable conceptual definition, the units are:

1. Meter   m   39.4 inches
2. Liter   L   1.06 quarts
3. Gram   g   0.0352 ounce
4. Celsius degree °C   Water boils at 100°C and freezes at 0°C.
5. Calorie   cal Energy required to heat 1 g of water 1 Celsius degree
6. Mole   mol Number of atoms in 12 g of carbon-12 isotope

Figures 2–8 through 2–11 along with Table 2–3 will aid the student in developing a "feel" for these units. Appendices A and B give further information about the SI system and precise definitions for some of the units commonly used in elementary chemistry courses. Appendices C and D offer a systematic approach to using these units to work some practical chemical problems.

TABLE 2–3 Recipe for Perfect Brownies in English and Metric Measurements

INGREDIENT	ENGLISH	METRIC (SI)
Unsweetened chocolate squares	2.0 oz	57 g
Butter or margarine	0.50 cup	120 mL
Sugar	1.0 cup	240 mL
Eggs	2	2
Vanilla	1.0 teaspoon	5 mL
Sifted enriched flour	0.50 cup	120 mL
Chopped walnuts	0.50 cup	120 mL
Oven	325°F	163°C
Pan	8 × 8 × 2 inches	20 × 20 × 5.0 cm

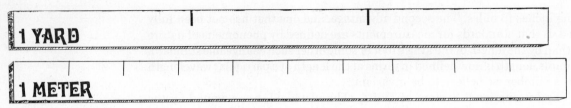

FIGURE 2–8 One meter = 39.4 inches.

While it is not the purpose of this text to require students to master a number of quantitative calculations, the authors do hope the reader will realize that the SI system requires no higher level of thinking than the English system and that problem solving is usually easier in the newer system. Consider the following examples:

EXAMPLE 1

1. English: How many feet are in 0.5000 miles?

   ? feet = 0.5000 miles × 5280 feet/mile = 2640 feet

2. SI: How many meters are in 2.0 km?
   The prefix *kilo-* means *1000 times,* so

   ? m = 2.0 km × 1000 m/km = 2000 m

FIGURE 2–9 The volume of a liter is slightly more than that of a quart. One liter = 1.06 quarts.

FIGURE 2–10
Thermometer scales are defined in terms of the expansion of common materials such as mercury and in terms of fixed reference points such as the changes of state of water and other common materials. It is only a matter of preference and convenience whether one scale or another is used.

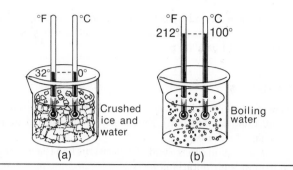

It is hardly worth the trouble to write anything down in this solution. One just thinks 2000 meters as one thinks one dollar for ten 10-cent items.

EXAMPLE 2

1. English: How many ounces are in 1.50 gallons?

    You might remember there are 32 ounces per quart and 4 quarts per gallon. Your solution then would be:

    ? ounces = 1.50 gallons × 4 quarts/gallon × 32 ounces/quart = 192 ounces

FIGURE 2–11   Units of measure are selected for convenience: the millimeter for the small parts of this tiny radio (above), and the kilometer for measuring intercontinental distances. The metric system simply adds the ease of using multiples of 10.

2. SI: How many milligrams are in a coin that weighs 5 g?

The prefix *milli-* means *one-thousandth of,* so there are 1000 mg in a gram. Five **g** then would be 5000 mg.

EXAMPLE 3

A typical piece of white bread contains 70 dietary calories (Cal). One dietary calorie is equal to 1000 small calories (cal). How many small calories are in a typical piece of white bread? 70,000 small calories.

(Note: You can see why the dietitians like the larger unit; with it, they can use smaller numbers in their notations.)

SELF-TEST 2–C

1. Consider the equation: $2 Na + 2 HCl \rightarrow H_2 + 2 NaCl$. Explain what is meant by the symbols:

a. Na _____

b. 2 Na _____

c. HCl _____

d. $\rightarrow$ _____

e. H_2 _____

f. 2 NaCl _____

2. Name three concepts that a chemical symbol can represent.

a. _____

b. _____

c. _____

3. A chemical formula gives what two pieces of information?

a. _____

b. _____

4. Which is proper to change when you balance a chemical equation: a coefficient or a subscript? _____

5. The SI system is more accurate than the English system of weights and measures. True ( ) or False ( )

6. How many fundamental physical units are required to express our present knowledge of the universe? _____

MATCHING SET

_____ 1. Produces a new type of matter

_____ 2. Air

_____ 3. Unchanged by further purification

_____ 4. Used to separate a solid from a liquid

a. Filtration
b. Chemical change
c. Element
d. Properties of pure substance
e. Mixture
f. Uses multiples of ten

_____ 5. Cannot be reduced to simpler substances

_____ 6. SI system

_____ 7. Symbol for iron

_____ 8. Mole of atoms

_____ 9. Molecule containing three oxygen
atoms

_____ 10. Carbon monoxide

_____ 11. 100 cm

_____ 12. Element symbol

_____ 13. Theoretical definition

_____ 14. Basic research

_____ 15. Plasma

_____ 16. Solution

g. O_3
h. CO
i. Fe
j. 1 meter
k. 6.022×10^{23} atoms
l. Based on idea of atoms
m. Information oriented
n. Made of charged particles
o. The element, an atom or a mole
p. Homogeneous mixture

QUESTIONS

1. Name as many materials as you can that you have used during the past day that were not chemically changed by artificial means.
2. In pottery making, an object is shaped and then baked. Which part of this process is chemical, and which part is physical?
3. Identify the following as physical or chemical changes. Justify in terms of the operational definitions for these types of changes.
   a. Formation of snowflakes
   b. Rusting of a piece of iron
   c. Ripening of fruit
   d. Fashioning a table leg from a piece of wood
   e. Fermenting grapes
   f. Boiling a potato
4. If physics is the study of matter and energy, why can it be said that the study of chemistry is a special case within the general study of physics?
5. Name a physical change food undergoes as you eat it. Name a chemical change.
6. Would it be possible for two pure substances to have exactly the same set of properties? Give reasons for your answer.
7. Chemical changes can be both useful and destructive to humanity's purposes. Cite a few examples of each kind with which you have had personal experience. Also give observed evidence that each is indeed a chemical change and not a physical change.
8. For many years water was thought to be an element. Explain.
9. Name two pure substances that are used at the dinner table. Identify each as an element or a compound.
10. Classify each of the following as a physical property or a chemical property. Justify in terms of operational definitions.
    a. Density
    b. Melting temperature
    c. Substance that decomposes into two elements upon heating
    d. Electrical conductivity of a solid
    e. A substance that does not react with sulfur
    f. Ignition temperature of a piece of paper
11. Classify each of the following as an element, a compound, or a mixture. Justify each answer.
    a. Mercury   e. Ink
    b. Milk   f. Iced tea
    c. Pure water   g. Pure ice
    d. A tree   h. Carbon
12. Which of the materials listed in Question 11 can be pure substances?
13. Explain how the operational definition of a pure substance allows for the possibility that it is not actually pure.
14. Why do theoretical definitions come after operational definitions in a particular concept?
15. Is it possible for the properties of iron to change? What about the properties of steel? Explain your answer.
16. Suggest a method for purifying water slightly contaminated with a dissolved solid.

17. Did most purification techniques arise from theory or practice?
18. Define a solution as a particular kind of mixture.
19. What is the most abundant element in the universe?
20. Name an element that is a solid at room temperature. A gas. A liquid.
21. Given the following sentence, write a chemical reaction using chemical symbols that convey the same information. "One nitrogen molecule containing two nitrogen atoms per molecule reacts with three hydrogen molecules, each containing two hydrogen atoms, to produce two ammonia molecules, each containing one nitrogen and three hydrogen atoms."
22. Aspirin is a pure substance, a compound of carbon, hydrogen, and oxygen. If two manufacturers produce equally pure aspirin samples, what can be said of the relative worth of the two products?
23. Is it possible to have a mixture of two elements and also to have a compound of the same two elements? Explain. Can you think of an example?
24. How is the salt content of the sea related to the purity of rain water? What method of purification does nature employ in the purification of rain water?
25. Name four forms of energy.
26. What is a difference between a chemical change and a nuclear change?
27. As you look up from this page, which are most abundant in your field of view: mixtures, compounds, or elements?
28. How is a plasma different from a gas?
29. What is the difference between applied research and basic research? Which is more likely to be done in a university laboratory?
30. Describe in words the chemical process that is summarized in the following equation:

$$2\,Na + Cl_2 \rightarrow 2\,NaCl$$

31. How many *atoms* are present in each of the following:
   a. One mole of He
   b. One mole of Cl_2
   c. One mole of O_3
32. Name ten elements and give the symbol for each one.

33. Describe the meaning of each symbol and number in the chemical equation:

$$2\,NO_2 \rightarrow N_2O_4$$

34. Would you think that tea in tea bags is a pure substance? Use the process of making tea to make an argument for your answer. How would your argument apply to instant tea?
35. Which contains more atoms?
   a. One mole of water, H_2O
   b. One mole of hydrogen, H_2
   c. One mole of oxygen, O_2
36. Do elements sometimes retain their physical properties in the formation of compounds? Give two examples to support your argument.
37. Find as many pure substances as you can in a kitchen.
38. What is the most fundamental assumption relative to structure and properties in chemical theory?
39. Consider the equation:

$$\underset{\substack{\text{HEPTANE}\\\text{(A COMPONENT}\\\text{OF GASOLINE)}}}{C_7H_{16}} + \underset{\text{OXYGEN}}{11\,O_2} \rightarrow \underset{\substack{\text{CARBON}\\\text{DIOXIDE}}}{7\,CO_2} + \underset{\text{WATER}}{8\,H_2O}$$

Write out in words as much of the information presented in this equation as you can decipher.
40. In the normal usage of the terms *atom* and *molecule*, which is composed of the other?
41. The number 12 is to a dozen, and 144 is to a gross, as 6.022×10^{23} is to a(n) _____.
42. If you had a mole of elephants, how many moles of elephant ears would you have? A mole of O_3 molecules contains how many moles of oxygen atoms?
43. Which English unit is closer to a liter? A meter?
44. How tall are you in meters?
45. What is your weight in kilograms?
46. Which is colder; $0\,°C$ or $0\,°F$?
47. Convert the distance of your last auto trip into kilometers.
48. Library assignment: Look up other purification methods such as sublimation, extractions, and zone refining, and tell in each case how the method is used in the separation of pure substances.

Chapter 3

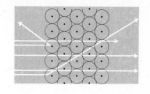

ATOMS — THE BUILDING BLOCKS ARE MADE OF BUILDING BLOCKS

As you focus on a dot over an "i," can you visualize the thousands and thousands of individual, very small atoms in the dot? Trying to fathom the minuteness of the atom is as deeply challenging to the human mind as trying to fathom the wholeness of the universe. One lures the mind to unseen smallness; the other to unseen largeness. If atoms are too small for us to observe, how do we know they exist? And if there were some way we could observe an individual atom, would there be some way we could probe inside it?

Answers to these questions are the focus of this chapter. Assembling the pieces of the puzzle has taken more than two thousand years, but much of the work has been done in the present era. What difference does it make what is inside atoms, or even if they exist? The drive to discover is a strong force in certain individuals and cannot be discounted. Simply, it is thrilling to discover what makes nature tick. A knowledge of the atom leads to a deep and enlightened understanding of bonding, chemical reactivity, light, and other intriguing phenomena. Perhaps the most powerful outcome of knowledge about atoms is the ability to predict accurately the properties of matter.

THE GREEK INFLUENCE

The ancient Greeks recorded the first theory of atoms. Leucippus and his student, Democritus (460–370 B.C.), argued for the concept of atoms. Democritus used the word **atom** (literally means "uncuttable") to describe the ultimate particles of matter, particles that could not be divided further. He reasoned that in the division of a piece of matter, such as gold, into smaller and smaller pieces, one would ultimately arrive at a tiny particle of gold that could not be further divided and still retain the properties of gold. The atoms that Democritus envisioned representing different substances were all made of the same basic material. His atoms differed only in shape and size.

Democritus used his concept of atoms to explain the properties of substances. For example, the high density and softness of lead could be caused by lead atoms packed very closely together like marbles in a box and moving easily one over another. Iron was known to be a less dense metal that is quite hard. Democritus argued that the properties of iron resulted from atoms shaped like corkscrews,

Democritus

Since the writings of Leucippus and Democritus have been destroyed, we know about their ideas only from recorded opposition to atoms and from a lengthy poem (55 B.C.) by the Roman poet, Lucretius.

atoms that would entangle in a rigid but relatively lightweight structure. Although his concept of the atom was limited, Democritus did explain in a simple way some well-known phenomena, such as the drying of clothes, how moisture appears on the outside of a vessel of cold water, how an odor moves through a room, and how crystals grow from a solution. He imagined the scattering or collecting of atoms as needed to explain the events he saw. All atomic theory has been built on the assumption of Leucippus and Democritus: atoms, which we cannot see individually, are the cause of the phenomena that we can see.

Plato (427–347 B.C.) and Aristotle (384–322 B.C.) led the arguments against the atom by asking to be shown atoms. They also argued that the idea of atoms was a challenge to God. If atoms could be used to explain nature, there would be no need for God. For centuries most of those in the mainstream of enlightened thought rejected or ignored the atoms of Democritus.

Although the idea was proposed three and a half centuries earlier by Roger Bacon, it was not until 1620 that Francis Bacon wrote his book, *New Organon,* which put experimental science in the most refined and scholarly terms and made it possible for other scholars to accept it.

Ideas about atoms drifted in and out of philosophical discussions for about 2200 years without playing a major role. Galileo (1564–1642) reasoned that the appearance of a new substance through chemical change involved a rearrangement of parts too small to be seen. Francis Bacon (1561–1626) speculated that heat might be a form of motion by very small particles. Robert Boyle (1627–1691) and Isaac Newton (1642–1727) used atomic concepts to interpret physical phenomena.

It was John Dalton (1766–1844), an English schoolteacher, who forcefully revived the idea of the atom. By Dalton's time experimental results had gained a position of greater respect than authoritative opinions. More clearly than any before him, Dalton was able to explain general observations, experimental results, and laws relative to the composition of matter. Dalton was particularly influenced by the experiments of two Frenchmen, Antoine Lavoisier (1743–1794) and Joseph Louis Proust (1754–1826). We shall look at the major contributions of these two experimentalists before we examine Dalton's theory.

ANTOINE-LAURENT LAVOISIER: THE LAW OF CONSERVATION OF MATTER IN CHEMICAL CHANGE

There are many reasons why Antoine-Laurent Lavoisier has been acclaimed the father of chemistry. He clarified the confusion over the cause of burning. He wrote an important textbook of chemistry, *Elementary Treatise on Chemistry.* He was the first to use systematic names for the elements and a few of their compounds. While he made still other contributions, his most notable achievement was to show the importance of very accurate weight measurements of chemical changes. His work began the process of establishing chemistry as a quantitative science.

Antoine-Laurent Lavoisier

Lavoisier weighed the chemicals in such changes as the decomposition of mercury oxide by heat into mercury and oxygen.

$$\underset{\text{MERCURY OXIDE}}{2\,HgO} \rightarrow \underset{\text{MERCURY}}{2\,Hg} + \underset{\text{OXYGEN}}{O_2}$$

Very accurate measurements showed that the total weight of all the chemicals involved remained constant during the course of the chemical change. Similar measurements on many other chemical reactions led Lavoisier to the summarizing statement now known as the **Law of Conservation of Matter:** *Matter is neither lost nor gained during a chemical reaction.* In other words, if one weighed all of the products of a chemical reaction — solids, liquids, and gases — the total would be the same as the weight of the reactants. Substances can be destroyed or created in a

chemical reaction, but matter cannot. In an atomic view, a chemical reaction was just a recombination of atoms. As a further example of the law of conservation of matter, consider Figure 3–1.

THE PERSONAL SIDE

With all of his success, Lavoisier had his problems and disappointments. His highest goal, that of discovering a new element, was never achieved. He lost some of the esteem of his colleagues when he was accused of saying the work of someone else was his own. In 1768, he invested half a million francs in a private firm retained by the French government to collect taxes. He used the earnings (about 100,000 francs a year) to support his research. Although Lavoisier was not actively engaged in tax collecting, he was brought to trial as a "tax-farmer" during the French Revolution. Lavoisier, along with his father-in-law and other tax-farmers, was guillotined on May 8, 1794, just two months before the end of the revolution.

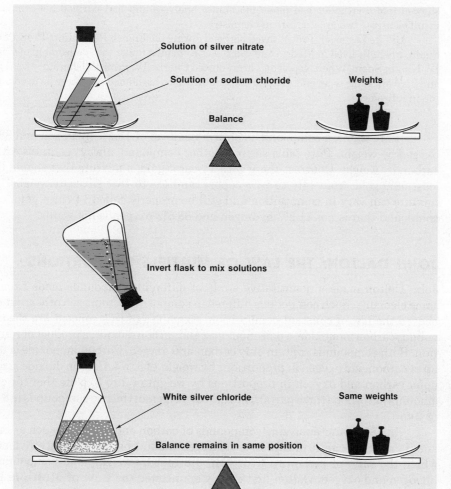

FIGURE 3–1 Mixing a solution of sodium chloride with a solution of silver nitrate produces a new substance, solid silver chloride, but the total weight of matter remains the same.

JOSEPH LOUIS PROUST: THE LAW OF CONSTANT COMPOSITION

Following the lead of Lavoisier, several chemists investigated the quantitative aspects of compound formation. One such study, made by Proust in 1799, involved copper carbonate. Proust discovered that, regardless of how copper carbonate was prepared in the laboratory or how it was isolated from nature, copper carbonate always contained five parts of copper, four parts of oxygen, and one part of carbon by weight. His careful analyses of this and other compounds led to the belief that a given compound has an unvarying composition. These and similar discoveries are summarized by the **Law of Constant Composition:** *In a compound, the constituent elements are always present in a definite proportion by weight.*

THE PERSONAL SIDE

Compounds have constant composition, while mixtures may have variable composition.

Proust's generalization has been verified many times for many compounds since its formulation, but its acceptance was delayed by controversy. Comte Claude Louis Berthollet (1748–1822), an eminent French chemist and physician, believed and strongly argued that the nature of the final product was determined by the amount of reacting materials one had at the beginning of the reaction. The running controversy between Proust and Berthollet reached major proportions, but more careful measurements supported Proust. Proust showed that Berthollet had made inaccurate analyses and had purified his compounds insufficiently — two great errors in chemistry.

Unlike Lavoisier, Proust saved his head during the French Revolution. Proust fled to Spain, where he lived in Madrid and worked as a chemist under the sponsorship of Charles IV, King of Spain. When Napoleon's army ousted Charles IV, Proust's laboratory was looted and his work came to an end. Later, Proust returned to his homeland, where he lived out his life in retirement.

Pure water, a compound, is always made up of 11.2% hydrogen and 88.8% oxygen by weight. Pure table sugar, another compound, always contains 42.11% carbon by weight. Contrast these with 14-carat gold, a mixture that should be at least 58% gold, from 14% to 28% copper, and 4% to 28% silver by weight. This mixture can vary in composition and still be properly called 14-carat gold, but a compound that is not 11.2% hydrogen and 88.8% oxygen is not water.

JOHN DALTON: THE LAW OF MULTIPLE PROPORTIONS

John Dalton

Methane is the main component of natural gas. Ethylene is the only component of polyethylene.

John Dalton made a quantitative study of different compounds made from the same elements. Such compounds differed in composition from each other, but each obeyed the law of constant composition. Examples of this concept are the compounds carbon monoxide, a poisonous gas, and carbon dioxide, a product of respiration. Both compounds contain only carbon and oxygen. Carbon monoxide is made up of carbon and oxygen in proportions by weight of 3 to 4. Carbon dioxide is made up of carbon and oxygen in proportions by weight of 3 to 8. Note that for equal amounts of carbon (three parts), the ratio of oxygen in the two compounds is 8 to 4, or 2 to 1.

In 1803, after analyzing compounds of carbon and hydrogen such as methane (in which the ratio of carbon to hydrogen is 3 to 1 by weight) and ethylene (in which the ratio of carbon to hydrogen is 6 to 1 by weight) and compounds of nitrogen and oxygen, Dalton first clearly enunciated the **Law of Multiple Proportions:** *In the formation of two or more compounds from the same elements, the*

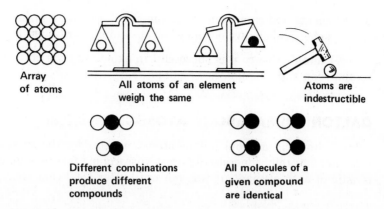

FIGURE 3–2
Features of Dalton's atomic theory. Isotopes have been ignored.

Array of atoms

All atoms of an element weigh the same

Atoms are indestructible

Different combinations produce different compounds

All molecules of a given compound are identical

weights of one element that combine with a fixed weight of a second element are in a ratio of small whole numbers (integers) such as 2 to 1, 3 to 1, 3 to 2, or 4 to 3.

DALTON'S ATOMIC THEORY

Why do the laws of conservation of matter, constant composition, and multiple proportions exist? How can they be explained? John Dalton employed the idea of atoms and endowed them with properties that enabled him to explain these chemical laws (Fig. 3–2).

THE PERSONAL SIDE

While Lavoisier is considered the father of chemical measurement, Dalton is considered the father of chemical theory. Dalton, a gentle man and a devout Quaker, gained acclaim because of his work. He made careful measurements, kept detailed records of his research, and expressed them convincingly in his writings. However, he was a very poor speaker and was not well received as a lecturer. When Dalton was 66 years old, some of his admirers sought to present him to King William IV. Dalton resisted because he would not wear the court dress. Since he had a doctor's degree from Oxford University, the scarlet robes of Oxford were deemed suitable, but a Quaker could not wear scarlet. Dalton, being colorblind, saw scarlet as gray, so he was presented in scarlet to the court but in gray to himself. This remarkable man was, in fact, the first to describe color blindness. He began teaching in a Quaker school when only 12 years old, discovered a basic law of physics, the law of partial pressure of gases, and helped found the British Association for the Advancement of Science. He kept over two hundred thousand notes on meteorology. Despite his accomplishments he shunned glory and maintained he could never find time for marriage.

The major points of Dalton's theory, presented in modernized statements, are:

* memorize

1. Matter is composed of indestructible* particles called atoms.
2. All atoms of a given element have the same properties such as size, shape, and weight,† which differ from the properties of atoms of other elements.

* Radioactive atoms are self-destructive. Dalton had no knowledge of this phenomenon.
† We now know that all of the atoms of the same element do not necessarily have the same weight. The idea of isotopes is introduced later in this chapter.

3. Elements and compounds are composed of definite arrangements of atoms, and chemical change occurs when the atomic arrays are rearranged.

Dalton's theory was successful in explaining the three laws of chemical composition and reaction. See Figure 3–3.

DALTON'S IDEA ABOUT ATOMIC WEIGHTS: THE FOLLOW-UP

John Dalton's idea about unique atomic weights for the atoms of the different elements naturally generated interest in searching for the atomic weight characteristic of each element. It was not until after 1860 that chemists developed a

Law	Statement of Law	Explanation of Law
Law of Conservation of Matter	Matter is neither lost nor gained in a chemical change.	A chemical change is the result of a new arrangement of the same atoms present initially; hence, the weight is the same before and after the change.

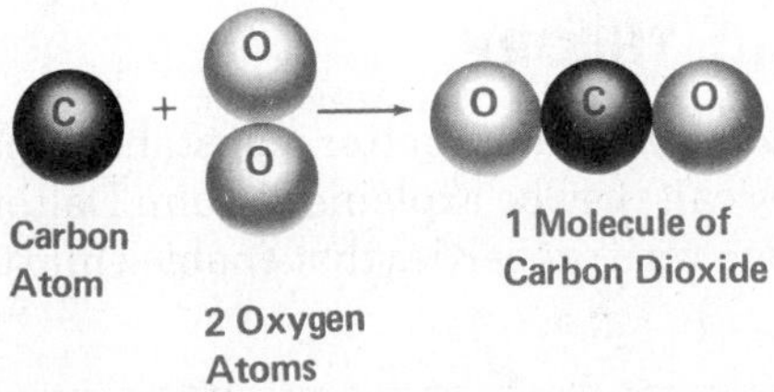

Carbon Atom + 2 Oxygen Atoms → 1 Molecule of Carbon Dioxide

Law	Statement of Law	Explanation of Law
Law of Constant Composition	When two or more elements combine to form a given compound, the ratio of the weights of the elements involved is always the same.	The smallest unit of a compound is a molecule. It has a fixed ratio of atoms, hence a fixed ratio of weights. Any larger sample of this compound would merely represent a multiple of the weights in the same ratio.

Carbon monoxide

Law	Statement of Law	Explanation of Law
Law of Multiple Proportions	In the formation of two or more compounds from the same elements, the weights of one element that combine with a fixed weight of a second element are in a ratio of integers such as 2:1, 3:1, 3:2, or 4:3.	If, for example, the first compound has a ratio of one atom of C to one atom of O (above), and a second compound has a ratio of one atom of C to two atoms of O (below), then for a fixed number of atoms of C, the ratio of atoms of O (and the weights of O) is a ratio of integers: 1:2.

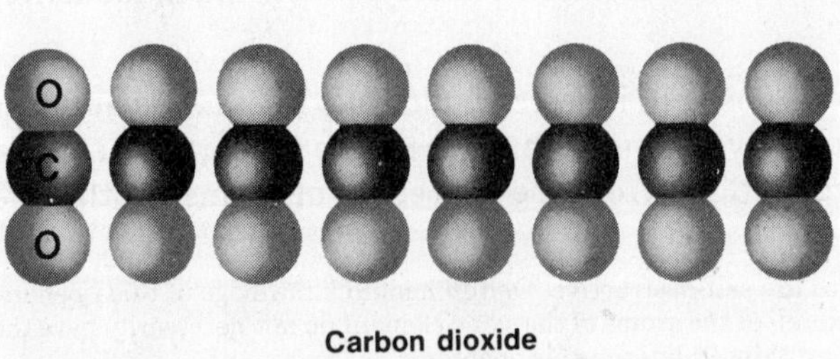

Carbon dioxide

FIGURE 3–3   John Dalton's explanation of three laws of chemistry in terms of atoms.

consistent set of atomic weights, although several notable attempts were made before that, including an early attempt by Dalton himself. In September 1860, many of the most brilliant minds in chemistry met in Karlsruhe, Germany, to discuss the inconsistencies in the atomic weights proposed at that time. There were differences of opinion on whether the formula of water was HO or H_2O, whether hydrogen gas was H_2 or H, and whether oxygen gas was O_2 or O. Water is 88.8% oxygen and 11.2% hydrogen by weight — a firmly established experimental fact by that time. If water is HO, as Dalton argued (based on his belief that the simplest formula is likely to be the correct one), then the weight of an oxygen atom should be about eight times that of a hydrogen atom:

$$\frac{\text{weight of an oxygen atom}}{\text{weight of a hydrogen atom}} = \frac{88.8}{11.2} = \frac{7.9}{1}$$

If the formula for water is H_2O, as the scientist Amedeo Avogadro (1776–1856) had proposed in 1811, then one oxygen atom is 88.8% of the molecule but two hydrogen atoms are 11.2%. Each hydrogen atom would be 1/2(11.2%), or 5.6%. With the formula H_2O, then, an oxygen atom would be about 16 times heavier than a hydrogen atom:

$$\frac{\text{weight of an oxygen atom}}{\text{weight of a hydrogen atom}} = \frac{88.8}{5.6} = \frac{15.9}{1}$$

Near the end of the meeting at Karlsruhe, Stanislao Cannizzaro (1826–1910) argued for the ideas of Avogadro. In spite of his arguments, which later proved to be correct, the confusion about atomic weights was not resolved during the conference. At the close of the meeting, however, a friend of Cannizzaro named Angelo Pavesi distributed copies of a paper written by Cannizzaro two years earlier. Several years later, chemists finally accepted H_2O as the formula of water, and a consistent set of atomic weights was generally agreed on and used.

The fact that an oxygen atom is about 16 times heavier than a hydrogen atom does not tell us the weight of either atom. These are relative weights in the same way that a grapefruit may weigh twice as much as an orange. This information gives neither the weight of the grapefruit nor that of the orange. However, if a specific number is *assigned* as the weight of any particular atom, this fixes the numbers assigned to the weights of all other atoms. The standard for comparison of relative atomic weights was for many years the weight of the oxygen atom, which was taken as 16.0000 atomic weight units. This allowed the lightest atom, hydrogen, to have an atomic weight of 1.008, or approximately 1.

The modern set of atomic weights (inside the back cover) is an outgrowth of the set of weights begun in the 1860s. The present atomic weight scale, adopted by scientists worldwide in 1961, is based on assigning the weight of a particular kind of carbon atom, the carbon-12 atom, as exactly 12 atomic weight units. On this scale, an atom of magnesium (Mg) with an atomic weight of about 24 has twice the weight of a carbon-12 atom. An atom of titanium (Ti) with an atomic weight of 48 has four times the weight of a carbon-12 atom.*

---

* What we are talking about here is really atomic mass. Atomic weight, an uncorrected misnomer from the past, persists today in, for example, the "Tables of Atomic Weights" published in most chemistry textbooks. If they were really atomic weights, they would change value wherever the force of gravity changes on earth (less at the equator, more at the poles). Instead, we have only one table for the whole world, which means atomic weights are really atomic masses. In this text, we shall use both terms: *atomic weight,* because it is a practice of chemistry and a term possibly more familiar to students, and *atomic mass,* where it seems necessary to clarify the thought.

Stanislao Cannizzaro (1826–1910), Italian chemist whose work resulted in the clarification of the atomic weight scale.

ATOMS ARE DIVISIBLE — DALTON AND THE GREEKS WERE WRONG

Dalton's concept of the indivisibility of atoms was severely challenged by the subsequent discoveries of radioactivity and cathode rays and was even in conflict with some previously known electrical phenomena such as static electrical charges.

Electrical charge was first observed and recorded by the ancient Egyptians, who noted that amber, when rubbed, attracted light objects. A bolt of lightning, a spark between a comb and hair in dry weather, and a shock on touching a doorknob are all results of the discharge of a buildup of electrical charge.

The two types of electrical charge had been discovered by the time of Benjamin Franklin (1706–1790). He named them positive (+) and negative (−) because they appear as opposites, in that they can neutralize each other. The existence and nature of the two kinds of charge, and their effects on each other, can be shown with a simple electroscope (Fig. 3–4). When a hard rubber rod is rubbed vigorously with silk and allowed to touch the lightweight balls, the balls spring apart immediately.

FIGURE 3–4 Effects of charged matter on other charged matter. Like charges repel. Unlike charges attract.

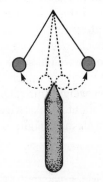

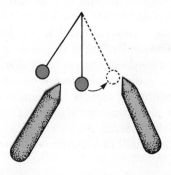

Neutral

Rubber rod rubbed with silk or wool and touched to pith balls.

Rubber rod rubbed with silk and touched to pith balls repels balls which have the same charge as the rod.

Rubber rod rubbed with wool attracts pith balls which have a charge opposite to the charge on the rod.

Marie Curie's laboratory. (From Weeks and Leicester: *Discovery of the Elements,* 7th ed. Easton, PA, Journal of Chemistry Education, 1968.)

The touching allowed the rod and the balls to share the same type of charge (positive). If the rod is then brought near one of the balls, the ball moves away from the rod. This movement indicates that *like charges repel.*

If the same rod is now rubbed vigorously with wool and brought near the charged balls, they move toward the rod. The opposite type of charge is now on the rod. The generalization is: *Unlike charges attract, and like charges repel.*

The discovery of natural radioactivity, a spontaneous process in which some natural materials give off very penetrating radiations, indicated that atoms must have some kind of internal structure. Henri Becquerel (1852–1908) discovered this property in natural uranium and radium ores in 1896. His student, Marie Curie (1867–1934), isolated the radioactive element radium and some of its pure compounds. It turns out that radioactivity is characteristic of the elements, not the compounds, and that about 25 elements are naturally radioactive.

Radioactive elements commonly emit alpha, beta, and gamma rays, as shown in Figure 3–5. Alpha and beta rays are composed of particles with charges and masses, while gamma rays have no detectable mass and are more like light. Alpha particles have a mass of 4 on the carbon-12 atomic weight scale, positive charge, and low penetrating power (they will not penetrate skin, for example). In

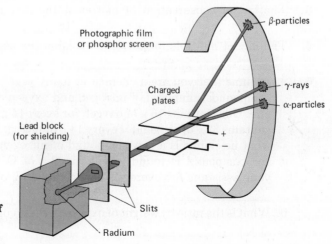

FIGURE 3–5 Separation of alpha, beta, and gamma rays by electrical field.

TABLE 3–1 Summary of Properties of Alpha Particles, Beta Particles, and Gamma Rays

	CHARGE	RELATIVE MASS	SYMBOLS
Alpha particle	Positive (+2)	4	α, $^4_2\alpha$, ^4_2He
Beta particle	Negative (−1)	0.0005	β, $^0_{-1}\beta$
Gamma ray	Neutral (0)	0	γ, $^0_0\gamma$

the arrangement shown in Figure 3–5, they are attracted toward the negatively charged plate. Beta particles have a mass of 0.0005 on the carbon-12 atomic weight scale, negative charge (they are attracted toward the positive plate), and enough penetrating power to go through kitchen-strength aluminum foil. Gamma rays are a type of electromagnetic energy like light and X rays, but more penetrating. Gamma rays can penetrate a considerable thickness of aluminum and even thin sheets of lead. They are not deflected at all by charged plates.

Cathode rays, as we shall see subsequently, are similar to beta rays in that both are composed of negatively charged particles, with identical charges and masses.

The discoveries of natural radiation, cathode rays, and electrical charge are evidence that atoms can be divided and may even divide spontaneously. The smallest atom is 1836 times more massive than the beta or cathode-ray particle. Therefore, beta (and cathode-ray) particles appear to be subatomic in origin. The properties of the three fundamental subatomic particles are summarized in Table 3–1.

SELF-TEST 3–A

1. Two Greek philosophers who were influential in advocating the concept of atoms were _____ and _____.

2. The Greek approach to the "discovery" of atoms can best be described as:
   a. Experimentation
   b. Philosophy (use of logic)
   c. Direct observation of atoms
   d. Consistent explanation of well-known, established laws of nature
   e. Deductive reasoning

3. The law of conservation of matter states that matter is neither lost nor _____ in a _____ reaction.

4. The law of multiple proportions explains the existence of compounds like _____ and _____.

5. a. Assume that you are a chemist of many years ago. Your field of study is compounds composed of nitrogen and oxygen only. You know about several. One contains 16 g of oxygen for every 14 g of nitrogen, while another contains 32 g of oxygen for every 14 g of nitrogen. Your assistant discovers what he claims is a new compound of nitrogen and oxygen. On analysis, the compound is found to contain 8 g of O for every 14 g of N. Has your assistant discovered a new compound, or is it one of the others?

   _____

   b. What is the ratio by weight of oxygen in these compounds for a given weight of N? _____

c. What fundamental law of chemistry is illustrated by a comparison of the compounds? _____

6. According to Dalton's atomic theory, what happens to atoms during a chemical change? Select one:
   a. Atoms are made into new and different kinds of atoms.
   b. Atoms are lost.
   c. Atoms are gained.
   d. Atoms are recombined into different arrangements.

7. According to Dalton's atomic theory, a compound has a definite percentage by weight of each element because
   a. All atoms of a given element weigh _____.
   b. All molecules of a given compound contain a definite number and kind of

   _____.

8. Like charges _____; unlike charges _____.

9. The three types of radiation from a radioactive element such as radium are _____, _____, and _____, of which _____ pass through an electrical field without being deflected.

THE ELECTRON — THE FIRST SUBATOMIC PARTICLE DISCOVERED

The first ideas about electrons came from experiments with cathode-ray tubes. A forerunner of neon signs, fluorescent lights, and TV picture tubes, a typical cathode-ray tube is a partially evacuated glass tube with a piece of metal sealed in each end (Fig. 3–6). The pieces of metal are called electrodes; the one given a negative charge is called the **cathode,** and the one given a positive charge is called the **anode.**

Cathode rays are streams of the negatively charged particles called *electrons*.

If a sufficiently high electrical voltage is applied to the electrodes, an electrical discharge can be created between them. This discharge appears to be a stream of particles emanating from the cathode. This cathode ray will cause gases and fluorescent materials to glow and will heat metal objects in its path to red heat. Cathode rays travel in straight lines and cast sharp shadows. Unlike light, however, cathode rays are attracted to a positively charged plate. This led to the conclusion that cathode rays are negatively charged.

Careful microscopic study of a screen that emits light when struck by cathode rays shows that the light is emitted in tiny, random flashes. Thus, not only are cathode rays negatively charged, but they also appear to be composed of particles, each one of which produces a flash of light upon collision with the material of the screen. The cathode-ray particles became known as **electrons.**

The charge and mass of an electron were determined by a combination of experiments by Sir Joseph John Thomson in 1897 and by Robert Andrews Millikan in 1911. Both scientists were awarded Nobel prizes, Thomson in 1906 and Millikan in 1923.

By using a specially designed cathode-ray tube (Fig. 3–7), Thomson applied electrical and magnetic fields to the rays. Using the basic laws of electricity and magnetism, he determined the **charge-to-mass ratio** of the electrons. He was able to measure neither the absolute charge nor the absolute mass of the electron, but he established the ratio between the two numbers and made it possible to calculate

Thomson discovered the charge-to-mass ratio of the electron.

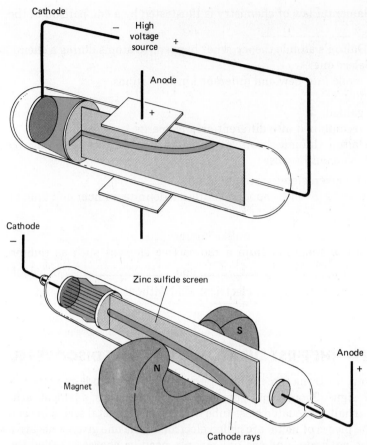

FIGURE 3–6 Deflection of a cathode ray by an electric field and by a magnetic field. When an external electric field is applied, the cathode ray is deflected toward the positive pole. When a magnetic field is applied, the cathode ray is deflected from its normal straight path into a curved path.

either one if the other could ever be measured. What Thomson did for the concept of the electron is like showing that a peach weighs 40 times more than its seed. What is the weight of the peach? What is the weight of the seed? Neither is known, but if it can be determined by other means that the peach weighs 120 g, then the weight of the seed, by ratio, must be 3 g.

Electrons are present in all of the elements.

An important part of Thomson's experimentation was his use of 20 different metals for cathodes and of several gases to conduct the discharge. Every combination of metals and gases yielded the same charge-to-mass ratio for the cathode rays. This led to the belief that electrons are common to all of the metals and gases used in the experiments, and probably to all atoms in general. Thus it appeared that the electron was an atomic building block.

THE PERSONAL SIDE

Sir Joseph John Thomson (1856–1940) was a scientist chiefly working in mathematics until he was elected Cavendish Professor at Cambridge University in 1884. He was not skilled at experimental techniques, but his ability to suggest experiments and interpret their results

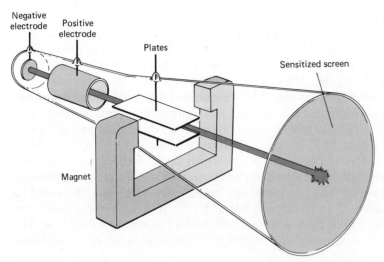

FIGURE 3–7 J. J. Thomson experiment. Electric field, applied by plates, and magnetic field, applied by magnet, cancel each other's effects to allow cathode ray (electron beam) to travel in straight line.

led to the discovery of the electron. In 1897 Thomson wrote in the *Philosophical Magazine*, "We have in the cathode rays a new state, a state in which the subdivision of matter is carried very much further than in the ordinary gaseous state — this matter being the substance from which the chemical elements are built up." Thomson won the Nobel prize in 1906 and was knighted in 1908. Seven of his research assistants later won Nobel prizes for their own research work. Thomson was buried in Westminster Abbey near the grave of Sir Isaac Newton.

Millikan measured the fundamental charge of matter — the charge on an electron. A simplified drawing of his apparatus is shown in Figure 3–8. The experiment consisted of measuring the electrical charge carried by tiny drops of oil that are suspended in an electrical field. By means of an atomizer, oil droplets were sprayed into the test chamber. As the droplets settled slowly through the air, high-energy X rays were passed through the chamber to charge the droplets negatively (the X rays caused air molecules to give up electrons to the oil). By using a beam of light and a small telescope, Millikan could study the motion of a single

Millikan measured the charge on an electron.

FIGURE 3–8 Millikan's oildrop experiment for determining the charge on an electron. The pull of gravity on the drop is balanced by the upward electrical force. Drawing does not include X-ray source for charging droplets.

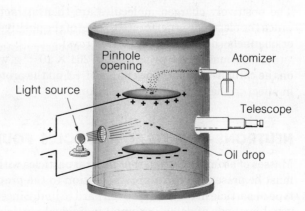

droplet. When the electrical charge on the plates was increased enough to balance the effect of gravity, a droplet could be suspended motionless. At this point, the gravitational force would equal the electrical force. Measurements made in the motionless state, when inserted into equations for the forces acting on the droplet, enabled Millikan to calculate the charge carried by the droplet.

Millikan found different amounts of negative charge on different drops, but the charge measured each time was always a whole-number multiple of a very small basic unit of charge. The *largest* common divisor of all charges measured by this experiment was 1.60×10^{-19} coulomb (the coulomb is a charge unit). Millikan assumed this to be the fundamental charge, which is the charge on the electron.

With a good estimate of the charge on an electron and the ratio of charge-to-mass as determined by Thomson, the very small mass of the electron could be calculated. The mass of an electron is 9.11×10^{-28} g. On the carbon-12 relative scale, the electron would have a weight of 0.000549 atomic weight units. The negative charge on an electron of -1.60×10^{-19} coulomb is set as the standard charge of -1.

PROTONS — THE ATOM'S POSITIVE CHARGE

The first experimental evidence of a fundamental positive particle came from the study of canal rays. A special type of cathode-ray tube produces canal rays (Fig. 3–9). The cathode is perforated, and the tube contains a gas at very low pressure. When high voltage is applied to the tube, cathode rays can be observed between the electrodes as in any cathode-ray tube. On the other side of the perforated cathode, a different kind of a ray is observed. These rays are attracted to a negative plate brought alongside the rays. The rays must therefore be composed of positively charged particles. Each gas used in the tube gives a different charge-to-mass ratio for the positively charged particles. When hydrogen gas was used, the largest charge-to-mass ratio was obtained, indicating that hydrogen provides the positive particles with the smallest mass. This particle was considered to be the fundamental positively charged particle of atomic structure, and was called a **proton** (from Greek for "the primary one").

Experiments on canal rays were begun in 1886 by E. Goldstein and further work was done later by W. Wien. The production of canal rays is caused by high-energy electrons moving from the negative cathode to the positive anode, hitting the molecules of gases occupying the tube. Electrons are knocked from some atoms by the high-energy electrons, leaving each molecule with a positive charge. The positively charged molecules are then attracted to the negative electrode. Since the electrode is perforated, some of the positive particles go through the holes or channels (hence the name *canal rays*).

The mass of the proton is 1.67261×10^{-24} g, which is 1.00727 relative weight on the carbon-12 scale. The charge of $+1$ on the proton is equal in size but opposite in effect to the charge on the electron.

NEUTRONS — NEUTRAL PARTICLES FOUND IN MOST ATOMS

Masses of atoms indicated that neutral particles with about the mass of the proton must be present in the atom in addition to the protons and electrons. This third type of particle in the atom proved hard to find. Since the particle has no charge, the usual methods of detecting small individual particles could not be used.

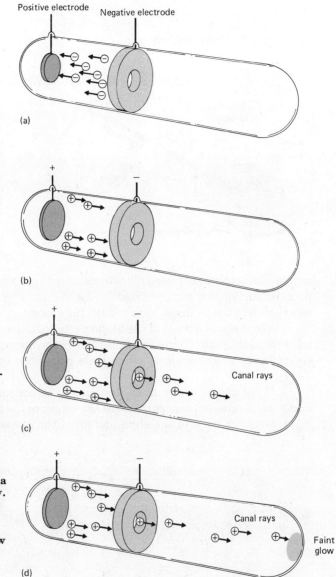

Positive electrode   Negative electrode

(a)

(b)

(c)   Canal rays

(d)   Canal rays   Faint glow

FIGURE 3–9 *(a)* **Electrons rush from the negative electrode to the positive electrode as a result of high voltage.** *(b)* **Electrons collide with gas molecules to produce positive ions, which are accelerated toward the negative electrode.** *(c)* **Some of the positive ions escape capture by the electrode and rush through the opening as a result of their kinetic energy.** *(d)* **Some of the positive ions in the positive ray collide with gas molecules to produce a characteristic glow and strike the end of the glass tube to produce a luminous spot.**

It remained for James Chadwick, in 1932, to devise a clever experiment that produced neutrons by a nuclear reaction and then detected them by having the neutrons knock hydrogen ions, a detectable species, out of paraffin.

A neutron has no electrical charge and has a mass of 1.67492×10^{-24} g, which is a relative weight of 1.00867 on the carbon-12 scale.

Paraffin is the hydrocarbon that seals home-canned strawberry preserves.

THE NUCLEUS — AN AMAZING ATOMIC CONCEPT

When Ernest Rutherford and his students directed alpha particles toward a very thin sheet of gold foil in 1909, they were amazed to find a totally unexpected result (Fig. 3–10). As they had expected, the paths of most of the alpha particles were only slightly changed as they passed through the gold foil. The extreme deflection of a few of the alpha particles was a surprise. Some even "bounced" back toward the

Alpha particles are scattered by the nuclei of the gold atoms.

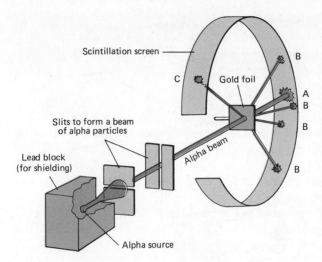

FIGURE 3–10
Rutherford's gold foil experiment. A cylindrical scintillation screen is shown for simplicity; actually, a movable screen was employed. Most of the alpha particles pass straight through the foil to strike the screen at point A. Some alpha particles are deflected to points B, and some are even "bounced" backward to points such as C.

source. Rutherford expressed his astonishment by stating that he would have been no more surprised if someone had fired a 15-inch artillery shell into tissue paper and then found it in flight back toward the cannon.

What allowed most of the alpha particles to pass through the gold foil in a rather straight path? According to Rutherford's interpretation, the atom is mostly empty space and, therefore, offers little resistance to the alpha particles (Fig. 3–11).

What caused a few alpha particles to be deflected? According to Rutherford's interpretation, concentrated at the center of the atom is a **nucleus** containing most of the mass of the atom and all of the positive charge. When an alpha

Alpha-particle scattering can be explained if the nucleus occupies a very small volume of the atom.

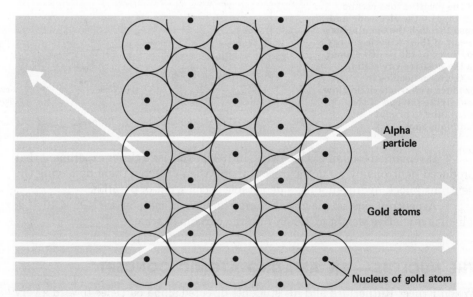

FIGURE 3–11   Rutherford's interpretation of how alpha particles interact with atoms in a thin gold foil line up? Actually, the gold foil was about 1000 atoms thick. For illustration purposes, points are used to represent the gold nuclei, and the path widths of the alpha particles are drawn much larger than scale.

particle passes near the nucleus, the positive charge of the nucleus repels the positive charge of the alpha particle; the path of the smaller alpha particle is deflected. The closer an alpha particle comes to a target nucleus, the more it is deflected. Those alpha particles that meet a nucleus head on bounce back toward the source as a result of the strong positive-positive repulsion, since the alpha particles do not have enough energy to penetrate the nucleus.

Rutherford's calculations, based on the observed deflections, indicate that the nucleus is a very small part of an atom. An atom occupies about a million million times more space than does a nucleus; the radius of an atom is about 10,000 times greater than the radius of its nucleus. Thus, if a nucleus were the size of a baseball, then the edges of the atom would be about one third of a mile away. And most of the space in between would be absolutely empty.

Since the nucleus contains most of the mass and all of the positive charge of an atom, the nucleus must be composed of the most massive atomic particles, the protons and neutrons. The electrons are distributed in the near-emptiness outside the nucleus.

Truly, Rutherford's model of the atom was one of the most dramatic interpretations of experimental evidence to come out of this period of significant discoveries.

Lord Rutherford (Ernest Rutherford, 1871–1937) was Professor of Physics at Manchester when he and his students discovered the scattering of alpha particles by matter. Such scattering led to the postulation of the nuclear atom. For this work he received the Nobel prize in 1908. In 1919, Rutherford discovered and characterized nuclear transformations.

ATOMIC NUMBER—EACH ELEMENT HAS A NUMBER

The **atomic number** of an element indicates the number of protons in the nucleus of the atom, which is the same as the number of electrons outside the nucleus. The two types of particles must be present in equal numbers for the atom to be neutral in charge. Note that the periodic table of the elements, inside the back cover, is an arrangement of the elements consecutively according to atomic number. Beginning with the atomic number 1 for hydrogen, there is a different atomic number for each element.

The lightest atom is the hydrogen atom.

$$\begin{pmatrix} \text{Number of} \\ \text{electrons} \\ \text{per atom} \end{pmatrix} = \begin{pmatrix} \text{Number of} \\ \text{protons} \\ \text{per atom} \end{pmatrix} = \begin{pmatrix} \text{Atomic number} \\ \text{of the} \\ \text{element} \end{pmatrix}$$

The **atomic mass** of a particular atom is the sum of the masses of the protons, neutrons, and electrons in that atom. Since an electron has such a small mass, the atomic mass is very nearly the sum of the masses of the protons and neutrons in the nucleus. Both protons and neutrons have masses of approximately 1.0 on the atomic weight scale. See Table 3–2. Hydrogen, with an atomic weight of 1, must be composed of one proton (and no neutrons) in the nucleus and one electron outside the nucleus. Helium has an atomic number of 2 and an atomic weight of 4. The atomic number of 2 indicates two protons and two electrons per

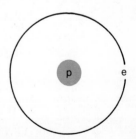

Hydrogen atom

TABLE 3–2   Summary of Properties of Electrons, Protons, and Neutrons

	RELATIVE CHARGE	RELATIVE MASS	LOCATION
Electron	−1	0.00055	Outside the nucleus
Proton	+1	1.00727	Nucleus
Neutron	0	1.00867	Nucleus

atom of helium. The atomic weight of 4 means that, in addition to the two protons in the nucleus, there are two neutrons.

$$\begin{pmatrix} \text{Approximate} \\ \text{number of} \\ \text{neutrons} \\ \text{per atom} \end{pmatrix} = \begin{pmatrix} \text{Atomic weight} \\ \text{of the} \\ \text{element} \end{pmatrix} - \begin{pmatrix} \text{Atomic number} \\ \text{of the} \\ \text{element} \end{pmatrix}$$

A notation frequently used to show the atomic mass (also called **mass number**) and atomic number of an atom uses subscripts and superscripts to the left of the symbol:

Atomic mass

$^{19}_{9}\text{F}$ ⟵——— Symbol of the element

Atomic number

For an atom of fluorine, $^{19}_{9}\text{F}$, the number of protons is 9, the number of electrons is also 9, and the number of neutrons is $19 - 9 = 10$.

Symbolism refers to individual isotopes.

ISOTOPES — DALTON NEVER GUESSED!

Many of the elements, when analyzed by a special type of canal-ray tube called a *mass spectrometer* (Fig. 3–12), are found to be composed of atoms of different masses (Fig. 3–13). Atoms of the same element having different atomic masses are called **isotopes** of that element.

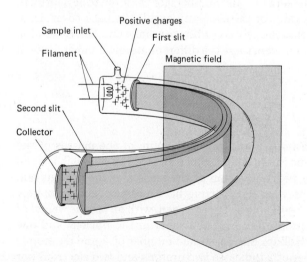

FIGURE 3–12   Mass spectrometer. Sample to be studied is injected near filament. Electrodes (not shown) subject sample to electron beam that ionizes a part of the sample by knocking electrons from neutral atoms or molecules. Electrodes are arranged to accelerate positive ions toward first slit. The positive ions that pass the first slit are immediately put into a magnetic field perpendicular to their path and follow a curved path determined by the charge-to-mass ratio of the ion. A collector plate, behind the second slit, detects charged particles passing through the second slit. The relative magnitudes of the electrical signals are a measure of the numbers of the different kinds of positive ions.

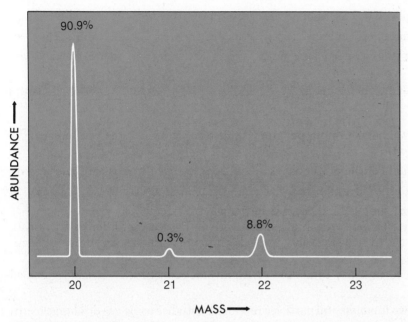

FIGURE 3–13 Mass spectrum of neon (+1 ions only). The principal peak corresponds to the most abundant isotope, neon-20. Percent relative abundance is shown. (From W. L. Masterton, and E. J. Slowinski: *Chemical Principles*. Philadelphia, Saunders College Publishing, 1973.)

The element neon is a good example to consider. A natural sample of neon gas is found to be a mixture of three isotopes of neon:

$^{20}_{10}$Ne $^{21}_{10}$Ne $^{22}_{10}$Ne

The fundamental difference between isotopes is the different number of neutrons per atom. All atoms of neon have 10 electrons and 10 protons; about 90% of the atoms have 10 neutrons, some have 11 neutrons, and others have 12 neutrons. Because they have different numbers of neutrons, they must have different masses. Note that all the isotopes have the same atomic number. They are all neon.

There are only 109 known elements, yet more than 1000 isotopes have been identified, many of them produced artificially (see Chapter 7). Some elements have many isotopes; tin, for example, has 10 natural isotopes. Hydrogen has three isotopes, and they are the only three that are generally referred to by different names: $^1_1$H is called protium, $^2_1$H is called deuterium, and $^3_1$H is called tritium. Tritium is radioactive. The natural assortment of isotopes, each having its own distinctive atomic mass, results in fractional atomic weights for many elements.

> Isotopes are atoms of the same element having different numbers of neutrons.

> The weighted average of the atomic weights of the isotopes in a natural mixture is the noninteger atomic weight of the element.

SELF-TEST 3–B

1. Isotopes of an element are atoms that have nuclei with the same number of _____ but different numbers of _____.
2. The nucleus of an atom occupies a relatively large (   ) or small (   ) fraction of the volume of the atom.

3. The positive charges in an atom are concentrated in its _____ .

4. The negatively charged particles in an atom are _____ ; the positively charged particles are _____ ; and the neutral particles are _____ .

5. In a neutral atom there are equal numbers of _____ and _____ .

6. The number of protons per atom is called the _____ number of the element.

7. The mass of the proton is _____ times the mass of the electron.

8. An atom of arsenic, $^{75}_{33}$As, has _____ electrons, _____ protons, and _____ neutrons.

9. Positive (canal) rays obtained with different gases are (different/identical), while the cathode rays obtained using different cathodes are (different/identical).

10. Cathode rays are composed of a universal constituent of matter named _____ .

11. The two fundamental particles revealed by studies using gas discharge (cathode-ray) tubes are the _____ and _____ .

12. All atoms of a given element are exactly alike. True ( ) or False ( )

WHERE ARE THE ELECTRONS?

Two major theories have been presented concerning the position, movement, and energy of electrons in an atom. The Bohr theory of the hydrogen atom was put forth in 1913 by Niels Bohr. This theory was extended and modified by Erwin Schrödinger, Werner Heisenberg, Louis de Broglie, and others in 1926. The newer theory is referred to as the quantum mechanical theory, the wave mechanical theory, or Schrödinger's theory.

The Bohr Model of the Atom

Energy of matter in motion is *kinetic* energy.

Energy stored in matter is *potential* energy.

Bohr assumed that an atom can exist only in certain energy states.

In Bohr's concept, electrons revolve around a nucleus in definite orbits, much as planets revolve around the sun. He equated classical mathematical expressions for the force tending to keep the electron traveling in a straight line and the force tending to pull the electron inward (the positive-to-negative attraction between a proton and an electron). The total energy of the atom is the kinetic energy of the electrons plus the potential energy due to the electron's separation from the nucleus. In a revolutionary sort of way, Bohr suggested that electrons stay in rather stable orbits and can have only certain energies within a given atom. According to Bohr, an electron can travel in one orbit for a long period or in another orbit some distance away for a long period, but it cannot stay for any measurable time between the two orbits. A rough analogy is provided by considering books in a bookcase. Books may rest on one shelf or on another shelf for very long periods but cannot rest between shelves. In moving a book from one shelf to another shelf, the potential energy of the book changes by a definite amount. When an electron moves from

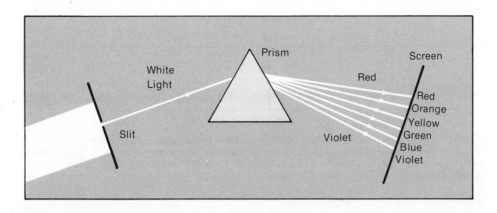

one orbit to another, its energy changes by a definite amount, called a **quantum** of energy.

Packets of light energy are called *photons*, or *quanta*.

Energy is required to separate objects attracted to each other. For example, energy is required to lift a rock from the earth, or to separate two magnets, or to pull a positive charge away from a negative charge. Bohr suggested that a very definite and characteristic amount of energy is required to move an electron from one energy level to another. The energy added to move an electron farther from its nucleus is stored in the system as potential energy (energy of position). Thus, the electron has more energy when it is in an orbit farther from the nucleus and less energy when it is in an orbit close to the nucleus. When the electron passes from an outer orbit to an inner orbit, energy is emitted from the atom, generally in the form of light energy.

Bohr used the idea of electrons moving up and down a "bookcase" of energy levels corresponding to orbits to explain the observable bright-line emission spectrum of hydrogen. A **spectrum** is the display produced when light is separated or dispersed into its component colors. The spectrum of white light is the rainbow display of separated colors shown in Figure 3-14. An **emission** spectrum is observed when the light emitted by atoms energized by a flame or an electric arc is allowed to pass through a narrow vertical slit and then through a prism of glass or quartz. If sunlight or light from a white-hot solid is dispersed, all of the colors of the rainbow are seen. This is a **continuous emission spectrum.** However, if the light from an energized gaseous element is dispersed, only a few colored lines are produced, the lines being separated by black spaces. This is a **bright-line emission spectrum** (Fig. 3-15). Each line is a pure color and is really an image of the slit in that particular color. A quantum of light of any given color has a characteristic energy that is different from the energy of a quantum of any other color.

According to Bohr, the light forming the lines in the bright-line emission spectrum of hydrogen comes from electrons moving toward the nucleus after having first been energized and pushed to orbits farther from the nucleus (Fig. 3-16). A movement between two particular orbits involves a definite quantum of energy. Each time an electron moves from one orbit to another orbit closer to the nucleus, energy loss occurs and a quantum of light having a characteristic energy (and color) is emitted. Transitions toward the nucleus between two outer orbits emit quanta having smaller characteristic energies. Transitions from an outer orbit to orbits near the nucleus emit quanta having larger characteristic energies. Each line corresponds to its own particular energy of the same-sized quanta of light.

Dispersed light produces a *spectrum.*

Niels Bohr (1885-1962) received his doctor's degree the same year that Rutherford announced his discovery of the atomic nucleus, 1911. After studying with Thomson and Rutherford in England, Bohr formulated his model of the atom. Bohr returned to the University of Copenhagen and, as Professor of Theoretical Physics, directed a program that produced a number of brilliant theoretical physicists. He received the Nobel prize for physics in 1922.

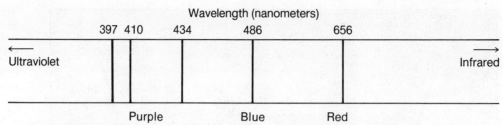

FIGURE 3–15   The visible portion of the hydrogen bright-line emission spectrum. Frequency and energy increase to the left; wavelength increases to the right. There are 7 more lines to the left in the ultraviolet range and 13 more lines to the right in the infrared range.

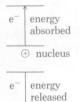

Not only could Bohr explain the cause of the lines in the bright-line emission spectrum of hydrogen, but he also calculated the expected wavelengths of the lines. He expressed the results of his calculations in the alternate view of the nature of light, its wave nature. The wave properties of light — wavelength, frequency, and speed — are considered in Figure 3–17 as they apply to any phenomenon possessing wave properties (water waves, sound waves from violin strings, radio waves, and light).

With brilliant imagination, Bohr applied a little algebra and some classical mathematical equations of physics to his tiny solar-system model of the hydrogen atom. The unprecedented requirement was that only a few allowable paths (quantized orbits) are available in which electrons can move stably around the nucleus. A further requirement was that energy differences (quanta) existed between any two

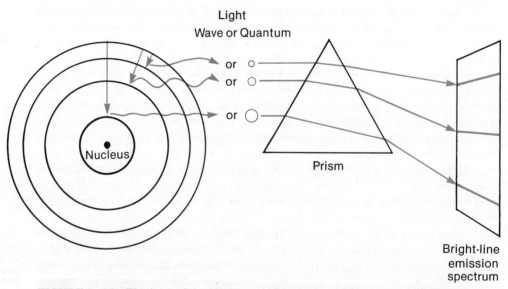

FIGURE 3–16   The formation of light according to the atomic theory. Electrons previously energized to higher energy levels make transitions back toward the nucleus (three transitions are shown). The decrease in potential energy during a transition is transformed into light. Transitions between different energy levels produce different-sized quanta of light. When the quanta are dispersed by being passed through a prism, the bright lines can be seen on film or on a spectroscope. Billions of transitions per second make each line bright enough (sufficient same-sized quanta emitted) for each line to be detected.

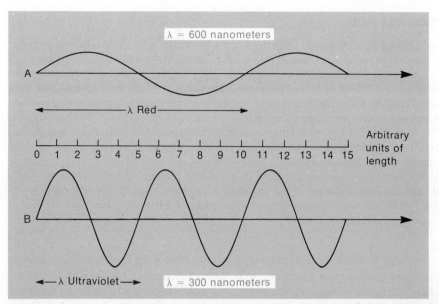

FIGURE 3–17 The wave theory of light considers light to be waves of wavelength λ (lambda) vibrating at right angles to their path of motion. Red light, *A*, completes one vibration or wave in the same distance and time that involve two complete waves of ultraviolet light, *B*. Ultraviolet light is of shorter wavelength than violet light. A nanometer is 10^{-9} meter.

orbits. Execution of the mathematics of the model produced the predicted wavelengths of the lines in the hydrogen spectrum, some of which are shown in Table 3–3. Note the close agreement between the measured values and the values predicted by the calculations of the Bohr theory. He calculated the wavelengths of all lines that had been observed with remarkably close agreement. Niels Bohr had tied the unseen (the interior of the atom) with the seen (the observable lines in the hydrogen spectrum) — a fantastic achievement.

Bohr's theoretical calculations for hydrogen spectral lines agreed amazingly well with experimental data.

The Bohr theory was accepted almost immediately after its presentation, and Bohr was awarded the Nobel prize in physics in 1922 for his contribution to the understanding of the hydrogen atom.

TABLE 3–3 Agreement Between Bohr's Theory and the Lines of the Hydrogen Spectrum*

CHANGES IN ENERGY LEVELS	WAVELENGTH PREDICTED BY BOHR'S THEORY (nm)	WAVELENGTH DETERMINED FROM LABORATORY MEASUREMENT (nm)	SPECTRAL REGION
2 → 1	121.6	121.7	Ultraviolet
3 → 1	102.6	102.6	Ultraviolet
4 → 1	97.28	97.32	Ultraviolet
3 → 2	656.6	656.7	Visible red
4 → 2	486.5	486.5	Visible blue-green
5 → 2	434.3	434.4	Visible blue
4 → 3	1876	1876	Infrared

* These lines are typical; other lines could be cited as well, with equally good agreement between theory and experiment. The unit of wavelength is the nanometer (nm), 10^{-9} meter.

THE PERSONAL SIDE

Bohr had a close call in his escape from Denmark when Hitler's forces ravished the country. Having done all he could to get Jewish physicists to safety, Bohr was still in Denmark when Hitler's army suddenly occupied the country in 1940. In 1943, to avoid imprisonment, he escaped to Sweden. There he helped to arrange the escape of nearly every Danish Jew from Hitler's gas chambers. He was later flown to England in a tiny plane, in which he passed into a coma and nearly died from lack of oxygen.

He went on to the United States, where until 1945 he worked with other physicists on the atomic bomb development at Los Alamos. His insistence upon sharing the secret of the atomic bomb with other allies, in order to have international control over nuclear energy, so angered Winston Churchill that he had to be restrained from ordering Bohr's arrest. Bohr worked hard and long on behalf of the development and use of atomic energy for peaceful purposes. For his efforts, he was awarded the first Atoms for Peace prize in 1957. He died in Copenhagen on November 18, 1962.

> Orbits are sometimes called shells, or just energy levels.

One way to decorate the interior of atoms other than hydrogen is to use the orbits devised by Bohr for hydrogen and insert the proper number of electrons. While this is at best a rough approximation, we shall find it is adequate in Chapter 5 for explaining such phenomena as some kinds of bonding and the formation of ions.

Atom Building Using the Bohr Model

Let us build up some atoms in Tinker Toy fashion. First, we need a few rules in order to play the game. Recall that the atomic number is the number of electrons (and protons) per atom of the element. Consistent with the ionization energies discussed in Chapter 5, the maximum number of electrons per orbit is $2n^2$, where n is the number of the orbit. Orbits are numbered with integers, beginning with 1 for the orbit closest to the nucleus. As practice, use the formula to check these numbers.

ORBIT	MAXIMUM NUMBER OF ELECTRONS
1	2
2	8
3	18
4	32
5	50

> The ground state is the lowest energy state for all of the electrons in an atom.

A general, overriding rule to the preceding numbers is that the outside orbit can have no more than eight electrons. When electrons are placed in orbits as close to the nucleus as possible, the electrons are said to be in their **ground state.**

You might like to follow along in Figure 3–18 (the Bohr Model column) as the building-up process is described. Hydrogen, with atomic number 1, has one electron. In its ground state, this electron is in the first orbit. The two electrons of helium are in its first orbit since the first orbit can have a maximum of two electrons.

For all atoms of other elements, two electrons are in the first orbit, and the other electrons of the atoms are assorted into higher-numbered energy levels. In atomic-number order, lithium through neon, two electrons are placed in the first orbit (which fill it), and into the second orbit are placed one, two, three, and so on to eight electrons (for Ne). Eight electrons fill the second orbit.

Placement of Electrons in Ground State

Element	Atomic Number	Bohr Model	Wave Mechanical Model
Hydrogen (H)	1	1p — 1)e	$1s^1$
Helium (He)	2	2p 2n — 2)e	$1s^2$
Lithium (Li)	3	3p 4n — 2)e 1)e	$1s^2 2s^1$
Beryllium (Be)	4	4p 5n — 2)e 2)e	$1s^2 2s^2$
Boron (B)	5	5p 6n — 2)e 3)e	$1s^2 2s^2 2p^1$
Carbon (C)	6	6p 6n — 2)e 4)e	$1s^2 2s^2 2p^2$ (or $2p_x^1 2p_y^1$)
Nitrogen (N)	7	7p 7n — 2)e 5)e	$1s^2 2s^2 2p^3$ (or $2p_x^1 2p_y^1 2p_z^1$)
Oxygen (O)	8	8p 8n — 2)e 6)e	$1s^2 2s^2 2p^4$ (or $2p_x^2 2p_y^1 2p_z^1$)
Fluorine (F)	9	9p 10n — 2)e 7)e	$1s^2 2s^2 2p^5$
Neon (Ne)	10	10p 10n — 2)e 8)e	$1s^2 2s^2 2p^6$
Sodium (Na)	11	11p 12n — 2)e 8)e 1)e	$1s^2 2s^2 2p^6 3s^1$
Magnesium (Mg)	12	12p 12n — 2)e 8)e 2)e	$1s^2 2s^2 2p^6 3s^2$
Aluminum (Al)	13	13p 14n — 2)e 8)e 3)e	$1s^2 2s^2 2p^6 3s^2 3p^1$
Silicon (Si)	14	14p 14n — 2)e 8)e 4)e	$1s^2 2s^2 2p^6 3s^2 3p^2$ (or $3p_x^1 3p_y^1$)
Phosphorus (P)	15	15p 16n — 2)e 8)e 5)e	$1s^2 2s^2 2p^6 3s^2 3p^3$ (or $3p_x^1 3p_y^1 3p_z^1$)
Sulfur (S)	16	16p 16n — 2)e 8)e 6)e	$1s^2 2s^2 2p^6 3s^2 3p^4$ (or $3p_x^2 3p_y^1 3p_z^1$)
Chlorine (Cl)	17	17p 18n — 2)e 8)e 7)e	$1s^2 2s^2 2p^6 3s^2 3p^5$
Argon (Ar)	18	18p 22n — 2)e 8)e 8)e	$1s^2 2s^2 2p^6 3s^2 3p^6$
Potassium (K)	19	19p 20n — 2)e 8)e 8)e 1)e	$1s^2 2s^2 2p^6 3s^2 3p^6 4s^1$
Calcium (Ca)	20	20p 20n — 2)e 8)e 8)e 2)e	$1s^2 2s^2 2p^6 3s^2 3p^6 4s^2$

FIGURE 3–18 Electron arrangements of the first 20 elements. The nuclear contents of a typical isotope are shown.

Sodium (Na), with 11 electrons, has the first two orbits filled with 2 and 8 electrons, respectively, and has 1 electron in the third orbit. Each succeeding element in atomic-number order, magnesium through argon, adds 1 more electron to the third orbit of its atoms.

At argon (Ar), the maximum of eight electrons in the outside orbit comes into play. When 19 electrons are present, as in an atom of potassium (K), the first orbit has 2 electrons, the second orbit has 8 electrons, and the third orbit could have the other 9 electrons (maximum of 18 electrons) if it were not the outside orbit. So to accommodate 19 electrons, there are two choices: 2-8-9 or 2-8-8-1. The first choice violates the requirement of no more than 8 in the outside orbit. The second is the proper choice. Calcium (Ca) with 20 electrons per atom has an electronic arrangement of 2-8-8-2.

Beginning with scandium (Sc), atomic number 21, and continuing through zinc (Zn), atomic number 30, 10 electrons are added to the third orbit to complete its maximum of 18. Zinc has the electronic arrangement 2-8-18-2.

You might pause in your reading here and predict the ground-state electronic arrangement of gallium (Ga), atomic number 31, and rubidium (Rb), atomic number 37, using this system.

The Wave Mechanical Model — A "Messy" Atom

Electrons are described by both particle and wave theories.

The Bohr model failed when applied to elements other than hydrogen because it could not account exactly for the line spectra of atoms with more than one electron. It was also weak in explaining why the periods (the horizontal rows) of the periodic table vary considerably in length.

After Bohr's work, a more modern, highly sophisticated mathematical theory of the atom was developed by Schrödinger, Heisenberg, Dirac, and others. In

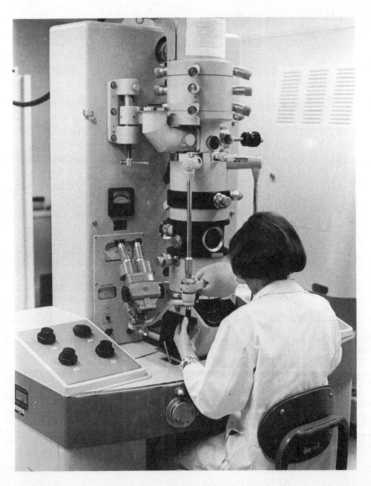

Electron microscope depends on the wave properties of electrons. (From L. H. Greenberg: *Physics for Biology and Pre-Med Students.* Philadelphia, W. B. Saunders Company, 1975, p. 554.)

this theory, electrons are treated as having both a particle and a wave nature. The locations of the electrons are treated as **probabilities,** without seeking to locate the exact spot for an electron at a given time. This approach suggested that the Bohr theory describing the electrons with fixed orbits sought more precision than nature would allow.

A Frenchman, Louis de Broglie, was the first to suggest (in 1924) that electrons and other small particles should have wave properties. In this respect, he said, electrons should behave like light, a suggestion that scientists of the time found hard to accept. However, in a few years separate experiments by George Thomson (son of J. J. Thomson) in England and Clinton Davisson in the United States justified de Broglie's hypothesis (Fig. 3–19). The electron microscopes found in many research laboratories today are built and operated on our understanding of the wave nature of the electron.

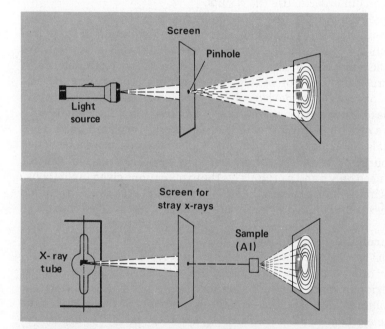

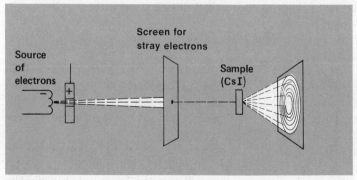

FIGURE 3–19 Similar patterns are shown by light, X rays, and electrons as each is diffracted. Diffraction is the bending and spreading of wave motion around edges. The effect is prominent when the wavelength is large compared to the size of the obstacle and small when the wavelength is short compared to the size of the obstacle. Similar effects from light, X rays, and electrons indicate a property common to all: each has a wave nature.

It should not be surprising to find that matter can be treated by both wave and particle theories (the duality of matter), since its convertible counterpart — energy — has been treated successfully by both theories for a long time. Keep in mind that we do not really know if matter or light is a wave or a particle. However, because there are limits on what we can visualize in our physical world, in talking about something like subatomic behavior we are forced to use physical models based on known behavior, rather than more sophisticated models that would describe some type of intermediate behavior with which we are unfamiliar in our macroscopic world.

The wave theory of the atom was developed in the 1920s, principally by Erwin Schrödinger. The most fundamental aspects of the theory are the mathematical wave equations used to describe the electrons in atoms. Solutions to the equations are called wave functions, or **orbitals.** Calculations involving the wave equations are complicated and time-consuming, but we do not need to do the elaborate calculations in order to use the results.

The principal result of the wave equation for an atom is a series of orbitals. Orbitals are different in their type, energy, and likely configuration in space (related to the probability of finding the electron there). The types of orbitals are distinguished by the letter **s, p, d,** and **f.** These letters were derived from terms in spectroscopy (sharp, principal, diffuse, and fundamental, respectively) and emphasize again that atomic theory developed very closely with atomic spectra.

The orbitals of the wave theory are actually subdivisions of the Bohr orbits.* Each orbit has an s orbital. Beginning with the second orbit, each orbit also has a set of three p orbitals. The third orbit and all orbits thereafter also have a set of five d orbitals. The fourth orbit and all orbits thereafter also have a set of seven f orbitals. We shall need only s and p orbitals for the explanations given in this book, since the electrons in the outer orbits of the various atoms are in the s and p orbitals of those atoms. These outer s and p electrons are the ones most involved with other atoms.

Only the **probability** of finding an electron in a given volume of space around the nucleus can be calculated from the orbital resulting from the Schrödinger equation. In order to portray the probabilities of finding an electron, usually the surface of a region in space (similar to the surface of a balloon) is plotted that will enclose the volume where the electron will be expected to be found 90% of the time. Actually, the electron structure of the atom is rather "messy" to describe in any definite way because of the uncertainty of locating its electron.

How are orbitals visualized?

As shown in Figure 3–20, an s orbital is always spherical. A p orbital is shaped like a dumbbell. Three different spatial orientations are possible for the p orbital, hence the designations p_x, p_y, and p_z. The d and f orbitals have more complicated shapes. The shapes of the orbitals are about the best that we can do to relate the wave theory of electron probabilities in a pictorial way, for we are trying to visualize where in space a given electron will be. The geometries associated with the orbitals help to explain the structures of molecules that result from the interaction and combinations of atoms.

Maximum of two electrons per orbital.

The first Bohr orbit is synonymous with the 1s orbital. (The 1 indicates the orbit; the s indicates the type of orbital.) According to the wave theory, an orbital may be occupied by a *maximum* of two electrons. Hence, the 1s orbital (and also the

---

* This shows that both the Bohr and Schrödinger theories can yield the same result in simple cases, as they should, since they are describing the same thing.

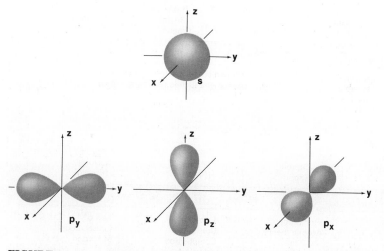

FIGURE 3–20 Spatial orientations of s and p orbitals for hydrogen-like atoms. The x, y, and z subscripts denote p orbitals with different orientations in space with reference to the imaginary axes.

first orbit) can have a maximum of two electrons. When the electron of a hydrogen atom is as close to its nucleus as it can stably be (in its **ground state**), the electron is in a 1s orbital, expressed as $1s^1$. The two electrons of helium can be in the same orbital and are in the 1s orbital in their ground state, $1s^2$.

Since the second Bohr orbit can have a maximum of eight electrons, four orbitals are required to accommodate the eight electrons. The four orbitals are a 2s orbital and a set of three 2p orbitals designated $2p_x$, $2p_y$, and $2p_z$ (Fig. 3–20). When the three electrons of a lithium atom are in their ground state, two electrons are in the 1s orbital and the other electron is in the 2s orbital, $1s^2 2s^1$ (Fig. 3–18). The five electrons of a boron atom are distributed with two in the 1s orbital, two in the 2s orbital, and one in the 2p orbital (either $2p_x$, $2p_y$, or $2p_z$, since these orbitals have the same energy unless the atom is in a strong magnetic field), $1s^2 2s^2 2p^1$.

How are electrons assorted into orbitals?

If there are two, three, or four electrons in a set of p orbitals, the electrons are spread among the orbitals as much as possible. This gives a more stable arrangement of electrons from spin and angular momentum considerations. What this means is that no orbital in a set will have two electrons until each orbital in the set has one electron. Thus, the six electrons of carbon distribute as $1s^2 2s^2 2p_x^1 2p_y^1$. The seven electrons of nitrogen have the configuration $1s^2 2s^2 2p_x^1 2p_y^1 2p_z^1$, and the eight electrons of oxygen are arranged in the ground state as $1s^2 2s^2 2p_x^2 2p_y^1 2p_z^1$.

In Figure 3–18, wave-mechanical electronic arrangements are shown through atomic number 20. Note the ground state order of energy for the orbitals: 1s 2s 2p 3s 3p 4s. How does the number of electrons in the 2s and 2p orbitals compare with the number of electrons in the second orbit? The 3s and 3p and the third orbit?

What, then, is an atom really like? The atomic concepts have changed over a long period of time (Fig. 3–21). We have Dalton's concept of an atom as a hard sphere similar to a small billiard ball. We have Bohr's concept of the atom as a small three-dimensional solar system with a nucleus and electrons in paths called orbits. In the modern theory, we have more detail in that orbits now have suborbits called orbitals, and we are given approximate spaces where electrons exert their greatest

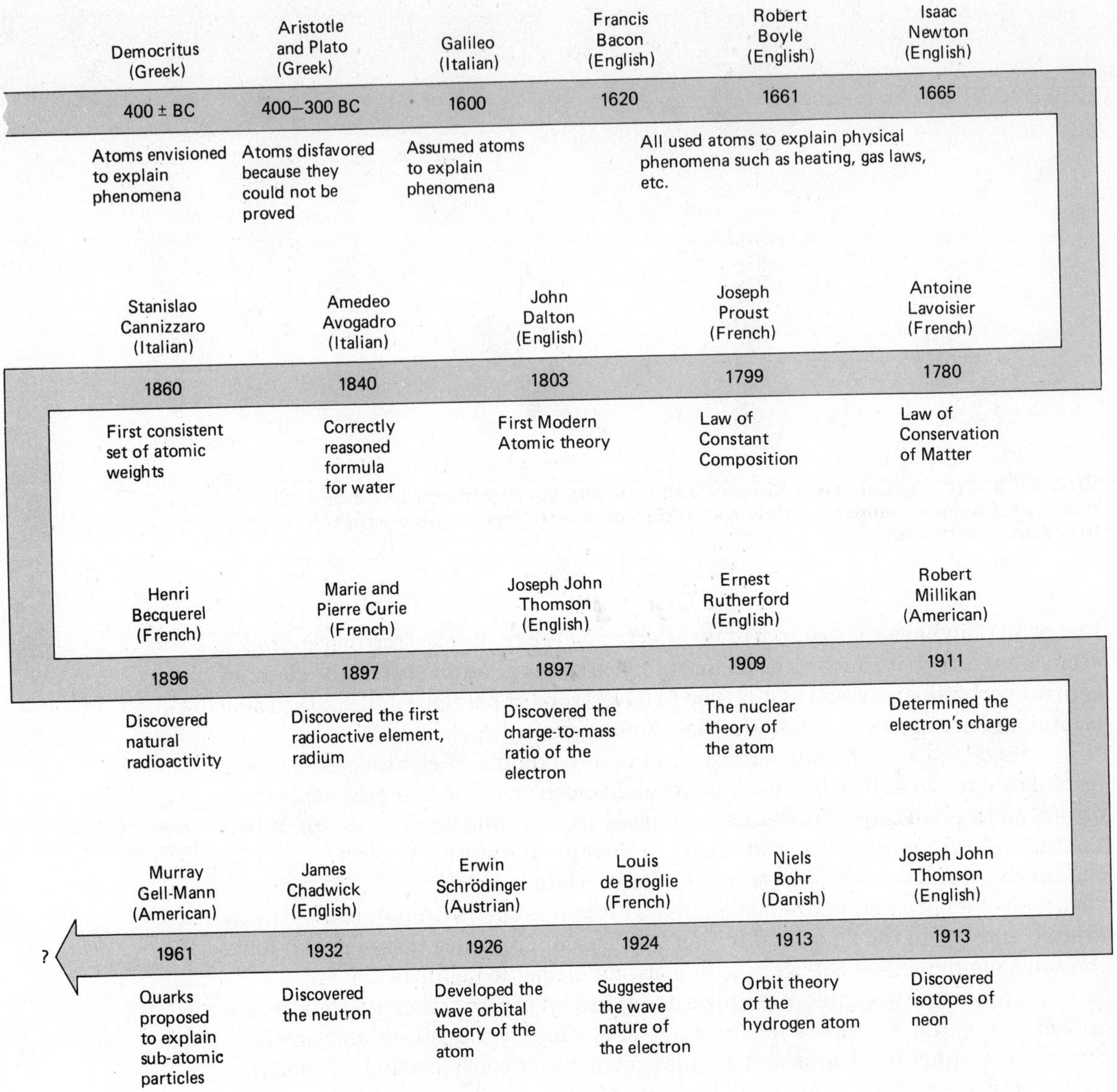

Democritus (Greek)	Aristotle and Plato (Greek)	Galileo (Italian)	Francis Bacon (English)	Robert Boyle (English)	Isaac Newton (English)
400 ± BC	400–300 BC	1600	1620	1661	1665
Atoms envisioned to explain phenomena	Atoms disfavored because they could not be proved	Assumed atoms to explain phenomena	All used atoms to explain physical phenomena such as heating, gas laws, etc.		

Stanislao Cannizzaro (Italian)	Amedeo Avogadro (Italian)	John Dalton (English)	Joseph Proust (French)	Antoine Lavoisier (French)
1860	1840	1803	1799	1780
First consistent set of atomic weights	Correctly reasoned formula for water	First Modern Atomic theory	Law of Constant Composition	Law of Conservation of Matter

Henri Becquerel (French)	Marie and Pierre Curie (French)	Joseph John Thomson (English)	Ernest Rutherford (English)	Robert Millikan (American)
1896	1897	1897	1909	1911
Discovered natural radioactivity	Discovered the first radioactive element, radium	Discovered the charge-to-mass ratio of the electron	The nuclear theory of the atom	Determined the electron's charge

Murray Gell-Mann (American)	James Chadwick (English)	Erwin Schrödinger (Austrian)	Louis de Broglie (French)	Niels Bohr (Danish)	Joseph John Thomson (English)
1961	1932	1926	1924	1913	1913
Quarks proposed to explain sub-atomic particles	Discovered the neutron	Developed the wave orbital theory of the atom	Suggested the wave nature of the electron	Orbit theory of the hydrogen atom	Discovered isotopes of neon

FIGURE 3–21   **Some of the important milestones in the development of atomic theory.**

influence in an atom. Why present all three theories? First, an understanding of the simpler Dalton and Bohr theories helps us to understand the more complicated, more detailed modern theory of the atom. Second, all three theories help us to understand the phenomena we observe. We simply use whatever detail is necessary to explain what we see. For example, the simpler Dalton concept adequately explains many properties of the gaseous, liquid, and solid states. Most bonding between atoms of the light elements can be explained by application of the orbits of Bohr. The shapes of molecules and the arrangement of atoms with respect to each other can best be explained by the orbital representations of the modern theory. In the explanations given in this text, we shall follow the principle that simplest is best.

SELF-TEST 3–C

1. Under some conditions light has properties of _____ and under other conditions exhibits the properties of _____.

2. When light is dispersed into the different colors composing the light, a _____ is produced.

3. According to Bohr's theory, light of characteristic wavelength is produced as an electron passes from an orbit closer to (   ) or farther from (   ) the nucleus to an orbit closer to (   ) or farther from (   ) the nucleus.

4. Which of the following led to the modern theory of the atom and was not included in the Bohr theory?
    a. Concept of the nucleus
    b. Quantum theory
    c. Particle nature of the electron
    d. Wave nature of the electron

5. According to de Broglie, every moving particle has not only mass and velocity but also a characteristic _____.

6. The maximum number of electrons in the $n = 3$ energy level is _____, and the maximum number in any orbital is _____.

7. Consider the meaning of the representations of the orbitals shown in Figure 3–20.
    a. Are the representations those of the paths of electrons? _____
    b. Are the representations the containers of electrons? _____
    c. Do the representations show where an electron is most likely to be found? _____

MATCHING SET

_____ 1.	Atomic mass	a. Attract
_____ 2.	Unlike electrical charges	b. Equal to number of protons in nucleus
_____ 3.	$2n^2$	c. Demonstrated wave nature of electron
_____ 4.	Nucleus	d. Cathode-ray particle
_____ 5.	Electron	e. Neutrons plus protons
_____ 6.	^{22}Ne and ^{20}Ne	f. A small, definite amount of energy
_____ 7.	Atomic number	g. Proton and an electron
_____ 8.	Quantum	h. Uncharged elementary particle
_____ 9.	Gamma ray	i. Contains most of the mass in an atom
_____ 10.	Particles in an H atom	j. Maximum number of electrons in an orbit
_____ 11.	Neutron	k. Predicted wave nature of electron
_____ 12.	G. P. Thomson C. J. Davisson	l. Probable location for electrons in an atom
_____ 13.	de Broglie	m. A form of radiant energy
_____ 14.	Orbital	n. Isotopes

QUESTIONS

1. What kinds of evidence did Dalton have for atoms that the early Greeks (Democritus, Leucippus) did not have?

2. How does Dalton's atomic theory explain:
   a. The law of conservation of matter?
   b. The law of constant composition?
   c. The law of multiple proportions?

3. The laws of chemical change presented in this chapter are often referred to as empirical laws. What does *empirical* mean? How does empirical differ from theoretical?

4. Although there may not be a very reliable way to check the conservation of matter in a large explosion of dynamite, what leads us to believe that the law of conservation of matter is obeyed?

5. Describe the potential energy and kinetic energy relationships as a rock tumbles off a cliff.

6. What experimental evidence indicates that
   a. Cathode rays have considerable energy?
   b. Cathode rays have mass?
   c. Cathode rays have charge?
   d. Cathode rays are a fundamental part of all matter?
   e. Two isotopes of neon exist?
   f. Atoms are destructible?

7. Describe in detail Rutherford's gold-foil experiment under the following headings:
   a. Experimental setup
   b. Observations
   c. Interpretations

8. Why was Thomson's charge-to-mass ratio determination for electrons very significant although he did not determine either the charge or the mass of the electron?

9. What part do electrons play in producing positive rays?

10. How do the following discoveries indicate that the Daltonian model of atoms is inadequate?
    a. Cathode rays
    b. Positive rays
    c. Nucleus
    d. Natural radioactivity
    e. Isotopes

11. Characterize the three types of emissions from naturally radioactive substances as to charge, relative mass, and relative penetrating power.

12. Explain what the following terms mean:
    a. Isotopes of an element
    b. Atomic number
    c. An alpha emitter

13. If electrons are a part of all matter, why are we not electrically shocked continually by the abundance of electrons about and in us?

14. There are more than 1000 kinds of atoms, each with a different weight. Yet there are only 109 elements. How does one explain this in terms of sub-atomic particles?

15. What is a practical application of cathode-ray tubes?

16. A common isotope of lithium (Li) has a mass of 7. The atomic number of lithium is 3. What are the constituent particles in its nucleus?

17. An element has 12 protons in its nucleus. How many electrons do the atoms of this element possess?

18. An isotope of atomic mass 60 has 33 neutrons in its nucleus. What is its atomic number, and what are the name and chemical symbol of the element?

19. An isotope of cerium (Ce) has 88 neutrons in its nucleus. How many protons plus neutrons does this nucleus contain?

20. The element iodine (I) occurs naturally as a single isotope of atomic mass 127; its atomic number is 53. How many protons and how many neutrons does it have in its nucleus?

21. An element with an atomic number of 8 is found to have three isotopes, with atomic masses of 16, 17, and 18. How many protons and neutrons are present in each nucleus? What is the element?

22. Suppose Millikan had determined the following charges on his oil drops:
    1.33×10^{-19} coulomb
    2.66×10^{-19} coulomb
    3.33×10^{-19} coulomb
    4.66×10^{-19} coulomb
    7.92×10^{-19} coulomb
    What do you think his value for the electron's charge would have been?

23. Suppose an isotope of aluminum has an atomic mass of 27.0. How many protons, neutrons, and electrons are in an atom of this isotope? What is the charge on the nucleus?

24. What is a quantum? What is a photon?

25. Discuss, in quantum terms, how a ladder works.

26. Distinguish between atomic number and atomic weight.

27. Distinguish between a continuous spectrum and a bright-line spectrum under the two headings:
    a. General appearance
    b. Source

28. How does the Bohr theory explain the many lines in the spectrum of hydrogen although the hydrogen atom contains only one electron?

29. Helium, neon, argon, krypton, xenon, and radon form a group of similar elements in that they form very, very few compounds. From their atomic

structures, suggest a reason for this similarity in relative inactivity.

30. Compare your view of a valley as you walk from a mountain top to the floor of the valley with the progression of atomic theory.

31. The law of conservation of matter is to a chemical change as the law of constant composition is to a chemical _____ .

32. True or False
   a. If compounds conform to the law of multiple proportions, they must necessarily conform to the law of constant composition.
   b. If compounds conform to the law of constant composition, they must necessarily conform to the law of multiple proportions.

33. What is constant about a compound?
   a. The weight of a sample of the compound
   b. The weight of one of the elements in samples of the compound
   c. The ratio by weight of the elements in the compound

34. If pure water is 88.8% oxygen and 11.2% hydrogen by weight,
   a. Is it likely to have *only* 88.8 g of oxygen in 110 g of water?
   b. Is it likely to have *exactly* 22.2 g of oxygen in 25.0 g of water?

35. In John Dalton's concept of atoms, were they more like billiard balls, cotton puff balls, tennis balls, or small solar systems?

36. How are fluorescent lights and TV picture tubes related to the study of the atom?

37. If you found the number of wheels received by an assembly plant to be twice the number of motors, what type of vehicle would you assume to be assembled there? Of what chemical law does this remind you?

38. Which is the empirical (observable) fact:
   a. Water is 88.8% oxygen by weight, or
   b. Water molecules contain one atom of oxygen each?

39. In recent years we have found that pure substances, such as some plastics, do vary in composition and that some elements can be decomposed (nuclear fission). What does this say to you about concepts and progress in science?

40. If different materials when heated give off different and characteristic colors of light, what can you assume about the structure and kinds of light?

41. Why is it impossible to produce a positive charge without producing a negative charge at the same time?

42. Krypton is the name of Superman's home planet and also that of an element. Look up the element krypton, and list its symbol, atomic number, atomic weight, and electronic arrangement.

43. Explain in your own words why alpha particles are deflected in one direction in an electrical field while beta particles are deflected in the opposite direction.

44. Read about lasers, and seek similarities between the explanation for the generation of laser light and the explanation for the production of bright-line elemental spectra.

45. Without looking at Figure 3–18 (except for checking later), write out the placement of electrons in their ground state
   a. Into orbits according to the Bohr theory for atoms having 6, 10, 13, and 20 electrons.
   b. Into orbitals according to the wave mechanical theory for atoms having 7, 13, 16, and 20 electrons.

Chapter 4

ELEMENTS IN USEFUL ORDER — THE PERIODIC TABLE

At this point in the story of chemistry, we have the first meaningful opportunity to discuss the **periodic table** (or **chart**), which you have probably noticed on the classroom wall. The careful study of the properties of the elements (defined and partially described in Chapter 2) led to the formulation of the periodic table, and the established periodic table furnishes evidence for atomic theory (Chapter 3).

What is there to know about the periodic table? Why is a so-called periodic table on a wall of most science classrooms and labs? Is it just a portrait of chemistry to adorn a wall, or is it useful? Why is the name "periodic" appropriate? Why is the table so arranged, and what are its important features? Does the table give order to the 109 known elements?

We shall discover in this chapter that the periodic table is important because it summarizes, correlates, and predicts a wealth of chemical information. In essence, the periodic table does bring order to 109 individual elements. Elements in an orderly arrangement provide the same benefits as your class notes arranged in a logical order, your room neatly and orderly arranged, or the goods arranged into departments in a store — ease of use and facilitation of understanding. From the standpoint of its logic, the periodic table can be of great help to a student of chemistry. As you read and study this chapter, look for both how and what chemical information is summarized, correlated, simplified, and predicted by the orderly arrangement of the elements in the periodic table.

Why arrange things in order?

ELEMENTS DESCRIBED

If we are to find and see order among the elements, we must have some general acquaintance with them. A few of the chemical and physical properties of 20 of the elements are summarized in Table 4–1. This format has been chosen so you can compare the properties more easily. The properties chosen for comparison are density, hardness, and relative reactivity.

The mass of a substance in a given volume is its **density.** For example, 1 mL of water at room temperature contains 1 g of matter; its density is 1 gram per milliliter (1 g/mL), while lithium has a density of 0.534 g/mL. A piece of lithium will float on water as it reacts with the water. Most gases have very low densities, about 0.001 g/mL, while many metals have densities much greater than that of water.

Hardness is a relative term used to describe solids. Diamond, a form of the element carbon, is one of the hardest substances known, while talc, like that used in talcum powder, is among the softest of the solids. **Reactivity** is often a useful term to describe an element. If an element is described as very reactive, that means it may react vigorously with air upon exposure, with moisture on your fingertips if you touch it, or with other elements. An element called unreactive may not even form *any* compounds!

Metals usually have high reflectivity (known as metallic luster), the ability to be bent and drawn into wire without shattering, and higher densities than nonmetals. Most metals conduct heat and electricity well and react with *nonmetals*. Since most metals conduct heat so well, in a cool environment metals feel colder to

TABLE 4–1 Some Properties of 20 Elements

ELEMENT	ATOMIC NUMBER	DESCRIPTION	COMPOUND FORMATION* With Cl (or Na)	With O (or Mg)
Hydrogen (H)	1	Colorless gas; reactive	HCl	H_2O
Helium (He)	2	Colorless gas; unreactive	None	None
Lithium (Li)	3	Soft metal; low density; very reactive	LiCl	Li_2O
Beryllium (Be)	4	Harder metal than Li; low density; less reactive than Li	$BeCl_2$	BeO
Boron (B)	5	Both metallic and nonmetallic; very hard; not very reactive	BCl_3	B_2O_3
Carbon (C)	6	Brittle nonmetal; unreactive at room temperature	CCl_4	CO_2
Nitrogen (N)	7	Colorless gas; nonmetallic; not very reactive	NCl_3	N_2O_5
Oxygen (O)	8	Colorless gas; nonmetallic; reactive	Na_2O, Cl_2O	MgO
Fluorine (F)	9	Greenish-yellow gas; nonmetallic; extremely reactive	NaF, ClF	MgF_2, OF_2
Neon (Ne)	10	Colorless gas; unreactive	None	None
Sodium (Na)	11	Soft metal; low density; very reactive	NaCl	Na_2O
Magnesium (Mg)	12	Harder metal than Na; low density; less reactive than Na	$MgCl_2$	MgO
Aluminum (Al)	13	Metal as hard as Mg; less reactive than Mg	$AlCl_3$	Al_2O_3
Silicon (Si)	14	Brittle nonmetal; not very reactive	$SiCl_4$	SiO_2
Phosphorus (P)	15	Nonmetal; low melting point; white solid; reactive	PCl_3	P_2O_5
Sulfur (S)	16	Yellow solid; nonmetallic; low melting point; moderately reactive	Na_2S, SCl_2	MgS
Chlorine (Cl)	17	Green gas; nonmetallic; extremely reactive	NaCl	$MgCl_2$, Cl_2O
Argon (Ar)	18	Colorless gas; unreactive	None	None
Potassium (K)	19	Soft metal; low density; very reactive	KCl	K_2O
Calcium (Ca)	20	Harder metal than K; low density; less reactive than K	$CaCl_2$	CaO

* The chemical formulas shown are lowest ratios. The molecular formula for $AlCl_3$ is Al_2Cl_6, and for P_2O_5 is P_4O_{10}.

the touch than do most nonmetals. The metal is conducting heat from your skin and you feel cooler.

Nonmetals are insulators; that is, they are extremely poor conductors of heat and electricity. Their crystals are brittle and tend to shatter easily. Therefore nonmetals cannot be drawn into wire or beaten into shapes like metals. Many nonmetals are gases at room temperature. Their densities are usually less than the densities of metals, and they react readily with metals and other nonmetals.

Although you probably know the properties of some of the elements listed in Table 4–1, our primary purpose is not to ask you to learn these properties, but rather to use them to search out any trends and similarities among the elements. Do you see any trends or similarities among the elements listed in the table? The elements in Table 4–1 are listed in atomic number order, but is there another, better arrangement for them?

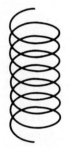

A helix is similar to a coiled spring. An early periodic table took this form.

Other spellings observed for Mendeleev's name in English: *Mendeleef* and *Mendeleyeff.*

THE PERIODIC LAW—THE BASIS OF THE PERIODIC TABLE

The periodic law did not occur to anyone until 1869, although considerable information was available concerning the then-known elements. Parts of the complete idea had occurred as early as 1817 when Johann Wolfgang Döbereiner saw trends and similarities among several groups of three elements each, which he called *triads.* By 1862, A. Beguyer de Chancourtois saw similarities in elements along vertical lines when the elements were arranged in order of their atomic weights along a helix. A most interesting insight occurred in 1866 when John Newlands arranged elements in the order of their atomic weights and observed that every eighth element had similar properties. Newlands coined the "Law of Octaves" for which he was harshly ridiculed by his peers. All of these early ideas were incomplete and gained no lasting support.

On the evening of February 17, 1869, at the University of St. Petersburg (now Leningrad) in Russia, a 35-year-old professor of general chemistry, Dmitri Ivanovich Mendeleev (1834–1907), was writing a chapter for his soon-to-be-famous textbook on chemistry. He had the properties of each element written on a separate card for each element. While he was shuffling the cards trying to gather his thoughts before writing his manuscript, Mendeleev realized that if the elements were arranged in the order of their atomic weights, there was a trend in properties that repeated itself several times! Thus the periodic law and table were born, although only 63 elements had been discovered by 1869 (the noble gases, He, Ne, Ar, Kr, Xe, and Rn, were not discovered until after 1893), and the clarifying concept of the atomic number was not known until 1913.

Within a month, Mendeleev had prepared a paper and had delivered it before the Russian Chemical Society. His idea and textbook achieved great success, and he rose to a position of prestige and fame as he continued to teach at St. Petersburg. In 1890, he resigned from the university during an episode of student unrest against the government, in which he sided with the students.

By 1871, Mendeleev published a more elaborate periodic table (Fig. 4–1). This version was the forerunner of the modern table currently seen in classrooms and textbooks.

Two features of the 1871 version were especially interesting. Empty spaces were left in the table, and there was a problem with the positions of tellurium (Te) and iodine (I).

	Group I R_2O RCl	Group II RO RCl_2	Group III R_2O_3 RCl_3	Group IV RO_2 RCl_4	Group V R_2O_5 RH_3	Group VI RO_3 RH_2	Group VII R_2O_7 RH	Group VIII RO_4
1	H = 1							
2	Li = 7	Be = 9.4	B = 11	C = 12	N = 14	O = 16	F = 19	
3	Na = 23	Mg = 24	Al = 27.3	Si = 28	P = 31	S = 32	Cl = 35.5	
4	K = 39	Ca = 40	— = 44	Ti = 48	V = 51	Cr = 52	Mn = 55	Fe = 56, Co = 59 Ni = 59, Cu = 63
5	(Cu = 63)	Zn = 65	— = 68	— = 72	As = 75	Sb = 78	Br = 80	
6	Rb = 85	Sr = 87	?Yt = 88	Zr = 90	Nb = 94	Mo = 96	— = 100	Ru = 104, Rh = 104 Pd = 106, Ag = 108
7	(Ag = 108)	Cd = 112	In = 113	Sn = 118	Sb = 122	Te = 125	I = 127	
8	Cs = 133	Ba = 137	?Di = 138	?Ce = 140	—	—	—	— —
9	(—)	—	—	—	—	—	—	— —
10	—	—	?Er = 178	?La = 180	Ta = 182	W = 184	—	Os = 195, Ir = 197 Pt = 198, Au = 199
11	(Au = 199)	Hg = 200	Tl = 204	Pb = 207	Bi = 208	—	—	
12	—	—	—	Th = 231	—	U = 240	—	— — —

FIGURE 4–1   An 1871 version of Mendeleev's periodic table. The formulas for simple oxides, chlorides, and hydrides are shown under each group heading. *R* represents the element in each group.

The empty spaces showed the genius and daring of Mendeleev. He left the empty spaces to retain the rationale of ordered arrangement based on periodic recurrence of the properties. For example, in atomic weight order are copper (Cu), zinc (Zn), and then arsenic (As). If As had been placed next to Zn, As would have fallen under aluminum (Al). But As forms compounds similar to those formed by phosphorus (P) and antimony (Sb), not Al. Mendeleev reasoned that two as yet undiscovered elements existed and moved As over two spaces to the position below P. The two missing elements were soon discovered: gallium (Ga) in 1875 and germanium (Ge) in 1886. The other gaps in his 1871 periodic table were later filled by discovered elements.

Mendeleev aided the discovery of the new elements by predicting their properties with remarkable accuracy, and he even suggested the geographical regions in which minerals containing the elements could be found. The properties of a missing element were predicted by consideration of the properties of its neighboring elements in the table. He had learned from Döbereiner, perhaps, that the density of an element is approximately the arithmetical average of the density of the lighter element above the missing element and the density of the heavier element just below. An example of Mendeleev's prediction of the properties of an undiscovered element is shown in Table 4–2. The term *eka* comes from Sanskrit, and means "one"; thus, *ekasilicon* means "one place away from silicon." He also predicted the properties of ekaboron (scandium) and ekaaluminum (gallium).

Dmitri Mendeleev (1834–1907). Born in Siberia, Mendeleev rose to Professor of Chemistry at St. Petersburg (now Leningrad) and then to director of the Russian Bureau of Weights and Measures. Although a prolific writer, a versatile chemist and inventor, and a popular teacher, the fame of this brilliant scientist rests on his discovery of the periodic law.

TABLE 4-2   Some of Mendeleev's Predicted Properties of Ekasilicon and the Corresponding Observed Properties of Germanium

	EKASILICON (Es)	GERMANIUM (Ge)
Atomic weight	72	72.6
Color of element	Gray	Gray
Density of element (g/mL)	5.5	5.36
Formula of oxide	EsO_2	GeO_2
Density of oxide (g/mL)	4.7	4.228
Formula of chloride	$EsCl_4$	$GeCl_4$
Density of chloride (g/mL)	1.9	1.844
Boiling point of chloride (°C)	Under 100	84

The empty spaces in the table and Mendeleev's predictions of the properties of missing elements stimulated a flurry of prospecting for elements in the 1870s and 1880s. As a result, gallium (Ga) was discovered in 1875; scandium (Sc), samarium (Sm), holmium (Ho), and thulium (Tm) in 1879; gadolinium (Gd) in 1880; neodymium (Nd) and praseodymium (Pr) in 1885; and germanium (Ge) and dysprosium (Dy) in 1886. Many of these elements are not even common today, yet they are important as ingredients in catalysts and color television screens.

If Mendeleev had followed the atomic weight order precisely, some elements with similar properties would not have been in the same column, or group. In the 1869 table, tellurium (Te) with an atomic weight of 128 was placed one position ahead of iodine, which has a lower atomic weight of 127. On the basis of its chemical properties, Te belonged with Sb, S, and O, and I belonged with F, Cl, and Br.

Mendeleev believed the atomic weight of tellurium was in error, but this was later shown not to be the case. In the 1871 table, the weight of Te had been changed from 128 to 125—an example of the unwise practice of changing data to fit a theory. The record is not clear as to why he changed the value.

Other reversed pairs in the modern periodic table are U before Np, Ar before K, Co before Ni, and Th before Pa. Upon realization of the atomic number concept in 1913, the question was resolved.

About nine months after Mendeleev delivered his paper before the Russian Chemical Society, Julius Lothar Meyer (1830–1895), a German physician and professor of chemistry at the University of Tübingen, prepared a table very similar to Mendeleev's. Apparently, both men were unaware of each other's work, yet both had left gaps for undiscovered elements. Meyer's table was based primarily on the repeatable trends in physical properties as the property is plotted against the atomic weight. Meyer grouped elements in subfamilies so, for example, zinc (Zn), cadmium (Cd), and mercury (Hg) with similar chemical properties could be separated from their chemical cousins magnesium (Mg), calcium (Ca), strontium (Sr), and barium (Ba). Mendeleev's table was superior because Meyer did not predict the properties of the undiscovered elements, and he did not rectify atomic-weight-position errors.

Building on the work of Mendeleev, Meyer, and others and using the clarifying concept of the atomic number, we are now able to state the modern periodic law: *When elements are arranged in the order of their atomic numbers, their chemical and physical properties show repeatable trends.*

Refer again to Table 4–1 and note how the trend in properties from lithium (Li) to neon (Ne) matches the trend from sodium (Na) to argon (Ar). The pattern in the properties of the elements, then, is *periodic;* hence the name *periodic law* or *table*. Other familiar periodic phenomena include the average daily temperature, which is periodic with time in a temperate climate. Low temperatures in January give way to high temperatures in July and low temperatures again in December. The trend repeats each year, not with exactly the same numbers but with the same pattern of change. Drowsiness follows a trend each 24 hours. Cash flow follows a cycle related to pay day. Hunger pains may be periodic several times each day. A shingled roof has the same pattern over and over and is, therefore, periodic.

So, to build up a periodic table according to the periodic law, line up the elements in a horizontal row in the order of their atomic numbers. Every time you come to an element with similar properties to one already in the row, start a new row. The columns, then, will contain elements with similar properties.

The properties of the elements are periodic functions of their atomic numbers.

FEATURES OF THE MODERN PERIODIC TABLE

A modern, popular version of the periodic table is shown in Figure 4–2. Note the following features.

The vertical columns are called **groups.**

The horizontal rows are called **periods.**

The letters A and B distinguish **families** of elements. For example, Group IA is the alkali metal family, and the closely related Group IB is sometimes called the coinage metal family.

The groups of elements are catalogued into four categories. The A groups are the **representative** elements. As we shall see, simple atomic theory represents these elements well. The B groups and Group VIII are the **transition** elements that link the two areas of representative elements. The **inner transition** elements are the lanthanide series, which fits between La and Hf, and the actinide series, which fits between Ac and Ku. The **noble gases** are unique and comprise a group to themselves.

SELF-TEST 4–A

1. In which group of the periodic table are Mg _____, Pd _____, Cl _____, Ga _____, Ag _____?

2. In which period of the periodic table are Li _____, Mo _____, Nd _____, U _____, Br _____?

3. Which are transition elements (T), which are representative elements (R), which are noble gases (N), and which are inner transition elements (I)?
   Be _____, P _____, Cr _____, Kr _____, Am _____

4. Who was primarily responsible for formulating the periodic table? _____

5. According to the periodic law, when the elements are arranged in the order of their _____, their properties show periodicity.

FIGURE 4-2 Periodic table of the elements. (Modified from Morris Hein, et al.: *Foundations of Chemistry in the Laboratory*, 4th ed. Belmont, CA, Dickenson Publishing Co., 1977.)

Legend:

- Atomic number — 11
- Name — Sodium
- Symbol — Na
- Electron structure — 2 8 1
- Atomic weight — 22.9998

Atomic weights are based on Carbon-12. Atomic weights in parentheses indicate the most stable or best-known isotope. Slight disagreement exists as to the exact electronic configuration of several of the high atomic-number elements.

Transition elements

Inner transition elements

Group IA

Period		
1	Hydrogen 1, H, 1.0079	
2	Lithium 3, Li, 2 1, 6.939	
3	Sodium 11, Na, 2 8 1, 22.9898	
4	Potassium 19, K, 2 8 8 1, 39.098	
5	Rubidium 37, Rb, 2 8 18 8 1, 85.47	
6	Cesium 55, Cs, 2 8 18 18 8 1, 132.905	
7	Francium 87, Fr, 2 8 18 32 18 8 1, (223)	

Group IIA

- Beryllium 4, Be, 2 2, 9.0122
- Magnesium 12, Mg, 2 8 2, 24.312
- Calcium 20, Ca, 2 8 8 2, 40.08
- Strontium 38, Sr, 2 8 18 8 2, 87.62
- Barium 56, Ba, 2 8 18 18 8 2, 137.34
- Radium 88, Ra, 2 8 18 32 18 8 2, (226)

IIIB
- Scandium 21, Sc, 2 8 9 2, 44.956
- Yttrium 39, Y, 2 8 18 9 2, 88.905
- *57 Lanthanum 57, La, 2 8 18 18 9 2, 138.91
- **89 Actinium 89, Ac, 2 8 18 32 18 9 2, (227)

IVB
- Titanium 22, Ti, 2 8 10 2, 47.90
- Zirconium 40, Zr, 2 8 18 10 2, 91.22
- Hafnium 72, Hf, 2 8 18 32 10 2, 178.49
- 104

VB
- Vanadium 23, V, 2 8 11 2, 50.942
- Niobium 41, Nb, 2 8 18 12 1, 92.906
- Tantalum 73, Ta, 2 8 18 32 11 2, 180.948
- 105

VIB
- Chromium 24, Cr, 2 8 13 1, 51.996
- Molybdenum 42, Mo, 2 8 18 13 1, 95.94
- Wolfram (Tungsten) 74, W, 2 8 18 32 12 2, 183.85
- 106

VIIB
- Manganese 25, Mn, 2 8 13 2, 54.938
- Technetium 43, Tc, 2 8 18 13 1, (99)
- Rhenium 75, Re, 2 8 18 32 13 2, 186.2
- 107

VIII
- Iron 26, Fe, 2 8 14 2, 55.847 | Cobalt 27, Co, 2 8 15 2, 58.933 | Nickel 28, Ni, 2 8 16 2, 58.71
- Ruthenium 44, Ru, 2 8 18 15 1, 101.07 | Rhodium 45, Rh, 2 8 18 16 1, 102.905 | Palladium 46, Pd, 2 8 18 18 0, 106.4
- Osmium 76, Os, 2 8 18 32 14 2, 190.2 | Iridium 77, Ir, 2 8 18 32 15 2, 192.2 | Platinum 78, Pt, 2 8 18 32 17 1, 195.09
- 108 | 109

IB
- Copper 29, Cu, 2 8 18 1, 63.546
- Silver 47, Ag, 2 8 18 18 1, 107.868
- Gold 79, Au, 2 8 18 32 18 1, 196.967

IIB
- Zinc 30, Zn, 2 8 18 2, 65.38
- Cadmium 48, Cd, 2 8 18 18 2, 112.40
- Mercury 80, Hg, 2 8 18 32 18 2, 200.59

IIIA
- Boron 5, B, 2 3, 10.811
- Aluminum 13, Al, 2 8 3, 26.9815
- Gallium 31, Ga, 2 8 18 3, 69.72
- Indium 49, In, 2 8 18 18 3, 114.82
- Thallium 81, Tl, 2 8 18 32 18 3, 204.37

IVA
- Carbon 6, C, 2 4, 12.0112
- Silicon 14, Si, 2 8 4, 28.086
- Germanium 32, Ge, 2 8 18 4, 72.59
- Tin 50, Sn, 2 8 18 18 4, 118.69
- Lead 82, Pb, 2 8 18 32 18 4, 207.19

VA
- Nitrogen 7, N, 2 5, 14.0067
- Phosphorous 15, P, 2 8 5, 30.9738
- Arsenic 33, As, 2 8 18 5, 74.922
- Antimony 51, Sb, 2 8 18 18 5, 121.75
- Bismuth 83, Bi, 2 8 18 32 18 5, 208.980

VIA
- Oxygen 8, O, 2 6, 15.9994
- Sulfur 16, S, 2 8 6, 32.064
- Selenium 34, Se, 2 8 18 6, 78.96
- Tellurium 52, Te, 2 8 18 18 6, 127.60
- Polonium 84, Po, 2 8 18 32 18 6, (210)

VIIA
- Fluorine 9, F, 2 7, 18.9984
- Chlorine 17, Cl, 2 8 7, 35.453
- Bromine 35, Br, 2 8 18 7, 79.904
- Iodine 53, I, 2 8 18 18 7, 126.904
- Astatine 85, At, 2 8 18 32 18 7, (210)

Noble Gases
- Helium 2, He, 2, 4.0026
- Neon 10, Ne, 2 8, 20.183
- Argon 18, Ar, 2 8 8, 39.948
- Krypton 36, Kr, 2 8 18 8, 83.80
- Xenon 54, Xe, 2 8 18 18 8, 131.30
- Radon 86, Rn, 2 8 18 32 18 8, (222)

Lanthanide series 6 (*)

- Cerium 58, Ce, 2 8 18 20 8 2, 140.12
- Praseodymium 59, Pr, 2 8 18 21 8 2, 140.907
- Neodymium 60, Nd, 2 8 18 22 8 2, 144.24
- Promethium 61, Pm, 2 8 18 23 8 2, (147)
- Samarium 62, Sm, 2 8 18 24 8 2, 150.35
- Europium 63, Eu, 2 8 18 25 8 2, 151.96
- Gadolinium 64, Gd, 2 8 18 25 9 2, 157.25
- Terbium 65, Tb, 2 8 18 27 8 2, 158.924
- Dysprosium 66, Dy, 2 8 18 28 8 2, 162.50
- Holmium 67, Ho, 2 8 18 29 8 2, 164.930
- Erbium 68, Er, 2 8 18 29 8 2, 167.26
- Thulium 69, Tm, 2 8 18 31 8 2, 168.934
- Ytterbium 70, Yb, 2 8 18 32 8 2, 173.04
- Lutetium 71, Lu, 2 8 18 32 9 2, 174.97

Actinide series 7 (**)

- Thorium 90, Th, 2 8 18 32 18 2, 232.038
- Protactinium 91, Pa, 2 8 18 32 20 9 2, (231)
- Uranium 92, U, 2 8 18 32 21 9 2, 238.03
- Neptunium 93, Np, 2 8 18 32 22 9 2, (237)
- Plutonium 94, Pu, 2 8 18 32 23 9 2, (242)
- Americium 95, Am, 2 8 18 32 25 8 2, (243)
- Curium 96, Cm, 2 8 18 32 25 9 2, (247)
- Berkelium 97, Bk, 2 8 18 32 26 9 2, (247)
- Californium 98, Cf, 2 8 18 32 28 9 2, (251)
- Einsteinium 99, Es, 2 8 18 32 29 9 2, (254)
- Fermium 100, Fm, 2 8 18 32 30 9 2, (253)
- Mendelevium 101, Md, 2 8 18 32 31 9 2, (256)
- Nobelium 102, No, 2 8 18 32 32 9 2, (254)
- Lawrencium 103, Lr, 2 8 18 32 32 9 2, (257)

6. When a phenomenon shows the same pattern over and over, we say the pattern is _____.

7. Elements that conduct heat and electricity well are classified as _____.

USES OF THE PERIODIC TABLE

The periodic table is very useful to chemists and students of chemistry in many ways. In addition to being a handy reference for atomic weights, atomic numbers, and whatever other information is printed in each square, the arrangement of the elements—the position of an element in the table—presents a wealth of useful information.

In the following sections, three major uses of the periodic table will be described: (1) elements in a group have similar properties, (2) elements in successive periods show repeating trends in their properties, and (3) the periodic table supports theory and relates theoretical concepts. These memory aids will be helpful to you as you study chemistry.

Elements in a Group Have Similar Properties

Refer back to Table 4–1 and note the properties of Li, Na, and K. They are soft metals, have low density, are very reactive, and form chlorides and oxides with formulas of MCl and M_2O. Now notice on the modern periodic table (Fig. 4–2) that rubidium (Rb), cesium (Cs), and francium (Fr) are also in the same group, Group IA. What properties would you expect Rb, Cs, and Fr to have? If you predicted soft metals, low density, very reactive, MCl and M_2O, you are right. Elements in a group have similar properties, but not the same properties.

In formulas MCl and M_2O, M represents an alkali metal.

Some properties of elements in a group differ by degree in a regular pattern. For example, the melting points beginning with Li and going down the column through Cs are (in °C) 179, 98, 64, 39, and 28, respectively. Lithium reacts slowly with water, sodium reacts faster, potassium still faster, and for the elements at the bottom of the group, just exposure to moist air produces an explosion.

Some properties differ by degree but not in a regular pattern. For example, the densities (in g/mL) of the solids Li through Cs are 0.53, 0.97, 0.86, 1.53, and 1.87, respectively.

Some properties are the same for every member of a group. Elements in a group generally react with other elements to form similar compounds. This is the most useful and powerful inference that can be made from the periodic table. For example, if the formula for the compound composed of lithium (Li) and chlorine (Cl) is LiCl, then probably there is a compound of Rb and Cl with the formula RbCl; the compound for Rb and Br (in Group VIIIA with Cl) would probably have the formula of RbBr. Likewise, if the formula Na_2O is known, then a compound with the formula K_2S predictably exists. This ability to predict formulas from the periodic table has limitations. For example, Na, K, Rb, and Cs all form superoxides (formula MO_2), but no superoxide with Li is known. The limitations do not prohibit the use of the periodic table for predicting formulas. In general, elements in the same group of the periodic table form some of the same types of compounds!

Four groups of elements in the periodic table are referred to by the name of the group. These four groups and some of their general properties are described below.

The **alkali metals** are Group IA (Li, Na, K, Rb, Cs, and Fr). The name *alkali* derives from an old word meaning "ashes of burned plants." When alkali metal oxides react with water, alkalis are formed. Alkalis (later in this book to be called bases) taste bitter (but don't try it!), feel slick when in solution, and neutralize or destroy the effects of acids. A common alkali is sodium hydroxide (NaOH), known commercially as lye. It is formed when sodium oxide (Na_2O) reacts with water.

$$Na_2O + H_2O \rightarrow 2NaOH$$

There are about 15 isotopes of francium, all of which are naturally radioactive.

The **alkaline earths** are Group IIA (Be, Mg, Ca, Sr, Ba, and Ra). They are metals harder and less reactive than the alkali metals. They form MCl_2 and MO. As the oxides, some react with water to form alkalis. For example, CaO (lime) reacts with water to form $Ca(OH)_2$ (slaked lime), a widely used alkali because of its low cost. Theatrical "limelight," a brilliant white light, got its name from calcium oxide being placed in an electric arc. In the Middle Ages, an *earth* was any solid substance that did not melt and was not changed by fire into some other substance. Under these conditions, many of the alkaline-earth compounds change into the oxides, which have high melting temperatures (in excess of 1900°C) and the same general white appearances of the original compounds. The high temperatures were not attainable by these early investigators to melt the oxides; hence, the name *earth* was applied and has stuck to this day.

The **halogens** are Group VIIA (F, Cl, Br, I, and At). Fluorine (F) and chlorine (Cl) are gases at room temperature, whereas bromine (Br) is a liquid and iodine (I) is a solid. In the elemental state, each of these elements exists as diatomic molecules (X_2). All isotopes of astatine (At) are naturally radioactive and disintegrate quickly. If you could accumulate enough astatine, it would be a solid at room temperature. The name *halogen* comes from a Greek word and means "salt-producing." The most famous salt involving a halogen is sodium chloride (NaCl), table salt. But there are many other halogen salts such as calcium fluoride (CaF_2), a natural source of fluorine; potassium iodide (KI), an additive to table salt that prevents goiter; and silver bromide (AgBr), the active photosensitive component of photographic film.

Aliases: noble gases, inert gases, rare gases.

The **noble gases** are He, Ne, Ar, Kr, Xe, and Rn. All are colorless monatomic gases at room temperature. They are referred to as noble because they generally lack chemical reactivity. The derivations of some of the names of these elements are consistent with their inactivity: argon (from Greek, *argon,* meaning "inactive"); xenon (from Greek, *xenon,* meaning "stranger"). Helium (Greek, *helios,* meaning "the sun") was discovered by analysis of the sun's light; later it was found on Earth. Neon (Greek, *neos,* meaning "new") is a common gaseous filler for "neon" lights. Neon glows red when excited in a discharge lamp. Other gases and painted tubes are used to give different colors. Radon is naturally radioactive.

The name *radon* comes from "radium" (Latin, *radius,* meaning "ray"). At first, radon was called niton (Latin, *nitens,* meaning "shining").

Until 1962, it was thought that all of the noble gases had absolutely no chemical reactivity. On some older periodic tables, the noble gas column was headed "inert gases." Many reasons were presented to explain why the noble gases were inactive and why they never would react. Beginning in 1962, the situation began to change. A Canadian, Neil Bartlett, prepared the compound O_2PtF_6.

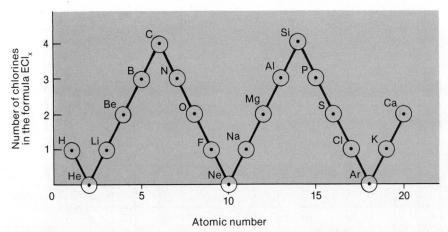

FIGURE 4–3   The number of chlorine atoms in the formula ECl_x, where E represents the element and x is the subscript on chlorine. The number of chlorine atoms is the combining power or valence of E.

Realizing that xenon might also form a similar compound, he discovered the first noble gas compound, $XePtF_6$. His discovery was followed quickly by the work of scientists at Argonne National Laboratory, who made some 30 compounds involving the heavier members of the noble gases combined with fluorine or oxygen. Some of the first prepared compounds were KrF_2, KrF_4, XeF_2, XeF_4, XeF_6, XeO_3, XeO_4, and RnF_4. No compounds with He, Ne, or Ar have yet been reported.

Hydrogen probably should be in a group by itself, although you may see H in both Group IA and Group VIIA in some periodic tables. Hydrogen forms compounds with formulas similar to those of the alkali metals, but with vastly different properties, such as NaCl and HCl; Na_2O and H_2O. Hydrogen also forms compounds similar to those of the halogens: NaCl and NaH (sodium hydride); $CaBr_2$ and CaH_2 (calcium hydride).

Hydrogen—the element without a home on the periodic table.

Similar Trends Occur in Successive Periods

What should be remembered about the periodic nature of certain properties is where the low values and high values occur. For example, from Table 4–1, the number of chlorine atoms combining with elements of the period Li through Ne is 1-2-3-4-3-2-1-0, respectively. The same trend occurs for Na through Ar (Fig. 4–3). These numbers correspond to the old term *combining power,* or **valence** (Latin, *valens,* meaning "strength"). If each chlorine has a combining power of one, then Li has a combining power of one (LiCl, one Li to one Cl), Be a combining power of two ($BeCl_2$), boron (B) a combining power of three (BCl_3), and so on. A modern, atomic-theory interpretation of combining power is discussed in the next chapter.

From left to right across each period, metallic character gives way to nonmetallic character (Fig. 4–4). The elements with the most metallic character are at the lower left part of the periodic table near francium (Fr). The elements with the most nonmetallic character are found near the upper right portion of the periodic table near fluorine (F).

The heavy line on the periodic table that begins at boron (B) and staircases down to astatine (At) roughly separates the metals and the nonmetals. Most of the

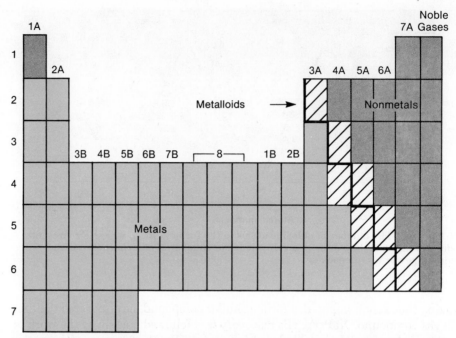

FIGURE 4–4   The location of metals, nonmetals, and semimetals (metalloids) in the periodic table. (Modified from Edward I. Peters: *Introduction to Chemical Principles*, 3rd ed. Philadelphia, Saunders College Publishing, 1982.)

Semiconductors conduct electricity less than metals such as silver and copper, but more than insulators such as sulfur; semiconductors are components of transistors.

elements (about 80%) lie to the left of this line and are considered metals. The elements positioned along the line are considered **semimetals** or **metalloids.** Their properties are intermediate between those of metals and nonmetals. For example, silicon (Si), germanium (Ge), and arsenic (As) are **semiconductors,** and boron (B) conducts electricity well only at high temperatures. It is these semiconductor elements that form the basic components of memory chips and computer logic.

Notice the periodic patterns of melting points of the elements when plotted versus atomic number (Fig. 4–5) and of boiling points when plotted versus atomic number (Fig. 4–6). The trends are not smooth, but a general periodic pattern is obvious. In which groups of the periodic table are the elements with the lowest

FIGURE 4–5   The periodic nature of the melting points of the elements when plotted versus atomic number.

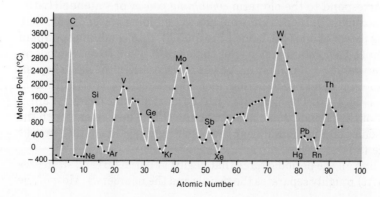

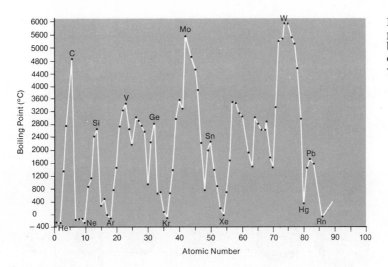

FIGURE 4–6 The periodic nature of the boiling points of the elements when plotted versus atomic number.

melting and boiling points? Which groups have the highest? Does this information correlate well with the general information given in Table 4–1?

Atomic volumes show periodicity with atomic number (Figs. 4–7 and 4–8). The volume of a mole of atoms in the solid state can be obtained by dividing the atomic weight (g/mole) by the density of the solid (g/mL). A plot of such atomic volumes versus atomic weight was a main exhibit in support of Meyer's periodic table. Figure 4–8 gives a better general perspective of the trends across and down the periodic table. Why do atoms get larger from top to bottom of a group? Do you suppose it has something to do with more layers making a larger onion? Yes, by analogy, the larger atoms simply have more energy levels (orbits) inhabited by electrons than do the smaller atoms.

Larger atoms have more orbits occupied by electrons.

Atomic volumes decrease across a period from left to right. You may see a paradox of adding electrons and getting smaller atoms. But, you are adding pro-

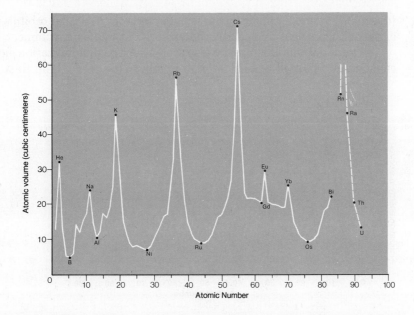

FIGURE 4–7 The periodic nature of the atomic volumes of the elements when plotted versus atomic number.

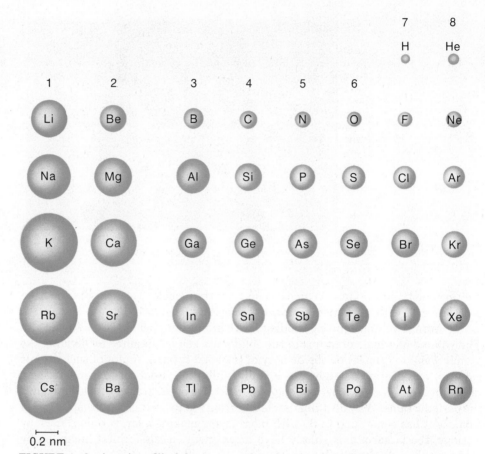

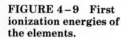

FIGURE 4–8   Atomic radii of the A group elements. Atomic radii increase as one goes down a group and in general decrease in going across a row in the periodic table. Hydrogen has the smallest atom, cesium the largest. (From Edward I. Peters: *Introduction to Chemical Principles,* 3rd ed. Philadelphia, Saunders College Publishing, 1982.)

tons, too. The greater nuclear charge pulls electrons in similar orbitals (same shell) closer to the nucleus and causes contraction of the atomic volume.

Periodic relationships are also seen when the first ionization energies of the elements are plotted against atomic number (Fig. 4–9). The **first ionization**

FIGURE 4–9   First ionization energies of the elements.

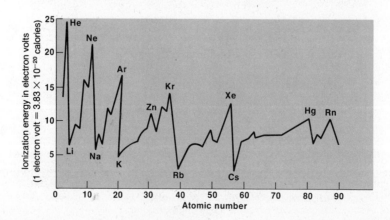

energy is the energy required to remove the first electron from an atom. The energy required to remove the second electron is the second ionization energy; removal of the third electron requires the third ionization energy, and similar terminology applies for the fourth and other electrons. Ionization energies can be determined experimentally for some elements by inserting the gaseous element into a cathode-ray tube and increasing the voltage until a surge of current occurs (the first ionization energy), increasing the voltage further until a second surge of current occurs (the second ionization energy), and so on. According to Figure 4–9, for which group of elements is it easiest to remove an electron? For which group of elements is it most difficult to remove an electron? Is it easier to remove electrons from metals or from nonmetals? If you answered Group IA, noble gases, and metals, respectively, you are correct. The theoretical reasons for your answers will be given in the next chapter, where ionization energies give a basis for understanding the use of electrons in bonding atoms to atoms.

In subsequent chapters, look for other trends in elemental properties and how they relate to the periodic table. Among these will be the types of bonding, electronegativity, and the number of valence electrons.

The Periodic Table and Atomic Theory Support Each Other

Atomic theory and the periodic table support the validity of each other. Atomic theory justifies the arrangement of the elements in the periodic table. The periodic table provides observational evidence for the growing understanding of trends in atomic structure. Chronologically, the periodic table was established prior to the development of our modern atomic theory and helped to make the general acceptance of modern atomic theory possible.

Why do elements in the same group in the periodic table have similar chemical behavior? The answer is because all of the elements in a group (particulariy the representative elements, the A groups, and the noble gases) have atoms with similar structural features. In Figure 4–2, the ground-state positions of electrons in orbits are given for each of the 109 atoms. The electronic structures for the first 20 atoms are repeated nearby in Figure 4–10. Note in Group IA that each

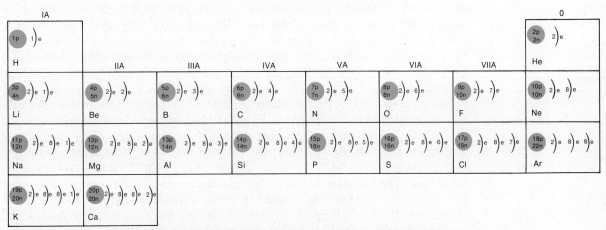

FIGURE 4–10 Electron arrangements of the first 20 elements. Above each column, or group, is the group number. The nuclear contents of a typical isotope are shown.

element has atoms with one and only one electron in the outermost occupied shell (called the **valence shell**). Group IIA elements all have two electrons in the valence shells of their atoms. Group IIIA elements and atoms all have three electrons in the valence shell, and the pattern continues through Group VIIA. Note that the group number is the number of electrons in the valence shell of each atom in the group.

The valence shell is the outermost occupied orbit.

What is the structural feature of the noble gases that results in their having little or no chemical reactivity? The noble gases have eight electrons in the valence shell of each atom (except for helium [He], which has a total of only two electrons). Eight electrons in the valence shell seem to provide a balanced, stable, structural arrangement that minimizes the tendency of an atom to react with other atoms.

Eight electrons in the valence shell provide a stable electronic arrangement.

Why are there repeatable patterns of properties across the periods in the periodic table? Again, it is because there is a repeatable pattern in atomic structure. Each period begins with one electron in the valence shell of the atoms of the elements in Group IA. Each period builds up to eight electrons in the valence shell, and the period ends. This pattern repeats across periods two through six. As more elements are made by nuclear accelerators, period seven may end someday. When it does, the periodic table and atomic theory predict the last element in period seven will be element number 118, with eight electrons in the valence shells of its atoms.

Nuclear accelerators are discussed in Chapter 7.

Atomic theory and the periodic table complement each other perfectly. One verifies the other, and vice versa.

THE PERIODIC TABLE IN THE FUTURE

The periodic table ties together well what is known about familiar elements, and it predicts accurately properties of unfamiliar elements. It is an indispensable memory aid. It makes intelligent and informal guessing easy, especially when it comes to predicting chemical formulas. All of these benefits are very important to students of chemistry.

Beyond these benefits, perhaps the most elegant contribution of the periodic table toward understanding nature is its stimulation of research. We are already aware of how Mendeleev's gaps stimulated the search for new elements. Going a step further, there is now active research under way to make elements beyond element 109. (There is a paradox here in using huge nuclear accelerators that cover many acres to try to shoot alpha particles and other very small particles into unseen, very small nuclei.) Part of the stimulus for this research is to see whether the prepared elements have the properties predicted by the periodic table. Without the periodic table, this reaching out would not be occurring. At least 19 elements have already been produced or discovered since 1939, bringing the number of known elements to 109. Predictions on elements 110 through 118 are that they will be very stable but still radioactive. Element 118 is expected to be a noble gas.

The periodic table will continue to stimulate the making of new compounds as it has in the past. An example from the past is sodium perbromate ($NaBrO_4$). Sodium perchlorate ($NaClO_4$) and sodium periodate ($NaIO_4$) had been known for some time. The fact that $NaBrO_4$ could not be prepared by the same methods was puzzling since Br is between Cl and I in Group VIIA of the periodic table, and all three elements should form similar compounds. All attempts failed (at least seven papers appeared in the chemical literature detailing why $NaBrO_4$ would never be made) until 1968, when another new compound, XeF_2, reacted with sodium bromate ($NaBrO_3$) in water to produce sodium perbromate for the first time. Faith in the correctness and predictive powers of the periodic table stimulated this research.

In the future, the groups in the periodic table will likely be labelled differently. Since 1959, the International Union of Pure and Applied Chemistry (IUPAC) has been considering the differences between the European and American practices of labelling A and B groups. For example, Group IIIA in European usage is Group IIIB in American practice. The IUPAC has decided to avoid the issue and is working a recommendation through approval procedures that would label the groups consecutively 1 through 18 from left to right.

"Similarities within groups" are "similarities within groups"; new labels will not change the integrity of the form of the table. We are using the older American system in this book because of the preference of distinguishing A and B groups when teaching atomic theory and because the new IUPAC recommendation was not approved when this edition went to press.

The periodic table is not a panacea for the chemist, but it is an important correlating unit for tying the properties and relationships of the elements together. For its place in *your* future, we propose this hypothesis: the periodic table will be your most lasting memory of the chemistry you study in this course. As you look at that portrait of chemistry on your classroom wall, what do you see now that you did not see a while ago? You now see what you may not have seen before—a requirement for the appreciation of art—and of science.

IUPAC resolves subjective issues related to chemistry.

"A rose is a rose regardless of what it is called."

SELF-TEST 4–B

1. What are the combining power and formula for the chloride of:
   Ga? _____ _____ , Ba? _____ , _____ ,
   Se? _____ _____ , I? _____ _____

2. Classify each of the following as a metal, nonmetal, or metalloid.
   Si _____ , Ce _____ , Cl _____ ,
   Cs _____ , Ca _____ , O _____ ,
   H _____ , Ge _____

3. How many electrons are in the valence shell of Na _____ ,
   Ca _____ , F _____ , Cl _____ ,
   O _____ , Al _____ , C _____ ?

4. The amount of energy required to remove an electron from an atom is called the _____ energy.

5. Which element in each pair has the greater ionization energy?
   He or O _____ , Na or F _____ , Ca or Br _____ ,
   K or S _____

6. Which element in each pair has the larger atoms? Li or K _____ ,
   F or Br _____ , Na or S _____ , B or In _____

7. In which groups of the periodic table would elements with the following electron configurations be found? 2-8-1 _____ , 2-8-4 _____ ,
   2-8-8-2 _____

MATCHING SET

_____	1.	Periodic	a. Electronic arrangement 2-8-2
_____	2.	Ionization energy	b. Generally a gas or a brittle solid
_____	3.	Larger atoms	c. Greater for Group VIIA than for Group IA
_____	4.	Two valence electrons	d. Eight valence electrons
_____	5.	A noble gas	e. At the bottom of a group
_____	6.	A metal	f. Praseodymium (Pr)
_____	7.	A nonmetal	g. Electronic arrangement 2-8-1
_____	8.	A halogen	h. Seven valence electrons
_____	9.	An inner transition element	i. Repeated pattern
_____	10.	Valence shell	j. Ruthenium (Ru)
			k. Outermost occupied orbit

QUESTIONS

1. State the periodic law.
2. How did the discovery of the periodic law lead to the discovery of elements?
3. How was the atomic weight 72, which appears in Mendeleev's periodic table of 1871 (Fig. 4–1), for a then-unknown element evaluated?
4. Omitting argon, write the formulas for a bromide of each of the elements with atomic numbers 11 through 20.
5. From their positions in the periodic table, predict which will be more metallic: (a) beryllium (Be) or boron (B); (b) beryllium (Be) or calcium (Ca); (c) calcium (Ca) or potassium (K); (d) arsenic (As) or germanium (Ge); (e) arsenic (As) or bismuth (Bi).
6. Use the information on the periodic chart to answer the following:
   a. The nuclear charge on cadmium (Cd)
   b. The atomic number of arsenic (As)
   c. The atomic mass (or mass number) of an isotope of bromine (Br) having 46 neutrons
   d. The number of electrons in an atom of barium (Ba)
   e. The number of protons in an isotope of zinc (Zn)
   f. The number of protons and neutrons in an isotope of strontium (Sr), atomic mass (or mass number) of 88
   g. An element forming compounds similar to those of gallium (Ga)
7. In a general way, how do average daily temperatures over the past three years at your location relate to properties and electronic structures of the elements when the elements are taken in the order of their atomic numbers?

8. Given one formula, based on the positions of the elements in the periodic table, predict the other formula.
   a. $BaCl_2$; formula for Sr and Br
   b. Na_2S; formula for K and Se
   c. Al_2O_3; formula for Ga and S
   d. NCl_3; formula for P and Br
9. Sodium reacts violently with water and forms hydrogen in the process. Magnesium will react with water only when the water is very hot. Copper does not react with water. Suppose you find a bottle containing a lump of metal in a liquid and a label, "Cesium (Cs)." Based on your knowledge of the periodic table, what danger is there, if any, of disposing of the metal by throwing it into a barrel of water?
10. How many elements are present in each period?
11. Write the symbols of the halogen family in the order of increasing size of their atoms.
12. Why does cesium (Cs) have larger atoms than lithium (Li)?
13. What similarities do you observe in the elements in Group IIA?
14. What general electronic arrangement is conducive to chemical inactivity?
15. How are the elements in a group related to each other?
16. Write the names and symbols of the alkaline earth elements.
17. Write the symbols for the family of elements that have three electrons in the valence shells of their atoms.
18. What is common about the electron structures of the alkali metals?

19. Pick the electron structures below that represent elements in the same chemical family.
    a. $1s^2 2s^1$
    b. $1s^2 2s^2 2p^4$
    c. $1s^2 2s^2 2p^2$
    d. $1s^2 2s^2 2p^6 3s^2 3p^4$
    e. $1s^2 2s^2 2p^6 3s^2 3p^6$
    f. $1s^2 2s^2 2p^6 3s^2 3p^6 4s^2$
    g. $1s^2 2s^2 2p^6 3s^2 3p^6 4s^1$
    h. $1s^2 2s^2 2p^6 3s^2 3p^6 3d^1 4s^2$

20. In how many different principal energy levels do electrons occur in period 1, period 3, and period 5?

21. Complete the following table:

ATOMIC NO.	NAME OF ELEMENT	ELECTRON STRUCTURE	PERIOD	METAL OR NONMETAL
6				
12				
17				
37				
42				
54				

22. How do the electronic structures of transition elements differ from those of "regular" (representative) elements?

23. How many electrons does the last element in each period have in its valence shell?

24. Answer this question without referring to the periodic table. Element number 55 is in Group IA, period 6. Describe its valence shell when all electrons are in the ground state. In how many orbits are there electrons?

25. If element 36 is a noble gas, in what groups would you expect elements 35 and 37 to occur?

26. Oxygen and sulfur are very different elements in that one is a colorless gas and the other a yellow crystalline solid. Why, then, are they both in Group VIA?

27. Suppose the popular press reports the discovery of a large deposit of pure sodium in northern Canada. What is your reaction as an informed citizen?

28. True or False. There are more nonmetallic elements than metallic elements.

29. List several common occurrences to you that are periodic.

30. Many elements are known to form compounds with hydrogen. Letting E be an element in any group, the following table represents the possible formulas of such compounds.

GROUP	IA	IIA	IIIA	IVA	VA	VIA	VIIA
	EH	EH_2	EH_3	EH_4	EH_3	H_2E	HE

Following the pattern in the table, write the formulas for the hydrogen compounds of (a) Na, (b) Mg, (c) Ga, (d) Ge, (e) As, (f) Cl.

31. Why do you suppose that Mendeleev did not predict the existence of the noble gases?

32. Write the symbol for an alkali metal, a lanthanide, an alkaline earth, a halogen, an actinide, and a transition metal (first series).

33. Below are some selected properties of lithium (Li) and potassium (K). Before looking up the numbers, estimate values for the corresponding properties of sodium (Na).

	LITHIUM	SODIUM	POTASSIUM
Atomic weight	6.9	—	39.1
Density (g/cm³)	0.53	—	0.86
Melting point (°C)	180	—	63.4
Boiling point (°C)	1330	—	757

34. Give the names and symbols for two elements most like selenium (Se), atomic number 34.

35. Predict some chemical and physical properties for the element francium (Fr, atomic number 87).

36. What is the likelihood of discovering another family of elements such as the noble gases?

37. Tin (II) chloride ($SnCl_2$) and tin (IV) chloride ($SnCl_4$) are known compounds of tin. From the positions of tin (Sn) and thallium (Tl) in the periodic table, predict the two expected chlorides of thallium.

38. Complete the following table by writing the predicted formula.

ELEMENT	F	O	Cl	S	Br	Se
Na						
K						
B						
Al						
Ga						
C						
Si						

39. Look up the properties of the other halogens and use them to predict the following properties of astatine: melting point, boiling point (1 atm), density in the gaseous, liquid, and solid states, valence(s) toward oxygen and toward hydrogen, solubility of NaAt in water at 25°C and 100°C.

40. If element 118 is ever produced, what will be its position in the periodic table?

41. Compare the first ionization energies of oxygen and xenon using Figure 4–9. Explain Neil Bartlett's reasoning when he expected xenon to form a compound with PtF_6 similar to O_2PtF_6.

Chapter 5

CHEMICAL BONDS — THE ULTIMATE GLUE

Chemical bonds hold atoms, molecules, and ions together.

What holds matter together? In other words, why does glue stick, or what causes pieces of hard candy to stick together? Why is a diamond so hard; why is wax soft? Or, in reverse, why do things break or fall apart? Why is table salt so brittle? Why does paint peel? Why do some substances melt at a rather low temperature, while others melt at higher temperatures?

These and similar questions can be answered logically and be consistent with experimental evidence if we think of matter as one atom bound to another. Granted, it is a little hard to consider the Empire State Building or the Washington Monument or a living organism as a conglomeration of atoms bonded one to the other. But large pieces of matter, even the Rocky Mountains, conform to the same fundamental principles of nature as a small crystal of sugar or salt.

Most of the reasons for matter bonding to matter (or atom to atom) can be summarized by two concise notions discussed in Chapter 3.

1. Unlike charges attract.
2. Electrons tend to exist in pairs.

Couple these two ideas (one empirical; one theoretical) with the proximity requirement that only the outer electrons of the atoms (the *valence electrons*) interact, and you have the basic concepts that explain how atoms in over 6 million compounds bond to each other. Just how the different atoms use these principles to bond atom to atom is the subject of this chapter. We shall see that the action of an atom in the formation of a bond is dictated by its atomic structure and generalized by its position in the periodic table.

Various interactions of the atoms cause the formation of five major types of chemical bonds. The types of bonds, along with some common materials in which they occur, are:

1. Ionic bonding   Salts, such as table salt (sodium chloride); and metal oxides, such as lime, iron rust, ruby, and sapphire

2. Covalent bonding   Molecular compounds, such as water, methane, and sugar; and polymers, such as polyethylene

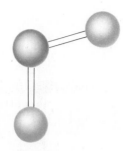

FIGURE 5–1 A water molecule is often represented by a ball-and-stick model. This model tells which atoms are bonded together and the angle involved but gives no information as to why the atoms are bonded in a particular pattern, the relative sizes of the atoms, or the actual distances between them.

3. Hydrogen bonding   Intermolecular bonding among molecules of water, ammonia, DNA, and proteins

4. Intermolecular London forces   Liquid helium and solid CO_2 (dry ice)

5. Metallic bonding   Metals and alloys

As always, it is the properties of the substances that dictate and verify the related theories. It is properties such as chemical reactivity, volatility (ability to pass into the gaseous state), melting point, electrical conductivity, and color that often give some indication of how atoms are bonded to each other. For example, since melting involves atoms or molecules becoming less firmly bound to their neighbors, a high melting point implies that a solid is held together by very stable chemical bonds. As we shall see shortly, compounds composed of a network of tightly bound ions or atoms tend to have relatively high melting points. The volatility of a substance also indicates how strongly molecules are attracted to each other. For example, in the case of carbon dioxide, CO_2, we must assume that the bonding between molecules (intermolecular bonding) is slight, since it takes relatively little energy to break up solid CO_2 (dry ice).

In the ensuing discussion of chemical bonds, major emphasis will be placed on accounting for the properties of a given substance by the bonds that hold that substance together.

CO_2 changes readily from a solid to a gas (sublimes).

Bonding theories must explain the observed behavior of chemicals.

IONIC BONDS

Ions and Ion Formation

A large category of compounds forms hard, brittle crystalline solids with relatively high melting points. When melted or in solution, they conduct electricity well, but they do not conduct when solid. If in solution or melted, these compounds often react quickly with each other. Compounds with these properties are known as **ionic compounds.** Examples of ionic compounds are sodium chloride (NaCl), magnesium fluoride (MgF_2), and calcium oxide (CaO).

All of the properties of these compounds can be explained if the compounds are assumed to be composed of charged atoms (called **ions**) rather than neutral atoms. X-ray and mass spectrographic studies of these kinds of compounds strongly confirm that ions exist.

How, then, do atoms become ions, and which atoms are most likely to form ions?

Since electrons constitute the outermost parts of the atom, it is reasonable to assume that electrons—and only electrons—are manipulated to form ions. Electrons can be *removed* from an atom, with the result that part of the positive

An ion is a charged atom or group of atoms.

There is a periodic table on p. 80 and on the inside of the back cover.

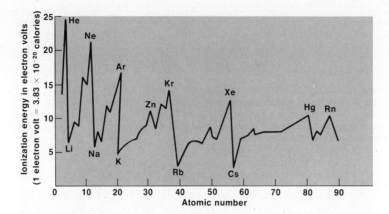

FIGURE 5-2 First ionization energies of the elements.

Ionization energies are measured experimentally and correspond to the reaction: atom → positive ion + electron.

Metals lose electrons to form positive ions.

The electrons easiest to remove are the outermost electrons, the valence electrons. Which is easier to remove, the peel or the seed of an orange?

Table 5-1 shows the amounts of energy needed to remove one or more electrons from various atoms.

nuclear charge is not neutralized and the atom becomes an ion with a positive charge. Electrons can also be *gained* by an atom, with the result that excess negative charge has been added to the atom, making it a negative ion. Metals form positive ions in the presence of nonmetals, and nonmetals form negative ions in the presence of metals.

Recall that representative metals constitute the left side of the periodic table and that they have one, two, or three valence electrons. Energy is required to remove these electrons from atoms. This is the ionization energy that was shown to be periodic in Chapter 4 (lower ionization energies on the left of the periodic table, higher on the right; see Fig. 5-2 for review). Ionization energy must be added to an atom, and the energy can come either from an outside source (heat or electricity, for example) or from the energy given off when nonmetals receive and pair electrons. Since metals have the lower ionization energies, their electrons are easier to remove than the electrons of nonmetals.

The number of electrons removed from metals to form positive ions is the number of valence electrons per atom, as confirmed by the ionization energies listed in Table 5-1. Atoms of Group IA metals (Li, Na, and K as well as Rb, Cs, and Fr) have only one electron easily removed. To remove a second electron requires at least seven times more energy than to remove one electron. Group IA metal atoms have one valence electron. According to Table 5-1, atoms of Group IIA metals (Be, Mg, and Ca shown plus Sr, Ba, and Ra) have two easier-to-remove electrons, and thus two valence electrons. Atoms of Group IIIA metals (B and Al, for example) have three easier-to-remove electrons, and hence three valence electrons.

The removal of electrons from metals and the consequent formation of positive ions can be depicted in varying degrees of detail. For Group IA, using sodium (Na, atomic number 11) as the example:

11p
12n) 2) 8) 1 + energy ⟶ 11p
12n) 2) 8 + 1 e⁻

Sodium atom
(neutral)

Sodium ion
(+1)

or

$$Na + energy \rightarrow Na^+ + e^-$$
2-8-1   2-8

TABLE 5–1 Ionization Energies of Selected Gaseous Atoms*

ATOMIC NUMBER	ATOM	IONIZATION ENERGIES (EV)							
		1st	2nd	3rd	4th	5th	6th	7th	8th
1	H	13.6							
2	He	24.6	54.4						
3	Li	5.4	75.6	122.4					
4	Be	9.3	18.2	153.9	217.7				
5	B	8.3	25.1	37.9	259.3	340.1			
6	C	11.3	24.4	47.9	64.5	392.0	489.8		
7	N	14.5	29.6	47.4	77.5	97.9	551.9	666.8	
8	O	13.6	35.1	54.9	77.4	113.9	138.1	739.1	871.1
9	F	17.4	35.0	62.6	87.2	114.2	157.1	185.1	953.6
10	Ne	21.6	41.1	64	97.2	126.4	157.9		
11	Na	5.1	47.3	71.7	98.9	138.6	172.4	208.4	264.2
12	Mg	7.6	15.0	80.1	109.3	141.2	186.9	225.3	266.0
13	Al	6.0	18.8	28.4	120.0	153.8	190.4	241.9	285.1
14	Si	8.1	16.3	33.5	45.1	166.7	205.1	264.4	303.9
15	P	10.6	19.7	30.2	51.4	65.0	220.4	263.3	309.3
16	S	10.4	23.4	35.0	47.3	72.5	88.0	281.0	328.8
17	Cl	13.0	23.8	39.9	53.5	67.8	96.7	114.3	348.3
18	Ar	15.8	27.6	40.9	59.8	75.0	91.3	124.0	143.5
19	K	4.3	31.8	46	60.9	82.6	99.7	118	155
20	Ca	6.1	11.9	51.2	67	84.4	109	128	147

* An electron volt (ev) is the energy acquired by an electron when accelerated by a potential difference of 1 volt. For each element, electrons must be removed to the heavy vertical line in order to attain a noble gas electronic configuration. The heavy lines separate regions of low and high ionization energies.

or, simply

$$Na \rightarrow Na^+ + e^-$$

Likewise for Group IIA metals, using Mg as the example:

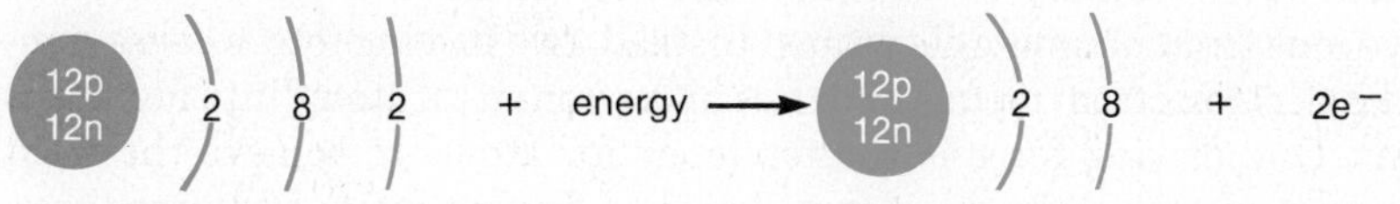

Magnesium atom (neutral)   Magnesium ion (+2)

or

$$\underset{2\text{-}8\text{-}2}{Mg} + energy \rightarrow \underset{2\text{-}8}{Mg^{2+}} + 2e^-$$

or, simply

$$Mg \rightarrow Mg^{2+} + 2e^-$$

In a similar fashion for Group IIIA metals, using Al as the example:

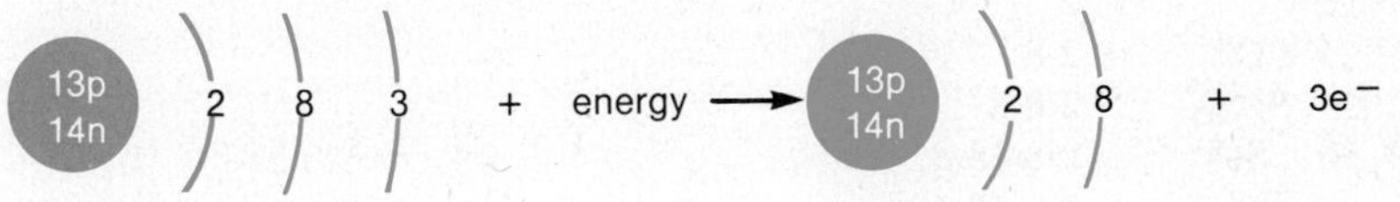

Aluminum atom (neutral)   Aluminum ion (+3)

or

$$Al + energy \rightarrow Al^{3+} + 3e^-$$
$$\text{2-8-3} \qquad\qquad \text{2-8}$$

or, simply

$$Al \rightarrow Al^{3+} + 3e^-$$

The more detailed depiction of positive ion formation points out an interesting coincidence for many, if not most, ions. For the ions formed from metals in Group IA, Group IIA, and Group IIIA (B and Al only), each ion has eight electrons in its new outermost orbit. This is exactly the number of valence electrons in the atoms of the noble gases (except He, of course). It appears that when some atoms achieve a noble gas electronic arrangement, further transfer of electrons is unlikely. Some positive ions with noble gas electronic configurations are shown in Table 5–2. Stable positive ions that do not conform to the noble gas electronic arrangement are most of the transition metal ions and positive ions formed from elements to the right of the transition metals in Periods 4, 5, and 6.

In summary, positive ions are formed when metal atoms lose one electron (Group IA), two electrons (Group IIA), or three electrons (Group IIIA). The resulting ions often have the same electronic arrangement as a noble gas.

Why not a +8 stable ion?

The difficulty of removing successive electrons from the same atom is verified by the ionization energies in Table 5–1. As more and more electrons are removed from a single atom such as a boron (B) atom, the net positive charge builds up with the loss of each electron. The unneutralized charge helps to hold the remaining electrons more securely. For this reason (no other stabilizing forces considered), a B^{3+} ion is more difficult to make and less likely to occur than a Li^+ ion.

Likewise, moving across a period in the periodic table, the lower ionization energies of the metals give way to the higher ionization energies of the nonmetals. It is much more difficult to remove electrons from nonmetals than from metals. The energies available in ordinary chemical changes are simply not great enough to form positive ions from nonmetallic atoms. Instead, two driving forces cause nonmetals to take electrons from metals and thus make nonmetal atoms into nonmetal negative ions. One driving force is the tendency for atoms to achieve the eight valence electrons of the noble gas atoms. Another driving force, and probably a more fundamental one, is the stable arrangement of a pair of electrons. Recall from Chapter 3 that two electrons occupy an orbital. The compatibility of two electrons in an orbital is thought to be caused by the mutual stability afforded by opposite

TABLE 5–2 Electronic Configurations of the Noble Gases and Ions with Identical Configurations

SPECIES	CONFIGURATION
He, Li^+, Be^{2+}, H^-	2
Ne, Na^+, Mg^{2+}, F^-, O^{2-}	2-8
Ar, K^+, Ca^{2+}, Cl^-, S^{2-}	2-8-8
Kr, Rb^+, Sr^{2+}, Br^-, Se^{2-}	2-8-18-8
Xe, Cs^+, Ba^{2+}, I^-, Te^{2-}	2-8-18-18-8

spins on the two electrons. A moving charge (rotation or otherwise) produces a magnetic field about it. When the moving charges spin in opposite directions, compatible magnetic fields are produced that attract the charges toward each other.

Electrons in an orbital spin in opposite directions.

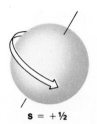

$s = +\frac{1}{2}$

A nonmetal atom, then, strives to pair all of its valence electrons to achieve a noble gas electronic arrangement of eight valence electrons. Thus a nonmetal can add to its five, six, or seven valence electrons three, two, or one electron(s) and have the stable eight valence electrons.

Consider an atom of fluorine (F, atomic number 9). Its electronic arrangement is 2-7 (or $1s^2\, 2s^2\, 2p_x^2\, 2p_y^2\, 2p_z^1$). A fluorine atom needs one electron to pair with its one unpaired electron, and this single electron completes the stable eight valence electrons. The gain of electrons by the nonmetals, like the loss of electrons by the metals, can be depicted in various degrees of detail.

$s = -\frac{1}{2}$

9p 10n) 2 7) + e$^-$ → 9p 10n) 2 8) + energy

Fluorine atom (neutral)   Fluoride ion (-1)

$$F + e^- \rightarrow F^- + energy$$
2-7   2-8

$$:\ddot{F}\cdot + e^- \rightarrow :\ddot{F}:^-$$

$$F + e^- \rightarrow F^-$$

Nonmetal atoms gain electrons to form negative ions.

Note that an electron is on the left side of each equation, and the electron is therefore gained. The third depiction is very informative, and involves **electron dot formulas,** suggested by G. N. Lewis in 1916. Only the valence electrons are represented around the symbol for the element. The electrons are placed at north, south, east, and west positions adjacent to the symbol. The method clearly shows the pairing of electrons as well as the attainment of eight valence electrons when the negative ion is formed.

Next consider how an oxygen atom becomes an oxide ion. The electronic arrangement of oxygen (O, atomic number 8) is 2-6 (or $1s^2\, 2s^2\, 2p_x^2\, 2p_y^1\, 2p_z^1$), which in electron dot representation is $\cdot\ddot{O}:$. How an oxygen atom gains two electrons to become an oxide ion (O^{2-}) can be depicted in several ways as before.

8p 8n) 2 6) + 2e$^-$ → 8p 8n) 2 8) + energy

Oxygen atom (neutral)   Oxide ion (-2)

$$O + 2e^- \rightarrow O^{2-} + energy$$
2-6   2-8

$$\cdot\ddot{O}: + 2e^- \rightarrow :\ddot{O}:^{2-}$$

$$O + 2e^- \rightarrow O^{2-}$$

Other common negative ions are listed in Table 5-2. The types of ions formed by the various groups of elements are shown on the periodic table in Figure

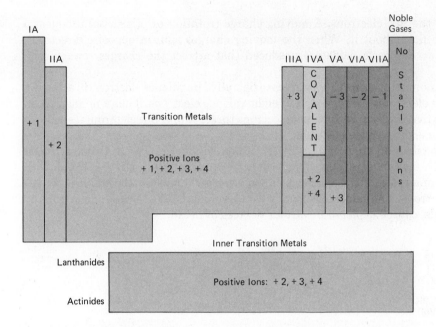

Electrical conductivity by
ionic substances is
illustrated in Figure 9-3
on p. 207.

FIGURE 5-3   The periodic table and the formation of ions.

5-3. Just as the difficulty of removing successive electrons from an atom of a metal increases, it is increasingly more difficult to add successive electrons to a nonmetal atom. The negative charge built up on a negative ion repels an incoming electron. For this reason, highly charged negative ions (-4, -5, -6, etc.) are not stable and do not exist in stable compounds. An atom with four valence electrons has no pronounced tendency to lose or gain electrons or to form ions.

In summary, nonmetals in the presence of metals tend to gain one, two, or three electrons to form negative ions, which have all valence electrons paired and have the stable eight-electron arrangement of noble gases.

PROPERTIES OF IONS, ATOMS, AND MOLECULES

Ions and their parent atoms differ in their properties. Besides the difference of being charged or neutral, there are other attendant, important differences. For example, sodium atoms react quickly with water to produce hydrogen gas. Sodium ions and water produce no hydrogen gas. Chlorine atoms are poisonous to the human body. Chloride ions are not considered poisonous. In fact, sodium chloride (Na^+ and Cl^- ions) is palatable on tomatoes, whereas Na and Cl atoms would blow up the tomato and poison the eater! Hydrogen gas is combustible; oxygen gas supports combustion; hydrogen and oxygen bonded into water molecules put out a fire. The bonding and charging of atoms change their nature. Compounds that contain ionic bonds (formed by the *transfer* of electrons) are classified as **salts,** in contrast to molecular compounds, in which all the atoms are held together by *sharing* electrons (covalent bonds, p. 101).

The *electrical conductivity* of melted ionic compounds is based on the movement of free ions to oppositely charged poles when an electrical field is imposed. The movement of the ions transports charge, or electric current, from one place to another. In a rigid solid, the immobile ions are not free to move, and the solid does not conduct electricity.

The *hardness* of ionic compounds is caused by the strong bonding between ions of unlike charge. The strong bonds require much energy to separate the ions and allow the freer movement of the melted state. Much energy means *higher melting points,* which are characteristic of ionic compounds.

Ionic compounds are *brittle* because the structure of the solid is a regular array of ions. Take, for example, the structure of sodium chloride (NaCl) (Fig. 5–4). Each Na^+ ion is surrounded by six Cl^- ions, and each Cl^- ion is, in turn, surrounded by six Na^+ ions. Each ion is attracted to all of the oppositely charged ions and repelled by all of the identically charged ions in the structure. If a plane of ions is shifted just one ion's distance in any direction, identically charged ions are now next to each other; there is repulsion — no attraction — and the crystalline solid breaks. Sodium chloride cannot be hammered into a thin sheet. It shatters instead.

There is no unique pairing of ions; hence ionic compounds have *no molecules.*

All chemical compounds are inherently *neutral.* For ionic compounds, this means there must be present as many unit positive charges as unit negative charges. This is a theoretical justification for the usual chemical formulas seen for ionic compounds. For example, a calcium ion (Ca^{2+}) requires two fluoride ions (F^-) to have exactly the same amount of each kind of charge. The chemical formula is CaF_2.

Ion sizes are different from parent atom sizes. Positive ions are smaller than the atoms from which they were made; negative ions are larger than their atoms

Ionic bonds are the attractions between ions of opposite charge.

An ionic structure is a regular geometrical array of ions.

Ionic compounds have no molecules.

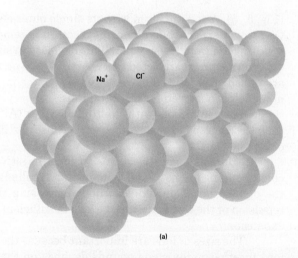

(a)

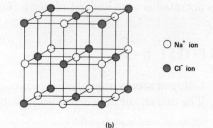

○ Na$^+$ ion

● Cl$^-$ ion

(b)

FIGURE 5–4 Structure of sodium chloride crystal. *(a)* **Model showing relative sizes of the ions;** *(b)* **ball-and-stick model showing cubic geometry.**

METALS

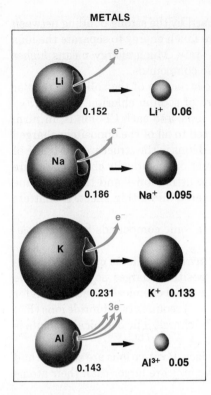

NONMETALS

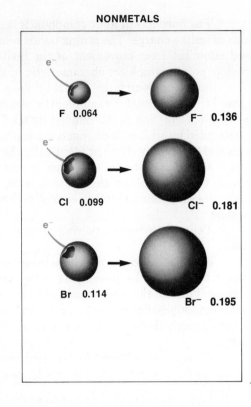

FIGURE 5–5
Relative sizes of selected atoms and ions. Numbers given are atomic or ionic radii in nanometers.

Chemical formulas of ionic compounds result from the number of electrons lost or gained by the reacting atoms and the neutrality of the compound, which requires equal numbers of positive and negative charges.

One nanometer, nm, is 10^{-9} meter.

(Fig. 5–5). A sodium atom with its single outer electron has a radius of 0.186 nm. One would expect that when this electron is removed (forming the Na^+ ion) the resulting ion would be smaller. This decrease in size results because there are now only ten electrons attracted to a charge of $+11$ on the nucleus, and these electrons are pulled closer to the nucleus by this charge imbalance. The same type of phenomenon is observed for all metal ions.

The metal ions with multiple charges (Al^{3+}, Fe^{2+}, etc.) are much smaller than the corresponding metal atom, because of still greater surplus positive charge on the nucleus.

Nonmetals gain electrons to form negative ions that are *larger* than the corresponding atoms. This phenomenon results from the addition of electrons to the outer orbit of an atom without increasing the charge on the nucleus. The repulsion of the electrons and the lack of sufficient charge on the nucleus cause the expansion.

The sizes of ions are important because the strength of the forces that hold ions together in ionic compounds depends on the sizes (and charges) of the ions involved. If two ions have the same charge, the smaller ion would have a more concentrated charge and get closer to another ion to form a stronger bond.

SELF-TEST 5–A

1. Charged atoms are called _____.
2. The attraction between positive and negative ions produces a(n)
   _____ bond.

3. A sodium atom loses _____ electron(s) in achieving a noble gas configuration.

4. What is the correct formula for calcium iodide (Ca^{2+} and I^-)? _____

5. Which ion gained an electron in its formation: Na^+ or Cl^-? _____

6. Electrons in the outer orbit may be called _____ electrons.

7. Positive ions are formed from neutral atoms by ( ) losing or ( ) gaining electrons.

8. Negative ions are formed from neutral atoms by ( ) losing or ( ) gaining electrons.

9. Predict the number of electrons lost or gained by the following atoms in forming ions. Indicate whether the electrons are gained or lost.

Rb _____   S _____ _____

Ca _____ _____   Mg _____ _____

K _____ _____   Br _____ _____

COVALENT BONDS

What holds together carbon monoxide (CO), methane (CH_4), water (H_2O), quartz (SiO_2), ammonia (NH_3), carbon tetrachloride (CCl_4), and molecules of about 5 million other compounds in which all of the elements are nonmetals? They are all very poor conductors of electricity in the melted state. Remember that all nonmetals have higher ionization energies, and none are prone to form positive ions to balance possible negative ions.

The driving forces of electron pairing and the stable eight-electron (**octet**) arrangement of the noble gases can be accommodated by *sharing* pairs of electrons between atoms of elements in Groups IVA, VA, VIA, and VIIA. The sharing of electrons between two atoms produces a **covalent bond.** The strength of the bond comes from interaction of an orbital of one atom with an orbital of another atom (Fig. 5–6). The shared electrons are held to each other by pairing forces, and the electron pairs are held to the two nuclei by attractions between unlike charges.

The drive to attain the stable eight valence electrons of a noble gas is known as the octet rule. The rule is particularly applicable to carbon and a few other nonmetals.

G. N. Lewis (1875–1946) proposed in 1916 that a *covalent bond* results from sharing an electron pair. He was a professor of chemistry at University of California, Berkeley. Many of his ideas about bonding are still applicable today.

Single Covalent Bonds

A single covalent bond is formed when two atoms share a single pair of electrons. The simplest examples are diatomic (two-atom) molecules such as H_2 (hydrogen), F_2 (fluorine), and Cl_2 (chlorine).

A hydrogen atom has one electron. If a hydrogen atom could share its electron with another atom that has an unpaired valence electron of opposite spin,

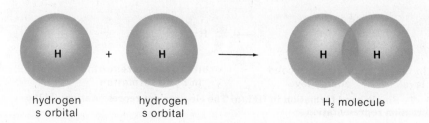

hydrogen
s orbital

hydrogen
s orbital

H_2 molecule

**FIGURE 5–6
Interaction of s orbitals
in H_2 molecule.**

To break a bond requires energy; when bonds are formed, energy is released.

a stable pairing of the two electrons can be achieved and the hydrogen atom can then have the electronic structure of helium, a noble gas. This arrangement can be achieved by two hydrogen atoms sharing their single electrons. The electron dot formula for the H_2 molecule is

$$2H\cdot \rightarrow \quad H\!:\!H \quad + energy$$
ATOMS   MOLECULE

Since each fluorine atom has one unpaired electron ($:\ddot{\ddot{F}}\cdot$) ($1s^2\, 2s^2\, 2p_x^2\, 2p_y^2\, 2p_z^1$), two fluorine atoms also can share an electron each and form a single covalent bond and a F_2 molecule.

$$2:\ddot{\ddot{F}}\cdot \rightarrow \quad :\ddot{\ddot{F}}\!:\!\ddot{\ddot{F}}: \quad + energy$$
ATOMS   FLUORINE MOLECULE

Only the pair of electrons represented between the two symbols (the two F's) are bonding electrons. The other six pairs of electrons are called **nonbonding** valence electrons, and they repel each other.

The ×'s and •'s distinguish the sources of the identical electrons.

Before reading any further, you might draw the electron dot structures of Cl_2, Br_2, and I_2.

When hydrogen ($H\times$) and fluorine ($\cdot\ddot{\ddot{F}}:$) combine to form HF ($H\!\overset{\times}{:}\!\ddot{\ddot{F}}:$), the s orbital of hydrogen interacts with a half-filled p orbital of fluorine (Fig. 5–7) to form a single covalent bond.

In a water molecule, two O—H single covalent bonds are formed. An oxygen atom has six valence electrons, of which two are unpaired ($\cdot\ddot{O}:$). It needs two more electrons to pair up its electrons and produce the stable octet of a noble gas. Two hydrogen atoms supply the two electrons.

$$:\ddot{O}\cdot + 2H\times \rightarrow :\ddot{O}\overset{\times}{\underset{\times}{:}}H + energy$$
$$\overset{\times}{\underset{H}{}}$$
WATER

The choice of H:Ö: over H:Ö:H will be explained later in this chapter.

An ammonia (NH_3) molecule has three N—H single covalent bonds. A nitrogen atom has five valence electrons, of which three are unpaired ($\cdot\ddot{N}\cdot$). The atom needs three more electrons to pair up its electrons and give it the stable eight. Three hydrogen atoms supply the three electrons to form the NH_3 molecule.

$$\cdot\ddot{N}\cdot + 3H\times \rightarrow H\overset{\times}{:}\ddot{N}\overset{\times}{\underset{\times}{:}}H + energy$$
$$\underset{H}{}$$
AMMONIA

A molecule of BF_3 has three B—F single covalent bonds. A boron atom has only three valence electrons ($\times\overset{\times}{B}\times$), and each fluorine atom has seven valence electrons, one unpaired ($\cdot\ddot{\ddot{F}}:$). When a boron atom and three fluorine atoms share

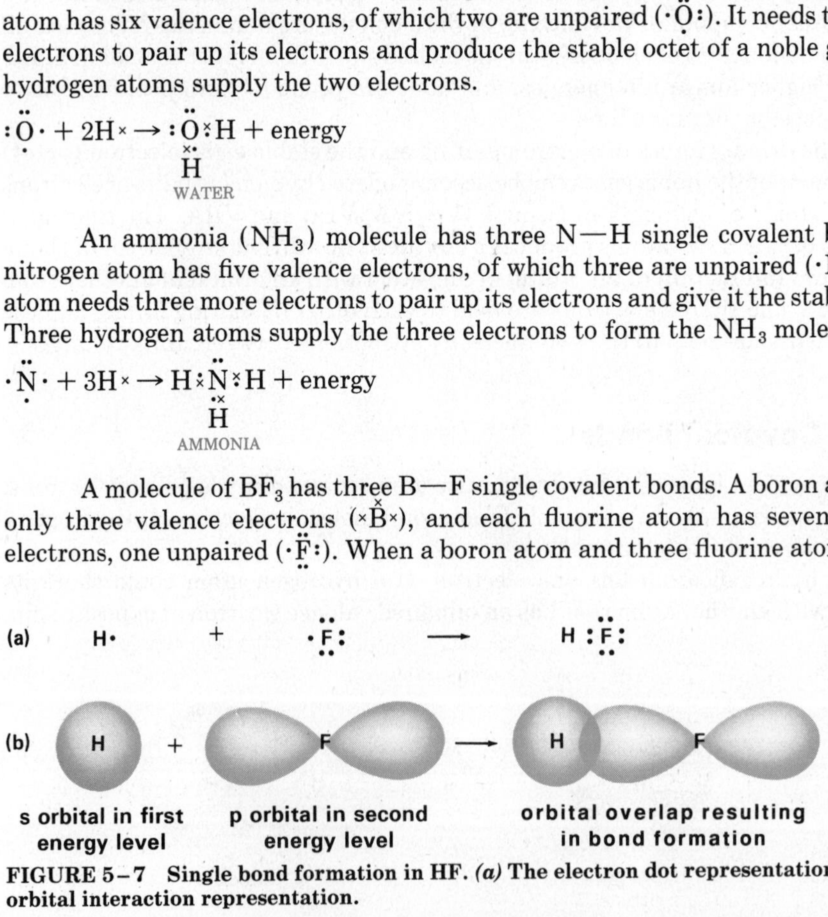

(a)   H·   +   ·F̈:   ⟶   H :F̈:

(b)   H   +   F   ⟶   H   F

s orbital in first   p orbital in second   orbital overlap resulting
energy level   energy level   in bond formation

FIGURE 5–7   Single bond formation in HF. *(a)* The electron dot representation; *(b)* the orbital interaction representation.

electrons to form a molecule of BF_3, the boron has all of its electrons paired, but it has only six electrons (three pairs) in its valence shell.

$$\times \overset{\times}{\underset{\times}{B}} \times + 3 \cdot \overset{\cdot\cdot}{\underset{\cdot\cdot}{F}} : \rightarrow : \overset{\cdot\cdot}{\underset{\cdot\cdot}{F}} \overset{\times}{\underset{\times}{B}} \overset{\times\overset{\cdot\cdot}{\overset{\cdot\cdot}{F}}\cdot\cdot}{} \overset{\cdot\cdot}{\underset{\cdot\cdot}{F}} : + \text{energy}$$

The octet rule has exceptions.

Since BF_3 gas can be isolated, the requirement to have all valence electrons paired must supersede the requirement to have eight electrons in the valence shell. Yes, pairing sometimes supersedes the octet rule, but the molecular examples are few. Carbon occurs in more than 5 million known compounds, covalently bonded and requiring the stable eight electrons in the valence shell of each carbon atom. This is reason enough to use the simple rule of eight, remembering that occasionally there are exceptions.

Before reading on, draw the electron dot structures for methane, CH_4, and carbon tetrachloride, CCl_4, remembering that the four valence electrons of carbon are unpaired ($\cdot \overset{\cdot}{C} \cdot$).

Have you ever seen the chemical formula for trisodium phosphate (TSP, Na_3PO_4) or for blue vitriol [copper (II) sulfate, $CuSO_4$]? These are common substances sold in hardware stores and elsewhere for cleaning floors (Na_3PO_4) and killing algae in ponds ($CuSO_4$). Both of these substances have the properties of ionic compounds. When Na_3PO_4 is dissolved in water, sodium ions (Na^+) and phosphate ions (PO_4^{3-}) are formed. Copper (II) sulfate forms copper ions (Cu^{2+}) and sulfate ions (SO_4^{2-}) in water. Since the PO_4^{3-} and SO_4^{2-} ions are composed of nonmetal atoms only, the P—O and S—O bonds are covalent bonds. If the P and S atoms are surrounded by the oxygen atoms in electron dot structures, the accepted structure is represented. Circles are used to represent the electrons transferred from the metal (Na, Cu) atoms to the PO_4^{3-} (addition of three electrons) and SO_4^{2-} (addition of two electrons). The bonds marked **coordinate covalent** are formed by one atom supplying both electrons for the shared bond.

The bonds between Na^+ and PO_4^{3-} and between Cu^{2+} and SO_4^{2-} are ionic bonds.

The phosphate ion and the sulfate ion are examples of **polyatomic** (many-atom) **ions,** which are held intact by covalent bonds. A few common examples are listed in Table 5–3.

TABLE 5–3   A Few Polyatomic Ions

Ammonium	NH_4^+	Hypochlorite	ClO^-	Chromate	CrO_4^{2-}
Acetate	$CH_3CO_2^-$	Chlorate	ClO_3^-	Silicate	SiO_3^{2-}
Nitrate	NO_3^-	Perchlorate	ClO_4^-	Phosphate	PO_4^{3-}
Nitrite	NO_2^-	Carbonate	CO_3^{2-}	Arsenate	AsO_4^{3-}
Hydroxide	OH^-	Sulfate	SO_4^{2-}		

Before continuing on, draw the electron dot structure for the perchlorate ion, ClO_4^-.

Multiple Bonding

One pair of electrons is *one* covalent bond.

When an atom has fewer than seven electrons in its valence shell, it can form covalent bonds in two ways. The atom may share a single electron with each of several other atoms, which can contribute a single electron each. This leads to **single** covalent bonds. But the atom can also share two (or three) pairs of electrons with a single other atom. In this case there will be two (or three) bonds between these two atoms. When two shared pairs of electrons join together the same two atoms, we speak of a **double bond,** and when three shared pairs are involved, the bond is called a **triple bond.** Examples of these bonds are found in many compounds such as those shown in Figure 5–8.

A *double bond* consists of two electron pairs shared between two atoms.

As we can see from these structures, molecules may contain several types of bonds. Thus, ethylene (Fig. 5–8) contains a double bond between the carbon atoms and single bonds between the hydrogen atoms and the carbon atoms. For convenience, an electron pair bond is often indicated by a dash as follows:

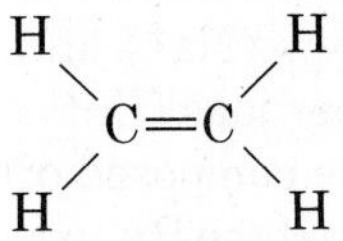

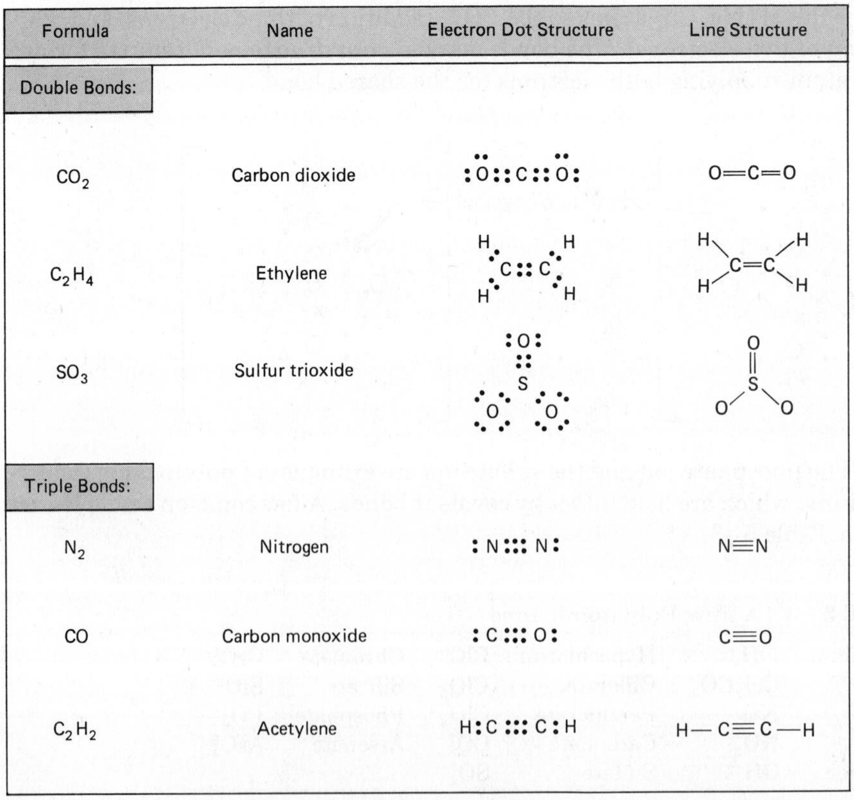

FIGURE 5–8
Electron dot structures of some molecules containing multiple bonds. Line structures are shown for comparison.

Formula	Name	Electron Dot Structure	Line Structure
Double Bonds:			
CO_2	Carbon dioxide		O=C=O
C_2H_4	Ethylene		
SO_3	Sulfur trioxide		
Triple Bonds:			
N_2	Nitrogen		N≡N
CO	Carbon monoxide		C≡O
C_2H_2	Acetylene		H—C≡C—H

TABLE 5–4   Some Bond Lengths and Bond Energies

Bond type	C—C	C=C	C≡C	N—N	N=N	N≡N
Bond length (nm)	0.154	0.134	0.120	0.140	0.124	0.109
Bond energy (kcal/mole)	83	146	200	40	100	225

kcal/mole = thousands of calories necessary to break 6.02×10^{23} bonds.

The H_2 molecule with a single bond is shown as H—H; ethylene, with a double bond, is shown as $H_2C=CH_2$; and diatomic nitrogen, with a triple bond, is shown as N≡N. Note that in each of these cases the stable octet rule is obeyed if the shared electrons can be counted as belonging to both atoms.

A line between two atoms, as in H—H, represents a bonding pair of electrons.

Single, double, and triple bonds differ in length and strength. Triple bonds are shorter than double bonds, which in turn are shorter than single bonds. Bond energies normally increase with decreasing bond length as a result of greater orbital interaction. **Bond energy** is the amount of energy required to break a mole of the bonds. Some typical bond lengths and energies are listed in Table 5–4.

Polar Bonds

In a molecule like H_2 or F_2, where both atoms are alike, there is equal sharing of the electron pair. Where two unlike atoms are bonded, however, the sharing of the electron pair is unequal and results in a shift of electric charge toward one partner. Recall that the more nonmetallic an element is, the more that element attracts electrons. This is due to the relatively large ratio of nuclear charge to atomic size. In effect, the larger this ratio, the more strongly an atom attracts its electrons and those shared with other atoms in covalent bonding.

In a *polar bond*, there is an unequal sharing of the bonding electrons.

The attraction for the electrons in a chemical bond can be expressed on a quantitative basis and is called **electronegativity.** Nonmetallic character increases across and up the periodic table toward fluorine (F), which has the largest electronegativity of the nonmetals. In 1932 Linus Pauling first proposed the concept of electronegativity based on bond energy differences. The currently accepted values for electronegativities of elements are shown in Figure 5–9. The most electronegative element is fluorine, with an electronegativity of 4.0. The electronegativities generally increase along a diagonal line drawn from francium (Fr) to fluorine. The values for other elements are between these two extremes. Although electronegativities show a periodic trend (Fig. 5–10), the pattern is not as regular as that for ionization energies (Fig. 4–9) or atomic radii (Fig. 4–8).

Linus Pauling (1901–   ) has been awarded two Nobel prizes: in 1954 for his work on molecular structure and in 1963 for his efforts on nuclear disarmament. Only he and Marie Curie have received two Nobel prizes.

When two atoms are bonded covalently and the electronegativities of the two atoms are the same, there is an equal sharing of the bonding electrons, and the bond is a **nonpolar** covalent bond. The bonds in H_2, F_2, and NBr_3 (N and Br have the same electronegativity, 3.0) are nonpolar.

The electronegativity of an atom is a measure of its ability to attract electrons to itself in a compound. The most electronegative atom is fluorine.

Two atoms with different electronegativities bonded covalently form a **polar** covalent bond. The bonds in HF, NO, SO_2, H_2O, CCl_4, and BeF_2 are polar.

In a molecule of HF, for example, the bonding pair of electrons is more under the control of the highly electronegative fluorine atom than of the less electronegative hydrogen atom (Fig. 5–11).

When covalent bonds join different atoms, the bonds are generally polar because one of the atoms has distorted the electron distribution toward itself as a

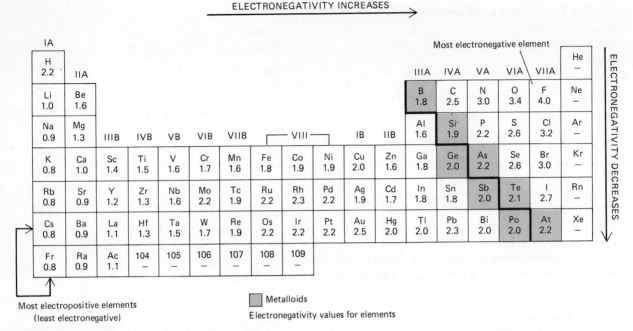

ELECTRONEGATIVITY INCREASES

Most electronegative element

Most electropositive elements
(least electronegative)

☐ Metalloids

Electronegativity values for elements

FIGURE 5–9 Some electronegativity values in a periodic table arrangement.

result of its greater electronegativity. Thus, polar covalent bonds occur in practically every molecule that has different kinds of covalently bonded atoms.

A very common bond that we shall discuss frequently in the remainder of this text is the C—H bond. Since carbon has an electronegativity of 2.5 and hydrogen 2.2, the C—H bond is only slightly polar. The arrangement of C—H bonds around a carbon atom generally makes —CH$_2$— and —CH$_3$ groups nonpolar.

The polar bonds in beryllium difluoride (BeF$_2$), water (H$_2$O), carbon tetrachloride (CCl$_4$), and chloroform (CHCl$_3$) are indicated in Figure 5–12 by arrows in the direction of the electron shift, pointing toward the more electronegative atom.

If a substance has polar bonds, it may have polar molecules or it may have nonpolar molecules. It all depends on the three-dimensional geometric shape of the molecule, discussed in the last section of this chapter. If the electron shifts within the molecule balance out (are symmetrical), the substance has **nonpolar molecules.** (See BeF$_2$ and CCl$_4$ in Fig. 5–12.) Or, to say it another way, if the centers of

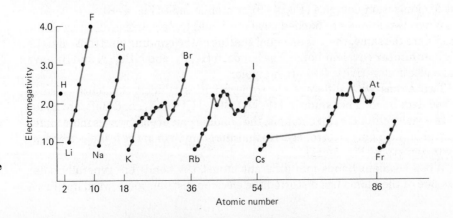

FIGURE 5–10 Periodic nature of the electronegativities when plotted versus atomic number.

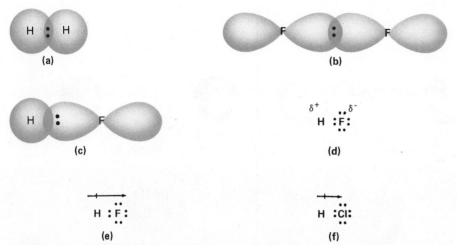

FIGURE 5–11 Polar bonds in HF and HCl. *(a)* Symmetrical distribution of electrons in H_2 results in the center of negative charge being identical with the center of positive charge. This is symbolized by the electron dots placed in the overlap area. *(b)* Overlap of p orbitals in F_2 also results in symmetrical distribution of charge. *(c)* In HF, the electron pair is displaced toward the fluorine nucleus since fluorine is more electronegative than hydrogen. Note the electron dots have been placed to the right of the overlap area to convey the idea of polarity (separation of charge). *(d)* δ^+ (delta positive, meaning fractional positive charge) and δ^- (delta negative, meaning fractional negative charge) are used to indicate poles of charge. In *(e)* and *(f)* an arrow is used to indicate electron shift, the arrow having a "plus" tail to indicate partial positive charge on the hydrogen atom. Note that the longer arrow in the HF structure indicates a greater degree of polarity than in HCl. This should not be confused with the greater bond length in HCl.

positive and negative charge coincide, the molecule is nonpolar. On the other hand, if the electron shifts within the molecule do not balance out (are asymmetrical), the substance has **polar molecules** (H_2O and $CHCl_3$ in Fig. 5–12). In other words, if the centers of positive and negative charge do not coincide, the molecule is polar, and the substance will have properties reflecting this polar nature.

Water is a polar molecule.

TABLE 5–5   Characteristics of Ionic, Polar Covalent, and Pure Covalent Compounds

TYPE OF BOND	IONIC	POLAR COVALENT	PURE COVALENT
Disposition of the electrons	Transferred from metal to nonmetal	Partially transferred	Shared
Elements involved	Groups IA, IIA, transition, inner transition metals with Groups VA, VIA, VIIA nonmetals	Nonmetals with nonmetals: IVA, VA, VIA, VIIA	Nonmetals with nonmetals: IVA, VA, VIA, VIIA
Electronegativity difference	Great (more than 2)	Small	None
Conductance of electricity as a solid	No	No	No
Conductance of electricity as a liquid (melted solid)	Yes	No	No
Molecules	No	Yes	Yes
Ions	Yes	No	No

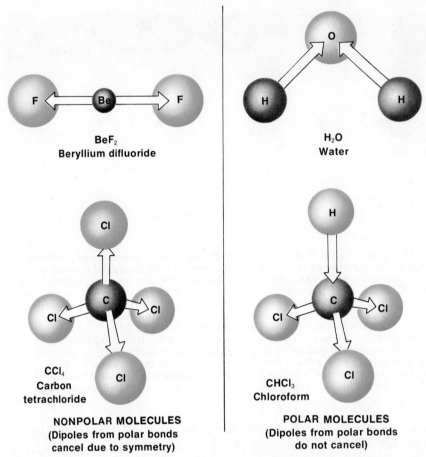

BeF$_2$
Beryllium difluoride

H$_2$O
Water

CCl$_4$
Carbon tetrachloride

CHCl$_3$
Chloroform

NONPOLAR MOLECULES
(Dipoles from polar bonds cancel due to symmetry)

POLAR MOLECULES
(Dipoles from polar bonds do not cancel)

FIGURE 5–12   Polar bonds may or may not result in polar molecules. The polar bonds in beryllium difluoride and carbon tetrachloride are arranged about the center atom in such a way as to cancel out the polar effect. In contrast, the polar bonds in water and chloroform molecules do not cancel as a result of the molecular shape but combine to give a polar molecule.

Whether a substance is polar or nonpolar can have a great effect on the chemical reactivity of the substance and its solubility in various liquids. For example, an old rule of thumb is that **like dissolves in like:** polar substances dissolve in polar liquids; nonpolar substances dissolve in nonpolar liquids. Therefore, if rubbing alcohol (2-propanol or isopropyl alcohol) will dissolve in polar water, rubbing alcohol must be a polar substance. Likewise, if gasoline will not dissolve in polar water, gasoline is nonpolar and will dissolve in nonpolar carbon tetrachloride.

An experimental method for detecting polar molecules is represented in Figure 5–13.

Features of ionic, polar covalent, and nonpolar or pure covalent compounds are summarized in Table 5–5.

Intermolecular bonds are bonds between molecules.

Hydrogen bonds can form between molecules or within a molecule. These are known as intermolecular and intramolecular hydrogen bonds, respectively.

INTERMOLECULAR FORCES

The attractive forces between molecules are collectively known as **van der Waals forces,** named after the Dutch physicist who won the Nobel prize in 1910 for his

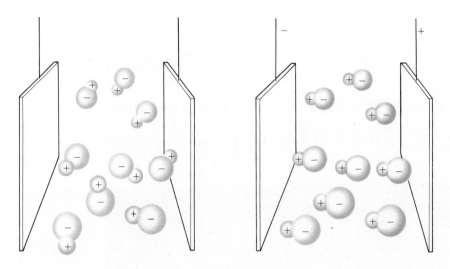

Field off   Field on

FIGURE 5–13   Physical evidence for both the existence of polar molecules and the degree of polarity is provided by a simple electrical capacitor. The capacitor is composed of two electrically conducting plates with nonconducting material (an electrical insulator) between the plates. The storage of charge by the capacitor is least when there is a vacuum between the plates; charge storage is improved when nonpolar substances are placed between the plates and is most effective with polar substances between the plates. Energy is required to orient the molecules, and so energy is stored in the molecules until the field is turned off. Then the energy is released as the dipoles become random again.

studies of intermolecular forces in gases and liquids. They include dipole–dipole forces and temporary dipole (London forces). **Hydrogen bonding** is a special case of dipole–dipole interactions, in which the hydrogen atom in a polar molecule interacts with an electronegative atom in either an adjacent molecule or the same molecule.

When a covalently bound hydrogen atom forms a second bond to an electronegative atom, the second bond is called a hydrogen bond.

Hydrogen Bonding

For a series of molecular substances with similar structures, the boiling points ordinarily increase as the molecular weights increase. For example, the boiling points of fluorine (F_2), chlorine (Cl_2), bromine (Br_2), and iodine (I_2) increase rather regularly with increasing molecular weight (Fig. 5–14).

The general relationship between boiling points and molecular weights also holds for hydrogen chloride (HCl), hydrogen bromide (HBr), and hydrogen iodide (HI). However, the boiling point of hydrogen fluoride (HF), the lightest member of this series of compounds, is abnormally high (Fig. 5–15). Irregularities similar to this are also found in other compounds in which hydrogen is bonded to fluorine, oxygen, and nitrogen. The increased attraction between molecules containing H—F, H—O, or H—N bonds is termed **hydrogen bonding.**

The explanation for hydrogen bonding is to be found in the extremely large electronegativity differences of the H—F, H—O, and H—N bonds, the polarity being due to the extreme electronegativity of fluorine, oxygen, and nitrogen. Consider the bonding in liquid HF. Since unlike-charged ends of these molecules should attract each other, we expect HF molecules to be associated with one another. This association is illustrated in Fig. 5–16. The increased association

For a substance to evaporate or boil, its molecules must gain enough energy to break loose from each other.

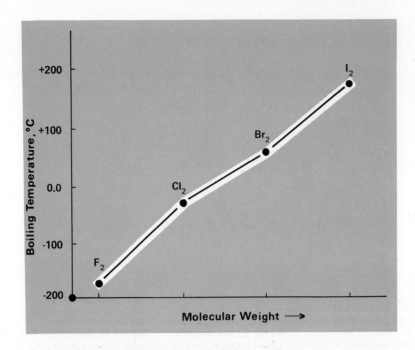

FIGURE 5–14
Boiling points of F$_2$,
Cl$_2$, Br$_2$, and I$_2$, as a
function of molecular
weight.

Water has an abnormally
high boiling point because
of hydrogen bonding
between the molecules.

between HF molecules compared with that found between HCl molecules offers a
ready explanation for the unusually high boiling point of HF.

Water provides another good example of hydrogen bonding. Figure 5–15
illustrates the boiling point of H$_2$O is about 200 degrees higher than would be
predicted if hydrogen bonding were not present. Water molecules are not linear but
rather are angular, with two nonbonding (unshared) electron pairs located toward
one end of the molecule and the partially positive hydrogen atoms located toward
the opposite end. The two polar bonds in this geometry result in a distinctly polar
molecule.

The δ^- and δ^+ represent
partial charges.

$$\ddot{\underset{\delta^+}{\text{H}}}\quad\overset{\overset{\delta^-}{\cdot\ddot{\text{O}}\cdot}}{}\quad\text{H}$$

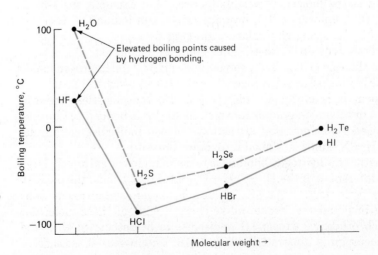

FIGURE 5–15
Boiling points of Group
VIA and Group VIIA
hydrides are plotted
against molecular
weights.

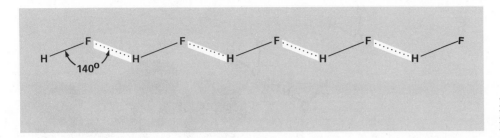

FIGURE 5-16
Hydrogen bonding in HF.

In liquid and solid water, where the molecules are close enough to interact, the hydrogen atom on one of the water molecules is attracted to the **nonbonding electrons** on the oxygen atom of an adjacent water molecule. This is possible because of the small size of the hydrogen atom. The result of this association is called hydrogen bonding because the slightly positive hydrogen atom acts as a sort of a bridge to hold two molecules together in much the same way as electron pairs hold atoms together in molecules. Since each hydrogen atom can form a hydrogen bond to an oxygen atom in another water molecule, each water molecule can form a maximum of four hydrogen bonds to four other water molecules (Fig. 5-17a). The result is a tetrahedral cluster of water molecules around the central water molecule. Imagine that liquid water is made up of clusters of hydrogen-bonded water molecules, and the extent of hydrogen bonding is a function of temperature.

In ice, hydrogen bonding is more extensive and the resulting three-dimensional hookup of tetrahedral clusters gives the open structure shown in Fig. 5-17b. Consequently, at ordinary pressures ice is *less* dense than water. The melting of ice breaks about 15% of the hydrogen bonds, and this collapses the structure shown in Fig. 5-17b to a more dense liquid. As the liquid is heated, more hydrogen bonds are broken and the clusters become smaller. Not all the hydrogen bonds are broken, however, and large aggregates of water molecules exist in liquid water even near 100°C. As water is heated, thermal agitation disrupts the hydrogen bonding until, in water vapor, there is only a small fraction of the number of hydrogen bonds that are found in liquid or solid water.

Although hydrogen bonds are much weaker than ordinary covalent bonds, hydrogen bonding plays a key role in the chemistry of life. Later chapters in the text will discuss hydrogen bonding in connection with the properties of a number of substances such as water, DNA, proteins, alcohols, cotton, and hair.

> Bond energies of hydrogen bonds are about 10% of normal covalent bond energies.

> Why does ice float?

> Hydrogen bonding plays a key role in the chemistry of life.

London Forces

The forces between molecules that cannot be explained by dipole–dipole attractions or hydrogen bonds are often due to a weak attraction known as **London forces.**

In helium, for example, the two electrons are nonbonding, and yet there is a slight attraction between two helium atoms. This slight attraction allows helium to become a solid at −272°C under high pressure.

London forces can arise from a variety of causes. For instance, as two atoms or molecules approach each other, intermolecular interactions will cause a temporary shifting of their electron clouds. An uneven electron distribution in an atom makes the atom itself temporarily polar, producing a dipole (meaning "two poles").

> These weak intermolecular forces are named after Fritz London, who proposed the temporary dipole concept in 1937.

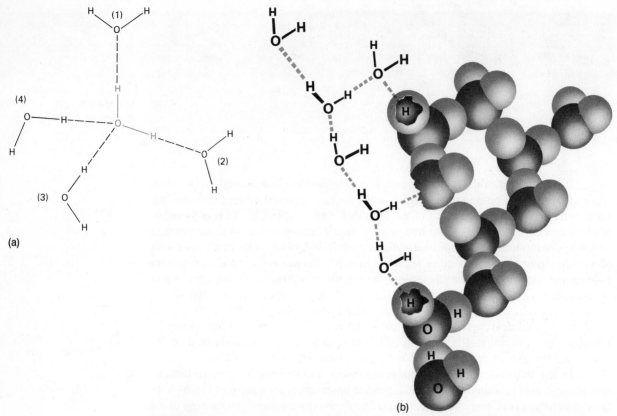

FIGURE 5-17   *(a)* Tetrahedral cluster of four water molecules around a fully hydrogen-bonded water molecule in the center. *(b)* Hydrogen bonding in the structure of ice. The hydrogen bonds are indicated by the dashed lines. In liquid water the hydrogen bonding is not as extensive as it is in ice.

The temporary poles on two adjacent atoms can interact with each other, resulting in a momentary attractive force (Fig. 5-18). The existence of the solid state of many nonpolar molecular substances (such as oxygen, nitrogen, and helium) can be explained by invoking London forces.

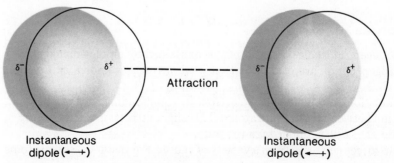

FIGURE 5-18   An illustration of London forces. One instantaneous dipole interacts with another in a neighboring atom.

METALLIC BONDING

Metals have some properties totally unlike those of other substances. For example, most metals are good electrical conductors; they are shiny solids and have relatively high melting points (with a few notable exceptions such as mercury). Any theory of the bonding of metal atoms must be consistent with these properties. Structural investigations of metals have led to the conclusion that metals are composed of regular arrays or lattices of metal ions in which the bonding electrons are loosely held. The loosely held electrons can be made to move rather easily through the lattice upon application of an electric field. In this way the metal acts as a conductor of electricity. As a consequence of this movement of bonding electrons, we cannot really write a satisfactory description of the bonding in a metal using an electron dot structure.

Loosely held electrons are found in metals. They can move freely and are not confined to the area between any particular pair of atoms.

SELF-TEST 5–B

1. a. An example of a molecule containing covalent bonding where the electrons are equally shared between the atoms is _____ ; b. one where they are unequally shared is _____ .
2. The number of electrons shared in a triple covalent bond is _____ .
3. There are _____ covalent bonds in an ammonia (NH_3) molecule.
4. Which atom cannot form a double bond, fluorine or sulfur? _____
5. a. How many valence (bonding) electrons are thought to be involved in molecules containing covalently bound atoms of period 2 and 3 elements? _____ b. This is known as the _____ rule. c. Is it true most of the time or all of the time? _____
6. Which is the most electronegative of all elements? _____
7. Hydrogen bonding very probably occurs when hydrogen is bound to atoms of _____ , _____ , or _____ .
8. Which molecule (H_2O, H_2, O_2, CCl_4) is a polar molecule? _____

MOLECULAR STRUCTURE

The modern theory of atomic structure is effective in explaining bonding and periodicity of the elements. It is also capable of explaining related chemical phenomena, such as molecular structure. The structure of a molecule is determined by the arrangement of the atoms in the molecule with respect to each other, that is, by the angles between bonds and the lengths of the bonds. The angle formed by two intersecting lines drawn from the two nuclei of the attached atoms through the nucleus of the central atom is called the **bond angle.** Molecular structures are established by many different experimental techniques. Here we shall see what molecular structures are predicted by atomic theory and what modifications are needed for theoretical and experimental findings to agree.

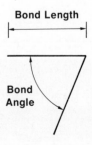

X-ray diffraction is the principal experimental method for the determination of crystal structures.

PREDICTING SHAPES OF MOLECULES

A simple, reliable method for predicting the shapes of molecules is the **Valence Shell Electron Pair Repulsion Theory (VSEPR),** which is based on the idea that electron pairs in the valence shell repel each other. In fact, this theory assumes that electron pairs in the valence shell of a central atom behave like a group of electrically charged balloons that are connected to a central point. If similarly charged, the balloons would tend to be as far apart as possible. The geometric shapes that give the maximum distance for two, three, four, five, and six electron pairs are linear, trigonal planar, tetrahedral, trigonal bipyramidal, and octahedral, respectively (Fig. 5–19).

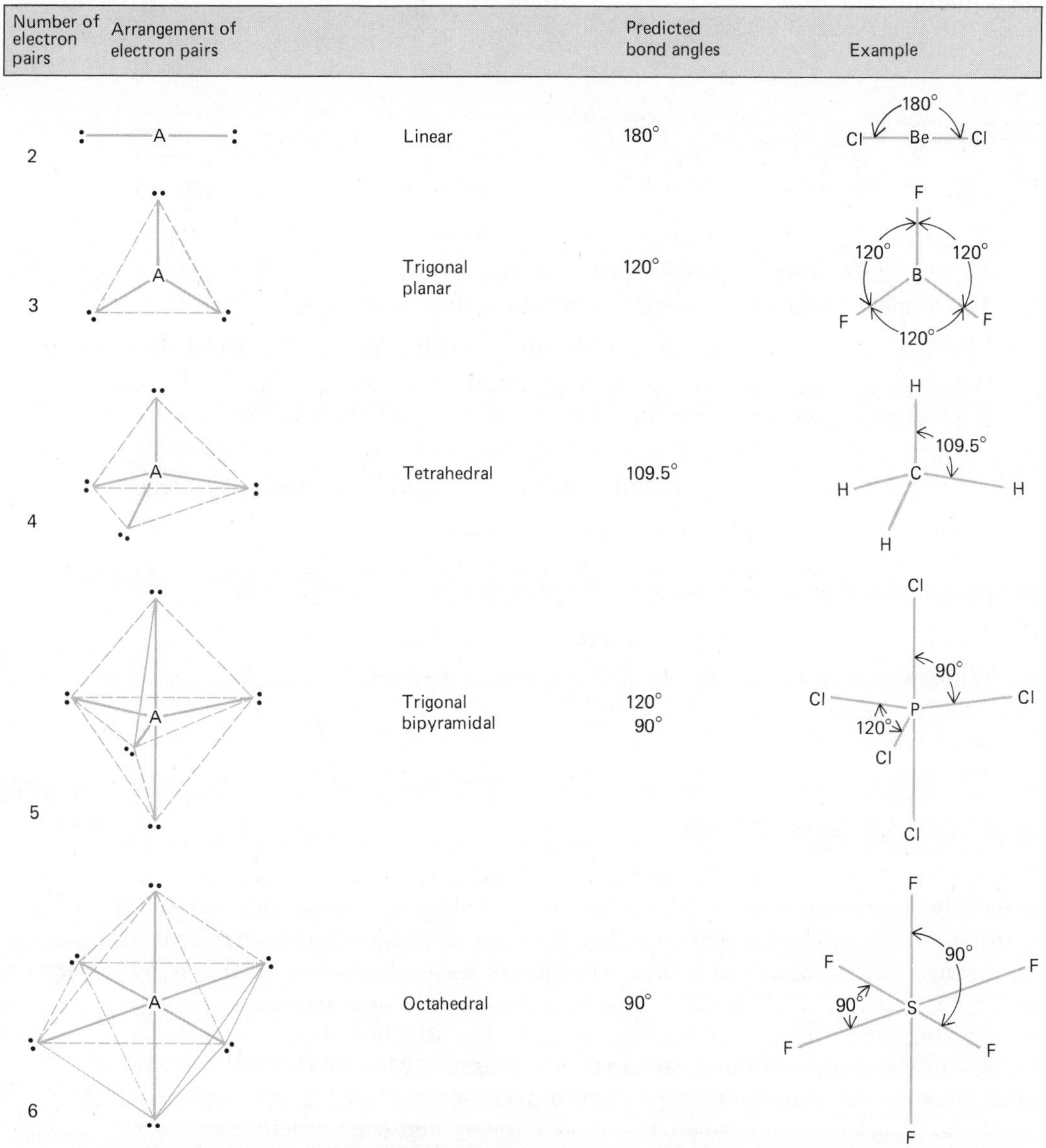

Number of electron pairs	Arrangement of electron pairs		Predicted bond angles	Example
2		Linear	180°	
3		Trigonal planar	120°	
4		Tetrahedral	109.5°	
5		Trigonal bipyramidal	120° 90°	
6		Octahedral	90°	

FIGURE 5–19 **Arrangements of electron pairs according to the Valence Shell Electron Pair Repulsion Theory.**

Since the noble gas configuration of eight valence electrons gives a stable configuration for many common nonmetals, the most common geometry is tetrahedral, with four electron pairs at the corners of a tetrahedron. Examples of molecules with four bonding pairs of electrons are silicon tetrachloride, $SiCl_4$ and methane, CH_4.

$SiCl_4$

Atomic silicon has four electrons in its outer orbit. When these are paired with the unpaired electrons of four chlorine atoms, $:\ddot{C}l\cdot$, four electron-pair bonds are formed. These four pairs of electrons in the valence shell will repel each other to form the predicted tetrahedral structure.

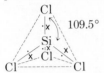

One of the advantages of VSEPR theory is the ability to predict shapes of molecules that contain both bonding and nonbonding pairs. Ammonia and water are two important examples. In both cases the central atom is surrounded by four pairs of electrons. Nitrogen in ammonia has one nonbonding pair and three bonding pairs, while oxygen in water has two nonbonding pairs and two bonding pairs. Figure 5–20 shows the tetrahedral representation of four electron pairs in CH_4, NH_3, and H_2O. The bonding angles in ammonia and water are predicted to be slightly smaller than the normal tetrahedral angle of 109.5 degrees. VSEPR theory attributes this to the larger volume occupied by nonbonding pairs compared to that occupied by bonding pairs. This spreads the nonbonding pairs farther apart and squeezes the bonding pairs closer together. Hence nonbonding pair–nonbonding pair repulsions are larger than bonding pair–bonding pair repulsions.

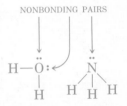

Regarding the shapes of molecules with nonbonding electrons, only the shape formed by the atoms in the molecule is stated since X-ray structure studies locate atoms, not electron pairs. You can visualize the shape of the molecule by simply ignoring the nonbonding pairs. For example, NH_3 is pyramidal and H_2O is angular.

It is often enough to know the formula of the molecule and the number of valence electrons of the central atom to predict shapes of other molecules even when the octet rule is not followed. For example, sulfur, phosphorus, and other nonmetals in the third or higher periods form several covalent molecules that have more than eight electrons in the valence shell. Two examples shown in Figure 5–19 are PCl_5 and SF_6. VSEPR predicts a trigonal pyramidal shape for PCl_5 (5 bonding pairs, 10 valence electrons) and an octahedral shape for SF_6 (6 bonding pairs, 12 valence electrons). These geometries are in agreement with the experimentally determined structures.

An example of the predictive power of VSEPR theory is the correct prediction of the shape of XeF_4 after it had been synthesized in 1962 at Argonne National Laboratory, but before its structure had been determined. Xenon is a noble gas, and as we mentioned in Chapter 4, noble gases were not expected to undergo reactions to form compounds since they have a stable octet of valence electrons. The isolation of XeF_4 created a real challenge for bonding theorists, but according to VSEPR theory, the number of bonding pairs would be four from the formation of four covalent bonds with fluorine atoms. This leaves four electrons or two nonbonding pairs of electrons. The total of six electron pairs leads to a prediction of an octahe-

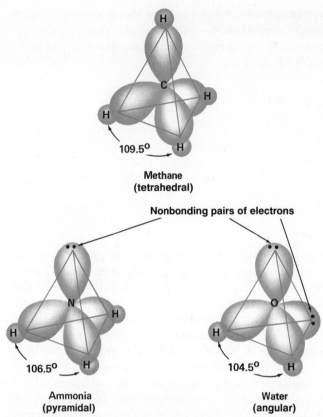

**Methane
(tetrahedral)**

Nonbonding pairs of electrons

**Ammonia
(pyramidal)**

**Water
(angular)**

FIGURE 5–20   **Methane molecules have only bonding pairs of valence electrons; water
and ammonia molecules have both bonding and nonbonding pairs. All three structures
have a tetrahedral arrangement of valence electron pairs around the central atom. Since
only atoms (not electron pairs) determine molecular structure, the different molecules
have different molecular structures.**

dral shape for the electron pairs, but where do you put the lone pairs? VSEPR
theory tells us to place them as far from one another as possible, so placing them at
opposite corners of an octahedron gives the maximum distance possible. This
leaves a square planar shape for the XeF_4 molecule (cover up the nonbonding pairs
to see this), and this was later shown by experiment to be the correct shape.

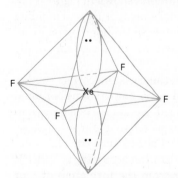

VSEPR theory can also be used to predict the shapes of molecules or ions
that contain double or triple bonds. The rule is that if you treat double or triple

bonds as though they were a single bond, your prediction of the geometry will be correct. For example, formaldehyde has the dot structure

Since the double bond would be treated the same as a single bond, the molecule would be predicted to have a trigonal planar shape based on three electron pairs. This is the correct shape for formaldehyde.

SELF-TEST 5–C

1. ( ) Bonding or ( ) nonbonding pairs of electrons repel more than ( ) bonding or ( ) nonbonding pairs of electrons.
2. Give the molecular shapes expected for the hypothetical molecules

    AX_2 _____, BZ_3 _____, CY_4 _____,

    DQ_6 _____. (There are no nonbonding valence electrons in these structures, and all of the bonds are single bonds.)
3. The water molecule has _____ bonding pairs and _____ nonbonding pairs of electrons.
4. The ammonia molecule has _____ bonding pair(s) and _____ nonbonding pair(s) of electrons.

MATCHING SET I

_____ 1. Double covalent bond	a. An electrically neutral arrangement of covalently bonded atoms
_____ 2. Ionic bonds	b. Shared electrons
_____ 3. Ionization energy	c. Requires O, F, or N
_____ 4. Metallic bonding	d. Positive ions attracted to negative ions
_____ 5. London forces	e. Ionic compound
_____ 6. Noble gas	f. Electrons free to move
_____ 7. Covalent bonds	g. Element with eight valence-shell electrons
_____ 8. NaCl	h. Covalent compound
_____ 9. Metal ion	i. Attraction between nonpolar neutral particles
_____ 10. Hydrogen bonds	j. Measures gaseous atom's hold on electron
_____ 11. NH_3	k. Smaller than parent atom
_____ 12. Molecule	l. Four electrons shared
_____ 13. Single covalent bond	m. Two electrons shared

MATCHING SET II (Shapes of Molecules)

_____ **1.** $BeCl_2$ (no nonbonding valence electrons)

_____ **2.** H_2O

_____ **3.** $SiCl_4$

_____ **4.** SF_6

_____ **5.** BCl_3

_____ **6.** XeF_4

_____ **7.** NH_3

a. Linear
b. Trigonal planar
c. Bent
d. Tetrahedral
e. Octahedral
f. Pyramidal
g. Square planar

QUESTIONS

1. Diamond (a form of carbon) has a melting point of 3500°C, whereas carbon monoxide (CO) has a melting point of −207°C. What does this suggest about the kinds of bonding found in these two substances?

2. Write the electronic configuration for the element potassium (atomic number 19). What will be the electronic configuration when a K^+ ion is formed?

3. Is Ca^{3+} a possible ion under normal chemical conditions? Why?

4. Write the symbols for the six elements with the highest ionization potentials, selecting one from each period of the periodic table.

5. Match the electronic configurations that would be expected to lead to similar chemical behavior. The numbers denote the numbers of electrons in the orbits.
    a. 2-2   d. 2-8-2
    b. 2-5   e. 2-1
    c. 2-8-8-1   f. 2-8-5

6. Fluorine (atomic number 9) has an electronic configuration of 2-7. How many electrons will be involved in the formation of a single covalent bond?

7. What kind of bond (ionic, pure covalent, polar covalent) is likely to be formed by the following pairs of atoms?
    a. A Group IA element with a Group VIIA element
    b. A Group VIA element with a Group VIIA element
    c. Two chlorine atoms
    d. An element with low electronegativity and an element with high electronegativity
    e. Two elements with about the same electronegativity
    f. Two elements with the same electronegativity

8. How many electrons would there be in an iodide ion, I^-?

9. Write the electron dot structures for the fluoride ion, F^-, the chloride ion, Cl^-, and the bromide ion, Br^-.

10. Draw the electron dot structure for water. Based on bonding theory, why is water's formula not H_3O?

11. Define the term _bond energy_.

12. Draw electron dot structures for the following molecules:
    a. NF_3 h. CH_3OH o. ClO_3^-
    b. CCl_4 i. Br_2 p. SO_3^{2-}
    c. C_2Cl_2 j. HCl q. NH_4^+
    d. OF_2 k. BCl_3 r. OH^-
    e. H_2S l. PH_3 s. AsO_4^{3-}
    f. CO m. SiH_4
    g. N_2H_4 n. IBr

13. The members of the nitrogen family, N, P, As, and Sb, form compounds with hydrogen: NH_3, PH_3, AsH_3, and SbH_3. The boiling points of these compounds are

    SbH_3 −17°C
    AsH_3 −55°C
    PH_3 −87.4°C
    NH_3 −33.4°C

    Comment on why NH_3 doesn't follow the downward trend of boiling points.

14. Match the following substances with the type of bonding responsible for holding units in the solid together.

    solid krypton (Kr)   ionic
    ice   covalent
    diamond   metallic
    CaF_2 hydrogen bonding
    iron   London forces

15. Predict the general kind of chemical behavior (i.e., loss, gain, or sharing of electrons) you would ex-

pect from atoms with the following electron arrangements:

a. 2-8-1

b. 2-7

c. 2-4

16. Show how two fluorine atoms can form a bond by the interaction of their half-filled p orbitals.

17. Select the *polar* molecules from the following list and explain why they are polar:
N_2, HCl, CO, NO

18. How many bonds join the two atoms in each of the following?
CN^-, Cl_2, S_2

19. Boron trichloride has the electron dot formula
:C̈l:
:C̈l×B̈×C̈l:. What does this tell you about the octet rule even for Period 2 elements?

20. Use your chemical intuition and suggest a reaction that might occur between boron trichloride (Question 19) and ammonia, H×N̈×H.
                                              H

21. Give the basic points of the valence-shell electron-pair repulsion theory.

22. Explain the 106.5° H—N—H bond angles of NH_3, the ammonia molecule.

23. How many atomic positions must be specified in order to define a bond angle?

24. $BeCl_2$ is a compound known to contain polar Be—Cl bonds, yet the $BeCl_2$ molecule is not polar. Explain.

25. If ionic bonds are represented by white, how would pure covalent bonds and polar covalent bonds be represented by black and grey?

26. How is an ionic bond formed? How is a covalent bond formed?

27. A compound will not conduct electricity when melted, and it melts at 46°C, a low melting point. What type of bond holds atom to atom in this compound?

28. What ions would probably be formed by Br, Al, Ba, Na, Ca, Ga, I, S, O, Mg, K, At, Fr, all Group IA metals, all Group VIIA nonmetals?

29. How are ionic solids held together?

30. A compound will conduct electricity when melted, but it is rather hard to melt. What type of bonds are in this compound?

31. A substance is composed of carbon only, or of two nonmetals. The substance has a high melting point and is very hard. What kind of bonds hold the atoms of the substance together?

32. Identify the following two statements as either empirical or theoretical.
a. Unlike charges attract.
b. Electrons tend to exist in pairs.

33. In which case would hydrogen bonding be most extensive: (a) liquid water, (b) water vapor, or (c) ice?

34. Why is water a liquid at room temperature?

35. Predict the formulas of compounds consisting of the following elements.

NONMETAL	METAL	Ba	Al	K
Cl				
O				
S				
N				
I				

36. Do you suppose any type of glue holds things together without chemical bonds?

37. Which is harder to break, an ordinary covalent bond or a hydrogen bond?

38. What is the direction of energy transfer in a
a. bond-making process?
b. bond-breaking process?

39. Liquid water consists of water molecules held together by covalent O—H bonds, and the water molecules are held together loosely by hydrogen bonds. When water boils, which type of bond breaks first?

40. Since energy is required to break bonds, do you expect water molecules to break apart at some elevated temperature above the normal boiling point of water?

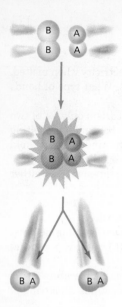

Chapter 6

SOME PRINCIPLES OF CHEMICAL REACTIVITY

Literally millions of chemical changes have been observed and described in the literature. A typical pre-medical student in organic chemistry will study the chemical changes involving approximately 2000 organic compounds in a two-semester course and, during this time, chemists will add many more than 2000 new organic compounds to the known list. Can there be any hope, then, that a liberal arts student like yourself, or even a chemist for that matter, can learn enough chemistry to find a sensibility that surrounds this science and a sense of its usefulness to help meet individual and societal needs? One helpful solution is to look for principles that are related to all of the chemical changes. Happily, there are only a few such generalizations that bring order to a multiplicity of facts. It is even more impressive when chemical principles, coupled with related theory, offer predictability for new chemical information. Some of the more important principles are presented in this chapter.

REACTANTS BECOME PRODUCTS

In all chemical reactions some pure substances disappear and others appear. A familiar example is the change of shiny steel to a red-brown iron rust (Fig. 6–1). This is a very important reaction since it has been estimated that the loss from corrosion of iron and steel in the United States is in excess of $24 billion per year.

Some other chemical changes are given in the following equations:

Reactants   *Products*

$$CaO + H_2O \rightarrow Ca(OH)_2$$
CALCIUM   WATER   CALCIUM
OXIDE   HYDROXIDE
(QUICKLIME)   (SLAKED LIME)

$$2\,Na + Cl_2(gas) \rightarrow 2\,NaCl$$
SODIUM   CHLORINE   SODIUM
                              CHLORIDE
                              (TABLE SALT)

$$H_2(gas) + I_2(gas) \rightarrow 2\,HI(gas)$$
HYDROGEN   IODINE   HYDROGEN
                              IODIDE

FIGURE 6–1 Chemical changes produce new substances, generally with properties very different from the starting substances. Bright, shiny nuts and bolts are changed to crumbly, dull rust by iron reacting with oxygen.

From these reactions we can note our first important point concerning chemical reactions:

> In chemical reactions, reactants become products; some substances are consumed and new ones appear.

As was pointed out in Chapter 2, one of the first laws established in chemistry is the conservation of matter in chemical change. Pure substances are destroyed and created, but the matter involved is preserved. The practicality of the conserved quantitative relationship between the amount of reactants consumed and products produced is explored in the following section.

WEIGHT RELATIONSHIPS IN CHEMICAL REACTIONS

An important question concerning weight relationships in chemical reactions could be: What weight of aluminum can be produced from one ton of aluminum oxide? Once the atomic weight scale was established, such important calculations could be accomplished readily in view of the conservation of matter involved. We shall return to this question in Example 4, after looking at simple relationships based on the balanced chemical equation.

Hydrogen reacts with chlorine to form hydrogen chloride. The weight of the reactants, hydrogen and chlorine, used in the reaction must equal the weight of the product, hydrogen chloride. The reaction can be written more concisely by using the chemical equation:

$$H_2 + Cl_2 \rightarrow 2\ HCl$$

In chemical reactions new substances are formed and old ones disappear.

All calculations involving chemical equations are based on the law of conservation of matter.

Accounting for atoms requires balancing an equation. To balance a chemical equation:

1. Place numbers (coefficients) only *before* formulas. ($2\ H_2$ means 2 molecules of H_2 *and* $2 \times 2 = 4$ atoms of H.)
2. Have same number of each kind of atom on each side of the arrow ($\rightarrow$).
3. Do not change subscripts. Each symbol represents one atom. H_2 means a molecule composed of two atoms.

Note that a coefficient, 2, has been placed in front of the hydrogen chloride so that two atoms of hydrogen and two atoms of chlorine are represented in both the reactants and the products; that is, none are gained or lost. What is the meaning of the symbolism of the equation? There are two alternative but equally meaningful ways to interpret the equation:

One molecule of hydrogen reacts with one molecule of chlorine to form two molecules of hydrogen chloride;

or,

One mole of hydrogen molecules reacts with 1 mole of chlorine molecules to form 2 moles of hydrogen chloride molecules.

A mole of any item is 6.02×10^{23} of that item.

Once the equation is balanced, the relative number of moles for each substance involved is given by the respective coefficients. Furthermore, since 1 mole weighs 1 gram molecular weight, a set of weights for all substances involved in the reaction can be obtained easily by adding up the atomic weights of the atoms represented by the formula.

For example, the molecular weights of the three molecules involved are:

	Atomic weight	$\times$	No. of atoms	$=$	Total weight
H_2	1	$\times$	2	$=$	2
Cl_2	35.5	$\times$	2	$=$	71
HCl (two elements)					
H	1	$\times$	1	$=$	1
Cl	35.5	$\times$	1	$=$	35.5
			Total for molecule $=$		36.5

Note that one molecule of HCl has a molecular weight that is one half of the sum of the molecular weights for the reactants, H_2 and Cl_2. When it is realized that two molecules of HCl are produced for each molecule of H_2 and/or Cl_2 reacted, the weight relationship is understood properly.

$$H_2 \quad + \quad Cl_2 \quad \rightarrow \quad 2\,HCl$$
$$(1+1) + (35.5+35.5) = 2(1+35.5)$$
$$2 \quad + \quad 71 \quad = \quad 73$$
$$73 \quad = \quad 73$$

Consider another example to illustrate the calculation of a molecular weight. One of the components of gasoline is octane, C_8H_{18}. The molecular weight is calculated as follows:

	Atomic weight	$\times$	No. of atoms	$=$	Total weight
C	12.0	$\times$	8	$=$	96.0
H	1.0	$\times$	18	$=$	18.0
		Molecular weight $=$			114.0

The molecular weight is simply the sum of all of the atomic weights in the formula. The gram molecular weight of C_8H_{18} is 114.0 g, and this is the weight of 6.02×10^{23} molecules or one mole of C_8H_{18}. It follows that two moles of C_8H_{18} molecules weigh 2 moles $\times$ 114 g per mole, or 228 g.

These facts allow several types of mole calculations to be made for chemical reactions. The following examples will illustrate some of these.

EXAMPLE 1

How many moles of nitrogen (N_2) are required to react with 6 moles of hydrogen (H_2) in the formation of ammonia (NH_3)?

1. Write and balance the equation:

    $$N_2 + 3\,H_2 \rightarrow 2\,NH_3$$

2. Since 1 mole of N_2 reacts with 3 moles of H_2, how many moles of N_2 will react with 6 moles of H_2? The answer is 2 moles of N_2. If the number of moles of H_2 is doubled, then the number of moles of N_2 must be doubled to keep the same ratio of nitrogen and hydrogen that react with each other.

Ammonia is used as a crop fertilizer.

EXAMPLE 2

How many moles of nitrogen dioxide (NO_2) will be produced by 4 moles of oxygen reacting with sufficient nitrogen oxide (NO)?

1. Write and balance the equation:

    $$2\,NO + O_2 \rightarrow 2\,NO_2$$

2. From the balanced equation we see that the number of moles of NO_2 produced is twice that of the oxygen reacting. Therefore 4 moles of oxygen would produce 8 moles of NO_2.

Nitrogen dioxide is a major air pollutant.

EXAMPLE 3

How many grams of carbon dioxide (CO_2) can be produced by burning 2650 g of gasoline (C_8H_{18})?

1. Write and balance the equation:

    $$2\,C_8H_{18} + 25\,O_2 \rightarrow 16\,CO_2 + 18\,H_2O$$

2. The balanced equation states that 2 moles of gasoline produce 16 moles of CO_2. Since the molecular weight of C_8H_{18} is 114, 2 moles would weigh 228 g. The molecular weight of CO_2 is 44, that is, $(1 \times 12) + (2 \times 16)$, so 16 moles would weigh $16 \times 44 = 704$ g. Thus, 228 g of gasoline would produce 704 g of CO_2; that is, the weight of CO_2 produced is about three times the weight of gasoline burned. This means that our 2650 g of gasoline should produce about 8000 g of CO_2. To be more exact:

    $$\text{grams of } CO_2 = 2650 \text{ g } C_8H_{18} \times \frac{704 \text{ g } CO_2}{228 \text{ g } C_8H_{18}}$$

    $$= 8182 \text{ g } CO_2$$

EXAMPLE 4

How many pounds of aluminum can be obtained from 1.00 ton of pure aluminum oxide, Al_2O_3? (This was the opening question in this section.)

1. Write and balance the chemical equation:

    $$2\,Al_2O_3 \rightarrow 4\,Al + 3\,O_2$$

2. The balanced equation states that 2 moles of Al_2O_3 produce 4 moles of Al. Add up the formula weight of Al_2O_3 [(27.0 g/mole $\times$ 2 moles Al/mole Al_2O_3) + (16.0 g/mole $\times$ 3 moles O/mole Al_2O_3) = 102 g/mole]. Thus, 2 moles would weigh 2 moles $\times$ 102 g/mole = 204 g. Four moles of Al weigh 4 moles $\times$ 27.0 g/mole = 108 g. Since 108 is

about half of 204, a ton of Al_2O_3 should produce about half a ton of Al, which is about 1000 pounds. To be more exact:

$$\text{pounds of Al} = 1.00 \text{ ton } \cancel{Al_2O_3} \times \frac{108 \cancel{\text{ g Al}}}{204 \cancel{\text{ g Al}_2O_3}} \times \frac{\text{pound Al}}{454 \cancel{\text{ g Al}}}$$

$$\times \frac{454 \cancel{\text{ g Al}_2O_3}}{\cancel{\text{pound Al}_2O_3}} \times \frac{2000 \cancel{\text{pounds}}}{\cancel{\text{ton}}}$$

$$= 1060 \text{ pounds of Al}$$

By observing this problem carefully, perhaps you can see a short cut. If the units given for Al_2O_3 and the units sought for Al are the same units (ton, for example), the numbers obtained from the chemical equation need no additional factors. (We did not need the two 454 g/pound factors in the preceding numerical solution; they cancel.) Thus, the numbers obtained from the chemical equation can have any weight units (tons, pounds, grams, kilograms, and so on) *as long as both weights have the same units*. In other words, the solution to the problem resolves simply into the following setup:

$$\text{pounds of Al} = 1.00 \text{ ton } \cancel{Al_2O_3} \times \frac{108 \cancel{\text{ tons Al}}}{204 \cancel{\text{ tons Al}_2O_3}} \times \frac{2000 \text{ pounds}}{\cancel{\text{ton}}}$$

$$= 1060 \text{ pounds of Al}$$

QUANTITATIVE ENERGY CHANGES IN CHEMICAL REACTIONS

Chemical reactions may produce heat, an **exothermic** process, or absorb heat, an **endothermic** process. Furthermore, the amount of heat energy involved in a chemical change is just as quantitative as the amounts of chemicals involved. Consider again the reactions given at the beginning of this chapter, but this time add the heat effect:

	Reactants		*Products*	*Heat Effect**

+ Heat means heat is liberated.
(EXOTHERMIC)

CaO + H_2O → $Ca(OH)_2$ + 15.6 kcal per mole of $Ca(OH)_2$
CALCIUM OXIDE (QUICKLIME)   WATER   CALCIUM HYDROXIDE (SLAKED LIME)

$2 Na$ + $Cl_2(gas)$ → $2 NaCl$ + 196.4 kcal (98.2 kcal per mole of NaCl)
SODIUM   CHLORINE   SODIUM CHLORIDE (TABLE SALT)

− Heat means heat is required.
(ENDOTHERMIC)

$H_2(gas)$ + $I_2(gas)$ → $2 HI(gas)$ − 12.4 kcal (−6.20 kcal per mole of HI)
HYDROGEN   IODINE   HYDROGEN IODIDE

In the first reaction, calcium oxide (quicklime) reacts with water to give calcium hydroxide (slaked lime) with the evolution of heat. In the second reaction, metallic sodium reacts with the greenish-yellow gas, chlorine, to give sodium chloride (table salt). If a piece of hot sodium is put into a flask containing chlorine, the sodium burns quickly, liberating a great deal of heat and light, to produce white crystals of sodium chloride. In the last reaction, gaseous hydrogen reacts with gaseous iodine to produce gaseous hydrogen iodide, with the absorption of heat.

Energy changes in chemical reactions are proportional to the amount of reactant (or product).

* Heat energy can be measured in calories (cal). A kilocalorie (kcal) is 1000 calories. A calorie is the amount of heat required to raise the temperature of 1 g of water 1°C.

These facts, along with similar ones, lead to a second generalization about chemical reactions:

A given amount of a particular chemical change corresponds to a proportional amount of energy change.

For example, the preparation of 1 mole of $Ca(OH)_2$ from CaO and H_2O releases 15.6 kcal; for 2 moles of $Ca(OH)_2$, 2×15.6, or 31.2, kcal of heat is released.

Sometimes energy changes in reactions are difficult to observe because of the very slow rate of reaction. An example is the rusting of iron. This is a very important reaction since it has been estimated that the loss from corrosion of iron and steel in the United States is slightly over $100 per person per year. The reaction involved is complicated, but we can represent it by the simplified equation:

$$4 \text{ Fe} + 3 \text{ O}_2 + 6 \text{ H}_2\text{O} \rightarrow 4 \text{ Fe(OH)}_3 + 788 \text{ kcal}[197 \text{ kcal per mole of Fe(OH)}_3]$$
IRON   MOIST AIR   IRON
                                    HYDROXIDE
                                  + HEAT (RUST)

Ordinarily, the rusting of iron occurs so slowly that the liberation of heat is perceptible only with the aid of special instruments. The total amount of heat evolved in rusting is considerable, but it typically takes place over a long period of time.

SELF-TEST 6 – A

1. In photosynthesis, carbon dioxide is combined with water to form the simple sugar, glucose, and oxygen:

   _____ CO_2 + _____ H_2O → _____ $C_6H_{12}O_6$ + _____ O_2

   a. Balance the equation.
   b. How many molecules of CO_2 are necessary to produce one molecule of sugar? _____
   c. How many moles of CO_2 are necessary to produce 1 mole of sugar?

   _____

   d. What is the molecular weight for CO_2? _____
      For $C_6H_{12}O_6$? _____
   e. How many grams of CO_2 are required to make 1 mole of sugar?

   _____

2. If 68 kcal of energy is released in the formation of 18 g (1 mole) of water, how much energy would be released in the formation of 36 g of water? _____

3. Balance the following equations:
   a. _____ Mg + _____ O_2 → _____ MgO
   b. _____ Si + _____ Cl_2 → _____ $SiCl_4$
   c. _____ Al + _____ O_2 → _____ Al_2O_3

4. An important source of hydrogen, used as a rocket fuel, is the decomposition of water by electrical energy. The reaction is:

   $$H_2O \xrightarrow{\text{electrical energy}} H_2 + O_2$$

a. Balance the equation.

_____

b. What weight of water is necessary to produce 2.0 g of hydrogen?

_____

c. How many grams of oxygen would be produced as a byproduct?

_____

d. How much water would be necessary to produce 2.0 tons of hydrogen?

_____

RATES OF CHEMICAL REACTION

How fast do you digest the food you eat? How fast can iron be made from iron ore? How fast does gasoline burn? The whole notion of how fast or how slow chemical reactions proceed can be put on a quantitative basis by the concept of *reaction rate*. The rate of a reaction is always defined in terms of the changes in the amounts of chemical substances still present per unit of time. Thus, if we consider the burning of sulfur to produce sulfur dioxide,

$$S + O_2 \rightarrow SO_2$$

Raising the temperature speeds up chemical reactions.

we can discuss the rate of the reaction in terms of the amount of SO_2 formed per minute or of the amount of S or O_2 consumed per minute. A number of factors affect chemical reaction rates.

Effect of Temperature on Reaction Rate

For many reactions, a temperature rise of 10°C doubles the rate.

It is possible to alter the rate of a chemical reaction by changing the temperature. If the temperature is raised, the rates of chemical reactions are increased; if the temperature is reduced, the rates are decreased. We make use of this principle in cooking foods (a roast will cook at a faster rate at a higher temperature) and in preserving foods (foods spoil less quickly if refrigerated). Figure 6–2 illustrates the effect of temperature on the reactions that take place in a slice of fruit exposed to air.

FIGURE 6–2 The biochemical processes of decomposition occur more rapidly at higher temperatures. Half the peach shown in the photograph was refrigerated, while the other half was kept warm. The refrigerated half on the right shows little discoloration, whereas the other one shows the typical signs of decay.

Effect of Concentration on Reaction Rate

It is also possible to alter the rate of a reaction by changing the concentrations of the reactants. For example, in the reaction of sulfur and oxygen given earlier, if air replaces oxygen, the reaction will proceed at a slower rate, since air is a mixture of about one part oxygen and four parts nitrogen. The rusting of iron can be retarded by painting or coating the surface of the metal to cut down on the concentration of the oxygen and moisture at the surface. A demonstration such as that shown in Figure 6-3 contrasts the reaction of iron with oxygen in relatively low and in relatively high concentrations.

Increasing the concentration of reactants speeds up a reaction.

A theoretical (molecular) explanation for both the concentration effect and the temperature effect on the rate of chemical reactions is illustrated in Figure 6-4. On a molecular basis, the reaction between sulfur and oxygen in air occurs more slowly than in pure oxygen because there are fewer oxygen molecules per volume of air to react (collide) with sulfur molecules.

An interesting and sometimes very dangerous aspect of the concentration effect in reaction rates is the state of subdivision of the chemical reactant. You would find it difficult to impossible to burn a sack of flour in an ordinary fireplace, as the flour would tend to smother the wood and block its contact with oxygen in the air. However, the flour will burn if you can keep it hot enough and keep the air flowing to it. It is not unlike trying to burn a stack of magazines. Would you be surprised to learn that the same flour as dust in the air forms an explosive mixture so powerful that it can literally blow concrete buildings apart? Indeed, this is sadly so, and many lives have been lost in such dust explosions. It is simply a matter of having the combustible particles in close contact with the oxygen of air. Reducing particle size is really a concentration effect in terms of bringing more of the reacting

FIGURE 6-3 Effect of concentration on reaction rate. Steel wool held in the flame of a gas burner is oxidized rapidly. It is in contact with air, which is 20% oxygen. When the red hot metal is placed in pure oxygen in the flask, it oxidizes much more rapidly.

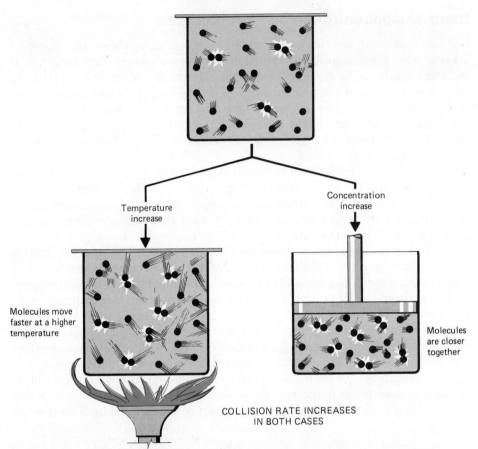

FIGURE 6–4   Effects of temperature and concentration on rates of chemical reactions. At the higher temperature, more collisions occur between molecules, and a greater percentage of the collisions produces a chemical reaction. (The molecules move faster at higher temperatures, and if kept in the same volume, they will strike more often. Moving faster, they will have more energy to cause structural changes to occur.) At the higher concentration (no temperature change), more collisions occur, but the percentage of effective collisions remains the same.

particles together in a given period. The surface concentration of the flour, the portion of the flour that is in contact with the oxygen, is increased as the particles get smaller and smaller in the same volume.

Effect of a Catalyst on Reaction Rate

Slow chemical reactions will often proceed at a much faster rate in the presence of a third chemical or group of chemicals. For example, in the manufacture of sulfuric acid, the number one chemical of commerce, it is necessary to convert sulfur dioxide to sulfur trioxide:

$$2 \, SO_2 + O_2 \rightarrow 2 \, SO_3$$

If the pure chemicals are mixed, the reaction is very slow, much too slow for a profitable industrial process under practical temperature and concentration conditions. However, if some oxides of nitrogen are introduced into the system, the desired reaction proceeds rapidly. Furthermore, the nitrogen compounds are not permanently changed in the process. Such a chemical or chemicals are referred to

Biological catalysts are called enzymes.

as catalysts. Obviously, a catalyst offers an alternative and easier pathway for the interacting atoms to achieve a resulting molecular structure.

> Catalysts are substances that increase the rate of a chemical reaction without being permanently consumed.

Later, in the study of biochemistry, it will be amazing to note the ability of biological systems to produce catalysts on demand and then destroy them after a particular chemical need is met.

REVERSIBILITY OF CHEMICAL REACTIONS

Most chemical processes can be reversed under suitable conditions. When a chemical reaction is reversed, some of the products are converted back into reactants. For example, heating calcium hydroxide will drive off water. This process is the reverse of adding water to quicklime (CaO):

$$Ca(OH)_2 + heat \rightarrow CaO + H_2O$$

Chemical reactions are capable of going forward or backward.

Other methods can be used to reverse chemical reactions. For example, if we put calcium hydroxide in a vacuum, there will soon be water vapor in the space around the solid.

It is easier to reverse chemical reactions when they are associated with small heat changes. For the compound hydrogen iodide (HI), it is possible to break the HI molecules apart into hydrogen and iodine:

$$2 HI + heat \rightarrow H_2 + I_2$$

at slightly elevated temperatures. On the other hand, water is decomposed into hydrogen and oxygen only by the use of considerable amounts of electrical energy. The reaction that is the reverse of the electrolytic decomposition of water is also shown in Figure 6–5. Hydrogen when burned in air produces water vapor, which can be condensed to liquid water, and much energy.

The fact that many reactions can be approached from either direction leads to the conclusion: *Chemical reactions are generally reversible.*

There are many reversible reactions important to human life. One of these is involved in the transport of atmospheric oxygen from the lungs to the various parts of the body. This task is carried out by hemoglobin, a complex compound found in the blood. This substance takes up oxygen while in the lungs to form oxyhemoglobin.

$$Hemoglobin + O_2 \leftrightarrows Oxyhemoglobin$$

The double arrows, $\leftrightarrows$, indicate a reversible reaction.

The oxyhemoglobin is then carried by the bloodstream to the various parts of the body, where it releases the oxygen for use in metabolic processes (Fig. 6–6).

CHEMICAL EQUILIBRIUM

Chemicals do not always react to form products with the complete extinction of the reactants. In theory, they never do. We may get the idea that all chemical reactions go to completion when we watch a piece of wood "burn up." However, nature quite often displays a reaction in which both reactants and products are present in the reaction medium at constant, but not necessarily the same, concentration levels. When reversible reactions reach the point where the forward reaction is proceeding at the same rate as the reverse reaction, the amount of chemicals present will

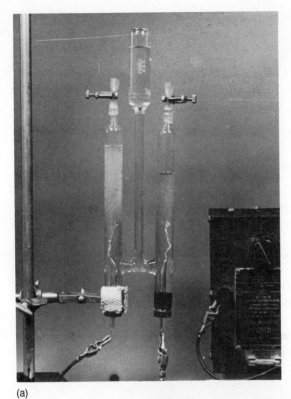

(a)

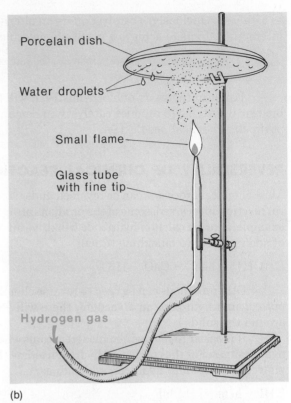

(b)

FIGURE 6-5 *(a)* **Electrical energy is required to decompose water into hydrogen** *(right tube)* **and oxygen** *(left tube)*. **Note that two volumes of hydrogen are produced for each volume of oxygen.** *(b)* **Hydrogen and oxygen burn to produce water in the gaseous state. The water is condensed on the cooler porcelain dish.**

remain constant because a particular chemical will be produced as fast as it is consumed. At this point, we have **chemical equilibrium.**

> A chemical change is at chemical equilibrium when products are produced at the same rate the products are consumed in reproducing reactants.

Consider the equilibrium among limestone ($CaCO_3$), lime (CaO), and carbon dioxide. If dry limestone is placed in a vacuum, carbon dioxide gas will soon appear in the container with the limestone. The reaction is:

$$CaCO_3 \rightleftarrows CaO + CO_2$$

After the carbon dioxide builds up to a certain concentration level, the system is at equilibrium and the carbon dioxide level will not rise further. We know that the reaction is proceeding in both directions when the system is at equilibrium, because we can introduce radioactive carbon in either the reactant or the product and trace it through the reaction in either direction with devices that detect radiation.

For every chemical reaction there is an **equilibrium constant,** which is expressive of the point of equilibrium. The equilibrium constant is equal to a function of the product concentrations divided by a function of the reactant concentrations. For a reaction that has more products than reactants at equilibrium, the constant is large. Conversely, a reaction that barely manufactures products in the presence of considerable amounts of reactants has a small equilibrium con-

Double arrows also indicate that the reaction proceeds in both directions at one time.

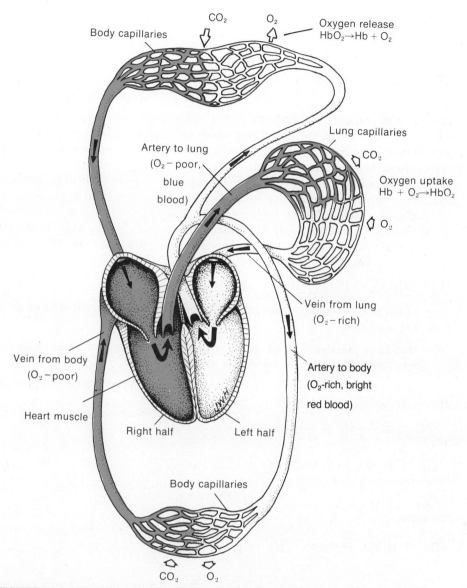

FIGURE 6-6 Simplified diagram of human circulation. The heart (shown in front view) is divided into two parallel halves. The right half pumps oxygen-poor blood to the lungs; the left half pumps oxygen-rich blood to the body. Hb = hemoglobin; HbO_2 = oxyhemoglobin.

stant. If the equilibrium constant is too large to be measured (it is there only in theory), the reaction is said to go to completion.

The idea of the equilibrium constant is complicated by the fact that it is a constant only for a particular temperature. Equilibrium constants are most often measured at normal laboratory temperature, and as you would guess, there is a quantitative relationship between the constants for a reaction at different temperatures and the amount of heat energy produced or consumed in the reaction. Returning to the limestone and carbon dioxide for a moment, the constant gets quite large and the reaction goes to completion in the conversion of limestone to lime (CaO) if the stone is heated in a kiln. This is how lime is made for the manufacture of cement.

REACTIONS BY GROUPS OF ATOMS

Under ordinary reaction conditions, chemical reactions proceed with minimal molecular alterations. The evidence is that molecules do not just become unglued completely and break up into atoms when entering into a chemical reaction. Rather, a relatively stable group of atoms enter into competition with another stable group for small particles, such as an electron, a proton, an atom, or a group of atoms. For example, sulfuric acid (H_2SO_4) dissolves in water to form a solution of ions. Evidence indicates that the sulfuric acid solution contains large amounts of hydronium ions (H_3O^+) and hydrogen sulfate ions (HSO_4^-) along with some sulfate ions (SO_4^{2-}). Apparently, the water removes hydrogen ions (H^+) from the sulfuric acid molecules without completely disrupting the sulfur-oxygen structure in the sulfate ion.

$$H_2O + H_2SO_4 \rightarrow H_3O^+ + HSO_4^-$$

This is an example of the **principle of minimal structural change** in chemical reactions:

Ordinary chemical change occurs with a minimum amount of change in the structures of the atoms, molecules, or ions involved.

Consider the following group of changes to note the apparent stability of the sulfate group.

$$\underset{\text{MAGNESIUM}}{Mg} + \underset{\substack{\text{SULFURIC} \\ \text{ACID}}}{H_2SO_4} \rightarrow \underset{\substack{\text{MAGNESIUM} \\ \text{SULFATE}}}{MgSO_4} + \underset{\text{HYDROGEN}}{H_2}$$

$$\underset{\text{CALCIUM}}{Ca} + H_2SO_4 \rightarrow \underset{\substack{\text{CALCIUM} \\ \text{SULFATE}}}{CaSO_4} + H_2$$

$$\underset{\text{STRONTIUM}}{Sr} + H_2SO_4 \rightarrow \underset{\substack{\text{STRONTIUM} \\ \text{SULFATE}}}{SrSO_4} + H_2$$

$$\underset{\text{BARIUM}}{Ba} + H_2SO_4 \rightarrow \underset{\substack{\text{BARIUM} \\ \text{SULFATE}}}{BaSO_4} + H_2$$

At this point, you may want to glance at some of the formulas in Chapter 21 on medicines and drugs. Our ability to take very complicated molecules and modify them slightly for desired properties both illustrates this important principle and challenges us in our efforts to control the chemical structures for our purposes.

The chemical principles presented in this chapter are fundamental and a beginning point for any student of the science. These principles have been selected because they are important to the presentation of this text and not because they are more important than other chemical principles that would be concurrently developed in a more formal study for students majoring in the physical sciences. It is important to realize that a physical principle is a concise summary statement of a multitude of related facts, and it is difficult to find a few words to express grand ideas that took many years to focus in human understanding. Do not be too hard on yourself and expect complete understanding on your introduction to some of these ideas. We will pick up on these ideas again and again as the chemical story unfolds.

Recall that an ion is a charged atom or group of atoms. Production of ions from neutral species is termed ionization.

Hydronium ions are hydrogen ions attached to water molecules. (See Chapter 9.)

All alkaline Earth metals react to form similar compounds.

SELF-TEST 6−B

1. Which factor affecting reaction rate—temperature or concentration—is most closely related to freezing foods to prevent spoilage? _____ Why? _____

2. If water can be produced by burning hydrogen in oxygen, what are the products of the decomposition of water? _____ and _____

3. Name two reversible chemical changes.

   _____

   _____

4. To what extent is a catalyst (a) used and (b) used up in a chemical reaction?

   a. _____

   b. _____

5. Are equal amounts of reactants and products necessary to achieve chemical equilibrium? _____

6. In what way is an equilibrium constant not a constant?

7. Which should burn faster: (a) a pound of flour in a sack or (b) the same flour in dust form in the air of a flour mill? _____

8. What chemical reaction can you name that is constantly reversed in your bloodstream? _____

9. Name two chemical reactions that are not easily reversed. _____ and _____

10. What is the principle of minimal structural change in chemical reactions?

   _____

   _____

   _____

MATCHING SET

_____ 1. Rate of chemical reaction

_____ 2. Corrosion

_____ 3. Lower temperature

_____ 4. Catalyst

_____ 5. Equilibrium

_____ 6. Balanced equation

_____ 7. 6.02×10^{23}

_____ 8. Gram molecular weight of P_4O_{10}

_____ 9. Weight of 1 mole of water molecules

_____ 10. Conserved during chemical reaction

_____ 11. Number of atoms in the formula $(NH_4)_3PO_4$

a. Atom count same on both sides of the arrow
b. Speeds up chemical reaction
c. Equal reaction rates
d. Atomic number ordering
e. Slower chemical reaction
f. Formation of rust
g. Amount of matter reacted in a given time
h. 18 g
i. 20
j. 1 mole
k. 284 g
l. Mass or weight
m. 18

QUESTIONS

1. What is a chemical catalyst?
2. How has the term *catalyst* been used in a social sense?
3. Give an example of a chemical reaction that does not go to completion.
4. Why is it necessary to balance a chemical equation before it can be used in making a calculation?
5. Identify four chemical reactions that we use in our daily lives in which energy plays an important role.
6. Give an example of a chemical reaction whose rate is fast and one whose rate is slow.
7. List three characteristics of all chemical reactions and illustrate each characteristic with an example.
8. Would you expect all of the chemical bonds in a molecule of aspirin to break as the chemical acts in your body?
9. In 1968 an Apollo spacecraft cabin fire killed three astronauts. The fact that pure oxygen was used as the cabin atmosphere contributed to the severity of the fire. How?
10. Is dust a safety factor to be considered in a grain-grinding mill?
11. Consider what the relative rates of iron rusting would be in a dry climate as opposed to a damp climate. What principle(s) of reaction rate is (are) involved?
12. Fires have been started by water seeping into bags in which quicklime (CaO) was stored. Why would this produce a fire?
13. Write and balance the chemical equation for the reaction between water and hot carbon to form gaseous hydrogen (H_2) and gaseous carbon monoxide (CO).
14. When iron rusts, heat energy is produced. Why does a rusty piece of iron not feel warm?
15. How could greater surface area (hot coal dust versus hot lump coal in the presence of air) affect the rate of a chemical reaction?
16. Balance the following equations:
    a. $SO_2 + O_2 \rightarrow SO_3$
    b. $NO_2 + H_2O \rightarrow HNO_3 + NO$
    c. $K + Br_2 \rightarrow KBr$
    d. $Mg_3N_2 + H_2O \rightarrow NH_3 + Mg(OH)_2$
    e. $P_4 + Cl_2 \rightarrow PCl_5$
    f. $PbO_2 \rightarrow PbO + O_2$
    g. $Al + O_2 \rightarrow Al_2O_3$
    h. $Fe + O_2 \rightarrow Fe_3O_4$
    i. $HgO \rightarrow Hg + O_2$
    j. $Fe + H_2O \rightarrow Fe_3O_4 + H_2$
    k. $Al + H_2SO_4 \rightarrow Al_2(SO_4)_3 + H_2$
    l. $CH_4 + O_2 \rightarrow CO_2 + H_2O$
    m. $C_8H_{18} + O_2 \rightarrow CO_2 + H_2O$
    n. $O_3 \rightarrow O_2$
    o. $C_{12}H_{22}O_{11} + O_2 \rightarrow CO_2 + H_2O$
17. If the term *endothermic* is used to describe a chemical reaction that absorbs energy as it proceeds, what adjective would be used to describe a reaction that produces energy?
18. In a balanced chemical equation, which is conserved: (a) molecules or (b) atoms?
19. The kinetic-molecular theory states that matter is made up of molecules and that molecules move faster at higher temperatures. Why should molecules move faster at elevated temperatures?
20. The electrolysis (electrical decomposition) of water is the reverse of what chemical reaction?
21. Firefighters use the methods of controlling the rate of a chemical reaction to combat a fire. Beside each of the following firefighting methods, give the rate-controlling factor that is being applied.
    a. Use of water
    b. Limiting the fuel supply
    c. Use of a fire blanket
    d. Carbon dioxide extinguisher
22. Which burns faster, a large log or the same log cut into small sticks of wood? What principle of chemical reactivity applies?
23. What principle of chemical reactivity applies to the storage of food in a freezer?
24. What can be said about the magnitude of an equilibrium constant for a reaction that appears to go to completion?
25. Add the atomic weights to determine the formula weights:
    a. H_2O_2 e. $C_6H_{12}O_6$
    b. H_3BO_3 f. Ag_2O
    c. $C_2H_4(OH)_2$ g. H_2SO_4
    d. Fe_2O_3 h. $Ca_3(PO_4)_2$
26. What is the weight (in grams) of 1 mole of each of the following?
    a. Xe   e. NH_3
    b. C_2H_5OH f. $Na_2S_2O_3 \cdot 5\ H_2O$
    c. H_2NCH_2COOH g. $MgSO_4 \cdot 7\ H_2O$
    d. Au   h. CF_2Cl_2
27. Write the chemical equation for the decomposition of limestone by heat.
28. What are some factors that drive reactions toward products?
29. What is the molecular weight of H_2O? of H_2SO_4?
30. How many moles of KCl can be made using 1 mole of potassium and 1 mole of chlorine?

    $2\ K + Cl_2 \rightarrow 2\ KCl$

31. How many moles of sulfur trioxide are needed to

form 100 moles of sulfuric acid? The equation is:

$$SO_3 + H_2O \rightarrow H_2SO_4$$

32. Hydrogen chloride (HCl) is produced by the action of sulfuric acid (H_2SO_4) on sodium chloride (NaCl):

$$NaCl + H_2SO_4 \rightarrow HCl + Na_2SO_4$$

How many moles of HCl can be made from 50.5 moles of H_2SO_4? Be sure to balance the equation first.

33. Chlorine can be made by the electrical decomposition of melted sodium chloride:

$$NaCl \xrightarrow{\text{energy}} Na + Cl_2$$

How many moles of products can be made from 1 mole of sodium chloride? Balance the equation, and be sure to include both products.

34. How many grams of hydrogen are liberated when 75.0 g of sodium metal react with excess water?

$$2\,Na + 2\,H_2O \rightarrow 2\,NaOH + H_2$$

35. Copper metal can be produced by heating copper sulfide with carbon and air:

$$CuS + O_2 + C \rightarrow Cu + SO_2 + CO_2$$

a. Balance the equation.
b. How many moles of oxygen are required for each mole of CuS?
c. How many grams of copper can be produced from 100 g of CuS?

36. How many grams of sulfur dioxide (SO_2) can be formed by burning 70.0 g of sulfur?

$$S + O_2 \rightarrow SO_2$$

37. Silver sulfide (Ag_2S) is the common tarnish on silver objects. What weight of silver sulfide can be made from 1.00 mg of hydrogen sulfide (H_2S) obtained from a rotten egg?

$$4\,Ag + 2\,H_2S + O_2 \rightarrow 2\,Ag_2S + 2\,H_2O$$

Chapter 7

NUCLEAR REACTIONS — ELECTRONS, PROTONS, NEUTRONS, AND MORE!

The 19th-century Daltonian atom pictured each element with its own type of characteristic, indestructible atom. The discovery of natural radioactivity by Henri Becquerel (Chapter 3) changed these ideas. The accepted theory now considers an atomic nucleus, which Rutherford first suggested, a collection of subatomic particles called neutrons and protons, with some nuclei being unstable. This means there are nuclear reactions involving nuclear structural changes just as there are chemical reactions resulting from changes in electronic structure.

In this chapter we look at nuclear reactions and their importance, and how they have led to an even better understanding of the structure of the atoms that compose the universe. As in some other areas of scientific discovery, the study of nuclear reactions has led to both beneficial and harmful results. For example, we can be both healed and injured by nuclear radiation. The heat from nuclear reactions can warm the homes and buildings we live in; however, the energy from uncontrolled nuclear reactions can virtually destroy our world. The control of these nuclear changes and the consequent good or ill associated with them depend first on our understanding of such changes and ultimately on judgments concerning their use. We will begin our study by looking at how nuclear particles are detected and how their properties are measured.

NUCLEAR PARTICLES AND REACTIONS

As in the case of atomic and molcular theory, we must depend on circumstantial evidence to establish the identity of the particles involved in nuclear reactions. The study is somewhat more difficult with nuclear reactions because many of the reactions of interest lead to products that can exist for only a very short period (sometimes as short as 10^{-8} second). For such nuclear reactions it is quite impossible to collect molar amounts of the reaction products and thereby deduce characteristic properties of the submicroscopic particles involved. It is evident then that successful methods for the study of nuclear reactions must involve rapid observations and the ability to record these observations for later study.

Several instruments have been designed specifically for the detection of high-energy particles or photons from nuclear transitions. These include the

Geiger counter, the scintillation counter, the cloud chamber, and the bubble chamber.

The **Geiger counter** is a modified cathode-ray tube with the electrical circuits needed to amplify the current across the electrodes in the tube so the current can be detected or recorded. Most Geiger tubes consist of a rugged metal case with a thin window of mica to allow the entry of radiation into the tube. Running through the center of the tube and insulated from the metal case is a charged wire (Fig. 7–1). The tube is evacuated and then partially filled with the inert gas argon. The wire is charged electrically to the extent that the tube is on the verge of discharge. If discharged by voltage alone, it would be a cathode-ray tube (Chapter 3). A high-energy particle entering through the mica window will cause one or more of the argon atoms to ionize.

$$\underset{\substack{\text{BETA} \\ \text{PARTICLE}}}{e^-} + Ar \rightarrow Ar^+ + 2\,e^-$$

The resulting charged particles (electrons and argon ions) cause other argon atoms to ionize in a cascade effect. The result of this one event is a sudden, massive electrical discharge that causes a current to flow through the tube. The number of discharges per second can be measured by the current output of the tube. Variations of the electrical circuit in the Geiger counter also allow individual pulses to be counted. The Geiger counter can detect products of nuclear decay and can measure their numbers as well. The Geiger counter is rugged and dependable, but it suffers from one disadvantage: in order to be detected by the counter, the particles must have sufficient energy to penetrate the mica window. Most alpha particles do not have sufficient energy for Geiger counter detection.

The **scintillation counter** is sensitive to "soft-nuclear" emissions, such as energies of most alpha particles. The outside of a glass window in a scintillation tube is covered with a phosphor coating. The minute flash of light produced by a relatively low-energy nuclear emission passes through the glass to a very sensitive photoelectric detector. The resulting electric current is amplified and either displayed or recorded.

Scientists were not able to characterize the electron, alpha particle, beta particle, and positive-ray particles until it was learned how to observe their behav-

Mica is a mineral that is found in relatively strong, thin sheets.

The scintillation counter is like a TV picture tube; a subatomic particle produces light when it strikes the phosphor coating.

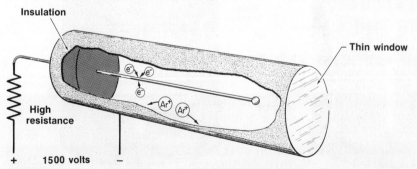

FIGURE 7–1 Schematic drawing of a Geiger tube. The voltage is adjusted to just below the discharge potential. Under these conditions high-energy particles that penetrate the thin window cause the argon atoms to ionize. The result is a cascade effect, in which newly formed charged particles, accelerated by the electric field, produce more ions, and a massive and sudden discharge occurs. The high resistance in series with the high-voltage electrodes prevents a large current for more than an instant, thus making the high-voltage device safe to handle.

ior in electric and magnetic fields (Chapter 3). With the charge-to-mass ratios determined by Thomson and the charge per particle determined by Millikan, investigators were quickly able to study those particles that could be produced (by electrical discharge or natural radioactivity) in relatively large and steady streams. At this point, it was reasonable to assume that other nuclear particles escaped detection because they were produced in insufficient quantities to be detected.

In 1911, C. T. R. Wilson invented the **cloud chamber,** which could actually "see" a single high-energy, nuclear particle in flight, the collision of such a particle with another nuclear particle, and the path of the products of such a reaction. A single nuclear event could be observed! Furthermore, when the cloud chamber is placed in a magnetic field, the charged particles will follow a curved path and the individual particles can be characterized.

The cloud track is somewhat analogous to the vapor trails of a high-flying jet airplane; even when the plane is too high to be seen itself, the condensed water from the exhaust clearly marks its pathway.

The structure of a simple cloud chamber is illustrated in Figure 7–2. Pressure is exerted on a closed system containing a nuclear particle source (such as an alpha emitter), air, and a layer of water or other liquid, such as ethanol. As pressure is exerted on this system, the concentration of water vapor is increased in the air space around the alpha emitter. Now, if the pressure is suddenly reduced, the temperature of the air drops, and the air will contain more water vapor than it can normally hold. Consequently, there will be a strong tendency for the water vapor to condense (precipitate). In such a supersaturated system the water molecules readily condense on charged particles. Now, the alpha particle, because of its high energy, ionizes air particles in its path. As a result, there will be a visible path of condensed water (a cloud track, Figure 7–3) tracing the alpha particle pathway. Any charged particles with sufficient energy to ionize the molecules of the air can thus be observed. It is a relatively simple matter to photograph such cloud tracks and record nuclear events, such as a collision between an alpha particle and a nitrogen molecule, for later study.

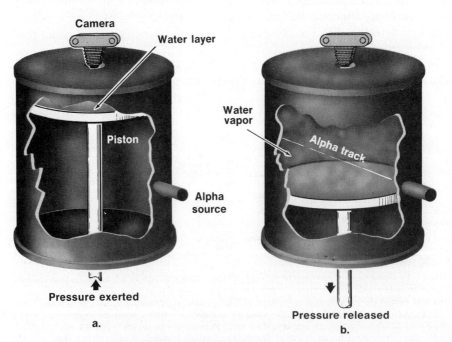

FIGURE 7–2   A simple Wilson cloud chamber.

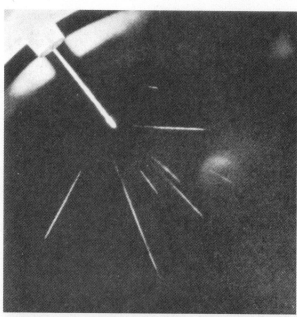

FIGURE 7–3   Alpha tracks photographed in a cloud chamber. Although the alpha particles do have different energies, owing to collisions with other particles in escaping the source, the apparent difference in path lengths is exaggerated because they are not all moving in the same plane. Under the same conditions proton tracks from the reaction, $^{14}_{7}N + ^{4}_{2}He \rightarrow ^{1}_{1}H + ^{17}_{8}O$, would tend to be considerably longer. (From J. H. Wood et al.: *Fundamentals of College Chemistry.* 1st ed. New York, Harper and Row, 1963.)

A device similar to the cloud chamber is the **bubble chamber,** invented by D. A. Glaser in 1952. In the bubble chamber, superheated helium or hydrogen is kept above its boiling point but held in the liquid state by applying pressure. If the pressure is released suddenly, the unstable state of the superheated liquid exists for a few seconds before boiling starts. Charged particles passing through this superheated liquid cause bubbles to be produced; the first tendency for the liquid to boil is at the path of the ionizing radiation. The bubble chamber may be recycled faster than the cloud chamber and is therefore more sensitive for "seeing" short-lived subatomic particles.

TRANSMUTATION IN NATURE

Armed with the ability to detect and characterize high-energy particles produced by nuclear reactions, nuclear scientists began collecting information concerning the nuclear reactions of naturally radioactive substances. There are many such reactions since all of the elements above bismuth (atomic number = 83) and a few below have one or more naturally occurring radioactive isotopes. Each of these isotopes is an emitter of alpha, beta, and/or gamma rays, and with each emission a nuclear reaction occurs spontaneously.

One of the dreams of the alchemists (1200–1700 A.D.) was to transmute base metals such as lead and iron into gold. The dream was discarded only after the acceptance of Dalton's indestructible atom.

The isotope of uranium with atomic mass 238 is an alpha emitter. The atomic number of uranium is 92, which means that $^{238}_{92}U$ has 92 protons and 146 (i.e., 238 − 92) neutrons in the nucleus. When the $^{238}_{92}U$ nucleus gives off an alpha particle, made up of two protons and two neutrons, four units of atomic mass and two units of atomic charge are lost. The resulting nucleus has a mass of 234 and a nuclear charge of 90. Now, atoms containing 90 protons in the nucleus are atoms of thorium, not uranium. This spontaneous nuclear reaction then has changed an atom of one element into an atom of another element and is an example of the **transmutation** of elements.

The decomposition of the $^{238}_{92}U$ nucleus is stated briefly by the following nuclear equation:

$$^{238}_{92}U \rightarrow \, ^4_2He + \, ^{234}_{90}Th$$

In this equation, the **mass number** (rounded off atomic mass) of the particle is given by the superscript and the **nuclear charge** (atomic number) is given by the subscript. If the characterized alpha emission was not proof enough that this reaction occurs, additional evidence is supplied by the fact that $^{234}_{90}Th$ is always found with $^{238}_{92}U$ in natural ore deposits and almost always in the concentration predicted by the rates of the reactions involved.

Thorium-234 is also radioactive. However, this nucleus is a beta emitter. This poses an interesting question: How can a nucleus containing protons and neutrons emit a beta particle, which is an electron? It has been established that an electron and a proton can combine outside the nucleus to form a neutron. Therefore, the reverse process is proposed to occur in the nucleus. A neutron decomposes, giving up an electron and changing itself into a proton:

$$^1_0n \rightarrow \, ^1_1H + \, ^{\,\,\,0}_{-1}e + Energy$$

Since the mass of the electron is essentially zero compared with that of the proton and neutron, the nucleus would maintain essentially the same mass, but the nucleus would now carry one more positive charge (a proton instead of one of the neutrons). This nucleus is no longer thorium, since thorium has only 90 protons in the nucleus; it is now a nucleus of element 91, protactinium (Pa). The reaction is the following:

$$^{234}_{90}Th \rightarrow \, ^{234}_{91}Pa + \, ^{\,\,\,0}_{-1}e + Energy$$

Gamma radiation may or may not be given off simultaneously with alpha or beta rays, depending on the particular nuclear reaction involved. Since gamma rays involve no charge and essentially no mass, the emission of a gamma photon cannot alone account for a transmutation event.

The decay of uranium-238 is extremely slow compared with the decay of thorium-234. The rate of decay can be represented by a characteristic **half-life.** A half-life represents the period required for half of the radioactive material originally present to undergo transmutation.

The half-life is independent of the amount and chemical form of radioactive material present and is determined only by the type of radioactive nucleus present in the sample.

For example, in the preceding reaction, the half-life of $^{234}_{90}Th$ is 24 days. This means that half of the thorium will remain unreacted after 24 days. In another 24 days half of the half (¼) will remain. This process continues indefinitely with half of the $^{234}_{90}Th$ remainder decaying each 24 days.

Figure 7–4 illustrates graphically how the concept of half-life works for a radioactive isotope. Some half-lives are extremely long and others are extremely short. The half-life for the $^{238}_{92}U$ alpha decay is 4.5 billion years. As one would expect,

----

Mass number
↓
$^{238}_{92}U$
↑
Nuclear charge (atomic number or number of protons)

No matter how much of a radioactive substance is present at the beginning, only half of it remains at the end of one half-life.

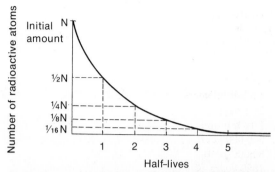

FIGURE 7–4 Half-life. The rate of decay for a radioactive atom depends in a very special way on the number of those atoms present. The rate is such that in a given period—the half-life for the species—half of the original number of atoms will be gone regardless of the number present at the start. In this graph the number of atoms remaining is plotted against time. At the end of one half-life period the original number, N, is reduced to $\frac{1}{2}$N. After two of these periods, the number is reduced to half of $\frac{1}{2}$N, or $\frac{1}{4}$N.

relatively large amounts of $^{238}_{92}$U can be found in nature while only trace amounts of $^{234}_{90}$Th occur.

The radioactive decay of $^{234}_{90}$Th into $^{234}_{91}$Pa is the second step in a series of nuclear decays that starts with $^{238}_{92}$U. After 14 decays the series ends with a stable, nonradioactive isotope of lead, $^{206}_{82}$Pb. This decay series is called the **uranium series** (Fig. 7–5). Table 7–1 gives the half-lives of the isotopes in the uranium series. Two other natural decay series exist that are similar to the uranium series,

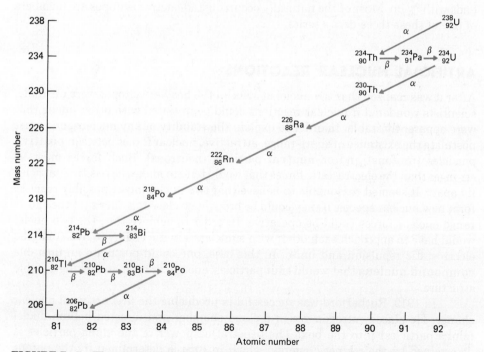

FIGURE 7–5 The uranium radioactive decay series. An alpha particle decay results in a decrease of two atomic charge units and four mass units. Beta particle decay results in no loss of mass and an increase in one atomic charge unit. Gamma emissions are not shown.

**TABLE 7–1 Half-Lives of the Naturally
Occurring Radioactive Elements
in the Uranium-238 ($^{238}_{92}$U) Series**

ISOTOPE	TYPE OF DISINTEGRATION	HALF-LIFE
^{238}U	α	4.5 billion years
^{234}Th	β	24.1 days
^{234}Pa	β	1.18 minutes
^{234}U	α	250,000 years
^{230}Th	α	80,000 years
^{226}Ra	α	1620 years
^{222}Rn	α	3.82 days
^{218}Po	α, β	3.05 minutes
^{214}Pb	β	26.8 minutes
^{214}Bi	α, β	19.7 minutes
^{210}Tl	β	1.32 minutes
^{210}Pb	β	22 years
^{210}Bi	β	5 days
^{210}Po	α	138 days
^{206}Pb	Stable	

Each decay series ends with an isotope of lead as the final product.

but they start out with a different isotope and proceed through a different set of radioactive decay products. The **thorium series** begins with $^{232}_{90}$Th (a different isotope from the two thorium isotopes that occur in the uranium series) and ends with stable $^{208}_{82}$Pb. A third series, called the **actinium series,** begins with $^{235}_{92}$U and ends with $^{207}_{82}$Pb. Most of the naturally occurring radioactive isotopes are members of one of these three decay series.

ARTIFICIAL NUCLEAR REACTIONS

After it was realized that the nuclei of some of the heavier isotopes were unstable, scientists wondered if nuclear reactions could be initiated with other nuclei that were apparently stable. In order to explain the stability of any nucleus, one must postulate the existence of short-range, attractive, nuclear forces between **positive** particles (protons) and/or **neutral** particles (neutrons). Such forces must be stronger than the electrostatic forces that would tend to make the positive particles fly apart. It seemed reasonable to believe that two nuclei might possibly react to form new nuclear species if they could be brought so close together that the short-range nuclear forces could be operative. In order to achieve this, the two nuclei would have to approach each other with sufficient kinetic energy to overcome the electrostatic repulsion and unite. In this case, one could postulate an unstable **compound nucleus** that would emit particles, energy, or both, in seeking a stable structure.

Recall that like charges repel and unlike charges attract.

In 1919, Rutherford was successful in producing the first artificial nuclear change. He placed nitrogen gas in a cloud chamber and directed helium nuclei (alpha particles) into the box. The penetrating power of the alpha particles is determined by their kinetic energy, which in turn is determined by the parent reaction producing them. Since the alpha particles from a given source all have about the same energy, the alpha tracks in a cloud chamber would be of essentially the same length (Fig. 7–3). In this experiment, Rutherford found some tracks that were much longer than the typical alpha track. Furthermore, these longer tracks did not appear to start at the origin of the alpha tracks but seemed to begin at the termination of an alpha track. When these tracks were studied in a magnetic field, their curvature indicated a particle with a charge-to-mass ratio identical to the value for the proton. Rutherford concluded that the tracks were produced by high-energy protons, which, because of their smaller size and charge, are more penetrating than an alpha particle for a given amount of energy. All of the results of the experiment could be explained if one assumed the nuclear reaction to be:

$$^{14}_{7}N + {}^{4}_{2}He \rightarrow [{}^{18}_{9}F] \rightarrow {}^{17}_{8}O + {}^{1}_{1}H$$

where $^{18}_{9}F$ is an unstable compound nucleus. Natural fluorine consists exclusively of the isotope $^{19}_{9}F$. Since both the $^{17}_{8}O$ and the hydrogen nuclei are stable, the products show no further tendency to undergo nuclear change.

Following Rutherford's original transmutation experiment, there was considerable interest in discovering new nuclear reactions. Many isotopes were subjected to beams of high-energy particles. As you might guess, numerous reactions were found; for example, bombardment of beryllium with alpha particles produced carbon.

$$^{9}_{4}Be + {}^{4}_{2}He \rightarrow {}^{13}_{6}C \rightarrow {}^{12}_{6}C + \underset{\text{(NEUTRON)}}{{}^{1}_{0}n}$$

Chadwick discovered the neutron by this reaction in 1932.

Although the $^{12}_{6}C$ produced in this reaction is stable, the neutron is given off with sufficient energy to provoke additional nuclear reactions in nuclei with which the neutron collides. It was just this nuclear reaction that was used by James Chadwick in 1932 to prove the existence of the previously postulated neutron.

Not all nuclear reactions produce stable isotopes. If $^{25}_{12}Mg$ is bombarded with an alpha source, a radioactive isotope of aluminum, $^{28}_{13}Al$, is produced that does not exist in nature:

$$^{25}_{12}Mg + {}^{4}_{2}He \rightarrow {}^{29}_{14}Si \rightarrow {}^{28}_{13}Al^* + {}^{1}_{1}H$$

An asterisk is often used to denote a radioactive isotope in nuclear equations. Radioactive isotopes, such as $^{28}_{13}Al$, have characteristic half-lives just as do the naturally occurring ones. The half-life of $^{28}_{13}Al$ is relatively short, only 2.3 minutes. The $^{28}_{13}Al$ nucleus emits a beta particle and becomes a stable isotope of silicon:

$$^{28}_{13}Al^* \rightarrow {}^{28}_{14}Si + {}^{0}_{-1}e$$

Nitrogen can be bombarded with neutrons from the Chadwick reaction to produce radioactive $^{14}_{6}C$:

$$^{14}_{7}N + {}^{1}_{0}n \rightarrow {}^{15}_{7}N \rightarrow {}^{14}_{6}C^* + {}^{1}_{1}H$$

An interesting question arises as to why the alpha particles were scattered by the gold foil in Rutherford's gold-foil experiment (Chapter 3), and yet the same alpha source can produce a nuclear change with a smaller atom such as $^{9}_{4}Be$ (Fig. 7–6). The answer lies in the fact that the charge on the gold nucleus is $+79$, whereas the charge on the beryllium nucleus is $+4$. Most of the alpha particles emitted from natural radioactive decay do not have enough energy to penetrate a heavy, positively charged nucleus such as that of gold. Therefore, if the artificial nuclear reactions are to be studied for the heavier elements, the kinetic energy of the subatomic projectile particles must be increased.

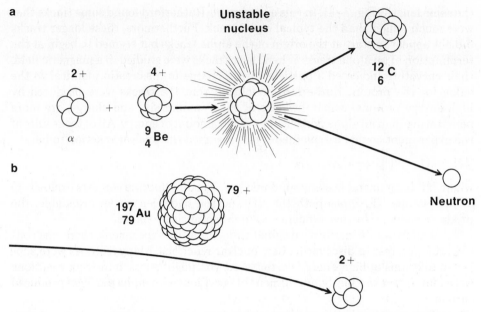

FIGURE 7–6 *(a)* A beryllium nucleus (Be) is struck by an alpha particle, which has sufficient energy to overcome the repulsions of like charges. A nuclear reaction occurs producing a carbon atom and a neutron. *(b)* In Rutherford's gold foil experiment (the experiment that suggested the nuclear atom), the alpha particles were not energetic enough to penetrate the gold nucleus (Au) and were deflected.

SELF-TEST 7–A

1. A nuclear radiation detection device utilizing a cloud of water vapor is called a

   _____.

2. When a $^{87}_{35}$Br nucleus emits a beta particle, the nuclear species that results is

   _____.

3. When a $^{216}_{84}$Po nucleus emits an alpha particle, the nuclear species that results

   is _____.

4. The half-life of $^{44}_{19}$K is 22 minutes. If a 1-g sample of $^{44}_{19}$K is taken, how much $^{44}_{19}$K

   will remain after three half-lives (66 min)? _____

5. In the following reaction, what is the compound nucleus?

   $^{7}_{3}$Li + $^{1}_{1}$H → _____ → $^{7}_{4}$Be + $^{1}_{0}$n

6. The scientist who discovered the neutron was _____.

   The process was _____.

7. High-energy particles from nuclear disintegrations cause gas molecules in a

   Geiger tube to _____. In the scintillation counter, the high-energy

   particles strike a phosphor that _____.

8. Which device is more likely to detect an alpha particle, the Geiger counter or

   the scintillation counter? _____

   Why? _____.

NUCLEAR PARTICLE ACCELERATIONS — NUCLEAR BULLETS

Since charged particles interact with magnetic and electrical fields (Chapter 3), these forces can be used to cause nuclear reactions that otherwise would not be observed in nature. The naturally occurring radiation was observed to cause some nuclear reactions. However, many more reactions could be provoked if the nuclear radiation particles could be accelerated, thereby giving them more kinetic energy. The devices used to produce these high-energy nuclear projectiles are called **particle accelerators.** We will look at a few of these, how they work, and their uses.

Although the construction of particle accelerators is often quite elaborate and different for each apparatus, the basic principles of operation are comparatively simple: (1) opposite charges attract, and (2) the path of a charged particle is curved as it passes through a magnetic field. When these effects are combined with the fact that electrical fields do not penetrate to the inside of a charged metal container (the fields are shielded by the "free" electrons in the metal), you have the fundamental facts necessary to explain how particle accelerators work.

Sufficient kinetic energy for a particle to penetrate the electrical fields of an atom and to enter a large nucleus cannot be gained readily by accelerating the particle in one step between two electrical poles. This would require an impossibly large potential difference of millions of volts. However, if the acceleration is done in several thousand steps, with readily obtainable potential differences of 2000 to 10,000 volts being applied to each step, sufficient energy can be imparted to the particle. The **linear accelerator** and the **cyclotron** illustrate the two primary ways the stepwise acceleration is accomplished.

The operation of the linear accelerator does not require a magnetic field and therefore is less complicated than that of the cyclotron. The principle of operation is outlined in Figure 7–7. A source of electrons or protons is provided by ionization (as in a canal-ray rube) at one end of the line of hollow metal cylinders. A target material is located at the other end. The entire device is enclosed in a near-perfect vacuum so that the accelerated particles may move in a straight path with little possibility of collisions with molecules in the air. Adjacent cylinders of opposite charge cause the travelling charged particles to be repelled by the previous tube and attracted by the upcoming tube. When a given particle enters a cylinder, the sign of the charge on the cylinder changes so the particle will be repelled as it enters the gap between cylinders. Simultaneously, the signs of the charge on all the other cylinders are also changed. As a result, the particle is successively attracted to the upcoming cylinder and repelled by the previous cylinder. At each gap, then, the particles increase their speed and their energy. Often moving near the speed of light, the accelerated particles strike the target atoms and enter their nuclei, producing nuclear changes.

The largest linear accelerator in the world went into operation in 1967 at Stanford, California. It is 2 miles long and accelerates electrons to energies of 20 to 40 billion electron volts (Bev).

The cyclotron was developed in 1931 by Ernest O. Lawrence (element 103, a synthetic element, is named in his honor) and M. S. Livingston. In addition to the electronic circuits and the huge magnets, the instrument consists of two hollow D-shaped metal containers enclosed in a vacuum as shown in Figure 7–8. Charged particles are formed near the center of the gap between the Ds and begin their acceleration to high energy by being attracted into one of the Ds. While the particles are in a D, the influence of the magnetic field causes their path to curve. By the

These accelerators are complicated, but their operating principles are simple.

An electron volt is a unit of energy; it is the amount of energy gained by an electron when it is accelerated by an electric field, the electric potential across the field being 1 volt.

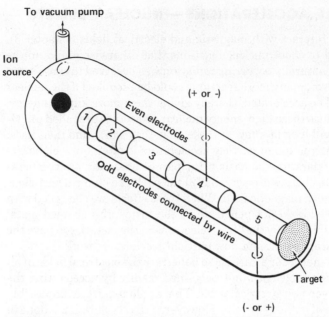

FIGURE 7–7   Linear accelerator diagram. Charged particles are produced at the ion source and are attracted toward an oppositely charged electrode, cylinder 1. When the charged particles are passed through cylinder 1, the charge on the cylinder is changed to the same sign as that on the particle being accelerated. At the same instant, the sign of the charge on cylinder 2 becomes opposite to that of the charge on the particle. Thus the particle is accelerated as it crosses the gap between cylinders. Note that the tubes become successively longer to accommodate increased speeds of the particle.

time the particles have completed the semicircle, the signs on the Ds have changed and the repulsion of the previous D and the attraction by the upcoming D accelerate the particles across the gap between the Ds. Each time the particles traverse the gap, they are accelerated to a greater speed. The increased speed causes the particles to move in a wider arc on each revolution. After many accelerations and

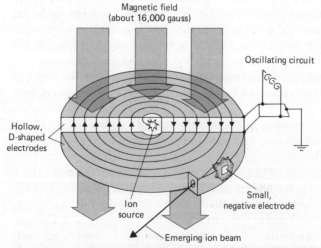

FIGURE 7–8   Schematic diagram illustrating the operation of the cyclotron. (Adapted from Linus Pauling: *College Chemistry*, 3rd ed. San Francisco, W. H. Freeman and Company, Copyright © 1964.)

revolutions, the arc is sufficiently large for the charged deflector plate to repel the particles through a window and onto the target sample. In some cyclotrons, the particles go too fast to be directed outward by the deflector plate. In this case, the target sample is placed inside one of the Ds.

The first successful model of the cyclotron had Ds with diameters of about 4 inches. Later models had diameters of about 2 feet. These models accelerated protons to energies of about 0.5 **million electron volts (Mev).** Modern cyclotrons have diameters of about 10 feet and accelerate protons to energies of about 250 Mev. While particle energies this high may seem large, much higher energies have been needed to probe the atom.

In May 1974, the world's largest particle accelerator was dedicated in Batavia, Illinois (Fig. 7–9). This installation is called the Fermi National Accelerator Laboratory, after Enrico Fermi, who produced the first nuclear chain reaction (see Chapter 8). This large proton accelerator is 4 miles in circumference, but even it will be dwarfed in the future by an accelerator under construction in Switzerland that will have a circumference of 17.1 miles. The Fermi Lab accelerator can produce particles with energies greater than 1 *trillion* electron volts. The larger, planned accelerator will produce even more energetic particles.

When a beam of such high-energy particles (electrons, protons, or deuterons) bombards nuclei, many different reactions take place. A massive nucleus, having captured a high-energy particle, will usually emit one or more subatomic particles before reaching a stable state. If the bombarding particles have energies less than about 100 Mev, the emitted particles can be described in terms of relatively simple subatomic particles, such as protons, neutrons, electrons, positrons, neutrinos, and antineutrinos. However, when the bombarding particles have energies of thousands of Mev or greater, new and strange particles are created; some of these are briefly described in Table 7–2.

Consider all of the particles produced when a carbon nucleus is struck by a high-energy proton (Fig. 7–10).

For a number of years theoreticians agonized over the lack of any reasonable explanation for the existence of these particles. In 1961 Murray Gell-Mann pro-

(a) (b)

FIGURE 7–9 Particle accelerator at the Fermi National Laboratory, near Batavia, Illinois. *(a)* Aerial view of main accelerator. *(b)* Magnets in interior of main accelerator, which is 4 miles in circumference and 1.27 miles in diameter. (Fermi Lab Photo.)

TABLE 7–2   Subatomic Particles Produced in Artificial Nuclear Reactions

Particles produced from collisions involving projectile particles having energies below 100 Mev include:

		MASS IN MULTIPLES OF THE MASS OF AN ELECTRON	CHARGE
e^-	Electron	1	-1
e^+	Positron	1	$+1$
p	Proton	1837	$+1$
n	Neutron	1838	0
v	Neutrino	0	0

Particles produced from collisions involving projectile particles having energies above 100 Mev include:

		MASS IN MULTIPLES OF THE MASS OF AN ELECTRON	CHARGE
Mesons			
μ mu (muon)		206	$+1$ or -1
$\pi^\pm$ pi		273	$+1$ or -1
π^* pi		264	0
κ kappa (kaon)		967	0 or $+1$ or -1
Baryons (or hyperons)			
Λ lambda		2183	0
Σ sigma		2330	0 or $+1$ or -1
Ξ cascade		2580	-1 or 0

posed the "eightfold" way, a scheme to relate the vast number of particles by their mathematical symmetries. He proposed that each of the newly discovered subatomic particles is composed of three more fundamental particles, which he named **quarks.** Each quark posesses a set of characteristics, much like the electron's spin, charge, and mass. Quarks have a property called color, which determines how they bind together. Another property is spin (up or down), and another is mass. All of these properties give a quark a "flavor." The three varieties of quarks used by

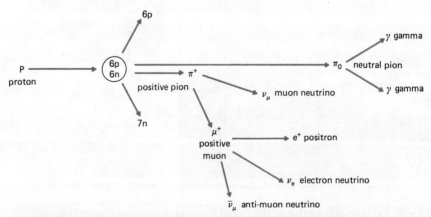

FIGURE 7–10   Particle production resulting from the collision of a high-energy proton with a carbon atom.

Gell-Mann were up, down, and strange. Quarks are assumed to have an electric charge of one third or two thirds (plus or minus). According to the theory, a proton consists of two up quarks ($+\frac{2}{3}$ each) and one down quark ($-\frac{1}{3}$). This theory seemed to be developing nicely until 1974, when a fourth variety of quark was discovered. By 1984, a sixth variety was discovered! The six varieties of quarks now known are up (u), down (d), charmed (c), strange (s), top (t), and bottom (b).

Where will these new discoveries and theories lead us in understanding chemistry? No one knows yet. Up to now chemists have been content with only three fundamental particles, the electron, proton, and neutron. But perhaps future scientists will be able to design chemistry experiments that may somehow help us to measure the properties of quarks, much as Chadwick did when he showed the existence of the neutron. Perhaps by controlling the composition of certain atoms in terms of more than three fundamental subatomic particles, we may be able to influence nuclear and chemical reactions. The applications might be as astounding as nuclear energy and the atomic bomb were in the application of early nuclear theory.

TRANSURANIUM ELEMENTS

The heaviest known element before 1940 was uranium. The invention of the cyclotron and other devices to obtain high-energy particles made it possible for these particles to react with heavy nuclei and to form even more massive nuclei. Thus, **transuranium** elements with atomic numbers greater than 92 were prepared.

In 1940 at the University of California, E. M. McMillan and P. H. Abelson prepared element 93, the synthetic element neptunium (Np). The experiment involved directing a stream of high-energy deuterons ($^2_1$H) onto a target of $^{238}_{92}$U. A deuteron is the nucleus of an isotope of hydrogen with one neutron as well as one proton. The initial reaction was the conversion of $^{238}_{92}$U to $^{239}_{92}$U.

The first synthetic element was prepared in 1940.

$$^{238}_{92}\text{U} + {}^2_1\text{H} \rightarrow {}^{239}_{92}\text{U} + {}^1_1\text{H}$$

Uranium-239 has a half-life of 23.5 minutes and decays spontaneously to the element neptunium by the emission of beta particles.

$$^{239}_{92}\text{U} \rightarrow {}^{239}_{93}\text{Np} + {}^{\ 0}_{-1}\text{e}$$

Neptunium is also unstable, with a half-life of 2.33 days; it converts into a second new element, plutonium.

$$^{239}_{93}\text{Np} \rightarrow {}^{239}_{94}\text{Pu} + {}^{\ 0}_{-1}\text{e}$$

Plutonium-239, like neptunium, is radioactive, with a half-life of 24,100 years. Because of the relative values of the half-lives, very little neptunium could be accumulated, but the plutonium could be obtained in larger quantities. The $^{239}_{94}$Pu is important as fissionable material since atomic bombs (see Chapter 8) can be made with it as well as with naturally occurring $^{235}_{92}$U. The names of neptunium and plutonium were taken from the mythological names Neptune and Pluto in the same sequence as the planets Uranus, Neptune, and Pluto.

Plutonium is used to make atomic bombs and is also one of the most toxic elements known.

Although Neptune and Pluto are the last of the known planets in the solar system, their namesakes are not the last in the list of elements. The rush of transuranium experiments that followed produced additional elements: americium (Am), curium (Cm), berkelium (Bk), californium (Cf), einsteinium (Es), fermium

TABLE 7–3   Nuclear Reactions Used to Produce Some Transuranium Elements*

ELEMENT	ATOMIC NUMBER	REACTION
Neptunium, Np	93	$^{238}_{92}U + ^{1}_{0}n \rightarrow ^{239}_{93}Np + ^{0}_{-1}e$
Plutonium, Pu	94	$^{238}_{92}U + ^{2}_{1}H \rightarrow ^{238}_{93}Np + 2^{1}_{0}n$
		$^{238}_{93}Np \rightarrow ^{238}_{94}Pu + ^{0}_{-1}e$
Americium, Am	95	$^{239}_{94}Pu + ^{1}_{0}n \rightarrow ^{240}_{95}Am + ^{0}_{-1}e$
Curium, Cm	96	$^{239}_{94}Pu + ^{4}_{2}He \rightarrow ^{242}_{96}Cm + ^{1}_{0}n$
Berkelium, Bk	97	$^{241}_{95}Am + ^{4}_{2}He \rightarrow ^{243}_{97}Bk + 2 ^{1}_{0}n$
Californium, Cf	98	$^{242}_{96}Cm + ^{4}_{2}He \rightarrow ^{245}_{98}Cf + ^{1}_{0}n$
Einsteinium, Es	99	$^{238}_{92}U + 15 ^{1}_{0}n \rightarrow ^{253}_{99}Es + 7 ^{0}_{-1}e$
Fermium, Fm	100	$^{238}_{92}U + 17 ^{1}_{0}n \rightarrow ^{255}_{100}Fm + 8 ^{0}_{-1}e$
Mendelevium, Md	101	$^{253}_{99}Es + ^{4}_{2}He \rightarrow ^{256}_{101}Mv + ^{1}_{0}n$
Nobelium, No	102	$^{246}_{96}Cm + ^{12}_{6}C \rightarrow ^{254}_{102}No + 4 ^{1}_{0}n$
Lawrencium, Lr	103	$^{252}_{98}Cf + ^{10}_{5}B \rightarrow ^{257}_{103}Lr + 5 ^{1}_{0}n$
—	104	$^{242}_{94}Pu + ^{22}_{10}Ne \rightarrow ^{260}_{104}? + 4 ^{1}_{0}n$
—	105	$^{249}_{98}Cf + ^{15}_{7}N \rightarrow ^{260}_{105}? + 4 ^{1}_{0}n$
—	106	$^{249}_{98}Cf + ^{18}_{8}O \rightarrow ^{263}_{106}? + 4 ^{1}_{0}n$
—	107	$^{209}_{83}Bi + ^{54}_{24}Cr \rightarrow ^{262}_{107}? + ^{1}_{0}n$
—	109*	$^{209}_{83}Bi + ^{58}_{26}Fe \rightarrow ^{266}_{109}? + ^{1}_{0}n$

* Element 109 was discovered in Germany in August 1982. It was prepared by a technique known as cold fusion. Only a single atom of this element was detected.

(Fm), mendelevium (Md), nobelium (No), lawrencium (Lr), and elements 104, 105, 106, 107, 108, and 109 (as yet unnamed). Obviously, the new elements were named after countries, states, cities, and people. Reactions employed in the production of most of the transuranium elements are given in Table 7–3. As accelerators with greater and greater energy capacities are produced, even more nuclear reactions should be available for study.

RADIOISOTOPE DATING OF THE UNIVERSE, MINERALS, AND ARTIFACTS

The concept of radioisotope half-life discussed earlier was almost immediately recognized as a useful tool for measuring the age of radioactive materials when reasonable assumptions were made. The assumptions for radioisotope dating are:

Assumptions are made in radio dating.

1. The nuclear decay is independent of the history of the isotope.
2. The decay is independent of the present chemical environment.
3. The rate of decay has always been constant.
4. There was definite initial isotope composition at the beginning of the radioactive decay process. (For example, no lead present initially in the uranium ore described subsequently.)

Three methods of radioactive dating have proved widely applicable and are discussed here. They are $^{238}_{92}U/^{206}_{82}Pb$, $^{40}_{19}K/^{40}_{18}Ar$, and $^{14}_{6}C$ dating.

Uranium/Lead Dating

The decay scheme for natural $^{238}_{92}U$ was presented in Figure 7–5. Since the decay of $^{238}_{92}U$ eventually results in the stable $^{206}_{82}Pb$ isotope, an analytical determination of the relative amounts of these two isotopes can provide an estimate of the age of the rock formations in which they are found. This assumes, of course, that no $^{206}_{82}Pb$ was present in the sample at the initial time and that all of the $^{206}_{82}Pb$ present has appeared through this known process. An estimate of age is possible since the half-life for each decay reaction is known.

A related method was suggested as early as 1905 by Rutherford while he was lecturing at Yale University. He suggested that the helium resulting from alpha decays in the uranium series of decay reactions could be measured as an indication of age. About the same time, Bertram Boltwood suggested that the $^{238}_{92}U/^{206}_{82}Pb$ ratio could be measured as a criterion for dating rocks. Boltwood dated a sample of uraninite ore taken from Spruce Pine, North Carolina, as 510 million years old. Modern instrumentation using the same basic method on the same ore has yielded a date of 344 to 385 million years as its age.

To understand how this method works, consider the fact that 1.00 gram of $^{238}_{92}U$ in its half-life of 4.5 billion years would leave 0.50 grams of $^{238}_{92}U$ and in the process produce 0.43 gram of $^{206}_{82}Pb$. The time required for all of the other decay steps after the breakdown of $^{238}_{92}U$ is relatively short. Hence, the rate of lead formation is controlled by this *rate-determining* step. The amount of $^{206}_{82}Pb$ can be calculated by using the fact that one $^{238}_{92}U$ atom is converted into one $^{206}_{82}Pb$ atom and by using the conversion

$$(207.21 \text{ g Pb}/238.07 \text{ g U}) \times 0.50 \text{ g U} = 0.43 \text{ g Pb}$$

Now, if the ratio of 0.50 gram $^{238}_{92}U$ to 0.43 gram of $^{206}_{82}Pb$ is found in a uranium ore, it would follow that the rock is 4.5 billion years old, the half-life of $^{238}_{92}U$.

Age determinations for various rocks taken from different parts of the world all indicate their ages to be in the neighborhood of 3 billion years. Some meteorites have been determined to be 4.5 billion years old. As a consequence, the age of the planets in the solar system is thought to be 4.5 billion years. Table 7–4 gives the ages of some lunar rocks and dust found at the Sea of Tranquility using $^{238}_{92}U$, $^{235}_{92}U$, and $^{232}_{90}Th$ radioisotope dating techniques.

The use of radioactive processes to determine the age of minerals was suggested by Rutherford in 1905.

Potassium/Argon Dating

The dating of mineral samples is possible because of the presence of a radioactive isotope of the element potassium, $^{40}_{19}K$. This isotope decays to a stable isotope of argon, $^{40}_{18}Ar$, by a process known as **electron capture** followed by gamma ray emission.

$$^{40}_{19}K + {}^{0}_{-1}e \rightarrow {}^{40}_{18}Ar^* \rightarrow {}^{40}_{18}Ar + {}^{0}_{0}\gamma$$

TABLE 7–4 Radioisotope Dating of Lunar Samples from the Sea of Tranquility by Three Different Methods

RADIOISOTOPE	HALF-LIFE (BILLION YEARS)	DECAY PRODUCT	AGE (BILLION YEARS)	
			Crystalline Sample	Lunar Dust
$^{232}_{90}Th$	13.9	$^{208}_{82}Pb$	3.6	4.5
$^{235}_{92}U$	0.71	$^{207}_{82}Pb$	3.9	4.7
$^{238}_{92}U$	4.5	$^{206}_{82}Pb$	3.8	4.7

Electron capture is one method by which an unstable nucleus can decrease its atomic number by capturing an orbital electron close to the nucleus. The electron combines with a proton in the nucleus to form another neutron, thereby decreasing the atomic number by one unit. The $^{40}_{18}Ar^*$ species is unstable (denoted by the *) and will radiate energy in the form of a gamma ray in going to a lower energy state.

The potassium-argon method of age determination depends on measuring the amount of $^{40}_{18}Ar$ trapped as a gas within the rock where the argon was produced. Subsequently, the amount of $^{40}_{19}K$ in the rock is determined. The total amount of all isotopes of potassium is determined (usually by an emission spectrophotometric technique), and the amount of the $^{40}_{19}K$ isotope is measured. The amounts of $^{40}_{18}Ar$ and $^{40}_{19}K$ are determined by using a mass spectrograph. Once the amounts of $^{40}_{18}Ar$ and $^{40}_{19}K$ are known for a given sample, the age of the rock can be determined by using a graph such as Figure 7–11 or by using a rather complex mathematical equation (not given here) that was used to determine the graph.

Reliability of the potassium-argon dating method rests heavily on the accuracy of measuring the very small amounts of $^{40}_{18}Ar$ and $^{40}_{19}K$ in a rock sample. $^{40}_{19}K$ decays not only by electron capture but also by beta emission.

$$^{40}_{19}K \rightarrow \, ^{0}_{-1}e + \, ^{40}_{20}Ca$$

For the calculation of age to be accurate, the two decay rates of $^{40}_{19}K$ must be known accurately. Repeated age determinations on different samples of the same material show that the method is about 98% accurate for ages of about 3 million years and about 99% accurate for ages of about 160 million years. This assumes that the rock sample contains all of the $^{40}_{18}Ar$ emitted by $^{40}_{19}K$ decay. The method, then, is limited to nonporous geologic materials that will retain argon completely. This rules out almost all sedimentary rocks.

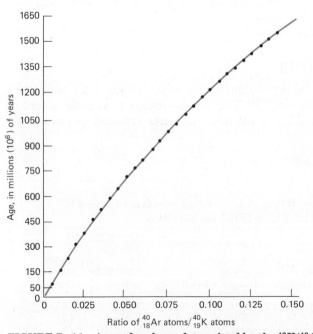

FIGURE 7–11   Ages of rocks as determined by the $^{40}_{19}K/^{40}_{18}Ar$ method of dating.

TABLE 7–5   Results of Several Potassium-Argon Age Determinations

SUBJECT	AGE (MILLION YEARS)
Rock from Olduvai Gorge, Tanzania (where L. S. B. Leakey found fossil remains of *Zinjanthropus,* an extinct primate)	1.75
Quartz monzonite from Marysvale, Utah	26.0
El Capitan, Yosemite, California	88.0
Volcanic rock in central Arizona	1,800

Table 7–5 lists some of the results of using the potassium-argon method of dating geologic materials.

Unlike the radioactive $^{14}_{6}C$ dating procedure (discussed next), which depends on the disappearance of $^{14}_{6}C$, the potassium-argon dating procedure depends on the appearance of the $^{40}_{18}Ar$ isotope. The clock is "set" then at time zero when geologic conditions were such that gaseous argon could be trapped. This usually occurs when a molten rock formation solidifies. If events such as reheating take place during a rock's history, it is likely to lose some argon and, hence, appear too young (not enough $^{40}_{18}Ar$ will be found).

This method cannot be used on sedimentary rock.

Carbon-14 Dating

Cosmic rays (interstellar radiation) are composed of many forms of very high-energy particles such as H^+ and He^{2+}. Many of these particles enter the earth's atmosphere every second. Cosmic rays undergo nuclear reactions with stable nuclei in the upper atmosphere to produce slow-moving neutrons. These neutrons can react with $^{14}_{7}N$ nuclei present in nitrogen molecules in the upper atmosphere to produce a radioactive isotope of carbon, $^{14}_{6}C$, which has a half-life of 5730 years.

$$^{1}_{0}n + {}^{14}_{7}N \rightarrow {}^{14}_{6}C^* + {}^{1}_{1}H$$

The $^{14}_{6}C$ decays by a beta emission.

$$^{14}_{6}C^* \rightarrow {}^{0}_{-1}e + {}^{14}_{7}N$$

Radioactive $^{14}_{6}C$ in the compound carbon dioxide mixes with the ordinary carbon dioxide in the atmosphere and, in turn, is incorporated into the structure of all living matter through natural food chains. Upon death of the organism, the intake of food ceases and the natural level of radioactive carbon present within the structure begins to decrease at the rate of 50% every 5730 years. The realization of this fact led Professor Willard F. Libby, at the University of Chicago, to postulate that radioactive $^{14}_{6}C$ could be used to date ancient artifacts derived from living matter such as parchment, cloth, and wood carvings. The $^{14}_{6}C$ remaining in the artifact would have to be compared with the normal isotopic ratio of $^{12}_{6}C$, $^{13}_{6}C$, and $^{14}_{6}C$.

Professor Libby received a Nobel Prize in 1960 for $^{14}_{6}C$ dating.

An important assumption is that the flow of $^{14}_{6}C$ into the biosphere is constant over time. According to several studies, radioactive carbon is indeed slowly mixed with its nonradioactive isotopes. The assumption is approximately true that the rate of production of $^{14}_{6}C$ has been essentially constant over the past several thousand years. Recently, growth rings on sequoia and bristlecone pine trees have been accurately measured and compared with $^{14}_{6}C$ dates. According to these experiments, $^{14}_{6}C$ production has fluctuated, particularly during the first millenium B.C. In

Uranium/lead dating— billions of years; Potassium/argon dating—millions of years; Carbon-14 dating— thousands of years.

TABLE 7–6   Comparison of Ages* of Various Artifacts over a Span of 3500 Years by Radiocarbon Dating and Other Methods†

MATERIAL	RADIOCARBON AGE	AGE BY ANOTHER METHOD‡
Mammalian remains from middle of an Inca temple	450 ± 150 years	444 ± 25 years
Sequoia tree ring	930 ± 100	880 ± 15
Wood from Roman ship	2,030 ± 200	1,990 ± 3
Charcoal from Etruscan tomb	2,730 ± 240	2,600 ± 100
Wood from Egyptian tomb of Zoser	3,979 ± 350	4,650 ± 75

* These ages are given as the age in 1950.
† This table is taken from the *McGraw-Hill Encyclopedia of Science and Technology,* Vol. 11. New York, McGraw-Hill, 1971, p. 291.
‡ Other methods include tree-ring dating and chronological methods.

other words, there is not perfect agreement between the tree-ring age and the $^{14}_{6}C$ found in each ring. Nevertheless, radioactive $^{14}_{6}C$ dates do compare reasonably well with dates obtained from other methods (Table 7–6). Figure 7–12 shows the relative uncertainties for samples of various ages using the carbon-14 dating method.

The various dating methods complement one another in terms of the time spans for which they are useful.

RADIATION DAMAGE

We are constantly bombarded by radiation from a number of sources. This radiation includes cosmic rays, medical X rays, radioactive fallout from countries that do nuclear testing, and naturally occurring, widespread radioisotopes. Fortunately,

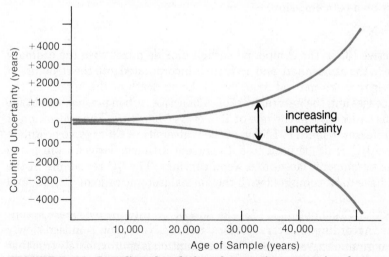

FIGURE 7–12   In radiocarbon dating, the counting uncertainty becomes increasingly important for older samples, resulting in very large uncertainties in ages beyond 30,000 years. Other factors, such as variabilities in cosmic rays over centuries, introduce still other uncertainties. (From *McGraw-Hill Encyclopedia of Science and Technology,* Vol. 11. New York, McGraw-Hill, 1971, p. 302.)

most radiation damage is too slight to be noticed immediately, although its very presence should be regarded as one of the hazards of everyday life.

As we have seen earlier, a radioisotope will disintegrate into a stable species, or it will become part of a decay series. In a sample of radioactive matter large enough to measure, there will be many disintegrations over a given time if the half-life is short or few disintegrations over the same interval if the half-life is long.

Three principal factors render a radioactive substance dangerous: (1) the number of disintegrations per second, (2) the half-life of the isotope, and (3) the type or energy of the radiation produced. In addition, radiation can be very damaging if the radioactive substance is of a chemical nature such that it can be incorporated into a food chain or otherwise enter a living organism.

Radioactive disintegrations are measured in **curies** (Ci; one Ci is 37 billion disintegrations per second). A more suitable unit is the microcurie (μCi), which is 37,000 disintegrations per second. One curie of a radioisotope is a potent sample if the energy per disintegration is large enough to cause a biochemical change.

Normal background radiation to the human body is 2 to 3 disintegrations per second.

The unit **roentgen** is used to measure the intensity of X rays or gamma rays. One roentgen is the quantity of X ray or gamma ray radiation delivered to 0.001293 g of air, such that the ions produced in the air carry 3.34×10^{-10} coulomb of charge. A single dental X ray represents about 1 roentgen.

The three types of natural radioactive emissions differ in their penetrating ability (Fig. 7–13), with gamma rays being by far the most penetrating.

Damage by radiation is due to ionization caused by the fast-moving particles colliding with matter and by the excitation of matter by gamma and X rays, which in turn produce ionization. Neutrons are produced in nuclear explosions, in nuclear reactions, and by background cosmic radiation. A neutron does not produce ionization per se but instead imparts its kinetic energy to atoms, which in turn may ionize or break away from the atom to which they are bonded. Neutrons render many engineering materials, such as plastics and metals, structurally weak over long periods as a result of the decay caused by breaking chemical bonds.

Biological tissue is easily harmed by radiation. A flow of high-energy particles may cause destruction of a vital enzyme, hormone, or chromosome needed for life of a cell. The radiation may also produce free radicals, which poison the cell. In

There is a normal background radiation from natural causes.

Wilhelm Roentgen discovered X rays in 1895 and was awarded the Nobel Prize for this work in 1901.

Neutrons can damage metals, causing structural failure. This is a severe problem in nuclear reactors.

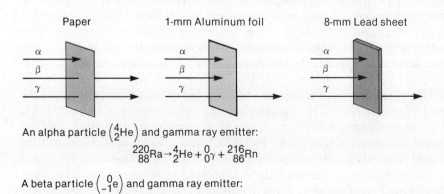

Paper   1-mm Aluminum foil   8-mm Lead sheet

An alpha particle $\left(^4_2\text{He}\right)$ and gamma ray emitter:

$$^{220}_{88}\text{Ra} \rightarrow\ ^4_2\text{He} +\ ^0_0\gamma +\ ^{216}_{86}\text{Rn}$$

A beta particle $\left(^{\ 0}_{-1}\text{e}\right)$ and gamma ray emitter:

$$^{131}_{53}\text{I} \rightarrow\ ^{\ 0}_{-1}\text{e} +\ ^0_0\gamma +\ ^{131}_{54}\text{Xe}$$

FIGURE 7–13 Penetrating ability of alpha (α), beta (β), and gamma (γ) radiation. Gamma rays even penetrate an 8-mm lead sheet. Skin will stop alpha rays but not beta rays.

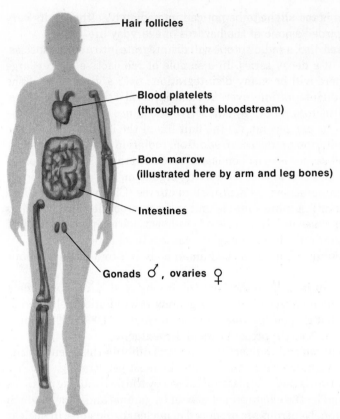

FIGURE 7–14 The fast-dividing cells within the body are the ones most harmed by radiation. These include cells in bone marrow, white cells, platelets of the blood, those lining the gastrointestinal tract, hair follicles, and gonads. In addition, the lymphocytes (cells producing the immune responses) are easily killed by radiation.

general, those cells that divide most rapidly are most easily harmed by radiation (Fig. 7–14).

Whole body radiation effects are divided into **somatic effects,** which are confined to the population exposed, and **genetic effects,** which are passed on to subsequent generations. A unit of measurement of radiation density is helpful in measuring the effect of radiation on tissue. The **rad** is defined as 100 ergs of energy imposed on a gram of tissue. Whole body doses of radiation of up to 150 rads produce scarcely any symptoms, whereas doses of 700 rads produce death. Intermediate doses produce vomiting, diarrhea, fatigue, and loss of hair. Often the somatic effects are delayed. Perhaps the best studied of the delayed effects are the incidences of cancer related to exposure to radiation. It has been estimated that 11% of all leukemia cases and about 10% of all forms of cancer are attributable to background radiations. Certainly an individual who is exposed to a higher than normal level of radiation over a considerable length of time increases the chances of cancer. The alteration of normal cells to cancerous cells caused by radiation is undoubtedly a series of changes, since in almost all cases the onset of cancer lags behind the exposure to radiation by an induction period of 5 to 20 years.

The genetic effects of radiation are the result of radiation damage to the germ cells of the testes (sperm) or the ovary (egg cells). Ionization caused by

One *rad* is roughly the energy absorbed by tissue exposed to one roentgen of gamma rays.

1 joule $= 10^7$ ergs
1 calorie $= 4.184 \times 10^7$ ergs

TABLE 7–7 **Average Dose of Radiation to Soft Tissues and Gonads from Surroundings***

SOURCE	DOSE TO GONADS PER YEAR (RAD)
Natural Background	
Cosmic rays	0.028
Local gamma rays	0.047
Radon in air	0.001
Potassium-40	0.019
Carbon-14	0.001
Other sources	0.002
Subtotal	0.098
Man-Made	
Medical X rays	0.100
Luminous watch dials	0.001
Occupational exposure	0.002
Television sets	0.001
Fallout from weapons test	0.001
Subtotal	0.105
	Total 0.203

Medical X rays account for about 50% of the average radiation dosage shown here.

* This table is taken from the *McGraw-Hill Encyclopedia of Science and Technology,* Vol. 11. New York, McGraw-Hill, 1971, p. 250.

radiation passing through a germ cell may break a DNA strand or cause it to be altered in some other way. When this damaged DNA is replicated (the process producing copies of the DNA structure during cell division, Chapter 15), the result may be the transmission of a new message to successive generations, a **mutation.** Every type of laboratory animal on which radiation damage experiments have been performed has responded by increased incidence of mutation. Therefore, the necessity of protecting the population of childbearing age from radiation should be apparent. Theoretically, at least, one photon or one high-energy particle can ionize a chromosomal DNA structure and produce a genetic effect that will be carried for generations. Table 7–7 shows average doses of radiation to soft tissue and the gonads from a variety of sources.

SELF-TEST 7–B

1. A particle having an energy of 1 Bev has an energy of _____ electron volts.
2. One type of particle accelerator that moves the charged particles in a circle is called a _____ .
3. The first transuranium element to be "made by man" is _____ .
4. The transuranium element of greatest atomic number to be "made by man" is element number _____ .
5. In $^{238}_{92}U$ dating of moon rocks, the final decay product measured is _____ .
6. What radiation dosage are you receiving each year from background radiation? _____

7. Which type of radiation dating would be useful for dating the oldest objects found in the universe? _____

8. Which has the longest half-life, neptunium-239 or plutonium-239? _____ _____ Relate this to the amount of each element that can be accumulated. _____

9. Describe the energies required to produce nuclear particles more fundamental than electrons, protons, and neutrons.

_____

_____

_____

USES OF NUCLEAR RADIATION

The damaging aspects of nuclear radiation must always be kept in mind, especially when the possibilities of accidental or unintended exposures are great. However, the harmful radiation from radioisotopes can be put to beneficial use. Consider the important application of killing harmful pests that would destroy our food during storage. In some parts of the world stored-food spoilage may claim up to 50% of the food crop. In our society, refrigeration, canning, and chemical additives lower this figure considerably.

Still, there are problems with food spoilage. Food protection costs amount to a sizable fraction of the final cost of food. Food irradiation using gamma-ray doses from sources such as ^{60}Co and ^{137}Cs is commonly used in European countries, Canada, and Mexico. The U.S. Food and Drug Administration (FDA) has been reluctant to allow this form of food preservation, but changes seem to be coming soon. Foods may be pasteurized by irradiation to retard the growth of organisms such as bacteria, molds, and yeasts. This irradiation prolongs shelf life under refrigeration much in the same way that heat pasteurization protects milk. Normally chicken has a three-day refrigerated shelf life. After irradiation, chicken may have a three-week refrigerated shelf life. The FDA may soon permit irradiation up to 100 kilorads for the pasteurization of foods.

There are 1000 kilorads in 1 megarad.

Radiation levels in the 1- to 5-megarad range sterilize; that is, every living organism is killed. Foods irradiated at these levels will keep indefinitely when sealed in plastic or aluminum-foil packages. The FDA is unlikely to approve irradiation sterilization of foods in the near future because of potential problems caused by as yet undiscovered, but possible, "unique radiolytic products." These would-be substances produced by the high-energy irradiation of foods might be harmful in some way. For example, it might produce a chemical substance that is capable of causing genetic damage. To prove or disprove the presence of these substances, animal feeding studies using irradiated foods are presently being conducted.

Ethylene dibromide (EDB) has been used widely in fumigating fruits. Now EDB is suspected to cause cancer and damage to human reproductive organs. Because of this toxicity, EDB has been banned by the U.S. Environmental Protection Agency.

Presently, over 40 classes of foods are irradiated in 24 countries. In the United States, only a small number of foods may be irradiated (Table 7–8).

Recent findings regarding the potentially harmful health effects of several common agricultural fumigants have indicated that irradiation of fruits and vegetables could be an effective alternative to some chemical fumigants. The agricultural products may be picked, packed, and readied for shipment. After that, the

TABLE 7–8   Examples of Irradiated Foodstuffs

FOOD	PURPOSE	STATUS
Potatoes	Retardation of sprouts	FDA approved
Wheat	Insect disinfection	FDA approved
Wheat flour	Insect disinfection	FDA approved
Spices	Retardation of microbe growth	FDA approved
Grapefruit	Mold control	For export
Strawberries	Mold control	For export
Fish	Microbe control	For export
Shrimp	Microbe control	For export

entire shipping container can be passed through a building containing a strong source of radiation (Fig. 7–15). This type of sterilization offers greater worker safety because it lessens chances of exposure to harmful chemicals (see Chapter 16) and protects the environment because it lessens chances of contamination of water supplies with these toxic chemicals (Chapter 17).

The radioisotope ^{60}Co, a gamma-ray emitter, has proved quite useful in the testing of metal castings in industry. Contained within an aluminum thimble, the cobalt radioisotope can be placed inside a casting after a piece of photographic film has been positioned on the outside of the object (Fig. 7–16). The gamma rays penetrate the metal part and make observable any structural flaws in the metal by exposing the photographic film. The intensity of the gamma rays passing through the flawed portion of the casting is different from the intensity passing through the rest of the metal. After development, the photographic film can be examined to detect and locate the presence of any flaws. Aviation safety has been increased by the use of radiation detection of flaws and structural weaknesses in structural members of aircraft.

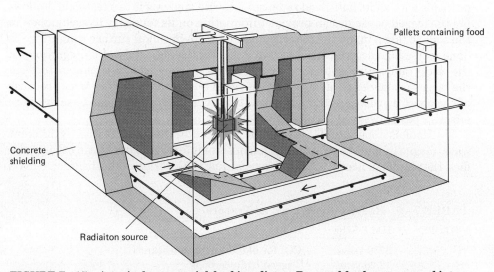

Pallets containing food

Concrete shielding

Radiaiton source

FIGURE 7–15   A typical commercial food irradiator. Boxes of food are conveyed into the shielded chamber and around the radiation source (center). When not in use, the source can be lowered into a pool of water below.

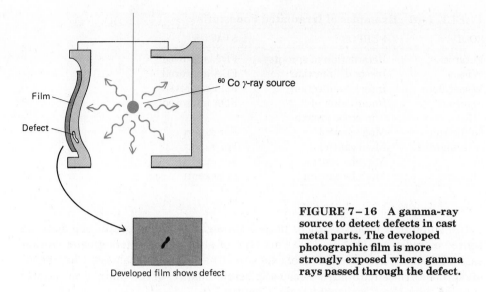

Film

Defect

^{60}Co γ-ray source

Developed film shows defect

FIGURE 7–16 A gamma-ray source to detect defects in cast metal parts. The developed photographic film is more strongly exposed where gamma rays passed through the defect.

Because radioisotopes act chemically in a manner almost identical to that of the nonradioactive isotopes of that element, chemists have been using radioactive isotopes as **tracers** in various chemical reactions since their use was discovered in 1945. Several of the more common radioisotopes used as tracers are listed in Table 7–9. For example, since plants are known to take up the element phosphorus from the soil through their roots, the use of the radioactive phosphorus isotope ^{32}P, a beta emitter, presents a way not only of detecting the uptake of phosphorus by a plant but also of measuring the speed of uptake under various conditions. Plant biologists can grow hybrid strains of plants that can absorb phosphorus quickly and then test this ability with the radiophosphorus tracer. This type of research leads to faster maturing crops, better yields per acre, and more food or fiber at less expense.

One can measure important characteristics of pesticides by tagging the pesticide with short half-lived radioisotopes and applying it to a test field. Following the tagged pesticide can provide information on its tendency to accumulate in the soil, be taken up by the plant, and accumulate in runoff surface water. This is done with a high degree of accuracy by counting the radioactive disintegrations of the tracer radioactive isotope. After these tests are completed, the radioisotopes in the tagged pesticides decay to a harmless level in a few days or a few weeks because of the short half-lives of these species. This type of research leads to safer, more effective pesticides (see Chapter 19).

Radioisotopes are also used in **nuclear medicine** in two distinctly different ways, diagnosis and therapy. In the diagnosis of internal disorders and other maladies, the physician needs information regarding the location of the disorder. This is

TABLE 7–9   Radioisotopes Used as Tracers

ISOTOPE	HALF-LIFE	USE
^{14}C	5730 years	CO_2 for photosynthesis research
^{3}H	12.26 years	Tagged hydrocarbons
^{35}S	86.7 days	Tagged pesticides, air flow
^{32}P	14.3 days	Phosphorus uptake by plants
^{131}I	8.05 days	Medical purposes

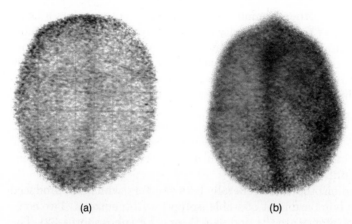

(a)   (b)

FIGURE 7–17   Brain scans using radioactive technetium-99m. *(a)* **Scan of a normal brain.** *(b)* **Scan of an abnormal brain showing an accumulation of radioisotope in a region of suspected tumor growth.**

done by **imaging,** a technique by which the radioisotope either alone or combined with some other chemical will accumulate at the site of the disorder. There, acting like a homing device, the radioisotope disintegrates and emits its characteristic radiation, which is detected. The detectors in modern medical diagnostic instruments are controlled by computers that not only determine where the radioisotope is located in the patient's body but also actually construct an image of the area within the body where radioisotopes are concentrated (Fig. 7–17).

Four of the most common diagnostic radioisotopes are given in Table 7–10. All of these are made by using a particle accelerator in which heavy charged nuclear particles are made to react with other radioisotopes or stable atoms. Each of these radioisotopes produces gamma radiation, which in low doses is less harmful to the tissue than ionizing radiations such as beta or alpha rays.

By the use of special carriers, these radioisotopes can be made to accumulate in specific areas of the body. For example, the pyrophosphate ion, $P_4O_7^{4-}$, a simple polyatomic ion, can bond to the technetium-99m radioisotope and together they accumulate in the skeletal structure where abnormal bone metabolism is taking place. Such investigations often pinpoint bone tumors.

The technetium-99m radioisotope is metastable (denoted by the letter m). Metastable isotopes lose energy by disintegrating to a more stable version of the same isotope.

$$^{99m}\text{Tc} \rightarrow {}^{99}\text{Tc} + \gamma$$

TABLE 7–10   Diagnostic Radioisotopes

RADIOISOTOPE	NAME	HALF-LIFE (HOURS)	USES
^{99m}Tc*	Technetium-99m	6	As TcO_4^- to the thyroid, brain, kidneys
^{201}Tl	Thallium-201	21.5	To the heart
^{123}I	Iodine-123	13.2	To the thyroid
^{67}Ga	Gallium-67	78.3	To various tumors and abscesses

* The technetium-99m isotope is the one most commonly used for diagnostic purposes. The m stands for "metastable," a term explained in the text.

Technetium-99m, like the other common diagnostic radioisotopes, has a short half-life, which means that the radioactivity does not linger for an unacceptably long period in the patient's body. When a diagnosis is to be derived with technetium-99m, the radioisotope is washed from a cartridge containing molybdenum-99 (half-life 66 hours) using a salt solution. The molybdenum radioisotope is constantly producing technetium isotopes.

$$^{99}Mo \rightarrow \,^{99m}Tc + \,_{-1}^{0}e + \gamma$$

The longer half-life of the molybdenum isotope makes it possible to ship the technetium in a form that will ensure the arrival of technetium-99m with sufficient strength to still be useful.

Therapeutic radioisotopes are generally beta emitters, which are produced in nuclear reactors by bombardment of stable isotopes with neutrons. Two common therapeutic radioisotopes are ^{131}I (iodine-131) and ^{32}P (phosphorus-32). The most common use of iodine-131 is in the treatment of thyroid cancers. A patient drinks a solution of ^{131}I ions as potassium iodide. The iodine, as iodide ion, then makes its way to the thyroid, where the beta rays produced by the ^{131}I radioisotopes destroy the cancerous thyroid cells. Of course, healthy cells are also destroyed, but not in sufficient numbers to destroy all of the healthy tissue.

Nuclear medicine, the use of radioisotopes for therapeutic and diagnostic purposes, has established itself in medical practice throughout the world. Over 2000 hospitals in the United States are licensed to use radioisotopes. Discoveries of new substances to carry radioisotopes to specific sites in the body offer one of the most promising areas of research. Another is the use of computers for imaging those sites in the body where the radioisotopes are concentrated.

Still another useful application of nuclear radiation in medicine is the heating effect caused by radiation. This heat can be used to generate electricity, which, in turn, can power a cardiac pacemaker. Pacemakers powered by plutonium-238 (half-life 89 years) have been used in humans in France since 1970 and in the United States since 1972. The electricity generated by the plutonium thermoelectric source is sent as a pulse directly to the ventricles of the heart at a preset rate. Because of the relatively long half-life of the plutonium source, these pacemakers can remain in the patient for longer periods without the need for additional surgery than can pacemakers powered by batteries.

It is hoped that, together, these and other beneficial uses of radioisotopes will allow nuclear science to save far more lives than nuclear bombs have or ever will have destroyed. Obviously, the science is neither good nor bad, as it can be used by human choice for good or ill.

SELF-TEST 7–C

1. Name three uses of radioactive isotopes. ——————, ——————, and ——————
2. Which radioisotope is used for examining metal castings? (a) ^{60}Co, (b) ^{32}P, (c) ^{67}Ga, (d) ^{99m}Tc
3. Name a radioisotope that might be useful as a tracer in agricultural research. (a) ^{32}P, (b) ^{14}C, (c) ^{3}H, (d) all of these
4. The process of concentrating a radioisotope at a particular site of the body in order to locate and measure the extent of a disorder is called: (a) radiotherapy, (b) imaging, (c) sterilization.

5. In the symbol for the radioisotope technetium-99m, the *m* stands for (a) middle, (b) mathematical, (c) metastable.

6. With its half-life of approximately 6 hours, how much technetium-99m would remain 18 hours after injection into a patient? (a) one eighth of the original dose, (b) one half of the original dose, (c) one sixth of the original dose, (d) one fourth of the original dose.

7. If two radioisotopes were available for diagnosis, worked equally well, and each decayed by giving off gamma rays, but one had a half-life of 13 hours while the other had a half-life of 6 hours, which one would you recommend? (a) 13-hour half-life isotope or (b) 6-hour half-life isotope

MATCHING SET

_____ 1. Somatic effect
_____ 2. 1 microcurie
_____ 3. $^{14}_{6}C$
_____ 4. $^{40}_{19}K$
_____ 5. Genetic effect
_____ 6. $^{238}_{92}U$ dating
_____ 7. Element 109
_____ 8. E. O. Lawrence
_____ 9. $^{218}_{84}Po$
_____ 10. James Chadwick
_____ 11. Half-life
_____ 12. C. R. T. Wilson
_____ 13. Bone marrow
_____ 14. Geiger counter
_____ 15. ^{99m}Tc
_____ 16. ^{60}Co

a. Intake stops when organism dies
b. Radiation effect on general population
c. First suggested by Rutherford
d. Developed the cyclotron
e. Radiation damage to DNA
f. Tissue easily damaged by radiation
g. Radioisotope used in medical diagnosis
h. Time required for half of the nuclei to disintegrate
i. 37,000 disintegrations per second
j. Used to detect defects in metal castings
k. Generates argon in a geologic clock
l. Can detect ionizing radiation
m. Latest synthetic element
n. Developed cloud chamber
o. Discovered neutron
p. Alpha decay product of $^{222}_{86}Rn$
q. Invented the bubble chamber
r. Half-life of 0.5 sec

QUESTIONS

1. Describe the operation of a Geiger counter.
2. Why is the cloud chamber (or the bubble chamber) more useful than other types of nuclear particle detectors?
3. What does the symbol $^{11}_{5}B$ mean?
4. In general, how have the synthetic transuranium elements been produced?
5. Complete or supply the following nuclear equations:
   a. $^{1}_{1}H + ^{35}_{17}Cl \rightarrow ^{4}_{2}He + ?$
   b. Beta emission of $^{60}_{27}Co$
   c. Alpha emission of $^{238}_{90}Th$
   d. $^{1}_{0}n + ^{60}_{28}Ni \rightarrow ? + ^{1}_{1}H$
   e. $^{2}_{1}H + ^{1}_{1}H \rightarrow ? + ^{0}_{0}\gamma$
   f. $^{238}_{92}U + ^{12}_{6}C \rightarrow ? + 4\,^{1}_{0}n$

6. We speak of "seeing" a nuclear event with a cloud or bubble chamber. Explain why we are more truthful in using the quotation marks about the word *seeing*.

7. Look up the origin of the word *mutation* and explain why the word *transmutation* was an apt word choice to describe the changing of one element into another.

8. What is the difference between a thermal neutron and a high-energy neutron resulting from cosmic radiation?

9. If a radium atom ($^{226}_{88}$Ra) loses one alpha particle per atom, what element is formed? What is its atomic weight? What is its atomic number?

10. What are the important assumptions made in radiocarbon dating?

11. Name two methods by which the age of rocks can be determined. What assumptions are made in these methods?

12. What errors would be introduced into the age determination of a piece of granite if it has been reheated at some point during its existence?

13. What error would be introduced into the age determination of a tree ring if the amount of cosmic rays had been double their present value at the time the tree grew that ring?

14. What is meant by "delayed somatic effect"? Give an example.

15. The ^{99}Mo (half-life, 67 hours) canisters used to prepare solutions of ^{99m}Tc have an effective life of about one week. Can you suggest a reason for this based on half-lives of radioisotopes?

16. Suggest a therapeutic use for the gamma-emitting radioisotope ^{60}Co.

17. Suggest an experiment using a radioisotope tracer in agriculture.

18. Iodine-123 (half-life, 13.2 hours) is used to measure iodine uptake by the thyroid gland. If 1 mg is injected into a patient's bloodstream, how long will it take for the radioisotope to be reduced to less than 1 μg?

19. Ask for an interview with a radiologist at your local hospital. Ask him or her to tell you about the greatest benefits and risks in the use of radioisotopes.

Chapter 8

ENERGY AND OUR SOCIETY

Today we are more aware than ever before of the need for energy. We feel its importance through increased prices of electricity, natural gas, gasoline, and consumer goods, which are produced and transported with energy. In our highly industrialized and appliance-oriented society, large amounts of energy are now a daily necessity for our way of life (Fig. 8–1).

Our dependence on the energy from coal, petroleum, and natural gas has been uncomfortably and abruptly realized several times since 1970, but shortages and high prices of energy have been on the horizon for several years. The power brownouts in the eastern United States in the late 1960s were a signal that something was amiss. In the early 1970s, gasoline shortages were caused by inadequate refining capacity, low return on investment capital, and a war in the Middle East. In 1973, the Organization of Petroleum Exporting Countries (OPEC) declared an

Why an energy shortage?

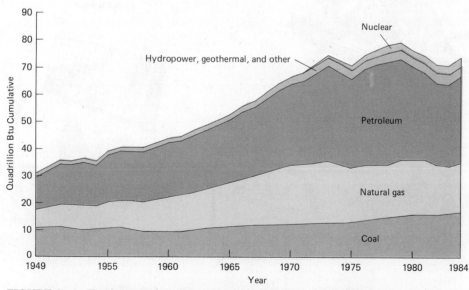

FIGURE 8–1 Energy consumption in the United States—the immediate view. Note that the burning of fossil fuels (coal, petroleum, and natural gas) furnishes nearly all of our present energy despite all the talk about hydroelectric and nuclear energy. *Btu* stands for British thermal unit (see Table 8–1). (Source: Energy Information Administration, *Annual Energy Review 1984*.)

embargo on oil shipments to all nations that supported Israel, demanding and getting political concessions. This made energy even more a matter of public concern. Since 1973, the price of oil has been raised several times, reaching a high of about $39 a barrel and then decreasing to less than $20 per barrel. Overproduction of oil and an improved efficiency in its use have contributed to the lower prices.

Who is using energy?

To put the increasing dependence on energy into perspective, consider Figure 8–2. As late as 1940, 80% of the world's population was living at the primitive or advanced agricultural level. Consequently, little energy was being used. Today, about 25% to 30% of the present world population lives in a highly industrialized society and uses 80% of the world's energy. In the United States, 5% of the world's population uses 30% of the world's energy production in any given year. Through conservation of energy in the United States and increased worldwide consumption of energy, our relative position in the use of the world's production of energy is declining. The advanced development of other countries causes serious worldwide competition for the types of energy sources currently available, and the supplies of the types being used are becoming depleted.

Purposes of this chapter.

Our primary purposes in this chapter are to examine our present sources of energy to see what chemicals are used and how energy is obtained from these

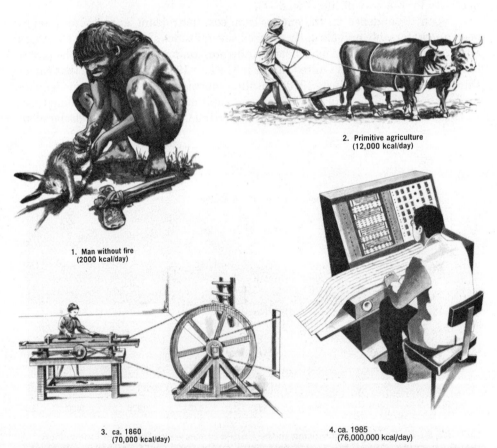

1. Man without fire
(2000 kcal/day)

2. Primitive agriculture
(12,000 kcal/day)

3. ca. 1860
(70,000 kcal/day)

4. ca. 1985
(76,000,000 kcal/day)

FIGURE 8–2 Energy use per capita based on various types of societies. (From A. Turk et al.: *Environmental Science,* 2nd ed. Philadelphia, Saunders College Publishing, 1978.)

chemicals. However, we shall first discuss some of the fundamental principles that guide and limit the extraction and use of energy.

FUNDAMENTAL PRINCIPLES OF ENERGY
Definitions of Energy and Power

Energy is the ability to move matter, that is, to do work. Some comparisons among units of energy and sources of energy are shown in Table 8–1. Types of energy include chemical, light, heat, sound, electrical, nuclear, and mechanical (a wound clock spring or moving car). All of these types of energy can move things. A closely related term that is often confused with energy is **power,** the rate at which energy is used. Power has the units of energy used per time, such as calories per second or joules per second (watts).

Law of Conservation of Energy

Also known as the first law of thermodynamics, this law asserts that energy is neither lost nor gained in all energy processes. In heating a beaker of water with a burner, all of the energy given off by the flame can be accounted for in the increased energy of the water and surroundings. No energy is lost or gained in the transformation of chemical energy to heat energy. Furthermore, the transformation is quantitative in that a certain amount of gas burned produces a certain amount of energy (see Table 8–1). In the changing of one kind of energy into another, the exchange rate is definite, reliable, and reproducible.

This law also implies that the total amount of energy in the universe is constant. The energy is being transformed regularly from one kind to another, but the total remains the same. Of interest to Earth travellers is the fact that the sun and the energy stored in the chemicals on Earth are what we have to use — that is all! No creation of energy is ongoing.

One last implication of the law applicable to this study is the limitation the law puts on perpetual motion. The law recognizes that a machine, by running, cannot produce enough energy to run itself, much less the creation of enough

Energy is the ability to move matter.

Power is the rate of using energy.

Thermodynamics is the movement of energy.

First statement of the first law of thermodynamics: "A force [translated: energy] once in existence cannot be annihilated." Julius Robert Mayer, a ship's doctor, 1840.

The law of conservation of energy has been extended by the discovery of Albert Einstein, who showed the interrelationship between mass and energy by $E = mc^2$.

The more general law is stated: The total amount of matter and energy in the universe is constant.

TABLE 8–1   A Handy Chart of Energy Units*

CUBIC FEET OF NATURAL GAS	BARRELS OF OIL	TONS OF BITUMINOUS COAL	BRITISH THERMAL UNITS (Btu's)	KILOWATT HOURS OF ELECTRICITY	JOULES	KILO-CALORIES†
1	0.00018	0.00004	1000	0.293	1.055×10^6	252
1000	0.18	0.04	1×10^6	293	1.055×10^9	0.25×10^6
5556	1	0.22	5.6×10^6	1628	5.9×10^9	1.40×10^6
25,000	4.50	1	25×10^6	7326	26.4×10^9	6.30×10^6
1×10^6	180	40	1×10^9	293,000	1.055×10^{12}	0.25×10^9
3.41×10^6	614	137	3.41×10^9	1×10^6	3.6×10^{12}	0.86×10^9
1×10^9	180,000	40,000	1×10^{12}	293×10^6	1.055×10^{15}	0.25×10^{12}
1×10^{12}	180×10^6	40×10^6	1×10^{15}	293×10^9	1.055×10^{18}	0.25×10^{15}

* Based on normal fuel heating values. 10^6 = 1 million, 10^9 = 1 billion, 10^{12} = 1 trillion, 10^{15} = quadrillion (quad).
† A food Calorie = 1000 cal = 1.000 kcal.

energy to be used elsewhere. At a minimum, the machine would have to create enough energy to move its parts and to overcome friction. Since this is a creation process, and not a transformation, the law says this is impossible.

Energy Is Conserved in Quantity But Not in Quality

This is but one of the many ways to state the second law of thermodynamics; another statement pertinent to this discussion will be given presently. But, first, what does it mean that energy is conserved in quantity but not in quality? Perhaps two of the many available examples will clarify the concept. Consider as a first example the commonly known facts that coal, petroleum, and wood, along with air, have energy stored in their chemical structures and that some of this energy is released during burning. It is also well known that the main products of the burning process, carbon dioxide and water, will not burn and release more energy. In the burning process, both matter and energy are conserved, which is required by the laws of conservation of matter and energy, respectively. However, the reactants and their stored energy are more useful in energetic terms than the products and their spent energy.

As a second example, consider an electrical motor. The electricity that runs the motor is more useful than the heat that comes from the warm motor. Again, energy is conserved in the process of running an electrical motor, yet the usable energy is not conserved.

In concept, the energy relationships in the second law of thermodynamics can be compared to the relationships among gross income, deductions, and net pay (or realizable income) in a paycheck. A certain amount of energy is available for the process considered; this is analogous to gross income. Some of the energy is not usable because of frictional losses, electrical shorts and drains, retention of some energy in the chemical products, or some other factor affecting efficiency; the energy that is not usable is represented in the analogy by the paycheck deductions. Finally, some of the energy is usable and is represented by net pay in the analogy. Both the energy and the money are accounted for as required by the first law of thermodynamics.

In all processes, then, some energy is wasted — not lost — by conversion into energy that is not usable in doing work. The wasted (or unusable) energy is represented by **entropy,** a measure of the disorder in a physical system. Entropy is not energy per se, but it is a function of energy with units of energy per degree, such as calories/degree.

Another statement of the second law of thermodynamics is based on entropy: In all natural processes, entropy is increased. Taken to its extreme, this means that the entropy of the whole universe is increasing at the expense of stars running down in usable energy at a tremendous rate. This is not a reason for worry because the universe is so vast that enough usable energy is there for all conceivable purposes for many billions of years. However, a source of usable energy that is not limitless is the so-called fossil fuels (coal, petroleum, and natural gas), which when gone are not easily restored. It would take eons for photosynthesis to regenerate the material for new fossil fuel deposits.

Why, then, is the energy used to increase entropy not usable energy? The derivation of the word *entropy,* meaning "disorder," explains. The ultimate fate of any change in energy is a form of heat energy caused by the random, disordered

Usable energy is not conserved.

No matter how we try, we can never convert all of the stored energy in a system into usable energy.

Entropy means disorder and measures nonuseful energy.

motion of molecules. Have you ever thought about what happens to the light energy coming from a light bulb, or what happens to the electrical energy once it is used to run an electrical motor, or what happens to the sometimes large amounts of energy that result from an explosion? All forms of energy, including sound, are converted eventually into random molecular motion — the molecules move faster and (or) farther apart. Molecules moving in all directions are not as useful in bringing about controlled change as are electrons, photons of light, or molecules when they are moving from one point to another in organized fashion. The type of energy is important if it is to be useful (moving molecules cannot run an electrical motor; moving electrons can), but for all types of usable energy the usefulness also comes from the organized, nonrandom movement, for example, of electrons from a generator or battery to the motor, of light from its source outward, and of molecules streaming from a gas jet. Useful energy involves organized flow.

The end of the line for energy is the random motion of molecules.

Let us summarize this brief encounter with the second law of thermodynamics by describing the energy coming from a burning match. Some of the energy is usable to ignite other objects, or to heat an object, or to provide light. This is the directional, organized energy. All the while that the usable energy is being used, some of the total energy emanating simply heats molecules in the vicinity and increases the entropy of the molecules. Eventually, all of the heat and light coming from the match will become increased random motion of the molecules.

What does the second law of thermodynamics mean to the informed citizen? Simply stated, when usable energy-rich chemicals such as coal and petroleum are consumed, the usable energy is lost to us forever.

When fossil fuels are gone, then what?

The Efficiency of Energy Use Is Low

In every energy process, the efficiency of the use of the energy for doing work is less than 100% — usually far less. Automobiles are about 20% to 25% efficient, that is, about 80% of the useful energy available to do work is lost and not applied to turning the wheels. Some fuel cells are about 70% efficient. The human body is about 45% efficient in converting the energy of glucose metabolism to muscle movement. Photosynthesis is 2% to 10% efficient; steam turbines for producing electricity are about 38% efficient; heating homes with electricity is about 38% efficient, whereas heating homes with natural gas is about 70% efficient. The efficiency is usually greater when using a **primary source** of energy on site (burning gas) than when using a **secondary source** (electricity). For example, it takes about 10,000 Btu's to produce 1 kilowatt-hour (kwh) of electricity. If this 1 kwh is then used for heating, only 4000 Btu's of heat are produced. Natural gas burned on site would be more efficient than if burned in a steam generator plant to produce the electricity.

Efficiency is
$$\frac{\text{used energy}}{\text{available energy}}.$$

Primary source of energy: one transformation on site (e.g., chemical → heat via combustion).
Secondary source: usually more than one transformation plus long-distance transport (e.g., chemical → heat via combustion → steam → mechanical → electricity).

Energy Not Lost Is Energy Gained

Energy can be transported through wires (electricity), stored in chemicals (batteries), and carried through the air (radio waves). On the other hand, energy can be prevented from moving by means of insulators. The insulation of houses has popularized the **R value** for heat insulators. The R value (the resistance) is inversely proportional to the conductivity of heat through a slab of material. A common unit of R value is (ft^2) (°F) (hr/Btu). The typical recommendation of an R value of 30 for the ceilings of single-family dwellings means that an average square foot of such a ceiling would lose heat by conduction at a rate of (1/30) Btu/hr for every 1°F difference in temperature. The higher the R value, the fewer Btu's escape per hour per square foot of ceiling. Some R values for 1-inch slabs of the material (in units of ft² °F hr/Btu) are air, 5.9; polyurethane foam, 5.9; rock wool, 3.3; fiberglass, 3.0; white pine, 1.3; and window glass, 0.14. Dry, still air has an insulating value (R value) as great as almost any building material. In fact, many commercial materials owe their heat-insulating ability to entrapped, isolated pockets of air.

Some Materials Have a Higher Energy Cost Than Other Materials

It costs more in energy terms to produce a ton of some substances than to produce the same amount of other substances (Table 8–2). Certain applications now using plastics or metals might more efficiently use ceramics or brick in order to conserve energy. Of course, other factors such as labor costs also influence the economic decisions involved.

Now let's consider the chemicals that provide energy, and see how energy is obtained from them.

FOSSIL FUELS

As indicated in Figure 8–1, nearly all of the present energy needs in the United States are being supplied by fossil fuels: petroleum, natural gas, and coal. This is also true on a worldwide basis. Although the chemistry and geology of fossil fuel formation are not thoroughly understood, it is generally agreed that buried plant material formed these fuels over millions of years. So far as we know, no coal and oil are being formed underground today.

The energy stored in fossil fuels came from sunlight, which was converted to chemical potential energy at an efficiency of 8% and stored in the plants. And what is the source of sunlight? Nuclear energy, as we shall see.

TABLE 8–2   Energy Requirements to Produce Some Common Products

PRODUCT	MILLIONS OF Btu/TON
Titanium	482
Aluminum	244
Copper	112
Polyethylene	100
Polystyrene	64
Polyvinylchloride	49
Plate glass	25
Steel slabs	24
Paper	22
Portland cement	8
Brick	4

(From *Chem Tech*, September 1980, p. 550.)

TABLE 8–3   Energy of Combustion of Fossil Fuels

FUEL	REACTION EQUATION	HEAT ENERGY EVOLVED*
Coal	$C + O_2 \rightarrow CO_2$	94 kcal/mole; 7.8 kcal/g
Natural gas†	$CH_4 + 2\ O_2 \rightarrow CO_2 + 2\ H_2O$	211 kcal/mole; 13.2 kcal/g
Petroleum‡	$2\ C_8H_{18} + 25\ O_2 \rightarrow 16\ CO_2 + 18\ H_2O$	1303 kcal/mole; 11.4 kcal/g

* Numbers are for the quantity of *fuel* (C, CH_4, or C_8H_{18}) specified.
† CH_4, methane, is the principal constituent in natural gas (up to 97%).
‡ C_8H_{18}, octane, is only one of many hydrocarbons present in petroleum.

When fossil fuels are burned, chemical energy is released. The amount of energy varies with the type of fuel (Table 8–3). In each case, the complete combustion products are carbon dioxide and water.

Petroleum

Vast deposits of petroleum were first discovered in the United States (Pennsylvania) in 1859 and in the Middle East (Iran) in 1908. Since that time petroleum has found wide use as an energy source, first as kerosene for lighting, then as gasoline and aircraft fuel for transportation, and more recently as fuel oil to produce electricity.

One barrel of petroleum contains 42 gallons.

The petroleum demand in the United States has declined over the last few years (a percentage point or two each year) and hovers around an average of 15 million barrels a day. In the first six months of 1985, petroleum supplied 40.3% of the energy for the United States; natural gas accounted for 25.7%, coal's share was 23.3%, and other sources supplied 10.7%. Domestic oil production is increasing slightly and averages around 9 million barrels a day. Gasoline demand has been increasing slightly, and the requirement is about 7 million barrels a day.

The overall U.S. oil supply is summarized in Figure 8–3. There were an estimated 200 billion barrels of oil to be recovered in the United States (including

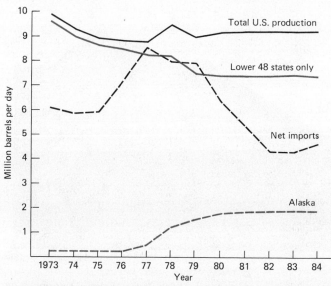

FIGURE 8–3   Oil supply to the United States for 1973 through 1984. Our daily use is the sum of the "Total U.S. Production" and the "Net Imports." (Source: Energy Information Administration, *Annual Energy Review 1984.*)

Alaska) when oil production first began. Half of this amount has now been produced.

Numbers of this type are often quoted, and a few qualifications should be noted. **Recoverable oil** is based on 30% recovery, that is, pumping 30% from the ground and leaving 70% behind, since to recover more would cost more and hence raise the cost of the average barrel of oil. It stands to reason that as supplies decrease and costs go up, it will prove economically feasible to "recover" more of the oil.

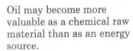

The definition of recoverable oil is based on economics.

A second estimate of U.S. reserves of oil can be based on as yet undiscovered oil under the continental shelf. This estimate is 400 billion barrels, thus doubling the amount of U.S. oil. Either way, the amount of petroleum is limited.

Oil supply, transport, and projected availability are presented in Figure 8–4 (international production of crude oil from 1960 through 1984), Figure 8–5 (a depiction of the origin and destination of the world's oil), and Figure 8–6 (how much of the world's oil supply has been used), respectively. The assumption underlying the projection in Figure 8–6 is that the rest of the world will continue in its lower rate of energy consumption. This is not a feasible assumption. Based on the present trends, a reasonable projection is that 80% of the world's petroleum will have been used in a 60- to 70-year period, ending at about the year 2025. Unless drastic changes are made at some point in the relatively near future, perhaps as early as the year 2000, petroleum will become an increasingly rare commodity.

Oil may become more valuable as a chemical raw material than as an energy source.

The fate of crude oil as it is refined is shown in Figure 8–7. The largest portion becomes gasoline. Only a very small portion is not used for combustion in transportation vehicles and in furnaces and heaters. However, from the small fraction that is not burned come thousands of chemicals — the petrochemicals — which we consider necessary for life today. Consider, for example, doing without plastics, some medicines, food additives, lubricants, most man-made fibers and

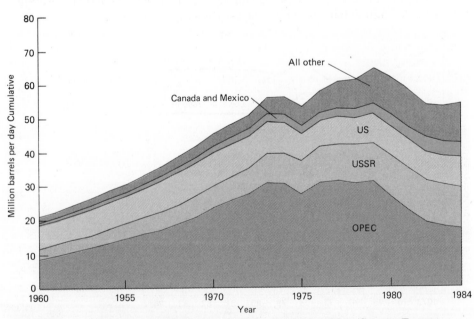

FIGURE 8–4 International production of crude oil, 1960–1984. (Source: Energy Information Administration, *Annual Energy Review 1984*.)

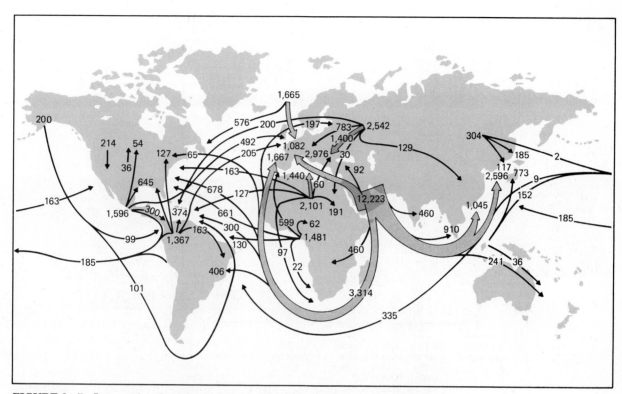

FIGURE 8–5 International crude oil flow in 1982. The numbers are thousands of barrels of oil per day. (Source: Energy Information Administration, *Annual Energy Review 1984*.)

cloth, and paints. These commodities are just the start of thousands of petrochemical-based consumer goods on the market today. In the future, the petrochemical industry may be required to produce basic foodstuffs. Edible fats were produced from petroleum in Germany during World War II. Glycerol is now made from petroleum on a commercial scale, and the process for making sugar from oil has been developed. A real chemical consideration that must come into focus in the

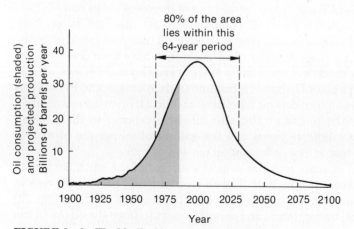

FIGURE 8–6 World oil use: past and projected.

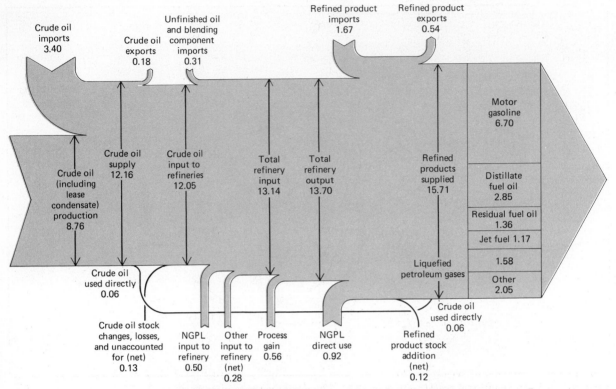

FIGURE 8–7 Petroleum flow in 1984 in the United States. The numbers are millions of barrels per day. NGPL (natural gas plant liquids) are hydrocarbon liquids separated from natural gas at processing plants. (Source: Energy Information Administration, *Annual Energy Review 1984*.)

future as the petroleum supply decreases is whether we want to burn petroleum for energy production or save it as a starting material for petrochemical products.

Oil Shale

Extracting oil from oil shale was not economical until increases in the cost of crude oil prompted at least a dozen major oil companies to get involved. It has been estimated that the United States has reserves of shale oil that could yield oil in the trillions of barrels. Three huge deposits of oil-shale rock have been discovered in the United States. The most developed one is in Utah and Colorado. A second huge oil-shale deposit lies in a giant U-shaped formation from Michigan and Pennsylvania to Alabama. This enormous deposit is believed to hold a trillion barrels of oil. In remote north-central Alaska lies a third vast oil-shale deposit. In the Alaskan deposit, geologic assays indicate yields of a few gallons of oil per ton of rock to saturated ores, which test at 102 gallons of oil per ton of rock.

Most oil locked in shale rock lies near the surface of the earth, and in its economic favor, you never hit a dry hole. The oil is obtained by heating the rock in the absence of air. One technique involves: (1) blasting the rock to break it up, (2) burning a part of the oil underground and using the heat to force the oil out of the rock, and (3) pumping the collected oil to the surface.

This vast source of oil appears not to be attractive economically as long as other sources of petroleum provide the oil at $30 or less per barrel.

Coal

Unlike petroleum, where discoveries are being made almost daily, geologists believe that all of the world's coal supplies have now been discovered. According to the Energy Information Administration, the world's reserves of coal are estimated to be 986.54 billion short tons, of which 28.7% is in the United States. Not all of this coal is mineable.

Mineable coal is defined as 50% of all coal that is in a seam at least 12 inches thick and within 4000 feet of the surface. In the United States the recoverable coal reserves are divided among anthracite (2%), bituminous (52%), subbituminous (38%), and lignite (8%). Some properties of these different kinds of coal are listed in Table 8–4.

Coal is a mixture and has no single molecular structure. However, continuing research has deduced a general type of structure for coal. For example, molecular weights range between 300 and 1000, and the structure is characterized by a profusion of rings of carbon atoms, some rings bonded to each other (fused) and other rings bonded into long chains. The chains are bonded to each other at various points. A representative portion of the combined ideas on the structure of coal is shown in Figure 8–8. The structure can be better understood after a study of Chapters 12 and 13 (organic chemistry) and will be useful in understanding an origin of benzo(alpha)pyrene, a carcinogen. By way of contrast with coal, the structure of petroleum has fewer rings of carbon atoms and less bonding of chains to chains.

Of the 890 million short tons of coal produced in the United States in 1984, the largest portion (74.6%) was burned to produce electricity. Other uses include exports (9.2%), production of coke (5.1%), residential and commercial heating (1.0%), and miscellaneous (10.2%).

The length of time mineable coal will supply our energy needs has been estimated. Figure 8–9 indicates that coal will be available as a fuel much longer than oil. If new mining techniques are developed, then more of the deposited coal might be termed mineable.

Two of coal's major drawbacks are that it is a relatively dirty fuel and difficult to handle. Coupled with the atmospheric pollution caused by sulfur-containing coal, these drawbacks were prime reasons for a major shift from coal to petroleum in electrical generating plants, particularly in the industrialized nations,

The largest supply of fossil fuel is in the form of coal.

A short ton is 2000 pounds. A long ton is 2200 pounds.

Most coal is burned to make electricity.

TABLE 8–4   Some Properties and Characteristics of Types of Coal

	TYPE OF COAL			
	Anthracite	Bituminous	Subbituminous	Lignite
Heat content	High	High	Medium	Low
Sulfur	Low	High	Low	Low
Hydrogen/carbon mole ratio	0.5	0.6	0.9	1.0
Major deposits	New York, Pennsylvania	Appalachian Mountains, the Midwest, Utah	Rocky Mountains	Montana

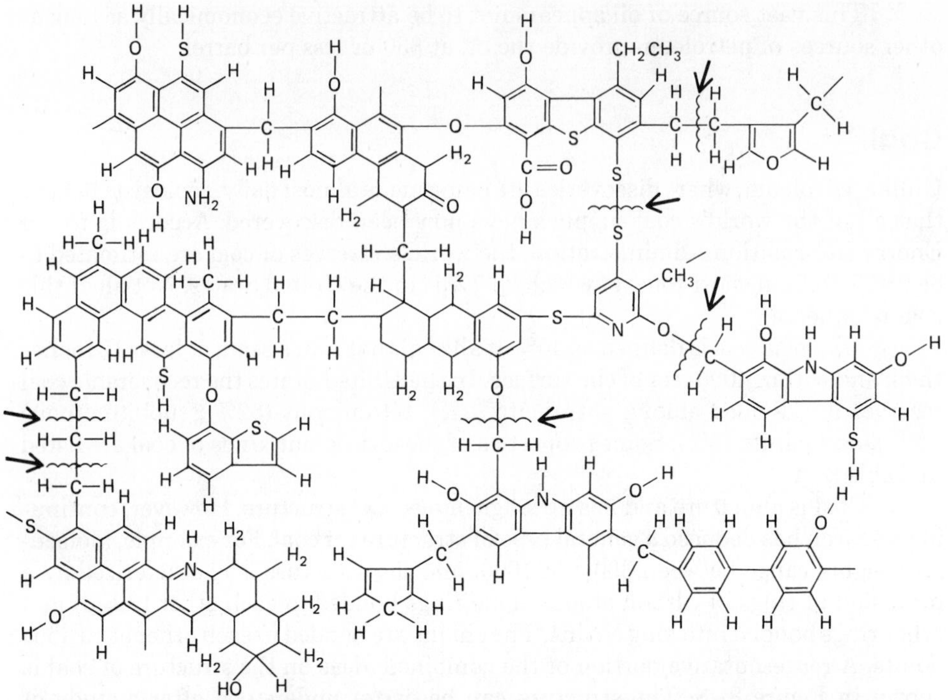

FIGURE 8–8 A partial molecular structure of coal. The arrows indicate weak bonds that may be broken easily during heating. The six-sided figures (hexagons) represent six carbon atoms in a ring. When the rings join on a side, those carbon atoms are in two rings. By way of contrast, the structure of petroleum has fewer rings of carbon atoms and less bonding of chains to chains.

late in the 1960s. The dangerous and unhealthful character of deep coal mining and the environmental disruption caused by strip mining contributed to the shift.

COAL GASIFICATION Coal can be converted into a relatively clean-burning fuel by a process known as **gasification** (Fig. 8–10). In this process, coal is made to react with a limited supply of either hot air or steam. In the reaction of coal with air, the product is a gaseous mixture known as **"power gas,"** and the reaction is exothermic.

Gasification can make coal cleaner to burn and easier to handle from supplier to user.

$$\text{Coal} + \text{Air} \rightarrow \underset{\text{POWER GAS}}{CO(g)} + H_2(g) + N_2(g) + 26.39 \text{ kcal/mole carbon}$$

Power gas contains up to 50% nitrogen by volume and is consequently a relatively poor fuel. In fact, power gas of this composition has only one sixth the heat content of methane.

If the coal is allowed to react with high-temperature steam, a mixture of carbon monoxide and hydrogen known as **synthesis gas** or **coal gas** is obtained. Unlike power gas, this mixture contains no nitrogen.

This reaction is endothermic.

$$\underset{\text{COAL}}{C} + \underset{\text{STEAM}}{H_2O(g)} \rightarrow \underset{\substack{\text{SYNTHESIS GAS} \\ \text{OR COAL GAS}}}{CO(g)} + H_2(g) - 31 \text{ kcal/mole C}$$

When air and steam are mixed in the correct proportions, the reaction of the mixture with coal can be self-sustaining, since the production of power gas is

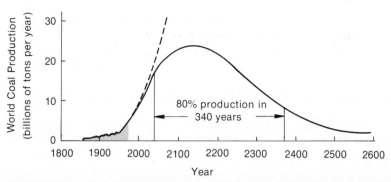

FIGURE 8–9 The coal mined to date *(shaded area)* represents only a small fraction of the mineable coal. The rate of increase in coal consumption *(dashed line)* is 4% per year. It is obvious that such an exponential rise cannot continue long after the year 2000. At the present usage (held constant) coal would last for many hundreds of years.

exothermic and produces enough energy to drive the endothermic production of coal gas.

In both power gas and coal gas mixtures, the CO and H_2 are burned by oxygen in the air to produce heat. The heat produced is about one third that of an equal volume of methane (natural gas).

$$2\ CO + O_2 \rightarrow 2\ CO_2 + 135.3\ \text{kcal (67.6 kcal/mole CO)}$$
$$2\ H_2 + O_2 \rightarrow 2\ H_2O + 115.6\ \text{kcal (57.8 kcal/mole } H_2)$$

These reactions are exothermic.

In a newer coal gasification process, high-energy methane is the end product. The process uses a catalyst (usually potassium hydroxide or potassium car-

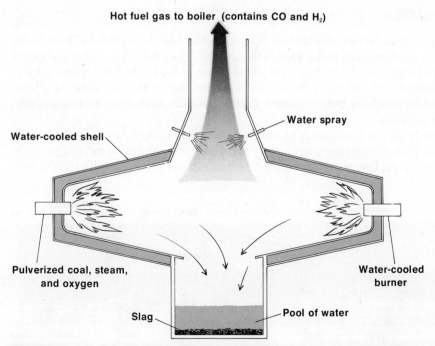

FIGURE 8–10 Schematic drawing of coal gasifier. A relatively cool combustion of powdered coal in a limited supply of oxygen produces a mixture of carbon monoxide and hydrogen along with other gases. The mineral content in the coal collects in the slag.

bonate) and is thermally neutral (neither exothermic nor endothermic) at 700°C, the temperature at which the process is usually run.

In the process, crushed coal is mixed with an aqueous catalyst; the mixture is then dried and sent to a gasifier chamber where CO and H_2 are added. The mixture is then heated to 700°C.

Reactions that occur in the gasifier are (numbers are values at 25°C):

$$2 \text{ C} + 2 \text{ H}_2\text{O} \rightarrow 2 \text{ CO} + 2 \text{ H}_2 - 64 \text{ kcal/2 moles C}$$
$$\text{CO} + \text{H}_2\text{O} \rightarrow \text{CO}_2 + \text{H}_2 + 8 \text{ kcal}$$
$$\text{CO} + 3 \text{ H}_2 \rightarrow \text{CH}_4 + \text{H}_2\text{O} + 54 \text{ kcal}$$

The overall (or net) reaction is:

$$2 \text{ C} + 2 \text{ H}_2\text{O} \rightarrow \text{CH}_4 + \text{CO}_2 - 2 \text{ kcal/mole CH}_4$$

Any unreacted CO and H_2 are cycled back through the gasifier. Recycled steam is used to help dry the coal before the coal enters the gasifier. The catalyst is recovered and reused.

This process converts solid, messy coal into easily transported, efficiently burned methane, the chief component of natural gas. The energy consumed by the process is small, and, best of all, the combustion of methane is environmentally clean.

Other opportunities exist in the area of coal modification, since liquid fuels can also be obtained from coal by hydrogenating the coal. Knowledgeable estimates indicate that we will not be able to rely on coal modification to supply our energy needs on an extensive scale before 1990.

Natural Gas

Natural gas approaches an ideal fuel. It burns with a high heat output (Table 8–3), with little or no residue, and is easily transported. The problem with natural gas is its limited supply. Federal price regulations begun in 1954 held down prices, stimulated demand, and decreased incentives for exploration of new gas deposits. These price regulations have been removed. It remains to be seen whether new gas will become available.

In 1973, the production of natural gas in the United States peaked at 22 trillion (10^{12}) cubic feet per year (Fig. 8–11). The world reserves of natural gas were estimated by the Energy Information Administration to be 3402 trillion cubic feet

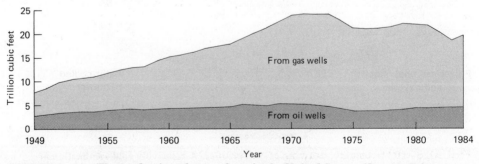

FIGURE 8–11   The production of natural gas in the United States. (Source: Energy Information Administration, *Annual Energy Review 1984.*)

in 1984. Most of these reserves are in the Middle East, Eastern Europe, and the U.S.S.R. Even if significant new deposits of natural gas are discovered in such locations as the outer continental shelves, the North American natural gas deposits are about 60% depleted. Importation of natural gas is complicated by difficulties encountered in condensing natural gas to a liquid for ocean transit and the danger of transporting and storing such concentrated and volatile fuels.

Even with higher prices and an impending depletion of natural gas, more homes are heated by burning natural gas than by any other means. In 1983, 55% of the homes in the United States were heated by natural gas, followed by electricity (18.5%), fuel oil (14.9%), wood (4.8%), and liquefied gas (butane, propane) (4.6%). Coal and kerosene come in at a low 0.5%.

POTENTIAL, NON-MAINSTREAM SOURCES OF ENERGY

Other energy sources offer more potential than they now supply. Some of these are available only in certain locales; some may be novel and may never be very useful. Into these categories falls energy of geothermal origin, from wind and ocean currents, and from burning garbage.

Geothermal Energy

Geothermal energy, or heat from the Earth, comes from the hot, molten rock that forms the Earth's interior. As the Earth's surface is penetrated, the temperature rises an average of 25°C with each kilometer (five eighths of a mile) of depth. Although 32 million quadrillion Btu's of energy simmer within 10 km of the surface of the United States, most of the Earth's heat is too deep to be extracted for practical use.

Hot springs and geysers are the most common manifestations of geothermal energy. Hot springs have heated homes in Boise, Idaho, since the 1890s. Electricity is produced from geothermal water on a commercial basis in California's Imperial Valley. Heat is extracted from dry, hot rock by circulating water through drill holes and man-made fractures in the rock.

In 1984, geothermal energy contributed 2.6×10^{13} Btu's to the national energy pool. By the year 2020, the contribution is expected to be 18.5 quads (1.85×10^{16} Btu's).

Energy from Wind Currents

Energy from wind currents is secondhand solar energy. The basic driving force is the unequal heating of the Earth and its atmosphere with the characteristic flow of the wind influenced by the Earth's rotation.

Your first thought of controlled energy from the wind is probably the picturesque windmills of Denmark and the Netherlands. However, almost a million windmills operated in the United States in the 1920s. Most were on farms and were used to pump water. Although fewer windmills dot the farms now, there is new experimentation with large wind turbines. The research was stimulated by a $900 million program resulting from the Wind Energy Systems Act of 1980.

Energy from Ocean Currents

An immense amount of energy is available in the ocean currents, if we can find practical means of channelling it. The major drawback is the slowness with which tides and ocean currents move.

Through the use of a property of the oceans other than its movement, enough energy is expected to be produced by the year 2000 to replace 400,000 barrels of oil per day. The process, called Ocean Thermal Energy Conversion (OTEC), capitalizes on the difference between the sun-warmed surface water and the cold ocean depths. The warm surface water vaporizes ammonia (NH_3). The gaseous ammonia drives a turbine to turn a generator and produce electricity. Cold water is pumped from a depth of 3000 feet to cool and condense the ammonia to liquid. The process is then repeated. A small OTEC project has operated near Hawaii for several years.

Energy from Garbage

An energy-producing plant that burns garbage is a solution to both an energy problem and the problem of what to do with our trash. Several such plants are in operation in several countries, including France and the United States. The Nashville Thermal Transfer Corporation in downtown Nashville, Tennessee, began operation in February 1974 (Fig. 8–12). The plant supplies steam and (or) cold water to 28 buildings and 30,000 people in the downtown area. The energy comes from burning residential garbage. From 200 to 300 truckloads arrive daily for four days each week and supply the 400 tons of garbage to be burned each day. Pipes laid under the city during urban renewal carry the steam and cold water to the various buildings. Electricity produced by steam-powered turbines cools the water, acting similar to a very large refrigerator. The ash from burning the garbage has to go to a landfill, but the ash is less than 10% of the volume and 30% of the weight of the original garbage. Oil and gas are available as back-up energy sources if needed, but they are used rarely.

Another use of garbage as a source of energy is at the world's largest garbage dump at Fresh Kills on Staten Island, New York. Underneath the huge mounds of

FIGURE 8–12 The Nashville, Tennessee, thermal energy plant.

garbage, bacteria turn old, buried garbage into methane. The Brooklyn Union Gas Company has tapped this gas, and enough can be provided to fuel 16,000 homes on Staten Island.

ELECTRICITY PRODUCTION

The major secondary source of energy in the United States is electricity. Secondary energy is made from a primary source (coal, petroleum, natural gas, nuclear, geothermal, winds, etc.) on the way to the end user. In 1984, 35% of all energy consumed was used in the production of electricity (Fig. 8–13). In addition to the 26.05 quads being put into the production of electricity, 18.27 quads of this energy are lost in the conversions and transmission of energy. At least part of this loss is expected by knowing the second law of thermodynamics (discussed earlier in this chapter), which states a natural process loses some energy to entropy (disorder).

 A longer view of the generation of electricity by type of primary source of fuel is shown in Figure 8–14. Coal has been, and probably will be for many years, the major primary source for generation of electricity.

 A familiar expression associated with electricity is **electrical power.** Appliances and other electrical devices are rated according to their electrical power (kilowatts). However, most consumers purchase electrical energy (kilowatt-hours). What is the difference? Energy is the ability to do work, and power is the *rate* of doing work (Table 8–5). Power has a time factor. For example, a 100-watt

When a power unit is multiplied by a time unit, the result is an equivalent energy unit.

energy units = watts × time, i.e., 1 kwh = 10^3 w × 1 hr = $\dfrac{10^3 \text{ joules}}{\text{second}}$ × $\dfrac{3600 \text{ sec}}{\text{hr}}$ × 1 hr = 3.6 × 10^6 joules.

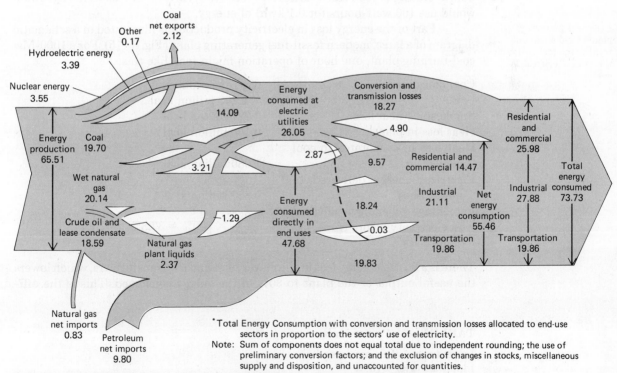

FIGURE 8–13 Total energy flow in the United States in 1984. The numbers are energies expressed in quadrillion (10^{15}) Btu's, otherwise known as quads. (Source: Energy Information Administration, *Annual Energy Review 1984.*)

*Total Energy Consumption with conversion and transmission losses allocated to end-use sectors in proportion to the sectors' use of electricity.
Note: Sum of components does not equal total due to independent rounding; the use of preliminary conversion factors; and the exclusion of changes in stocks, miscellaneous supply and disposition, and unaccounted for quantities.

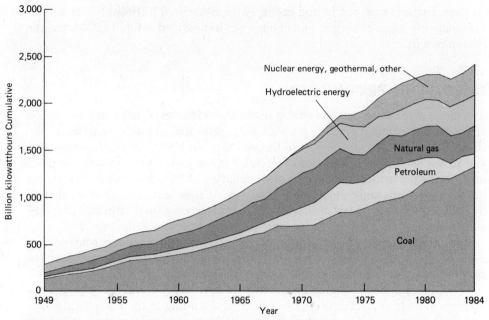

FIGURE 8–14   The generation of electricity by type of energy source. (Source: Energy Information Administration, *Annual Energy Review 1984*.)

bulb operating at 100 volts would draw 1 ampere of current. In 1 hour, the light bulb would use 100 watt-hours (or 0.1 kwh) of energy.

Electrical generating plants yield about one third of the fuel energy in the form of electrical energy.

        Part of the energy loss in electricity production is illustrated in a schematic diagram of a large, modern fossil-fuel generating plant (Fig. 8–15). For a 1000-Mw coal-burning plant, one hour of operation might look like this:

Coal consumed	696 tons producing 2.270 billion kcal
Smokestack heat loss	0.227 billion kcal
Heat loss in plant	0.106 billion kcal
Heat loss in evaporator to cool condenser	1.080 billion kcal
Electrical energy delivered to power lines	0.857 billion kcal
Percentage of energy delivered as electricity before transmission losses	$\dfrac{0.857}{2.27} \times 100\% = 37.8\%$

There is a further energy loss in the power lines and the transformers, which lowers the useful output of the plant to 30% of the energy consumed. This is the **effi-**

TABLE 8–5   Power Units

1 watt (w) = 1 joule per second
1 kilowatt (kw) = 10^3 w
1 megawatt (Mw) = 10^3 kw = 10^6 w
1 watt = 1 volt $\times$ 1 ampere

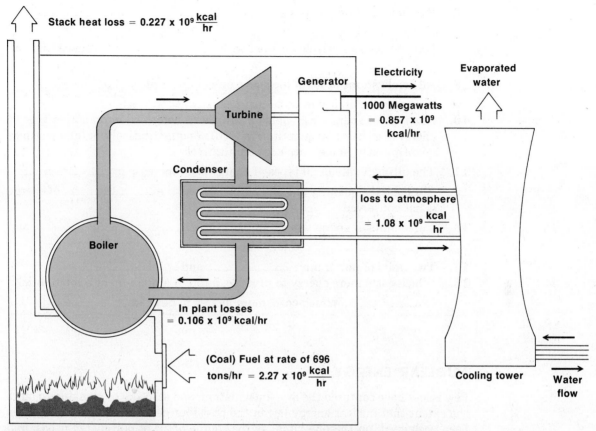

FIGURE 8–15   The heat balance of a 1000-megawatt coal burning electrical generating plant. Note that the 969 tons of coal burned per hour furnish 2.27×10^9 kcal of heat energy, but only 0.857×10^9 kcal of energy, or 38%, is converted to electricity. Note also the large amounts of heat energy lost to the cooling water and atmosphere.

ciency figure for the overall operation. It is important to note that we pay for 300 kcal of heat energy in the form of coal or fuel oil but receive less than 100 kcal of energy in the form of electricity. Obviously, it requires much less fuel to heat homes with the fuel itself than with electricity made from the fuel.

SELF-TEST 8–A

1. In 1984, what energy source provided the most energy for U.S. consumption? See Figure 8–1. _____

2. Which furnishes the most heat energy per gram — coal, petroleum, or natural gas? _____

3. How many gallons of oil are there in one barrel? _____

4. Is the composition of "power gas" obtained from coal gasification of CO, H_2, N_2 or CO, H_2? _____

5. The typical efficiency of an electrical generating plant is about (100, 50, 33, 10) percent. _____

6. Examples of fossil fuels are _____, _____, and _____.

7. Natural gas and petroleum react with _____ to produce CO_2 and _____.

8. All combustions of fossil fuels give off energy. True ( ) or False ( )

9. Energy is the ability to do _____.

10. One type of energy (e.g., light) is always transformed into another type of energy (e.g., heat) (a) quantitatively, (b) not quantitatively, or (c) sometimes quantitatively, sometimes not quantitatively.

11. The ultimate fate of all types of energy is an increase in _____.

12. Although the quantity of energy is conserved, the _____ of energy is not conserved.

13. Three units of energy are _____, _____, and _____.

14. Two units of power are _____ and _____.

15. Which costs more energy to produce, a ton of aluminum or a ton of brick? _____ Which costs more money to buy? _____

NUCLEAR ENERGY

Few issues have captured the awe, imagination, and scrutiny of mankind to quite the extent that nuclear energy has in the past four decades. Nuclear energy has been acclaimed, on the one hand, as the source of all of our energy needs, and accused, on the other hand, of being our eventual destroyer.

Part of the interest in nuclear power is the tremendous amount of energy generated by a relatively small amount of fuel. The mechanics of nuclear reactions are described in Chapter 7; in this section, we focus on the energy that accompanies nuclear reactions.

Methane is the major component of natural gas.

The vast amounts of energy are released when heavy atomic nuclei split, the **fission** process, and when small atomic nuclei combine to make heavier nuclei, the **fusion** process. Consider the energy contrast between combustion of a fossil fuel and a nuclear fusion reaction. When one mole (6.02×10^{23} molecules, or 16 g) of methane is burned, over 200 kcal of heat are liberated:

$$CH_4 + 2\,O_2 \rightarrow CO_2 + 2\,H_2O + 211 \text{ kilocalories (kcal/mole } CH_4)$$

Energy changes associated with nuclear events may be many thousands of times larger than those associated with chemical events.

In contrast, a lithium nucleus can be made to react with a hydrogen nucleus to form two helium nuclei in a nuclear reaction. The energy released per mole of lithium in this reaction is 23,000,000 kcal. This means that 7 g of lithium and 1 g of hydrogen produce 100,000 times more energy through fusion of nuclei than 16 g of methane and 64 g of oxygen produce by electron exchange.

$$^{7}_{3}Li + ^{1}_{1}H \rightarrow 2\,^{4}_{2}He + 23{,}000{,}000 \text{ kcal/mole of } ^{7}_{3}Li$$

Atomic mass, Atomic number $^{7}_{3}Li$

Realizing that nuclear changes could involve giant amounts of energy relative to chemical changes for a given amount of matter, Otto Hahn, Fritz Strassman, Lise Meitner, and Otto Frisch discovered in 1938 that $^{235}_{92}U$ is fissionable. Subsequently the dream of controlled nuclear energy became a reality, followed by the bomb and nuclear power plants. In the 1950s it was hoped that nuclear energy would soon relieve the shortage of fossil fuels. To date this has not been accomplished, although the production of nuclear energy has grown very rapidly in recent years. The use of nuclear energy to generate electricity is much more advanced in Europe than in the United States. For example, nuclear reactors supply 65% of electricity in France versus 16% in the United States.

Fission Reactions

Fission can occur when a thermal neutron (with a kinetic energy about the same as that of a gaseous molecule at ordinary temperatures) enters certain heavy nuclei with an odd number of neutrons ($^{235}_{92}$U, $^{233}_{92}$U, $^{239}_{94}$Pu). The splitting of the heavy nucleus produces two smaller nuclei, two or more neutrons (an average of 2.5 neutrons for $^{235}_{92}$U), and much energy. Typical nuclear fission reactions may be written:

$$^{235}_{92}U + ^{1}_{0}n \rightarrow ^{141}_{56}Ba + ^{92}_{36}Kr + 3\,^{1}_{0}n + energy$$

$$^{235}_{92}U + ^{1}_{0}n \rightarrow ^{103}_{42}Mo + ^{131}_{50}Sn + 2\,^{1}_{0}n + energy$$

Note that the same nucleus may split in more than one way. The fission products, such as $^{141}_{56}$Ba and $^{92}_{36}$Kr, emit beta particles ($_{-1}^{0}$e) and gamma rays ($^{0}_{0}\gamma$) until stable isotopes are reached.

$$^{141}_{56}Ba \rightarrow \,_{-1}^{0}e + ^{0}_{0}\gamma + ^{141}_{57}La$$

$$^{92}_{36}Kr \rightarrow \,_{-1}^{0}e + ^{0}_{0}\gamma + ^{92}_{37}Rb$$

The products of these reactions emit beta particles, as do their products. After several such steps, stable isotopes are reached: $^{141}_{59}$Pr and $^{90}_{40}$Zr, respectively.

The neutrons emitted can cause the fission of other heavy atoms if they are slowed down by a moderator, such as graphite. For example, the three neutrons emitted in the first preceding reaction could produce fission in three more uranium atoms, the nine neutrons emitted by those nuclei could produce nine more fissions, the 27 neutrons from these fissions could produce 81 neutrons, the 81 neutrons could produce 243, the 243 neutrons could produce 729, and so on. This process is called a **chain reaction** (Fig. 8–16), and it occurs at a maximum rate when the uranium sample is large enough for most of the neutrons emitted to be captured by other nuclei before passing out of the sample. Sufficient sample in a certain volume to sustain a chain reaction is termed the **critical mass.**

In the atomic bomb the critical mass is kept separated into several smaller subcritical masses until detonation, at which time the masses are driven together by an implosive device. It is then that the tremendous energy is liberated and everything in the immediate vicinity is heated to temperatures of 5 to 10 million degrees. The sudden expansion of hot gases literally explodes everything nearby and scatters the radioactive fission fragments over a wide area. In addition to the movement of gases, there is the tremendous vaporizing heat that makes the atomic bomb so devastating.

There is no danger of an atomic explosion in the uranium mineral deposits in the Earth for two reasons. First, uranium is not found pure in nature — it is found only in compounds, which in turn are mixed with other compounds. Second, less than 1% of the uranium found in nature is fissionable $^{235}_{92}$U. The other 99% is $^{238}_{92}$U, which is not fissionable by thermal neutrons. In order to make nuclear bombs or

Note in nuclear reactions that the sum of the atomic numbers on the left side of the equation equals the sum of the atomic numbers on the right side of the equation. Likewise for the atomic masses.

Fission is the breakup of heavy nuclei.

$^{1}_{0}$n represents a neutron.

$_{-1}^{0}$e or $_{-1}^{0}\beta$ represents a beta particle.

A low-energy neutron will disrupt some large nuclei.

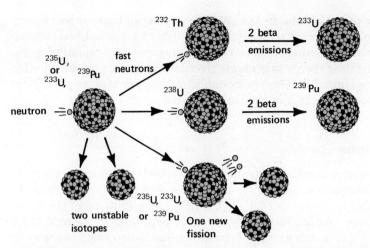

FIGURE 8–16   A chain reaction. A thermal neutron collides with a fissionable nucleus and the resulting reaction produces three additional neutrons. These neutrons can either convert nonfissionable nuclei such as $^{232}_{90}$Th to fissionable ones or cause additional fission reactions. If enough fissionable nuclei are present, a chain reaction will be sustained.

nuclear fuel for electrical generation, a purification enrichment process must be performed on the uranium isotopes, thus increasing the relative proportion of $^{235}_{92}$U atoms in a sample. Ordinary uranium such as that found in ores is only 0.711% $^{235}_{92}$U.

It is interesting to note that fission products can be found in the Gabon Republic of West Africa, which indicate that a uranium ore deposit "went critical" about 150,000 years ago. At that time the natural uranium-235 content would have been higher than it is now.

Mass Defect — The Ultimate Nuclear Energy Source

What is the source of the tremendous energy of the fission process? It ultimately comes from the conversion of mass into energy, according to Einstein's famous equation, $E = mc^2$, where E is energy that results from the loss of an amount of mass, m, and c^2 is the speed of light squared. If separate neutrons, electrons, and protons are combined to form any particular atom, there is a loss of mass called the **mass defect.** For example, the calculated mass of one $^4_2$He atom from the masses of the constituent particles is 4.032982 amu:

$2 \times 1.007826 = 2.015652$ amu, mass of two protons and two electrons
$2 \times 1.008665 = \underline{2.017330}$ amu, mass of two neutrons
total $= 4.032982$ amu, calculated mass of one $^4_2$He atom

Since the measured mass of a $^4_2$He atom is 4.002604 amu, the mass defect is 0.030378 amu:

4.032982 amu
−4.002604 amu
0.030378 amu, mass defect

Because the atom is more stable than the separated neutrons, protons, and electrons, the atom is in a lower energy state. Hence, the 0.030378 amu lost per atom would be released in the form of energy if the $^4_2$He atom were made from separate

---

Separation of uranium isotopes had to precede the control of atomic energy.

amu = atomic mass unit

$6.02 \times 10^{23} \dfrac{\text{amu}}{\text{gram}}$

or

$\dfrac{1 \text{ mole amu}}{\text{gram}}$

The mass that is lost leaves in the form of energy: $E = mc^2$.

protons, electrons, and neutrons. The energy equivalent of the mass defect is called the **binding energy.** The binding energy is analogous to the earlier concept of bond energy, in that both are a measure of the energy necessary to separate the package (nucleus or molecule) into its parts.

Atoms with atomic numbers between 30 and 63 have a greater mass defect per nuclear particle than very light elements or very heavy ones, as shown in Figure 8–17. This means the most stable nuclei are the middle-weight ones found in the atomic number range from 30 to 63.

Because of the relative stabilities, it is in the intermediate range of atomic numbers that most of the products of nuclear fission are found. Therefore, when fission occurs and smaller, more stable nuclei result, these nuclei will contain less mass per nuclear particle. In the process, mass must be changed into energy. This energy gives the fission process its tremendous energy. It takes only about 1 kg of $^{235}_{92}$U or $^{239}_{94}$Pu undergoing fission to be equivalent to the energy released by 20,000 tons (20 kilotons) of ordinary explosives like TNT. The energy content in matter is further dramatized when it is realized that the atomic fragments from the 1 kg of nuclear fuel weigh 999 g, so only one tenth of 1% of the mass is actually converted to energy. The fission bombs dropped on Japan during World War II contained approximately this much fissionable material.

Separated nuclear particles have more mass than when combined in a nucleus.

Intermediate-sized nuclei tend to have the greatest nuclear stability.

Controlled Nuclear Energy

The fission of a $^{235}_{92}$U nucleus by a slow-moving neutron to produce smaller nuclei, extra neutrons, and large amounts of energy suggested to Enrico Fermi and others that the reaction could proceed at a moderate rate if the number of neutrons could be controlled. If a neutron control could be found, the concentration of neutrons could be maintained at a level sufficient to keep the fission process going but not

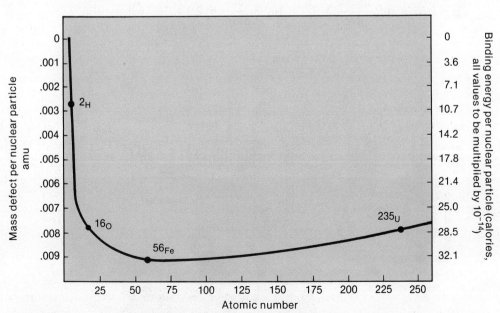

FIGURE 8–17 Mass defect for different nuclear masses. The most stable nuclei center around $^{56}_{26}$Fe, which has the largest mass defect per nuclear particle.

high enough to allow an uncontrolled explosion. It would then be possible to drain the heat away from such a reactor on a continuing basis to do useful work. In 1942, Fermi, working at the University of Chicago, was successful in building the first atomic reactor, called an **atomic pile.**

An atomic reactor has several essential components. The charge material (fuel) must be fissionable or contain significant concentrations of a fissionable isotope such as $^{235}_{92}U$, $^{239}_{94}Pu$, or $^{233}_{92}U$. Ordinary uranium, which is mostly the nonfissionable $^{238}_{92}U$, cannot be used since it has a small concentration of the $^{235}_{92}U$ isotope. A moderator is required to slow the speed of the neutrons produced in the reactions without absorbing them. Graphite, water, and other substances have been used successfully as moderators. A substance that will absorb neutrons, such as cadmium or boron steel, is present in order to have a fine control over the neutron concentration. Shielding, to protect the workers from dangerous radiation, is an absolute necessity. Shielding tends to make reactors heavy and bulky installations. A heat-transfer fluid provides a large and even flow of heat away from the reaction center.

Once the heat is produced in a nuclear reactor and safety measures are employed to protect against radiation, conventional technology allows this energy to be used to generate electricity, to power ships, or to operate any device that uses heat energy. A system for the nuclear production of electricity is illustrated in Figure 8–18.

What are the fuel requirements in nuclear fission energy production? In a typical fission event such as

$$^{1}_{0}n + ^{235}_{92}U \rightarrow ^{93}_{37}Rb + ^{141}_{55}Cs + 2\,^{1}_{0}n + 200\ Mev$$

the energy release, 200 Mev, is equivalent to 7.7×10^{-12} calorie per atom of $^{235}_{92}U$, or 4.64×10^9 kcal/mole. Since 1 g of pure $^{235}_{92}U$ contains 2.56×10^{21} atoms, the total energy release for 1 g of uranium-235 undergoing fission would be

$$1\ g \times 2.56 \times 10^{21}\ \frac{atoms}{1\ g} \times 7.7 \times 10^{-12}\ \frac{cal}{atom} = 2.0 \times 10^{10}\ cal$$

Atomic pile:

1. Carefully diluted fissionable material;
2. Moderator to control fission reaction;
3. Coolant to control heat;
4. Shielding to limit radiation.

The first one was piled together at the University of Chicago in 1942.

1 million electron volts (Mev) = 3.827×10^{-14} cal.

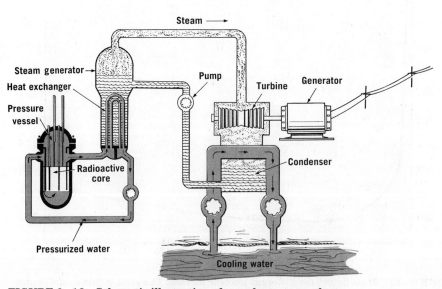

FIGURE 8–18   Schematic illustration of a nuclear power plant.

This is the amount of energy that would be released if 5.95 tons of coal were burned, or if 13.7 barrels of oil were burned to produce heat to power a boiler. This means that about 3 kg of $^{235}_{92}$U fuel per day would be required for a 1000-Mw electric generator. The fuel used, however, is not pure $^{235}_{92}$U, but **enriched** uranium containing up to 3% $^{235}_{92}$U.

When 1 g of ^{235}U undergoes fission, it provides the same energy as burning about 6 tons of coal.

Nuclear energy is mostly used for electricity production. In 1965, when nuclear energy usage began, and until about 1974, more energy was produced in the United States from burning firewood than from nuclear energy. In 1984, 4.8% of the U.S. energy supply came from nuclear sources; by 1990, almost 10% of our energy need is expected to be met by nuclear sources. On January 1, 1986, there were 98 operable U.S. nuclear reactors and 32 on order but not yet operating (Fig. 8–19). The United States is producing less uranium now, down from a high of 21,850 short tons of U_3O_8 in 1980. The trend was downward to 7500 short tons in 1984. Estimates of the reserves in 1983 were 885,000 short tons U_3O_8 (reasonably assured) to 4,394,000 short tons U_3O_8 (speculated and estimated).

It is possible to convert the nonfissionable $^{238}_{92}$U and $^{232}_{90}$Th into fissionable fuels by using a **breeder reactor.** In such a reactor, a blanket of nonfissionable material is placed outside the fissioning $^{235}_{92}$U fuel (Fig. 8–20), which serves as the source of neutrons in the breeder reactions. The two breeder reaction sequences are

^{235}U is in very short supply.

$$^{238}_{92}\text{U} + ^{1}_{0}\text{n} \rightarrow ^{239}_{92}\text{U} \xrightarrow{\beta} ^{239}_{93}\text{Np} \xrightarrow{\beta} ^{239}_{94}\text{Pu}$$

$$^{232}_{90}\text{Th} + ^{1}_{0}\text{n} \rightarrow ^{233}_{90}\text{Th} \xrightarrow{\beta} ^{233}_{91}\text{Pa} \xrightarrow{\beta} ^{233}_{92}\text{U}$$

The products of the breeder reactions, $^{233}_{92}$U and $^{239}_{94}$Pu, are both fissionable with slow neutrons, and neither is found in the Earth's crust.

The Clinch River Breeder Reactor project died in 1983, but research continues at Argonne National Laboratory outside Chicago. The breeder concept came

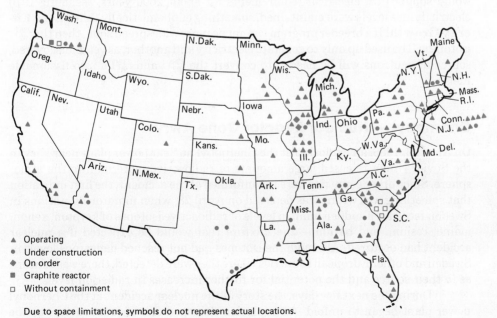

▲ Operating
● Under construction
◆ On order
■ Graphite reactor
□ Without containment

Due to space limitations, symbols do not represent actual locations.

FIGURE 8–19 Status of nuclear reactors in the United States on January 1, 1986.

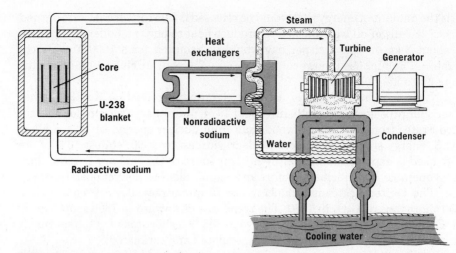

FIGURE 8–20   Schematic diagram of a fast breeder reactor and steam-turbine power generator.

$^{239}_{94}$Pu is toxic from a radiation as well as a chemical point of view.

from Walter Zinn, Argonne's first director. The first nuclear reactor of any kind to produce electricity was a research breeder reactor at Idaho Falls, Idaho. Operation began in 1951. An updated version (EBR-2, experimental breeder reactor) has been in operation for more than 20 years.

Breeder reactors present many technological problems, not the least of which is the potential of a disaster caused by mishandling of the $^{239}_{94}$Pu isotope, which is extremely toxic and can also be fabricated into a fission bomb. Nevertheless, the expected benefit from the breeder program is massive amounts of energy. For example, if all the uranium used for electrical generation were used in breeder reactors, instead of running out of uranium fuel in several decades, the breeder fuels would supply U.S. electrical requirements for about 2600 years, assuming 1970 electricity-use levels were maintained, something euphemistically called **Zero Energy Growth!** If a breeder program cannot be more widespread soon, then the $^{235}_{92}$U isotope will be used up only to generate electricity and another, as yet undiscovered, source of neutrons will be needed to convert the $^{238}_{92}$U and $^{232}_{90}$Th into fissionable fuels.

Chernobyl — Nuclear Reactor Gone Awry

The figure on page 191 shows a schematic of the impact of a reactor meltdown on the environment.

On April 26, 1986, an explosion at the Chernobyl nuclear power plant near Kiev in the Soviet Union released large amounts of radioactive material into the atmosphere. Since other countries were not informed of the accident, the first indication that something was wrong was detected on April 28, when monitoring stations in Sweden recorded a sudden jump in levels of radioactive isotopes of krypton, xenon, iodine, cesium, and barium — the mixture that would be expected if a nuclear accident had occurred. Although the isotopes had not reached dangerous levels in Sweden and other European countries where they were detected, there was concern as to their source and the potential for further increases in radiation levels.

During the next few days, the story of the nuclear accident at the Chernobyl power plant began to unfold. An explosion damaged one of four reactors as the Chernobyl facility was going into a planned shutdown. Although the cause of the explosion is not known at the time of this writing, it is known that the Chernobyl

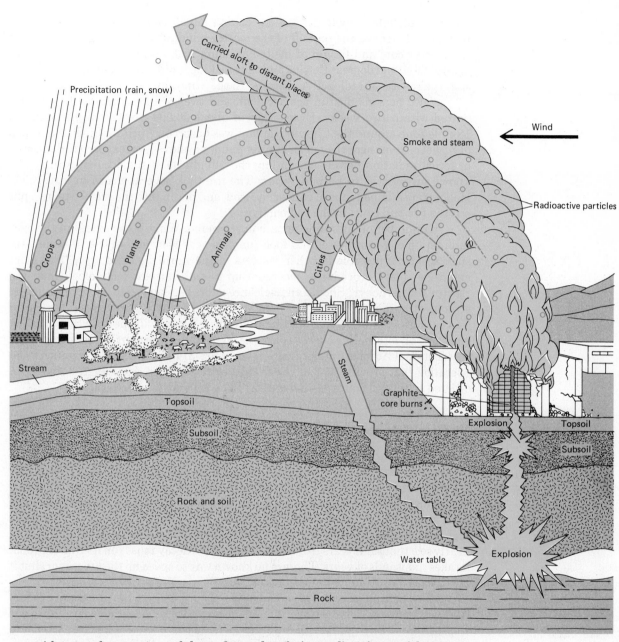

Precipitation (rain, snow)

Carried aloft to distant places

Wind

Smoke and steam

Radioactive particles

Crops

Plants

Animals

Cities

Steam

Stream

Graphite
core burns

Topsoil

Explosion

Topsoil

Subsoil

Subsoil

Rock and soil

Water table

Explosion

Rock

After a nuclear reactor meltdown, fire, and explosion, radioactive particles are broadcast from the site by explosive forces, rising hot air, and steam. Such an explosion is chemical, *not nuclear*. With current reactors, the steam in contact with zirconium or other active metals produces hydrogen gas, which explodes. If the explosion and/or fire produces cracks to the water table, groundwater carries radioactive particles throughout the region. Radioactivity incorporated into plants or absorbed by streams on the surface finds its way up the food chain into human beings. Either through ingested radioactive food or through inhaled radioactive particles, human beings and animals become susceptible to cancer.

plant used U-235 as fuel, graphite as moderator, and water as coolant. The most likely series of events is (1) overheating of the reactor core because of the loss of water coolant; (2) melting of uranium fuel and the zirconium alloy that encased the fuel; (3) reaction of molten zirconium and uranium with superheated steam to give off large amounts of hydrogen gas; (4) detonation of hydrogen gas, destroying the reactor and blowing the roof off of the reactor building; (5) burning of graphite in the reactor core and fission of melted uranium fuel, keeping the temperature at several thousand degrees Celsius and thus making it difficult to put out the fire and stop the release of radioactive material to the atmosphere.

Two persons died in the accident, and twenty additional deaths were reported within two weeks. Evacuation of about 50,000 Soviet citizens from an 18-mile zone around the Chernobyl plant was begun on April 27. Since thousands of Soviet citizens near the nuclear power plant may have been exposed to damaging or lethal radiation before evacuation, the full effects of the accident will not be known for months or perhaps years. The fire took several weeks to extinguish. Helicopters dumped tons of sand, clay, lead, and boron into the reactor to help put out the fire and contain the radioactivity.

A number of questions remain unanswered. What is the extent of the contamination of groundwater and food supplies near Kiev and in the Ukraine? Are such accidents possible in other reactors in the world that generally have more extensive safety procedures and more up-to-date nuclear technology than the reactor at Chernobyl? Should all nuclear reactors be required to have containment structures, as did the reactor at Three Mile Island in the United States, as a way of preventing radiation leakage to the atmosphere? How will the growth of nuclear energy as an alternative energy source be affected by Chernobyl? According to the International Atomic Energy Agency, there are 374 nuclear power plants now in operation that provide 15% of the world's electric power. Another 157 have been ordered as several countries move away from dependency on fossil fuels.

Nuclear Waste — A Problem with No Apparent Solution

Radioactive isotopes are produced in nuclear fission reactors. The radiation levels produced by many of these isotopes are dangerous to life. Compounding the danger is the fact that some of the half-lives extend the active lifetime of the radioactive wastes into thousands of years. If all nuclear power and weapons research had come to an abrupt halt after the Three Mile Island incident in 1978, most of the nation's high-level nuclear waste stockpiled up to that point would be as deadly today, and nearly so a thousand years from now. Radioactive decay must run its course even if it takes thousands of years. There is no known way to speed up the process or shut it off. This is a problem with no apparent solution.

The persistent half-lives are not the only problem. In the process of producing nuclear energy, a large quantity of radioactive waste is made. According to the Department of Energy, 71 million pounds of radioactive waste were discharged into the air, water, and ground from 1946 to 1983 at seven facilities, three at Oak Ridge, Tennessee, one at Paducah, Kentucky, and three in Ohio (Fernald, Piketon, and Ashtabula). Additional wastes were discharged from other facilities.

The waste gases were mostly radioactive isotopes of the noble gases, such as krypton-85, with a half-life of 10.76 years. Tritium, an isotope of hydrogen ($^{3}_{1}H$, half-life of 12.26 years), is also discharged in gaseous form, principally as water vapor. As the hydrogen part of a water molecule, tritium may be taken up into biological food chains.

Radioactive isotopes in liquid wastes are generally converted to solid form through precipitation, thereby decreasing the volume to one tenth or less of the original volume. Then, the solid is stored. The 51 million pounds of radioactive wastes buried in the ground at the Y-12 plant at Oak Ridge are buried in unlined trenches and covered with dirt — no casks, no containers. It is not known the extent to which ground water will spread the radioactive materials over the long term of hundreds of years.

If we cannot control radioactive decay and we have these huge buildups of radioactive wastes, what can be done? Proposed solutions to these problems center on the theme, "one person's solution is another person's pollution (problem)." The premise is to get the wastes as far away from us as possible. Fanciful schemes, such as rocketing the nuclear waste into the Sun or deep space or burying it in the deep oceans, have been largely dismissed as too expensive and risky. The Department of Energy is committed to putting on-line in 1998 underground storage chambers carved in salt beds, clay, or rock. According to the plan, the waste will first go into specially built containers designed to withstand shock and corrosion for thousands of years. Scientists believe this is the best of the few alternatives available.

Even if the storage method planned by the Department of Energy is adequate, there is still another problem — that of transporting the radioactive wastes from the end-user site to the storage site. In some cases, this requires transporting the wastes across most of the United States. If a truck, train, or ship is used, how can the transporting be done safely? Some of the transporting is through heavily populated centers.

The danger in transporting nuclear wastes is the spread of dangerous radioactivity, not a nuclear explosion.

Of the myriad problems associated with nuclear waste, there is a unique challenge to mankind. It is not the size of the problem in terms of the amount of wastes to be stored or the transportation of the wastes, but rather the complete lack of control of the rate of radioactive decay. The chemical theory and technology for abating air pollution, water pollution, and recycling chemical wastes are known today, despite the enormity and complexity of these problems. But nothing can be done to change the rate at which radioactive decay goes away. Outside of health problems, this may be the most perplexing and frustrating problem to be faced by mankind today.

Fusion Reactions

When very light nuclei, such as H, He, and Li, are combined, or **fused,** to form an element of higher atomic number, energy must be given off consistent with the greater stability of the elements in this intermediate atomic number range (Fig. 8–17). This energy, which comes from a decrease in mass, is the source of the energy released by the Sun and by hydrogen bombs. Typical examples of fusion reactions are:

Fusion is the combination of very light nuclei.

$$4\,{}^1_1\text{H} \rightarrow {}^4_2\text{He} + 2\,{}^0_{+1}\text{e} + 26.7 \text{ Mev for four } {}^1_1\text{H fused}$$

$${}^2_1\text{H} + {}^2_1\text{H} \rightarrow {}^3_2\text{He} + {}^1_0\text{n} + 3.2 \text{ Mev}$$

$${}^2_1\text{H} + {}^2_1\text{H} \rightarrow {}^3_1\text{H} + {}^1_1\text{H} + 4.0 \text{ Mev}$$

$${}^3_1\text{H} + {}^2_1\text{H} \rightarrow {}^4_2\text{He} + {}^1_0\text{n} + 17.6 \text{ Mev}$$

${}^2_1\text{H}$ = Deuterium
${}^3_1\text{H}$ = Tritium
${}^0_{+1}\text{e}$ = Positron

The net reaction for the last three reactions given here is:

$$5\,{}^2_1\text{H} \rightarrow {}^4_2\text{He} + {}^3_2\text{He} + {}^1_1\text{H} + 2\,{}^1_0\text{n} + 24.8 \text{ Mev for five } {}^2_1\text{H fused}$$

Deuterium is a relatively abundant isotope — out of 6500 atoms of hydrogen in sea water, for example, one is a deuterium atom. What this means is that the oceans are a potential source of fantastic amounts of deuterium. There are $1.03 \times$

Materials for fusion reactions are available in enormous amounts.

10^{22} atoms of deuterium in a single liter of sea water. In a single cubic kilometer of sea water, therefore, there would be enough deuterium atoms with enough potential energy to equal the burning of 1360 billion barrels of crude oil, and this is approximately the total amount of oil originally present in this planet.

Fusion reactions occur rapidly only when the temperature is of the order of 100 million degrees or more. At these high temperatures atoms do not exist as such; instead, there is a **plasma** consisting of unbound nuclei and electrons. In this plasma nuclei merge or combine. In order to achieve the high temperatures required for the fusion reaction of the hydrogen bomb, a fission bomb (atomic bomb) is first set off.

One type of hydrogen bomb depends on the production of tritium (^3_1H) in the bomb. In this type, lithium deuteride ($^6_3Li^2_1H$, a solid salt) is placed around an ordinary $^{235}_{92}U$ or $^{239}_{94}Pu$ fission bomb. The fission is set off in the usual way. A ^6_3Li nucleus absorbs one of the neutrons produced and splits into tritium, ^3_1H, and helium, ^4_2He.

$$^6_3Li + ^1_0n \rightarrow ^3_1H + ^4_2He$$

The temperature reached by the fission of $^{235}_{92}U$ or $^{239}_{94}Pu$ is sufficiently high to bring about the fusion of tritium and deuterium:

$$^3_1H + ^2_1H \rightarrow ^4_2He + ^1_0n + 17.6 \text{ Mev}$$

A 20-megaton bomb usually contains about 300 pounds of lithium deuteride, as well as a considerable amount of plutonium and uranium.

The enhanced radiation weapon (ERW) commonly known as the **neutron bomb** is a modified hydrogen bomb (Fig. 8–21). The weapon is not a bomb in the sense that it is delivered aboard a bomber and then dropped over a target. Rather it is small enough to be fired from field artillery. One bomb would deliver a destructive force of roughly 1 kiloton, which is about one twentieth the destructive power of the 20-kiloton fission bomb dropped on Hiroshima, Japan, in August 1945.

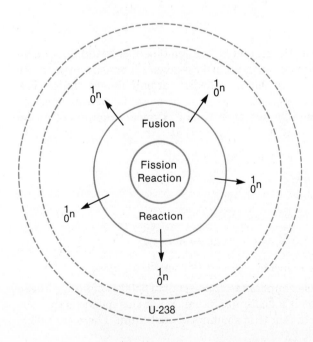

FIGURE 8–21 The general arrangement within a neutron bomb with the U-238 shield of the hydrogen bomb shown in its place. The neutron bomb (or enhanced radiation weapon [ERW]) does not have a U-238 shield.

The major modification of a hydrogen bomb into a neutron bomb requires removal of the U-238 shield. In the hydrogen bomb, the shield enables a larger buildup of pressure and hence a larger explosion. The shield reduces the velocity of the neutrons and is a source of harmful radioactivity. With the shield absent, as in neutron bombs, 80% of the energy released is in the form of high-speed neutrons. The neutrons readily penetrate iron and steel, so armored tanks provide no protection from the effects of the neutrons. Since there is minimal heat and minimal blast from the neutron bomb, little destruction is done to inanimate matter. However, people and animals within a 1000-foot radius would be paralyzed within 5 minutes and dead in two days. Other living beings within a 1000- to 2000-foot radius would be dead in four to six days, although those in concrete shelters would be safe. There is virtually no residual radiation or fallout.

CONTROLLED FUSION There are three critical requirements for controlled fusion. First, the temperature must be high enough for ignition to occur. For the deuterium-tritium combination given earlier, a temperature of about 10^8 to 10^9 degrees is needed. Second, the plasma must be confined long enough to release a net output of energy. Third, the energy must be recoverable in some usable form.

Containment is one of the biggest problems in developing controlled fusion.

As yet the fusion reactions have not been "controlled." No physical container can contain the plasma without cooling it below the critical fusion temperature. Magnetic "bottles," enclosures in space bounded by a magnetic field, have confined the plasma, but not for long enough periods. Recent advances suggest that, with further development, these "bottles" may hold the plasma long enough for the fusion reaction to occur.

Tokamaks and Stellarators are magnetic bottle research devices.

Thermal energy conversion (Fig. 8–22) could be used to take the power from a deuterium-tritium–fueled fusion reaction. Liquid lithium would be used to ab-

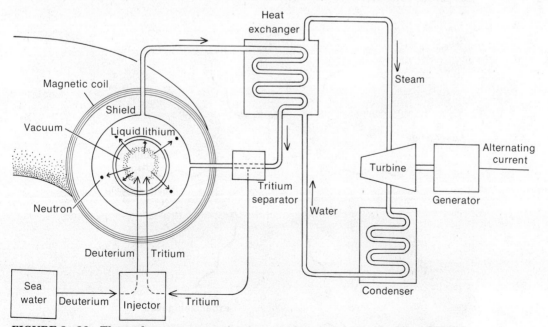

FIGURE 8–22 Thermal energy conversion from nuclear fusion. Such a scheme, based on the deuterium-tritium fuel cycle, relies on the energy of highly energetic neutrons. A liquid lithium shield absorbs these neutrons and is heated. The lithium then exchanges this heat with water to generate steam.

sorb kinetic energy of fast-moving neutrons and then exchange this heat with water to drive a steam turbine, thus producing electricity. This system is actually a breeder reactor, as some of the lithium is converted into fuel — tritium, ^3_1H — by neutron absorption.

A newer confinement method is based on a laser system that simultaneously strikes tiny hollow glass spheres called **microballoons,** which enclose the fuel, consisting of equal parts of deuterium and tritium gas at high pressures (Fig. 8–23).

Two other attempts to achieve fusion are aneutronic fusion and particle beam fusion accelerator (PBFA).

Aneutronic fusion, also called migma (Greek word for "mixture"), was presented theoretically by Bogdan Maglich in 1973. Since this process does not involve neutrons either as products or reactants, a penetrating, hard-to-capture, potentially damaging particle is eliminated. Fusion is achieved by accelerating nuclei (such as deuterons, ^2_1H) in linear accelerators to an energy of 0.7 million electron volts, which is equivalent to a temperature of 7 billion degrees centigrade. The high-energy ions are directed on a lithium target. The fusion of a deuteron and a lithium nucleus produces a helium nucleus, two protons, and energy.

In the PBFA process, electrical charge is stored in capacitors and discharged in 40-nanosecond pulses. The energy accelerates lithium ions to a kinetic energy of between 1 and 2 million joules. The lithium ions impinge on a target of deuterium (^2_1H) and tritium (^3_1H). Lithium nuclei fuse with one or the other of the hydrogen isotopes and produce energy.

There is hope that controlled fusion will be demonstrated during the next decade, but it appears that fusion will not furnish any significant fraction of the world's energy needs before the turn of the century.

Controlled fusion energy should result in a rather limited production of dangerous radioactivity. The lighter elements involved can be radioactive enough

A nanosecond is one billionth of a second.

Controlled fusion might end many of the world's energy problems.

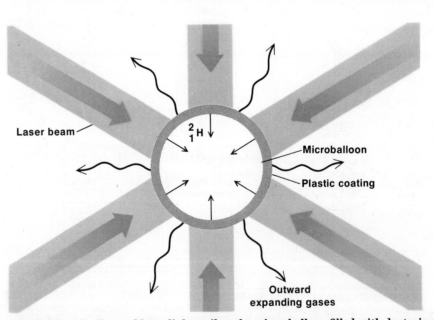

FIGURE 8–23 Focused laser light strikes the micro-balloon filled with deuterium and tritium, causing a plastic outer layer to burn off (or ablate). The outwardly expanding gases from the plastic material drive the glass sphere and its fuel contents inward. The high density and high temperatures produced might result in fusion.

to be a serious hazard, but only for a short period of time; the half-lives of these isotopes are short. Storage and then return to the environment would be quite satisfactory.

SOLAR ENERGY

Earth's ultimate source of energy is the Sun, and this energy is generated by nuclear fusion reactions. Although the Earth receives only about three ten-millionths (0.0000003) of the total energy emitted from the Sun, this amount of energy is enormous—about 2×10^{15} kcal/min, or 2.0 cal/cm²/min. Owing to reradiation from the atmosphere, and the absorption and scattering of radiant energy by molecules in the lower portions of the atmosphere, the amount of radiation actually reaching the surface of this planet is about 1 cal/cm²/min. The actual value depends on location, season, and weather conditions. Even this is a large amount of energy. For example, the roof of an average-sized house will receive about 10^8 calories/day when the radiation level is 1 cal/cm²/min. This is equivalent to the heat energy derived from burning about 32 pounds of coal per day. This is also equal to 120 kilowatt-hours of electrical energy per day—more than enough to heat an average American home on a winter day.

Only about 1% of the solar radiation used by plants in photosynthesis ends up as stored chemical energy such as foodstuffs. After fossil fuels are depleted, could photosynthesis be considered a viable source of energy outside of food materials? Much of the energy used in photosynthesis is for "fixing" carbon in the form of cellulose and carbohydrates. Recently it has been shown that certain blue-green algae, *Anabaena cylindrica,* can convert sunlight and water into hydrogen and oxygen. A colony of such algae, coupled with a fuel cell utilizing hydrogen and oxygen, could be a source of electricity during sunlight hours (Fig. 8–24).

Solar energy, used efficiently, could solve many energy problems. For example, the world's present energy requirements per year are about 10^{19} kcal. An area of

Solar energy is transmitted nuclear energy.

Some catalysts reported in late 1982 for the decomposition of water in the presence of sunlight are indium phosphide (InP), phosphorus-doped silicon coated with Pt or Ni, and p-type iron oxide semiconductor. All are presently too expensive and (or) inefficient (12%, 12%, and 0.05%, respectively) to compete with hydrogen produced by the reaction of coal with steam.

$C + H_2O \rightarrow CO + H_2$

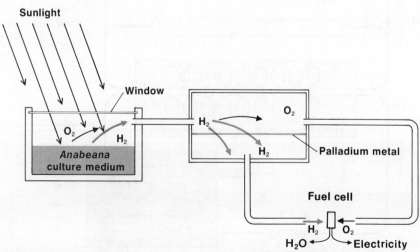

FIGURE 8–24 Schematic diagram of an electricity-producing photosynthesis process. H₂ and O₂ produced by the *Anabaena* are separated by palladium metal, which is permeable to H₂ but not to O₂. The H₂ and O₂ are then combined in the fuel cell to produce electricity.

desert of about 28,000 square miles with little cloud cover or dust, near the equator, would receive about 10^{17} kcal/year of solar radiation. Such deserts exist in northern Chile, and could, if the need were great enough to justify the costs, supply a large portion of the world's energy needs.

The solar energy could heat water to steam, which in turn could generate electricity (Fig. 8–25), which could electrolyze water to hydrogen and oxygen. The hydrogen could be piped to where the energy is needed and then converted to electricity. Such an arrangement would give rise to a **hydrogen economy,** one that has many advantages over present energy sources such as fossil fuels.

Another approach to the direct utilization of solar energy is the **solar battery,** known as a photovoltaic device. The solar battery converts energy from the sun into electron flow. Solar batteries are about 13% to 14% efficient and are capable of generating electrical power from sunlight at the rate of at least 90 watts per square yard of illuminated surface. They are now used in space flight applica-

Hydrogen can be burned in most devices that now burn natural gas.

Some hydrogen-powered buses and cars are now operating on an experimental basis.

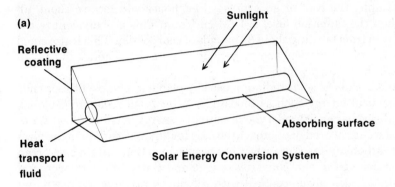

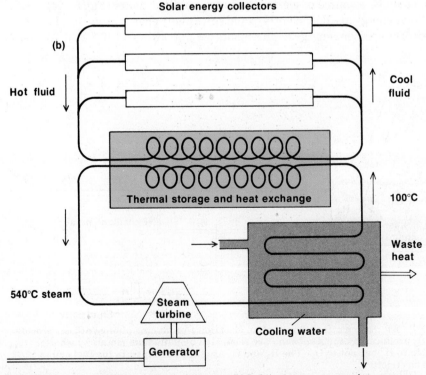

FIGURE 8–25 *(a)* **Solar energy collector.** *(b)* **Solar energy conversion system.**

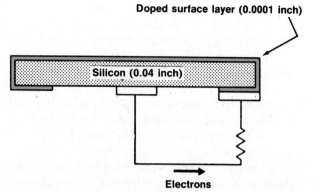

Doped surface layer (0.0001 inch)

Silicon (0.04 inch)

Electrons

FIGURE 8–26 Silicon photodiode (Bell solar battery).

Solar calculators are powered by solar cells.

tions and communication satellites and in Israel, India, Pakistan, South Africa, and Azerbaijan SSR to obtain electrical power.

One type of solar battery consists of two layers of almost pure silicon (Fig. 8–26). The lower, thicker layer contains a trace of arsenic and the upper, thinner layer a trace of boron. Silicon has four valence electrons and forms a tetrahedral, diamond-like, crystalline structure. Each silicon atom is covalently bonded to four other silicon atoms. Arsenic has five valence electrons. When arsenic atoms are included in the silicon structure, only four of the five valence electrons of arsenic are used for bonding with four silicon atoms; one electron is relatively free to roam (Fig. 8–27). Boron has three valence electrons. When boron atoms are included in

The goal is to capture and use the solar energy "on the run" without upsetting the energy flow into (light side) and away from (dark side) the Earth.

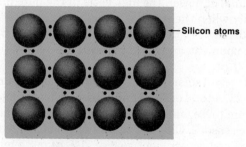

Silicon atoms

Perfect crystal

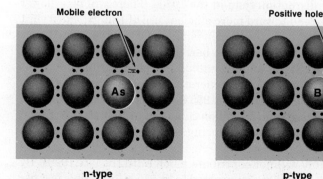

Mobile electron

Positive hole

As

B

n-type

p-type

FIGURE 8–27 Schematic drawing of semiconductor crystal layers derived from silicon. (From W. L. Masterton and E. J. Slowinski: *Chemical Principles*. Philadelphia, Saunders College Publishing, 1977.)

the silicon structure, there is a deficiency of one electron around the boron atom; this creates "holes" in the boron-enriched layer. Even without sunlight, the "extra" electrons of arsenic diffuse into the holes in the boron. The driving force is the strong tendency to pair electrons. Externally applied potentials as high as 1000 volts are incapable of reversing the flow. The negative charge built up in the boron layer would hinder the flow of electrons into that layer and eventually stop the flow. The opposing factors — repulsion between free electrons and the drive to pair electrons — finally bring about an equilibrium.

When sunlight strikes the boron layer, the equilibrium is disturbed. If the wafer is connected as an ordinary battery would be, electrons flow from the arsenic layer through the circuit and back into the boron layer. The fact that electrons enter the circuit from the arsenic layer can be explained if sunlight unpairs electrons in the boron layer, and the freed electrons are repelled to the arsenic layer and into the circuit. To complete the circuit, electrons enter the boron layer, where there are holes for electrons. If the sun is still shining on the cell, unpaired electrons will continue to be repelled to the positive arsenic layer. Since there are no places for the incoming electrons in the structure of the arsenic layer, the electrons go into the external circuit, where there is less opposition to their flow. An analogy to the operation of the solar cell would be the benches on which people rest in a crowded amusement park (the boron layer). The flow is toward the benches, but they are too hot to sit on. The flow is continued by those who do not know (or forget) that the benches are too hot.

The advantage of the solar battery is that it has no moving parts, no liquids, and no corrosive chemicals — it just keeps on generating electricity indefinitely while exposed to sunlight. The drawbacks of the solar battery are the large area required for large amounts of power, the high costs of the pure materials, and the fact that they work only when the sun is shining. Since the first practical use of solar batteries in 1955 to power eight rural telephones in Georgia, they have undergone a great deal of development, and much more is expected because of their great potential use.

Solar batteries are still very expensive.

The four major types of solar devices for heating water or air in buildings are illustrated in Figure 8–28. A total of 16.83 million square feet of solar collectors was sold in 1983. Energy from these and all other solar systems depends on changes in the cloud cover and the seasons. Cloudy and cold days considered, solar energy is expected to meet the official government goal of 20% of our energy needs by the year 2000.

What else can be expected energywise in the year 2000? We can expect to be less vulnerable to oil supply disruption than in the 1970s. There will be a slower growth of energy demand, a trend toward more efficient use of energy, and a shift from oil to other sources. Hence, a reduced dependence on OPEC is expected. The total oil demand is predicted to be 16.3 million barrels per day, with 10.5 million barrels supplied by the United States and 5.8 million barrels imported.

In the year 2000, coal (for electricity generation primarily) is expected to increase from the 900 million short tons used in 1985 to 1700 million short tons, and natural gas is expected to decrease from 20 trillion cubic feet in 1980 to 17.5 trillion cubic feet. Nuclear power may grow from producing one tenth of the electricity to one fourth.

Synthetic fuels probably will not contribute much until the late 1990s. By the year 2000, the production of coal gas is expected to be 255 billion cubic feet per year. Shale oil may contribute 515 thousand barrels per day. Energy from biomass

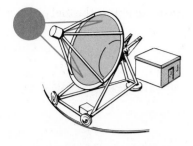

Tracking concentrators follow
the sun, boil water, produce
steam, and generate electricity.

Flat plate collectors, mounted
on a roof facing south, heat
water or air in pipes by the
greenhouse effect.

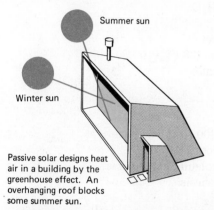

Summer sun

Winter sun

Passive solar designs heat
air in a building by the
greenhouse effect. An
overhanging roof blocks
some summer sun.

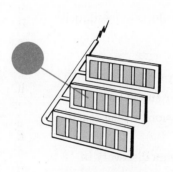

Photovoltaic (or solar)
cells produce electricity.

FIGURE 8–28   **Solar devices for heating water and air in buildings and for producing electricity.**

is estimated to be equivalent to 65 thousand barrels of oil per day. This will be energy primarily from alcohol fuels produced from crops, wood, and waste.

Jot these numbers down, and check them when you are about 30 to 35 years old. We hope the energy situation provides a good lifestyle for us all in the year 2000.

SELF-TEST 8–B

1. The splitting of an unstable nucleus to produce energy is termed (a) *fission* or (b) *fusion.*
2. A sufficiently large sample to sustain a chain reaction in a fission process is called the _____.
3. Of the two major isotopes of uranium, which is fissionable, uranium-238 or uranium-235? _____
4. When light nuclei combine to form heavy nuclei and energy, the process is (a) fission or (b) fusion.
5. The major problem with obtaining fusion energy is (a) containment of reactants at high temperature, (b) enough fuel, or (c) costs.
6. When and if fusion is used to produce energy in a breeder reaction, the fuel produced will be _____.

7. When fission is used to produce energy in a breeder reaction, the fuel produced will be _____ and _____ .

8. Complete the nuclear reactions:

a. $^{85}_{34}\text{Se} \rightarrow {}^{0}_{-1}\text{e} + $ _____

b. $^{246}_{96}\text{Cm} \rightarrow {}^{4}_{2}\text{He} + $ _____

MATCHING SET

_____	1.	User of 35% of world's energy
_____	2.	Fossil fuels
_____	3.	Combustion products of fossil fuels
_____	4.	Mineable coal
_____	5.	Synthesis gas
_____	6.	Fissionable isotope
_____	7.	Basis of nuclear energy
_____	8.	Product of a fission breeder reactor
_____	9.	Deuterium
_____	10.	Date of petroleum discovery in United States
_____	11.	Tritium
_____	12.	Source of deuterium
_____	13.	Used to confine fusion fuel
_____	14.	One use of solar radiation
_____	15.	Approximate efficiency of a solar battery

a. 1859
b. CO_2 and H_2O
c. Uranium-235 ($^{235}_{92}U$)
d. Mass defect
e. Plutonium-239 ($^{239}_{94}Pu$)
f. United States
g. $^{2}_{1}H$
h. Sea water
i. 90%
j. Within 4000 feet of surface
k. Microballoons
l. Coal, petroleum, natural gas
m. Photosynthesis
n. CO and H_2
o. 10–14%
p. $^{3}_{1}H$
q. China
r. 1740

QUESTIONS

1. What is your attitude toward using up the fossil fuels within a few decades? Do we owe future generations a supply of these resources? Would you agree to give up air-conditioning, private cars, and power tools, to mention a few examples, and to limit heating and cooking if necessary to share these fuels with your grandchildren?

2. Which theoretically yields the greatest energy per mole?
a. The burning of gasoline
b. The fission of uranium-235
c. The fusion of hydrogen

3. Which is the more efficient use of energy: burning coal in a house to heat it or heating the house electrically with energy produced in a coal-burning power plant?

4. Give three examples of systems that contain chemical energy that can be used as a source of heat energy.

5. Is the electrical energy where you live produced by burning fossil fuels? If not, what is the energy source? Are there pollution problems associated with the generation of the electrical power?

6. What produces the tremendous energy of a fission reaction?

7. What was the original source of energy that is tied up in fossil fuels?

8. Why is it difficult to fuse two $^{52}_{24}Cr$ nuclei? Explain.

9. What is meant by a chain reaction?

10. What major problem is associated with harnessing the energy from a fusion reaction?

11. Suggest several ways solar energy might be harnessed.

12. Name two sources of energy not specifically mentioned in this chapter.

13. Explain how useful energy might be obtained from garbage.

14. Which is more fundamental—a supply of energy or a supply of food? Explain.

15. The energy consumption of the United States in 1970 was 2×10^{13} kilowatt-hours. What is this amount of energy expressed in kilocalories? In Btu's?

16. Assume the world population to be 4.5 billion and calculate the Earth's energy needs if everyone used as much energy as is used in the United States.

17. Which fuel has the greatest energy content per gram of fuel burned: coal, natural gas, or petroleum? Is this factor more important to you, the consumer, than is economics or pollution?

18. Define energy.

19. List three so-called fossil fuels. Why are they called fossil fuels?

20. Which fuel—coal, petroleum, or natural gas—burns naturally with the least amount of pollution?

21. Describe what is meant by coal gasification. How is it accomplished?

22. Do you pay for electricity as electrical power (kilowatts) or electrical energy (kilowatt-hours)? What is the difference?

23. How do mass defect and binding energy relate? How do they arise?

24. How do the neutron bomb and the hydrogen bomb differ? How are they alike?

25. What is meant by an insulator R value of 30?

26. What are some constructive applications of radioactive radiation?

27. What is the principal element in the photoelectric diode? What two elements are used in trace quantities in a photocell?

28. What are two dangerous properties of plutonium-239?

29. What problems are encountered in dealing with nuclear waste? What is your preference as a solution to the nuclear waste problem?

30. What gases are contained in both coal gas and power gas?

31. Which is the more efficient transport of energy: gas through pipes or electricity through wires?

32. If solar energy is so clean, why are we so slow in moving to its use?

33. Do you think the United States should go forward with the use of the breeder reactor? Why?

34. How much has the price of oil changed during your lifetime?

35. If the mass of a proton is 1.007275 amu, the mass of an electron is 0.000551 amu, and the mass of a neutron is 1.008665 amu, what is (a) the calculated mass of one atom, and (b) the mass defect when the following measured masses are given?
    (1) $^{7}_{3}$Li, 7.01601 amu
    (2) $^{9}_{4}$Be, 9.01219 amu
    (3) $^{18}_{8}$O, 17.99916 amu

36. Is the energy crisis a crisis of quality or quantity of energy?

37. Is electricity a primary or secondary source of energy? Explain your answer.

38. As a project, update the energy situation in the United States and the world by consulting the most recent edition of the *Annual Energy Review* (published by the Department of Energy, Energy Information Administration), a copy of which is probably in your library.

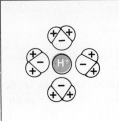

Chapter 9

ACIDS AND BASES — CHEMICAL OPPOSITES

When we discover in nature a large group of useful, related compounds, we have simplified categories to study rather than many isolated compounds to characterize. When we learn that the two large groups of compounds react with each other and, as antagonists, ultimately can neutralize (or pacify) the effects each group had originally, we have an additional intriguing reason to find out about these substances. Two such groups of compounds are *acids* and *bases*.

Are acids and bases useful? We cannot live even a minute without them. In the following list are but a few of the routine encounters we have with acids and bases.

Household	Cooking—baking powder (base)
	baking soda (base)
	vinegar (acid)
	Lye and drain cleaners (bases)
	Citrus fruits such as lemons and oranges (acids)
	Acid skin and pimples (acids)
	Toilet bowl cleaners (bases)
Soil	Lime added to "sweeten the soil" (bases)
Automobile	Battery acid
	If antifreeze is too acidic, radiators corrode.
Acid rain	
Acid mine drainage	
Streams	Fish die if acidity is too high.
Medicine	Antacids alleviate indigestion (bases).
Body functions	Acidity must be controlled carefully to preserve health and life; this is the every-minute application.
	Some acids and bases are very toxic to the human body.

Beyond our routine encounters with acids and bases, the control of acidity is necessary in many procedures for analyzing chemicals and in many industrial processes. Some of these applications will be discussed later in this textbook. The purpose of this chapter is to provide the fundamentals that will help you understand the world of acids and bases.

First of all, what is an acid? The word *acid* comes from the Latin *acidus*, meaning "sour" or "tart," since in water solutions, acids have a sour or tart taste.

Acids in water react with metals such as zinc and magnesium to liberate hydrogen, react with bases to produce a salt and water, and change the color of litmus, a vegetable dye, from blue to red. These properties are produced by the release of hydrogen ions, H^+, in water.

A common acidic substance known since antiquity is vinegar, the sour constituent from the fermentation of apple cider. The acid of vinegar is acetic acid, $HC_2H_3O_2$.

Classically, a base is a substance capable of liberating hydroxide ions, OH^-, in water. Some of the most common bases of this type are the hydroxides of the alkali metals — NaOH and KOH — and of the alkaline earth metals — $Ca(OH)_2$ and $Mg(OH)_2$.

Water solutions of bases, which are called alkaline solutions or basic solutions, taste bitter, are slippery or soapy to the touch, change litmus from red to blue, and react with acids to form a salt and water.

In the reaction of an acid with a classic base, the acid supplies hydrogen ions, H^+, which react with hydroxide ions, OH^-, from the base to form water, HOH.

$$H^+ \quad + \quad OH^- \rightarrow HOH$$
<small>(SUPPLIED   (SUPPLIED
BY ACID)   BY BASE)</small>

Since the common properties of acids and bases and the reactions between acids and bases occur in water solutions, we shall begin with a discussion of the general properties of aqueous solutions.

FORMATION OF LIQUID SOLUTIONS

Recall from Chapter 2 that a solution is a homogeneous mixture. A liquid solution, then, is a uniform distribution of one substance in another, with the mixture having the properties of a liquid.

How many liquid solutions are familiar to you? How about sugar or salt dissolved in water, or oil paints dissolved in turpentine, or grease dissolved in gasoline? In each of these solutions, the substance present in the greater amount, the liquid, is the *solvent,* and the substances dissolved in the liquid, the ones present in smaller amounts, are the *solute(s).* For example, in a glass of tea, water is the solvent, and sugar, lemon juice, and the tea itself are solutes. A theoretical concept of a solution of sugar in water pictures a collection of sugar molecules evenly dispersed among the water molecules (Fig. 9–1).

In this presentation of acid-base reactions, most of the chemistry studied will be in water or aqueous solutions, where water is the solvent. Generally, one of the species exchanging hydrogen ions is the solute. In addition to being the solvent,

Hydrogen ions are protons. Because of their high charge-to-size ratio, hydrogen ions in solution are bonded to one or more water molecules.

Litmus is but one acid-base indicator. Another, phenolphthalein, is colorless in acid and pink in base.

Solution: homogeneous (uniform) mixture of atoms, ions, or molecules.

Aqueous solutions are water solutions.

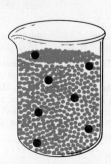

● **Sugar molecule**

· **Water molecule**

FIGURE 9–1   A schematic illustration at the molecular level of sugar solution in water. Large circles represent the sugar molecules and the small circles water. The size of the container and the size of the particles are not to scale.

water molecules can and do exchange hydrogen ions with solute particles under suitable conditions.

IONIC SOLUTIONS (ELECTROLYTES) AND MOLECULAR SOLUTIONS (NONELECTROLYTES)

Solute particles may be ions or molecules.

Solutes in aqueous solutions can be classified by their ability or inability to render the solution electrically conductive. When aqueous solutions are examined to see whether they conduct electricity, we find that solutions fall into one of two categories: **electrolytic** solutions, which conduct electricity, and **nonelectrolytic** solutions, which do not. A simple apparatus such as that shown in Figure 9-2 can be used to determine into which classification a given solution falls.

Ionic dissociation is the separation of ions of a solute when the substance is dissolved.

The conductance of electrolytic solutions is theoretically explained by the solute particles in such solutions being ions rather than molecules. Recall that sodium chloride crystals are composed of sodium ions, which are positively charged, and chloride ions, which are negatively charged. When sodium chloride dissolves in water, **ionic dissociation** occurs (see Fig. 9-7). The resulting solution (Fig. 9-3a) contains positive sodium ions and negative chloride ions dispersed in water. Of course, the solution as a whole is neutral, since the total number of positive and negative charges are equal.

$$Na^+Cl^- \xrightarrow{\text{water}} Na^+_{(aq)} + Cl^-_{(aq)}$$

(SOLID)   (AQUEOUS)   (AQUEOUS)
             SODIUM   CHLORIDE
              ION   ION

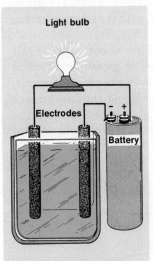

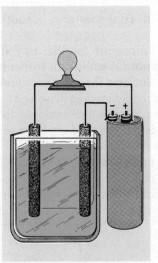

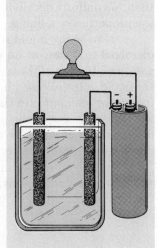

(a) **Solution of table salt**
(an electrolytic solution)

(b) **Solution of table sugar**
(a nonelectrolytic solution)

(c) **Pure water**
(a nonelectrolyte)

FIGURE 9-2 A simple test for an electrolytic solution. In order for the light bulb to burn (*a*), electricity must flow from one pole of the battery and return to the battery via the other pole. To complete the circuit, the solution must conduct electricity. A solution of table salt, sodium chloride, results in a glowing light bulb. Hence, sodium chloride is an electrolyte. In (*b*), the light bulb does not glow. Hence, table sugar is a nonelectrolyte. In (*c*), it is evident that the solvent, water, does not qualify as an electrolyte since it does not conduct electricity in this test.

The random motions of the sodium and chloride ions are not completely independent. The charges on the particles prevent all of the sodium ions from going spontaneously to one side of the container while all of the chloride ions are going to the other side. However, a net motion of ions occurs when charged electrodes are placed in an electrolytic solution (Fig. 9–3b). If the negative ions give up electrons to one electrode while the positive ions receive electrons from the other electrode, a flow of electrons or electricity is maintained.

Ions migrate toward oppositely charged electrodes in an electric field.

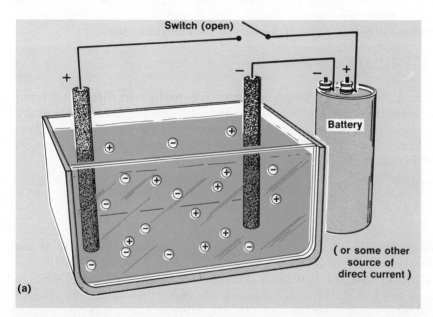

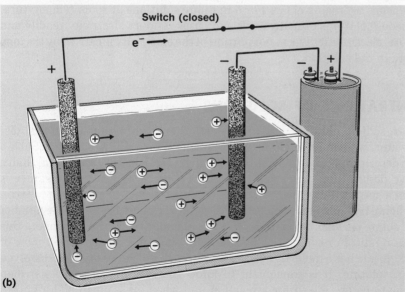

FIGURE 9–3 Conductance of electricity by ionic solution. (*a*) The hydrated ions are randomly distributed throughout the salt solution; the net charge is zero. (*b*) Negative electrode attracts positive ions; positive electrode attracts negative ions. If electrons are transferred from negative electrode to positive ions and from negative ions to positive electrode, the circuit is complete, and electricity will flow through the circuit.

Nonelectrolytic solutions composed of solute molecules dispersed throughout solvent molecules are insensitive to negatively and positively charged electrodes unless the voltage is so great that it breaks the molecules into ions.

Sometimes ionic solutions arise when a molecular substance dissolves in water. For example, hydrogen chloride, HCl, is a gas composed of covalent, diatomic molecules, each having one hydrogen atom and one chlorine atom. When hydrogen chloride dissolves in water an **ionization** reaction occurs, producing ions from molecules. The resulting solution is composed of hydrogen ions and chloride ions dispersed among the water molecules; consequently it is a conducting solution, and hydrogen chloride in water is properly termed an electrolyte.

> *When a molecular solute dissolves in water to produce ions, the process is called ionization.*
>
> *The symbol $H^+_{(aq)}$ means a hydrogen ion dissolved in water, an aqueous system.*

$$\underset{\text{MOLECULE}}{HCl} \xrightarrow{\text{water}} \underset{\text{IONS}}{H^+_{(aq)} + Cl^-_{(aq)}}$$

The hydrogen ion in aqueous systems is not free to roam. Recall that water molecules are polar. A free proton (isolated hydrogen ion) could not exist in such a medium; it becomes attached to the negative end of one of the water dipoles. In fact, the attraction of water dipoles for the polar HCl molecule probably causes its ionization in the first place.

$$H^+ + H\!:\!\overset{\cdot\cdot}{\underset{H}{O}}\!: \;\rightarrow\; \left[H\!:\!\overset{\cdot\cdot}{\underset{\underset{\text{HYDRONIUM ION}}{H}}{O}}\!:\!H \right]^+$$

> *H_3O^+ is the hydronium ion.*

Thus, the hydrogen ion in water is **hydrated** and is often referred to as the **hydronium** ion, H_3O^+ or $H^+(H_2O)$. When one considers hydrogen bonding between water molecules, it is very likely that other water molecules are attached to the molecule to which the proton is attached. The best representation we can give for the hydrogen ion in water then is $H^+(H_2O)_n$, where n is a constantly changing number, perhaps averaging about 4 or 5 in dilute solutions at room temperature.

Aqueous solutions of many acids and bases conduct electricity readily and are, therefore, electrolytic. The major portion of the electricity is carried by the ions of the solute in the solution.

CONCENTRATIONS OF SOLUTIONS

When sugar, sodium chloride, alcohol, or any other readily soluble material dissolves in water, we can have either a concentrated or a dilute solution. Such a qualitative description of concentration is much less satisfactory and useful than a quantitative description, which tells us just how much of a given substance is dissolved in a specified volume.

> *Molar concentration: number of moles of a substance per liter of solution.*

In chemistry, concentrations of solutions are often expressed in the number of *moles* of solute *per liter* of solution. Molar and molarity are used to denote this concentration unit. For example, 1 liter of a 1-molar solution contains 1 mole of solute. If a solution has a molarity of six, the solution has 6 moles of solute dissolved in 1 liter of solution. It is convenient to know the molarity of solutions when different solutions are mixed for a reaction and when we want to add a known number of moles of each reactant.

Solutions of known concentration are prepared using volumetric flasks. These are glass vessels with the stems precisely marked to indicate specific volumes, such as 1.000 liter. The procedure involves the steps shown in Figure 9–4.

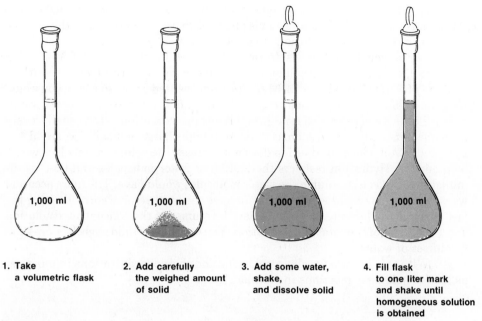

1. Take
   a volumetric flask

2. Add carefully
   the weighed amount
   of solid

3. Add some water,
   shake,
   and dissolve solid

4. Fill flask
   to one liter mark
   and shake until
   homogeneous solution
   is obtained

FIGURE 9–4   Laboratory procedure for the preparation of a solution of known concentration.

To show how concentrations are determined, let us consider a case where a 25-g sample of NaCl is carefully weighed, then transferred to a 1-liter volumetric flask and dissolved in water. The next step is to add water to the flask until the solution has a total volume of 1 liter. In order to determine the concentration of such a solution, we need to know the number of moles of NaCl present. The formula weight of NaCl is 23.0 + 35.5, or 58.5. Thus, if we have 58.5 g of NaCl, we have 1 mole. We have 25 g, so the concentration is

$$\frac{25\ \text{g}}{58.5\ \text{g/mole}} = 0.43 \text{ mole of NaCl per liter of solution}$$

We usually indicate this as 0.43 M NaCl, where M stands for moles of solute per liter of solution and is read as "molar."

Suppose we have 86 g of sucrose (table sugar: $C_{12}H_{22}O_{11}$, molecular weight 342) dissolved in a volume of 500 mL. What is the concentration of sugar in this solution? Since we have $^{86}\!/_{342}$ mole of sugar, or 0.25 mole, dissolved in $^{500}\!/_{1000}$ of a liter, the concentration of the sugar solution is

$$\frac{\dfrac{86\ \text{g}}{342\ \text{g/mole}}}{\dfrac{500\ \text{mL}}{1000\ \text{mL/liter}}} = 0.50 \text{ M sucrose}$$

When a more dilute solution is needed, how is the solution prepared? For example, hydrochloric acid is sold commercially as a 12-molar (12-M) solution. If you wanted a 1-M solution of the acid, this solution needs to be one twelfth as concentrated as the commercial product. One volume of the concentrated acid (a cup or a milliliter or any volume will do) diluted with 11 (not 12) volumes of water

Molarity = M =

$\dfrac{\text{moles of solute}}{\text{liter of solution}}$

will result in the acid being in a medium 12 times larger than the acid was prior to dilution. Hence, the diluted solution is one twelfth as concentrated as the original solution, and the diluted solution is now 1-M hydrochloric acid.

Because of the heat generated when some acids are mixed with water, *acids should be added to water* to distribute the heat better. This is particularly important when mixing sulfuric acid with water.

Commercially available sulfuric acid is 18 M. If you wanted a 4-M solution of the acid, you would take 4 volumes of the concentrated acid and mix this with 14 volumes of water. If you had a 14-M solution and wanted a 5-M solution, you would measure 5 volumes of the concentrated solution and mix it into 9 volumes of water.

The method described for dilution is an approximation — but usually a very good one. Since some ions, such as those with high charge or small size, are highly hydrated, water added for dilution does not increase the volume as much as would be predicted. Hydration restricts the motion of water molecules and thereby diminishes the effective volume the water molecules would have. The more accurate way of diluting is to add enough water to have the total volume correct. Using the last example to illustrate, you would have 5 volumes of the concentrated solution and 14 volumes of the total dilute solution. You may have to add slightly more than 9 volumes of water.

With this knowledge about solutions and their concentrations in mind, let us look at acid-base reactions in solution.

THE HAPPY MEDIUM BETWEEN ACIDS AND BASES — NEUTRALIZATION

In 1923, J. N. Brønsted and T. M. Lowry defined acids and bases as they are generally recognized by chemists today.

Brønsted-Lowry acid: a chemical species that can *donate* hydrogen ions (also called protons or H^+ ions) is an acid.

Brønsted-Lowry base: a chemical species that can *accept* hydrogen ions is a base.

To illustrate these definitions, we again consider the reaction between gaseous hydrogen chloride (HCl) and water:

$$HCl(gas) + H_2O \rightarrow H_3O^+ + Cl^-$$

ACID   BASE   Hydronium Ion   Chloride Ion

HYDROCHLORIC ACID

Examination of the preceding reaction shows that the HCl molecule has donated a hydrogen ion (H^+) to the water molecule. This transfer of a hydrogen ion is understandable when electronegativity differences between the bonded atoms are considered. First, polar bonds are formed since chlorine (Cl) is more electronegative than hydrogen (H) and oxygen (O) is more electronegative than hydrogen. The electrical poles formed by the shift of electron pairs toward the more electronegative atom are shown in the following reaction by a delta positive (δ^+) for the positive pole and and delta negative (δ^-) for the negative pole.

$$H \!:\! \overset{..}{\underset{\delta^+ \;\; \overset{|}{H}}{O}} \!:\! {}^{\delta^-} + \overset{\delta^+}{H} \!:\! \overset{..}{\underset{..}{Cl}} \!:\! {}^{\delta^-} \rightarrow \left[H \!:\! \overset{..}{\underset{\overset{|}{H}}{O}} \!:\! H \right]^+ + \; : \! \overset{..}{\underset{..}{Cl}} \! : ^-$$

BASE   ACID

Why are hydrogen ions transferred from HCl to water rather than from water to HCl? Experimentally, water is more polar than HCl. Part of water's greater polarity is due to oxygen having greater electronegativity than chlorine. However, polarity is determined by other factors such as the arrangement of atoms in the molecule. Whatever the causes, water does win the battle for the hydrogen ions, and HCl relinquishes its hydrogen ions to the more polar water.

The reaction of HCl with water is practically complete. Almost all of the HCl is converted to H_3O^+ and Cl^-. A concentrated (about 12-M) solution of hydrogen chloride in water is mostly a solution of hydronium (H_3O^+) ions, chloride (Cl^-) ions, and water molecules, with relatively few dissolved HCl molecules, which give the concentrated solution its characteristic odor.

If the ionic solid sodium oxide, Na_2O, is dissolved in water, a vigorous reaction produces a solution containing sodium ions (Na^+) and hydroxide ions (OH^-). In this process, the oxide ion (O^{2-}) reacts with a polar water molecule to form the hydroxide ion. In this, as in other such aqueous reactions, it is understood that the ions are hydrated (i.e., water molecules are bonded to the ions on a transitory basis).

Reaction is:
$$Na_2O + H_2O \rightarrow 2\,Na^+ + 2\,OH^-$$

$$:\ddot{O}:^{2-} + H:\ddot{O}: \rightarrow :\ddot{O}:H^- + :\ddot{O}:H^-$$
$$\overset{\ddot{}}{H}$$

BASE   ACID   HYDROXIDE IONS

There are many other bases that take a hydrogen ion from a water molecule in this way. For example:

$$\underset{\substack{\text{SULFIDE ION} \\ \text{BASE}}}{S^{2-}} + \underset{\text{ACID}}{H_2O} \rightarrow \underset{\substack{\text{HYDROGEN} \\ \text{SULFIDE ION}}}{HS^-} + \underset{\substack{\text{HYDROXIDE} \\ \text{ION}}}{OH^-}$$

$$\underset{\substack{\text{CYANIDE ION} \\ \text{BASE}}}{CN^-} + \underset{\text{ACID}}{H_2O} \rightarrow \underset{\substack{\text{HYDROGEN} \\ \text{CYANIDE}}}{HCN} + \underset{\substack{\text{HYDROXIDE} \\ \text{ION}}}{OH^-}$$

According to the Brønsted-Lowry definition, water acts as an acid in these reactions and donates a hydrogen ion to the other molecule or ion, which acts as a base. A species such as water that can either donate or accept hydrogen ions is called **amphiprotic.** The existence of amphiprotic species implies that acid-base reactions possess a reciprocal nature; an acid and a base react to form another acid (to which a hydrogen ion has just been added) and another base (from which a hydrogen ion has just been removed). Because water is the most commonly used solvent, it is also the most usual reference compound for acid-base reactions.

Amphiprotic species can be either an acid or a base.

Also, one water molecule can transfer a hydrogen ion to another water molecule.

$$2\,H_2O \rightleftharpoons H_3O^+ + OH^-$$

When equilibrium is reached, the reaction has occurred to only a very small extent, as indicated by arrows of unequal length. In neutral water at 25°C, the concentrations of H_3O^+ and OH^- are the same, 0.0000001 M (or 10^{-7} M). Since H_3O^+ and OH^- are produced in equal amounts when only water is present, pure water is neither acidic nor basic, but is described as *neutral.*

Pure water is neutral.

A chemical species in water solution is commonly spoken of as an acid if it donates hydrogen ions to water and increases the concentration of H_3O^+ or $H^+_{(aq)}$. Similarly, a base in water solution is commonly described as a compound whose addition to water increases the concentration of OH^-. Since water is not the only possible solvent, these concepts are too narrow for general scientific use; they have been extended by the preceding definitions, which focus on the essential feature of

Acids form H^+ ions in water; bases form OH^- ions in water.

When an acid neutralizes
a base, acid and base
properties are suppressed.

such acid-base behavior—that is, the donation or acceptance of a hydrogen ion
(H^+) in a reaction.

When acids react with bases, the properties of both species disappear. The
process involved is called **neutralization.** To get a more precise picture of acid-
base neutralization reactions, we shall consider what happens when a solution of
hydrochloric acid is mixed with a solution of sodium hydroxide. The hydrochloric
acid contains H_3O^+ and Cl^- ions; the sodium hydroxide solution contains Na^+ and
OH^- ions. When these two solutions are mixed, a reaction occurs between H_3O^+
and OH^-.

$$Na^+ + Cl^- + \underset{\text{ACID}}{H_3O^+} + \underset{\text{BASE}}{OH^-} \rightarrow H_2O + H_2O + Na^+ + Cl^-$$

If we have an equal number of H_3O^+ and OH^- ions, they will react to produce a
neutral solution, with the hydronium ions (H_3O^+) donating their hydrogen ions to
the hydroxide ions (OH^-), forming molecules of water. Such reactions are called
neutralization reactions because the acids and bases neutralize each other's properties. If we have more H_3O^+ ions than OH^- ions, the extra H_3O^+ will make the
resulting solution **acidic.** If we have more OH^- ions than H_3O^+ ions, only a
fraction of the OH^- ions will be neutralized, and the extra OH^- ions will make the
resulting solution **basic.**

KINSHIP OF SOME ACIDS AND BASES

When an acid ionizes, it produces a hydronium ion plus a species called the **conjugate base** of that acid. For example:

Conjugate means
"kindred" or "kin to."

$$\underset{\substack{\text{NITRIC} \\ \text{ACID}}}{HNO_3} + \underset{\text{WATER}}{H_2O} \rightarrow \underset{\substack{\text{HYDRONIUM} \\ \text{ION}}}{H_3O^+} + \underset{\substack{\text{NITRATE ION, THE} \\ \text{CONJUGATE BASE OF} \\ \text{THE ACID HNO}_3}}{NO_3^-}$$

In the same manner we speak of nitric acid, HNO_3, as being the **conjugate acid** of
the nitrate ion, NO_3^-, a base. Conjugate acids and bases differ by one hydrogen ion.
The conjugate acid has one more hydrogen ion than its conjugate base.

SELF-TEST 9 – A

1. When ammonia dissolves in water, the resulting solution conducts electricity.
   Ammonia in water is therefore a(n) _____.
2. A chemical species that can accept a hydrogen ion is a(n) _____.
   A chemical species that can donate a hydrogen ion is a(n) _____.
3. A compound HA is found to undergo a reaction forming a product H_2A^+.
   Therefore HA is a(n) ( ) acid or ( ) base. If compound HA reacted to form
   A^-, then HA would be a(n) ( ) acid or ( ) base.
4. If the aforementioned compound HA undergoes both reactions described, then
   HA is termed _____.
5. A solution that contains equal concentrations of OH^- and H^+ ions is termed

   _____.

6. The word *aqueous* means _____.

7. In a neutralization reaction, a(n) _____ reacts with a(n)

_____ .

8. In order to make a 5-M solution from a 17-M solution, you would mix five
volumes of the concentrated solution with _____ volumes of water.

THE STRENGTHS OF ACIDS AND BASES

Because the water molecule is itself a weak base, the strongest acid (i.e., the best
hydrogen ion donor) that can exist in water is the hydronium ion (H_3O^+). If strong
acids such as sulfuric acid (H_2SO_4) or nitric acid (HNO_3) are added to water, they
donate their hydrogen ions to water molecules, and the resulting solutions contain
H_3O^+ and HSO_4^-, and H_3O^+ and NO_3^-, respectively. All of these reactions are
reversible, at least in principle, but when strong acids are dissolved in a large excess
of water to form dilute solutions, virtually all of the strong acid is converted to
products.

Reactants   *Products*
$$HCl + H_2O \rightarrow H_3O^+ + Cl^-$$
$$HNO_3 + H_2O \rightarrow H_3O^+ + NO_3^-$$
$$H_2SO_4 + H_2O \rightarrow H_3O^+ + HSO_4^-$$

CONJUGATE

ACID   BASE   ACID   BASE

Since many ions are present, the solution conducts electricity well; HCl, HNO_3,
and H_2SO_4 are **strong electrolytes** (Fig. 9–5).

> Strong electrolytes
> dissociate completely
> in water.

Not all acids lose hydrogen ions as readily to water as do nitric acid and
sulfuric acid. Some negative ions are capable of competing with water for the
hydrogen ion being exchanged. The result of this competition is the establishment
of an **equilibrium** (or balance) between neutral acid molecules and hydronium
ions in water solution.

> Under equilibrium con-
> ditions, the concentra-
> tions of the species in so-
> lution remain unchanged
> even though reactions in
> the forward and reverse
> directions continue.

Acetic acid, found in vinegar, is a weak acid; that is, its conjugate base,
acetate ion, competes well for a hydrogen ion. The molecular structure of acetic
acid ($HC_2H_3O_2$) is

$$\begin{array}{c} H \\ | \\ H-C-C \\ | \quad \diagdown \\ H \quad\quad O-H \end{array} \diagup O$$

ACIDIC HYDROGEN ATOM

The hydrogen atom bonded to highly electronegative oxygen in the molecule is the
only hydrogen donated to a base in water solution; for that reason, it is designated
an acidic hydrogen in the preceding formula. The other hydrogen atoms in the
acetic acid molecule are not acidic because the C—H bonds are almost nonpolar
since the electronegativities of carbon and hydrogen are nearly the same value.

When acetic acid is dissolved in water, some ions are produced. However,
most of the acetic acid molecules do not donate hydrogen ions to water molecules.
The result is a mixture containing a few H_3O^+ and $C_2H_2O_2^-$ ions and many
$HC_2H_3O_2$ molecules. The reaction is:

$$\underset{\text{ACETIC ACID}}{HC_2H_3O_2} + \underset{\text{WATER}}{H_2O} \rightleftarrows \underset{\substack{\text{HYDRONIUM} \\ \text{ION}}}{H_3O^+} + \underset{\substack{\text{ACETATE} \\ \text{ION}}}{C_2H_3O_2^-}$$

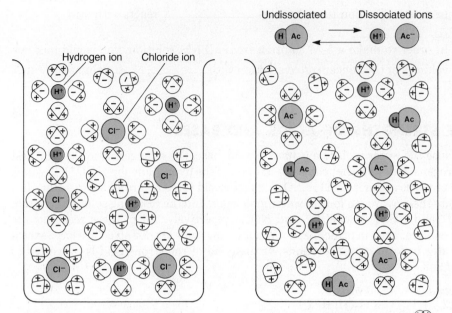

FIGURE 9–5 An illustration of what it may be like in an aqueous (water = 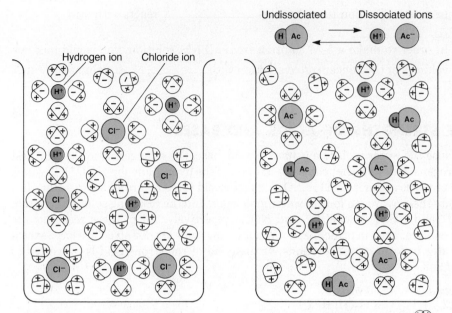)
solution of a strong acid (HCl) and an aqueous solution of a weak acid (acetic acid, HAc).
The strong acid is practically all ions; very few molecules. The weak acid is mostly
molecules with only a few ions.

The relatively few ions in an acetic acid solution do not conduct electricity very
effectively; consequently, acetic acid is a **weak electrolyte** (Fig. 9–5). (A dilute
solution of acetic acid would barely conduct in the apparatus shown in Fig. 9–2.)

Another way of looking at this reaction is to realize that there are two bases
in the mixture: the water molecule H_2O and the acetate ion $C_2H_3O_2^-$. Since the
reverse reaction dominates, the acetate ion must be a stronger base than the water
molecule.

The same kind of considerations can be made for other bases. Ammonia
dissolved in water is a weak base. The resulting reaction produces relatively few
ions; ammonia is mostly in the molecular form:

$$NH_3 \; + \; H_2O \; \rightleftarrows \; NH_4^+ \; + \; OH^-$$

AMMONIA   WATER   AMMONIUM   HYDROXIDE
                          ION   ION

Consequently, the relatively few ions present do not conduct electricity well, and
ammonia may thus be called a weak electrolyte.

Table 9–1 gives some common acids and bases ranked according to their
relative strengths.

THE pH SCALE

Because water solutions of dilute acids and bases are used so extensively and are so
important in biology and medicine, it is convenient to have a simple way of desig-

Ammonia is the number
three commercial chemical
in quantity produced,
sodium hydroxide is num-
ber seven, and nitric acid
is number eleven. (See
inside front cover.)

TABLE 9–1   Relative Strengths of Some Acids and Bases

Increasing Acid Strength ↑

ACID							CONJUGATE BASE
Perchloric acid	$HClO_4$	+	$H_2O \rightleftharpoons H_3O^+$	+	ClO_4^-		Perchlorate ion
Hydrochloric acid	HCl	+	$H_2O \rightleftharpoons H_3O^+$	+	Cl^-		Chloride ion
Nitric acid	HNO_3	+	$H_2O \rightleftharpoons H_3O^+$	+	NO_3^-		Nitrate ion
Hydronium ion	H_3O^+	+	$H_2O \rightleftharpoons H_3O^+$	+	H_2O		Water
Hydrofluoric acid	HF	+	$H_2O \rightleftharpoons H_3O^+$	+	F^-		Fluoride ion
Acetic acid	$HC_2H_3O_2$	+	$H_2O \rightleftharpoons H_3O^+$	+	$C_2H_3O_2^-$		Acetate ion
Water	H_2O	+	$H_2O \rightleftharpoons H_3O^+$	+	OH^-		Hydroxide ion
Hydroxide ion	OH^-	+	$H_2O \rightleftharpoons H_3O^+$	+	O^{2-}		Oxide ion

Increasing Base Strength ↓

nating the acidity or basicity of these solutions. The pH scale was devised for this purpose; it furnishes a number that describes the acidity of a solution. The pH is defined as the negative logarithm of the concentration of the hydrogen ion, $pH = -\log[H_3O^+]$. The brackets mean moles per liter of hydronium ions. For practical purposes, the pH scale runs from 0 to 14. A pH of 7 indicates a neutral solution (such as pure water), a pH below 7 indicates an acidic solution, and a pH above 7 indicates a basic solution.

The pH number is related to the concentration of the hydrogen ions in an aqueous solution expressed as the negative power of 10 (logarithm). A hydrogen ion concentration of 0.00001 or 1×10^{-5} M corresponds to a pH of 5. When the pH is 7, as in pure water, the hydrogen ion concentration would have to be 1×10^{-7} M. Note that for each unit decrease in the pH number, there is a tenfold increase in the concentration of the hydrogen ion (that is, the acidity).

At pH = 7, the number of H_3O^+ ions equals the number of OH^- ions.

To understand why the pH of pure water is 7, we shall need to reexamine the very slight acid-base reaction between water molecules:

$$H_2O + H_2O \rightleftharpoons H_3O^+ + OH^-$$

This ionization reaction produces ions, which should cause water to conduct electricity. But we have said that water is a nonelectrolyte. Actually, with sensitive electrical equipment, pure water can be shown to conduct electricity *very slightly* because of the small amount of ionization. Laboratory measurements reveal that 0.0000001, or 1×10^{-7}, mole per liter of water is present as H_3O^+ and OH^- ions at 25°C. Pure water contains 55.6 moles of water per liter. Consequently, we can see that the actual amount of the water that ionizes is very small compared to the total amount of water present. Pure water, then, which is defined as being neutral, contains 1×10^{-7} mole of hydrogen ions (hydronium ions) per liter, and the pH is 7. In pure water the concentrations of hydronium ions (H_3O^+) and hydroxide ions (OH^-) must be equal, since each time a hydronium ion (H_3O^+) is produced in the ionization reaction a hydroxide ion (OH^-) is also produced. Therefore, in pure water the concentration of the hydroxide ion (OH^-) is also 1×10^{-7} mole per liter.

One liter of water weighs 1000 g and is 55.6 molar:

$$\frac{\dfrac{1000 \text{ g}}{\text{liter}}}{\dfrac{18 \text{ g}}{\text{mole}}} = \frac{55.6 \text{ moles}}{\text{liter}}$$

Figure 9–6 graphically displays the relationship between the pH number and the concentration of the hydrogen ion. It also gives the approximate pH values for common solutions.

A close examination of Figure 9–6 reveals that basic solutions such as aqueous ammonia have hydrogen ion concentrations less than that of pure water.

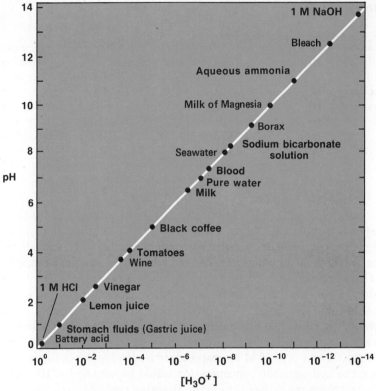

FIGURE 9–6 A plot
of pH versus hydrogen
ion concentration
$[H_3O^-]$. Note that the
pH *increases* as the
$[H_3O^-]$ decreases. The
pH values of some
common solutions are
given for reference. (A
solution in which
$[H_3O^-] = 1$ M has a pH
of 0 since $1 = 10^0$.)

Note that the pH of a typical sample of aqueous ammonia is 11. This is equivalent
to a hydrogen ion concentration of 1×10^{-11}, or 0.00000000001, mole per liter.

ACIDITY UNDER CONTROL — BUFFERS

Buffers control pH.

In response to other chemicals in solution, a buffer absorbs or releases hydrogen
ions and thereby maintains a steady pH. If a base is added, the buffer releases
hydrogen ions to neutralize the added base. If an acid is added, the buffer responds
by absorbing hydrogen ions.

Buffers are composed chemically of a conjugate acid-base pair. Appreciable
amounts of both members of the pair must be present in solution for the most
effective buffering action. In fact, equal amounts of the conjugate acid-base pair
produce optimum buffering capacity. Some commonly used conjugate acid-base
pairs are H_2CO_3 and HCO_3^-, $H_2PO_4^-$ and HPO_4^{2-}, $HC_2H_3O_2$ and $C_2H_3O_2^-$, and NH_3
and NH_4^+.

The control of pH is necessary in many industrial and natural processes. For
example, we can become very ill or even die unless the pH of our blood is controlled
within ± 0.1 unit around a pH of 7.4. Buffers are the controlling agents.

Several buffering agents buffer the blood. One of the buffering agents is the
$H_2PO_4^-$ and HPO_4^{2-} conjugate acid-base pair. Let us see how the phosphate system
might aid in maintaining a constant pH in the blood.

As you are aware, carbon dioxide (CO_2) is a product of the metabolism of food in the body. Carbon dioxide leaves the cell and is transported by the blood to the lungs, where CO_2 is exhaled. In the blood, CO_2 reacts with water to form a weak acid solution.

$$CO_2 + H_2O \rightleftarrows HCO_3^- + H^+$$

The hydrogen ions would lower the pH unless groups such as HPO_4^{2-} are present to take up the excess hydrogen ions.

$$HPO_4^{2-} + H^+ \rightleftarrows H_2PO_4^-$$

The $H_2PO_4^-$ ion is a weak acid, which means $H_2PO_4^-$ clings to its newly acquired hydrogen ion, keeping the hydrogen ion concentration down. In the lungs, the two reactions are reversed, and CO_2 is exhaled.

SALTS — PRODUCTS OF ACID-BASE NEUTRALIZATION
Preparation of Salts

The chemical compounds known as **salts** play a vital role in nature, in plant and animal growth and life, and in the manufacture of various chemicals for human use. They can be formed as the products of acid-base neutralizations, as in the following example:

$$(K^+ + OH^-) + (H_3O^+ + Cl^-) \rightarrow \underline{K^+ + Cl^-} + 2H_2O$$

| Potassium Hydroxide in Water BASE | Hydrochloric Acid in Water ACID | Crystallize the Salt by Removal of Solvent, Water |

KCl (Solid)
SALT

Most salts contain ions held together by **ionic bonding** (see Chapter 5). Solid potassium chloride, for example, is composed of an equal number of K^+ ions and Cl^- ions arranged in definite positions with respect to one another in an ionic structure or lattice (see Fig. 5–4). Since the salt crystal must be electrically neutral, it can have neither an excess nor a deficiency of positive or negative charge.

Salts are ionic compounds.

Let us imagine that we have at our disposal the ions in the following list, and let us see what salts could result.

IONS		SOME POSSIBLE SALTS	SALT NAME
Na^+	Sodium	NaCl	Sodium chloride
Ca^{2+}	Calcium	$NaNO_3$	Sodium nitrate
Cl^-	Chloride	Na_2SO_4	Sodium sulfate
NO_3^-	Nitrate	$NaC_2H_3O_2$	Sodium acetate
SO_4^{2-}	Sulfate	$CaCl_2$	Calcium chloride
$C_2H_3O_2^-$	Acetate	$Ca(NO_3)_2$	Calcium nitrate
		$Ca(C_2H_3O_2)_2$	Calcium acetate
		$CaSO_4$	Calcium sulfate

In the examples just given, notice that in order to attain an electrically neutral lattice, it is necessary to balance the charges of the ions. A sodium (Na^+) ion requires just one chloride ion (Cl^-), and the NaCl lattice contains an equal number of Na^+ and Cl^- ions. A sulfate ion (SO_4^{2-}) with two negative charges must have its negative charge balanced by two positive charges. This may be done by using two Na^+ ions

$$2\ Na^+ + SO_4^{2-} \rightarrow Na_2SO_4$$

or one Ca^{2+} ion

$$Ca^{2+} + SO_4^{2-} \rightarrow CaSO_4$$

In the formula of a salt, the positive and negative charges are equal.

It is possible to form many solid salts by mixing water solutions of different soluble salts with each other. For example, both lead acetate and sodium chloride are soluble in water. If we prepare solutions of these salts and then mix the solutions, we find the insoluble salt, lead(II) chloride, precipitates from the mixture and may be removed by filtration.

$$\underbrace{2\ Na^+ + 2\ Cl^-}_{\substack{\text{SODIUM CHLORIDE} \\ \text{IN SOLUTION}}} + \underbrace{Pb^{2+} + 2\ C_2H_3O_2^-}_{\substack{\text{LEAD(II) ACETATE IN} \\ \text{SOLUTION}}} \rightarrow \underbrace{PbCl_2}_{\substack{\text{SOLID} \\ \text{LEAD(II)} \\ \text{CHLORIDE}}} + \underbrace{2\ Na^+ + 2\ C_2H_3O_2^-}_{\substack{\text{SODIUM ACETATE} \\ \text{IN SOLUTION}}}$$

Sodium acetate may be recovered by evaporating the water.

This reaction illustrates an important principle in solubility. If the component ions of a compound of low solubility are mixed in solution in great enough concentrations, the compound containing those ions will precipitate from solution.

Salts in Solution

Although some salts are very soluble in water, others are quite insoluble. Salts are found with a wide range of water solubilities.

An important property of many salts is their solubility in suitable solvents. The amount of a salt that will dissolve in a given quantity of solvent tells us the salt's solubility in that solvent. The preparation of lead(II) chloride just shown was made possible by the differences in solubilities of different salts in the same solvent. As a result of these differences in solubilities, we can make roads out of calcium carbonate (limestone), which is insoluble in water, but not calcium chloride, which is water soluble.

The two most abundant ions in ocean water are Cl^- and Na^+ ions, sufficient to recover about 27 g of NaCl per kilogram of sea water. To put it another way, there are about 128 million tons of NaCl per cubic mile of sea water.

Consider what happens at the ionic level when a sodium chloride crystal is placed in contact with water. We know that sodium chloride is soluble in water. This means that most, if not all, of the attractive forces between the ions in the crystal lattice are somehow overcome in the solution process.

Because of its polar nature, the water molecule is ideally suited to interact with ions.

The surface of the salt crystal appears calm when the crystal is placed in water, but on the ionic level there is a great deal of agitation. Water molecules have sufficient polarity to interact strongly with the ions and bond with them. Once this occurs, the ion is less strongly bound in the lattice and so can be removed from the crystal. The crystal lattice now has a gap in it where the ion was removed. Ions that are bonded by solvent molecules are termed solvated ions, and the process of ion-solvent interaction is called **solvation.** But just how are these ions solvated? What causes this solvation?

The answers lie in the structure of the water molecule and its ability to interact with ions. As we saw in Chapter 5, water is a *polar* compound. The negative ends of its molecules are attracted to positive ions, and the positive ends are attracted to negative ions. As a result, several water molecules will interact with each ion. Figure 9–7 shows several water molecules solvating a Na^+ ion. The

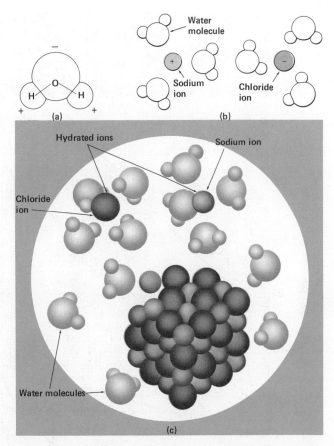

FIGURE 9–7 Dissolution of sodium chloride in water. (*a*) Geometry of the polar water molecule. (*b*) Solvation of sodium and chloride ions due to interaction (bonding) between these ions and water molecules. (*c*) Dissolution occurs as collisions between water molecules and crystal ions result in the removal of the crystal ion. In the process the ion becomes completely solvated.

positive end of the water molecule will tend to interact with a negative ion; this is shown for the Cl⁻ ion in Figure 9–7.

Every salt has what may be termed a solubility limit for a given solvent: at a given temperature, a certain number of grams of salt, and *no more*,* will dissolve in a certain quantity of solvent. A solution that contains all the dissolved salt that it can hold is termed a **saturated solution.** One might ask just why this type of solubility limit is found in nature. The reason for this can be understood if we remember that the oppositely charged ions of the salt solution actually attract one another. If one crowds the solution with solvated ions to too great an extent and "ties up too many solvent molecules," the ions will begin re-forming the crystal lattice. This is **crystallization,** or solvation in reverse. When undissolved salt is in contact with a saturated solution of that salt, a dynamic equilibrium is established, with the salt crystal being broken down at one point while being formed at another. The effect of temperature on the solubility of some common salts is shown in Figure 9–8.

Generally, for a given solvent, a salt will dissolve to a greater extent in the hot solvent than in the cold solvent.

* Supersaturated solutions can be formed under special conditions but are not stable in the presence of the solid solute. The presence of the solid solute causes the excess dissolved solute to crystallize, and a saturated solution is formed.

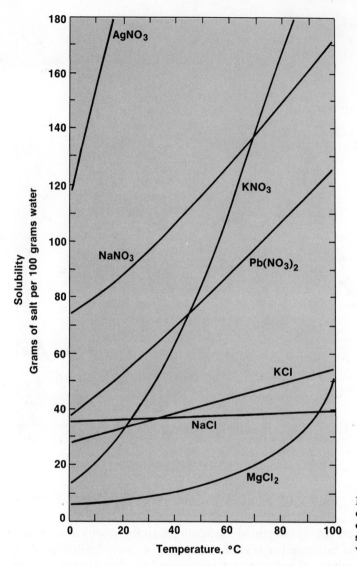

FIGURE 9-8 The effect of temperature on the solubility of some common salts in water.

SELF-TEST 9-B

1. Complete the matching set.

   pH 10   a. acidic
   pH 7   b. basic
   pH 3   c. neutral

2. If the solubility of KI is 140 g per 100 g of water at 20°C, how much KI will dissolve in 1000 g of water? _____

3. Write a chemical reaction showing water to be an acid. Use ammonia (NH_3) as the base.

   $H_2O + NH_3 \rightarrow$ _____ + _____

4. Write a chemical reaction showing water to be a base. Use HCl as the acid.

$$H_2O + HCl \rightarrow \underline{\hspace{3cm}} + \underline{\hspace{3cm}}$$

5. Which is more acidic, a pH of 6 or a pH of 2? _____

6. Is a pH of 8 more basic than that of water? _____

7. The pH of the blood is maintained at 7.4 by agents generally known as _____ .

8. What gas, when dissolved in blood, would lower blood's pH? _____

9. High pH means ( ) high or ( ) low hydrogen ion concentration.

10. Low pH means ( ) high or ( ) low hydrogen ion concentration.

MATCHING SET

_____ 1. Conjugate base of H_2A

_____ 2. Conjugate acid of A^{2-}

_____ 3. M

_____ 4. pH of pure water

_____ 5. Solution

_____ 6. Strong acid

_____ 7. Alkaline solution

_____ 8. Weak conjugate acid

_____ 9. Acid definition

_____ 10. Base definition

_____ 11. Weak acid

_____ 12. Electrolyte

_____ 13. Buffer

a. 7
b. Molarity; moles of solute per liter of solution
c. Hydrogen ion donor
d. Maintains pH
e. Hydrogen ion acceptor
f. Homogeneous mixture of atoms, molecules, and/or ions
g. $HC_2H_3O_2$, acetic acid
h. Strong conjugate base
i. Causes solution to conduct electricity
j. NaOH in HOH
k. HA^-
l. H_2SO_4
m. A^{2-}
n. 3

QUESTIONS

1. Define acid-base reactions in terms of hydrogen ions.

2. Indicate the solute and solvent in (a) a cup of coffee, (b) a 5% solution of alcohol in water, (c) a 5% solution of water in alcohol, and (d) a solution of 50% alcohol and 50% water.

3. What is the one test that all aqueous electrolytes must pass?

4. Describe a test to determine whether a solution is a weak acid or a strong acid.

5. Give an example of ionic dissociation. Give an example of ionization. What is the difference between the two?

6. What mobile units must be present in electrolytic solutions?

7. Describe a test to determine whether a solution is acidic or basic.

8. Why can boric acid (H_3BO_3) be used in eyewashes while hydrochloric acid (HCl) is not safe to use?

9. Distinguish between the hydrogen ion and the hydronium ion.

10. Write a neutralization reaction between lye (NaOH) and muriatic acid (HCl).

11. Write the equation for a chemical reaction in which water acts as a Brønsted-Lowry acid; as a Brønsted-Lowry base.

12. What is the main distinction between water solutions of strong and weak electrolytes?

13. What is the difference between ionization and dissociation in the production of mobile ions?

14. a. What is the pH of a neutral solution?
b. Which pH is more acidic, a pH of 5 or a pH of 2?
c. Which pH is more basic, a pH of 5 or a pH of 10?

15. Classify each of the following as acids or bases, using the Brønsted-Lowry definitions: H_2SO_4, CO_3^{2-}, Cl^-, HCO_3^-, O^{2-}, H_2O.

16. Would liquid Ajax be more likely to have a pH greater or less than 7? (It has a strong smell of ammonia.)

17. Two solutions contain 1% acid. Solution A has a pH of 4.6, and solution B has a pH of 1.1. Which solution contains the stronger acid?

18. Predict the formulas of salts formed with the following pairs of ions:
  Na^+ and SO_4^{2-}
  Ca^{2+} and I^-
  Mg^{2+} and NO_3^-
  Ca^{2+} and PO_4^{3-}
   K^+ and Br^-

19. Moist baking soda is often put on acid burns. Why? Write an equation for the reaction assuming the acid to be hydrochloric (HCl).

20. Hydrochloric acid is the acid present in the human stomach. Is this a strong or a weak acid?

21. What is the pH of a bicarbonate of soda (sodium bicarbonate) solution?

22. From a practical standpoint and for safety reasons, why should you know if the acid you are using is strong or weak?

23. If a drug consumes 37 times its weight in excess stomach acid, the reaction is called _____ and the drug must be a(n) _____

24. A special case of solvation is the hydration of ions by polar water molecules. What does this suggest about other polar solvents such as ethyl alcohol?

25. Describe what happens when an ionic solid dissolves in water.

26. Describe vividly the scenario of HCl molecules being added to water. Compare and contrast this scenario with what happens when acetic acid (HAc) molecules are added to water.

27. What ions are present in water solutions of the following salts: Na_2SO_4, $CaBr_2$, $Mg(NO_3)_2$?

28. In terms of the hydrogen ion (H^+) concentration, when is a solution acidic and when is it basic?

29. What is meant by an alkaline solution?

30. Which solution do you expect to be more acidic, a 1-M solution of HCl or a 1-M solution of $HC_2H_3O_2$? Explain.

31. Identify the conjugate acid-base pairs in the following equations:
a. $HCl + H_2O \rightarrow H_3O^+ + Cl^-$
b. $NH_3 + H_3O^+ \rightarrow NH_4^+ + H_2O$
c. $HC_2H_3O_2 + H_2O \leftrightarrows H_3O^+ + C_2H_3O_2^-$
d. $HC_2H_3O_2 + OH^- \rightarrow H_2O + C_2H_3O_2^-$
e. $CN^- + HC_2H_3O_2 \leftrightarrows HCN + C_2H_3O_2^-$
f. Ionization of sulfuric acid (2 steps)
  $H_2SO_4 + H_2O \rightarrow H_3O^+ + HSO_4^-$
  $HSO_4^- + H_2O \leftrightarrows H_3O^+ + SO_4^{2-}$

32. Write the electron dot structure for (a) the chloride ion and (b) the hydroxide ion.

33. Explain the following statement in terms of ionization and chemical bonding: When solutions of hydrogen chloride in water (polar) and in benzene (nonpolar) are prepared, the water solution is found to conduct electricity, but the benzene solution does not conduct an electric current.

34. An aqueous solution of methanol, CH_3OH, does not conduct electricity, but an aqueous solution of sodium hydroxide, NaOH, does. What does this information tell us about the -OH group in the alcohol?

35. Why does molten sodium chloride conduct electricity when solid NaCl, though ionic, does not?

36. Explain how a buffer such as the $H_2PO_4^- - HPO_4^{2-}$ system consumes hydrogen ions released into the bloodstream.

37. What is the function of a buffer?

38. Take at least two different applications of the word *buffer* as used in fields other than science, and relate these uses to how buffer is used with acids and bases.

39. On a simplified basis, how much water will have to be mixed with the volume of concentrated solution given in order to prepare the desired dilute solution? Fill in the blanks in the table below.

CONCENTRATED SOLUTION		DILUTE SOLUTION		
Concentration	Volume to Be Used (mL)	Concentration	Volume of Water to Be Used (mL)	TOTAL VOLUME (mL)
a. 18 M	5 mL	5 M	_____	_____
b. 18 M	_____	7 M	_____	36
c. 15 M	12 mL	6 M	_____	_____
d. 14 M	_____	4 M	_____	28
e. 12 M	_____	3 M	_____	48

40. A mole of ethyl alcohol weighs 46 g. How many grams of ethyl alcohol would be required to make 1 liter of 1.5-M solution?

41. Which has more grams of solute dissolved per liter of solution, a 0.50-M solution of sucrose (molecular weight 342) or a 0.50-M solution of sodium chloride (molecular weight 58.5)?

42. What is the molar (M) concentration of a solution containing 12.0 g of NaCl dissolved in 500 mL of solution?

43. If a hydrochloric acid solution is 0.1 M, how many grams of HCl are dissolved in 1 liter of this solution?

Chapter 10

OXIDATION-REDUCTION: ELECTRON TRANSFER CHEMISTRY

Equally important as the hydrogen ion transfer in acid-base chemistry are the processes called oxidation and reduction. As we shall find out later, oxidation and reduction reactions always occur together, so they are often named together as **oxidation-reduction.** In this chapter we will look at oxidation-reduction from both the theory, which seeks to explain what is going on at the molecular level, and the applications, which are of vast importance in our lives.

Oxidation got its name from the chemical change associated with the element oxygen combining with other elements. In fact, oxygen combines with every element except helium, neon, and argon. Prior to the discovery of the electron, oxidation was considered a simple combination of two elements that produced a compound called an **oxide.** Recall from Chapter 3 that Antoine Lavoisier used the decomposition of mercuric oxide (HgO), along with other reactions, to develop the Law of Conservation of Matter.

When oxygen combines with another element, heat is almost always produced. If this energy (as heat) is given off rapidly enough, the oxidation is called *combustion,* or *burning.* An example of rapid combustion is shown in Figure 10–1.

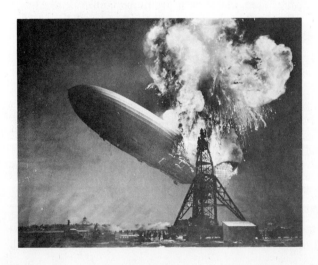

FIGURE 10–1 Photo of the explosion on May 6, 1937, of the German airship Hindenburg. (The Bettmann Archive, Inc.)

FIGURE 10–2 Photo of a large brush fire, common in the arid western United States. (The Bettmann Archive, Inc.)

Neither oxidation nor combustion is limited to oxygen combining with just elements. Compounds may be oxidized as well. Automobile engines burn hydrocarbon fuels (Chapter 12) and produce the oxides of hydrogen (water) and carbon (carbon monoxide and carbon dioxide). Oxides of nitrogen are produced as well. These nitrogen oxides come from the oxidation of some of the nitrogen in the air that is mixed with the fuel and ignited in the combustion chamber. Most oxidation is controlled, such as the combustion of fuels in engines, furnaces, fireplaces, and stoves, but some oxidation, such as rusting and forest and house fires, is not easily controlled, is unwanted, and may be life threatening (Fig. 10–2).

Combustion is always accompanied by heat and light.

Elemental oxygen makes up 21% by volume of our atmosphere. Most of the remainder of our atmosphere is nitrogen. Because of its many commercial uses, oxygen is extracted from the air in large quantities by liquefaction followed by distillation.

If the atmosphere were composed of a greater concentration of oxygen, then fires could more readily get out of control; rates of chemical reactions are related to the concentrations of the reactants.

Some properties of oxygen are summarized as follows:

Formula	O_2
Molecular weight	32.00
Melting point	$-218.4°C$
Boiling point	$-183.0°C$
Description	Colorless and odorless gas
Solubility in water	48.9 mL per liter of water at 0°C

WHAT IS OXIDATION?

Whenever oxygen combines with another element or compound, the chemical reaction is *oxidation,* and the products of the reaction are called *oxidation products.*

Most metals react readily with oxygen to form oxides. However, metals like gold and platinum, which do not readily oxidize, can form oxides using indirect means. When iron, an easily oxidized metal, reacts with oxygen, a red-brown oxide forms.

$$4 \, Fe + 3 \, O_2 \rightarrow 2 \, Fe_2O_3$$

A hydrate is a stable molecular or ionic substance associated with water.

In the presence of moisture, usually found in the air, a *hydrate* of iron oxide forms. This iron oxide hydrate is known as **rust.**

$$4 \, Fe + 3 \, O_2 + xH_2O \rightarrow 2 \, Fe_2O_3 \cdot xH_2O$$

In the formula for rust, the x represents a varying number of water molecules.

Oxygen also combines with nonmetals to form oxides. Carbon burns to form carbon monoxide and carbon dioxide.

$$2 \, C + O_2 \rightarrow 2 \, CO$$
$$C + O_2 \rightarrow CO_2$$

The formation of carbon monoxide when carbon dioxide could be formed is called **incomplete combustion.** In a limited supply of oxygen, carbon monoxide is the likely product. The carbon monoxide can be further oxidized to carbon dioxide.

$$2 \, CO + O_2 \rightarrow 2 \, CO_2$$

Carbon monoxide formed by the incomplete combustion of hydrocarbon fuels is a major component of urban air pollution.

Another name for "town gas" was "water gas." This name was based on the reaction used to manufacture the mixture, $H_2O + C \rightarrow CO + H_2$

Before natural gas became commonly used as a fuel, a mixture of carbon monoxide and hydrogen was piped into factories and homes as a product called "town gas." Both carbon monoxide and hydrogen will burn in air and produce enough heat to make this mixture an adequate fuel.

$$H_2 + 2 \, CO + 2 \, O_2 \rightarrow 2 \, CO_2 + H_2O + 193 \, kcal$$

The major drawback to town gas was the toxicity of the carbon monoxide. When town gas was used, carbon monoxide poisoning would occur when victims breathed unburned gas. Today, natural gas has largely supplanted town gas as a fuel. The principal component of natural gas is methane (CH_4), which is oxidized to carbon dioxide and water.

$$CH_4 + 2 \, O_2 \rightarrow CO_2 + 2 \, H_2O$$

While providing about 90% of all the energy needs for our society through the combustion of fuels, oxygen combines with other elements either in the air or in the fuels themselves to produce air pollutants (see Chapter 18).

SELF-TEST 10-A

1. A type of oxidation that produces heat and light is _____ .
2. When gasoline burns in plentiful air, the principal oxidation products are _____ and _____ .
3. When oxygen combines with an element to form a compound, that compound is often known as a(n) _____ .
4. When iron oxidizes in the presence of moisture, the product of the reaction is called _____ .

5. The products of the complete combustion of methane (CH_4) are _____ and _____.

6. The incomplete combustion of carbon produces _____.

OXIDATION-REDUCTION DEFINED FROM THE ELECTRON POINT OF VIEW

Oxidation can be defined in three ways, each of which explains what oxidation actually is. The theorist can even explain the first two definitions in terms of the third.

The first definition of oxidation, already given, is in terms of oxygen combining with some element or compound. Oxygen is said to be the **oxidizer,** and the other reactant gets **oxidized** to form oxidation products.

Oxidation is the gain of oxygen.

Glucose ($C_6H_{12}O_6$) is oxidized in living cells to the oxidation products carbon dioxide and water.

$$C_6H_{12}O_6 + 6\ O_2 \rightarrow 6\ CO_2 + 6\ H_2O + 669.5\ \text{kcal}$$

This reaction provides the energy of life for almost all living cells.

The carbon in carbon dioxide is more oxidized than the carbon in carbon monoxide. In general, when elements form several different compounds with oxygen, there is a **degree of oxidation.** Elements that are highly oxidized are often themselves capable of causing oxidation to occur. One name used for the compounds of these highly oxidized elements is **oxidizing agent.** Table 10–1 shows several of these oxidizing agents. Note the oxygen in their formulas.

A second definition of oxidation involves the loss of hydrogen atoms in organic molecules. The loss of hydrogen is not the cause of the oxidation but merely one way to recognize when oxidation has occurred in organic molecules. For example, ethanol is oxidized to the compound acetaldehyde in the liver with the aid of enzymes.

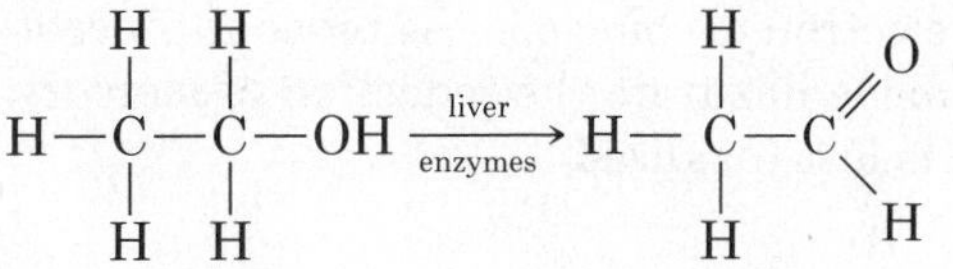

Oxidation is the loss of hydrogen atoms.

TABLE 10–1 Some Oxidizing Agents and Their Uses

NAME	FORMULA	USES AS OXIDIZING AGENT
Potassium dichromate	$K_2Cr_2O_7$	Tests for alcohol in breath
Potassium nitrate	KNO_3	Gunpowder
Calcium hypochlorite	$Ca(OCl)_2$	Bleach, swimming pool disinfectant
Lead dioxide	PbO_2	Lead storage batteries
Manganese dioxide	MnO_2	Batteries
Hydrogen peroxide	H_2O_2	Disinfectant, antiseptic
Potassium peroxydisulfate	$K_2S_2O_8$	Denture cleansers

Oxidation is the loss of electrons.

The acetaldehyde is more highly oxidized than ethanol because of the loss of the two hydrogen atoms. Other examples of reactions of this type will be discussed in Chapters 12 and 13 on the chemistry of organic compounds and in Chapter 15 on biochemistry.

The third and most general definition of oxidation involves **electron loss.** An element is said to be oxidized when it loses electrons. When a neutral atom becomes a positive ion, it has lost electrons and has been oxidized. Sodium is oxidized by chlorine to produce sodium ions and chloride ions.

$$2\,Na + Cl_2 \rightarrow 2\,Na^+ + 2\,Cl^-$$

These ions, in equal numbers, form sodium chloride, table salt.

In addition to oxygen, the elements fluorine and chlorine combine with elements and compounds in ways that can be called oxidation. Fluorine is rather exotic in its applications, but chlorine is commonly used in oxidation applications such as disinfecting water supplies and in bleaches and cleaning compounds.

REDUCTION — THE OPPOSITE OF OXIDATION

Reduction always accompanies oxidation. When something is oxidized, something else is reduced. When something gains oxygen and gets oxidized, something else loses oxygen and gets reduced. When something loses hydrogen and gets oxidized, something else gains hydrogen and gets reduced. When something loses electrons and gets oxidized, something else gains electrons and gets reduced. As an oxidizing agent causes oxidation, a **reducing agent** causes reduction.

Oxidation is always accompanied by reduction and vice versa. One cannot occur without the other.

	OXIDATION	REDUCTION
In terms of oxygen:	Gain of oxygen	Loss of oxygen
In terms of hydrogen:	Loss of hydrogen	Gain of hydrogen
In terms of electrons:	Loss of electrons	Gain of electrons

By observing the hydrogen and oxygen exchanges, one is limited to the compounds that contain H and O. The electron exchange is the broader concept. Lavoisier's experiments with mercuric oxide illustrate the reduction of mercury. When mercuric oxide is heated, oxygen is lost (oxidized).

$$2\,HgO \rightarrow 2\,Hg + O_2$$

The mercury is reduced.

The chemistry in the blast furnace is discussed in Chapter 11.

Most metal ores consist of metal oxides. Iron ore contains the oxide hematite ($Fe_2O_3 \cdot xH_2O$). When iron ore is fed into a blast furnace, the iron oxide is reduced, and metallic iron is one of the products. This reduction reaction occurs with the aid of the reducing agent coke, a form of carbon produced from coal. The overall reaction is

$$2\,Fe_2O_3 \cdot xH_2O + 3\,C \rightarrow 4\,Fe + 3\,CO_2 + 2x\,H_2O$$

Oxides like carbon monoxide can be reduced to organic compounds in the presence of catalysts, with hydrogen as the reducing agent.

$$CO + 2\,H_2 \rightarrow \underset{\text{METHANOL}}{CH_3OH}$$

This reaction is currently used for making alcohol fuels.

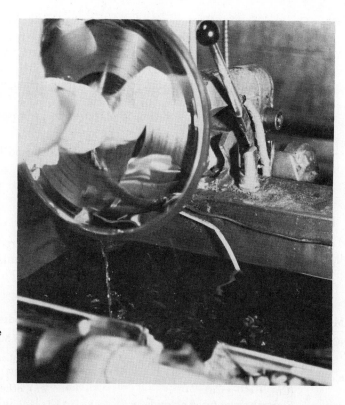

FIGURE 10–3
Stamping masters for making high-quality phonograph records are made by electroplating nickel onto a plastic record finely coated with silver.

When a silver compound like silver nitrate ($AgNO_3$) dissolves in water, silver ions and nitrate ions are distributed in the solution. If two oppositely charged electrodes are placed into the solution, the silver ions will be attracted to the surface of the negatively charged electrode. There, the silver ions will be reduced by gaining electrons.

An electrode can be any electrically conducting object.

$$Ag^+ + e^- \rightarrow Ag$$

A coating of silver atoms will be built up slowly on the surface of the negative electrode. In this way, metals may be plated on a variety of objects for both decoration and protection (Fig. 10–3).

Oxidation-reduction reactions are so important it is hard to imagine a moment of life or an event in our world that does not involve oxidation or reduction in some way.

Now let us turn our attention to some specific practical applications of oxidation and reduction chemistry. Other examples will be discussed in later chapters of this text.

The electrode where reduction takes place is called the cathode. Oxidation occurs at the anode.

SELF-TEST 10–B

1. In the oxidation of glucose in living cells, the oxidizing agent is _____.
2. When ethanol is oxidized in the liver, the oxidation product is _____. This type of oxidation is usually described as (a) oxygen addition, (b) electron loss, (c) hydrogen loss.

3. Conversion of carbon monoxide to methanol by the addition of hydrogen is an example of carbon monoxide being (a) oxidized, (b) reduced.

4. When a sodium atom loses a valence electron to become a sodium ion, the sodium atom is (a) reduced, (b) oxidized, (c) neither.

5. When coke (a form of carbon) reacts with iron ore in a blast furnace, the coke is the (a) oxidation product, (b) reducing agent, (c) oxidizing agent.

6. When a metal ion in solution is plated onto an object, the metal ion is (a) oxidized, (b) reduced.

7. The loss of electrons is called _____, and the gain of electrons is called _____.

REDUCTION OF METALS

One of the most practical applications of the oxidation-reduction principle is the separation of metals from their ores. The majority of metals are found in nature as compounds; that is, the metals are in an oxidized state. To be obtained, the metal

TABLE 10-2   Some Methods Used to Separate Metals from Their Ores by Reduction

METAL	OCCURRENCE	REDUCTION PROCESS	USES OF METAL
Cu	Cu_2S, chalcocite	Air blown through melted ore $Cu_2S + O_2 \rightarrow 2\ Cu + SO_2\uparrow$ $(Cu^+ + e^- \rightarrow Cu)$	Electrical wiring, boilers, pipes, brass (Cu 85%, Zn), bronze (Cu 90%, Sn, Zn), other alloys
Na	NaCl, rock salt	Electrolysis of fused chloride $2\ NaCl \rightarrow 2\ Na + Cl_2$ $(Na^+ + e^- \rightarrow Na)$	Coolant in nuclear reactors, yellow street lights, making ethyl gasoline
Mg	Mg^{2+}, sea water	Electrolysis of fused chloride $MgCl_2 \rightarrow Mg + Cl_2$ $(Mg^{2+} + 2\ e^- \rightarrow Mg)$	Light alloys such as duralumin (0.5% Mg, Al), Dowmetal H (90.7% Mg, Al, Zn, Mn), flares, some flash bulbs
Ca	$CaCO_3$, chalk, limestone, marble	Thermal reduction of chloride obtained from carbonate $3\ CaCl_2 + 2\ Al \xrightarrow{\text{heat}} 3\ Ca + 2\ AlCl_3$ $(Ca^{2+} + 2\ e^- \rightarrow Ca)$	Bearing metal alloys (0.7% Ca, Pb, Na, Li), storage battery electrodes
Al	$Al_2O_3 \cdot H_2O$, bauxite	Electrolysis in fused cryolite, Na_3AlF_6, at 800–900°C $2\ Al_2O_3 \rightarrow 4\ Al + 3\ O_2$ $(Al^{3+} + 3\ e^- \rightarrow Al)$	Packaging, airplane and automobile parts, alloys, roofing, siding
Ag	Ag_2S, argentite	Reduction by a more active metal $2\ Ag(CN)_2^- + Zn \rightarrow Zn(CN)_4^{2-} + 2\ Ag\downarrow$ $(Ag^+ + e^- \rightarrow Ag)$	Jewelry, tableware, plating, electrical wiring, photography
Zn	ZnS, zinc blende	Roasting in air with carbon $ZnS + 2\ O_2 + C \rightarrow Zn + SO_2\uparrow + CO_2\uparrow$ $(Zn^{2+} + 2\ e^- \rightarrow Zn)$	Galvanizing iron and steel, die-cast auto parts
Hg	HgS, cinnabar	Roasting in air $HgS + O_2 \xrightarrow{\text{heat}} Hg + SO_2\uparrow$ $(Hg^{2+} + 2\ e^- \rightarrow Hg)$	Some thermometers, electrical switches, blue-green street lights, amalgams in dentistry, fluorescent lighting
Fe	Fe_2O_3, hematite	Reduction by carbon monoxide $Fe_2O_3 + 3\ CO \rightarrow 3\ CO_2 + 2\ Fe$ $(Fe^{3+} + 3\ e^- \rightarrow Fe)$	As cast and forged iron, alloyed with C (0.1–1.5%) to make steels (stainless steel has 8% or more Cr)

must be reduced from the oxidized state to the neutral elemental state. This requires a gain of electrons. In addition to electricity, a variety of chemicals can supply the electrons for the reduction process. Some ways of reducing metals from their ores are given in Table 10-2.

ELECTROLYSIS

Several metals either are separated from their ores or are purified afterward by electrolysis, as noted in Table 10-2. **Electrolysis** is a type of chemical reaction caused by the application of electrical energy.

The suffix -*lysis* means "splitting" or "decomposition"; electrolysis is decomposition by electricity.

The principal parts of an electrolysis apparatus are shown in Figure 10-4. Electrical contact between the external circuit and the solution is obtained by means of electrodes, which are often made of graphite or metal. The electrode at which electrons enter an electrolysis cell is termed the **cathode,** and this is the electrode at which reduction takes place. The electrode at which the electrons leave the cell is the **anode.** At the anode, oxidation takes place.

The battery or generator produces a current of electrons, which flow toward one electrode and make it negatively charged, and away from the other electrode and make it positively charged. When the switch is closed, the positive ions in solution migrate toward the cathode. Soon, a chemical reaction is evidenced at the electrodes. Depending on the substances present in the solution, gases may be evolved, metals deposited, or ionic species changed at the electrodes. The ions that migrate to the electrodes are not necessarily the species undergoing reaction at the electrodes, because sometimes the solvent undergoes reaction more easily. Whatever happens, the chemical reactions occurring at the cathode and anode are due to electrons going into and coming out of the solution. The chemical reaction at the cathode furnishes electrons to solution species (reduction). At the anode, electrons

Reduction at cathode
Oxidation at anode

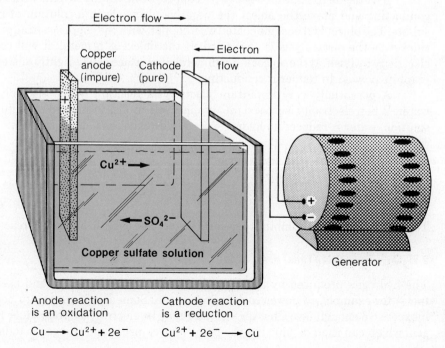

Electron flow ⟶

Copper anode (impure)

Cathode (pure)

⟵ Electron flow

Cu^{2+} ⟶

⟵ SO_4^{2-}

Copper sulfate solution

Generator

Anode reaction is an oxidation

$Cu \longrightarrow Cu^{2+} + 2e^-$

Cathode reaction is a reduction

$Cu^{2+} + 2e^- \longrightarrow Cu$

FIGURE 10-4
Electroplating from a copper sulfate solution.

are taken from species in solution, so the chemical reaction at the anode gives up electrons (oxidation).

The electroplating of copper is illustrated in Figure 10–4. Such an electrolysis can be used either to plate an object with a layer of pure copper or to purify an impure sample of copper metal; copper is transferred from the positive electrode into the solution and eventually to the negative electrode. If the positive electrode is impure copper to be purified, electrolysis deposits the copper as very pure copper on the negative electrode.

Now let us examine how the electrolysis transfers the copper from the positive electrode to the negative electrode. Electrons flow out of the negative terminal of the generator through the wire and into the negative electrode. Somehow this negative charge must be used up at the surface of the electrode.

Consider what happens when the electrons build up on the negative electrode. The positive copper ions nearby will be attracted to the surface and will take the electrons. Thus, the Cu^{2+} ions are reduced:

$$Cu^{2+} + 2\,e^- \rightarrow Cu \qquad \text{(CATHODE REACTION)}$$

In a similar way the negative sulfate ions migrate to the positive electrode (anode). However, it is easier to get electrons from the copper metal of the electrode than from the sulfate ions. As each copper atom gives up two electrons, the copper ion passes into solution:

$$Cu \rightarrow Cu^{2+} + 2\,e^- \qquad \text{(ANODE REACTION)}$$

In effect, then, the copper of the positive electrode is oxidized (the anode reaction) and passes into solution; the copper ions in solution migrate to the negative electrode, are reduced (the cathode reaction), and plate out as copper metal. Large amounts of copper are purified in this way each year. Silver and gold can be purified similarly.

If we desire to plate an object with copper, we have only to render the surface conducting and make the object the negative electrode in a solution of copper sulfate. The object will become coated with copper, with the copper coating growing thicker as the electrolysis is continued. If the object is a metal, it will conduct electricity by itself. If the object is a nonmetal, its surface can be lightly dusted with graphite powder to render it conducting.

A potentially very important electrolysis reaction is the electrolysis of water. When electricity is passed into graphite electrodes immersed in a dilute salt solution, water is reduced to hydrogen and hydroxide ions at the cathode:

$$2\,H_2O + 2\,e^- \rightarrow H_2\,\text{(gas)} + 2\,OH^- \qquad \text{(CATHODE REACTION)}$$

At the anode, water is oxidized to oxygen and hydrogen ions:

$$2\,H_2O \rightarrow O_2 + 4\,H^+ + 4\,e^- \qquad \text{(ANODE REACTION)}$$

The OH^- and H^+ ions combine to re-form water. The overall, or net, cell reaction is:

$$2\,H_2O \xrightarrow{\text{electricity}} 2\,H_2\,\text{(gas)} + O_2\,\text{(gas)}$$

The hydrogen produced by the reduction of water can be stored and used as a fuel — for example, to power rockets into space. Someday, if electricity becomes inexpensive enough (see Chapter 8), water may be electrolyzed to produce hydrogen, which can then be piped to the point of use, just as natural gas is today.

Copper can be plated onto an object by making that object the negative electrode in a cell containing dissolved copper salts.

RELATIVE STRENGTHS OF OXIDIZING AND REDUCING AGENTS

When a piece of metallic zinc is placed in a solution containing hydrated copper ions (Cu^{2+}), an oxidation-reduction reaction occurs:

$$Zn + Cu^{2+} \rightarrow Zn^{2+} + Cu$$

Evidence for this reaction is the deposit of copper on the zinc. The gradual decrease in the intensity of the blue color of the solution indicates removal of the Cu^{2+} ions.

The oxidation of zinc by copper ions can be thought of as a competition between zinc ions (Zn^{2+}) and copper ions (Cu^{2+}) for the two electrons. Since the reaction proceeds almost to completion, the Cu^{2+} ions obviously win out in the competition. Other metals can compete similarly for electrons.

The **activity** of a metal is a measure of its tendency to lose electrons. Zinc is a more active metal than copper on the basis of the experiment just described. This means that given an equal opportunity, the first reaction will take place to a greater extent:

A copper ion has a greater attraction for electrons than does a zinc ion.

$Zn \rightarrow Zn^{2+} + 2\ e^-$ (more likely to occur)

$Cu \rightarrow Cu^{2+} + 2\ e^-$ (less likely to occur)

Experiments of this type with various pairs of metals and other reducing agents yield an **activity series** of the elements, which ranks each oxidizing and reducing agent according to its *strength* or *tendency* for the electron transfer to take place. An iron nail will be partly dissolved in a solution of a copper salt containing Cu^{2+} ions, with copper being deposited on the nail that remains. From this, it is determined that iron, like zinc, is more active than copper. The reaction that occurs is

The active metals lose electrons more easily; hence, these free metals are not found in nature.

$$Fe + Cu^{2+} \rightarrow Fe^{2+} + Cu$$

Now, which is more active, zinc or iron? This question can be answered by placing an iron nail in a solution containing Zn^{2+} ions and, in a separate container, a strip of zinc in a solution containing Fe^{2+} ions. The zinc strip is found to be eaten away in the solution containing Fe^{2+} ions. The reaction, then, is

$$Zn + Fe^{2+} \rightarrow Fe + Zn^{2+}$$

Nothing happens to the iron nail in the solution of Zn^{2+} ions. We deduce that Zn loses electrons more readily than Fe.

Such an activity series can be extended to include other metals and even nonmetals. The concentrations of the ions in solution and other factors often must be considered for accurate work, but for our purposes these will be ignored. Table 10–3 is an activity series of some oxidizing and reducing agents.

The activity series can be used to predict whether a reaction will occur. Thus, a reducing agent in the table (right column) is able to reduce the oxidized form of any species below it. For example, magnesium can reduce Cu^{2+} to Cu:

Activity: $Zn > Fe > Cu$

$$Mg + Cu^{2+} \rightarrow Cu + Mg^{2+}$$

Magnesium can also reduce Ag^+ to Ag:

$$Mg + 2\ Ag^+ \rightarrow Mg^{2+} + 2\ Ag$$

Zinc can also reduce silver ions:

$$Zn + 2\ Ag^+ \rightarrow 2\ Ag + Zn^{2+}$$

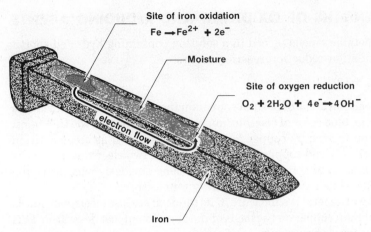

Site of iron oxidation
$Fe \rightarrow Fe^{2+} + 2e^-$

Moisture

Site of oxygen reduction
$O_2 + 2H_2O + 4e^- \rightarrow 4OH^-$

electron flow

Iron

FIGURE 10–5   The site of iron oxidation may be different from the point of oxygen reduction owing to the ability of the electrons to flow through the iron. The point of oxygen reduction can be located with an acid-base indicator because of the OH^- ions produced.

Zinc cannot reduce calcium ions, since calcium is above zinc in the series:

$Zn + Ca^{2+} \nrightarrow$ *No reaction*

The series also arranges oxidizing agents in the order of their effectiveness. Fluorine, F_2, can oxidize water, silver, iron (II) ion, or any species above it in the series; copper (II) ion, Cu^{2+}, can oxidize H_2, Fe, Zn, and the metals above Cu in the table, because the Cu^{2+} ion has a greater tendency to take on electrons than do the ions that are formed. Thus:

$Cu^{2+} + Fe \rightarrow Fe^{2+} + Cu$

$Cu^{2+} + Mg \rightarrow Mg^{2+} + Cu$

$Cu^{2+} + Ag \nrightarrow$ *No reaction*

CORROSION — UNWANTED OXIDATION-REDUCTION

In the United States alone, more than $10 billion is lost each year to corrosion. Much of this corrosion is the rusting of iron and steel, although other metals may oxidize as well. The problem with iron is that its oxide, rust, does not adhere strongly to the metal's surface once the rust is formed. Because the rust flakes off or is rubbed off easily, the metal surface becomes pitted. The continuing loss of surface iron by rust formation eventually causes structural weakness.

The corrosion of metals involves oxidation and reduction. The driving forces behind corrosion are the activity of the metal as a reducing agent and the strength of the oxidizing agent. Whenever a strong reducing agent (the metal) and a strong oxidizing agent (like oxygen) are together, a reaction between the two substances is likely. Factors governing the rates of chemical reaction such as temperature and concentration will affect the rate of corrosion as well. Consider the corrosion of an iron spike (Fig. 10–5). The surface of the iron is far from perfect. There are tiny microcrystals composed of loosely bound iron atoms on the surface of the metal. The iron can readily ionize into any water present on the surface of the metal.

$Fe \rightarrow Fe^{2+} + 2\,e^-$ (OXIDATION)

TABLE 10–3   Relative Strengths of Some Oxidizing and Reducing Agents: The Activity Series

	OXIDIZING AGENTS		REDUCING AGENTS	
	$Na^+ + e^-$	$\rightleftharpoons$	Na	
	$Ca^{2+} + 2\,e^-$	$\rightleftharpoons$	Ca	
	$Mg^{2+} + 2\,e^-$	$\rightleftharpoons$	Mg	
	$Zn^{2+} + 2\,e^-$	$\rightleftharpoons$	Zn	
	$Fe^{2+} + 2\,e^-$	$\rightleftharpoons$	Fe	
	$2\,H^+ + 2\,e^-$	$\rightleftharpoons$	H_2	
	$Cu^{2+} + 2\,e^-$	$\rightleftharpoons$	Cu	
	$Fe^{3+} + e^-$	$\rightleftharpoons$	Fe^{2+}	
	$Ag^+ + e^-$	$\rightleftharpoons$	Ag	
	$O_2 + 4\,e^- + 4\,H^+$	$\rightleftharpoons$	$2\,H_2O$	
	$F_2 + 2\,e^-$	$\rightleftharpoons$	$2\,F^-$	

Increasing Strength of Oxidizing Agent (left margin, upward arrow)

Increasing Strength of Reducing Agent (right margin, upward arrow)

The ionization of iron atoms into Fe^{2+} ions is an oxidation process and a result of the position of iron in the activity series (Table 10–3). Iron is a fairly active metal, that is, it tends to give its electrons up rather easily. Since iron is a good conductor of electricity, the electrons produced at this site can migrate to some point where they can reduce something. If these electrons did not migrate, the corrosion of iron would come to an abrupt halt as a result of a buildup of excessive negative charge. One location on the surface of the iron where electrons can be used would be any tiny drop of water containing dissolved oxygen. Here, the oxygen gains the electrons, forming hydroxide ions.

Oxidation cannot occur without reduction.

$$2\,O_2 + 2\,H_2O + 4\,e^- \rightarrow 4\,OH^- \qquad \text{(REDUCTION)}$$

This reduction of oxygen occurs so readily that when Fe^{2+} ions are encountered, they are further oxidized to Fe^{3+} ions. This happens to the dissolved Fe^{2+} ions in the water on the surface of the metal. The reaction is

$$4\,Fe^{2+} + O_2 + 2\,H_2O \rightarrow 4\,Fe^{3+} + 4\,OH^-$$

Finally, the Fe^{+3} ions combine with hydroxide ions to form the iron oxide we call rust.

$$2\,Fe^{3+} + 6\,OH^- \rightarrow Fe_2O_3 \cdot 3\,H_2O$$

The rate of rusting is enhanced by salts, which dissolve in the water on the surface of the iron and act like tiny salt bridges of an electrochemical cell (discussed in the next section). The hydroxide ions and Fe^{2+} and Fe^{3+} ions migrate more easily in the ionic solutions produced by the presence of the dissolved salts. Automobiles rust out more quickly when exposed to road salts in wintery climates. If road salts are used in your driving area, it's a good idea after snowy seasons to wash the undersides of automobiles to remove the accumulated salts.

Rusting can be prevented by protective coatings such as paint, grease, oil, enamel, or some corrosion-resistant metal like chromium. Some metals are more active than iron, but when these metals corrode, they form adherent oxide coatings. Coatings with these metals provide corrosion protection. One of these metals is zinc. Zinc coating of iron and steel is called **galvanizing** and may be done by dipping the object into a molten bath of zinc metal or by electroplating zinc onto the surface of an iron or steel object. In galvanized objects in which the zinc coating is

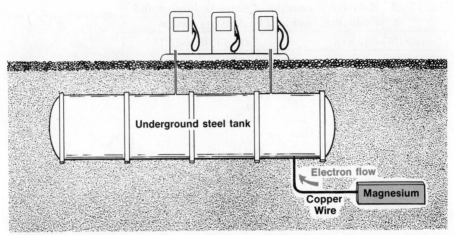

FIGURE 10–6 Cathodic protection. If magnesium is connected to the steel tank to be protected, the magnesium is more easily oxidized than the iron or copper connecting wire. The magnesium serves as a sacrificial anode. Hence, the cathode is protected with no points of oxidation occurring on its surface. The anode is the electrode where oxidation occurs; reduction occurs at the cathode. When the magnesium is used up, it is replaced by another block. The replacement is much easier and cheaper than replacing the tank.

exposed to air and water, a thin film of zinc oxide that protects the zinc from further oxidation forms. Galvanizing is a type of **cathodic protection.** As the name implies, a cathode is protected by using a more active metal in good electrical contact with the metal to be protected. The electrons for the reduction of oxygen,

$$O_2 + 2\ H_2O + 4\ e^- \rightarrow 4\ OH^-$$

are supplied by the more active metal. Thus a more active metal, electrically connected to a piece of iron, would be oxidized before the iron is oxidized.

Some cathodic protection relies on the cathode being sacrificed. An important application is the cathodic protection of underground steel storage tanks (Fig. 10–6) that hold gasoline and other hazardous liquids. These tanks must be protected since leakage would contaminate groundwater supplies. Beginning in 1986, these tanks must be cathodically protected under new federal regulations designed to protect groundwater.

The importance of ground-water purity is discussed in Chapter 17.

SELF-TEST 10–C

1. When electrons enter an electrode and cause a chemical reaction, that reaction is called _____ .
2. When electrons are removed from a chemical at an electrode, that reaction is called _____ .
3. What three reactants are necessary for rusting? _____ , _____ , and _____ .
4. According to the activity series shown in Table 10–3, could copper be used to protect iron cathodically? _____
5. According to the activity series, can magnesium reduce copper ions? _____

6. According to the activity series, which is the stronger oxidizing agent, fluorine or oxygen? _____

7. When water is electrolyzed, at the reduction electrode (cathode), _____ is produced. _____ is produced at the oxidation electrode (anode).

BATTERIES

One of the most useful applications of oxidation-reduction reactions is the production of electrical energy. A device that produces an electron flow (current) is called an **electrochemical cell.** Although a series of such cells is a **battery,** the term *battery* is commonly used even for single cells such as those we will describe.

Consider the reaction between zinc atoms and copper ions that was discussed previously. If zinc is placed in a solution containing Cu^{2+} ions, the electron transfer takes place between the zinc metal and the copper ions, and the energy liberated simply causes a slight heating of the solution and the zinc strip. If the zinc could be separated from the copper solution, and the two connected in such a way to allow current flow, the reaction can proceed, but now the electrons are transferred though the connecting wires. Figure 10–7 shows a battery that can be constructed to make use of the oxidation-reduction involved in the reaction of Zn with Cu^{2+}.

The anode reaction is the oxidation of zinc to Zn^{2+} ions.

$$Zn \rightarrow Zn^{2+} + 2\ e^-$$

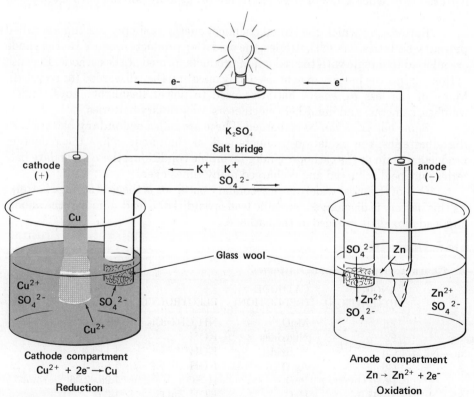

FIGURE 10–7   A simple battery involving the oxidation of zinc metal and the reduction of Cu^{2+} ions.

Historic battery used in the early telegraph.

The electrons flow from the Zn electrode through the connecting wire, light the lamp in the circuit, and then flow into the copper cathode where reduction of Cu^{2+} ions occurs:

$$Cu^{2+} + 2\,e^- \rightarrow Cu$$

The copper is deposited on the copper cathode.

In commercial batteries, the salt bridge is often replaced by a porous membrane.

This flow of electrons (negative charge) from the anode to the cathode compartment in the battery must be neutralized electrically. This is done by using a "salt bridge" provided to connect the two compartments. The salt bridge contains a solution of a salt such as K_2SO_4. Its purpose is to keep the two solutions neutral. Around the cathode the deposition of positive copper ions (Cu^{2+}) would tend to cause the solution to become negative owing to the presence of excess negative sulfate ions (SO_4^{2-}). Two actions can keep the solution around the cathode neutral: either positive potassium ions (K^+) pass into the solution, or negative sulfate ions pass out of the solution and into the salt bridge. Actually, both processes occur. Similarly, around the anode, the solution would tend to become positive because positive zinc ions (Zn^{2+}) are put into the solution. Two actions can keep the solution around the anode neutral: either negative sulfate ions pass into the solution or positive zinc ions pass out of the solution and into the salt bridge. In all of this exchange, it is necessary for the solution to maintain the same number of positive charges as negative charges. The reaction between zinc atoms and copper ions continues until one or the other is consumed.

Many different oxidation-reduction combinations are used in commercial batteries to produce a flow of electrons. A few of the more popular ones are listed in Table 10–4.

Batteries in which the stored chemical energy is simply used up are called **primary** batteries. In such batteries, the oxidation products produced at the anode are allowed to mingle with the reduction products formed at the cathode. Because of this mixing, the battery may be used only once and then discarded (or recycled). Many of the less expensive batteries used to power flashlights, toys, radios, watches, cameras, and hand-held calculators are primary batteries.

The mercury batteries (Table 10–4), which are so popular for small electronic devices, must be discarded carefully. If heated, these hermetically sealed batteries will rupture explosively due to expanding vapors within the package.

Some batteries can be recharged. These are called **secondary** batteries. In these batteries, the oxidation products stay at the anode, while the reduction products remain at the cathode. Under favorable conditions, these secondary batteries may be discharged and recharged many times over.

One of the most widely used secondary batteries is the lead storage battery. As this battery is discharged, metallic lead is oxidized to lead sulfate at the anode, and lead dioxide is reduced at the cathode.

TABLE 10–4 Characteristics of Some Batteries

SYSTEM	ANODE (OXIDATION)	CATHODE (REDUCTION)	ELECTROLYTE	TYPICAL OPERATING VOLTAGE PER CELL
Dry cell	Zn	MnO_2	NH_4Cl-$ZnCl_2$	0.9–1.4
Edison storage	Fe	Ni oxides	KOH	1.2–1.4
Nickel-cadmium—NiCd	Cd	Ni oxides	KOH	1.1–1.3
Silver cell	Cd	Ag_2O	KOH	1.0–1.1
Lead storage	Pb	PbO_2	H_2SO_4	1.95–2.05
Mercury cell	Zn(Hg)	HgO	KOH-ZnO	1.30
Alkaline cell	Zn(Hg)	MnO_2	KOH	0.9–1.2

Anode: $Pb + SO_4^{2-} = \underline{PbSO_4} + 2\,e^-$

Cathode: $PbO_2 + 4\,H^+ + SO_4^{2-} = \underline{PbSO_4} + 2\,H_2O$

The lead sulfate formed at both electrodes is an insoluble compound so it stays on the electrode surface. Since sulfuric acid is used in both the anode and the cathode reactions, the concentration of the sulfuric acid electrolyte decreases as the battery discharges. A measurement of the density of this battery acid gives a measure of the state of charge of the battery. The lower the density of the battery acid, the lower the state of charge.

A formula underlined like $\underline{PbSO_4}$ indicates that the substance is insoluble.

Recharging a secondary battery requires reversing the electrical current flow through the battery. When this occurs, the anode and cathode reactions are reversed.

At the negative electrode: $Pb + SO_4^{2-} \underset{\text{charge}}{\overset{\text{discharge}}{\rightleftharpoons}} \underline{PbSO_4} + 2\,e^-$

At the positive electrode: $PbO_2 + SO_4^{2-} + 4H^+ + 2\,e^- \underset{\text{charge}}{\overset{\text{discharge}}{\rightleftharpoons}} \underline{PbSO_4} + 2\,H_2O$

Normal charging of an automobile lead storage battery occurs during driving. The voltage regulator senses the output from the alternator, and when the alternator voltage exceeds that of the battery, the battery is charged. During the charging cycle in most batteries, some water is reduced at the cathode, while water is oxidized at the anode.

Oxidation of water: $2\,H_2O \rightarrow O_2 + 4\,H^+ + 4\,e^-$ (anode)

Reduction of water: $4\,H_2O + 4\,e^- \rightarrow 2\,H_2 + 4\,OH^-$ (cathode)

These reactions produce a mixture of hydrogen and oxygen in the atmosphere in the top of the battery. If this mixture is accidentally sparked, an explosion results. It is a good idea always to open a battery carefully and not introduce any sparks or open flames near a lead storage battery.

During starts, especially during extremely cold weather, the battery works very hard. This means that the battery must be recharged. Recharging often causes elongated crystals to grow on the electrode surfaces as the lead and lead oxide are redeposited on the negative and positive electrodes. Often these crystals of lead and lead oxide grow between the electrodes, causing internal short circuits. Usually, when this happens, the battery is "dead" and must be replaced. If electrolyte fluid runs low, the electrode surfaces dry, tending to make the surfaces not recharge properly. All in all, the lead storage battery is relatively inexpensive, reliable, and relatively simple and has an adequate life. Its high weight is its major fault. Newer secondary batteries have found use in some applications such as electronics, but none of these newer batteries can perform like the lead storage battery does for its cost.

FUEL CELLS

Fuel cells, like batteries, have a cathode and anode separated by an electrolyte. Unlike batteries, which are energy storage devices, fuel cells are energy conversion devices. Most fuel cells convert the energy of oxidation-reduction reactions of gaseous reactants directly into electricity. They are a special application of oxida-

Fuel cells are energy conversion devices. Batteries are energy storage devices.

tion-reduction chemistry. The most popularized application of fuel cells has been in the space program on board the Gemini, Apollo, and Space Shuttle missions.

Consider the reaction between hydrogen and oxygen to produce water and energy.

$$2\,H_2 + O_2 \rightarrow 2\,H_2O + \text{energy}$$

As mentioned earlier in this chapter, if a mixture of hydrogen and oxygen is sparked, the energy is released suddenly in the form of a violent explosion. In the presence of a platinum gauze, these gases will react at room temperature, slowly heating the catalytic surface to incandescence. In a fuel cell (Fig. 10–8), the oxidation of hydrogen by oxygen takes place in a controlled manner, with the electrons lost by the hydrogen molecules flowing out of the fuel cell and back in again at the electrode where oxygen is reduced. This electron flow powers the electrical needs of the spacecraft, or whatever else is connected to the fuel cell. The water produced in the fuel cell can be purified for drinking purposes.

Because of their light weight and their high efficiencies compared to batteries, fuel cells like the one shown in Figure 10–8 have proved valuable in the space program. Beginning with Gemini 5, alkaline fuel cells have logged over 10,000 hours of operation in space. The fuel cells used aboard the Space Shuttle deliver the same power that batteries weighing ten times as much would provide. On a typical seven-day mission, the Shuttle fuel cells consume 1500 pounds of hydrogen and generate 190 gallons of potable water.

Potable is a term used to describe drinkable water.

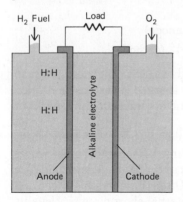

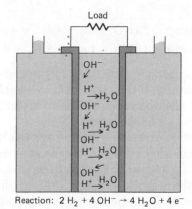

Reaction: $2\,H_2 + 4\,OH^- \rightarrow 4\,H_2O + 4\,e^-$

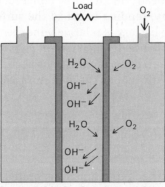

Reaction: $2\,H_2O + O_2 + 4\,e^- \rightarrow 4\,OH^-$

FIGURE 10–8 How an alkaline fuel cell works. The net reaction is $2\,H_2 + O_2 = 2\,H_2O$. Hydrogen is oxidized at the anode, and oxygen is reduced at the cathode. The water molecules produced are discharged into the alkaline electrolyte solution, which consists of potassium hydroxide (KOH) and water.

Other types of fuel cells that have been developed use air as the oxidizer and less pure hydrogen or carbon monoxide as the fuel. Ultimately, it is hoped that fuel cells capable of direct air oxidation of cheap gaseous fuels such as natural gas might be developed. These fuel cells might compete with gasoline engines and not produce air pollutants such as carbon monoxide and nitric oxide.

MATCHING SET

_____ 1. Chemicals in automobile lead storage battery

_____ 2. Most active metal in Table 10–3

_____ 3. Ore of mercury

_____ 4. Oxidizing agent used to purify drinking water

_____ 5. Product of incomplete combustion of carbon

_____ 6. Most oxidized form of carbon

_____ 7. Oxidation

_____ 8. Reduction

_____ 9. Gain of hydrogen

_____ 10. Pollutant from oxidation of sulfur in fuels

_____ 11. Pollutant from oxidation of nitrogen in air

_____ 12. Electrolyte

_____ 13. Strongest oxidizing agent

_____ 14. Secondary battery

_____ 15. A definitive name for oxidation

_____ 16. Fuel cell reaction

_____ 17. Product of ethanol oxidation in the liver

a. Fluorine
b. $2\,H_2 + O_2 \rightleftarrows 2\,H_2O$
c. Nitric oxide
d. Nickel-cadmium
e. Reduction
f. Carbon monoxide
g. Combustion
h. Pb, PbO_2, H_2SO_4
i. Causes solutions to conduct electricity
j. Loss of electrons
k. Cinnabar
l. Carbon dioxide
m. Acetaldehyde
n. Gain of electrons
o. Na
p. SO_2
q. Chlorine
r. Methane
s. $Fe_2O_3xH_2O$
t. Reduction

QUESTIONS

1. Write equations for the following reactions:
   a. The oxidation of methanol to formaldehyde
   b. The reduction of copper oxide (CuO) with hydrogen
   c. The reduction of copper ions (Cu^{2+}) to metallic copper at a cathode
   d. The burning of ethyl alcohol (C_2H_5OH) to form carbon dioxide and water
2. Give three examples of undesired oxidation.
3. Write formulas for four metal oxides and four nonmetal oxides.

4. Why would rusting of automobiles be less of a problem in Arizona than in Chicago?
5. Which do you think is more highly oxidized?
   a. MnO_2 or MnO
   b. PbO or PbO_2
   c. H_2O or H_2O_2
   d. K_2SO_4 or $K_2S_2O_8$
6. Give three uses for common oxidizing agents.
7. Which is the oxidized form of carbon, CO or C?
8. When iron is converted to rust, what is the iron said to be?

9. When iron ore is converted to iron, what is the iron said to be?

10. In electrolysis, reduction takes place at the negative electrode. What is this electrode called?

11. If an electrode is negatively charged, it will tend to attract ions of what charge?

12. Look at Table 10-2, which shows several methods for obtaining metals from their ores. What is the reducing agent in
    a. Making magnesium from sea water?
    b. Making sodium from rock salt?
    c. Making silver from silver ore?
    d. Making iron from iron ore?

13. When water is electrolyzed, what are the products of oxidation and reduction?

14. Using Table 10-3, which gives relative strengths of oxidizing and reducing agents, predict whether the following reaction would be expected to occur.
    a. $2 Na + Fe^{2+} \rightarrow 2 Na^+ + Fe$
    b. $Ca + F_2 \rightarrow Ca^{2+} + 2 F^-$
    c. $2 H^+ + Cu \rightarrow Cu^{2+} + H_2$
    d. $Zn^{2+} + Fe \rightarrow Fe^{2+} + Zn$

15. What is the purpose of a salt bridge in a battery?

16. Name two other examples of metallic corrosion besides rusting.

17. What is the effect of salts in the rusting process?

18. Would you expect an iron object to rust on the surface of the moon? Explain.

19. Describe two kinds of cathodic protection — one in which the cathode is sacrificed and one in which it is not.

20. What is the name given to coating steel with zinc?

21. State a general difference between a primary and a secondary battery.

22. What does the density of the electrolyte solution in a lead storage battery tell about the state of charge of the battery?

23. What is one cause of short circuits in lead storage batteries?

24. How is a fuel cell similar to most batteries?

25. How is a fuel cell different from most batteries?

26. What is the fuel in the type of fuel cells used on board spacecraft? What is the oxidant?

27. Explain why the spacecraft fuel cells are expensive to operate.

28. In many fuel cells the oxidant is pure oxygen. What would be a less expensive oxidant to use in fuel cells?

29. Besides electricity, what is produced in the NASA fuel cells?

Chapter 11

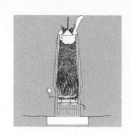

CHEMICAL RAW MATERIALS AND PRODUCTS FROM THE EARTH, SEA, AND AIR

The long view from space has dramatized what we already knew — the crust of the Earth is a very unusual environment, uniquely suited, at least in this solar system, for the production and support of life forms. Note again from Figure 2–6 that the chemical composition of the Earth's crust differs dramatically from that of the universe or even the composition of the Earth as a whole. Oxygen, in the air, water, and rocks, dominates at about 50%. Silicon, at about 25%, is a major part of silicate rocks, clays, and sand. Then come the major metals, iron, calcium, sodium, potassium, and magnesium, found mostly in mineral deposits and to some extent dissolved in the sea. Hydrogen, at less than 1% by weight, is ninth in abundance, and carbon, the central element in all life forms, is present in little more than trace amounts (Fig. 11–1). It is apparent, then, that life forms are a very small part of the whole and are clinging to the edge of the spaceship Earth.

Our environment is also quite heterogeneous in nature. Mixtures abound; everywhere we look the elements and compounds are almost lost in the complicated array of mixtures that have resulted from natural forces over very long periods of time.

Throughout most of history, we had not developed the power to alter our environment significantly. Most of the materials used, such as the wooden hammer or plow, were only physically changed from the natural material. Then came the chemical reduction of copper from its ores, followed by iron, and now a flood of new chemicals are produced each year. We have now developed, beyond question, the power to change the chemical mixtures that surround us. It is quite possible that we can control human-produced changes for the advantage of life forms, sterilize the Earth and destroy life, or do something in between these extremes. How we will address these controls is the burning question of this text, presented in the belief that it will take all of us, including the social, biological, and physical scientists, working together to address the problem adequately.

In this chapter we shall look at the chemistry of some major industrial operations on which our culture depends. However, the story only begins here, for the remainder of this text will deal with an ever-increasing number of chemical changes that are made at the expense of the natural mixtures of the Earth, sea, and air around us.

It is very important for the student to understand that the human knowledge is presently at hand to protect our natural environment. With the exception of radioactive nuclear wastes, we can, if we wish to pay the cost, cycle and recycle our natural resources through countless uses. We are learning that we do not have to use our iron and aluminum just once and then haul them to the dump. Perhaps the most striking aspect of natural chemistry is the cycling of the elements through a countless number of like uses.

We shall begin with the study of the chemistry of the major metals used in our society.

METALS AND THEIR REDUCTION

A continual search is under way for new ore deposits.

Metals occur mostly as compounds in the crust of the Earth, though some of the less active metals such as copper, silver, and gold can be found also as free elements. Fortunately, the distribution of elements in the crust is not uniform. Some elements that are not particularly abundant are familiar to us because they tend to occur in very concentrated, localized deposits, called **ores,** from which they can be extracted economically. Examples of these are lead, copper, and tin, none of which is among the more abundant elements in the crust of the Earth (Fig. 11–1). Other elements that actually form a much larger percentage of the crust are almost unknown to us because concentrated deposits of their ores are less commonly found or the metal is difficult to extract from its ore. An example is titanium, the tenth most abundant element in the crust of the Earth. Although the ores of titanium, rutile (mostly TiO_2), and ilmenite ($FeTiO_3$) are common, the use of the metal is rare because it is difficult to reclaim from the ores.

Some common minerals useful as ores are listed in Table 11–1.

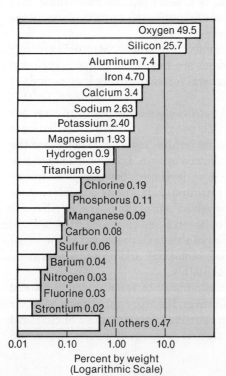

FIGURE 11–1 Abundance of the elements in the crust of the Earth (percentage by weight).

TABLE 11–1   Some Common Metals and Their Minerals

METAL	CHEMICAL FORMULA OF COMPOUND OF THE ELEMENT	NAME OF MINERAL
Aluminum	$Al_2O_3 \cdot xH_2O$	Bauxite
Calcium	$CaCO_3$	Limestone
Chromium	$FeO \cdot Cr_2O_3$	Chromite
Copper	Cu_2S	Chalcocite
Iron	Fe_2O_3	Hematite
	Fe_3O_4	Magnetite
Lead	PbS	Galena
Manganese	MnO_2	Pyrolusite
Tin	SnO_2	Cassiterite
Zinc	ZnS	Sphalerite
	$ZnCO_3$	Smithsonite

The preparation of metals from their ores, **metallurgy,** involves chemical reduction (Chapter 10). Indeed, the concept of oxidation and reduction developed from metallurgical operations. Iron in the ore iron oxide (Fe_2O_3) is in the form of Fe^{3+}. If we *reduce* Fe^{3+} ions to Fe atoms, we must find a source of electrons. (Recall that oxidation is the loss of electrons and that reduction is the gain of electrons.) Sometimes the desired metal is in solution (e.g., magnesium in the sea), where it exists in the oxidized form (Mg^{2+} ions). To obtain free metallic magnesium, we must add electrons to these ions (reduction) to produce neutral atoms.

Reduction of magnesium: $Mg^{2+} + 2\,e^- \rightarrow Mg$

Iron and Steel

The sources of most of the world's iron are large deposits of the iron oxides in Minnesota, Sweden, France, Venezuela, Russia, Australia, and England. In nature these oxides are frequently mixed with impurities, so the production of iron usually incorporates steps to remove such impurities. Iron ores are then reduced to the metal by using carbon, in the form of coke, as the reducing agent.

Iron ore is reduced in a blast furnace (Fig. 11–2). The solid material fed into the top of the blast furnace consists of a mixture of an oxide of iron (Fe_2O_3), coke (C), and limestone ($CaCO_3$). A blast of heated air is forced into the furnace near the bottom. Much heat is liberated as the coke burns, and the heat speeds up the reaction, which is important in making the process economical. The reactions that occur within the blast furnace are

Iron ores are iron compounds. To get iron from the ores, the iron in the compounds must be reduced.

$$\underset{\text{CARBON}}{2\,C} \;+\; \underset{\text{OXYGEN}}{O_2} \;\rightarrow\; \underset{\substack{\text{CARBON} \\ \text{MONOXIDE}}}{2\,CO} \;+\text{heat}$$

$$\underset{\substack{\text{IRON} \\ \text{OXIDE}}}{Fe_2O_3} + \underset{\substack{\text{CARBON} \\ \text{MONOXIDE}}}{3\,CO} \;\rightarrow \underset{\text{IRON}}{2\,Fe} + \underset{\substack{\text{CARBON} \\ \text{DIOXIDE}}}{3\,CO_2} + \text{heat}$$

Limestone (calcium carbonate) is added to remove the silica (SiO_2) impurity.

$$\underset{\substack{\text{CALCIUM} \\ \text{CARBONATE}}}{CaCO_3} \;\xrightarrow{\text{heat}}\; \underset{\substack{\text{CALCIUM} \\ \text{OXIDE}}}{CaO} + \underset{\substack{\text{CARBON} \\ \text{DIOXIDE}}}{CO_2}$$

$$\underset{\substack{\text{CALCIUM} \\ \text{OXIDE}}}{CaO} + \underset{\substack{\text{SILICON} \\ \text{DIOXIDE}}}{SiO_2} \rightarrow \underset{\substack{\text{CALCIUM} \\ \text{SILICATE}}}{CaSiO_3}$$

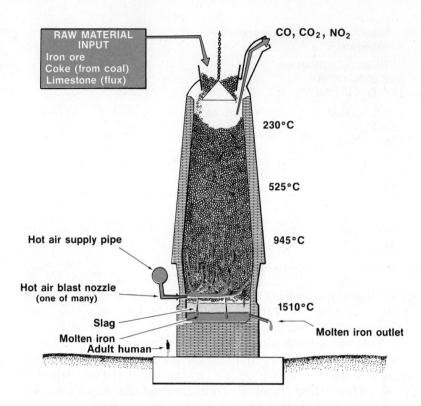

**FIGURE 11-2
Diagram of a blast
furnace.**

Alloy: a metal consisting
of two or more elements.

The calcium silicate, or **slag,** exists as a liquid in the furnace. Consequently, as the blast furnace operates, two molten layers collect in the bottom. The lower, denser layer is mostly liquid iron that contains a fair amount of dissolved carbon and often smaller amounts of other impurities. The upper, lighter layer is primarily molten calcium silicate with some impurities. From time to time the furnace is tapped at the bottom, and the molten iron is drawn off. Another outlet somewhat higher in the blast furnace can be opened to remove the liquid slag.

As it comes from the blast furnace, the iron contains too much carbon for most uses. If some of the carbon is removed, the mixture becomes structurally stronger and is known as **steel.** Steel is an **alloy** of iron with a relatively small amount of carbon (less than 1.5%); it may also contain other metals. In order to convert iron into steel, the excess carbon is burned out with oxygen.

There are several techniques for burning the excess carbon (see Figs. 11-3 and 11-4). A recent development that has been very widely adopted is the basic oxygen process (Fig. 11-4). In this process pure oxygen is blown into molten iron through a refractory tube (oxygen gun), which is pushed below the surface of the iron. At elevated temperatures, the dissolved carbon reacts very rapidly with the oxygen to give gaseous carbon monoxide and carbon dioxide, which then escape.

After the carbon content has been reduced to a suitable level, the molten steel is formed into desired shapes. During processing the steel is subjected to carefully regulated heat treatment to ensure that the steel has a uniform crystallinity, which in turn determines its pliability, toughness, and other useful mechanical properties.

All of the processes in steelmaking, from the blast furnace to the final heat treatment, use tremendous quantities of energy, mostly in the form of heat. In the production of a ton of steel, approximately one ton of coal or its energy equivalent is consumed.

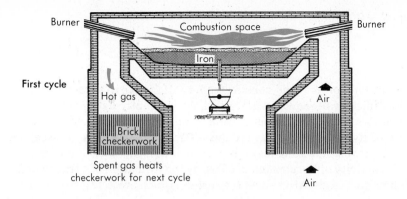

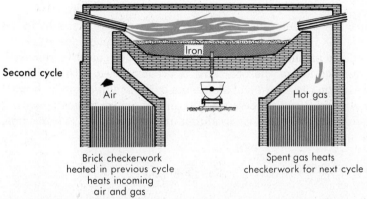

FIGURE 11–3   The open hearth furnace for the conversion of iron to steel. Air and fuel are blown in first in one direction and then the other in order that heat, which is costly, might be stored in the brick checkerwork. (From G. Lee, H. O. Van Orden, and R. O. Ragsdale: *General and Organic Chemistry.* Philadelphia, Saunders College Publishing, 1971.)

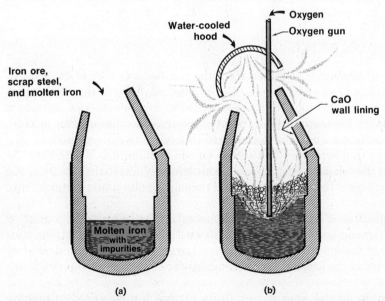

FIGURE 11–4   The basic oxygen process furnace. Much of the steel manufactured today is refined by blowing oxygen through a furnace charged with ore, scrap, and molten iron.

Aluminum

When aluminum was first made, it was very expensive and rare. A bar of aluminum was displayed next to the Crown Jewels at the Paris Exposition in 1855.

Aluminum, in the form of Al^{3+} ions, constitutes 7.4% of the Earth's crust. However, because of the difficulty of reducing Al^{3+} to Al, only recently have we learned to isolate and use this abundant element. Aluminum metal is soft and has a low density. Many of its alloys, however, are quite strong. Hence, it is an excellent choice when a lightweight, strong metal is required. In structural aluminum, the high chemical reactivity of the element is offset by the formation of a transparent, hard film of aluminum oxide, Al_2O_3, over the surface, which protects it from further oxidation:

$$4\,Al + 3\,O_2 \rightarrow 2\,Al_2O_3$$

The principal ore of aluminum contains the mineral bauxite, a hydrated aluminum oxide, $Al_2O_3 \cdot xH_2O$. Because impurities such as iron oxides in the ore have undesirable effects on the properties of aluminum, these must be removed, generally by the purification of the ore. This is accomplished with the Bayer process, which is based on the reaction of aluminum oxide or aluminum hydroxide with strong bases. In the Bayer process the mixture of oxides is treated with a sodium hydroxide solution, which dissolves aluminum oxide and leaves iron oxide, which is insoluble in the solution:

$$\underset{\text{(SOLID)}}{Al_2O_3 \cdot xH_2O} + \underset{\text{(SOLID)}}{Fe_2O_3} \xrightarrow[\text{solution}]{\text{NaOH}} \underset{\text{(SOLUTION)}}{Al(OH)_4^- + Na^+} + \underset{\text{(SOLID)}}{Fe_2O_3}$$

The mixture is filtered; $Al(OH)_3$ is then carefully precipitated out of the clear solution by the addition of carbon dioxide (an acid), which makes the solution less basic:

$$CO_2 + Al(OH)_4^- \rightarrow Al(OH)_3\downarrow + HCO_3^-$$

The aluminum hydroxide is heated to transform it into pure anhydrous aluminum oxide:

$$2\,Al(OH)_3 \xrightarrow{\text{heat}} Al_2O_3 + 3\,H_2O$$

Metallic aluminum is obtained from the purified oxide by electrolysis in molten cryolite (Fig. 11–5 and Table 10–2). Cryolite, Na_3AlF_6, has a melting point of 1000°C; the molten compound dissolves considerable amounts of aluminum oxide, which in turn lowers the melting point of the cryolite solution. This mixture of cryolite and aluminum oxide is electrolyzed in a cell with carbon anodes and a carbon cell lining that serves as the cathode on which aluminum is deposited. As the operation of the cell proceeds, the molten aluminum sinks to the bottom of the cell. From time to time the cell is tapped and the molten aluminum is run off into molds.

The top of the Washington Monument is a casting of aluminum made in 1884.

About ten times more energy is needed to produce a ton of aluminum than a ton of steel.

Aluminum is used both as a structural metal and as an electrical conductor in high-voltage transmission lines. It competes with copper as an electrical conductor because of the lower cost of aluminum. Larger diameter aluminum wires must be used to offset the lower electrical conductivity of aluminum compared with copper.

Can you list all the ways energy in its various forms is used to prepare aluminum from its ore?

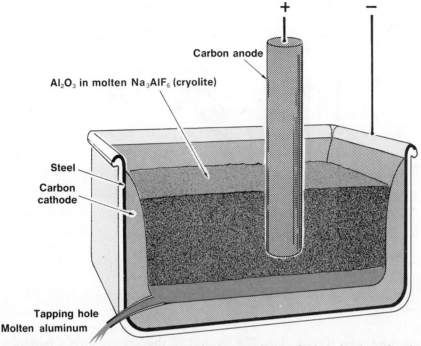

Carbon anode

Al$_2$O$_3$ in molten Na$_3$AlF$_6$ (cryolite)

Steel

Carbon cathode

Tapping hole
Molten aluminum

FIGURE 11−5 **Schematic drawing of a furnace for producing aluminum by electrolysis of a melt of Al$_2$O$_3$ in Na$_3$AlF$_6$. The molten aluminum collects at the bottom of the carbon cathode container.**

Copper

Although copper metal occurs in the free state in some parts of the world, the supply available from such sources is quite insufficient for the world's need. The majority of the copper obtained today is from various copper sulfide ores, most of which must be concentrated prior to the chemical processes that produce the metal. These minerals include $CuFeS_2$ (chalcopyrite), Cu_2S (chalcocite), and CuS (covellite). Because the copper content of these ores is around 1% to 2%, the powdered ore is first concentrated by the flotation process (Fig. 11−6).

In the flotation process, the powdered ore is mixed with water and a frothing agent such as pine oil. A stream of air is blown through to produce froth. The **gangue** in the ore, which is composed of sand, rock, and clay, is easily wet by the water and sinks to the bottom of the container. In contrast, a copper sulfide particle is hydrophobic — it is not wet by the water. The copper sulfide particle becomes coated with oil and is carried to the top of the container in the froth. The froth is removed continuously, and the floating copper sulfide minerals are recovered.

The preparation of copper metal from a copper sulfide ore involves roasting in air to oxidize some of the copper sulfide and any iron sulfide present:

$$2\ Cu_2S + 3\ O_2 \rightarrow 2\ Cu_2O + 2\ SO_2\uparrow$$
$$2\ FeS + 3\ O_2 \rightarrow 2\ FeO + 2\ SO_2\uparrow$$

Subsequently the mixture is heated to a higher temperature, and some copper is produced by the reaction:

$$Cu_2S + 2\ Cu_2O \rightarrow 6\ Cu + SO_2\uparrow$$

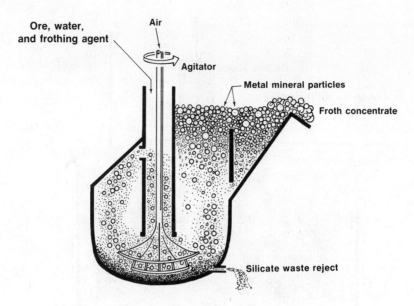

FIGURE 11-6
Apparatus for flotation concentration.

The iron oxides then form a slag. The product of this operation is a **matte,** a mixture of copper metal and sulfides of copper, iron, other ore constituents, and slag. The molten matte is heated in a converter with silica materials. When air is blown through the molten material in the converter, two reaction sequences occur. First, the iron is converted to a slag:

$$2\,FeS + 3O_2 \rightarrow 2\,FeO + 2\,SO_2$$
$$FeO + SiO_2 \rightarrow FeSiO_3$$
$$\text{(MOLTEN SLAG)}$$

The remaining copper sulfide is converted to copper metal:

$$2\,Cu_2S + 3\,O_2 \rightarrow 2\,Cu_2O + 2\,SO_2$$
$$Cu_2S + 2\,Cu_2O \rightarrow 6\,Cu + SO_2$$

The copper produced in this manner is crude or "blister" copper and is purified electrolytically.

In the electrolytic purification of copper, the crude copper is first cast into anodes; these are placed in a water solution of copper sulfate and sulfuric acid. The cathodes are made of pure copper. As electrolysis proceeds, copper is oxidized at the anode, moves through the solution as Cu^{2+} ions, and is deposited on the cathode (Chapter 10). The voltage of the cell is regulated so that more active impurities (such as iron) are left in the solution and less active ones are not oxidized at all. These less active impurities include gold and silver, and they collect as "anode slime," an insoluble residue. The anode slime is subsequently worked up to recover these rarer metals.

The copper produced by the electrolytic cell is 99.95% pure and is suitable for use as an electrical conductor. Copper for this purpose must be pure because very small amounts of impurities, such as arsenic, considerably reduce the electrical conductivity of copper.

Magnesium from the Sea

Magnesium, with a density of 1.74 g/mL, is the lightest structural metal in common use. For this reason it is most often used in alloys designed for light weight and great strength. It is a relatively active metal chemically because it loses electrons easily. Magnesium "ores" include sea water, which has a magnesium concentration of 0.13%, and dolomite, a mineral with the composition $CaCO_3 \cdot MgCO_3$. Because there are 6 million tons of magnesium present as Mg^{2+} salts in every cubic mile of sea water, the sea can furnish an almost limitless amount of this element.

There are about 328 million cubic miles of sea water.

The recovery of magnesium from sea water (Fig. 11–7) begins with the precipitation of magnesium hydroxide by the addition of lime to sea water:

$$CaO + H_2O \rightarrow Ca^{2+} + 2\ OH^-$$
$$Mg^{2+} + 2\ OH^- \rightarrow Mg(OH)_2\downarrow$$

The magnesium hydroxide is removed by filtration and then neutralized with hydrochloric acid to form the chloride:

$$Mg(OH)_2 + 2\ H^+ + 2\ Cl^- \rightleftharpoons Mg^{2+} + 2\ Cl^- + 2\ H_2O$$

The water is evaporated; this is followed by the electrolysis of molten magnesium chloride in a huge steel pot that serves as the negative electrode, or cathode (Fig. 11–8). Graphite bars serve as the positive electrodes, or anodes.

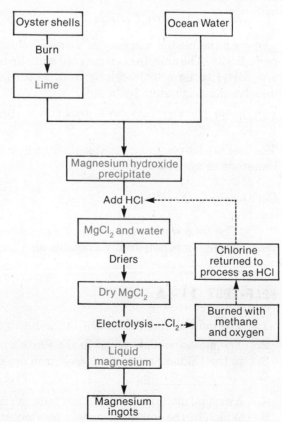

FIGURE 11–7 Flow diagram showing how magnesium metal is produced from sea water.

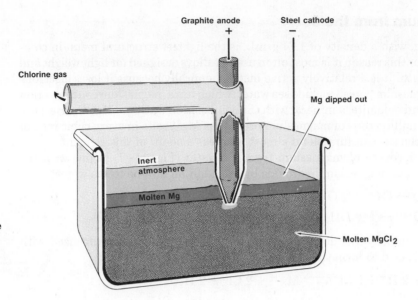

FIGURE 11–8 A cell for electrolyzing molten $MgCl_2$. The magnesium metal is formed on the steel cathode and rises to the top where it is dipped off periodically. Chlorine gas is formed on the graphite anode and is piped off.

To make magnesium one can use sea water, lime from oyster shells, methane from natural gas, and electricity.

$$Mg^{2+} + 2\ Cl^- \xrightarrow{\text{electrolysis}} Mg + Cl_2\uparrow$$
(MELTED)

REDUCTION
AT THE CATHODE: $Mg^{2+} + 2\ e^- \rightarrow Mg$

OXIDATION
AT THE ANODE: $2\ Cl^- \rightarrow Cl_2\uparrow + 2\ e^-$

As the melted magnesium forms, it floats to the surface and is removed periodically. The chlorine is recovered and reacts with air and natural gas (methane, CH_4) to form hydrochloric acid, which in turn is used to neutralize and dissolve the magnesium hydroxide:

$$4\ Cl_2 + \underset{\text{METHANE}}{2\ CH_4} + O_2 \rightarrow 2\ CO + 8\ HCl$$

The lime used to precipitate the magnesium as the hydroxide is obtained by heating limestone or oyster shells:

$$CaCO_3 \xrightarrow{\text{heat}} \underset{\text{LIME}}{CaO} + CO_2$$

The total world production of magnesium is only about 250,000 tons per year, although it is potentially available on a larger scale.

SELF-TEST 11–A

1. The most abundant element in the Earth's crust is _____.
2. The most abundant metal in the Earth's crust is _____.
3. In the United States, the largest iron ore deposits are found in the state of

   _____.

4. A natural material that is almost pure calcium carbonate is _____.
5. Which of the following metals may occur in the free or metallic state in mineral deposits — iron, copper, aluminum, or magnesium? _____

6. Which of the following metals — iron, copper, aluminum, and magnesium — are either produced or purified using electricity? _____

7. Another name for calcium silicate as it applies to production of iron is _____ .

8. In order for most metals to be prepared from their ores, they must be ( ) oxidized or ( ) reduced.

9. In an electrical refining process for metals, the purest metal will always be found at the ( ) anode or ( ) cathode.

10. Which metal is sufficiently concentrated in the oceans to be extracted commercially — magnesium, aluminum, copper, or iron? _____

11. Most metals are found in the Earth as ( ) neutral atoms, ( ) positive ions, or ( ) negative ions.

THE FRACTIONATION OF AIR

The atmosphere of the Earth is a fantastically large source of the elements nitrogen and oxygen and certain of the noble gases, including argon, neon, and xenon (see Table 11–2).

Before pure oxygen and nitrogen can be obtained from the air, water vapor and carbon dioxide must be removed first. This is usually done by precooling the air by refrigeration or by using silica gel to absorb water and lime to absorb carbon dioxide. Afterward, the air is compressed to a pressure exceeding 100 times normal atmospheric pressure, cooled to room temperature, and allowed to expand into a chamber. This expansion produces a cooling effect (the Joule-Thompson effect) due to breaking of weak attractive van der Waals bonds between the gaseous molecules. Recall that breaking bonds requires energy, so the expanding gas absorbs energy from the surroundings, thus cooling the surroundings and the gas itself. If this expansion is repeated and controlled properly, the expanding air actually cools to the point of liquefaction (Fig. 11–9). The temperature of the *liquid air* is usually well below the boiling points of nitrogen ($-195.8°C$), oxygen ($-183°C$), and argon ($-189°C$). This liquid air is then allowed to vaporize partially again, and since nitrogen is more volatile than oxygen or argon (N has a lower boiling point), the liquid becomes more concentrated in oxygen and argon. This process, known as the Linde process, produces high-purity nitrogen (99.5 + %) and oxygen with a purity of 99.5%. Further processing produces pure argon, neon

TABLE 11–2 The Atmospheric Composition of the Earth at Sea Level

GAS	PERCENTAGE BY VOLUME
Nitrogen	78.084
Oxygen	20.948
Argon	0.934
Carbon dioxide	0.033*
Neon	0.00182
Helium	0.00052
Methane	0.0002

* Estimated for 1986.

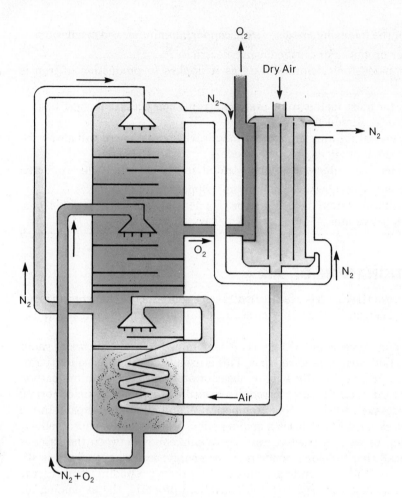

FIGURE 11–9
Diagram of a fractionating column for separating oxygen and nitrogen in an air supply.

(boiling point −246°C), and even helium (boiling point −268.9°C), but most helium used in the United States is produced from natural gas wells.

Most oxygen produced by the fractionation of liquid air is used in steel-making, although some is used in rocket propulsion (to oxidize hydrogen) and in controlled oxidation reactions of other types. Liquid oxygen (LOX) can be shipped and stored at its boiling temperature of −183°C at atmospheric pressure. Substances this cold are called **cryogens** (from Greek *kryos,* meaning "icy cold"). They represent special hazards since contact produces instantaneous frostbite, and structural materials such as plastics, rubber gaskets, and some metals become brittle and fracture easily at these temperatures. Liquid oxygen can accelerate oxidation reactions to the point of explosion because of the high oxygen concentration. For this reason, contact between liquid oxygen and substances that will ignite and burn in air must be prevented.

Special cryogenic containers holding liquid oxygen are actually huge vacuum-walled bottles much like those used to carry hot soup or hot coffee. These containers can be seen outside hospitals or industrial complexes, on highways and railroads, and even aboard ocean-going vessels (Fig. 11–10).

Liquid nitrogen is also a cryogen. It has uses in medicine (cryosurgery), for example in cooling a localized area of skin prior to removal of a wart or other unwanted or pathogenic tissue. Since nitrogen is so chemically unreactive, it is used as an inert atmosphere for certain applications such as welding, and liquid nitrogen

FIGURE 11-10 Photo of a cargo tanker capable of carrying several thousand gallons of liquefied, cryogenic oxygen, at a temperature of -183°C. Although it is extremely cold, its high concentration in the liquid state makes liquid oxygen exceptionally reactive with anything that can burn.

is a convenient source of high volumes of the gas. Because of its low temperature and inertness, liquid nitrogen has found wide use in frozen food preparation and preservation during transit. Containers of nitrogen atmospheres, such as railroad boxcars or truck vans, present health hazards since they contain little (if any) oxygen to support life, and workers have died when they entered such areas without breathing apparatus.

SELF-TEST 11-B

1. Name the two elements that are commercially prepared by the fractionation of the air. _____ and _____

2. LOX is the industrial abbreviation for _____.

3. What does the word *cryogen* mean? _____

4. Give four uses of the products from liquid air, and identify each with a particular element.

_____

_____

_____

_____

SILICON MATERIALS — OLD AND NEW

Silicon and oxygen make up 75% of the crust of the Earth, and it is the bonding between these two elements in clays and rocks that literally holds together the skin of the Earth. The chemical structures involved are many and complex. However, a few typical molecular structures related to some of the most important materials in our society, from glass to the computer chip, will be of significant value to any student of chemistry.

Glass

Silicon dioxide, SiO_2 (also called silica), occurs naturally in large amounts in rocks and sand, or more rarely in much larger crystals (quartz) (Fig. 11–11). It has a melting point of 1710°C. If the melted material is cooled rapidly, a noncrystalline solid is obtained. Crystalline quartz consists of an extended structure in which each silicon atom is bonded tetrahedrally to four oxygen atoms (Fig. 11–12a), and each oxygen atom is bonded to two silicon atoms. The bonding thus extends throughout the crystal (Fig. 11–12b). When silica is melted, some of the bonds are broken and the units move with respect to each other. When the liquid is cooled, the re-formation of the original solid requires a reorganization that is hard to achieve because of the difficulty the groups experience in moving. The very viscous liquid structure is thus partially preserved on cooling to give the characteristic feature of a **glass,** which is an apparently solid material (pseudo-solid) with some of the randomness in structure characteristic of a liquid. This random structure accounts for one of the typical properties of a glass: it breaks irregularly rather than splitting along a plane like a crystal.

By the addition of metal oxides to silica, the melting temperature of the mixture can be reduced from 1710°C to about 700°C. The oxides most often added are sodium oxide (added as Na_2CO_3, soda ash) and calcium oxide (added as $CaCO_3$). The metal ions form ionic bonds, which are nondirectional, with oxygen atoms that previously had been bonded rigidly to specific Si atoms. As a result, the so-called soda-lime glass has a lower melting temperature and viscosity than pure SiO_2 and can be produced and fabricated more easily.

Soda-lime glass will be clear and colorless only if the purity of the ingredients has been controlled carefully. If, for example, too much iron oxide is present,

Sodium carbonate (Na_2CO_3) is the number 10 commercial chemical. Calcium oxide is number 4. See inside the front cover.

Viscosity is the resistance to flow.

QUARTZ CRYSTALS

FIGURE 11–11
Quartz. (Courtesy of McGraw-Hill.)

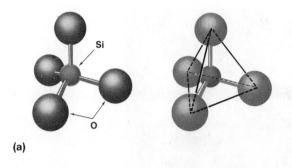

(a)

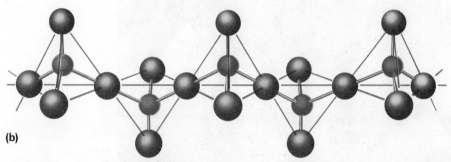

(b)

FIGURE 11–12 (*a*) **Tetrahedral structure of silicon and oxygen in silicates** (*b*) **Chain of tetrahedra showing that an oxygen is common at each point of contact between tetrahedra.**

the glass will be green. Other metal oxides produce other colors (see Table 11–3). To some extent, one color can counteract another.

The substances are melted together in a gas- or oil-fired furnace. As they react, bubbles of CO_2 gas are evolved.

$$CaCO_3 + SiO_2 \rightarrow CaSiO_3 + CO_2\uparrow$$
$$Na_2CO_3 + SiO_2 \rightarrow Na_2SiO_3 + CO_2\uparrow$$

The mixture is heated to about 1500°C to remove the bubbles of CO_2. At this temperature the viscosity is low, and the bubbles of entrapped gas escape easily. The mixture is cooled somewhat and then is blown into bottles by machines, drawn into sheets, or molded into other forms (Fig. 11–13).

Most glasses are made from the oxides of Si, Na, and Ca, which are melted together. Colored glass is produced by addition of other metal oxides.

TABLE 11–3   Substances Used in Colored Glasses

SUBSTANCE	COLOR
Copper (I) oxide	Red, green, or blue
Tin (IV) oxide	Opaque
Calcium fluoride	Milky white
Manganese (IV) oxide	Violet
Cobalt (II) oxide	Blue
Finely divided gold	Red, purple, or blue
Uranium compounds	Yellow, green
Iron (II) compounds	Green
Iron (III) compounds	Yellow

FIGURE 11–13
Craftsman working
with molten glass.
(Courtesy of Corning
Glass Company.)

It is possible to incorporate a wide variety of materials into glass for special purposes. Some examples are given in Table 11–4.

Ceramics

Ceramic materials have been made since well before the dawn of recorded history. They are generally fashioned from clay or other natural earths at room temperature and then permanently hardened by heat. Clays with a wide variety of properties are found in a considerable range of ceramic materials, from bricks to table china. The techniques developed with natural clay have been applied to a wide range of other inorganic materials in recent years. The result has been a considerable increase in the kinds of ceramic materials available. One can now obtain ceramic magnets as well as ceramics suitable for rocket nozzles — both were developed from mixtures of inorganic oxides by the use of ceramic technology, which includes as an indispensable process the heating of the materials to make them hard and resistant to wear.

The three basic ingredients of common pottery are silicate minerals: clay, sand, and feldspar. The term **clay** includes materials with a wide range of chemical compositions that are produced from the weathering of granite and gneiss rocks.

TABLE 11–4 Special Glasses

SPECIAL ADDITION OR COMPOSITION	DESIRED PROPERTY
Large amounts of PbO with SiO_2 and Na_2CO_3	Brilliance, clarity, suitable for optical structures: crystal or flint glass
SiO_2, B_2O_3, and small amounts of Al_2O_3	Small coefficient of thermal expansion: borosilicate glass. "Pyrex," "Kimax," and others
One part SiO_2 and four parts PbO	Ability to stop (absorb) large amounts of X rays and gamma rays: lead glass
Large concentrations of CdO	Ability to absorb neutrons
Large concentrations of As_2O_3	Transparency to infrared radiation
Suspended Se particles	Red color

Feldspars are aluminosilicates containing potassium, sodium, and other ions in addition to silicon and oxygen. An approximation of the weathering process can be made if we write feldspar as a mixture of oxides: $K_2O \cdot Al_2O_3 \cdot 6\ SiO_2$; this includes, among other reactions, the reaction of the mineral with water containing dissolved carbon dioxide to form clay.

$$K_2O \cdot Al_2O_3 \cdot 6\ SiO_2 + 2\ H_2O + CO_2 \rightarrow$$
$$\underbrace{Al_2O_3 \cdot 2\ SiO_2 \cdot 2\ H_2O}_{\text{A CLAY}} + 4\ SiO_2 + 2\ K^+ + CO_3^{2-}$$

The essential feature of the clay mineral is that it occurs in the form of extremely minute platelets, which, when wet, are plastic and can easily be shaped. When dry, the clay platelets are rigid; if heated to an elevated temperature, they become permanently rigid and are no longer subject to easy dispersion in water. When these clays are mixed with feldspars and silica, heating them produces a mixture of crystals held together by a matrix of glasslike material. The clays can be used by themselves to make bricks, flowerpots, and clay pipe, but finer quality ceramic materials contain purified clays and other ingredients in carefully controlled proportions. The clay is mixed with potter's flint (a form of silica) and feldspar in various proportions. The clay makes the mixture pliable; the silica decreases the amount of shrinkage that occurs after drying and firing in a kiln; the feldspar lowers the temperature needed for adequate firing. The feldspar acts as a glasslike material to hold the grains of clay and flint together. Clay must usually be dried before being fired; otherwise the rapid loss of water from the surface and the slower loss from the interior of the object will cause cracks in the clay.

In general, natural clays are extremely complex mixtures. If these are used in ceramics without treatment, the finished materials have a color and physical properties characteristic of the impurities present. The first pieces of fine Oriental chinaware arrived in Europe during the late Middle Ages, and European potters envied and admired the obviously superior product. This led to the beginning of systematic studies on the effect of composition on the nature of the ceramic produced and to a keen appreciation of the role of the purity of the clay in determining the color and potential decorative development of the piece (Fig. 11–14).

Alchemists made notable achievements in this area. One of these, Johann Friedrich Bottger, worked from about 1705 to 1719 for King Augustus of Saxony, who kept him almost as a prisoner. The king hoped to gain power from the alchemist's discoveries. Bottger succeeded in developing several novel ceramic mate-

FIGURE 11–14   **Fired pottery. Different colors and surface textures can be achieved by fusing the coating material with the clay itself. The coating ceases to be a "coat" and becomes a part of the whole, the entire structure being held together by covalent bonds. (Courtesy of the Robinson-Ransbottom Pottery Company.)**

rials, of which the most important was the first white glazed porcelain made in Europe (in 1709). Bottger devoted the rest of his life to the perfection of the manufacture and decoration of this material, in which he enjoyed considerable success. The china was made in Meissen and was both glazed and vitrified. The glazing was accomplished by coating the pieces with a material that melted and produced an impermeable layer on the surface. Vitrification was produced by firing the clay at a temperature sufficient to melt a portion of the material and, in effect, produce an impermeable glass that held the remaining particles together.

In the past few decades new ceramic materials have been developed and used on an increasingly wide scale. Nearly pure alumina (Al_2O_3) and zirconia (ZrO_2) are now used as bases for ceramic materials, which are excellent electrical or thermal insulators. Magnetic ceramics, which contain iron compounds, are used as memory elements in computers.

In recent years a new class of materials, the glass ceramics, has been discovered; these have unusual but very valuable properties. Normally glass breaks because once a crack starts, there is nothing to stop it from spreading. It was discovered that if glass is treated by heating until many tiny crystals have developed in it, the resulting material, when cooled, is much more resistant to breaking than normal glass. The process must be controlled carefully to obtain the desired properties. The materials produced in this way are generally opaque and are used for cooking utensils and kitchen ware. They include materials marketed under the name *Pyroceram* (Fig. 11–15). The initial manufacturing process is similar to that of other glass objects, but once the materials have been formed into their final shapes, they are heat treated to develop their special properties.

Portland Cement and Concrete

A cement is a material used to bind other materials together. Portland cement contains calcium, iron, aluminum, silicon, and oxygen in varying proportions. It has a structure somewhat similar to that described earlier for glass, except that in cement some of the silicon atoms have been replaced by aluminum atoms. Cement reacts in the presence of water to form a hydrated colloid of large surface area, which subsequently undergoes recrystallization and reaction to bond to itself and to bricks or stone. Cement is made by roasting a powdered mixture of calcium carbonate (limestone or chalk), silica (sand), or aluminosilicate mineral (kaolin,

FIGURE 11–15 Cookware made of Pyroceram. (Courtesy of Corning Glass Company.)

clay, or shale) and iron oxide at a temperature of up to 870°C in a rotating kiln (Fig. 11–16). As the materials pass through the kiln, they lose water and carbon dioxide and ultimately form a "clinker," in which the materials are partially fused. The "clinker" is then ground to a very fine powder after the addition of a small amount of calcium sulfate (gypsum). The composition of Portland cement is 60% to 67% CaO, 17% to 25% SiO_2, 3% to 8% Al_2O_3, up to 6% Fe_2O_3, and small amounts of magnesium oxide, magnesium sulfate, and potassium and sodium oxides.

The reactions that occur during the setting of cement are quite complex. The various constituents react with water and subsequently at the surface with the carbon dioxide in air. The initial reaction of cement with water gives a sticky gel that results from the hydrolysis of the calcium silicates. This sticks to itself and to the other particles (sand, crushed stone, or gravel). The gel has a very large surface area and is responsible for the strength of concrete. The setting process also involves the formation of small, densely interlocked crystals after the initial solidification of the wet mass. This continues for a long time after the initial setting and increases the compressive strength of the cement. Water is required since the setting reactions involve hydration. For this reason, freshly poured concrete is kept moist for several days. Over 400 million tons of cement are manufactured each year, most of which is used to make concrete. Concrete, like many other materials containing Si—O bonds, is highly noncompressible but lacks tensile strength. If concrete is to be used where it is subject to tension, it must be reinforced with steel.

(a)

(b)

FIGURE 11–16   (a) A cement kiln. Note the rollers on the supports, which allow the giant cylinder to rotate. As the kiln turns, the powder moves down the cylinder because it is at a slight angle. Intense heat is produced by the combustion of gaseous fuels. The powder loses volatile materials as it moves along, and the finished product is discharged from the lower end. (b) The first kiln used for making cement in the United States was constructed by David Saylor over a century ago in Coplay, Pennsylvania. The kiln still stands as pictured here. (Courtesy of the Portland Cement Northwestern States Company, Mason City, Iowa.)

"Pure" Silicon and "The Chip"

Silicon of about 98% purity can be obtained by heating silica and coke at 3000°C in an electric arc furnace.

$$SiO_2 + 2\,C \rightarrow Si + 2\,CO$$

Silicon of this purity is alloyed with aluminum and magnesium to increase their hardness and durability and is used in making silicone polymers.

High-purity silicon can be prepared by reducing $SiCl_4$ with magnesium.

$$SiCl_4 + 2\,Mg \rightarrow Si + 2\,MgCl_2$$

The magnesium chloride, being water soluble, is then washed from the silicon. The final purification of the silicon takes place by a melting process called **zone-refining** (Fig. 11–17), which produces silicon containing less than one part per *billion* of impurities such as boron, aluminum, and arsenic.

One outstanding property of silicon in a high state of purity is its electrical conductivity. Unlike a metal, which easily conducts electricity, and unlike a nonmetal, which fails to conduct electricity, silicon is a **semiconductor.** That is, it fails to conduct until a certain electrical voltage is applied, but beyond that it conducts moderately. By placing other atoms in a crystal of pure silicon, a process known as **doping,** experimenters have found that its conductivity properties can be changed. Doping a silicon crystal with a Group V element such as arsenic produces a crystal with extra electrons. (Arsenic has five valence electrons, whereas silicon has four.) This is known as an n-doped semiconductor. Doping silicon with a Group III element such as gallium produces a p-doped semiconductor, since gallium has only three valence electrons and looks positive in the silicon lattice.

In 1947 an electrical device called the **transistor** was invented. The simplest device used layers of n-p-n– or p-n-p–doped silicon. Germanium, a Group IV element just below silicon in the periodic table, was also used. Later, scientists used electrical fields to control conductivity in silicon transistors. These **field-effect transistors** have been put to good use by engineers designing low-noise amplifiers, receivers, and other forms of electronic equipment.

The extra electrons simulate a negatively charged material, hence the name n-doped semiconductor.

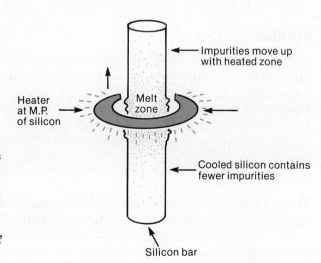

FIGURE 11–17 Zone refining. The hot zone moves upward on the silicon bar. As the silicon melts, impurities become mobile and move with the hot molten zone. Repeated passes of the heater produce a crystalline silicon bar with fewer than one part impurities per billion parts of silicon (1 ppb).

Impurities move up with heated zone

Heater at M.P. of silicon

Melt zone

Cooled silicon contains fewer impurities

Silicon bar

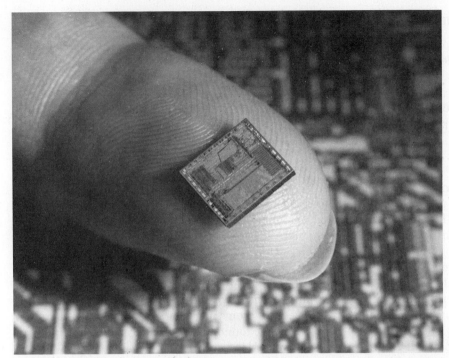

FIGURE 11–18 A tiny microcomputer (Intel 8748) fabricated from a single piece of highly purified silicon. Such computers are capable of many millions of computations per second. Their speed and small size have revolutionized computers and their applications. (Courtesy of Intel Corp.)

The most revolutionary application of silicon's semiconductor properties has been the design of **integrated electrical circuits,** computer memories, and even whole computers called **microprocessors** on tiny chips of silicon scarcely larger than a millimeter or so in diameter. As mentioned in Chapter 1, these devices have begun to permeate our whole society. You will find them in calculators, cameras, watches, toys, coin changers, cardiac pacemaker devices, and many other products. Truly, silicon is both the world we walk on and at the same time our constant companion in communications and electronic controls (Fig. 11–18).

SULFURIC ACID

Sulfur in underground mineral deposits is brought to the surface by the Frasch process (Fig. 11–19), which utilizes the fact that sulfur can be melted by superheated steam. The molten sulfur is raised to the surface of the Earth by means of compressed air and is then allowed to cool in large vats.

Sulfur is converted to sulfuric acid (H_2SO_4) by means of four steps, called the contact process. In the first step the sulfur is burned in air to give mostly sulfur dioxide:

$$S + O_2 \rightarrow SO_2(g)$$

The gaseous SO_2 is then converted to SO_3 by passing SO_2 over a hot catalytically active surface, such as platinum or vanadium pentoxide:

$$2\,SO_2 + O_2 \xrightarrow{\text{Catalyst}} 2\,SO_3(g)$$

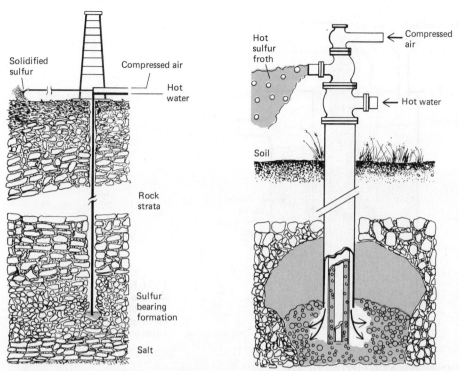

FIGURE 11–19   Diagram of the mining of sulfur by the Frasch process, which uses melting and pressure. Three concentric pipes are directed into the sulfur deposit. Compressed air and hot water are sent down two of the pipes, and molten sulfur is forced up the third pipe.

Although SO_3 can be converted directly into H_2SO_4 by passing SO_3 into water, the enormous amount of heat released in the reaction causes the formation of a stable fog of H_2SO_4. This is avoided by passing the SO_3 into H_2SO_4:

$$SO_3 + \underset{\text{SULFURIC ACID}}{H_2SO_4} \rightarrow \underset{\text{PYROSULFURIC ACID}}{H_2S_2O_7}$$

and then diluting the $H_2S_2O_7$ with water:

$$H_2S_2O_7 + H_2O \rightarrow 2\,H_2SO_4$$

Several million tons of sulfuric acid are prepared by this method each year.

Sulfuric acid is the number 1 chemical, 80 billion pounds per year!

The acid is used in huge quantities in the manufacture of fertilizers, in the petroleum industry, and in the production of steel. Also, sulfuric acid plays an important role in the manufacture of organic dyes, plastics, drugs, and many other products. The cost of sulfuric acid, about 1¢ per pound, has not changed much in 300 years, a tribute to improving technology.

SODIUM HYDROXIDE, CHLORINE, AND HYDROGEN CHLORIDE

Sodium hydroxide (NaOH), hydrogen, and chlorine (Cl_2) are prepared simultaneously by the electrolysis of a concentrated solution of sodium chloride in water (Fig. 11–20). There are several variations in the basic process that improve its efficiency and the purity of the products. The basic reactions, which occur at

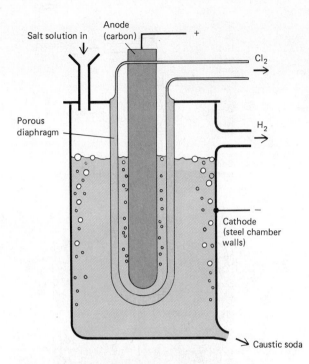

FIGURE 11–20 Cross section of cell for the electrolysis of brine to yield chlorine, hydrogen, and caustic soda (NaOH).

nonreactive solid electrodes, are:

Anode: $2\,Cl^- \rightarrow Cl_2(g) + 2\,e^-$ (OXIDATION)
Cathode: $2\,H_2O + 2\,e^- \rightarrow 2\,OH^- + H_2$ (REDUCTION)

The reaction replaces the Cl^- of the NaCl with OH^-.

     Purer sodium hydroxide is produced when the process is carried out using a mercury cathode. In this case, Na^+ is reduced to Na metal, which dissolves in the mercury electrode:

Cathode: $Na^+ + e^- \xrightarrow{\text{Hg}} Na(Hg)$

The sodium-mercury amalgam* is a liquid. The amalgam is removed from the cell continuously and reacted with pure water:

$$2\,Na(Hg) + 2\,H_2O \rightarrow 2\,Na^+ + 2\,OH^- + H_2(g)$$

The chlorine gas is collected from the anode compartment, compressed into tanks, and sold for further use. The hydrogen gas is also collected and sold. An important reaction between hydrogen and chlorine is the production of hydrogen chloride (HCl) gas and hydrochloric acid:

$$H_2(g) + Cl_2(g) \rightarrow 2\,HCl(g)$$

This process can be used to prepare hydrogen chloride, which is quite pure. Hydrochloric acid is a solution of hydrogen chloride in water. The use of mercury in this process is a potential hazard, as mercury contamination of streams results when the waste solutions from this process are discarded (see Chapter 1).

     Because electrical energy is quite expensive, an electrochemical process is economical only if all the products are sold. In recent years the demand for chlorine has grown very rapidly, but the demand for sodium hydroxide has not. Consequently it became necessary to devise some way of disposing of relatively large

---

* An amalgam is an alloy of metal with mercury; it can be either liquid or solid.

amounts of sodium hydroxide. This has been accomplished, in part, by transforming NaOH into sodium hydrogen carbonate:

$$Na^+ + OH^- + CO_2(g) \rightarrow Na^+ + HCO_3^- \rightarrow NaHCO_3$$

Because of the surplus of NaOH, this process has largely replaced older processes for the manufacture of $NaHCO_3$.

An alternate method for making HCl involves heating NaCl and H_2SO_4. The reaction occurs in two stages and can be driven to completion by heating:

$$NaCl + H_2SO_4 \rightarrow HCl(g) + NaHSO_4$$
$$NaCl + NaHSO_4 \rightarrow HCl(g) + Na_2SO_4$$

In principle, chlorine can be prepared by several reactions in which chloride is oxidized to Cl_2:

$$2\,Cl^- \rightarrow Cl_2(g) + 2\,e^-$$

Several oxidizing agents, including O_2 and hydrogen peroxide (H_2O_2), are suitable for this purpose.

PHOSPHORIC ACID

Phosphoric acid (H_3PO_4) or its salts are used in the manufacture of many materials encountered in our daily lives. These include baking powder, carbonated beverages, detergents, fertilizers, and fire-resistant textiles. Several processes are available for the manufacture of phosphoric acid, but the basic raw material is usually the calcium phosphate that occurs naturally in apatite minerals of the general formula $Ca_5X(PO_4)_3$, where X may be F^-, Cl^-, or OH^-.

An electric furnace is used to heat a mixture of the phosphate ore ($Ca_5F[PO_4]_3$), silica (SiO_2), and coke (C). At an elevated temperature a reaction occurs in which elemental phosphorus vapor (P_4) is produced:

$$4\,Ca_5F(PO_4)_3 + 18\,SiO_2 + 15\,C \rightarrow 18\,CaSiO_3 + 2\,CaF_2 + 15\,CO_2\uparrow + 3\,P_4\uparrow$$

The phosphorus is then condensed from the gaseous vapor and purified. The low melting point of phosphorus (44.1°C) allows easy storage and handling as a liquid if protected from air, in which phosphorus ignites spontaneously.

Elemental phosphorus is transformed into phosphoric acid by oxidation with air to give P_4O_{10}:

$$P_4(gas) + 5\,O_2 \rightarrow P_4O_{10}(gas)$$

which is then hydrated by absorption into hot phosphoric acid containing about 10% water:

$$P_4O_{10} + 6\,H_2O \xrightarrow{H_3PO_4} 4\,H_3PO_4$$

Arsenic, when present in the original ore, is carried through to the phosphoric acid produced, at which point arsenic can be precipitated by treatment with H_2S. This is important if the phosphoric acid is to be used in the manufacture of food products.

SODIUM CARBONATE

As noted in our discussion of glass making, sodium carbonate (Na_2CO_3) is a major ingredient of glass. Sodium carbonate is also important as a cheap alkaline material in the production of numerous chemicals and in the paper industry. It is especially useful to the homemaker as a water softener since the carbonate ion will precipitate the metal ions of iron, calcium, and magnesium, which cause water hardness (Chapter 17).

Trona ore, an impure form of sodium carbonate, is insufficient as a source of supply for this important chemical. Several chemical methods have been employed in the production of sodium carbonate. The Solvay process dominates. In this process, ammonia from the Haber process (Chapter 19) and carbon dioxide from the lime kiln are dissolved in a concentrated brine (NaCl) solution. Sodium hydrogen carbonate ($NaHCO_3$) precipitates from this solution:

$$NH_3 + CO_2 + H_2O \rightleftharpoons NH_4^+ + HCO_3^-$$
$$NH_4^+HCO_3^- + Na^+Cl^- \rightleftharpoons NaHCO_3\downarrow + NH_4^+Cl^-$$

On gentle heating, the dry sodium hydrogen carbonate yields the anhydrous sodium carbonate:

$$2\ NaHCO_3 \rightarrow Na_2CO_3 + H_2O\uparrow + CO_2\uparrow$$

Anhydrous sodium carbonate is known industrially as soda ash, while washing soda is the decahydrate $Na_2CO_3 \cdot 10\ H_2O$, which is prepared by crystallization directly from water solution.

The precipitation of sodium hydrogen carbonate or sodium carbonate from water solutions is an application of the acid-base chemistry presented in Chapter 9. Recall that a base requires protons or hydrogen ions at the expense of an acid. At a pH of 8 in the presence of the weak base ammonia, the bicarbonate ion HCO_3^- dominates over the carbonate ion CO_3^{2-} and will precipitate sodium as the bicarbonate:

$$Na^+ + HCO_3^- \rightleftharpoons NaHCO_3\downarrow$$

However, if the strong base NaOH is added to raise the pH to 12 or above, the bicarbonate ion loses its proton to the strong base and the carbonate ion then dominates in solution. Hence, the sodium carbonate will precipitate at the higher pH:

$$HCO_3^- + OH^- \rightarrow H_2O + CO_3^{2-}$$
$$2\ Na^+ + CO_3^{2-} + 10\ H_2O \rightarrow Na_2CO_3 \cdot 10\ H_2O\downarrow$$

SOURCES OF NEW MATERIALS

For most of history, we have been limited to the materials found on the surface of the Earth and to relatively few chemical transformations. Developments in mining techniques from technological advances and the control of chemical change through basic research have radically changed the materials available for human use.

Four sources of new materials are promised by recent advances: (1) Thousands of new compounds will be produced, described, and catalogued each year as a continuation of ongoing basic and applied research efforts. (2) New stable elements appear possible. If elements are made with atomic numbers up to 150 as predicted, some of the elements should be stable and should result in a multitude of new pure substances and mixtures. (3) Our reach through space programs comes ever closer to the exploitation of extraterrestrial materials from the moon, meteors, and nearby planets. Indications now are that the elements are the same, but the mineral configurations are often different for space materials. (4) Recent experiments make it clear that chemicals formed and purified in a gravity-free environment are significantly different from the "same" chemicals formed and purified in the presence of Earth's gravity.

In addition to the enhanced effects of purified copper in electrical conduction and ultrapure silicon in the control of electrical circuits and in the processing of information, one has to wonder what new materials and applications are in the offing. For example, consider again the making of glass. One of the biggest problems in controlling the purity of a particular glass is contamination from the container in which it is made. In space, no container is needed! Applications in making glass for improved optical fibers are presently exciting to researchers seeking still another breakthrough in the communications industry.

Light fibers have successfully carried 20 billion laser pulses per second, equivalent to 300,000 simultaneous conversations over a 42-mile distance. Under-ocean and intercity systems are presently being developed.

SELF-TEST 11–C

1. The two principal nonmetals in glass are _____ and _____ .
2. Flint, or crystal, glass contains a large amount of a compound of what metal? _____
3. What element is associated with the miniaturization of electronic components? _____
4. What three silicate minerals are used to make common pottery materials? _____ , _____ , and _____
5. Roasting a mixture of powdered limestone, clay, sand, and iron oxide produces what important commercial building material? _____
6. Which element is pumped from underground deposits as a molten material mixed with hot water? _____
7. The electrolysis of brine produces what element at the anode where oxidation occurs? _____
8. Phosphorus can be moved about as a liquid because of its low melting point provided it is kept out of contact with the _____ .
9. What is the chemical name of washing soda? _____ What related chemical is likely to be on the shelf in the kitchen? _____

MATCHING SET

_____	1. Copper	a.	Reduced in a blast furnace
_____	2. Aluminum	b.	Calcium fluoride added to glass
_____	3. Milk glass	c.	Contains phosphorus and oxygen
_____	4. Magnesium	d.	Used in making transistors
_____	5. Oyster shells	e.	$NaHCO_3$
_____	6. Sodium bicarbonate	f.	Mined with superheated water
_____	7. Sulfur	g.	A limitless supply in sea water
_____	8. Iron	h.	Supply calcium hydroxide for magnesium production
_____	9. Phosphate	i.	Purified electrolytically
_____	10. Silicon	j.	The most abundant metal in the Earth's crust
		k.	Sodium carbonate
		l.	An element distilled from air

QUESTIONS

1. Name three metals that you would expect to find free in nature. Name three that you would not.

2. What is the primary reducing agent in the production of iron from its ore?

3. Why is CaO necessary for the production of iron in a blast furnace?

4. What is the chemical difference between iron and steel?

5. Natural materials are most likely to be elements, compounds, or mixtures?

6. What metal is most used in industry?

7. What metal is recovered from sea water in industrial quantities?

8. Describe the solution used in a commercial cell for the electrolytic reduction of aluminum.

9. Why is it so important to purify industrial quantities of copper electrolytically to a level above 99.9% pure?

10. What chemical is obtained from oyster shells in the production of magnesium from sea water? What is the role of this chemical in the process?

11. Explain how the structure of glass and a liquid are similar.

12. What oxide is the main ingredient in glass?

13. Give reactions involved in the preparation of:
    $Ca(OH)_2$ from $CaCO_3$
    NH_3 from N_2 and H_2
    Iron from iron oxide
    Sulfuric acid from sulfur
    Phosphoric acid from phosphorus

14. A typical soda-lime glass has a composition reported as 70% SiO_2, 15% Na_2O, and 10% CaO. What ratio of weights of silica (SiO_2), sodium carbonate (Na_2CO_3), and calcium carbonate ($CaCO_3$) must be melted together to make this glass? The carbonates are decomposed by heat to evolve carbon dioxide.

15. What is the maximum weight (in pounds) of magnesium that can be obtained from 100 pounds of sea water? (See p. 251.)

16. Which will precipitate at the lower pH, sodium carbonate or sodium bicarbonate? Explain.

17. What is the purpose of using sodium carbonate in the family wash?

18. What is the purpose of oxidizing phosphorus in air after it has been reduced from phosphate rock in a furnace?

19. Two elements and a compound are produced in the electrolysis of brine. Name them and write the reactions involved in their production.

20. Why is it necessary to have two oxidation steps in the production of sulfuric acid from elemental sulfur?

21. Some argue that atomic energy is the most significant scientific and technical event of the 20th century. Others say the silicon chip has the greatest impact for change in our society. Should genetic engineering also be mentioned in this context? What do you think? Give reasons for your answer.

22. What chemicals are in Portland cement? Give the source of each.

23. When clay is fired, a rigid glasslike framework is established that is not attacked by water. Where do you think the "glass" comes from in the fired clay?

24. Is glass a mixture or a compound? Explain.

25. Is the recovery of oxygen from the air a chemical or a physical process? Give a reason for your answer.

26. Name two commercial sources for lime (CaO), and give the equations involved. Also, name two uses for this important chemical.

27. Two elements discussed in this chapter have special electrical properties only when they are in a high state of purity. What are they, and what are the applications for the pure materials?

Chapter 12

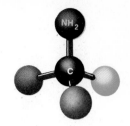

THE UBIQUITOUS CARBON ATOM—AN INTRODUCTION TO ORGANIC CHEMISTRY

The importance of carbon compounds to life on Earth cannot be overestimated. Consider what the world would be like if all the carbon and carbon compounds were removed suddenly. The result would be somewhat like the barren surface of the moon! Many of the little everyday things often taken for granted would be quite impossible without this versatile element. In an ordinary pencil, for example, the "lead" (made from graphite, an elementary form of carbon), the wood, the rubber in the eraser, and the paint on the surface are all carbon or carbon compounds. The paper in this book, the cloth in its cover, and the glue holding it together are also made of carbon compounds. All of the clothes one wears, including the leather in shoes, would not exist without carbon. If carbon compounds were removed from the human body, there would be nothing left except water and a small residue of minerals, and the same is true of all forms of living matter. Fossil fuels, foods, and most drugs are essentially made of carbon compounds. In addition, many carbon compounds such as plastics and detergents, which are not directly connected with the life processes, play an important role in our lives.

> Carbon and its compounds are vital to life on this planet.

Several million carbon compounds have been studied and described in the chemical literature, and thousands of new ones are reported every year. Although there are 88 other naturally occurring elements, there are many more known carbon compounds than known compounds that contain no carbon. The very large and important branch of chemistry devoted to the study of carbon compounds is **organic chemistry.** The name *organic* is actually a relic of the past, when chemical compounds produced from once-living matter were called "organic" and all other compounds were called "inorganic."

> Organic chemistry is the study of the nonmineral compounds of carbon.

WHY ARE THERE SO MANY ORGANIC COMPOUNDS?

The enormous number of organic compounds has intrigued chemists for over a hundred years. The atomic theory, as developed earlier for all atoms, describes a structure for the carbon atom that explains this multiplicity of carbon compounds. The peculiar structure of this atom allows it *to form covalent bonds with other carbon atoms in a seemingly endless array of possible combinations.* A simple organic

> The large number of carbon compounds is due to:
> 1. Stability of chains of carbon atoms
> 2. Occurrence of isomers
> 3. Reactivity of functional groups

molecule may contain a single carbon-carbon bond, whereas a complex molecule may contain literally thousands of such bonds. A few other elements are capable of forming stable bonds between like atoms. These include such elements as nitrogen (N_2), oxygen (O_2), and sulfur (S_8), to name a few. But only S, Sn, Si, and P can form long-chain molecules.

An additional factor in the large number of carbon compounds lies in the stability of carbon chains. The carbon chains are not normally subject to attack by water or, at ordinary temperatures, by oxygen. Chains formed by atoms of other elements undergo reaction with either water or oxygen, or both, much more easily than do carbon chains.

Isomers are two or more different compounds with the same number of each kind of atom per molecule.

A further reason for the large number of organic compounds is the ability of a given number of atoms to combine in more than one molecular pattern and, hence, produce more than one compound. Such compounds, each of which has molecules containing the same number and kinds of atoms, but arranged differently relative to each other, are called **isomers.** For example, the molecular structure represented by A—B—C is different from the molecular structure A—C—B, as is C—A—B; these three species are isomers. If we consider the number of possible ways the digits one through nine can be ordered to make nine-digit numbers, we can begin to imagine how a single group of atoms could possibly form hundreds of different molecules. Carbon, with its ability to bond to other carbon atoms, is especially well suited to form isomers.

A dash in a formula represents a single bond; two electrons are shared. The double dash represents a double bond; four electrons are shared.

A final factor explaining the large number of organic compounds is the ability of the carbon atom to form strong covalent bonds with atoms of numerous other elements, such as nitrogen, oxygen, sulfur, chlorine, fluorine, bromine, iodine, silicon, boron, and even many metals. As a result there are large classes of organic compounds. A **functional group,** a particular combination of atoms, appears in each member of a class. For example, all **organic acids** have a carboxyl group attached to another carbon atom.

$$\left(-C\overset{\displaystyle O}{\underset{\displaystyle OH}{}} \right)$$ CARBOXYL GROUP, A FUNCTIONAL GROUP

CHAINS OF CARBON ATOMS — THE HYDROCARBONS

Carbon is intermediate among the elements in its ability to attract electrons to it in a covalent bond. Thus, a carbon atom is unable to remove electrons completely from metals, and even fluorine (the best attractor of electrons among the elements) is unable to remove an electron completely from a carbon atom. As a result, carbon atoms tend not to form ionic bonds but rather to share electrons in the formation of covalent bonds.

Carbon has an intermediate electronegativity.

In addition to the tendency for carbon to form covalent bonds with many other atoms, there is also a remarkable inclination for carbon atoms to form relatively strong covalent bonds with each other. The carbon atoms join to give long chains up to thousands of carbon atoms long. The chains can also form branches or rings. Only two elements, hydrogen and carbon, are needed to form stable molecules with these branch, chain, or ring structures. As a result, **hydrocarbons** (compounds of hydrogen and carbon) are the largest class of organic compounds. The hydrocarbons can be divided into four groups: **alkanes** (contain C—C bonds),

alkenes (contain one or more C=C bonds), **alkynes** (contain one or more C≡C bonds), and **aromatic hydrocarbons** (compounds containing ring systems similar to those found in benzene).

Alkanes

The simplest alkane is methane (CH_4), the principal component of natural gas. Methane molecules are tetrahedral and have four C—H bonds, as shown in Figure 12–1.

The next member of the alkane family is ethane, C_2H_6, a hydrocarbon with two carbon atoms. The bonding in ethane is illustrated by the following formulas:

$$
\begin{array}{cccc}
& H \;\; H & & H \;\; H \\
& \ddots \;\; \ddots & & | \;\; | \\
H & :\!C\!:\!C\!: H & \text{or} \quad H & -\!C\!-\!C\!-H \\
& \ddots \;\; \ddots & & | \;\; | \\
& H \;\; H & & H \;\; H
\end{array}
$$

Even though the molecule is represented as flat, in reality the bonds to each carbon atom are in a tetrahedral arrangement, as shown in Figure 12–2. Also shown in Figure 12–2 is the rotation that takes place around the carbon-carbon single bond. Rotation of this type is a common feature in many molecules and becomes important as the molecular size increases.

By applying what we have learned, it is a simple matter to extend the concept of carbon-carbon bonding to a three-carbon molecule such as that of propane (C_3H_8). In Figure 12–3, note that the three carbon atoms in propane do not lie in a straight line because of the tetrahedral bonding about each carbon atom. Also, because of the rotation about the two C—C single bonds, the molecule is "flexible."

It is apparent that these bonding concepts can be extended to a four-carbon molecule and to a limitless number of larger hydrocarbon molecules. Actually, many such compounds are known; some, such as natural rubber, are known to contain over a thousand carbon atoms in a chain.

The first ten straight-chain alkanes, or saturated hydrocarbons, are listed in Table 12–1. Notice that each succeeding formula is obtained by adding CH_2 to the previous formula. Alkanes are an example of a **homologous series** — a series of

> Alkanes are **saturated hydrocarbons** because they contain the maximum amount of hydrogen allowed by the carbon bonding capacity.

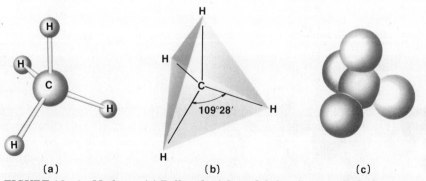

FIGURE 12–1 Methane. (*a*) Ball-and-stick model showing tetrahedral structure. (*b*) Geometry of regular tetrahedron. (*c*) Model of methane, CH_4, showing relative size of atoms in relationship to interatomic distances.

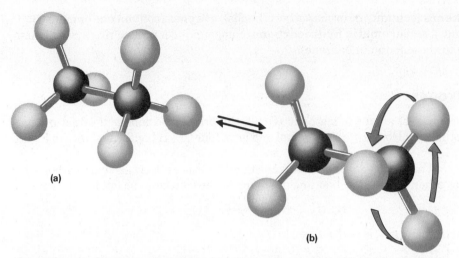

FIGURE 12−2 Two possible rotational forms of the ethane molecule. The hydrogen atoms in the methyl (CH_3) groups may be in an eclipsed position (*a*), staggered (*b*), or in any intermediate position. In ethane the two methyl groups can rotate easily about the carbon-carbon bond.

compounds of the same chemical type that differ only by a fixed increment. In this case the fixed increment is CH_2. The alkane homologous series can be represented by the general formula C_nH_{2n+2}, where *n* is the number of carbon atoms for a member of the series.

Structural isomers have the same molecular formulas but a different pattern of bonds.

STRUCTURAL ISOMERS: STRAIGHT OR BRANCHED-CHAIN VARIATIONS When we try to write the structure for butane, C_4H_{10}, we soon discover that two structures are possible.

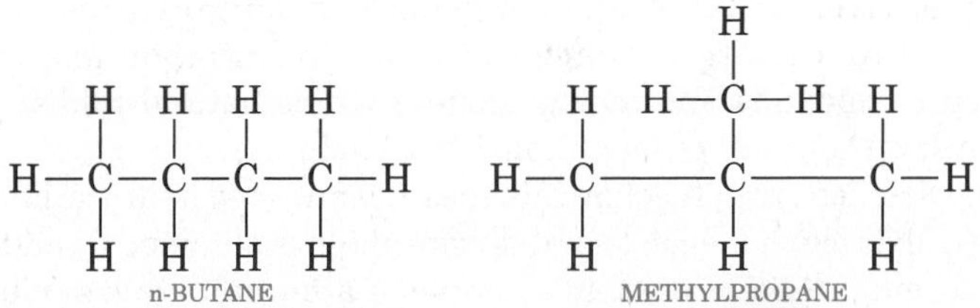

n-BUTANE		METHYLPROPANE (ISOBUTANE)
MELTING POINT	−138.3°C	−160°C
BOILING POINT	− 0.5°C	− 12°C
DENSITY (at 20°C)	0.579 g/mL	0.557 g/mL

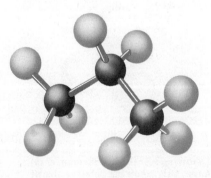

Propane

FIGURE 12−3 Ball-and-stick model of propane.

TABLE 12–1   The First Ten Straight-Chain Saturated Hydrocarbons

NAME	FORMULA	BOILING POINT, °C	STRUCTURAL FORMULA	USE
Methane	CH_4	−162	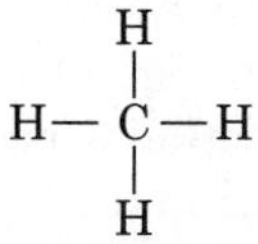	Principal component in natural gas
Ethane	C_2H_6	−88.5	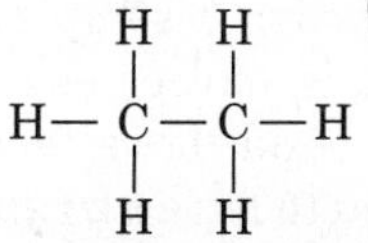	Minor component in natural gas
Propane	C_3H_8	−42	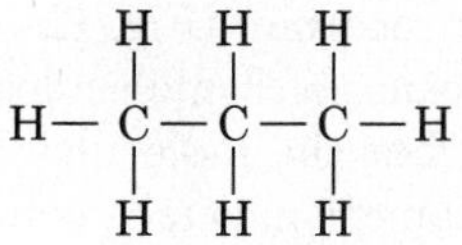	Bottled gas for fuel
n-Butane	C_4H_{10}	0	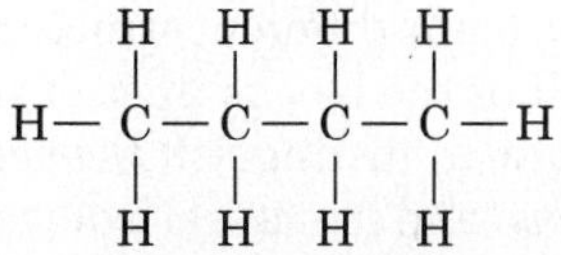	
n-Pentane	C_5H_{12}	36	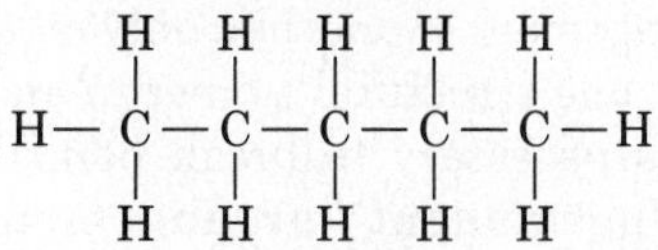	
n-Hexane	C_6H_{14}	69	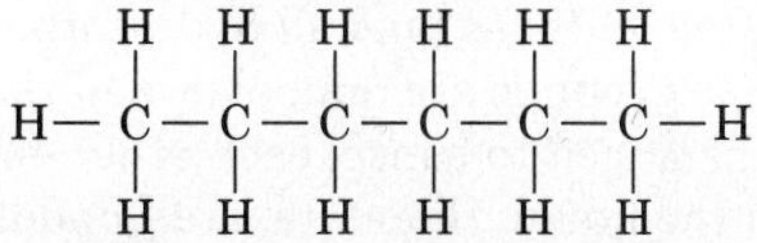	Some components of gasoline
n-Heptane	C_7H_{16}	98	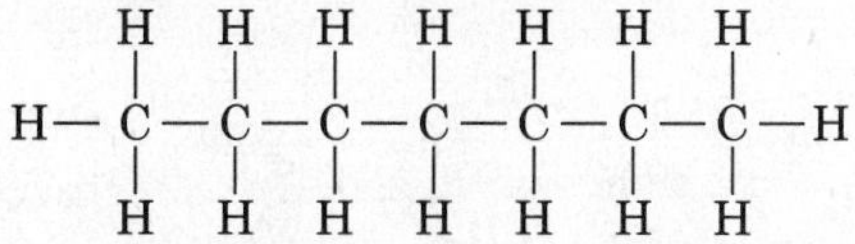	
n-Octane	C_8H_{18}	126	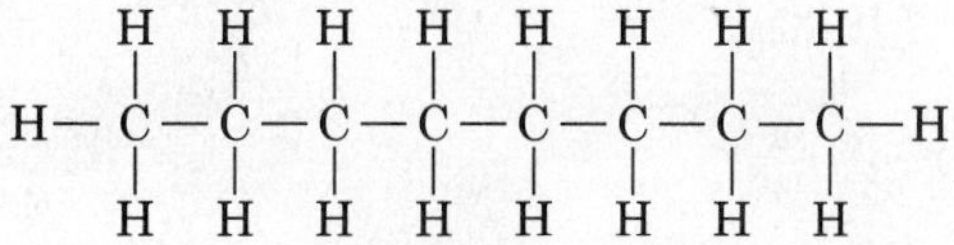	
n-Nonane	C_9H_{20}	151	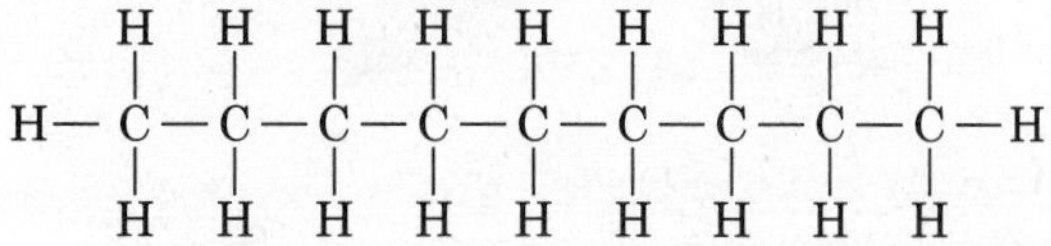	
n-Decane	$C_{10}H_{22}$	174	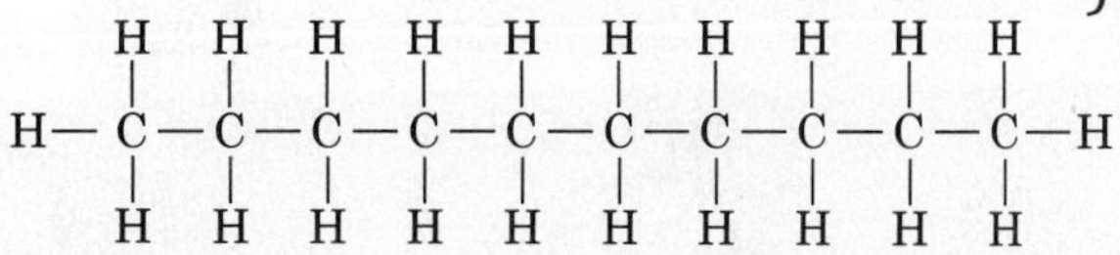	Found in kerosene

(Handwritten note across the structural formula column: "Memorize names, Formula, and structures for exams 2")

The two formulas represent two distinctly different compounds. Both are well known, each with its own particular set of properties. We must conclude, then, that molecular formulas such as C_4H_{10} are sometimes ambiguous and that structural formulas are necessary. Since these are different compounds, they have different names, whose derivation will be explained later in this chapter. However, all hydrocarbons that have four carbon atoms can be generally referred to as **butanes**.

If no carbon atom is attached to more than two other carbon atoms, the carbon chain is said to be a **straight-chain** structure. Actually, as shown in Figure 12-4, the carbon chain is bent (109.5°) at each carbon atom, but it is called a straight chain because the carbon atoms are bonded together in succession one after the other. You might note that many molecular shapes are possible for n-butane because of the possible rotational motions about the single bonds. These arrangements (called **conformations**) do not constitute different molecules. Because of the ease of bond rotation, a sample of a single pure hydrocarbon contains all of its conformations, that is, molecules rotated into all of their possible shapes.

In all branched-chain structures, at least one carbon atom is bonded to three or four other carbon atoms.

If one carbon atom is bonded to either three or four other carbon atoms in a molecule, the molecule is said to have a **branched chain.** Isobutane is an example of a branched-chain hydrocarbon (Fig. 12-4). Isobutane and n-butane are called **structural isomers** because both molecules contain exactly the same number and kinds of atoms, C_4H_{10}, but the molecules have different atom-to-atom bonding sequences. Structural isomerism can be compared to the results you might expect from a child building many different structures with the same collection of building blocks, and using all of the blocks in each structure.

Rotational forms resulting from the twisting around C—C single bonds are called conformations; they are not isomers.

It is important to distinguish between different bond rotational arrangements *(conformations)* and structural isomers *(configurations)*. To change from one rotational arrangement to another, only motion about a bond is required. However, to change from one structural isomer to another (for example, from isobutane to n-butane), it is necessary to break bonds and to form new ones. **Structural isomers are "permanent" arrangements; conformations are transient.**

The two butanes (and all hydrocarbon molecules) are essentially nonpolar, since the C—C bonds are nonpolar, and the slightly polar C—H bonds are symmetrically arranged to cancel each other out. The forces holding these molecules together in the liquid, therefore, are London forces, which depend on the surface

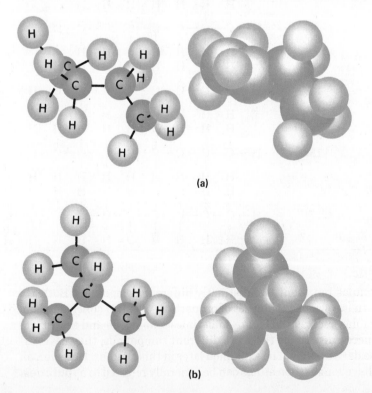

(a)

(b)

FIGURE 12-4 The isomeric butanes C_4H_{10}. (*a*) Normal butane, usually written n-butane. (*b*) Methyl-propane (isobutane).

area of a molecule and the closeness of approach of the molecules to each other. In general, a branched-chain isomer has a lower boiling point than a straight-chain isomer, since the branched-chain isomer does not permit intermolecular distances as short as those of a straight chain and has less surface area. Both of these mean less intermolecular attraction. Melting points of isomers generally do not follow the same pattern, since they also depend on the ease with which the molecules fit into a crystalline array.

Consider the isomeric pentanes, C_5H_{12}. There are three of these:

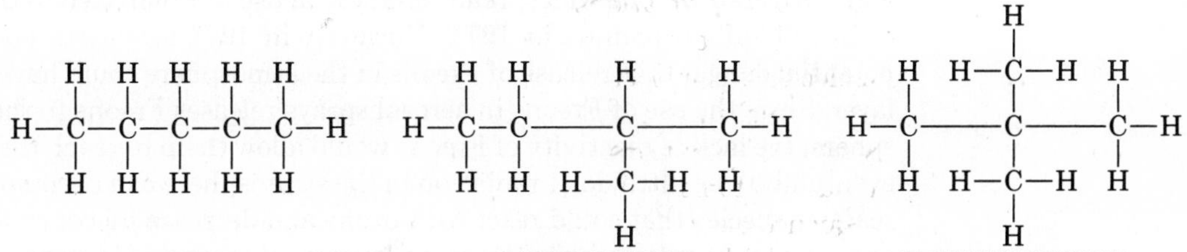

	n-PENTANE	2-METHYLBUTANE (ISOPENTANE)	2,2-DIMETHYLPROPANE (NEOPENTANE)
MELTING POINT	−130°C	−160°C	−17°C
BOILING POINT	36°C	28°C	9.5°C
DENSITY (at −20°C)	0.626 g/mL	0.62 g/mL	0.613 g/mL

All three isomers are predicted by bonding theory, and the theory predicts no other possible isomers for C_5H_{12}, since there are no other ways to unite the 17 atoms and have all valence electrons paired. This is the octet rule for carbon. These three isomers of pentane are well known, and no others have ever been found.

Table 12–2 gives the number of isomers predicted for some larger molecular formulas, starting with C_6H_{14}. Every predicted isomer, *and no more,* has been isolated and identified for the C_6, C_7, and C_8 groups. However, not all of the C_{15}'s and C_{20}'s have been produced, but there is sufficient belief in the theory to presume that if enough time and effort were spent, all of the isomers could eventually be produced. Structural isomerism certainly helps to explain the vast number of carbon compounds.

According to the octet rule, each atom (except H) shares or controls eight valence shell electrons. Hydrogen shares only two electrons.

SUBSTITUTION REACTIONS OF ALKANES Alkanes undergo **substitution reactions** in which the hydrogen atoms are replaced by other atoms. The reaction of methane with chlorine in the presence of light is an example of a substitution reaction.

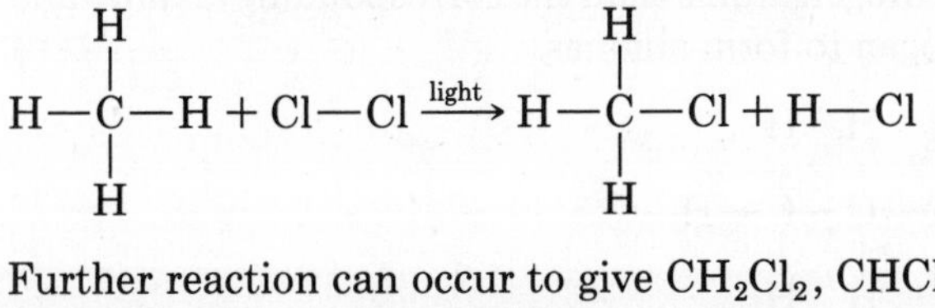

Further reaction can occur to give CH_2Cl_2, $CHCl_3$, or CCl_4.

TABLE 12–2 Structural Isomers of Some Hydrocarbons

FORMULA	ISOMERS PREDICTED	FOUND
C_6H_{14}	5	5
C_7H_{16}	9	9
C_8H_{18}	18	18
$C_{15}H_{32}$	4,347	—
$C_{20}H_{42}$	366,319	—
$C_{30}H_{62}$	4,111,846,763	—

Fluoride salts such as CoF_3 react with alkanes to give fully substituted hydrocarbons, known as perfluorocarbons. The high solubility of oxygen in perfluorocarbons has led to the use of perfluorocarbons as temporary blood substitutes.

Alkanes that contain both fluorine and chlorine are commonly known as Freons. Freons are unreactive, nontoxic, nonflammable, noncorrosive, odorless gases or liquids. These properties led to the use of Freon-11, CCl_3F, and Freon-12 (CCl_2F_2) as propellants in aerosol spray cans and as refrigerants in air conditioners and refrigerators. The widespread commercial use of Freons led to the production of over 1 billion pounds in 1974. However, in 1974 scientists pointed out the potential danger that release of Freons in the atmosphere could have on the ozone layer. Since the use of Freons in aerosol sprays releases Freons to the lower atmosphere, the lack of reactivity of Freons would allow them to reach the stratosphere eventually. The ultraviolet radiation in the stratosphere can decompose Freons to reactive species that could react with ozone and decrease its concentration in the ozone layer. Any decrease in the ozone layer would cause an increase in the ultraviolet radiation reaching the Earth's surface. Such a change would increase the incidence of skin cancer. As a result of these studies, the Environmental Protection Agency has limited the emissions of Freons.

> A detailed discussion of chlorofluorocarbons in the atmosphere is given in Chapter 18.

Alkenes

Molecules of alkenes have one or more carbon-carbon double bonds ($C=C$). The general formula for alkenes with one double bond is C_nH_{2n}. The first two members of the homologous alkene series are ethene (C_2H_4), and propene (C_3H_6) and their structural formulas are:

> Ethylene and propylene are common names that predate IUPAC nomenclature.

ETHENE OR ETHYLENE   PROPENE OR PROPYLENE

The structural formulas illustrate why alkenes are said to be **unsaturated hydrocarbons.** They contain fewer hydrogen atoms than the corresponding alkanes and can be made to react with hydrogen to form alkanes.

UNSATURATED ETHENE   SATURATED ETHANE

Ethene is the most important raw material used in the organic chemical industry. It ranks sixth in the top 50 chemicals (see table on inside of front cover) behind the inorganic chemicals sulfuric acid, nitrogen, ammonia, and lime and oxygen and is the number-one organic chemical. Over 30 billion pounds were produced in 1985 for use in making polyethylene, antifreeze (ethylene glycol), ethyl alcohol, and other chemicals.

In the United States ethene is produced by the **thermal cracking** of ethane, which is separated from natural gas.

$$C_2H_6 \xrightarrow{\text{heat}} H-\underset{\underset{}{|}}{\overset{\overset{H}{|}}{C}}=\underset{\underset{}{|}}{\overset{\overset{H}{|}}{C}}-H + H_2$$

In Europe and Japan, **catalytic cracking** of naphtha, a petroleum fraction, is the source of ethene since the natural gas in those countries contains relatively little ethane. Propene, the number-12 chemical, is produced by procedures similar to those used for ethene.

Naphtha is another name for the low boiling petroleum fraction of C_4 to C_{10} hydrocarbons (see Fig. 12–9).

Thermal cracking and catalytic cracking are discussed on p. 293.

Alkynes

The alkynes have one or more triple bonds ($-C\equiv C-$) per molecule and are very reactive unsaturated hydrocarbons. Alkynes have the general formula C_nH_{2n-2}. The simplest one is ethyne, commonly called acetylene (C_2H_2).

$$H-C\equiv C-H$$

A mixture of acetylene and oxygen burns with a flame hot enough to cut steel (3000°C).

ADDITION REACTIONS OF ALKENES AND ALKYNES Alkenes and alkynes are more reactive than alkanes because of the presence of double or triple bonds that can add halogens, hydrogen, or other molecules to form alkanes.

$$H-\overset{\overset{H}{|}}{C}=\overset{\overset{H}{|}}{C}-H + Br_2 \rightarrow H-\underset{\underset{Br}{|}}{\overset{\overset{H}{|}}{C}}-\underset{\underset{Br}{|}}{\overset{\overset{H}{|}}{C}}-H$$

ETHENE   1,2-DIBROMOETHANE

$$H-C\equiv C-H + 2\,HCl \rightarrow H-\underset{\underset{Cl}{|}}{\overset{\overset{H}{|}}{C}}-\underset{\underset{Cl}{|}}{\overset{\overset{H}{|}}{C}}-H$$

ETHYNE   1,2-DICHLOROETHANE

Addition reactions are used to prepare some polymers, an important class of compounds discussed in Chapter 14.

COMBUSTION: A COMMON REACTION FOR ALL HYDROCARBONS

Saturated and unsaturated hydrocarbons do undergo some similar reactions. One type shared by all hydrocarbons is **combustion reactions,** in which the hydrocarbon reacts with oxygen, that is, burns in air. Energy is produced, and if the amount of air is sufficient, the products are carbon dioxide (CO_2) and water (H_2O).

$$2\,CH_3-CH_3 + 7\,O_2 \rightarrow 4\,CO_2 + 6\,H_2O + 373 \text{ kcal/mole } C_2H_6$$
$$CH_2=CH_2 + 3\,O_2 \rightarrow 2\,CO_2 + 2\,H_2O + 337 \text{ kcal/mole } C_2H_4$$
$$2\,HC\equiv CH + 5\,O_2 \rightarrow 4\,CO_2 + 2\,H_2O + 311 \text{ kcal/mole } C_2H_2$$

MORE ON STRUCTURAL ISOMERISM: ALKENES AND ALKYNES

Structural isomers can also exist in molecules with double and triple carbon-carbon bonds. For example, two of the six isomers of C_4H_8, 1-butene and *trans*-2-butene, have the following structures:

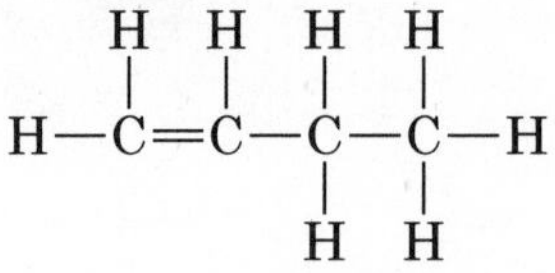

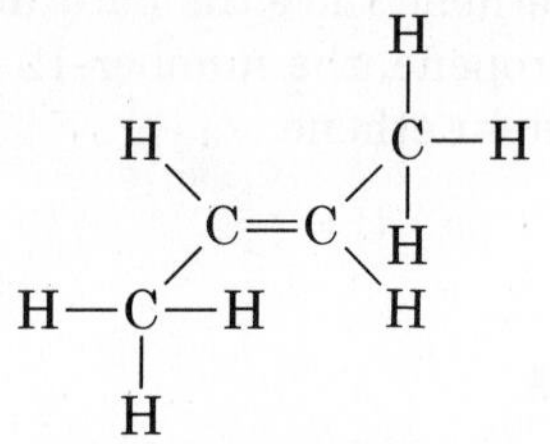

	1-BUTENE	TRANS-2-BUTENE
MELTING POINT	−185.4°C	−106.0°C
BOILING POINT	−6.3°C	1.0°C
DENSITY (AT −20°C)	0.641 g/ml	0.649 g/ml

The number placed before the name butene indicates the position number of the double bond. *Trans*-2-butene is an example of a geometric isomer that results from double bonds within a molecule. Geometric isomers will be considered later. Note that the properties of 1-butene and 2-butene definitely indicate two *different* compounds. 1-butyne and 2-butyne illustrate structural isomerism due to the positioning of triple bonds.

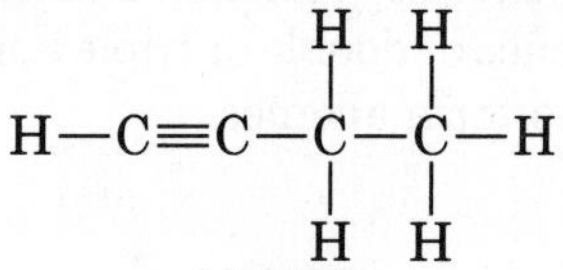

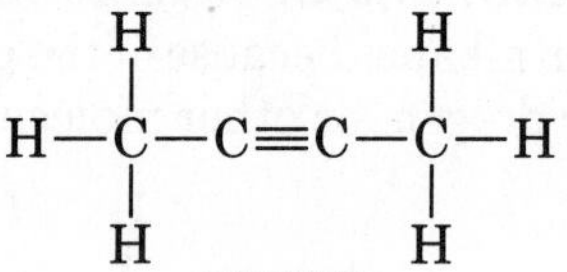

	1-BUTYNE	2-BUTYNE
MELTING POINT	−125.8°C	−32.2°C
BOILING POINT	8.1°C	27°C
DENSITY (at −20°C)	0.65 g/ml	0.69 g/ml

NOMENCLATURE — A SYSTEM OF NAMES

With so many organic compounds, a system of common names quickly fails owing to the shortage of unique names. As organic chemistry grew in complexity, a system of nomenclature developed that made use of numbers as well as names. Much attention has been given to the problems of naming organic compounds, and several international conventions have been held to work out a satisfactory system that can be used throughout the world. The International Union of Pure and Applied Chemistry has given its approval to a very elaborate nomenclature system (*IUPAC* system), which is now in general use.

A few of these IUPAC names will be needed for our discussion, and an appreciation of the basic simplicity of the approach in naming organic compounds is desirable. Inscrutable names, such as some of those encountered on medicine bottles, are replaced by systematic names that are descriptive of the molecules involved. The names of a few hydrocarbons are presented in Table 12–1.

Branched-Chain Hydrocarbons

For branched-chain hydrocarbons, it becomes necessary to name submolecular groups. The —CH_3 group is called the methyl group; this name is derived from

Can you draw another isomer of C_4H_8? (There are four others.)

The -*yl* ending indicates an attached group such as —CH_3, methyl.

methane by dropping the *-ane* and adding *-yl*. Any of the nine other hydrocarbons listed in Table 12–1 can give rise to a similar group. For example, the propyl group would be $—C_3H_7$. As an illustration of the use of the group names, consider this formula:

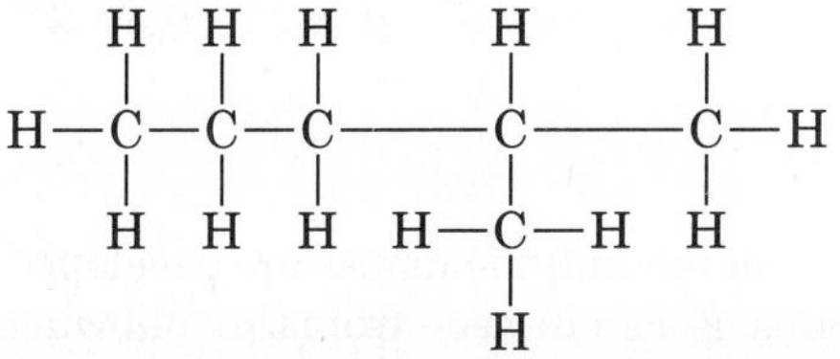

The *longest* carbon chain in the molecule is five carbon atoms long; hence, the root name is pentane. Furthermore, it is a methylpentane (written as one word) because a methyl group is attached to the pentane structure. In addition, a number is needed because the methyl group could be bonded to either the second or third carbon atom.

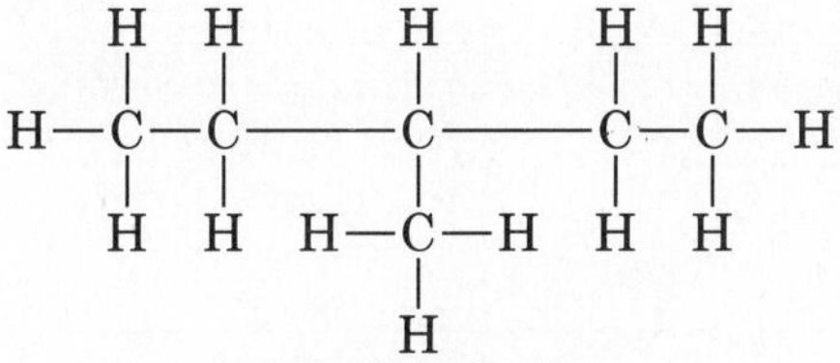

3-METHYLPENTANE 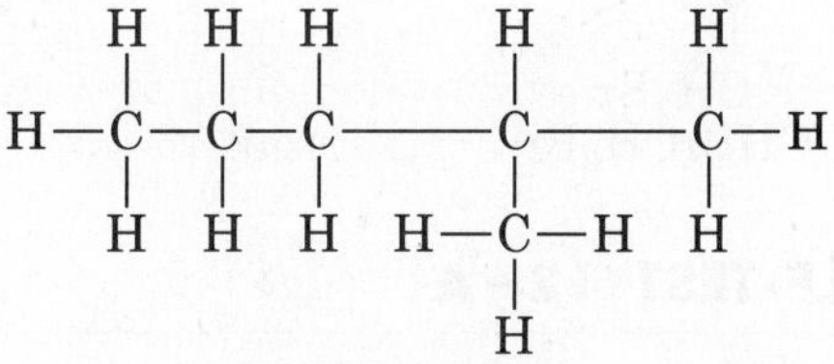   2-METHYLPENTANE

Note that 2-methylpentane is the same as 4-methylpentane since the latter would be the same molecule turned around; the accepted rule requires numbering from the end of the carbon chain that will result in the smallest numbers. Therefore, 2-methylpentane is the correct name.

Any number of substituted groups can be handled in this same fashion. Consider the name and the formula for 3,3,4,6-tetramethyl-5,5-diethyloctane.

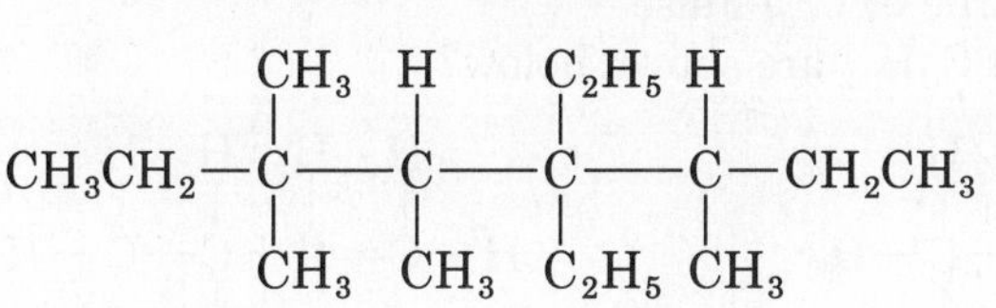

Alkenes

If a double bond appears in a hydrocarbon, then the root name, which indicates the number of carbon atoms, must be modified to reflect the double bond structure and its position. Changing *-ane* to *-ene* indicates the presence of the double bond, and a number is used to indicate its position. For example:

$CH_2{=}CH_2$ ETHENE (COMMON NAME: ETHYLENE)
$CH_2{=}CHCH_3$ PROPENE
$CH_2{=}CHCH_2CH_3$ 1-BUTENE
$CH_3CH{=}CHCH_3$ 2-BUTENE

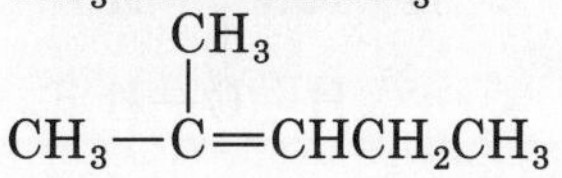

$$CH_3{-}\overset{\overset{\textstyle CH_3}{|}}{C}{=}CHCH_2CH_3$$ 2-METHYL-2-PENTENE

Margin notes:

$—CH_3$ is the same as

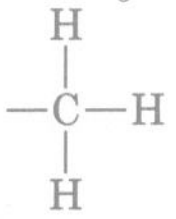

An *aliphatic compound* is any hydrocarbon that has an open chain of single-bonded carbon atoms.

Position of group on chain
↓
2-*methylpentane*
↑   ↖
group   longest
attached   chain of
to chain   C atoms

Positional numbers of groups should have the smallest possible sum.

No number is necessary for ethene or propene because there is only one possible position for the double bond.

Alkynes

If a triple bond is present, the *-ane* is changed to *-yne*. Examples:

$H—C≡C—H$ ETHYNE (COMMON NAME: ACETYLENE)
$CH_3CH_2C≡CH$ 1-BUTYNE
$CH_3—C≡C—CH_3$ 2-BUTYNE

Other Groups

When other groups are present in an organic compound, the names are developed on the same basis as those of the hydrocarbons, as can be seen from the following examples:

In formulas such as $CH_3CH_2CH_2Br$, elements directly following a carbon are bonded to that carbon. $CH_3CH_2CH_2Br$ is the same as

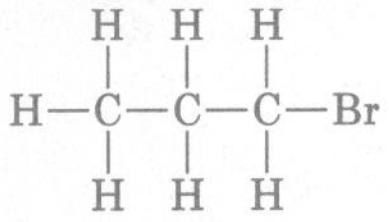

$$H—\overset{\displaystyle H}{\underset{\displaystyle H}{C}}—\overset{\displaystyle H}{\underset{\displaystyle H}{C}}—\overset{\displaystyle H}{\underset{\displaystyle H}{C}}—Br$$

$CH_3CH_2CH_2Br$ 1-BROMOPROPANE
$CH_3CHBrCH_3$ 2-BROMOPROPANE
$BrC≡CH$ BROMOETHYNE

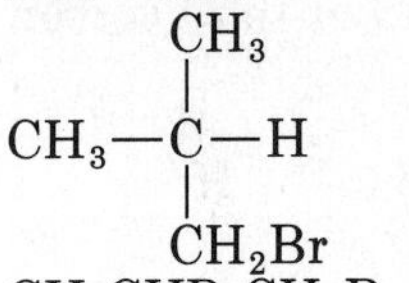

$$CH_3—\overset{\displaystyle CH_3}{\underset{\displaystyle CH_2Br}{C}}—H$$ 1-BROMO-2-METHYLPROPANE

$CH_3CHBrCH_2Br$ 1,2-DIBROMOPROPANE

SELF-TEST 12 – A

1. The branch of chemistry that deals with compounds of carbon is known as _____ chemistry.

2. The structure of the CH_4 molecule is described as _____.

3. How many covalent bonds are in a molecule of butane? _____ Does it make a difference which butane is considered? _____

4. A straight-chain hydrocarbon, such as pentane, actually has all of its carbon atoms in a straight line. ( ) True or ( ) False

5. How many different isomers of C_5H_{12} are shown below?

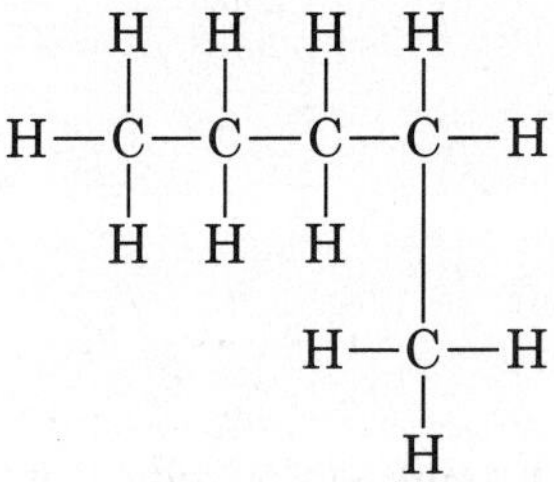

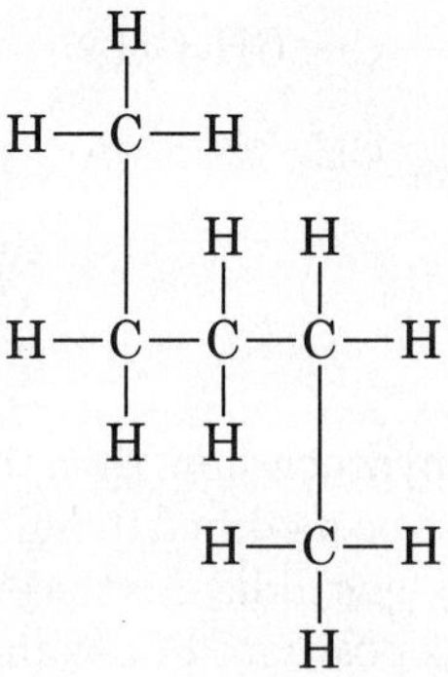

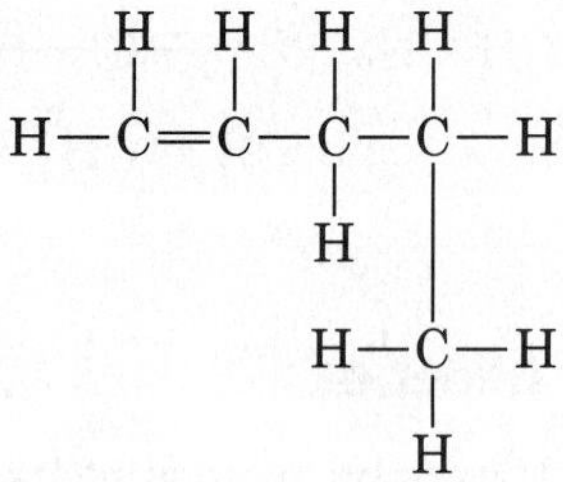

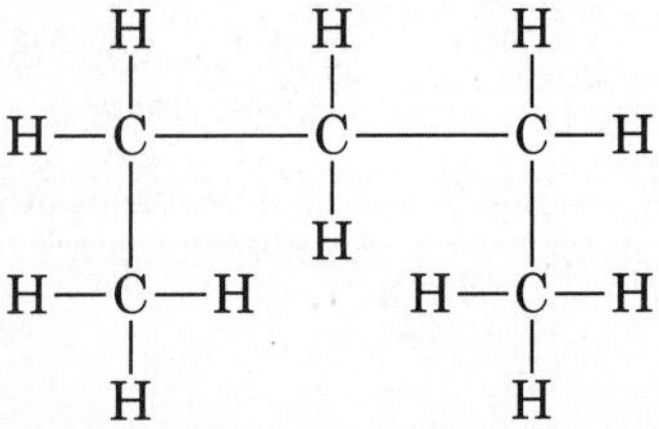

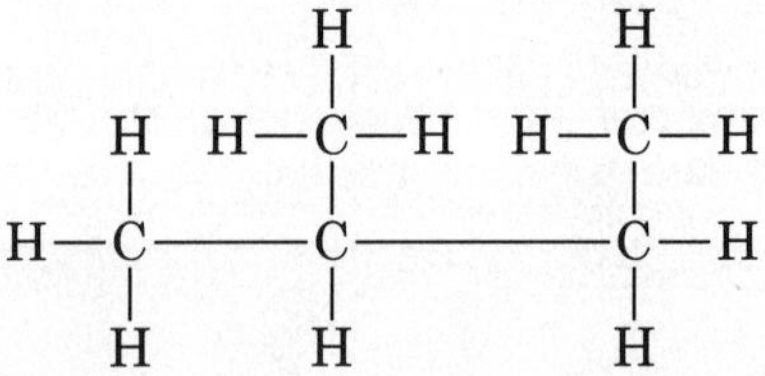

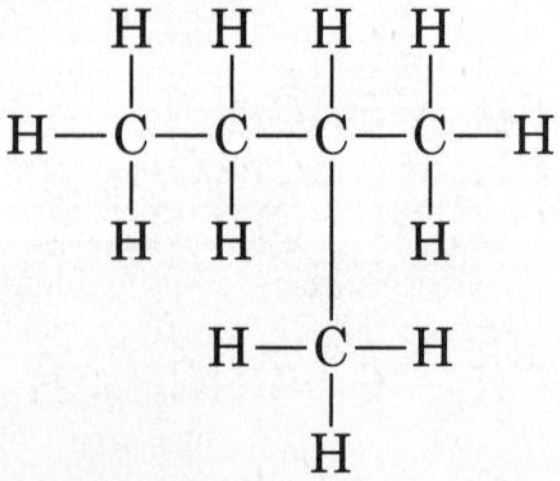

6. When the name of a compound ends in *-ene* (for example, *butene*), what structural feature is indicated? _____

7. Name the compound shown on the right:

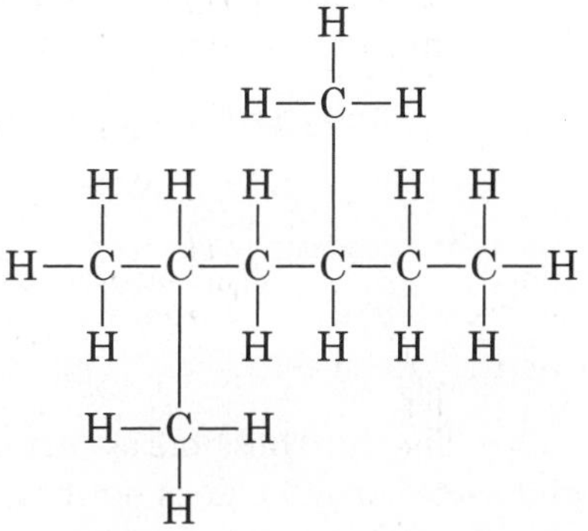

$$
\begin{array}{c}
\quad\quad\quad\quad\quad\quad\quad\quad H \\
\quad\quad\quad\quad\quad\quad H-C-H \\
\; H \;\; H \;\; H \quad\; | \quad H \;\; H \\
H-C-C-C-C-C-C-H \\
\; H \quad | \quad H \;\; H \;\; H \\
\quad\quad H-C-H \\
\quad\quad\quad | \\
\quad\quad\quad H
\end{array}
$$

8. The formula for the ethyl group is _____ .

MATCHING SET (Multiple answers are possible)

_____ **1.** Alkane

_____ **2.** Alkene

_____ **3.** Alkyne

a. Most unsaturated hydrocarbons bond
b. Will not undergo addition reaction
c. Contains carbon-carbon double bonds
d. Will react with Br_2 in an addition reaction
e. Will react with O_2 to produce heat

GEOMETRIC ISOMERS IN ALKENES

In the description of chain and branch structural isomers, it was pointed out that in order to change one *structural* isomer into another, at least two carbon atoms have to change the atoms to which they are bonded. However, it is possible for some sets of atoms to form two isomeric molecules, both of which have the same atoms bonded to each other. One type of this kind of isomerism is **geometric isomerism.**

Where carbon-carbon double bonds exist in a molecule, geometric isomerism is possible. A double bond between carbon atoms does not allow free rotation, and this lack of rotation provides a structural basis for geometric isomerism.

Consider the compound ethene, C_2H_4. Its six atoms lie in the same plane, with bond angles of approximately 120 degrees:

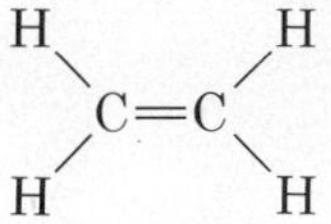

$$
\begin{array}{c}
H \quad\quad\; H \\
\backslash \quad\quad / \\
C=C \\
/ \quad\quad \backslash \\
H \quad\quad\; H
\end{array}
$$

If two chlorine atoms replace two hydrogen atoms, one on each carbon atom of ethene ($H_2C=CH_2$), the result is $CHCl=CHCl$. Experimental evidence confirms the existence of two compounds with this general arrangement. If the two chlorine atoms are close together, this is characteristic of one isomer (the **cis** isomer), and if they are far apart, another isomer (the **trans** isomer) is indicated. Both compounds

Stereoisomers have the same atoms and the same bonds, but the atoms are arranged in space differently.

To have geometric isomers, each carbon connected by the double bond must have two unlike groups attached.

are called 1,2-dichloroethene (the 1 and 2 indicate that the two chlorine atoms are attached to different carbon atoms). The two arrangements are distinguished from each other by the prefixes *cis* and *trans*. Note that the two isomeric compounds have significant differences in their properties.

	CIS-1,2-DICHLOROETHENE	*TRANS*-1,2-DICHLOROETHENE
MELTING POINT	−80.5°C	−50°C
BOILING POINT	60.1°C	48.4°C
DENSITY (AT 15°C)	1.291 g/ml	1.265 g/ml

The third possible isomer, 1,1-dichloroethene (a structural isomer of the *cis* and *trans* isomers), does not have *cis* and *trans* structures.

As a general rule, *trans* isomers have higher melting points than *cis* isomers because of the greater ease with which the *trans* molecules can fit into its solid structure and form strong intermolecular bonds.

Draw the structural isomer 1,1-dichloroethane.

When there is a carbon-carbon double bond in an organic molecule, the possibility exists for *cis* and *trans* isomers. Sometimes several such bonds can be found in the same molecule, giving rise to numerous isomeric compounds.

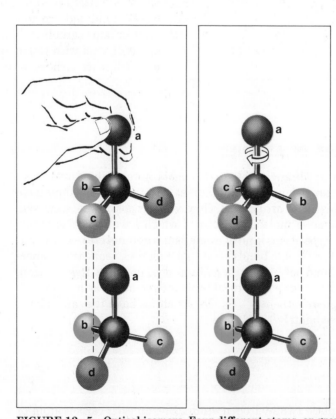

FIGURE 12–5 Optical isomers. Four different atoms, or groups of atoms, are bonded to tetrahedral center atoms so that the upper isomeric form cannot be turned in any way and exactly match the lower structure. The upper structure and the lower structure are nonsuperimposable mirror images. (See also Fig. 12–6.)

OPTICAL ISOMERS

Optical isomerism is possible when a molecular structure is **asymmetric** (without symmetry). One common example of an asymmetric molecule is one containing a tetrahedral carbon atom bonded to four *different* atoms or groups of atoms. Such a carbon atom is called an asymmetric carbon atom; an example is the carbon atom in the molecule CBrClIH.

Figure 12–5 shows the two ways to arrange four different atoms in the tetrahedral positions about the central carbon atom. These result in two nonsuperimposable, mirror-image molecules that are optical isomers.

There are many examples of nonsuperimposable mirror images in the macroscopic world. Consider your hands or right- and left-hand gloves, for instance. They are mirror images of one another and are nonsuperimposable.

All amino acids except glycine can exist as one of two optical isomers. In Figure 12–6, the mirror image relationship is shown for optical isomers of alanine, an amino acid with a tetrahedral carbon atom surrounded by an amino group (—NH$_2$), a methyl group (—CH$_3$), an acid group (—COOH), and a hydrogen atom. Note that the carbon atoms in the methyl and acid groups are not asymmetric since these atoms are not bonded to four different groups.

The properties of some optical isomers are almost identical. Different compounds whose molecules are mirror images of one another have the same melting point, the same boiling point, the same density, and many other identical physical and chemical properties. However, they always differ in one physical property: they rotate the plane of **polarized** light in opposite directions. According to the wave theory of light, a light wave traveling through space vibrates at right angles to its path (Fig. 12–7). A group of such rays traveling together vibrate in random directions, all of which are at right angles to the path of travel. If such a group of waves is passed through a polarizing crystal, such as Iceland spar (a form of CaCO$_3$), or through a sheet of Polaroid material, the light is split into two rays and the waves emerging along the incoming axis will vibrate in only one plane perpendicular to

Optical isomers are another type of stereoisomers.

All amino acids have an amine group (—NH$_2$) and an acid group (—COOH).

The formula of glycine is H$_2$NCH$_2$COOH. Why doesn't glycine have optical isomers?

D- and L- simply indicate that two structures are possible around an asymmetric C atom. The D- and L- notations do not indicate which way the substance will rotate the plane-polarized light.

$$\begin{array}{c} \text{COOH} \\ | \\ \text{C} \\ \text{H} \diagup \ | \diagdown \text{OH} \\ \text{CH}_3 \end{array}$$
D-LACTIC ACID

$$\begin{array}{c} \text{COOH} \\ | \\ \text{C} \\ \text{HO} \diagup \ | \diagdown \text{H} \\ \text{CH}_3 \end{array}$$
L-LACTIC ACID

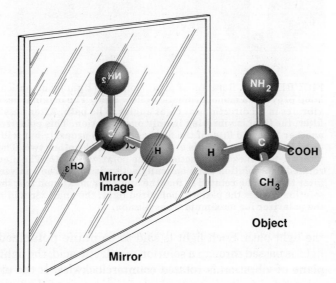

FIGURE 12–6 Optical isomers of the amino acid, alanine, 2-amino-propionic acid.

$$\begin{array}{c} \text{COOH} \\ | \\ \text{H}_2\text{N}-\text{C}-\text{H} \\ | \\ \text{CH}_3 \end{array}$$

The D-form is the non-superimposable mirror image of the L-form. (See also Fig. 12–5).

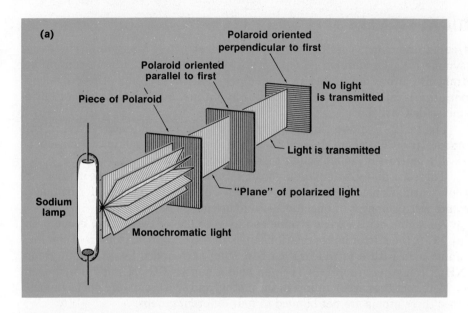

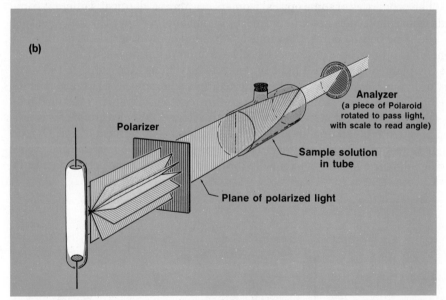

FIGURE 12–7   Rotation of plane-polarized light by an optical isomer. (*a*) A sodium
lamp provides a monochromatic yellow light. The original beam is nonpolarized; it
vibrates in all directions at right angles to its path. After passing through a Polaroid
filter, the light is vibrating in only one direction. This polarized light will pass through
another Polaroid filter if the filter is lined up properly but will not pass through the
third Polaroid filter if it is at right angles to the other two. The direction of the Polaroid
filters determines the direction of the polarization. (*b*) The plane of polarized light is
rotated by a solution of an optically active isomer. The analyzer can be a second Polaroid
filter that can be rotated to find the angle for maximum transmission of light. If the
solution rotates the plane of polarized light, the analyzer will not be at the same angle as
the polarizer for maximum transmission.

the light path. Such light is said to be **plane polarized.** When plane-polarized
light is passed through a solution of D-lactic acid, the light is still polarized, but the
plane of vibration is rotated counterclockwise. If the other lactic acid isomer is
substituted (L-lactic acid), clockwise rotation of the light is obtained.

Optical isomers can also differ in biological properties. An example is the hormone adrenalin (or epinephrine). Adrenalin is one of a pair of optical isomers. C* designates the asymmetric carbon atom. Only the isomer that rotates plane-polarized light to the left is effective in starting a heart that has stopped beating momentarily, or in giving a person unusual strength during times of great emotional stress. The other isomer is inactive.

All optically active amino acids in proteins are left-handed (L-isomers). Nature's preference for left-handed amino acids has provoked much discussion and speculation among scientists since Pasteur's discovery of optical activity in 1848 from studies of crystals of tartaric acid salts. However, no satisfactory explanation has been found for this "handedness" of life.

Enzymes, the catalysts for biochemical reactions, also have a handedness and, like a glove, will bind to only one of the optical isomers. For example, during contraction of muscles the body produces only the L-form of lactic acid and not the D-form.

Large organic molecules may have many asymmetric carbon atoms within the same molecule. At each such carbon atom there exists the possibility of *two* arrangements of the molecule. The total number of possible molecules, then, increases exponentially with the number of asymmetric centers. With two asymmetric carbon atoms there are 2^2, or four, possible structures; for three, there are 2^3, or eight, possible structures. It should be emphasized that each of the eight isomers can be made from the *same* set of atoms with the *same* set of chemical bonds. Glucose, a simple blood sugar also known as dextrose, contains four asymmetric carbon atoms per molecule. Thus, there are 2^4 (16) isomers in the family of stereoisomers to which D-glucose belongs. However, of the 16 possible isomers, only 3 are important. These are D-glucose, D-mannose, and D-galactose. Of these, D-glucose is by far the most common isomer.

ADRENALIN
(EPINEPHRINE)

Is there another set of life forms in another setting that may be "right-handed"?

The concentration of lactic acid in the blood is associated with the feeling of tiredness, and a period of rest is necessary to reduce the concentration of this chemical by oxidation.

D-GLUCOSE (C* = ASYMMETRIC CARBON ATOM)

D-MANNOSE

D-GALACTOSE

THE CYCLIC HYDROCARBONS

Hydrocarbons can form rings as well as straight chains and branched chains. The cyclic hydrocarbons include cycloalkanes (all single bonds), cycloalkenes (one carbon-carbon double bond), cycloalkynes (one carbon-carbon triple bond), and the aromatics (a unique combination of single bonds and electron delocalization around the ring).

Cycloalkanes

The simplest cycloalkane is cyclopropane, a highly strained ring compound:

Symbols like △ are just more chemical shorthand.

The bonds are strained because of the 60-degree angles in the ring; angles above 90 degrees show a much greater stability. Cyclopropane is an anesthetic. However, cyclopropane is highly flammable, and extreme caution must be taken when cyclopropane is used in surgery. Cyclopropane is an isomer of C_3H_6, an alkene. Although isomers of alkenes, the cyclic isomer is called a cycloalkane because all the bonds in the molecule are single bonds.

Some cyclic alkanes will add atoms in a similar way to the alkenes.

     The cycloalkanes are commonly represented by a polygon. Each corner represents a carbon atom, and the lines represent C—C bonds. The C—H bonds are not shown but are understood. Other common homologous cycloalkanes include cyclobutane, cyclopentane, and cyclohexane. These are represented as

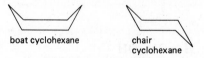

cyclobutane   cyclopentane   cyclohexane

Draw the structural formula for cyclohexane, showing all of the atomic symbols and bonds.

Cyclohexane exists in two conformations referred to as the "boat" and the "chair" forms. Since the end groups are farther apart in the chair form, steric (space) repulsions between the end groups will be less than in the boat form. As a result, the chair form is more stable. Cyclohexane is the prototype of the six-membered ring found in glucose and other sugars (Chapter 15).

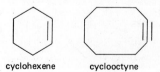

boat cyclohexane   chair cyclohexane

Cycloalkenes and Cycloalkynes

Examples of the unsaturated cyclic hydrocarbons include cyclohexene, which is used as a stabilizer in high-octane gasoline, and cyclooctyne, a synthetic achievement with no known practical use.

cyclohexene   cyclooctyne

Aromatic Compounds

All of the hydrocarbons we have discussed up to this point have localized electronic structures; that is, the bonding electrons are essentially fixed between two atomic centers as in C—C or C=C bonds. For a large group of organic compounds known as **aromatic** compounds, this type of complete electron localization is not found. Rather, these compounds have some of their bonding electrons spread over several atoms (delocalized) or even the entire molecule, a feature that leads to some interesting chemical properties.

The simplest aromatic compound is benzene (C_6H_6). The molecular structure of benzene is a ring of carbon atoms in a plane, with one hydrogen atom bonded to each carbon atom. The bonds between these carbon atoms are shorter than single bonds, but longer than double bonds. The measured bond angles are 120 degrees. The benzene structure is sometimes written with alternating double bonds.

Benzene is the number 16 commercial chemical.

However, the six electrons in the three double bonds are not localized but spread over the ring. Figure 12–8 illustrates the delocalization of six electrons above and below the plane of the ring. In other words, all six C—C bonds are equivalent, and a better representation of benzene is:

or

when hydrogen atoms are not shown. The delocalization of electrons around the ring accounts for the greater stability of aromatic compounds relative to that of unsaturated compounds, which contain double or triple bonds.

Many aromatic compounds related to benzene contain various atoms, hydrocarbon subgroups, and functional groups replacing ring hydrogen atoms. These compounds exhibit a wide variety of properties and differ greatly in their chemical reactivity. Some interesting examples of isomerism are also possible. Consider the

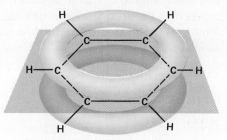

FIGURE 12–8 Bonding in an aromatic compound, benzene (C_6H_6).

Delocalized electrons above and below ring

three different compounds with the formula C_8H_{10} found in some coal tars. Several names are given to these isomers:

Xylenes rank number 26 among commercial chemicals.

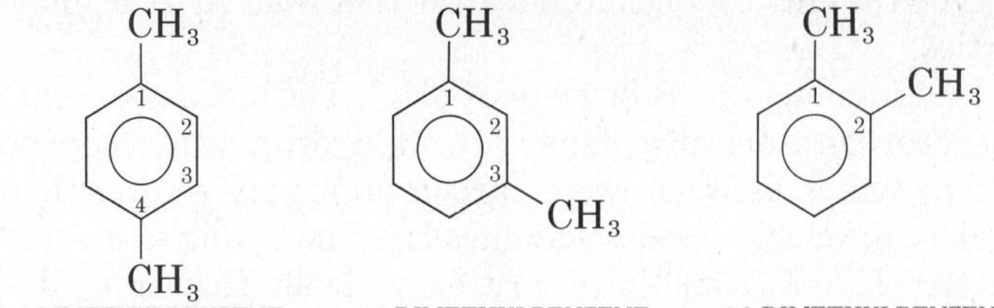

1,4-DIMETHYLBENZENE (*PARA*-XYLENE) mp 13.3°C

1,3-DIMETHYLBENZENE (*META*-XYLENE) mp −47.9°C

1,2-DIMETHYLBENZENE (*ORTHO*-XYLENE) mp −25°C

Each of these isomers has two methyl groups substituted for hydrogen atoms on the ring. The prefixes ***para-***, ***meta-***, and ***ortho-*** are used if there are only two groups on the benzene ring.

If more than two groups occur, a number system is most useful. Consider the following compounds:

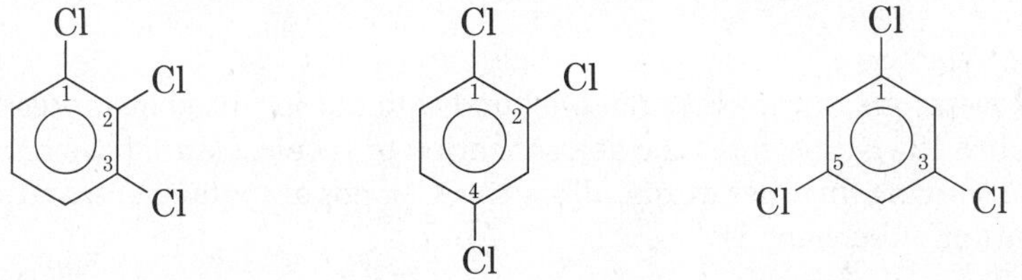

1,2,3-TRICHLOROBENZENE 1,2,4-TRICHLOROBENZENE 1,3,5-TRICHLOROBENZENE

There are no other ways of drawing these three isomers, and only three trichlorobenzenes have been isolated in the laboratory.

Nomenclature in aromatic compounds:
Ortho- groups are on adjacent carbon atoms (1,2 positions)
Meta- groups have one carbon atom between them (1,3 positions)
Para- groups have two carbon atoms between them (1,4 positions)

Some rings have nitrogen, oxygen, or sulfur atoms in place of a few carbon atoms in the ring. Only aromatic ring structures incorporating nitrogen will be encountered later in the text. Examples are pyridine and pyrimidine.

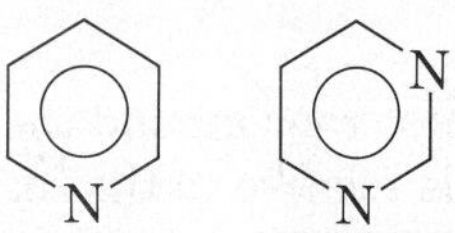

PYRIDINE PYRIMIDINE

SUBSTITUTION REACTIONS OF AROMATIC COMPOUNDS When the aromatic compound benzene reacts with chlorine in the presence of a suitable catalyst such as iron (III) chloride, a **substitution reaction** takes place in which one of the benzene hydrogen atoms is replaced by a chlorine atom. The mechanism for this reaction is similar to that of addition, except that a positively charged chlorine ion reacts with the aromatic benzene ring, which is electron rich. A positive hydrogen ion leaves the aromatic ring and becomes associated with the negative chloride ion as HCl.

$$Cl_2 + FeCl_3 \rightarrow FeCl_4^- + Cl^+$$

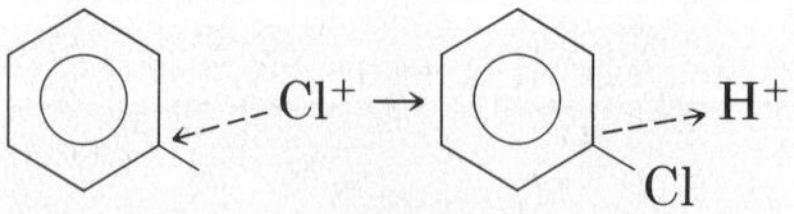

$$H^+ + FeCl_4^- \rightarrow FeCl_3 + HCl$$

The overall reaction can be written as:

$$\text{BENZENE} + Cl_2 \xrightarrow[\text{(catalyst)}]{FeCl_3} \text{CHLOROBENZENE} + HCl$$

Chlorobenzene is manufactured on an enormous scale and is used widely in industrial processes to prepare other organic compounds, such as *para*-dichlorobenzene, which is used with naphthalene in mothballs.

WHERE DO HYDROCARBONS COME FROM?

Complex mixtures of hydrocarbons, compounds containing only carbon and hydrogen, occur in enormous quantities in nature as natural gas, petroleum, and coal. These materials were formed from organisms that lived millions of years ago. After their death, they became covered with layers of sediment and ultimately were subjected to high temperatures and pressures in the depths of the Earth's crust. In the absence of free oxygen, these conditions converted once-living tissue into petroleum and coal. Hence, they are known as *fossil fuels*. After petroleum and natural gas are brought to the surface of the Earth, they can be separated into various fractions with different boiling points by the use of fractional distillation (Fig. 12–9). Once distilled, the fractions are refined further and made to undergo various types of chemical change in order to make desirable substances.

By heating coal or wood in the absence of air (destructive distillation), coal tar can be separated from the coke or charcoal (carbon). A ton of a typical soft coal produces about 140 pounds of coal tar, which is about one half pitch, the rest being composed of such aromatics as naphthalene, benzene, phenol, cresols, toluene, and xylenes.

The naturally occurring compounds not only are important in themselves but also serve as starting materials for making numerous organic compounds that do not occur in nature. From the simplest hydrocarbons come such diverse consumer products as plastic dishes, acrylic and polyester fibers for textiles, vinyl and latex paints, neoprene rubber, and such industrial products as Teflon and cattle feed.

Coal is one natural source of aromatic compounds.

Natural Gas

The natural gas found in North America is a mixture of methane (60–80%), ethane (5–9%), propane (3–18%), and butane and pentane (2–14%). However, in Europe and Japan the natural gas is essentially all methane. This difference affects how natural gas is used in different parts of the world. In either case, the natural gas is an important fuel.

Energy is obtained by burning the constituents of natural gas in air:

$$CH_4 + 2\ O_2 \rightarrow CO_2 + 2\ H_2O + 213\ \text{kcal/mole}\ CH_4$$
$$2\ C_2H_6 + 7\ O_2 \rightarrow 4\ CO_2 + 6\ H_2O + 372.8\ \text{kcal/mole}\ C_2H_6$$

The energy released in these reactions can be used to heat homes, run electric power plants, or power special internal combustion engines.

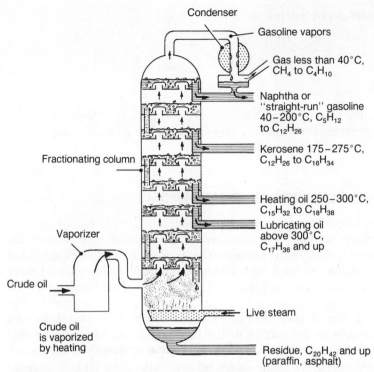

FIGURE 12–9   A diagram of a fractionating column for distilling petroleum. Boiling range and composition of each fraction are given to the right of the column. Notice that the higher boiling substances condense at the lower levels, and the lower boiling substances do not condense until the higher, cooler levels. (From J. I. Routh, D. P. Eyman, and D. J. Burton: *A Brief Introduction to General, Organic and Biochemistry.* **Philadelphia, Saunders College Publishing, 1971.)**

"Bottled gas" is actually stored in the liquid state under pressure. In order for bottled gas to be burned in stoves or furnaces, it must be vaporized.

In North America, propane and butane, the principal components of bottled gas, can be separated from natural gas and liquefied by the use of moderate pressure and cooling. Because propane is more volatile than butane, bottled gas is prepared so that it contains more propane in colder climates and more butane in warmer climates.

Refining Petroleum

From the time petroleum was first discovered in the United States in 1859 until the automobile became popular in 1900, most oil was refined to yield kerosene, a mixture of hydrocarbons in the C_{12} to C_{16} range. This liquid was used principally as a fuel for lamps.

Our supplies of coal, and especially petroleum, are limited.

The internal combustion engines used in early automobiles were designed to burn a more volatile mixture of hydrocarbons, the C_6 to C_{10} fraction, which became known as **gasoline.** The vapors of these lower molecular weight hydrocarbons mix readily with air in simple carburetors and burn fairly completely.

Mixtures of hydrocarbons in the C_5 to C_{12} range are known as *gasolines.*

With the increasing popularity of the automobile, petroleum refiners had to shift the output of a barrel of crude oil from a reasonably large fraction of kerosene to almost no kerosene and a much greater fraction of gasoline (Table 12–3). This dramatic increase in the amount of gasoline from a barrel of crude oil was accom-

TABLE 12–3   Division of a Barrel of Crude Oil

	1920 %	1967 %	1984 %
Gasoline	26.1	44.8	42.7
Kerosene	12.7	2.8	0.9
Jet fuel	—	7.6	7.5
Heavy distillates	48.6	22.2	19.1
Other (asphalt, road oil, residual fuel oil, etc.)	12.6	22.6	29.8

plished by the discovery of chemical processes that convert nongasoline molecules into ones that burn well in an automobile engine.

A barrel of crude oil is 42 gallons.

Thermal cracking involves heating saturated hydrocarbons under pressure in the absence of air. The hydrocarbons break into shorter-chain hydrocarbons—both alkanes and alkenes, some of which will be in the gasoline range.

Cracking breaks larger molecules into smaller ones.

$$C_{16}H_{34} \xrightarrow[\text{heat}]{\text{pressure}} C_8H_{18} + C_8H_{16}$$

AN ALKANE   AN ALKANE   AN ALKENE
IN THE GASOLINE RANGE

Later, **catalytic cracking** was carried out in units such as the one shown in Figure 12–10. These units use catalysts, which allow the processes to proceed at lower pressures and result in even higher yields of gasoline. Today, these catalysts

A catalyst increases the speed of a reaction without being consumed in the reaction.

FIGURE 12–10   Catalytic cracking unit. (Courtesy of Ashland Oil Company.)

include specially processed clays. Refiners of petroleum use their own special methods, which offer different advantages in cost and type of crude oil handled. The hydrocarbon molecules produced are much the same regardless of the methods used.

Cracking of petroleum fractions brought about not only an increase in the quantity of gasoline available from a barrel of crude oil, but also an increase in *quality*. That is, gasoline from a cracking process can be used at higher efficiency (in a high-compression engine) than can "straight run" gasoline, because the molecular structures of the hydrocarbons in the cracked gasoline allow them to oxidize more smoothly at high pressure. When the burning of the gasoline-air mixture is too rapid or irregular, ignition occurs in the combustion chamber too early, resulting in a small detonation, which is heard as a "knock" in the engine. This knocking can spell trouble to the owner of an automobile because it will eventually lead to the breakdown of the internal parts of the automobile's engine.

Knocking in gasoline
engines is a sign of
improper combustion.

An arbitrary scale for rating the relative knocking properties of gasolines has been developed. Normal heptane, typical of straight-run gasoline, knocks considerably and is assigned an octane rating of 0:

$$CH_3CH_2CH_2CH_2CH_2CH_3 \quad \text{(octane rating} = 0)$$
n-HEPTANE

whereas 2,2,4-trimethylpentane (isooctane) is far superior in this respect and is assigned an octane rating of 100:

$$
\begin{array}{ccccc}
& CH_3 & & CH_3 & \\
& | & & | & \\
CH_3- & C & -CH_2- & C & -CH_3 \quad \text{(octane rating} = 100)\\
& | & & | & \\
& CH_3 & & H &
\end{array}
$$
2,2,4-TRIMETHYLPENTANE

The octane rating of a gasoline is determined by first using the gasoline in a standard engine and recording its knocking properties. This is compared to the behavior of mixtures of n-heptane and isooctane, and the percentage of isooctane in the mixture with identical knocking properties is called the octane rating of the gasoline. Thus, if a gasoline has the same knocking characteristics as a mixture of 9% n-heptane and 91% isooctane, it is assigned an octane rating of 91. This corresponds to a regular grade of gasoline. Since the octane rating scale was established, fuels superior to isooctane have been developed, so the scale has been extended well above 100.

The octane scale measures
the ability of a mixture to
burn without knocking in
a gasoline engine.

Table 12–4 lists octane ratings for some hydrocarbons and octane enhancers. The "straight-run" gasoline fraction obtained from the fractional distillation of petroleum has an octane rating of 50 to 55, which is too low for use as a fuel in vehicles. From Table 12–4 we can see that the octane rating of a gasoline can be increased either by increasing the percentage of branched chain and aromatic hydrocarbon fractions or by adding octane enhancers (or a combination of both).

The **catalytic reforming** process is used to produce branched chain and aromatic hydrocarbons. Under the influence of certain catalysts, such as finely divided platinum, straight-chain hydrocarbons with low octane numbers can be re-formed into their branched-chain isomers, which have higher octane numbers.

$$CH_3CH_2CH_2CH_2CH_3 \xrightarrow[\text{heat}]{\text{platinum}} CH_3CH_2CHCH_3$$
$$\hspace{6cm} |$$
$$\hspace{6cm} CH_3$$

n-PENTANE   2-METHYLBUTANE
OCTANE NUMBER 62   94

TABLE 12–4   Octane Numbers of Some Hydrocarbons and Gasoline Additives

NAME	OCTANE NUMBER
n-heptane	0
n-hexane	25
n-pentane	62
2,4-dimethylhexane	65
1-pentene	91
1-butene	97
Tertiary-butyl alcohol	98
2,2,4-trimethylpentane (isooctane)	100
Benzene, technical grade	106
Ortho-xylene	107
Ethanol	112
Methanol	116
Para-xylene	116
Meta-xylene	118
Toluene, technical grade	118

Catalytic reforming is also used to produce aromatic hydrocarbons such as benzenes, toluene, and xylenes by using different catalysts and petroleum mixtures. For example, when the vapors of naphtha, kerosene, and light oil fractions are passed over a copper catalyst at 650°C, a high percentage of the original material is converted into a mixture of aromatic hydrocarbons from which benzene, toluene, xylenes, and similar compounds may be separated by fractional distillation. For example, n-hexane is converted into benzene

The hydrogen produced here can be used in the synthesis of ammonia by the Haber process. (See Chapter 19.)

$$CH_3CH_2CH_2CH_2CH_2CH_3 \rightarrow \bighexagon + 4\ H_2$$

n-HEXANE   BENZENE

and n-heptane is changed into toluene.

$$CH_3CH_2CH_2CH_2CH_2CH_2CH_3 \rightarrow \overset{CH_3}{\bighexagon} + 4\ H_2$$

n-HEPTANE   TOLUENE

The octane number of a given blend of gasoline can also be increased by adding "antiknock" agents or octane enhancers. Prior to 1975, the most widely used antiknock agent was tetraethyllead, $(C_2H_5)_4Pb$. The addition of 3 g of $(C_2H_5)_4Pb$ per gallon increases the octane rating by 10 to 15, and before the Environmental Protection Agency (EPA) required reductions in lead content, both regular and premium gasoline contained an average of 3 g of $(C_2H_5)_4Pb$ or $(CH_3)_4Pb$ per gallon. To prevent lead deposits in the engine, 1,2-dibromoethane or 1,2-dichloroethane was added to leaded gasoline. These combined with lead to form lead halides, which were then emitted from the auto exhaust.

Lead compounds are extremely toxic (see Chapter 16), and the low boiling point of tetraethyllead constitutes an additional hazard. However, levels of lead in the environment have been reduced drastically as a result of two actions on the part of the EPA. First, the decision to use a platinum-based catalytic converter to reduce emissions of carbon monoxide and nitrogen oxides required lead-free gasolines, since lead will deactivate the platinum catalyst. Beginning in 1975, new automobiles were required to use lead-free gasoline to protect the catalytic converter and to decrease the amount of airborne lead.

EPA regulations initiated in the 1970s also required a gradual reduction in the maximum lead content of leaded gasoline. Scheduled reductions in the lead content of gasoline have led to as little as 0.1 g of lead per gallon, and the goal is lead-free gasoline.

With the decreased use of tetraethyllead, other octane enhancers are being added to gasoline to increase the octane rating. These include toluene, 2-methyl-2-propanol (also called **tertiary**-butyl alcohol), methyl-**tertiary**-butyl ether (MTBE), methanol, and ethanol. In 1985, the most popular octane enhancer was MTBE, which joined the top-50 chemical list for the first time in 1984 (number 47) and moved up to number 44 in 1985.

Gasoline blends that contain methanol and ethanol are also being used as fuels. The EPA and all U.S. car manufacturers have approved the use of ethanol-gasoline blends up to 10% ethanol (known as **gasohol** when introduced in the 1970s). However, methanol is receiving much attention because it offers several advantages as an octane enhancer. When properly blended, methanol is more economical, has a higher octane rating, and can reduce emission levels of particulates, hydrocarbons, carbon monoxide, and nitrogen oxides. However, the biggest disadvantage of methanol relates to moisture. Small amounts of moisture destabilize the methanol-gasoline mixture, and metal corrosion of the engine becomes a serious problem.

The methanol moisture problem is solved by using another alcohol (ethanol, propanols, butanols) as a co-solvent in methanol blends. The EPA has approved several methanol blends that meet the vehicle emission standards and provide a high-octane gasoline. Most methanol blends contain about 2.5% methanol, 2.5% tertiary-butyl alcohol, 95% gasoline, and a corrosion inhibitor. Although 309,000 tons of methanol were blended into gasoline in 1984, auto makers still disagree about the use of methanol-gasoline blends. AMC/Renault, Ford, and General Motors have approved the use of methanol blends. However, a recent owner's manual for Chrysler cars warns that use of gasolines containing methanol could void the car's warranty. About half of the major foreign car makers do not recommend methanol blends either.

Both opponents and proponents of methanol blends agree there should be uniform labelling requirements for pumps that dispense methanol blends. Not all states require that the pump be labelled with the specific type of alcohol; only the word *alcohol* is required. As a result, consumers in many states are not aware that the gasoline they are buying contains methanol. Most assume the label *alcohol* refers to ethanol or gasohol, since ethanol was the first alcohol used in gasoline blends.

TERTIARY-BUTYL ALCOHOL

MTBE

Alcohols and ethers are discussed in Chapter 13.

Methanol is more soluble in water (recall hydrogen bonding) than in hydrocarbons. Hence, water will extract the methanol into a two-layered system.

SUMMARY

No wonder carbon is ubiquitous; there are over five million recorded compounds containing carbon. These millions of carbon compounds exist for the following reasons:

1. The ability of carbon to form covalent bonds to other carbon atoms almost without limit

2. The ability of a given number of carbon atoms to combine in more than one molecular pattern — isomers

3. The ability of carbon to form stable covalent bonds to a large number of other atoms — functional groups

4. The stability of carbon chains in the presence of substances such as oxygen and water, which usually destroy chains of other atoms

There are some specific reasons for the large number of carbon compounds.

Structural differences explain how two compounds can have the same chemical composition by weight, and yet have different physical and chemical properties. Some order can be brought out of chaos with an understanding of isomerism and a systematic approach to the possible molecular structures and their names.

SELF-TEST 12-B

1. In order to have optical isomers in carbon compounds, a carbon atom must have _____ different groups attached.

2. In what physical property do optical isomers that are mirror images differ? _____

3. How many optical isomers would be possible if five asymmetric carbon atoms were contained in a structure? _____

4. The benzene ring has both localized electrons and _____ electrons.

5. How many atoms does the symbol ⬡ represent? _____

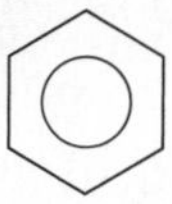

6. Name the following compound. _____

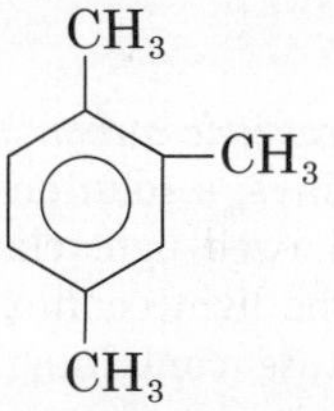

$$CH_3$$
$$\text{—}CH_3$$
$$CH_3$$

7. All ring structures contain only carbon atoms in the ring and have delocalized electrons. ( ) True or ( ) False

8. List four gasoline additives that will increase the octane rating of "straight-run" gasoline. _____

9. What is the problem in adding methanol to the "gasoline at the pump" in order to increase the octane rating? _____

10. Which of the following hydrocarbons would be expected to have the highest octane rating? _____

$$CH_3CH_2CH_2CH_2CH_2CH_2CH_3$$
a.

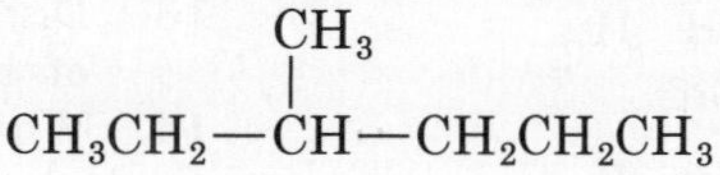

$$CH_3CH_2\text{—}\overset{\overset{\displaystyle CH_3}{|}}{CH}\text{—}CH_2CH_2CH_3$$
b.

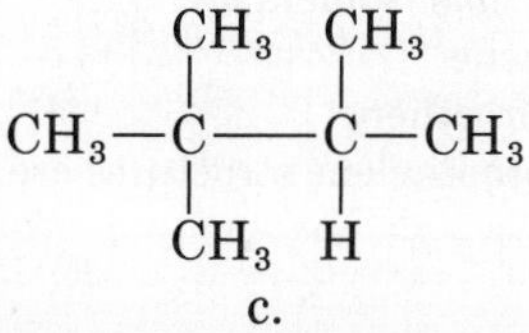

$$CH_3\text{—}\overset{\overset{\displaystyle CH_3}{|}}{\underset{\underset{\displaystyle CH_3}{|}}{C}}\text{——}\overset{\overset{\displaystyle CH_3}{|}}{\underset{\underset{\displaystyle H}{|}}{C}}\text{—}CH_3$$
c.

MATCHING SET

_____ **1.** Organic chemistry

_____ **2.** Isomers

_____ **3.** Cracking

_____ **4.** Hydrocarbon

_____ **5.** Methyl group

_____ **6.** Asymmetric carbon atom

_____ **7.** Delocalized electrons

_____ **8.** Octane rating

a. Breaks hydrocarbons into smaller molecules
b. Found in benzene
c. Compound such as C_3H_8
d. Chemistry of nonmineral carbon compounds
e. Same number and kinds of atoms differently arranged
f. $-CH_3$
g. Has four different groups attached
h. Measures knocking behavior in gasoline

QUESTIONS

1. *Saturated hydrocarbons* are so named because they have the maximum amount of hydrogen present for a given amount of carbon. The saturated hydrocarbons have the general formula C_nH_{2n+2}, where n is a whole number. What are the names and formulas of the first four members of this series of compounds?

2. What is the simplest aromatic compound?

3. Using the periodic table and electron dot formulas, illustrate the bonding in the compound cyclopropane, C_3H_6.

4. Draw the structural formula for each of the five isomeric hexanes, C_6H_{14}.

5. Write the structural formulas for:
   a. 2-methylbutane
   b. Ethylpentane
   c. 4,4-dimethyl-5-ethyloctane
   d. Methylbutane
   e. 2-methyl-2-hexene

6. Give the names for:

a.
$$H-\overset{\overset{\displaystyle H}{|}}{\underset{\underset{\displaystyle H}{|}}{C}}-\overset{\overset{\displaystyle H}{|}}{\underset{\underset{\displaystyle H-\overset{\overset{\displaystyle H}{|}}{\underset{\underset{\displaystyle H}{|}}{C}}-H}{|}}{C}}-\overset{\overset{\displaystyle H}{|}}{\underset{\underset{\displaystyle H}{|}}{C}}-H$$

b.
$$H-\overset{\overset{\displaystyle H}{|}}{\underset{\underset{\displaystyle H}{|}}{C}}-\overset{\overset{\displaystyle H}{|}}{\underset{\underset{\displaystyle H}{|}}{C}}-\overset{\overset{\displaystyle H}{|}}{\underset{\underset{\displaystyle H-\overset{\overset{\displaystyle H}{|}}{\underset{\underset{\displaystyle H}{|}}{C}}-H}{|}}{C}}-\overset{\overset{\displaystyle H}{|}}{\underset{\underset{\displaystyle H}{|}}{C}}-\overset{\overset{\displaystyle H}{|}}{\underset{\underset{\displaystyle H}{|}}{C}}-H$$

c.
$$H-\overset{\overset{\displaystyle H}{|}}{\underset{\underset{\displaystyle H}{|}}{C}}-\overset{\overset{\displaystyle H-\overset{\overset{\displaystyle H}{|}}{\underset{}{C}}-H}{|}}{C}=\overset{\overset{\displaystyle H-\overset{\overset{\displaystyle H}{|}}{\underset{}{C}}-H}{|}}{C}-\overset{\overset{\displaystyle H}{|}}{\underset{\underset{\displaystyle H}{|}}{C}}-H$$

d.
$$H-\overset{\overset{\displaystyle H}{|}}{\underset{\underset{\displaystyle H}{|}}{C}}-C\equiv C-\overset{\overset{\displaystyle H}{|}}{\underset{\underset{\displaystyle H}{|}}{C}}-H$$

7. How can optical isomers be distinguished from each other experimentally?

8. a. Which arrangement has a mirror image that is nonsuperimposable?
   b. Use the two structures

$$\underset{A\diagdown_{D}\diagup A}{\overset{B}{\underset{|}{C}}} \qquad \underset{A\diagdown_{D}\diagup E}{\overset{B}{\underset{|}{C}}}$$

   to explain the term *asymmetric carbon atom.*

9. If for a pair of optical isomers, a solution of the D-isomer rotates plane-polarized light clockwise and the L-isomer rotates the light counterclockwise, predict how a mixture containing equal amounts of the two isomers would affect the light.

10. How many optical isomers can there be for a molecular structure containing eight asymmetric carbon atoms?

11. What unique bond is present in an alkyne hydrocarbon?

12. Which do you think would be better to use for medicinal purposes, pure adrenalin obtained from natural products (found in nature) or the pure compound as synthesized in the laboratory? Why?

13. Distinguish between the classical and modern use of the word organic in chemistry.

14. What structural feature characterizes aromatic compounds?

15. Draw the *cis* and *trans* isomers for:
   a. 1,2-dibromoethene
   b. 1-bromo-2-chloroethene

16. How many trichloroethylene structures are possible?

17. Draw the structure of the compound 1,1,1-trichloroethane.

18. Define *ubiquitous,* and explain how this word is descriptive of the carbon atom.

19. Among the possibilities of structural, geometrical, and (or) optical isomers, what type(s) of isomers can isoprene form? (Isoprene is the fundamental structural unit of rubber; refer to the index.)

20. a. Carbon and hydrogen have almost the same electronegativities. On the basis of this information and the tetrahedral structure around each carbon atom in a hydrocarbon such as octane (C_8H_{18}), would octane be polar or nonpolar?
    b. Would octane be likely to dissolve in water?

21. When you see the symbol ⬡ in a chemistry book, what does it represent?

22. What two elements compose hydrocarbons?

23. Use the index to locate structures of glucose, epinephrine, dimethyldichlorosilane, histidine, and norepinephrine. Which of these substances have asymmetric carbon atoms and, therefore, can be one of a pair of optical isomers?

24. Are more organic or inorganic compounds known?

25. Why is carbon the "central" element in organic compounds?

26. How many carbon atoms are in a molecule of heptane?

27. What does the word *asymmetric* mean?

28. Describe the bonding in benzene.

29. What are the elements in sugar?

30. How many trichloroheptane structures are possible?

31. What is the structural formula for 1-pentene?

32. What type of bond is always in an alkene?

33. Draw the three dichlorobenzene structures and name them.

34. Draw the three trichlorobenzene structures and name them.

35. Biphenyl, ⬡—⬡ , is a compound that can be chlorinated to make chlorinated biphenyl. The family of chlorinated biphenyls is called polychlorinated biphenyls, or PCBs. Draw the structures of several polychlorinated biphenyls. There are 209 possible structures.

36. Do you think any hydrocarbons can exist as optical isomers? Give reasons for your answer.

37. Would you expect $CH_3CH_2CH_2CH_2CH_2CH_2CH_2CH_2CH_3$ to be an important useful constituent of bottled gas? Explain your answer. Would it be useful in gasoline? Explain.

38. Hydrocarbons are generally separated by what purification technique?

39. What hydrocarbons are gases in their natural state?

40. What is the use of tetraethyllead in gasoline?

41. Describe two ways to prevent knocking in gasoline.

42. The chair and boat forms of cyclohexane do not contain the same amount of structural molecular energy; one is a more strained molecule than the other. Which do you think contains more energy? Why?

43. If the temperature of the ordinary laboratory allows an equilibrium to exist between the chair and the boat form of cyclohexane, do you think this is one compound or two? Explain. What does this consideration do to the definition of a "pure substance" as given in Chapter 2?

Chapter 13

ORGANIC CHEMICALS OF MAJOR IMPORTANCE

The preparation of new and different organic compounds through chemical reactions is called **organic synthesis.** Millions of organic compounds have been synthesized in the laboratories of the world during the past 150 years. Prior to 1828, it was widely believed that chemical compounds synthesized by living matter could not be made without living matter — a "vital force" was necessary for the synthesis. In 1828, a young German chemist, Friedrich Wöhler, destroyed the vital force myth and opened the door to modern organic syntheses. Wöhler heated a solution of silver cyanate and ammonium chloride, neither of which had been derived from any living substance. From these he prepared urea, a major animal waste product found in urine.

$$AgOCN + NH_4Cl \rightarrow AgCl + NH_4OCN$$

SILVER CYANATE   AMMONIUM   SILVER CHLORIDE   AMMONIUM
               CHLORIDE   (PRECIPITATE)   CYANATE

$$NH_4OCN \xrightarrow{heat} H_2N\overset{\overset{\displaystyle O}{\|}}{C}NH_2$$

AMMONIUM CYANATE   UREA

The notion of a mysterious vital force declined as other chemists began to synthesize more and more organic chemicals without the aid of a living system. Soon it was shown that chemists could do more than imitate the products of living tissue; they could form unique materials of their own design.

Advances in understanding the structure of organic compounds gave organic synthesis a tremendous boost. Knowing the structure of compounds, the organic chemist could predict by analogy with simpler molecules what reactions might take place when organic reagents were used. Very elegant and reliable schemes of synthesis could then be constructed.

This chapter emphasizes some important organic compounds that not only are useful in themselves but also are necessary for the synthesis of many other organic compounds. The dependence of synthesis on a knowledge of structure will be pointed out from time to time.

Friedrich Wöhler (1800–1882) was Professor of Chemistry at the University of Berlin and later at Göttingen. His preparation of the organic compound urea from the inorganic compound ammonium cyanate did much to overturn the theory that organic compounds must be prepared in living organisms. He was one of the first to study the properties of aluminum, the first to isolate the element beryllium, and also is known for many other outstanding contributions to chemistry.

FUNCTIONAL GROUPS

The millions of organic compounds include classes of compounds that are obtained by replacing hydrogen atoms of hydrocarbons with atoms or groups of atoms known as **functional groups.** The important classes of compounds that result from attaching functional groups to a hydrocarbon framework are shown in Table 13–1. The "R" attached to the functional group represents the hydrocarbon framework with one hydrogen atom removed for each functional group added. The name

TABLE 13–1   Classes of Organic Compounds Based on Functional Groups*

GENERAL FORMULAS OF CLASS MEMBERS	CLASS NAME	TYPICAL COMPOUND	COMPOUND NAME	COMMON USE OF SAMPLE COMPOUND
R—X	Halide	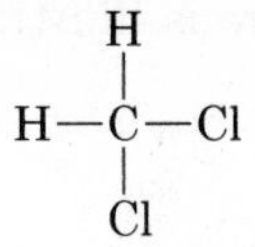 H—C(—Cl)(Cl), H bonded	Dichloromethane (methylene chloride)	Solvent
R—OH	Alcohol	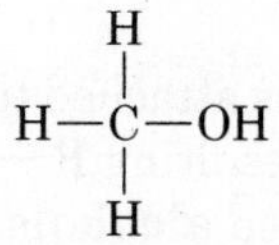 H—C(H)(H)—OH	Methanol (wood alcohol)	Solvent
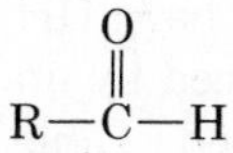 R—C(=O)—H	Aldehyde	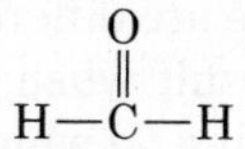 H—C(=O)—H	Methanal (formaldehyde)	Preservative
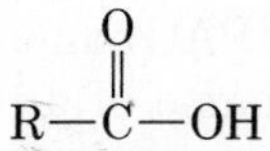 R—C(=O)—OH	Carboxylic acid	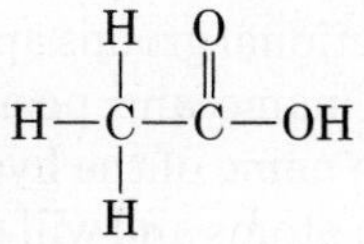 H—C(H)(H)—C(=O)—OH	Ethanoic acid (acetic acid)	Vinegar
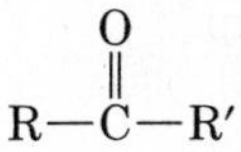 R—C(=O)—R′	Ketone	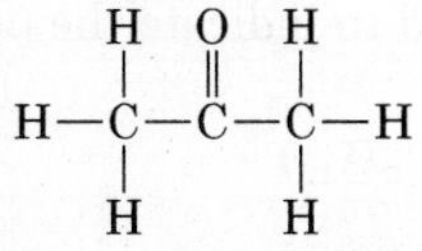 H—C(H)(H)—C(=O)—C(H)(H)—H	Propanone (acetone)	Solvent
R—O—R′	Ether	C_2H_5—O—C_2H_5	Diethyl ether (ethyl ether)	Anesthetic
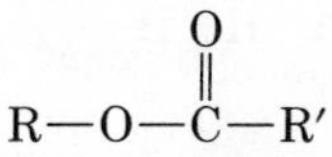 R—O—C(=O)—R′	Ester	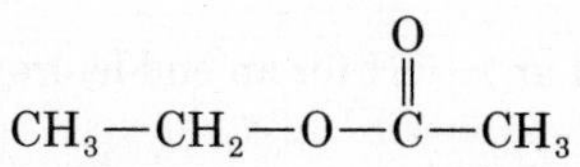 CH_3—CH_2—O—C(=O)—CH_3	Ethyl ethanoate (ethyl acetate)	Solvent in fingernail polish
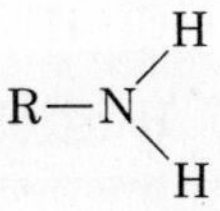 R—N(H)(H)	Amine	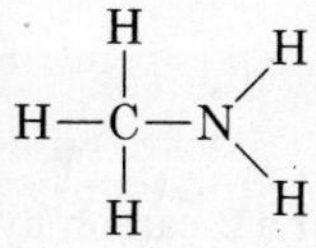 H—C(H)(H)—N(H)(H)	Methylamine	Tanning (foul odor)
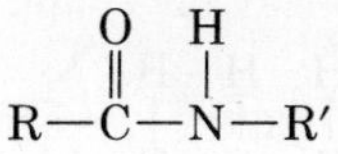 R—C(=O)—N(H)—R′	Amide	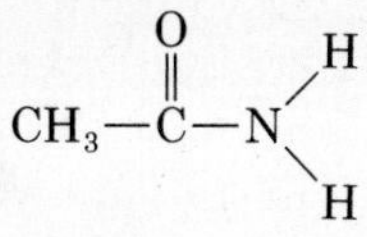 CH_3—C(=O)—N(H)(H)	Acetamide	Plasticizer

Handwritten annotations: "MEMORIZE", "FOR EXAM", arrow pointing to Aldehyde row.

* R stands for an H or a hydrocarbon group such as —CH_3 or —C_2H_5. R′ could be a different group from R.

for this R group is obtained by removing *-ane* from the parent hydrocarbon and adding *-yl.*

Parent Compound	*R Group*	*Functional Group*	*Substituted Hydrocarbon*

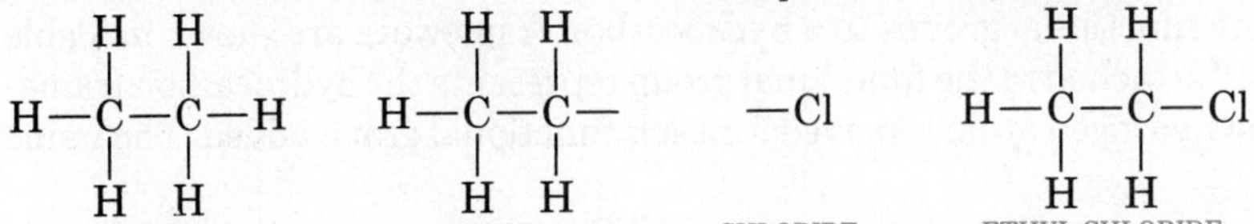

ETHANE   ETHYL   CHLORIDE   ETHYL CHLORIDE

Alcohols are compounds with the structure R—O—H, where R is a hydrocarbon group.

The alkyl halides, the first class of compounds listed in Table 13–1, were discussed in Chapter 12. This chapter describes both useful and naturally occurring compounds as representatives of the other functional group classes listed in Table 13–1. The importance of structure is illustrated by the classification of organic compounds by functional group.

Alcohols

When a hydroxyl (—OH) group is attached to a nonaromatic carbon skeleton of a hydrocarbon (an R group), the resulting R—OH compound has properties common to a class of compounds called **alcohols.** A single hydrocarbon molecule can give rise to several alcohols if there are different isomeric positions for the —OH group. Three different alcohols result when a hydrogen atom is replaced by an —OH group in n-pentane, depending on which hydrogen atom is replaced (Table 13–2).

When the —OH is attached directly to the aromatic ring, the compounds are known as phenols. These are considered as a separate class because phenols are much more acidic than alcohols due to the electron-withdrawing properties of the aromatic ring.

When one or more functional groups appear in a molecule, the IUPAC name reveals the functional group name and position. For example, the name of an alcohol will use the root of the name of the hydrocarbon to which it corresponds to indicate the number of carbon atoms and will use the suffix *-ol* to denote an alcohol. As before, a number is used to indicate the position of the alcohol group.

TABLE 13–2   Alcohols Derived from Pentane (C_5H_{12})

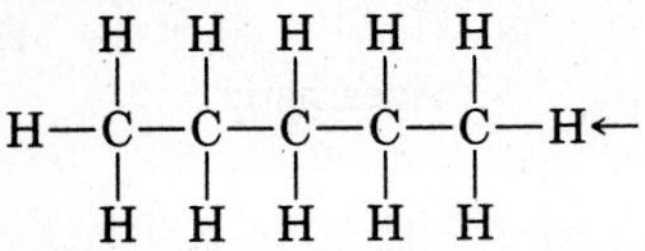

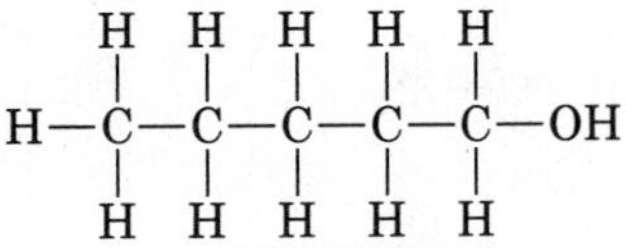

Substitution of an —OH for an end hydrogen   1-PENTANOL

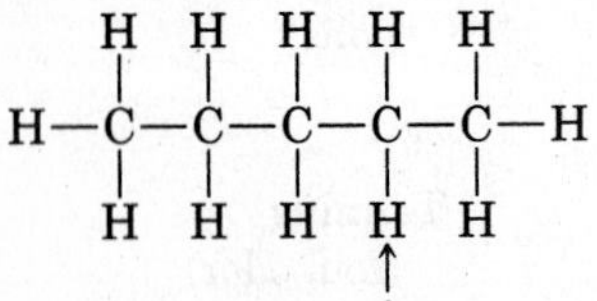

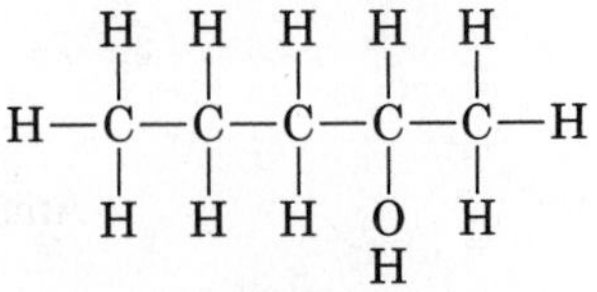

Substitution of an —OH for a 2-carbon hydrogen   2-PENTANOL

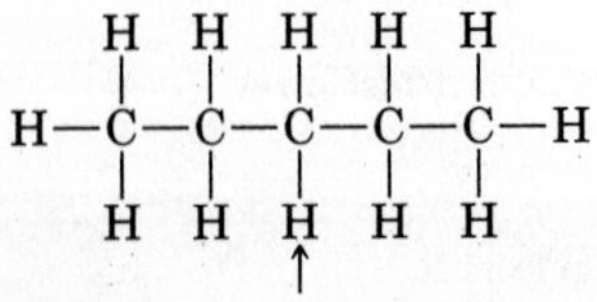

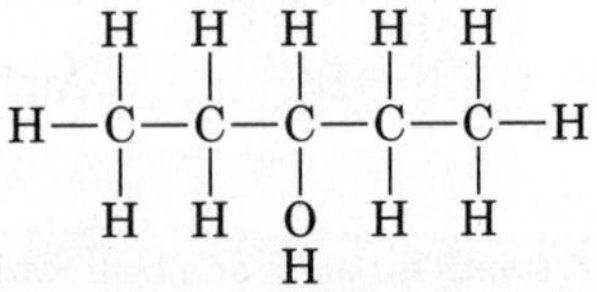

Substitution of an —OH for a 3-carbon hydrogen   3-PENTANOL

Alcohols are classified as primary, secondary, and tertiary based on the number of other carbons bonded to the —C—OH carbon.

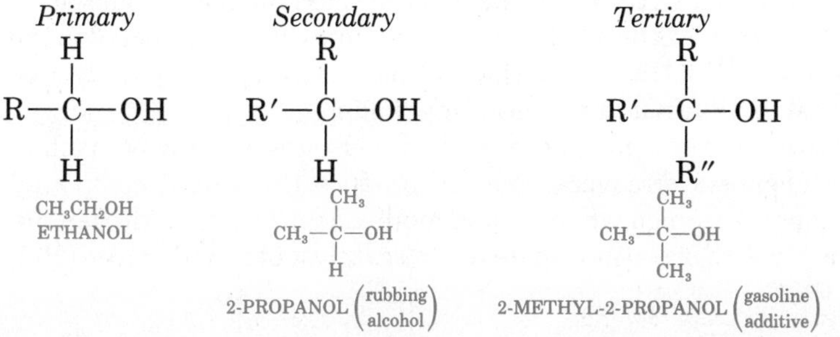

Primary

$$H$$
$$|$$
$$R—C—OH$$
$$|$$
$$H$$

CH₃CH₂OH
ETHANOL

Secondary

$$R$$
$$|$$
$$R'—C—OH$$
$$|$$
$$H$$

$$CH_3$$
$$|$$
$$CH_3—C—OH$$
$$|$$
$$H$$

2-PROPANOL $\left(\begin{smallmatrix}\text{rubbing}\\\text{alcohol}\end{smallmatrix}\right)$

Tertiary

$$R$$
$$|$$
$$R'—C—OH$$
$$|$$
$$R''$$

$$CH_3$$
$$|$$
$$CH_3—C—OH$$
$$|$$
$$CH_3$$

2-METHYL-2-PROPANOL $\left(\begin{smallmatrix}\text{gasoline}\\\text{additive}\end{smallmatrix}\right)$

The use of R, R′, and R″ indicates all R groups can be different.

The common name of 2-propanol is isopropyl alcohol.

The common name of 2-methyl-2-propanol is tertiary-butyl alcohol.

In Table 13–2, 1-pentanol is a primary alcohol, and 2-pentanol and 3-pentanol are secondary alcohols. Formulas and names of other important alcohols are given in Table 13–3.

TABLE 13–3   Some Important Alcohols

FORMULA	IUPAC NAME	COMMON NAME
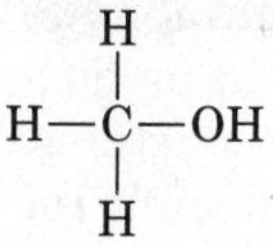 H—C—OH (with H above and H below)	Methanol	Methyl alcohol (wood alcohol)
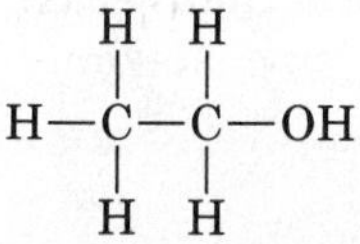 H—C—C—OH (ethane skeleton)	Ethanol	Ethyl alcohol (grain alcohol)
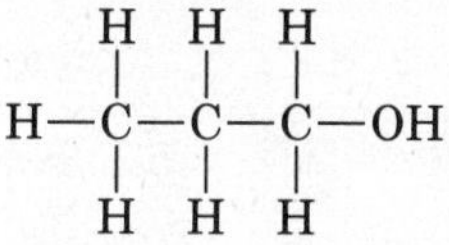 H—C—C—C—OH (propane skeleton)	1-propanol	n-propyl alcohol
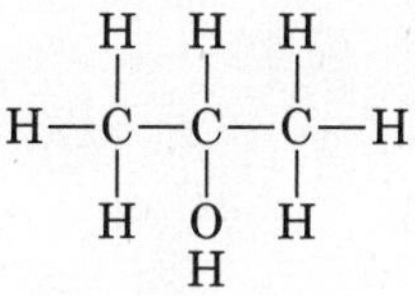 H—C—C—C—H (with O—H on middle carbon)	2-propanol	Isopropyl alcohol (rubbing alcohol)
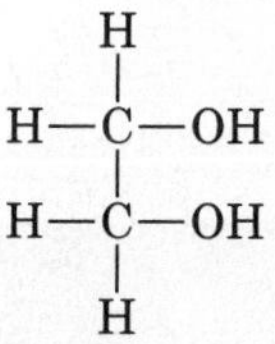 H—C—OH / H—C—OH	1,2-ethanediol	Ethylene glycol (permanent antifreeze)
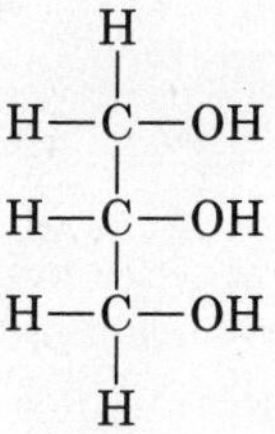 H—C—OH / H—C—OH / H—C—OH	1,2,3-propanetriol	Glycerol (glycerin)

METHANOL (METHYL ALCOHOL)   Methanol was originally called wood alcohol since it was obtained by the destructive distillation of wood. It is the simplest of all alcohols and has the formula CH_3OH. In the older method for the production of wood alcohol, hardwoods such as beech, hickory, maple, and birch are heated in the absence of air in a retort (Fig. 13–1). Methanol that is 92% to 95% pure can be obtained by fractional distillation of the resulting liquid.

Methanol is number 22 in the list of commercial chemicals.

In 1923, the price of wood alcohol in the United States was 88¢ per gallon. That year German chemists discovered how to produce this useful compound synthetically. Methanol is formed when carbon monoxide, CO, and hydrogen are heated at a pressure of 200 to 300 atmospheres over a catalyst of mixed oxides (90% ZnO – 10% Cr_2O_3).

Recall from Chapter 10 that this is a reduction of CO by H_2.

$$CO + 2\ H_2 \xrightarrow[\text{300°C pressure}]{\text{ZnO—Cr}_2\text{O}_3} CH_3OH$$

As a result of this synthetic process, German industrialists were able to sell pure methanol at 20¢ per gallon. Even a high tariff could not save the wood distillers and their outdated operations. The synthetic product soon dominated the market in the United States.

Industrial chemists are continually searching for less costly ways to prepare important chemicals such as methanol.

The production of synthetic methanol in the United States is over 8 billion pounds per year. About half of this is used in the production of **formaldehyde** (used in plastics, embalming fluid, germicides, and fungicides), 30% in the production of other chemicals, and smaller amounts for jet fuels, antifreeze mixtures, solvents, and as a denaturant (a poison added to ethanol to make it unfit for beverages). Methanol is a *deadly poison;* it causes blindness in less than lethal doses. Many deaths and injuries have resulted when this alcohol was mistakenly substituted for ethanol in beverages.

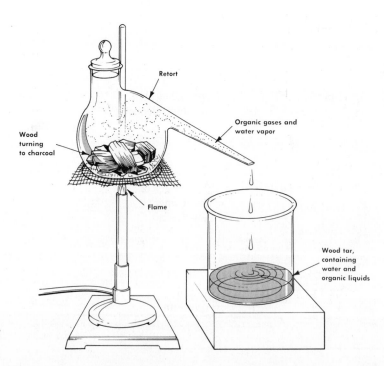

Retort

Organic gases and water vapor

Wood turning to charcoal

Flame

Wood tar, containing water and organic liquids

FIGURE 13–1   Destructive distillation of wood.

ETHANOL (ETHYL ALCOHOL) Ethanol (ethyl alcohol), which is used in alcoholic beverages, is called grain alcohol because it can be fractionally distilled from the fermented mash made from corn, rice, barley, and other grains. Fermentation is a breakdown of starch in the grain by means of enzymes. Enzymes are catalysts that are complex organic molecules produced by living cells. If the enzyme diastase is mixed with ground grain and water, and the mixture is allowed to stand at 40°C for a period of time, the starch in the grain will be changed into maltose.

$$2(C_6H_{10}O_5)_n + nH_2O \xrightarrow{\text{diastase}} nC_{12}H_{22}O_{11}$$
STARCH   MALTOSE (A SUGAR)

The subscript n in the formula for starch indicates that starch is made up of many $C_6H_{10}O_5$ units.

Brewers call the resulting mixture of maltose and water the **wort**. The wort is diluted and mixed with yeast and held at a temperature of 30°C for 40 to 60 hours. The living yeast cells secrete the enzymes maltase and zymase. The maltase causes the sugar, maltose, to hydrolyze into a simple sugar, glucose:

$$C_{12}H_{22}O_{11} + H_2O \xrightarrow{\text{maltase}} 2\,C_6H_{12}O_6$$
MALTOSE   GLUCOSE

The glucose, in turn, is converted by zymase to ethanol and carbon dioxide:

$$C_6H_{12}O_6 \xrightarrow{\text{zymase}} 2\,CO_2 + 2\,C_2H_5OH$$
GLUCOSE   ETHANOL

A solution of 95% ethanol and 5% water can be recovered from the mash by fractional distillation.

Synthetic ethanol is produced on a large scale for industrial use. The direct chemical addition of water to ethylene accounts for more than 80% of all ethanol production. Under high pressure in the presence of a catalyst and a large excess of water vapor, ethylene produces ethanol:

Over a billion pounds of ethanol are produced synthetically in the United States each year.

Ethylene is number six in the list of commercial chemicals.

ETHYLENE   ETHANOL

Pure ethanol is 200 proof (that is, twice the percentage of alcohol). Apart from the alcoholic beverage industry, ethanol is used widely in solvents and in the preparation of chloroform, ether, and many other organic compounds.

Proof $= 2 \times$ (% alcohol)

Some of the most commonly encountered alcoholic beverages and their characteristics are presented in Table 13–4.

TABLE 13–4   Common Alcoholic Beverages

NAME	SOURCE OF FERMENTED CARBOHYDRATE	AMOUNT OF ETHYL ALCOHOL	PROOF
Beer	Barley, wheat	5%	10
Wine	Grapes or other fruit	12% maximum, unless fortified*	20–24
Brandy	Distilled wine	40–45%	80–90
Whiskey	Barley, rye, corn, etc.	45–55%	90–110
Rum	Molasses	~45%	90
Vodka	Potatoes	40–50%	80–100

* The growth of yeast is inhibited at alcohol concentrations over 12%, and fermentation comes to a stop. Beverages with a higher concentration are prepared either by distillation or by fortification with alcohol that has been obtained by the distillation of another fermentation product.

TABLE 13–5   Alcohol Blood Level and Effect

BLOOD-ALCOHOL LEVEL (PERCENTAGE BY VOLUME)	EFFECT
0.05–0.15	Lack of coordination
0.15–0.20	Intoxication
0.30–0.40	Unconsciousness
0.50	Possible fatality

Different flavors of beverages are due to the presence of small amounts of other alcohols, aldehydes, and ketones, which form during fermentation.

Nearly 100 denaturation mixtures are approved for use in the denaturation of alcohol.

The federal tax on alcoholic beverages is about $20 per gallon. Since the cost of producing ethanol is only about $1 per gallon, ethanol intended for industrial use must be **denatured** to avoid the beverage tax. **Denatured alcohol** contains small amounts of a toxic substance, such as methanol or gasoline, which cannot be removed easily by chemical or physical means.

Although ethanol is not as toxic as methanol, one pint of pure ethanol, rapidly ingested, would kill most people. Ethanol is a depressant. The effects of different blood levels of alcohol are shown in Table 13–5. Rapid consumption of two 1-ounce "shots" of 90-proof whiskey or of two 12-ounce beers can cause the alcohol blood level to reach 0.05%. The breathalyzer test used to detect drunken drivers is based on the color change that occurs when ethanol is oxidized to acetic acid by dichromate anion ($Cr_2O_7^{2-}$) in acidic solution.

$$16\,H^+ + \underset{\text{YELLOW-ORANGE}}{2\,Cr_2O_7^{2-}} + 3\,CH_3CH_2OH \rightarrow 3\,CH_3COOH + \underset{\text{GREEN}}{4\,Cr^{3+}} + 11\,H_2O$$

Ethanol is oxidized by liver enzymes to acetaldehyde.

Ethanol is quickly absorbed by the blood and metabolized by enzymes produced in the liver. The rate of detoxification is about 1 ounce of pure alcohol per hour. The ethanol is oxidized to acetaldehyde, which is further oxidized to acetic acid, and eventually CO_2 and H_2O are produced.

ETHANOL   ACETALDEHYDE

A metabolic change that accompanies detoxification of ethanol is the synthesis of fat, which is deposited in liver tissue. Excessive drinking causes deterioration of the liver. Cirrhosis of the liver is eight times more common among alcoholics than among nonalcoholics. In 1974, cirrhosis of the liver rose above arteriosclerosis, influenza, and pneumonia to become the seventh leading cause of death. Alcoholics also tend to suffer from malnutrition and cardiovascular disease.

PROPANOLS (PROPYL ALCOHOLS)   When one considers the possible structures for propyl alcohol, it is apparent that two isomers are possible.

1-PROPANOL   2-PROPANOL
n-PROPYL ALCOHOL   ISOPROPYL ALCOHOL

Of the two propanols, 1-propanol is the more expensive; it is prepared by the oxidation of simple hydrocarbons. It finds uses as a solvent and as a raw material in the manufacture of other organic compounds.

The hydration of propylene yields the less expensive 2-propanol (isopropyl alcohol), which is rubbing alcohol. It has greater germicidal activity than the other simple alcohols and is used as an antiseptic.

Propylene is number twelve in the list of commercial chemicals.

$$CH_3-\overset{\overset{\displaystyle H}{|}}{C}=CH_2 + H_2O \xrightarrow{H_2SO_4} CH_3-\underset{\underset{\displaystyle H}{\underset{\displaystyle |}{\overset{\displaystyle O}{\overset{\displaystyle |}{}}}}}{\overset{\overset{\displaystyle H}{|}}{C}}-CH_3$$

PROPYLENE   2-PROPANOL

Ethylene Glycol and Glycerol (Glycerin)

More than one alcohol group (—OH) can be present in a single molecule. Glycerol and ethylene glycol, the base of permanent antifreeze, are examples of such compounds. Ethylene glycol is made in a two-step synthesis, starting with ethene.

Permanent antifreeze is ethylene glycol.

In organic reactions, catalysts, reagents, and reaction conditions are often listed with the arrow.

$$CH_2=CH_2 + \tfrac{1}{2}O_2 \xrightarrow[300°C]{Ag} CH_2\underset{\underset{\displaystyle O}{\diagdown\diagup}}{-}CH_2 + H_2O \xrightarrow{acid} \begin{matrix} H \\ | \\ H-C-OH \\ | \\ H-C-OH \\ | \\ H \end{matrix}$$

ETHENE   ETHYLENE   1,2-ETHANEDIOL
                    OXIDE   ETHYLENE GLYCOL

$$\begin{matrix} H \\ | \\ H-C-OH \\ | \\ H-C-OH \\ | \\ H-C-OH \\ | \\ H \end{matrix}$$

1,2,3-PROPANETRIOL
GLYCEROL (GLYCERIN)

Glycerol is a byproduct in the manufacture of soaps. Because of its moisture-holding properties, glycerol has many uses in foods and tobacco as a digestible and nontoxic humectant (gathers and holds moisture), and in the manufacture of drugs and cosmetics. It is also used in the production of nitroglycerin and numerous other chemicals. Perhaps the most important compounds of glycerol are its natural esters (fats and oils), which we shall discuss later in this chapter.

Hydrogen Bonding in Alcohols

The physical properties of water, methanol, ethanol, the propanols, ethylene glycol, and glycerol offer another interesting example of the effects of **hydrogen bonding** between molecules in liquids. In Table 13–6 the boiling points for these compounds are listed.

See Chapters 5 and 17 for a discussion of hydrogen bonding in water.

Since boiling involves overcoming the attractions between liquid molecules as they pass into the gas phase, a higher boiling point indicates stronger intermolecular forces holding the molecules together. Another factor is also present: as the molecules become larger, higher boiling points result because more energy is required to change the longer-chain molecules from the liquid to the gaseous phase, owing in part to the larger London forces. A graph showing the boiling points of the normal alcohols (straight carbon chains with the —OH group on an end carbon) as a function of chain length is given in Figure 13–2.

Methanol, like water, has an —OH group, and some hydrogen bonding is to be expected, as shown in Figure 13–3. Hydrogen bonding explains why methanol

Hydrogen bonding is responsible for the fact that an alcohol has a higher boiling point than its parent hydrocarbon.

TABLE 13-6   Boiling Points for Some —OH Compounds

Water	HOH	100°C
Methanol	CH_3OH	65.0°
Ethanol	CH_3CH_2OH	78.5°
1-Propanol	$CH_3CH_2CH_2OH$	97.4°
2-Propanol	$CH_3CHOHCH_3$	82.4°
Ethylene glycol	CH_2OHCH_2OH	198°
Glycerol	$CH_2OHCHOHCH_2OH$	290°

Note: The parent hydrocarbon of methanol is methane (bp −164°C); of ethanol and ethylene glycol, ethane (bp −88.6°C); and of the propanols and glycerol, propane (bp −42.1°C).

(molecular weight 32) is a liquid, whereas propane (C_3H_8, molecular weight 44), an even heavier molecule, is a gas at room temperature. Methanol has only one hydrogen through which it can hydrogen bond, while water can hydrogen bond from either of its two hydrogen atoms. Thus, water, with more extensive intermolecular bonding, has the higher boiling point even though water has lighter molecules.

Both methanol and ethanol can be used as antifreeze, but they tend to distill out of the coolant at the temperatures of a hot internal combustion engine. Protection against freezing is then lost over a period of time. Ethylene glycol is equally effective (molecule for molecule) in lowering the freezing point of water, and its high boiling point (198°C) makes it a permanent antifreeze. This property makes ethylene glycol more desirable, even though it takes almost twice as much ethylene glycol by weight as methanol to provide the same amount of protection for a car's cooling system. Ethylene glycol with suitable additives to protect the radiator system is sold under a number of brand names. The higher boiling point of ethylene glycol is readily explained in terms of the two —OH groups per molecule and the enhanced possibility for hydrogen bonding. Glycerol, with three —OH groups per molecule, has an even higher boiling point, as well as a very high viscosity (resistance to flow).

Since the —OH group in an organic molecule causes at least that area of the molecule to be polar, such molecules will be attracted to other polar molecules such

Methanol and ethanol oxidize easily to acids, and so corrode the radiator.

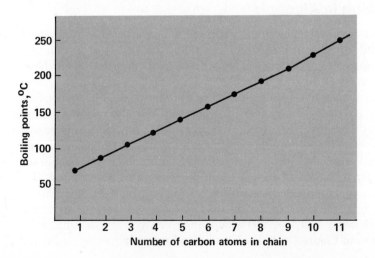

FIGURE 13-2 Boiling points of straight-chain alcohols (—OH group on an end carbon).

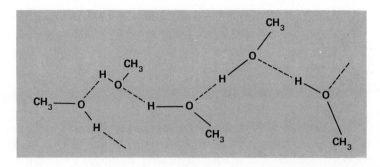

FIGURE 13–3
Hydrogen bonding in methanol.

as water molecules. As a result of these attractions, the lower molecular weight alcohols are quite soluble in water. In the higher molecular weight alcohols, the nonpolar hydrocarbon chain decreases significantly their solubility in water (see Table 13–7).

Synthesis with Alcohols

Alcohols can serve as the starting substances for the synthesis of many other types of organic compounds. Oxidation of alcohols may yield aldehydes, ketones, or acids, depending on the starting compound and the amount of oxygen added.

If the starting compound is a primary alcohol, the first oxidation product will be an aldehyde. Secondary alcohols are oxidized to give ketones. For example, the oxidation of ethanol, a primary alcohol, can be used to make acetaldehyde and acetic acid:

In organic compounds, oxidation is often accompanied by the loss of hydrogen or the addition of oxygen to a molecule.

$$
\underset{\text{ETHANOL}}{\overset{\displaystyle H}{\underset{\displaystyle H}{CH_3\overset{|}{\underset{|}{C}}-OH}}} \xrightarrow{\text{oxidation}} \underset{\text{ACETALDEHYDE}}{\overset{\displaystyle H}{CH_3\overset{|}{C}=O}} \xrightarrow{\text{oxidation}} \underset{\text{ACETIC ACID}}{\overset{\displaystyle O}{CH_3\overset{\|}{C}-OH}}
$$

TABLE 13–7   Water Solubilities of Some Alcohols

IUPAC NAME	FORMULA	SOLUBILITY (g/100 g H_2O)	COMMENT
Methanol	CH_3OH	∞*	Soluble in all proportions
Ethanol	CH_3CH_2OH	∞	Same
1-Propanol	$CH_3CH_2CH_2OH$	∞	Same
2-Propanol	$CH_3CHOHCH_3$	∞	Effect of —OH still strong
1-Butanol	$CH_3CH_2CH_2CH_2OH$	7.9	Hydrocarbon chain effect now apparent
2-Methyl-1-propanol	$(CH_3)_2CHCH_2OH$	10.0	Shorter overall chain length
2-Butanol	$CH_3CH_2CHOHCH_3$	12.5	Effect of —OH group stronger in this position
2-Methyl-2-propanol	$(CH_3)_3COH$	∞	Nonpolar effect is diminished
1-Pentanol	$CH_3CH_2CH_2CH_2CH_2OH$	2.3	Effect of the long hydrocarbon chain

* This symbol means infinitely "soluble," or that there is no limit to solubility.

Workable oxidants are hot copper oxide (CuO), potassium dichromate ($K_2Cr_2O_7$) with sulfuric acid, or potassium permanganate ($KMnO_4$).

The oxidation of 2-propanol, a secondary alcohol, provides the ketone, acetone:

$$\underset{\substack{| \\ \text{OH} \\ \text{2-PROPANOL}}}{\overset{\substack{\text{H} \\ |}}{CH_3CCH_3}} \xrightarrow{\text{oxidation}} \underset{\substack{\| \\ \text{O} \\ \text{ACETONE}}}{CH_3CCH_3}$$

Alcohols dehydrate to form alkenes or ethers. Two important dehydration reactions of ethanol illustrate how temperature can be used to determine the product. Sulfuric acid is the dehydrating agent.

Dehydration reactions involve the formation and removal of water molecules.

$$2 \underset{\text{ETHANOL}}{CH_3CH_2OH} \xrightarrow[\text{H}_2\text{SO}_4]{140°\text{C}} \underset{\text{DIETHYL ETHER}}{CH_3CH_2OCH_2CH_3} + H_2O$$

$$\underset{\text{ETHANOL}}{CH_3CH_2OH} \xrightarrow[\text{H}_2\text{SO}_4]{180°\text{C}} \underset{\text{ETHENE}}{H-\overset{\substack{\text{H} \\ |}}{C}=\overset{\substack{\text{H} \\ |}}{C}-H} + H_2O$$

Substitution reactions produce alkyl halides from simple alcohols. In the reaction shown, Br replaces an OH group on 2-propanol.

$$\underset{\substack{| \\ \text{OH} \\ \text{2-PROPANOL}}}{\overset{\substack{\text{H} \\ |}}{CH_3CCH_3}} + HBr \rightarrow \underset{\substack{| \\ \text{Br} \\ \text{2-BROMOPROPANE}}}{\overset{\substack{\text{H} \\ |}}{CH_3CCH_3}} + H_2O$$

Organic Acids

Organic acids are compounds of the type
$$R-\overset{\substack{O \\ \|}}{C}-OH.$$ They are generally weak acids.

Earlier an acid was defined as a species that has a tendency to donate hydrogen ions. We shall now consider the carboxylic acids, which contain the carboxyl group,

$$-C\overset{\displaystyle O}{\underset{\displaystyle OH}{}}$$. The electronegative character of the C=O group tends to drain electron density away from the region between the oxygen and hydrogen atoms. The partial positive charge assumed by the hydrogen then makes it possible for polar water molecules to remove hydrogen ions from some of the carboxyl groups. The strength of an organic acid depends on the group that is attached to the carboxyl group. If the attached group has a tendency to pull electrons away from the carboxyl group, the acid is a stronger acid. For example, trichloroacetic acid is much stronger than acetic acid:

TRICHLOROACETIC ACID   ACETIC ACID

As shown by the arrow, the highly electronegative chlorine atoms withdraw electron density from the region of the carboxyl group and make the loss of the hydrogen ion even easier than in acetic acid. However, trichloroacetic acid is still weaker than a strong mineral acid such as sulfuric acid.

TABLE 13–8   Carboxylic Acids and Odors

IUPAC NAME	FORMULA	ODOR
Methanoic acid	HCOOH	Sharp, irritating
Ethanoic acid	CH_3COOH	Sharp, irritating (vinegar)
Propanoic acid	CH_3CH_2COOH	Swiss cheese
Butanoic acid	$CH_3CH_2CH_2COOH$	Rancid butter
Pentanoic acid	$CH_3CH_2CH_2CH_2COOH$	Manure
Hexanoic acid	$CH_3CH_2CH_2CH_2CH_2COOH$	Goat

The ionization of carboxylic acids in water is simply:

$$R\overset{\overset{\displaystyle O}{\|}}{-}C-OH + H_2O \rightleftharpoons R\overset{\overset{\displaystyle O}{\|}}{-}C-O^- + H_3O^+$$

Carboxylic acids are **neutralized** by bases to form salts.

$$R\overset{\overset{\displaystyle O}{\|}}{-}C-OH + Na^+ + OH^- \longrightarrow R\overset{\overset{\displaystyle O}{\|}}{-}C-O^- Na^+ + H_2O$$

A SALT

The unequal double arrows ($\longleftarrow\!\!\!\longrightarrow$) indicate the equilibrium favors the unionized acid molecule; that is, the acid is weak.

The first six carboxylic acids and their odors are listed in Table 13–8. Many carboxylic acids have an unpleasant odor. Propanoic acid causes the major part of the odor of Swiss cheese, while butanoic acid is the foul smell of rancid butter. Butanoic acid is also one of the components of body odor. Odors worsen with the next acid in the homologous series, pentanoic acid, which smells like manure, followed by hexanoic acid, which smells like goats. Longer-chain carboxylic acids do not smell as bad, partly because they are less volatile. As Table 13–9 illustrates, the esters derived from these acids have very pleasant odors, and many of these esters are the major components of fruit and flower odors.

METHANOIC ACID (FORMIC ACID)   The simplest organic acid is methanoic acid, also called formic acid, in which the carboxyl group is attached directly to a hydrogen atom.

$$H-C\overset{\displaystyle O}{\underset{\displaystyle O-H}{\diagup}}$$

METHANOIC ACID
FORMIC ACID

This acid is found in ants and other insects and is part of the irritant that produces itching and swelling after a bite.

Formic acid may be prepared from its sodium salt, which is readily prepared by heating carbon monoxide (CO) with sodium hydroxide (NaOH):

$$CO + NaOH \xrightarrow[\text{6-10 min}]{200°\,C} HCOO^-Na^+$$

SODIUM FORMATE

Weak bases such as ammonia, NH_3, neutralize formic acid and are used in the treatment of insect bites.

If the resulting salt is mixed with a mineral acid, formic acid can be distilled from the mixture:

$$HCOO^-Na^+ + H_3O^+ + Cl^- \rightarrow HCOOH + Na^+Cl^- + H_2O$$

SODIUM FORMATE	HYDROCHLORIC ACID	FORMIC ACID	SODIUM CHLORIDE

ETHANOIC ACID (ACETIC ACID) Ethanoic (acetic) acid is the most widely used of the organic acids. It is found in vinegar, an aqueous solution containing 4% to 5% acetic acid. Flavor and colors are imparted to vinegars by the constituents of the alcoholic solutions from which they are made. Ethanol in the presence of certain bacteria and air is oxidized to acetic acid:

Oxidation of ethanol to acetic acid by oxygen in air is responsible for the souring of wine.

$$CH_3CH_2OH + O_2 \xrightarrow{\text{bacteria}} CH_3COOH + H_2O$$

$$\underset{\text{ETHANOL}}{} \quad \underset{\text{OXYGEN}}{} \quad \underset{\substack{\text{ETHANOIC ACID} \\ \text{(ACETIC ACID)}}}{} \quad \underset{\text{WATER}}{}$$

The bacteria, called mother of vinegar, form a slimy growth in a vinegar solution. The growth of bacteria can sometimes be observed in a bottle of commercially prepared vinegar after it has been opened to the air.

Acetic acid is an important starting substance for making textile fibers, vinyl plastics, and other chemicals and is a convenient choice when a cheap organic acid is needed.

FATTY ACIDS A fatty acid contains a carboxyl group attached to a long hydrocarbon chain. The chains often contain only carbon-carbon bonds but may contain carbon-carbon double bonds as well. Examples are stearic acid, palmitic acid, and oleic acid.

Fatty acids may be obtained from animal and vegetable fats or oils. The carbon chain in fatty acids is generally 8 to 18 carbon atoms in length.

$$CH_3CH_2CH_2CH_2CH_2CH_2CH_2CH_2CH_2CH_2CH_2CH_2CH_2CH_2CH_2CH_2C\overset{\displaystyle O}{\underset{\displaystyle OH}{}}$$

STEARIC ACID, $CH_3-(CH_2)_{16}-COOH$

$$CH_3CH_2CH_2CH_2CH_2CH_2CH_2CH_2CH_2CH_2CH_2CH_2CH_2CH_2C\overset{\displaystyle O}{\underset{\displaystyle OH}{}}$$

PALMITIC ACID, $CH_3-(CH_2)_{14}-COOH$

$$CH_3CH_2CH_2CH_2CH_2CH_2CH_2CH_2CH=CHCH_2CH_2CH_2CH_2CH_2CH_2CH_2C\overset{\displaystyle O}{\underset{\displaystyle OH}{}}$$

OLEIC ACID, $CH_3(CH_2)_7CH=CH(CH_2)_7COOH$

Stearic acid is obtained by the hydrolysis of animal fat, palmitic acid results from the hydrolysis of palm oil, and oleic acid is obtained from olive oil. These reactions are given in the next two sections. Stearic and palmitic acids are especially important in the manufacture of soaps.

SELF-TEST 13-A

1. The major component of permanent antifreeze is _____ .
2. Name the following compounds:

   a. CH_3OH _____

   b. CH_3CH_2OH _____

   c. $HCOOH$ _____

   d. $CH_3CH_2CHCH_3$ _____
   $\quad\quad\quad\quad\quad |$
   $\quad\quad\quad\quad\quad OH$

   e. CH_3COOH _____

   f. CH_2-CH_2
   $\quad\; |\quad\quad |$
   $\quad\; OH\quad OH$ _____

   g. CH_3CHO _____

   h. CH_3CHCH_3 _____
   $\quad\quad\quad |$
   $\quad\quad\quad Br$

3. Two methods by which ethanol is made on a large scale are:

   a. _____

   b. _____

4. An example of a carboxylic acid that contains a long hydrocarbon chain is

   _____ .

5. Identify the functional groups present in each of the following molecules:

   a. R—OH _____

   $$\text{c. } R-\overset{\overset{\displaystyle O}{\|}}{C}-H \text{ _____}$$

   $$\text{b. } R-\overset{\overset{\displaystyle O}{\|}}{C}-OH \text{ _____}$$

   $$\text{d. } R-\overset{\overset{\displaystyle O}{\|}}{C}-R' \text{ _____}$$

6. a. Write the structural formulas for ethane, ethanol, ethanal, ethanoic acid, diethyl ether, and ethyl amine.

   b. Give common names where possible.

   c. What R group is present in these compounds? _____

7. Why do high molecular weight acids have relatively little odor in comparison to pentanoic acid, which has a foul smell? _____

Esters

In the presence of strong mineral acids, organic acids react with alcohols to form compounds called **esters**. For example, when ethyl alcohol is mixed with acetic acid in the presence of sulfuric acid, ethyl acetate is formed. This reaction is a dehydration in which sulfuric acid acts as a catalyst and dehydrator.

Organic esters are compounds of the type $R-O-\overset{\overset{\displaystyle O}{\|}}{C}-R'$ formed by the reaction of organic acids and alcohols.

$$CH_3CH_2O\underset{-----}{\overline{+H + HO+}}\overset{\overset{\displaystyle }{}}{C}CH_3 \underset{}{\overset{H_2SO_4}{\rightleftharpoons}} CH_3CH_2O\overset{\overset{\displaystyle }{}}{C}CH_3 + H_2O$$

ETHYL ACETATE

Ethyl acetate is a common solvent for lacquers and plastics and is often used as fingernail polish remover.

Some odors of common fruits are due to the presence of mixtures of volatile esters (Table 13–9). In contrast, esters of higher molecular weight often have a distinctly unpleasant odor.

Fats and Oils

Fats and oils are esters of glycerol (glycerin) and a fatty acid. R, R', and R" stand for the hydrocarbon chains of the acids in the following equation:

Fats and oils are esters of fatty acids and glycerol. Fats are solids, and oils are liquids.

$$
\begin{array}{ccccc}
CH_2-OH & & HO-\overset{\overset{\displaystyle O}{\|}}{C}-R & & CH_2-O-\overset{\overset{\displaystyle O}{\|}}{C}-R \\
| & & & & | \\
CH-OH & + & HO-\overset{\overset{\displaystyle O}{\|}}{C}-R' & \rightleftharpoons & CH-O-\overset{\overset{\displaystyle O}{\|}}{C}-R' & + & 3\,H_2O \\
| & & & & | \\
CH_2-OH & & HO-\overset{\overset{\displaystyle O}{\|}}{C}-R'' & & CH_2-O-\overset{\overset{\displaystyle O}{\|}}{C}-R''
\end{array}
$$

GLYCEROL (ONE MOLECULE) FATTY ACID (THREE MOLECULES THAT MAY OR MAY NOT BE THE SAME) FAT OR OIL (ONE MOLECULE) WATER (THREE MOLECULES)

TABLE 13–9   Some Alcohols, Acids, and Their Esters

ALCOHOL	ACID	ESTER	ODOR OF THE ESTER

$CH_3CHCH_2CH_2OH$
    $\mid$
   CH_3
ISOPENTYL ALCOHOL
	CH_3COOH
	
	ACETIC ACID
		$CH_3CHCH_2CH_2-O-\overset{\overset{\textstyle}{\displaystyle}}{C}-CH_3$
		    $\mid$ $\parallel$
		   CH_3 O
		ISOPENTYL ACETATE
			Banana

$CH_3CHCH_2CH_2OH$
    $\mid$
   CH_3
ISOPENTYL ALCOHOL
	$CH_3CH_2CH_2CH_2COOH$
	
	PENTANOIC ACID
		$CH_3CHCH_2CH_2-O-\overset{}{C}-CH_2CH_2CH_3$
		    $\mid$ $\parallel$
		   CH_3 O
		ISOPENTYL PENTANOATE
			Apple

$CH_3CH_2CH_2CH_2OH$
 	
n-BUTYL ALCOHOL
	$CH_3CH_2CH_2COOH$
	
	BUTANOIC ACID
		$CH_3CH_2CH_2CH_2-O-\overset{}{C}-CH_2CH_2CH_3$
		              $\parallel$
		              O
		BUTYL BUTANOATE
			Pineapple

CH_3CHCH_2OH
    $\mid$
   CH_3
ISOBUTYL ALCOHOL
	CH_3CH_2COOH
	
	PROPIONIC ACID
		$CH_3CHCH_2-O-\overset{}{C}-CH_2CH_3$
		    $\mid$ $\parallel$
		   CH_3 O
		ISOBUTYL PROPIONATE
			Rum

CH_3CHCH_2OH
    $\mid$
   CH_3
ISOBUTYL ALCOHOL
	$HCOOH$
	
	FORMIC ACID
		$CH_3CHCH_2-O-\overset{}{C}-H$
		    $\mid$ $\parallel$
		   CH_3 O
		ISOBUTYL FORMATE
			Raspberry

⬡$-CH_2-OH$
 	
BENZYL ALCOHOL
	$CH_3CH_2CH_2COOH$
	
	BUTANOIC ACID
		⬡$-CH_2-O-\overset{}{C}-CH_2CH_2CH_3$
		           $\parallel$
		         O
		BENZYL BUTANOATE
			Roses

Lipids are soluble in fats and oils.

The term *fat* is usually reserved for solid glycerol esters (butter, lard, tallow) and *oil* for liquid esters (castor, olive, linseed, tung, and so forth). The term *lipid* includes fats, oils, and fat-soluble compounds.

Saturation (all single bonds with maximum hydrogen content) in the carbon chain of the fatty acids is usually found in solid or semi-solid fats, whereas unsaturated fatty acids (containing one or more double bonds) are usually found in oils. Hydrogen can be catalytically added to the double bonds of an oil to convert it into a semi-solid fat. For example, liquid soybean and other vegetable oils are hydrogenated to produce cooking fats and margarine.

Consumers in Europe and North America have historically valued butter as a source of fat. As the population increased, the advantages of a substitute for butter became apparent, and efforts to prepare such a product began about a hundred years ago. One problem that arose was the fact that common fats are almost all *animal* products with very pronounced tastes of their own. Analogous compounds from vegetable oils, which are bland or have mixed flavors, were generally *unsaturated* and consequently *oils*. A solid fat could be made from the much cheaper vegetable oils if an inexpensive way could be discovered to add hydrogen across the double bonds. After extensive experiments, many catalysts were found, of which finely divided nickel is among the most effective. The nature of the process can be illustrated by the reaction

Catalytic hydrogenation can convert a liquid oil into a solid fat.

$$H_2C-O-\underset{\underset{O}{\|}}{C}-(CH_2)_7CH=CH(CH_2)_7CH_3 \qquad H_2C-O-\underset{\underset{O}{\|}}{C}-(CH_2)_7CH_2CH_2(CH_2)_7CH_3$$

$$HC-O-\underset{\underset{O}{\|}}{C}-(CH_2)_7CH=CH(CH_2)_7CH_3 \xrightarrow[200\,C]{H_2\,Ni} HC-O-\underset{\underset{O}{\|}}{C}-(CH_2)_7CH_2CH_2(CH_2)_7CH_3$$

$$H_2C-O-\underset{\underset{O}{\|}}{C}-(CH_2)_7CH=CH(CH_2)_7CH_3 \qquad H_2C-O-\underset{\underset{O}{\|}}{C}-(CH_2)_7CH_2CH_2(CH_2)_7CH_3$$

TRIOLEIN, A LIQUID OIL   TRISTEARIN, A SOLID FAT

Oils commonly subjected to this process include those from cottonseed, peanuts, corn germ, soybeans, coconuts, and safflower seeds. In recent years, as it became apparent that saturated fats may encourage diseases of the heart and arteries, soft margarines and cooking oils (which still contain some of the unhydrogenated fatty acid) have been placed on the market.

DIETARY FATS AND ESSENTIAL FATTY ACIDS

Most diets in the United States gain 40% to 50% of their calories from fats or oils. This is rather high when compared with diets in most other parts of the world. Natural fats and oils are generally mixtures of various esters of glycerol with more than one kind of fatty acid. In our diets, most fatty acids are **saturated** fatty acids (Table 13–10). Such fatty acids can be (1) used as a source of energy if the body converts them to CO_2 and H_2O, (2) stored for possible future use in fat cells, or (3) used as starting materials for the synthesis of other compounds needed by the body.

TABLE 13–10   Ratio of Saturated and Unsaturated Fatty Acids from Common Fats and Oils*

OIL OR FAT	PERCENTAGE OF TOTAL FATTY ACIDS BY WEIGHT		
	Saturated	Monounsaturated	Polyunsaturated
Coconut oil	93	6	1
Corn oil	14	29	57
Cottonseed oil	26	22	52
Lard	44	46	10
Olive oil	15	73	12
Palm oil	57	36	7
Peanut oil	21	49	30
Safflower oil	10	14	76
Soybean oil	14	24	62
Sunflower oil	11	19	70

* *Saturated* means full complement of hydrogen (no C=C double bonds); *monounsaturated* means one C=C double bond per fatty acid molecule; *polyunsaturated* means two or more C=C double bonds per molecule of fatty acid. The chief unsaturated fatty acid is linoleic acid. Although derived from vegetable rather than animal fats, both coconut oil and peanut oil have been associated recently with hardening of the arteries when combined with a high cholesterol intake.

An ordinary scientific calorie is the amount of heat required to raise 1 g of water 1°C. A food calorie is 1000 scientific calories (or 1 kcal).

Fats are the most concentrated source of food energy in our diets, as they furnish about 9000 cal/g when burned for energy as compared with about 3800 cal/g for glucose. The human body can make some fats from carbohydrates and carries out such processes to store the excess energy furnished in the diet.

A high intake of dietary fat has been implicated as one of the factors that can give rise to **atherosclerosis,** a complex process in which the walls of the arteries suffer damage and ultimately develop scar tissue and fatty deposits. Atherosclerosis is generally considered to be a precursor to certain types of heart disease and strokes. Atherosclerosis may also be related to the amount of cholesterol in the diet, but the relationship of both dietary fat and cholesterol intake to atherosclerosis does not appear at this time to be a simple one.

It has been known for about 60 years that the human body has a small requirement for certain types of fatty acids (called **essential fatty acids**), and in recent years the basis for the need for these essential fatty acids has been determined.

The essential fatty acids are **linoleic, linolenic,** and **arachidonic** acids.

$$CH_3CH_2CH_2CH_2CH_2CH=CHCH_2CH=CHCH_2CH_2CH_2CH_2CH_2CH_2C \overset{O}{\underset{OH}{\diagup}}$$

LINOLEIC ACID $(C_{18}\Delta_{9,12})$

$$CH_3CH_2CH=CHCH_2CH=CHCH_2CH=CHCH_2CH_2CH_2CH_2CH_2CH_2C \overset{O}{\underset{OH}{\diagup}}$$

LINOLENIC ACID $(C_{18}\Delta_{9,12,15})$

$$CH_3CH_2CH_2CH_2CH_2CH=CHCH_2CH=CHCH_2CH=CHCH_2CH=CHCH_2CH_2CH_2C \overset{O}{\underset{OH}{\diagup}}$$

ARACHIDONIC ACID $(C_{20}\Delta_{5,8,11,14})$

Δ indicates the positions of the double bonds.

The presence of essential fatty acids in the diet permits the body to synthesize a very important group of compounds, the prostaglandins. The key compound here is linoleic acid, which the body cannot make from more saturated fatty acids. If linoleic acid is available, the body can make arachidonic acid and linolenic acid.

Prostaglandins are synthesized from the essential fatty acids. Even in very small amounts, prostaglandins have powerful effects on the human body.

Prostaglandins are a group of more than a dozen related compounds with potent effects on physiological activity such as blood pressure, relaxation and contraction of smooth muscle, gastric acid secretion, body temperature, food intake, and blood platelet aggregation. Their potential use as drugs is currently under widespread investigation. Two of the prostaglandins that have been characterized are prostaglandin E_1 (used to induce labor to terminate pregnancy) and prostaglandin E_2.

PROSTAGLANDIN E_1 $(C_{20}H_{34}O_5)$

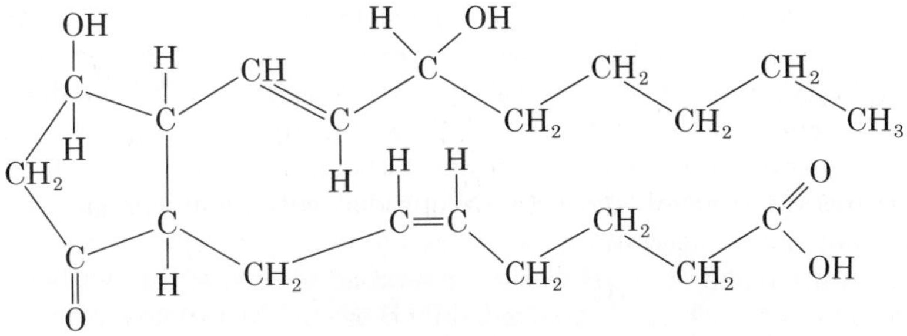

PROSTAGLANDIN E_2 ($C_{20}H_{32}O_5$)

Note that both of these prostaglandins contain exactly the same number of carbon atoms as arachidonic acid.

USEFUL PRODUCTS FROM ORGANIC SYNTHESIS REACTIONS

Many chemists are engaged in the synthesis of organic compounds. In educational and industrial laboratories throughout the world, they prepare new and different compounds on a small scale. If the new compound has commercial value, the preparation is subsequently adapted for full-scale plant operations.

The thousands of chemical changes required to synthesize the many known organic compounds have a few characteristics in common:

1. Several chemical changes are usually required to synthesize a single organic compound. Each chemical change is called a step, and each step produces an intermediate compound that is used in the next step. The final step produces the desired end product.
2. Normally only one functional group undergoes change in each step. The rest of the molecule remains intact and unchanged. This is known as the **principle of minimum structural change.**
3. From one principal starting substance can come many diverse products. The kind of product obtained depends on the reactants and the conditions imposed.
4. The more steps in the synthesis, the lower the percentage yield of the final product. The starting substance and each intermediate are only partially converted to the next intermediate because of equilibrium considerations, side reactions, or both, which convert some of the starting substance into undesirable products. If an intermediate product must be removed and purified before proceeding to the next step, some additional material is lost. The principal purification methods are recrystallization and extraction for solids and distillation for liquids.

Organic syntheses are usually carried out so that only one part of the molecule changes in each reaction step.

In our discussion of organic chemistry, we have given examples of several organic synthesis reactions that are run on a large scale. Many of these are single-step reactions. Examples include the addition reactions of ethene to form ethanol or 1,2-dichloroethane; substitution reactions of benzene such as the preparation of chlorobenzene; and the formation of methanol from carbon monoxide and hydrogen. An important multistep synthesis is the preparation of ethylene glycol from ethene (p. 307). Since multistep syntheses are often used to make important commercial products, another example will be given here for the synthesis of aspirin.

Aspirin, acetylsalicyclic acid, was first synthesized for medical use in 1893 by Felix Hofmann, a German chemist working for the F. Bayer Company. Aspirin is still the leading pain killer and the standard treatment to reduce fever and swelling. Over 30 million pounds of aspirin, or 150 tablets per person, are consumed in the United States each year, and worldwide use exceeds 100,000 tons per year.

The starting point for the synthesis of aspirin is benzene, which is obtained from coal tar. The steps in the synthesis are shown in Figure 13–4. Only the principal organic substance is shown for each step. Other products such as sodium chloride (a coproduct with phenol) and water (a coproduct with sodium phenoxide) are sometimes important in the synthesis because they have to be removed to avoid interference with subsequent steps. However, in a broad outline of the synthetic process, the coproducts are generally omitted; only those products made in a previous step and required for subsequent steps are included. In Figure 13–4 the step-by-step structural changes can be followed by noting groups in color. Conditions and additional reactants for each step are written with the arrow. These conventions are generally used to summarize organic syntheses.

Some intermediates are useful compounds in their own right. Phenol, commonly called carbolic acid, is used to prepare plastics such as Bakelite, drugs, dyes, and other compounds. Phenol also has medical application as a topical anesthetic for some types of lesions and in the treatment of mange and colic in animals. Methyl salicylate, or oil of wintergreen, is used as a flavoring agent and as a component of rubbing alcohol for sore muscles.

The conversion of phenol into sodium phenoxide is an acid-base reaction. Phenol, with an acidic hydrogen in the hydroxyl group, reacts with a base, sodium hydroxide, to give a salt, sodium phenoxide, and water. The reaction of salicylic acid to form oil of wintergreen is an **esterification.** The organic acid reacts with an alcohol in the presence of strong mineral acid to produce the ester, methyl salicylate, and water.

Obviously, it is beyond the scope of this text to give an extensive overview of organic synthesis. You can be assured, however, that if the "miracle cancer drug" is found or whenever any other new and useful compounds are discovered, their

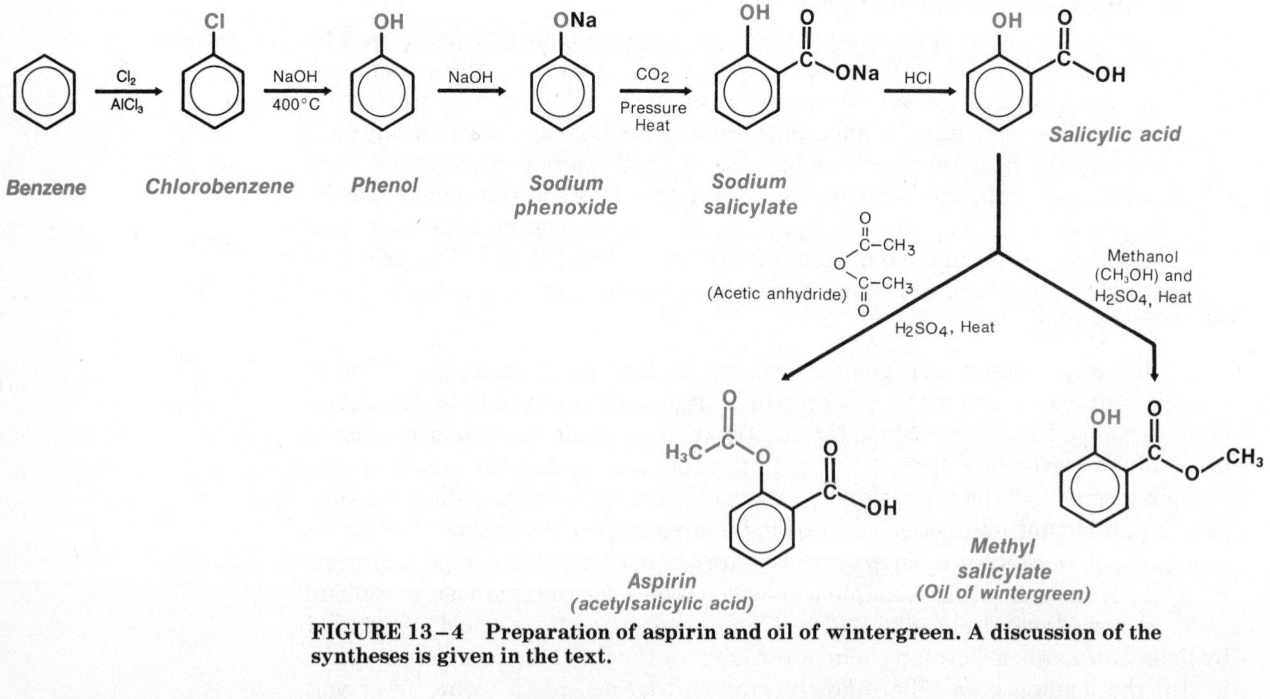

Any given organic compound can usually serve as the starting substance for the synthesis of many other organic compounds.

FIGURE 13–4   Preparation of aspirin and oil of wintergreen. A discussion of the syntheses is given in the text.

production will involve step-by-step molecular modifications to produce the desired product. An exciting aspect of organic synthesis is the prediction of desired properties of new molecular arrangements and then the testing of the theoretical properties. With so many possibilities yet to be discovered, one can only guess at the potential power of organic synthesis.

POSTSCRIPT

Organic compounds were obtained originally from plants, animals, and fossil fuels (coal, petroleum, gas), and these are still direct sources for many important chemicals such as sucrose from sugarcane or ethanol from fermented grain mash. However, the development of organic chemistry led to cheaper methods for the synthesis of both naturally occurring substances and new substances. The classification of organic compounds by functional groups and the study of the chemical reactions of functional groups are the bases of synthetic organic chemistry. They lead to the synthesis of new and useful compounds or to the more economical synthesis of known compounds; both are a central function of modern organic chemistry.

In this chapter we have given examples of useful compounds in some major functional group classes. The flow diagram in Figure 13–5 reviews the general reactions for obtaining compounds with these functional groups. A knowledge of these reactions along with an ability to recognize the functional groups in illustrations of organic molecules will help you understand both organic chemistry and biochemistry.

Some naturally occurring organic compounds are illustrated on the next several pages. See if you can identify the functional groups in these molecules.

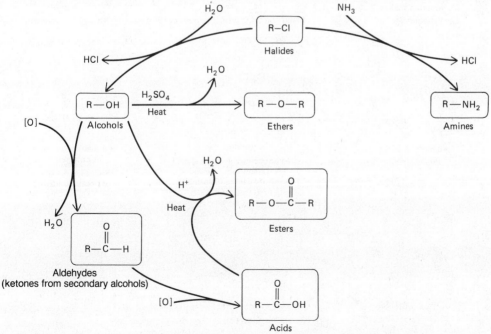

FIGURE 13–5   Flow diagram of the synthesis of compounds with different functional groups.

NATURALLY OCCURRING ORGANIC COMPOUNDS

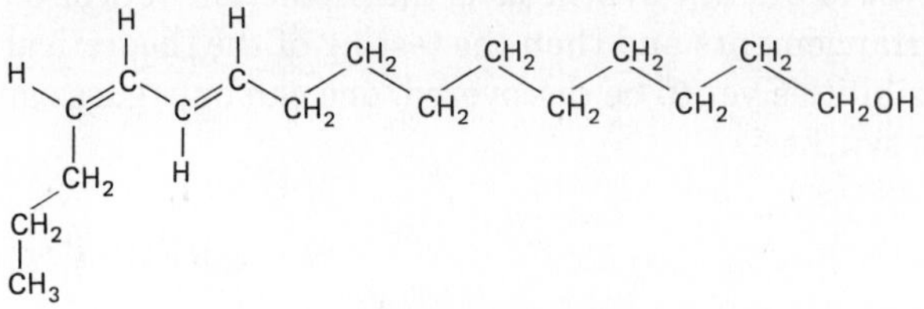

bombykol
(10-*trans*, 12-*cis*-hexadecadien-1-ol;
sex attractant of silkworm moth, physiologically
active at concentrations of 10 μg per ml of air)

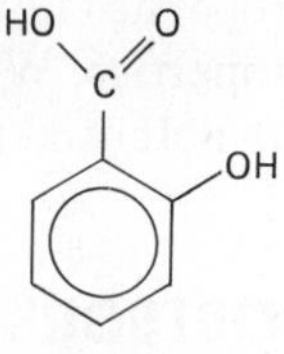

salicylic acid
(from the bark of willow trees;
reduces fever and swelling)

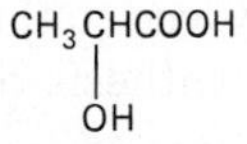

$CH_3CHCOOH$
   $|$
   OH

lactic acid
(contributes to
sore muscles)

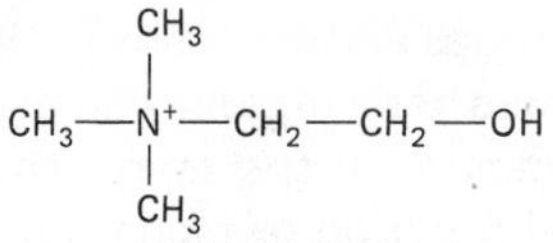

choline
(building block for
the neurotransmitter
acetylcholine)

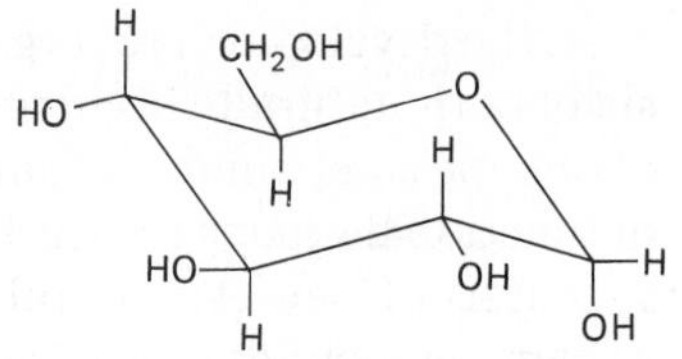

α-D-glucose
(a simple sugar)

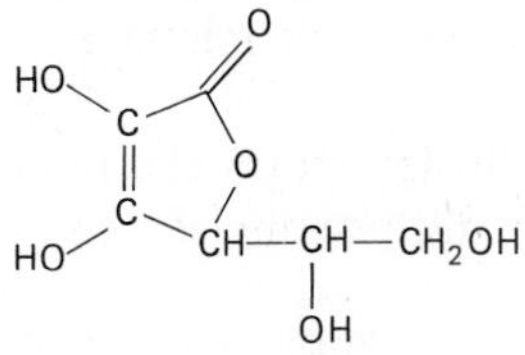

ascorbic acid
(vitamin C)

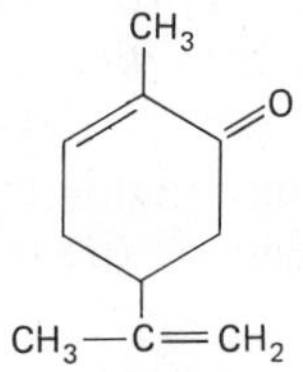

carvone
(chief component of spearmint oil)

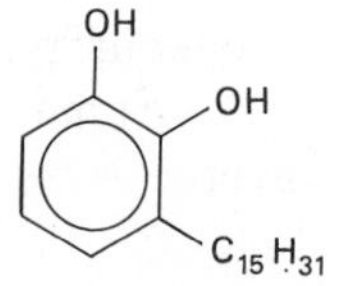

urushiol
(irritant in poison ivy)

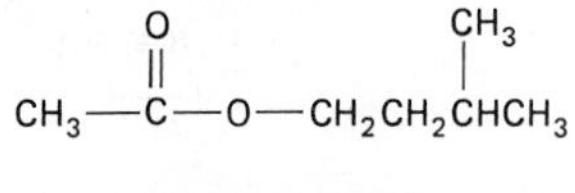

$CH_3 - \overset{O}{\overset{\|}{C}} - O - CH_2CH_2CHCH_3$
                                          $|$
                                         CH_3

isopentyl acetate
(banana oil)

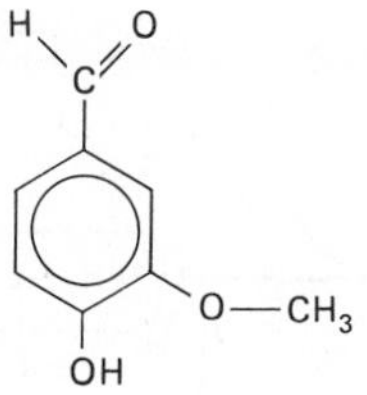

vanillin
(fragrant component of vanilla
bean, responsible for the
flavor of "vanilla flavoring"
and of extract of vanilla)

$H_2NCH_2CH_2CH_2CH_2NH_2$

**1,4-diaminobutane,
or putrescine**

(responsible for odor
of decaying fish)

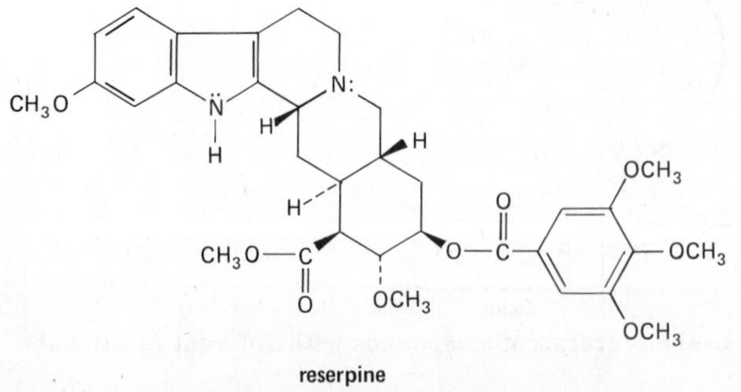

reserpine
(from Indian snakeroot; used as tranquilizer)

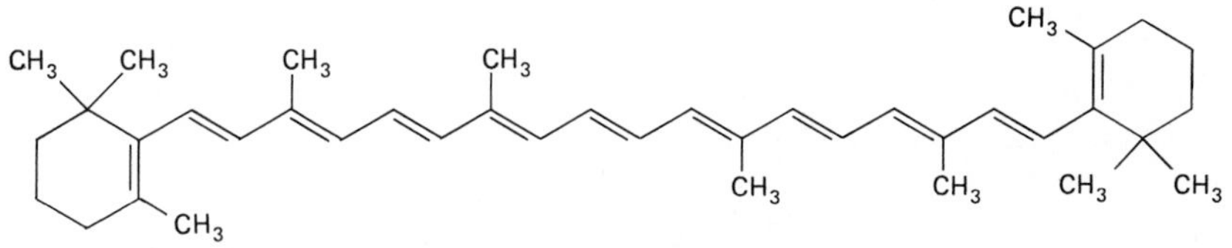

β-carotene
(precursor of vitamin A)

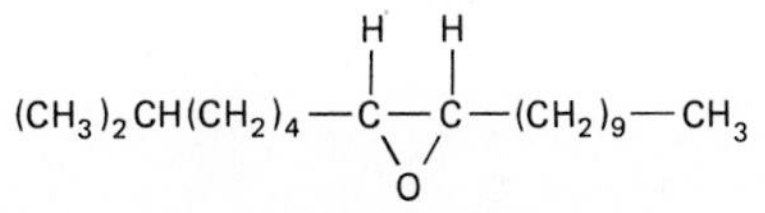

$(CH_3)_2CH(CH_2)_4$—C—C—$(CH_2)_9$—CH_3

disparlure
(gypsy moth sex attractant)

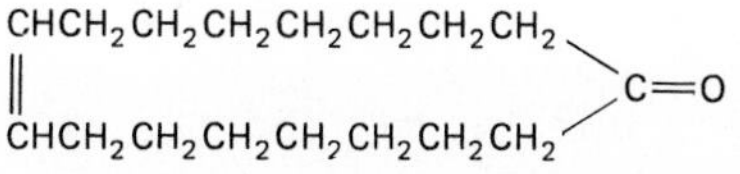

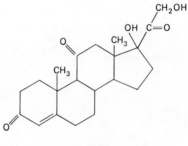

civetone
(secretion of the civet cat, used in perfume)

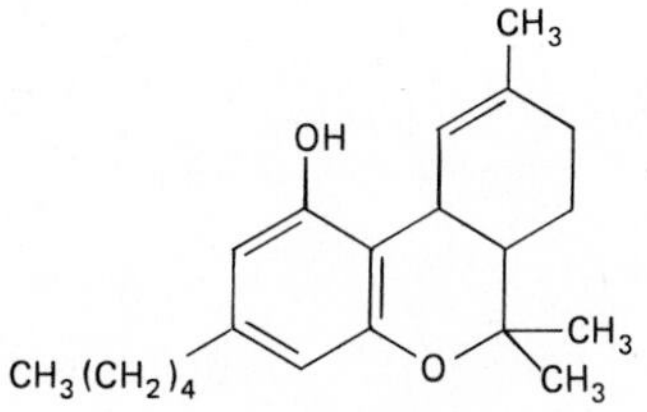

tetrahydrocannabinol
(marihuana)

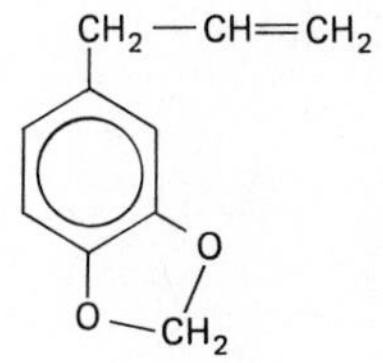

saffrole
(oil of sassafras)

cortisone
(hormone; regulation of carbohydrate
and protein metabolism; used to
reduce inflammation)

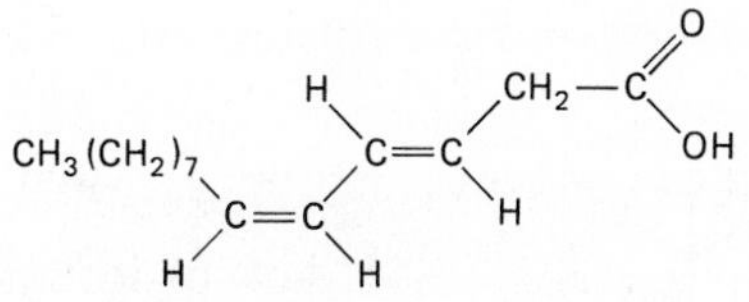

3-*trans*-5-*cis*-tetradecadienoic acid
(sex attractant of black carpet beetle)

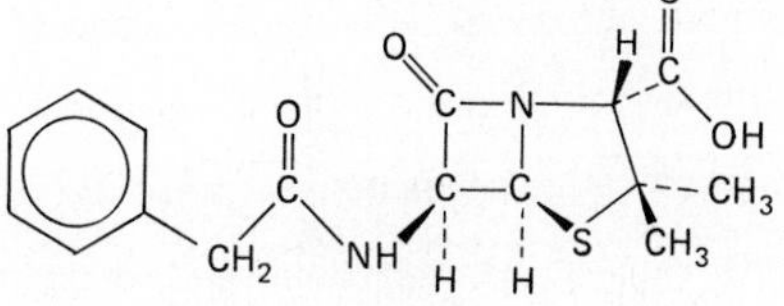

penicillin G
(secretion of a blue-green mold; antibiotic)

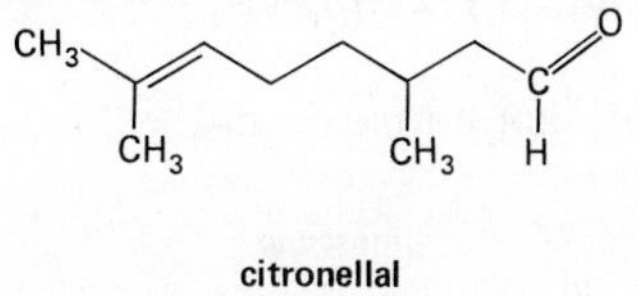

citronellal
(oil of lemon)

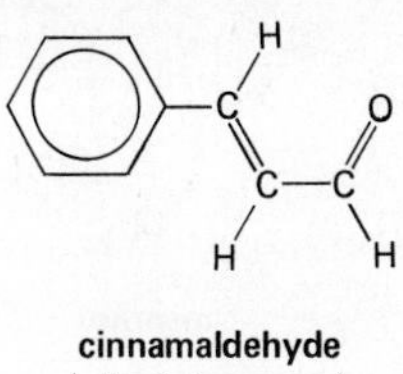

cinnamaldehyde
(oil of cinnamon)

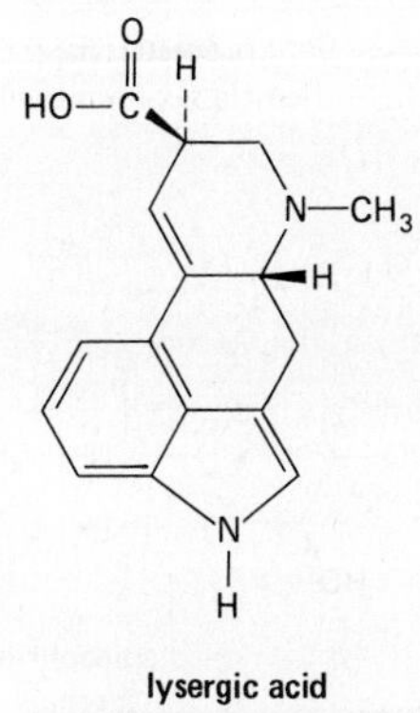

lysergic acid
(from ergot, a fungus that
grows on grain and hops)

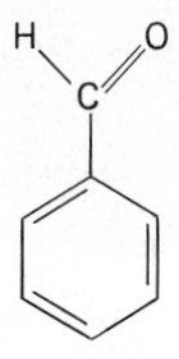

benzaldehyde
(oil of bitter almond)

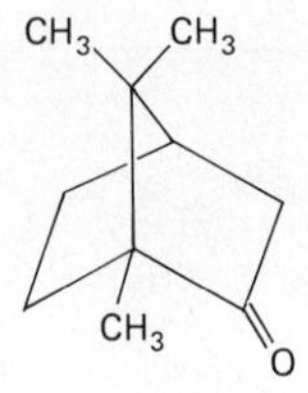

camphor
(oil of camphor tree)

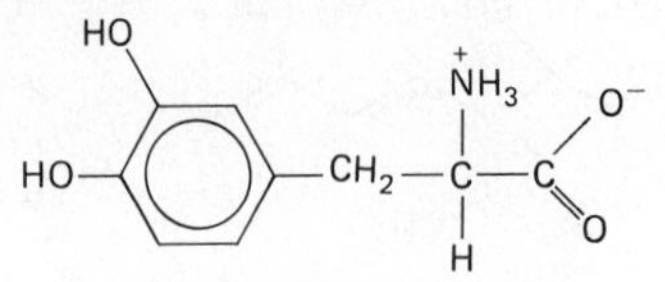

DOPA (3,4-dihydroxyphenylalanine)
(an amino acid—the L form is produced
in the adrenal gland and nerve terminals;
used in treatment of Parkinson's disease)

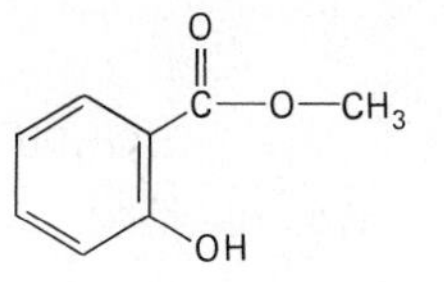

methyl salicylate
(oil of wintergreen)

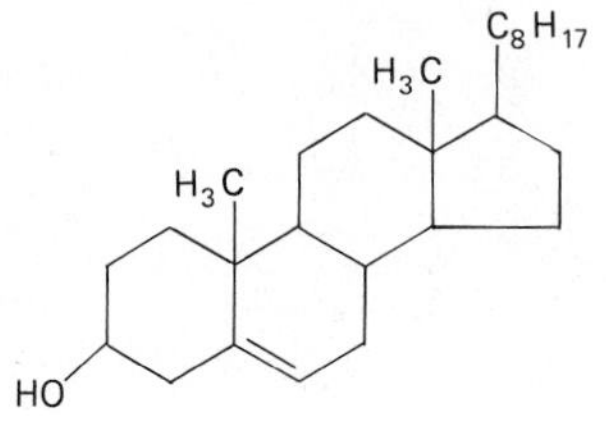

cholesterol

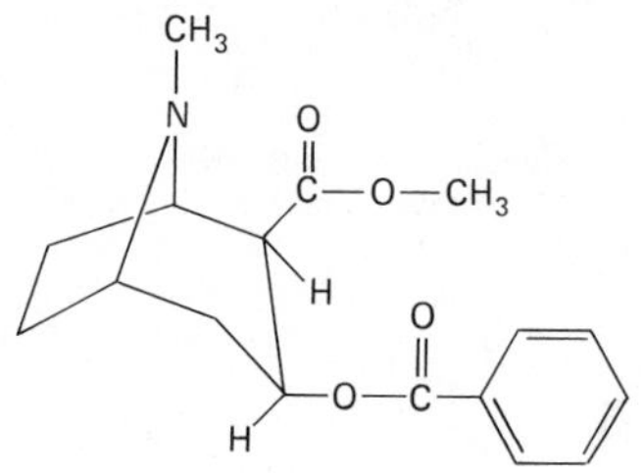

cocaine
(from leaves of coca plant)

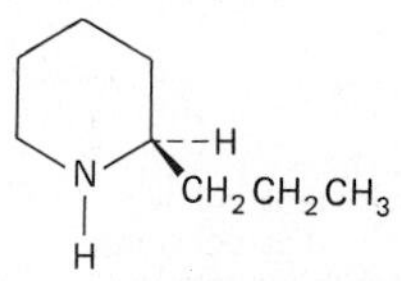

coniine
(poison that Socrates drank;
from hemlock plant)

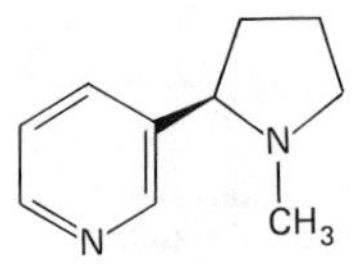

nicotine
(poison; from tobacco plant)

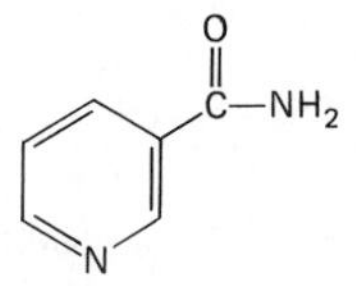

nicotinamide
(a B vitamin)

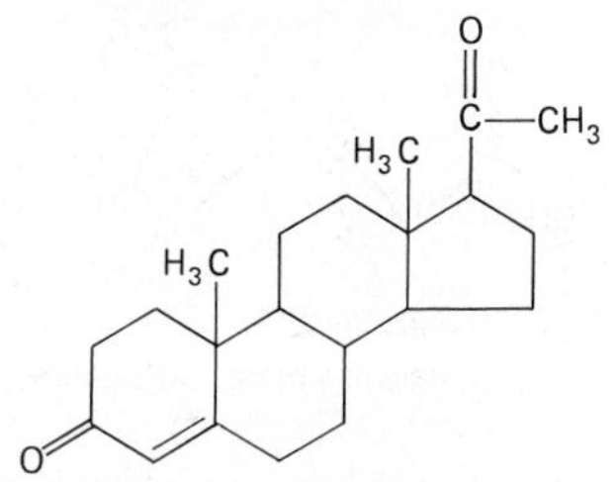

progesterone
(female sex hormone)

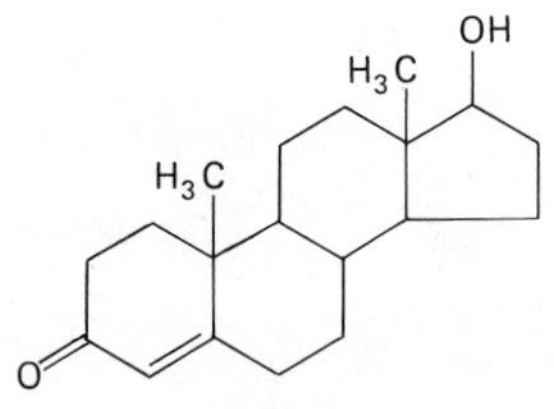

testosterone
(male sex hormone)

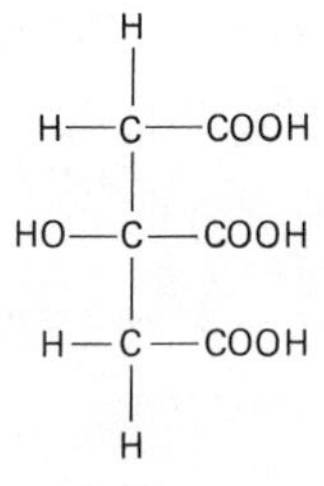

citric acid
(occurs in citrus fruits)

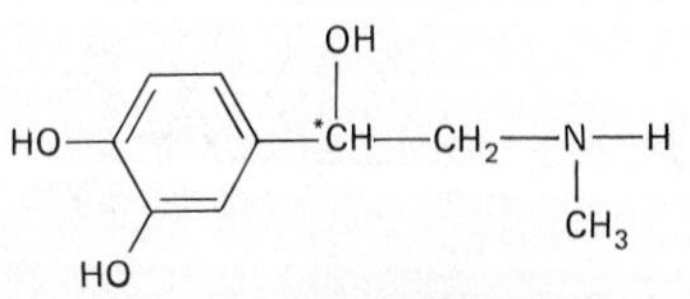

epinephrine
(adrenalin)

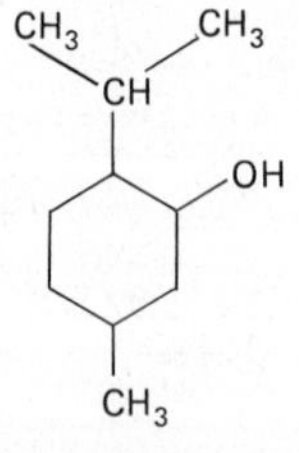

menthol

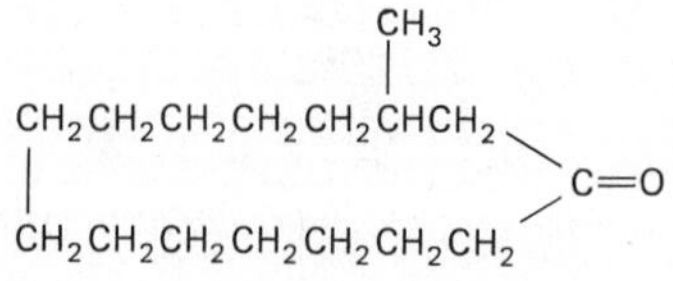

muscone
(gland of male musk deer, used in perfume)

SELF-TEST 13–B

1. Complete the following equation:

    $$CH_3CH_2CH_2OH + CH_3CH_2COOH \xrightarrow{H_2SO_4}$$

2. a. When referring to edible lipids, what is the difference between a fat and an oil? _____
    b. How can the melting points of most edible oils be increased?

    _____

3. In a typical synthesis of organic compounds, the starting compounds must be stripped to their atoms before new compounds can be made. ( ) True or ( ) False

4. Aspirin and oil of wintergreen
    a. are structurally similar
    b. are both acids
    c. can be made from chlorobenzene

5. Phenol is the same as ( ) carbonic or ( ) carbolic acid.

MATCHING SET

_____ 1. Synthesized from NH_4OCN by Wöhler

_____ 2. RCOOH

_____ 3. R—OH

_____ 4. $R-O-\overset{\displaystyle\parallel}{\underset{\displaystyle O}{C}}-R'$

_____ 5. Prostaglandins

_____ 6. RCHO

_____ 7. Linoleic acid

_____ 8. $R-NH_2$

_____ 9. Sodium stearate

a. Made from essential fatty acids
b. An unsaturated fatty acid
c. Amine
d. Aldehyde
e. Carboxylic acid
f. Soap
g. Urea
h. Alcohol
i. Ester

QUESTIONS

1. Name the following compounds:
    a. $CH_3CH_2CH_2COOH$
    b. $CH_3CH_2CH_2OH$

2. Write structural formulas for butanoic acid, aminomethane, 2-butanol, and 3-aminopentane.

3. Indicate what products would be formed in the reaction of
    a. Methanol and acetic acid
    b. 1-propanol and stearic acid
    c. Ethylene glycol and acetic acid

4. Explain how hydrogen bonding could play a significant role in fixing the boiling point of acetic acid.

5. Write structural formulas for the four alcohols with the composition C_4H_9OH.

6. Wood alcohol is a deadly poison that can be made from what deadly gas?

7. Draw a structural formula for each of the following:
    a. An alcohol   c. An ester
    b. An organic acid   d. Glycerol

8. Give an example of:
   a. An alkane
   b. An amine
   c. A carboxylic acid
   d. An ether
   e. An ester
   f. An alkene
   g. An alkyne
   h. An alcohol
   i. A ketone

9. What is meant by each of the following terms?
   a. Proof rating of an alcohol
   b. Denatured alcohol

10. Beginning with petroleum, outline the steps and write the chemical equations for the production of ethyl acetate.

11. Write structural formulas for two compounds that can have each of the molecular formulas listed:
   a. $C_5H_{12}O$
   b. C_3H_6O
   c. $C_5H_{10}O_2$

12. How do primary, secondary, and tertiary alcohols differ?

13. What reaction is used for the breathalyzer test?

14. How is ethylene glycol prepared?

15. Would you use the sodium salt of a fatty acid or the sodium salt of an inorganic acid to wash your hands? Explain.

16. Which would you expect to boil at a higher temperature, $CH_3CH_2CH_2CH_3$ or CH_3CH_2OH? Why?

17. Would you say that hydrogen bonding is stronger or weaker in alcohol when contrasted with that in water?

18. What ester smells like roses?

19. Draw structural formulas for:
   a. Aspirin
   b. Oil of wintergreen
   c. Phenol

20. What ester has the smell of bananas?

21. What chemical reactions can be used to distinguish between:
   a. C_2H_5OH and CH_3COOH
   b. $CH_3COOC_2H_5$ and CH_3COOH

22. What would be the product if *para*-dichlorobenzene were heated with NaOH at 400°C?

23. Tell how you could prepare methanol.

24. The product of oxidation of a primary alcohol is a(n) _____, which can be oxidized to a(n) _____.

25. Indicate the functional groups present in the following molecules:
   a. $CH_3CH_2CH_2COOH$
   b. $CH_3CH_2NH_2$
   c. $CH_3CHCH_2CH_2COOH$
      $\quad\quad |$
      $\quad\quad NH_2$
   d. CH_3CHCH_2COOH
      $\quad\quad |$
      $\quad\quad OH$
   e. $CH_3CCH_2CH_2COOH$
      $\quad\quad \|$
      $\quad\quad O$
   f. CH_3CHCH_2OH
      $\quad\quad |$
      $\quad\quad NH_2$

26. Pure ethyl alcohol is what proof?

27. Prostaglandins belong to what group of organic compounds?

28. Draw a structural formula for a molecule containing:
   a. Alcohol and amine functional groups
   b. Ether and ester functional groups
   c. A double bond and a triple bond
   d. A triple bond and an amine group

29. Explain the common names for methanol and ethanol.

30. Why is a fatty acid so named?

31. Is pure synthetic ethanol different from pure grain alcohol? Explain.

32. Which propanol is used as rubbing alcohol?

33. How many hydrogen bonds are possible per molecule of methanol? Of ethylene glycol? Of glycerol?

34. Would you expect glycerin to be water soluble? Why?

35. Which structural group hydrogen bonds most readily: alcohol, carboxylic acid, or ester?

36. What is the primary chemical in permanent antifreeze?

37. What structural features and properties make ethylene glycol a desirable antifreeze agent?

38. What is the acid in vinegar?

39. After reading this chapter and the previous one, what new thoughts do you have when you view a lump of coal or a drop of petroleum?

40. Consult a medical dictionary and determine the difference between atherosclerosis and arteriosclerosis.

41. What functional groups are found in glucose?

Chapter 14

MAN-MADE GIANT MOLECULES — THE SYNTHETIC POLYMERS

It is impossible for most Americans to get through a day without using a dozen or more materials based on synthetic **polymers.** Many of these materials are *plastics* of one sort or another. Examples of these include plastic dishes and cups, combs, automobile steering wheels and seat covers, telephones, pens, plastic bags for food and wastes, plastic pipes and fittings, plastic water-dispersed paints, false eyelashes and wigs, a wide range of synthetic fibers for clothing, synthetic glues, and flooring materials. In fact, these materials are so widely used they are usually taken for granted. All these materials are composed of **giant molecules.**

> A plastic is a substance that will flow under heat and pressure, and hence is capable of being molded into various shapes. All plastics are polymers, but not all polymers are plastic.

This "flood of plastic objects"did not arise accidentally; it slowly became necessary over a period of 30 or 40 years because (1) natural resources dwindled and so many natural materials became scarce, (2) with rising labor costs, many items could be made less expensively by molding than by whittling, shaving, sawing, and gluing, and (3) the new materials were so superior in properties that they did the job better.

Some of our most useful polymer chemistry has resulted from copying giant molecules found in nature. Rayon is remanufactured cellulose. Synthetic rubber is copied from natural latex rubber. As useful as they may be, however, polymer chemistry is not restricted to nature's models. Nylon, Dacron, and polycarbonates are a few examples of synthetic molecules that do not have exact duplicates in nature. We have gone to school on nature and extended our knowledge to new situations.

The purpose of this chapter is to investigate the structural chemistry of polymers to see just why they have such useful properties. Are these properties the result of stronger bonds, or groups of molecules acting together, or is there some other explanation? As we shall see, giant molecules were observed first in nature, and then copied and improved on by the chemist. In the next chapter we shall study some of nature's polymers and their functions.

WHAT ARE GIANT MOLECULES?

Many chemists were reluctant to accept the concept of giant molecules, but in the 1920s a persistent German chemist, Hermann Staudinger (1881–1965; Nobel prize, 1953), championed the idea and introduced a new term, ***macromolecule,*** for

> A *macromolecule* is a molecule with a very high molecular weight.

these giant molecules. Staudinger devised experiments that yielded accurate molecular weights, and he synthesized "model compounds" to test his theory. One of his first model compounds was prepared from styrene, a chemical made from ethylene and benzene.

Styrene is the number 20 commercial chemical.

$$H_2C\!=\!CH$$

STYRENE

Under the proper conditions, styrene molecules use the "extra" electrons of the double bond to undergo a **polymerization** reaction to yield polystyrene, a material composed of giant molecules. The word **polymer** means "many parts" (Greek, *poly* meaning "many," *meros* meaning "parts"). The molecules of styrene are the **monomers** (Greek, *mono* meaning "one"); they provide the recurring units in the giant molecule analogous to identical railroad cars coupled together to make a long train.

A polymer has molecules composed of a large number of similar units.

The macromolecule polystyrene is represented as a long chain of monomer units bonded to each other. Each unit is bonded to the next by a strong covalent bond. The polymer chain is not an endless one; some polystyrenes made by Staudinger were found to have molecular weights of about 600,000, corresponding to a chain of about 5700 styrene units. The polymer chain can be indicated as

$$R\!-\!CH_2\!-\!CH\!-\!\!\left(\!CH_2\!-\!CH\!-\!\right)_{\!n}\!CH_2\!-\!CH\!-\!R$$

where R represents some terminal group, often an impurity, and n is a large number.

Polystyrene is a clear, hard, colorless solid at room temperature. Since it can be molded easily at 250°C, the term ***plastic*** has become associated with it and similar materials. Polystyrene has so many useful properties that its commercial production, which began in Germany in 1929, today exceeds 5 billion pounds per year. It is used to make combs, bowls, toys, electrical parts, and many other items such as tough plastics used for radio and TV cabinets and synthetic rubber for tires.

Synthetic polymers are commonly called *plastics* when in a solid form.

There are two broad categories of plastics. One, when heated repeatedly, will soften and flow; when it is cooled, it hardens. Materials that undergo such reversible changes when heated and cooled are called **thermoplastics;** polystyrene is one example. The other type is plastic when first heated, but when heated further it forms a set of interlocking bonds. When reheated, it cannot be softened and re-formed without extensive degradation. These materials are called **thermosetting plastics** and include such familiar names as Bakelite and rigid-foamed polyurethane, a polymer that is finding many new uses as a construction material.

Thermoplastic polymers can be repeatedly softened by merely heating.

Thermosetting polymers form cross-linking bonds when heated and then become rigid.

In order to gain a better understanding of polymers, we must look at representative examples of the different types of polymerization processes.

ADDITION POLYMERS

In the previous section it was noted that some polymers, such as polystyrene, are made by adding monomer to monomer to form a polymer chain of great length. Perhaps the easiest addition reactions to understand chemically are those involving monomers containing double bonds. The simplest monomer of this group is ethene, C_2H_4. When ethene (ethylene) is heated under pressure in the presence of oxygen, polymers with molecular weights of about 30,000 are formed. In order to enter into reaction, the double bond of an ethene molecule must be broken. This forms **reactive sites** composed of unpaired electrons at either end of the molecule.

Ethylene (ethene) is the number five commercial chemical.

$$
\begin{array}{c}
\text{H} \quad \text{H} \\
| \quad | \\
\text{C}=\text{C} \\
| \quad | \\
\text{H} \quad \text{H}
\end{array}
\xrightarrow{\text{energy}}
\begin{array}{c}
\text{H} \quad \text{H} \\
| \quad | \\
\cdot\text{C}-\text{C}\cdot \\
| \quad | \\
\text{H} \quad \text{H}
\end{array}
$$

REACTIVE SITE

The partial breaking of the double bond can be accomplished by physical means such as heat, ultraviolet light, X rays, and high-energy electrons. This *initiation* of the polymerization reaction can also be accomplished with chemicals such as organic peroxides. These initiators, which are very unstable, break apart into pieces with unpaired electrons. These fragments (called **free radicals**) are ravenous in trying to find a "buddy" for their unpaired electrons. They react readily with molecules containing carbon-carbon double bonds.

An organic peroxide, RO—OR′, produces free radicals, RO·, each with an unpaired electron.

$$
\text{CH}_2 \text{:} \text{CH}_2 \xrightarrow[\text{a}]{\cdot\text{OR}} \cdot\text{CH}_2-\text{CH}_2\text{OR} \xrightarrow{n\text{CH}_2=\text{CH}_2} \text{(CH}_2-\text{CH}_2\text{)}_{n+1}\text{OR}
$$

peroxide free radical

n is a very large number, 40,000 or more

The extension of the polyethylene chain shown above comes from the unpaired electron bonding to an electron in an unreacted ethylene molecule. This leaves another unpaired electron to bond with yet another ethylene molecule. For example,

$$
\text{ROCH}_2-\text{CH}_2\cdot + \text{CH}_2\text{:}\text{CH}_2 \rightarrow \text{ROCH}_2-\text{CH}_2-\text{CH}_2-\text{CH}_2\cdot
$$

Polyethylenes formed under various pressures and catalytic conditions have different molecular structures and hence different physical properties. For example, chromium oxide as a catalyst yields almost exclusively the linear polyethylene shown in the margin. Actually, a methyl group is attached to about every eighth or tenth carbon in the chain. If ethylene is heated to 230°C at a pressure of 200 atmospheres, irregular branches result. Under these conditions, free radicals undoubtedly attack the chain at random positions, thus causing the irregular branching.

$$
-\text{CH}_2-\text{CH}_2-\text{CH}_2-\text{CH}_2- \rightarrow -\text{CH}_2-\text{CH}-\text{CH}_2-\text{CH}_2- + \text{H}\cdot
$$

$$
\begin{array}{c}
| \\
\text{CH}_2 \\
| \\
\text{CH}_2 \\
| \\
\text{R}
\end{array}
$$

$$\text{RCH}_2\text{CH}_2\cdot$$

BRANCHED POLYMER CHAINS

A portion of a linear polyethylene molecule. Ethylene is unsaturated; polyethylene is saturated.

TABLE 14–1   **Ethylene Derivatives That Undergo Addition Polymerization**

FORMULA	MONOMER NAME	POLYMER NAME	USES
$CH_2{=}CH_2$	Ethylene	Polyethylene	Coats milk cartons, wire insulation, bread wrappers, toys, films
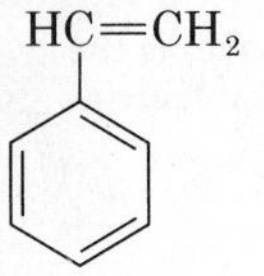 $HC{=}CH_2$ (phenyl ring)	Styrene	Polystyrene	Synthetic rubber, combs, toys, bowls, packaging, appliance parts
$CH_2{=}CHCl$	Vinyl chloride (number 19 commercial chemical)	Polyvinylchloride (PVC)	As a vinyl acetate copolymer in phonograph records, credit cards, rain wear, pipes, adhesives, films
$CH_2{=}CH$ — O—C—CH_3 ($\parallel O$)	Vinyl acetate	Polyvinylacetate	Latex paint
$CH_2{=}CH$ — CN	Acrylonitrile	Polyacrylonitrile (PAN) (Plexiglas modifier)	Rug fibers, high-impact plastics
$CH_2{=}CH{-}CH{=}CH_2$	Divinyl (1,3-butadiene)	Buna rubbers	Tires and hoses
$CH_2{=}C$—C (CH_3, O, $O{-}CH_3$)	Methyl methacrylate	Polymethyl methacrylate (Plexiglas, Lucite)	Transparent objects, lightweight "pipes"
$CF_2{=}CF_2$	Tetrafluoroethylene	Polytetrafluoroethylene (TFE) (Teflon)	Insulation, bearings, nonstick fry pan surfaces

(handwritten annotation next to Acrylonitrile: "very tough/durable")

(handwritten annotation next to Buna rubbers: "(car (automobile)")

The molecules in linear polyethylene can line up with one another very easily, yielding a tough, high-density crystalline compound that is useful in making toys, bottles, and structural parts. The polyethylene with irregular branches is less dense, more flexible, and not nearly as tough as the linear polymer, since the molecules are generally farther apart and their arrangement is not as precisely ordered. This material is used for trash bags, squeeze bottles, and other similar applications.

There is a large group of ethylene derivatives that undergo addition polymerization. Table 14–1 summarizes some information on these materials.

Synthetic "Natural" Rubber — A Tailor-Made Addition Polymer

Rubber is vulcanized by heating it with sulfur, which forms links between the polymer chains.

A very interesting application of stereochemical control over polymerization is the manufacture of synthetic rubber. When several structures are possible and only one is desired, stereochemical control must be exercised.

Natural rubber, a product of the *Hevea brasilieusis* tree, is a hydrocarbon with the composition C_5H_8, and when it is decomposed in the absence of oxygen it yields the monomer isoprene:

$$CH_2{=}\underset{\underset{\text{ISOPRENE}}{|}}{\overset{\overset{\displaystyle CH_3}{|}}{C}}{-}CH{=}CH_2$$

Natural rubber occurs as latex (an emulsion of rubber particles in water) that oozes from rubber trees when they are cut. Precipitation of the rubber particles yields a gummy mass that is not only elastic and water-repellent but also very sticky, especially when warm. In 1839, after 10 years' work on this material, Charles Goodyear (1800–1860) discovered that heating gum rubber with sulfur produced a material that was no longer sticky, but still elastic, water-repellent, and resilient.

Vulcanized rubber, as Goodyear called his product, contains short chains of sulfur atoms that bond together the polymer chains of the natural rubber and reduce its unsaturation (Fig. 14–1). The sulfur chains help to align the polymer chains, so the material does not undergo a permanent change when stretched but springs back to its original shape and size when the stress is removed. Substances that behave this way are called **elastomers.**

In later years chemists searched for ways to make a synthetic rubber so we would not be completely dependent on imported natural rubber during emergencies, such as during the first years of World War II. In the mid-1920s, German chemists polymerized butadiene (obtained from petroleum and structurally similar

VULCANIZATION:

SULFUR, OR S_8 MOLECULE

Stereoregulating: Controlling the arrangement of monomer units in the polymer.

$$CH_2{=}CH{-}CH{=}CH_2$$
BUTADIENE

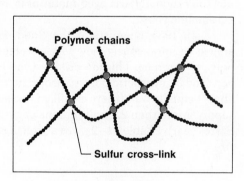

a. Before stretching

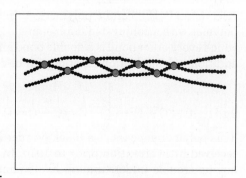

FIGURE 14–1 Stretched vulcanized rubber will spring back to its original structure, an elastomeric property.

b. Stretched

to isoprene, but without the methyl group side chain). The product was buna rubber, so named because it was made from butadiene (Bu—) and catalyzed by sodium (—Na).

The behavior of natural rubber (polyisoprene), it was learned later, is due to the specific arrangement within the polymer chain. We can write the formula for polyisoprene with the CH_2 groups on opposite sides of the double bond (the *trans* arrangement)

$$-CH_2 \diagdown_{CH_3} C{=}C \diagup^{H}_{CH_2-CH_2} \diagdown_{H}^{CH_3} C{=}C \diagup^{CH_2-CH_2}_{CH_3} \diagdown_{CH_2-}^{H} C{=}C$$

POLY-*TRANS*-ISOPRENE (THE —CH_2—CH_2— GROUPS ARE *TRANS*)

or with the CH_2 groups on the same side of the double bond (the *cis* arrangement, from Latin meaning "on this side").

$$-CH_2 \diagdown_{CH_3} C{=}C \diagup^{CH_2-CH_2}_{H} \diagdown^{CH_3}_{H} C{=}C \diagup^{CH_2-CH_2}_{H} \diagdown^{CH_2-}_{H} C{=}C$$

POLY-*CIS*-ISOPRENE (THE —CH_2—CH_2— GROUPS ARE *CIS*)

Natural rubber is poly-*cis*-isoprene.

Natural rubber is poly-*cis*-isoprene. However, the *trans* material also occurs in nature, in the leaves and bark of the sapotacea tree, and is known as *gutta-per-cha*. It is used as a thermoplastic for golf ball covers, electrical insulation, and other such applications. Without an appropriate catalyst, polymerization of isoprene yields a solid that is like neither rubber nor gutta-percha. Neither the *trans* polymer nor the randomly arranged material is as good as natural rubber *(cis)* for making automobile tires.

In 1955, chemists at the Goodyear and Firestone companies discovered, almost simultaneously, how to use stereoregulation catalysts to prepare synthetic poly-*cis*-isoprene. This material is, therefore, structurally identical to natural rubber. Today, synthetic poly-*cis*-isoprene can be manufactured cheaply and is used almost equally well (there is still an increased cost) when natural rubber is in short supply. More than 2.4 million tons of synthetic rubber are produced in the United States yearly. Table 14–2 gives a typical rubber formulation as it might be used in a tire.

POLYACETYLENE, A PLASTIC THAT CONDUCTS ELECTRICITY

Acetylene, C_2H_2, can be polymerized in the presence of a catalyst to produce a polymer with a conjugated double-single bond system, a chain of carbon atoms in which every other bond is a double bond and every other bond is a single bond. This addition polymerization reaction can be represented as follows:

$$2n\ H{-}C{\equiv}C{-}H \rightarrow \left[\begin{matrix} H & H & H & H \\ | & | & | & | \\ -C{=}C{-}C{=}C{-} \end{matrix} \right]_n$$

This polymer appears as a black powder in the usual laboratory preparation and received little attention prior to 1970. In that year a Korean university student, having trouble understanding his Japanese instructor, Hideki Shirakawa, prepared the polymer using an excessive amount of the catalyst. The result was a silver

TABLE 14–2 A Rubber Formulation

INGREDIENT	NAME	PERCENTAGE	FORMULA	FUNCTION
Rubber	Poly-*cis*-isoprene	62.0	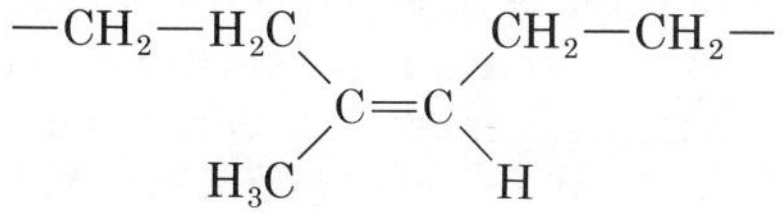 $\begin{array}{ccc} -CH_2-H_2C & & CH_2-CH_2- \\ & C=C & \\ H_3C & & H \end{array}$	Elastomer
Activators	Zinc oxide stearic acid	2.7 0.6	ZnO $C_{17}H_{35}COOH$	Activates vulcanizing agents; stearic acid acts as a lubricant in processing
Vulcanizing agent	Sulfur	1.5	S_8	Crosslinks polymer chains
Filler	Carbon black	30.5	C	Provides strength and abrasion resistance
Accelerator	Dibenzthiozole disulfide	1.1	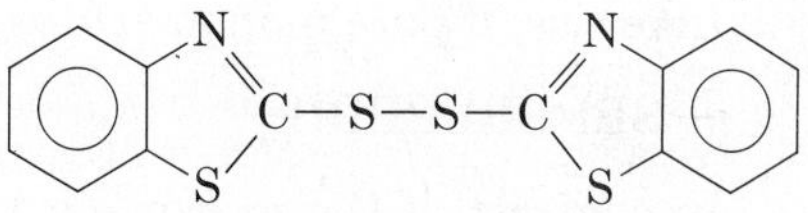 [benzothiazole]$C-S-S-C$[benzothiazole]	Catalyzes vulcanization
Antioxidant	Alkylated diphenylamine	1.1	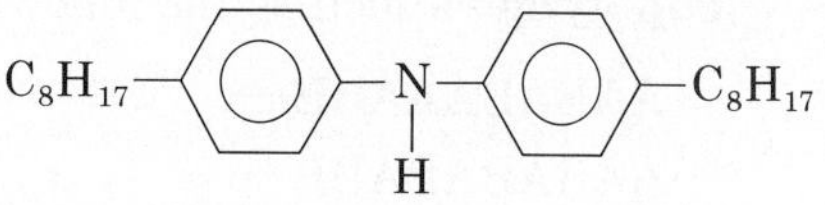 C_8H_{17}—[ring]—N(H)—[ring]—C_8H_{17}	Inhibits attack by oxygen or ozone in the air
Processing oil	Hydrocarbon oil	0.5	C_nH_{2n+2}	Plasticizer

film that looked more like a metal than anything else. Furthermore, the film conducted electricity, which was a first for plastic materials.

The metal-like polyacetylene can be explained in terms of very long conjugated polymer molecules that fit nicely together in crystalline structure. Also, it has long been known that conjugated carbon systems are adept at passing electrical charge from atom to atom along the system.

In 1975 at the University of Pennsylvania, Alan MacDiarmid began a systematic study of this new form of polyacetylene. It was soon learned that the electrical conductivity of the plastic could be increased a trillionfold (10^{12}) by the introduction of iodine into the polymer during its formation. The resulting conductivity rivaled that of metals! Recall that iodine atoms have an attraction for one additional electron per atom. If an iodine atom removed an electron from a double bond at one end of the polyacetylene molecule, the entire molecule would simply pass negative charge along the conjugated system and into the "positive hole" if an electron were available at the other end of the molecule. This flow of electrical charge is electrical conduction. An applied electrical potential could subsequently remove the newly received electron from the iodine atom, and the process could be repeated over and over again.

Following polyacetylene, many similar plastics with useful electrical properties can be conceived theoretically, and many are now being made. Obviously, the applications are very exciting to the many workers in this field. For example, lightweight batteries with plastic components have been made and appear to be commercially exploitable. Think what this might mean since the one great fault of the electric car has been the weight of the lead electrodes in batteries. Another area of intense research is the conversion of sunlight into electrical energy; the plastics may offer a much cheaper conversion surface where the energy transfer takes place. Motors, generators, wires, magnets, and electronic parts are all possible applications of the new materials. Will plastic conductors replace metal conductors as synthetic fibers have partially replaced silk and wool? Time will tell.

COPOLYMERS

After examining Table 14–1, you might well wonder what would happen if a mixture of two monomers is polymerized. If we polymerize pure monomer A, we get a **homopolymer,** poly A:

—AAAAAAAAAA—

Likewise, if pure monomer B is polymerized, we get a homopolymer, poly B:

—BBBBBBBBBB—

A copolymer is made by polymerizing two or more different monomers together.

In contrast, if the monomers A and B are mixed and then polymerized, we get **copolymers** such as the following:

—AABABAAABB—

—AABABABABB—

—BABABBAABA—

In such polymers the order of the units is often completely random, in which case the properties of the copolymer will be determined by the ratio of the amount of A to the amount of B and the reaction conditions during polymerization.

It is possible to produce copolymers that have long chains of similar monomers in their structures. These are called **block copolymers** and can be represented as:

—AAAAAABBBBBBBBBBAAAAAA—

To overcome brittleness in polypropylene, both random and block copolymers with ethylene are made. The block copolymers are more resistant to impact. They are molded into articles such as toys by a process called injection molding, in which the molten plastic flows under pressure into the mold. Random copolymers are more transparent, whereas the homopolymer is used for filaments in such applications as carpets and rope.

A copolymer can have useful properties that are different from and often superior to those of the polymers of its pure constituents. As an example, consider synthetic rubbers again. During World War II it was apparent to our military planners that we would be hard-pressed if our rubber supplies from Asia were cut off by Japan. A crash program was begun to develop synthetic rubber that would be as good as natural rubber. The Germans had earlier polymerized styrene, but this is a hard thermoplastic with little elasticity. They had also polymerized butadiene to make the first synthetic rubber (buna rubber), although it was not very serviceable. American chemists found, however, that a 1 to 6 copolymer of styrene and butadiene possessed properties closer to those of natural rubber.

$$CH_2=CH + CH_2=CH-CH=CH_2 \xrightarrow[\text{polymerization}]{\text{addition}}$$

STYRENE   BUTADIENE

$$-CH_2CH=CHCH_2CH_2CHCH_2CH=CHCH_2CH_2CH=CHCH_2-$$

SBR COPOLYMER (STYRENE-BUTADIENE RUBBER)

The double bonds remaining in the polymer chain allow them to undergo vulcanization like natural rubber polymer chains.

Copolymers are also used extensively in the plastics industry. For example, high-impact material for radio cabinets, tools, handles, and anything that might be dropped or struck is made of a copolymer of acrylate, butadiene, and styrene. Fibrous fillers and reinforcing agents add strength. A pure form of styrene-butadiene rubber has even replaced the latex in chewing gum.

The largest group of synthetic polymers is used to make rubber—about 10 billion pounds per year. About 40% is used for tires.

ADDITION POLYMERS AND PAINTS

All paints involve polymers in one form or another (Fig. 14–2). Popular latex and acrylic paints contain addition polymers that serve as **binders.** A paint binder forms a molecular network to hold the **pigment** (coloring agent) in place and to hold the paint to the painted surface. In oil-based paints, a drying oil (such as linseed oil) or a resin is the binder. All paints also have a volatile solvent or thinner; this is water in water-based paints and turpentine (or mineral spirits, or both) in oil-based paints.

Early latex paints were emulsions of partly polymerized styrene and butadiene in water (Fig. 14–3). Some type of emulsifying agent (such as soap) was present to keep the small drops of nonpolar styrene and butadiene dispersed in the polar water.

Immediately after the application of a latex paint, the water begins to evaporate. When some of the water is gone, the emulsion breaks down, and the remaining water evaporates quickly, leaving the paint film. Further polymerization of the styrene and butadiene follows slowly, but the paint appears to be dry in a few minutes. The pigment is trapped in the network of the polymer. If the paint is white, the pigment is probably titanium dioxide, TiO_2, which has replaced the poisonous compound "white lead," $Pb(OH)_2 \cdot 2\,PbCO_3$, that was used in older paints.

The styrene-butadiene resin is the least expensive binder material used, but it has a relatively long curing period, relatively poor adhesion, and a tendency to yellow with age. Polyvinylacetate is only a little more expensive and is an improvement over the styrene-butadiene resin. It quickly captured 50% of the latex market for interior paints. Another type with rapidly growing popularity, though about one third more expensive, includes the acrylic resins and the "acrylic latex" paints. These are more washable and much more resistant to light damage. They are especially useful as exterior paints.

In 1970, nearly 830 million gallons of coatings were sold by U.S. companies. The $3 billion sales represented 5.5% of sales of all chemicals and allied products during 1970.

The first commercial water-based latex paint was Glidden's Spred Satin, introduced in 1948.

Water-based latex paints reduce fire hazards and air pollution associated with the handling and application of oil-based paints.

Monomer of polyvinylacetate:

FIGURE 14–2   Photographers photographing a painter painting—in this picture the John Quincy Adams birthplace in Quincy, Massachusetts. The Sears Great American Home Series of advertisements illustrates the importance of surface coatings as preservatives. (Courtesy of Sears, Roebuck and Co., Chicago, Illinois.)

Acrylic polymers have a sheen that allows latex paint to compete in the exterior gloss market traditionally monopolized by oil-based coatings. Acrylics adhere well and control corrosion. Acrylics are polymers of acrylonitrile,

H   CN
  \   /
   C = C
  /   \
H   H

See Tables 14–1 and 14–4.

Mineral spirits are petroleum fractions of moderate volatility.

The fluoropolymers, similar to Teflon, are especially promising as surface coatings because of their great stability. Fluorine atoms are substituted for hydrogen atoms in the organic structure. Metals covered with polyvinylidene fluoride carry up to a 20-year guarantee against failure from exposure.

In the past few years, paint manufacturers have begun to blend linseed oil emulsions with latex emulsions in order to take advantage of the penetrating ability of the triglyceride molecules in the linseed oil. Some "latex" paints now contain as much as 75% linseed oil emulsion but still have the desirable characteristic of latex. Table 14–3 summarizes various additives to emulsion paints and the rationale for their use.

The drying of modern oil-based paints involves much more than the evaporation of the mineral spirits or turpentine solvent. The chemical reaction between a

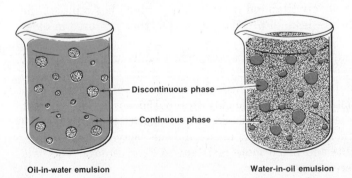

Oil-in-water emulsion   Water-in-oil emulsion

FIGURE 14–3   Two kinds of emulsions. An emulsion is composed of two immiscible liquids, one dispersed as tiny droplets in the other. An emulsifying agent is required to stabilize an emulsion.

TABLE 14–3 Additives Used in Emulsion Paints

Dispersing agents for pigments	Example: tetrasodium pyrophosphate ($Na_4P_2O_7$). The principle of like-charged particles repelling.
Protective colloids and thickeners	A thicker paint is slower to settle and drips and runs less. A protective colloid tends to stabilize the organic-water interface in the emulsion. Examples: sodium polyacrylates, carboxymethylcellulose, clays, gums. (Same mechanism as soap dispersing oil in water, Chapter 22.)
Defoamers	Foaming presents a serious problem if not corrected. Chemicals used: tri-n-butylphosphate, n-octyl alcohol and other higher alcohols, silicone oil.
Coalescing agents	As the water evaporates and the paint dries, an agent is needed to stick the pigment particles together. As the resin film forms, the agent evaporates. Coalescing agents must volatilize very slowly. Examples: hexylene glycol and ethylene glycol.
Freeze-thaw additives	Freezing will destory the emulsion. Antifreezes such as ethylene glycol are used.
pH controllers	The effectiveness of the ionic or molecular form of the emulsifier depends on the acid or alkaline conditions (pH). The wrong pH will break down the emulsion. Most paints tend to be too acidic. Ammonia, NH_3, is added to neutralize the acid.

Typical molecule in linseed oil.

drying oil and oxygen from the air completes the drying process. Common drying oils are soybean, castor, coconut, and linseed oils; the most widely used is linseed, which comes from the seed of the flax plant. All of these oils are glyceryl esters of fatty acids, as discussed in Chapter 13. Hydrolysis of a typical linseed oil yields the following assortment of fatty acids:

4% to 7%	Palmitic acid (16 C atoms)	(saturated)
2% to 5%	Stearic acid (18 C atoms)	(saturated)
9% to 38%	Oleic acid (18 C atoms)	(unsaturated)
3% to 43%	Linoleic acid (18 C atoms)	(unsaturated)
25% to 58%	Linolenic acid (18 C atoms)	(unsaturated)

The chemical action of oxygen on a drying oil is to replace a hydrogen atom on a carbon atom next to a $C=C$ double bond in an unsaturated fatty acid chain. When oxygen reacts with two fatty acids on two oil molecules, the result is cross-linking between the two molecules.

● Carbon atom
○ Oxygen atom

$$-CH_2-CH_2-CH=CH-CH_2-$$

$$+ O_2 \rightarrow$$

$$-CH_2-CH_2-CH=CH-CH_2-$$
part of another molecule

$$-CH_2-CH-CH=CH-CH_2-$$
$$|$$
$$O \text{ ether linkage}$$
$$|$$
$$-CH_2-CH-CH=CH-CH_2-$$

$$+ H_2O$$

The polymeric network produced by the crosslinking hardens the paint, traps the pigment, and secures the paint in the crevices of the painted surface.

Metal ions such as Zn^{2+}, Co^{2+}, Fe^{3+}, Mn^{2+}, and Ca^{2+} are added to oil-based paints to catalyze the drying process. These ions decompose peroxides (compounds containing the $-O-O$ group) formed during the crosslinking process and precipitate free acids as salts of the ions.

CONDENSATION POLYMERS

Polyesters

A chemical reaction in which two molecules react by splitting out or eliminating a small molecule is called a **condensation reaction.** For example, acetic acid and ethyl alcohol will react, splitting out a water molecule, to form ethyl acetate, an **ester.**

$$CH_3C\overset{O}{\overset{\|}{}}\!\!-OH + HOCH_2CH_3 \xrightarrow[\text{catalyst}]{H^+} CH_3C\overset{O}{\overset{\|}{}}\!\!-OCH_2CH_3 + H_2O$$

ACETIC ACID   ETHANOL   ETHYL ACETATE (AN ESTER)

This important type of chemical reaction does not depend on the presence of a double bond in the reacting molecules. Rather, it requires the presence of two kinds of functional groups on two different molecules. If each reacting molecule has *two* functional groups, both of which can react, it is then possible for condensation reactions to lead to a long-chain polymer. If we take a molecule with two carboxyl groups, such as terephthalic acid, and another molecule with two alcohol groups, such as ethylene glycol, each molecule can react at each end. The reaction of one acid group of terephthalic acid with one alcohol group of ethylene glycol initially produces an ester molecule with an acid group left over on one end and an alcohol group left over on the other:

TEREPHTHALIC ACID   ETHYLENE GLYCOL

(AN ESTER)

Subsequently, the remaining acid group can react with another alcohol group, and the alcohol group can react with another acid molecule. The process continues until an extremely large polymer molecule, known as a **polyester,** is produced with a molecular weight in the range of 10,000 to 20,000.

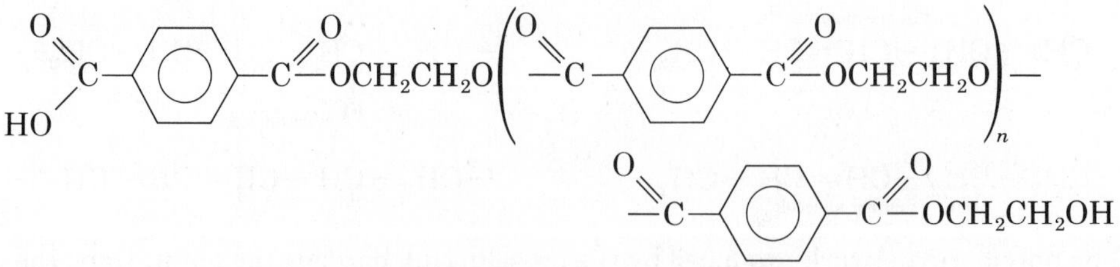

POLY(ETHYLENE GLYCOL TEREPHTHALATE)

Poly(ethylene glycol terephthalate) is used in making polyester textile fibers marketed under such names as *Dacron* and *Terylene* and films such as *Mylar.* The film material has unusual strength and can be rolled into sheets one thirtieth the thickness of a human hair. In film form this polyester is often used as a base for magnetic recording tape and for packaging frozen food. Dacron (and Teflon) tubes

---

Margin notes:

In a condensation polymerization, molecules are linked when they react to split out a small molecule such as water. The backbone of the polymer contains functional groups.

The esterification of a dialcohol and a diacid involves two positions on each molecule.

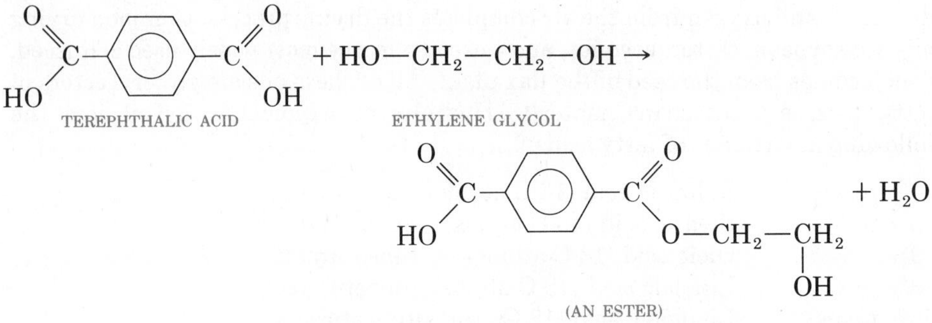

Terephthalic acid is the number 21 commercial chemical.

A typical polyester is produced from a dialcohol and a diacid.

substitute for human blood vessels in heart bypass operations. The inert, nontoxic, nonallergenic, noninflammatory, non–blood-clotting natures of these polymers make them excellent substitutes.

Condensation Polymers and Baked-On Paints

If you have ever had a car repainted, perhaps you have seen the baking oven in which the paint is dried (Fig. 14–4). Automobile finishes and those on major appliances (such as refrigerators, washing machines, and stoves) require very tough, adherent paints in order to withstand abuse. The tough coating is produced by extensively crosslinked condensation polymers.

A popular type of baked-on paint is the **alkyd** variety. The term comes from a combination of the words *alc*ohol and *acid.* Alkyds, then, are polyesters with extensive crosslinking. One of the simpler alkyds is formed from the diacid, phthalic acid, and the trialcohol, glycerol.

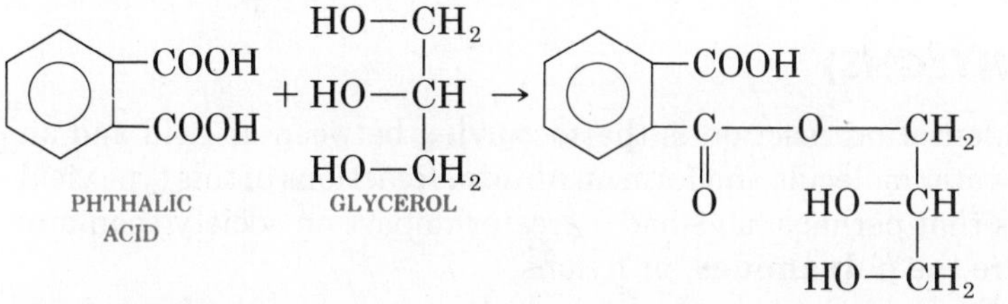

When General Motors lacquered the 1923 Oakland with a nitrocellulose lacquer, the protective coatings industry first began its expansion into the use of a wide variety of materials instead of a few naturally occurring oils and minerals.

The —OH and —COOH groups continue to react with more and more reactant molecules until extensive crosslinking occurs. Heating to about 130°C for about 1 hour causes maximum crosslinking. A portion of the resin's structure is shown as follows:

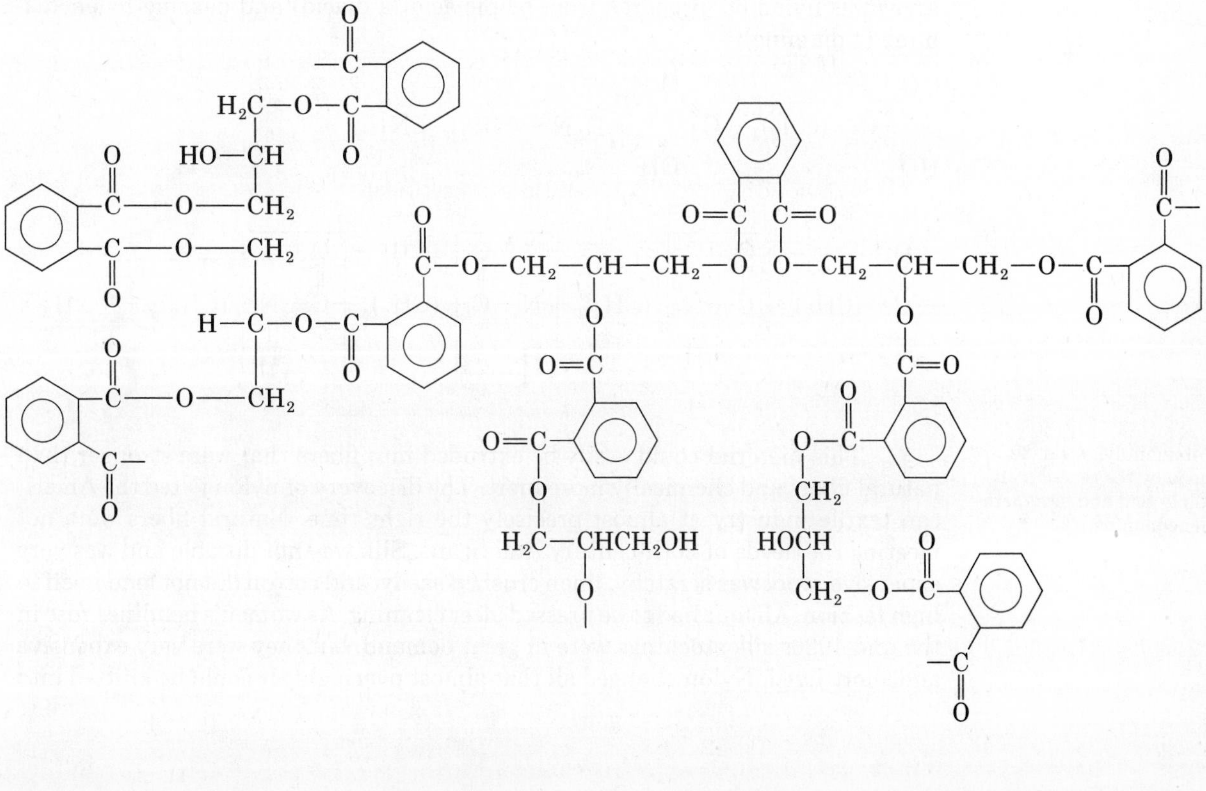

FIGURE 14–4   The high temperatures in a drying oven cause numerous crosslinking reactions to take place, which increase the surface strength of an alkyd paint.

POLYAMIDES (NYLONS)

Another useful condensation reaction is that occurring between an acid and an amine to split out a water molecule and form an **amide.** Reactions of this type yield a group of polymers that perhaps have had a greater impact on society than any other type. These are the **polyamides,** or nylons.

In 1928, the Du Pont Company embarked on a program of basic research headed by Dr. Wallace Carothers (1896–1937), who came to Du Pont from the Harvard University faculty. His research interests were high molecular weight compounds, such as rubber, proteins, and resins, and the reaction mechanisms that produced these compounds. In February 1935, his research yielded a product known as nylon 66, prepared from adipic acid (a diacid) and hexamethylenediamine (a diamine):

$$\underset{\text{ADIPIC ACID}}{\text{HO}\overset{\overset{\text{O}}{\parallel}}{\text{C}}-(CH_2)_4-\overset{\overset{\text{O}}{\parallel}}{\text{C}}\text{OH}} + \underset{\text{HEXAMETHYLENEDIAMINE}}{H_2N-(CH_2)_6-NH_2} \rightarrow$$

$$-\overset{\overset{\text{O}}{\parallel}}{\text{C}}-(CH_2)_4-\boxed{\overset{\overset{\text{O}}{\parallel}}{\text{C}}-\underset{\text{H}}{\text{N}}}-(CH_2)_6-\boxed{\text{N}-\overset{\overset{\text{O}}{\parallel}}{\text{C}}}-(CH_2)_4-\boxed{\overset{\overset{\text{O}}{\parallel}}{\text{C}}-\underset{\text{H}}{\text{N}}}-(CH_2)_6- + xH_2O.$$

NYLON 66
(The amide groups are outlined for emphasis.)

This material could easily be extruded into fibers that were stronger than natural fibers and chemically more inert. The discovery of nylon jolted the American textile industry at almost precisely the right time. Natural fibers were not meeting the needs of 20th-century Americans. Silk was not durable and was very expensive, wool was scratchy, linen crushed easily, and cotton did not lend itself to high fashion. All four had to be pressed after cleaning. As women's hemlines rose in the mid-1930s silk stockings were in great demand, but they were very expensive and short-lived. Nylon changed all that almost overnight. It could be knitted into

Common nylon can be made by the reaction of adipic acid and hexamethylenediamine.

the sheer hosiery women wanted, and it was much more durable than silk. The first public sale of nylon hose took place in Wilmington, Delaware (the hometown of Du Pont's main office), on October 24, 1939. The stockings were so popular they had to be rationed. World War II caused all commercial use of nylon to be abandoned until 1945, as the industry turned to making parachutes and other war materials. Not until 1952 was the nylon industry able to meet the demands of the hosiery industry and to release nylon for other uses as a fiber and as a thermoplastic.

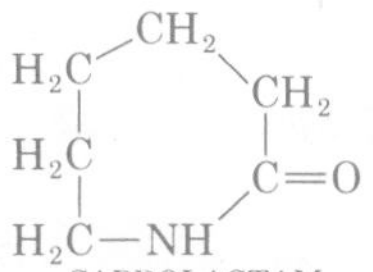

     Many kinds of nylon have been prepared and tried on the consumer market, but two, nylon 66 and nylon 6, have been most successful. Nylon 6 is prepared from caprolactam, which comes from aminocaproic acid. Notice how aminocaproic acid contains an amine group on one end of the molecule and an acid group on the other end.

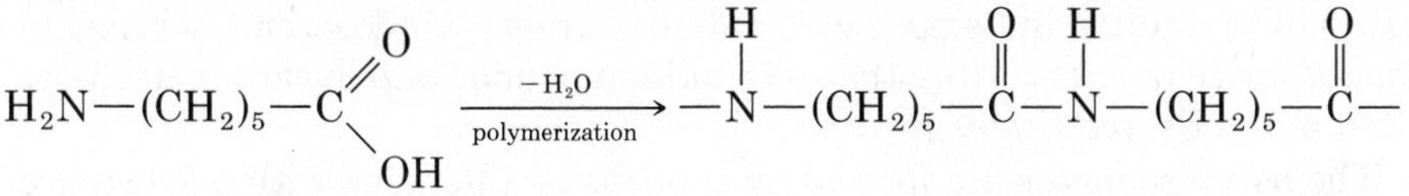

CAPROLACTAM

$$H_2N-(CH_2)_5-C\overset{O}{\underset{OH}{<}} \xrightarrow[\text{polymerization}]{-H_2O} -\overset{H}{\underset{|}{N}}-(CH_2)_5-\overset{O}{\overset{||}{C}}-\overset{H}{\underset{|}{N}}-(CH_2)_5-\overset{O}{\overset{||}{C}}-$$

AMINOCAPROIC ACID   PORTION OF NYLON 6

     Figure 14–5 illustrates another facet of the structure of nylon — **hydrogen bonding.** This type of bonding explains why the nylons make such good fibers. In order to have good tensile strength, the chains of atoms in a polymer should be able

Hydrogen bonding is important in determining the properties of nylon fibers.

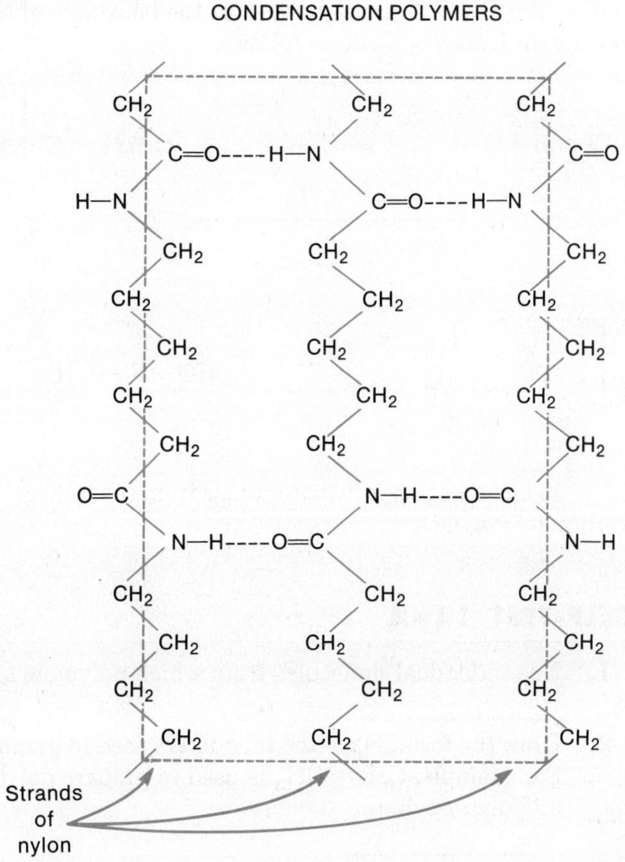

CONDENSATION POLYMERS

FIGURE 14–5 Structure and hydrogen bonding in nylon 6.

Strands of nylon

to attract one another, but not so strongly that the plastic cannot be initially extended to form the fibers. Ordinary covalent chemical bonds linking the chains together would be too strong. Hydrogen bonds, with a strength about one tenth that of an ordinary covalent bond, link the chains in the desired manner. We shall see later that this type of bonding is also of great importance in protein structures.

POLYCARBONATES

The tough, clear polycarbonates constitute another important group of condensation plastics. One type of polycarbonate, commonly called Lexan or Merlon, was first made in Germany in 1953. It is as "clear as glass" and nearly as tough as steel. A 1-inch sheet can stop a .38-caliber bullet fired from 12 feet away. Such unusual properties have resulted in Lexan's use in "bullet-proof" windows and as visors in astronauts' space helmets. More than 115 million pounds of polycarbonates were produced in the United States in 1980.

The polycarbonates are formed by condensing phosgene with a substance containing two phenol structures. A molecule of HCl is condensed out in the formation of a C—O bond to complete the ester ($-\overset{\overset{\textstyle O}{\|}}{C}-O-$) linkage. Other chlorines then react with other alcohol groups to give a polymer chain containing $-O-\overset{\overset{\textstyle O}{\|}}{C}-O-$ functional groups in the backbone of the chain. A representative portion of Lexan is made as follows:

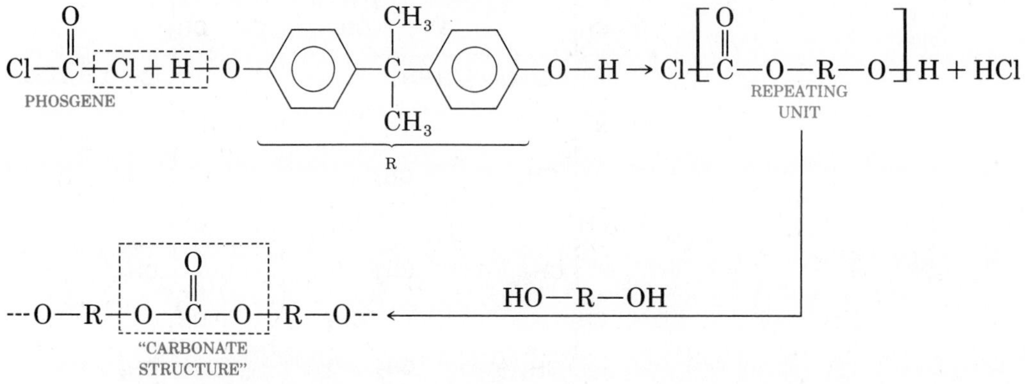

PHENOL

$$Cl-\overset{\overset{\textstyle O}{\|}}{C}\underset{\text{PHOSGENE}}{+Cl+H}+O-\bigcirc-\underset{\underset{\textstyle CH_3}{|}}{\overset{\overset{\textstyle CH_3}{|}}{C}}-\bigcirc-O-H \rightarrow Cl+\left[\overset{\overset{\textstyle O}{\|}}{C}-O-R-O\right]H + HCl$$

R

REPEATING UNIT

$$\cdots-O-R+\overset{\overset{\textstyle O}{\|}}{O-C-O}+R-O\cdots \longleftarrow \qquad HO-R-OH$$

"CARBONATE STRUCTURE"

The name *polycarbonate* comes from the linkage's similarity to an inorganic carbonate ion, CO_3^{2-}.

SELF-TEST 14–A

1. The individual molecules from which polymers are made are called

   _____.

2. Draw the formulas of the monomers used to prepare the following polymers. For example, $CH_2\!\!=\!\!CH_2$ is used to prepare polyethylene.
   a. Polypropylene

   _____

b. Polystyrene

_____

c. Teflon

_____

3. Many molecules of a carboxylic diacid reacting with many molecules of a dialcohol produce a _____.

4. Natural rubber is a polymer of _____.

5. When styrene and butadiene are polymerized together, the product is a type of _____.

6. Nylon is an example of a _____ polymer.

7. Polyamides are formed when _____ is split out from the reaction of many organic acid groups and many amine groups.

8. When an acid such as terephthalic acid $\underset{\text{HO}}{\overset{\text{O}}{\|}}\text{C}\text{—}\bigcirc\text{—}\underset{\text{OH}}{\overset{\text{O}}{\|}}\text{C}$ reacts with ethylene glycol, $HOCH_2CH_2OH$, the structure of the resulting polymer is: _____

9. Polyesters are formed by ( ) addition or ( ) condensation reactions.

10. In paints, the components are dispersed in a liquid called the _____, and the color is supplied by the _____.

11. The drying oil used most often in paints is _____ oil.

12. Water-based or "latex-type" paints are _____ because two liquids that are ordinarily incompatible are stabilized.

13. Baked-on paints are usually alkyds. This means that they are usually made from polyfunctional _____ and _____.

SILICONES

The element silicon, in the same chemical family as carbon, also forms many compounds with numerous Si—Si and Si—H bonds, analogous to C—C and C—H bonds. However, the Si—Si bonds and the Si—H bonds are reactive toward both oxygen and water; hence, there are no useful silicon counterparts to most hydrocarbons. However, silicon does form stable bonds with carbon, and especially oxygen, and this fact gives rise to an interesting group of condensation polymers containing silicon, oxygen, carbon, and hydrogen (bonded to carbon).

Silane, SiH_4, is structurally like methane, CH_4, in that both are tetrahedral.

In 1945, E. G. Rochow, at the General Electric Research Laboratory, discovered that a silicon-copper alloy will react with organic chlorides to produce a whole class of reactive compounds, the **organosilanes**.

$$\underset{\substack{\text{METHYL}\\\text{CHLORIDE}}}{2\,CH_3Cl} + \underset{\substack{\text{SILICON-}\\\text{COPPER ALLOY}}}{Si(Cu)} \rightarrow \underset{\substack{\text{DIMETHYLDICHLORO-}\\\text{SILANE}}}{(CH_3)_2SiCl_2} + Cu$$

The chlorosilanes readily react with water and replace the chlorine atoms with hydroxyl (—OH) groups. The resulting molecule is similar to a dialcohol.

$$(CH_3)_2SiCl_2 + 2\,H_2O \rightarrow (CH_3)_2Si(OH)_2 + 2\,HCl$$

Silicones are polymers
held together by a series
of covalent Si—O bonds.

Two dihydroxysilane molecules undergo a condensation reaction in which a water molecule is split out. The resulting Si—O—Si linkage is very strong; the same linkage holds together all the natural silicate rocks and minerals. Continuation of this condensation process results in polymer molecules with molecular weights in the millions:

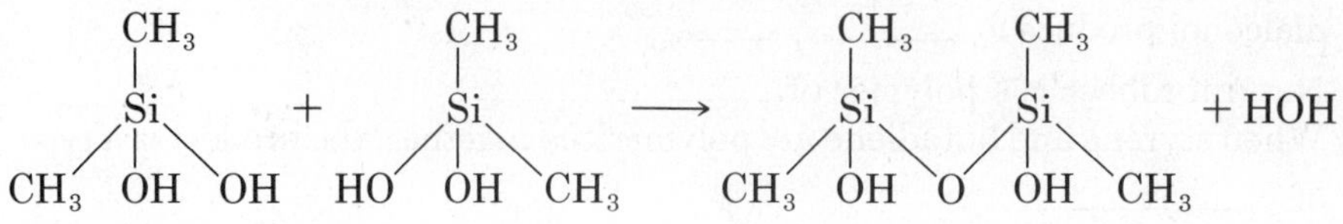

$$
\begin{array}{ccc}
\text{CH}_3 & \text{CH}_3 & \text{CH}_3 \quad \text{CH}_3 \\
| & | & | \quad\quad | \\
\text{Si} \quad + \quad \text{Si} & \longrightarrow & \text{Si} \quad\quad \text{Si} \quad + \text{HOH} \\
/\ \backslash \quad\quad /\ \backslash & & /\ \backslash \quad /\ \backslash \\
\text{CH}_3 \ \text{OH} \ \text{OH} \ \text{HO} \ \text{OH} \ \text{CH}_3 & & \text{CH}_3 \ \text{OH} \ \text{O} \ \text{OH} \ \text{CH}_3
\end{array}
$$

Further reaction yields:

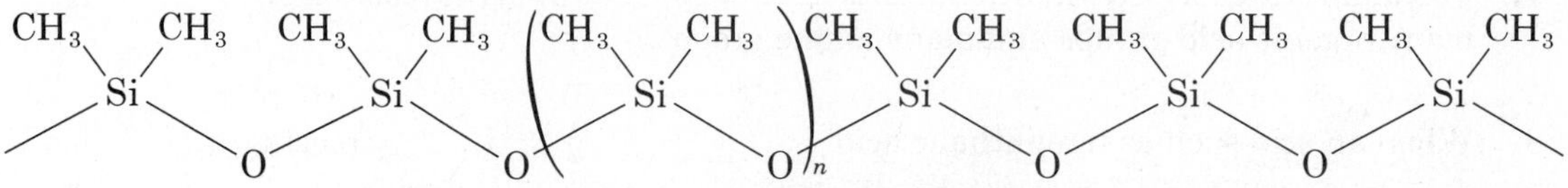

By using different starting silanes, polymers with different properties result. For example, two methyl groups on each silicon atom result in **silicone oils,** which are more stable at high temperatures than hydrocarbon oils and also have less tendency to thicken at low temperatures.

Silicone rubbers are very high molecular weight chains crosslinked by Si—O—Si bonds. Room-temperature-vulcanizing silicone rubbers are commercially available; they contain groups that readily crosslink in the presence of atmospheric moisture. The —OH groups are first produced, and then they condense in a crosslinking "cure" similar to the vulcanization of organic rubbers.

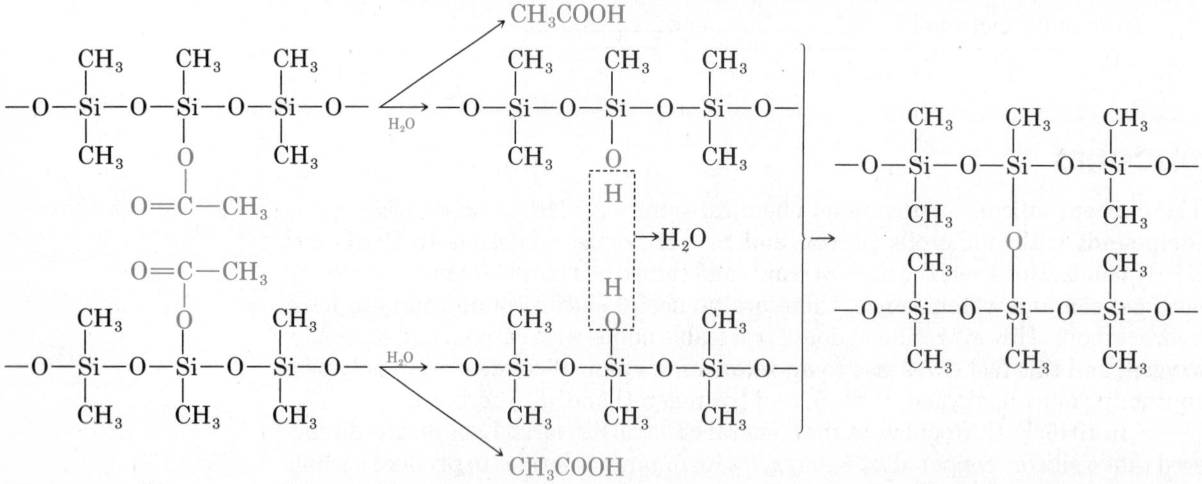

Over 3 million pounds of silicone rubber are produced each year in the United States. The uses include window gaskets; o-rings; insulation; sealants for buildings, space ships, and jet planes; and even some wearing apparel. The first footprints on the moon were made with silicone rubber boots, which readily withstood the extreme surface temperatures (Fig. 14–6).

Silicone oils and rubbers
find many medicinal uses.

"Silly Putty," a silicone widely distributed as a toy, is intermediate between silicone oils and silicone rubber. It is an interesting material with elastic properties

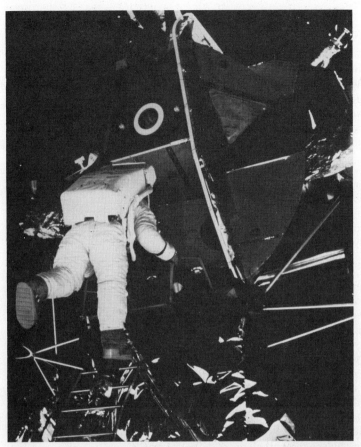

FIGURE 14–6 Examples of the use of silicone in the space program. Soles of lunar boots worn by the Apollo astronauts were made of high-strength silicone rubber. A silicone compound was also used for the air-tight seal of the lunar module hatch from which Astronaut Edwin E. Aldrin, Jr., has just emerged in this photo of the first manned landing on the moon on July 20, 1969. (*Chem. Tech.*, September 1980, p. 550.)

on sudden deformation, but its elasticity is quickly overcome by its ability to flow like a liquid when allowed to stand.

REARRANGEMENT POLYMERS

Some molecules polymerize by **rearrangement** reactions to yield very useful products. Molecules containing the isocyanate group (—NCO), for example, will react with almost any other molecule containing an active hydrogen atom (such as in an —OH or —NH$_2$ group) in a rearrangement process. An example is the reaction of hexamethylene diisocyanate and butanediol. The urethane linkage

$$\left(\begin{array}{c} -\text{N}-\text{C}-\text{O}- \\ | \quad \| \\ \text{H} \quad \text{O} \end{array} \right)$$ is produced by a shift (or rearrangement) of a hydrogen atom,

moving it from the alcohol (butanediol) to a nitrogen atom on the isocyanate group; the linkage (—N—C—O—) is similar to, but not the same as, the amide bond (—N—C—C—) in nylons.

TABLE 14–4   Composition of Some Trade-Name Polymers*

TRADE NAME	COMPOSITION	MONOMERS
Acrilan	85% acrylonitrile plus vinyl acetate or vinyl pyridine	$CH_2{=}C(OC_2H_5)_2$ ACETAL
Acrylic	At least 85% acrylonitrile	
Arnel	Cellulose triacetate	
Bakelite	Phenol-formaldehyde condensation product	$CH_2{=}CH(CONH_2)$ ACRYLAMIDE
Caprolan	Nylon 6	
Creslan	Copolymer of acrylonitrile and acrylamide	
Dacron	Ethylene glycol-terephthalic acid condensation product	
Delrin	Polyacetal	
Dynel	60% vinyl chloride plus 40% acrylonitrile	
Epoxy	Phenol-acetone-epichlorohydrin condensation product	
Formica	Phenol-formaldehyde condensation product	
Fortrel	Polyester similar to Dacron	
Herculon	Polypropylene	
Kodel	Polyester; terephthalic acid plus 1,4-cyclohexane dimethanol	
Lucite	Methyl methacrylate	
Melmac	Melamine-formaldehyde condensation product	
Mylar	Polyethylene terephthalate	
Neoprene	2-chlorobutadiene	
Nylon 501	Nylon 66	
Nytril	At least 85% vinylidene dinitrile	
Orlon	Originally pure acrylonitrile; now up to 14% of another monomer	
Plexiglas	Methyl methacrylate	
Polythene	Polyethylene	
Saran	Vinylidene chloride-vinyl chloride addition product	
Spandex	Polyurethane; ethylene glycol-diisocyanate condensation product	
Teflon	Polytetrafluoroethylene	
Terylene	Polyester similar to Dacron	
Vectra	Polypropylene	
Velon	Vinylidene chloride-vinyl chloride addition product; see Saran	
Zantrel	Rayon fiber	

Monomer structures (right column):

$CH_2{=}C(OC_2H_5)_2$ ACETAL

$CH_2{=}CH(CONH_2)$ ACRYLAMIDE

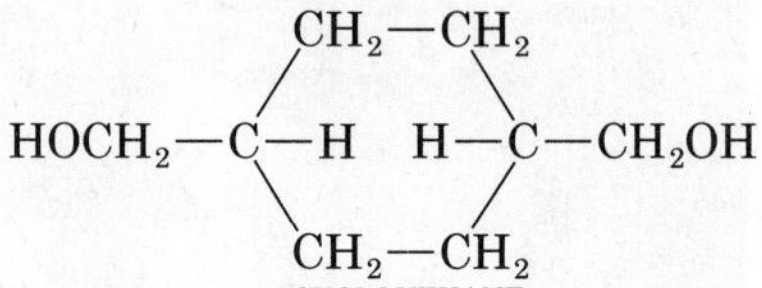

1,4-CYCLOHEXANE DIMETHANOL:

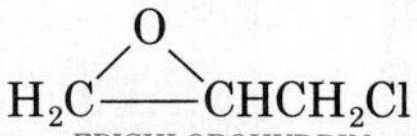

$$HOCH_2{-}C{-}H \quad H{-}C{-}CH_2OH$$

with $CH_2{-}CH_2$ above and $CH_2{-}CH_2$ below forming the cyclohexane ring.

EPICHLOROHYDRIN:

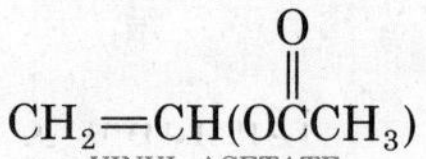

$$H_2C\overset{O}{\overbrace{\qquad}}CHCH_2Cl$$

$CH_2{=}CH(O\overset{O}{\overset{\|}{C}}CH_3)$ VINYL ACETATE

$CH_2{=}C(CN)_2$ VINYLIDENE DINITRILE

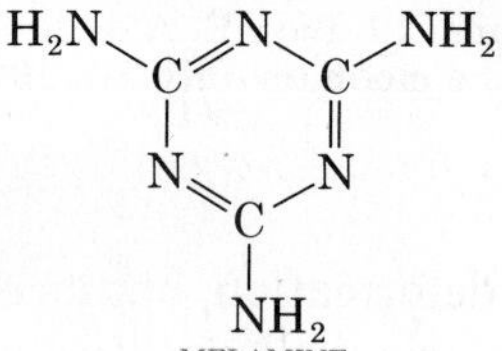

MELAMINE (triazine ring):

$H_2N{-}C$, NH_2, with NH_2 substituents on the ring.

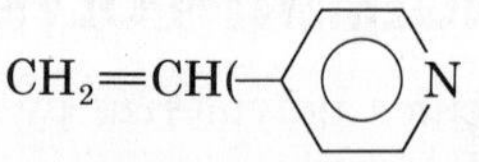

VINYL PYRIDINE:

$CH_2{=}CH({-}\bigcirc N)$

* Structures are given for those monomers not found elsewhere in the text. Consult the index for formulas not given.

$$OCN(CH_2)_6NCO + HO(CH_2)_4OH \rightarrow OCN(CH_2)_6\overset{\displaystyle H}{-}\overset{\displaystyle O}{N}\overset{\displaystyle \parallel}{-}C-O-(CH_2)_4OH$$

HEXAMETHYLENE   1,4-BUTANEDIOL   PRODUCT MOLECULE (A URETHANE)
DIISOCYANATE

rearrangement

The continued reaction of the other groups gives rise to a polymer chain—a polyurethane.

$$-\overset{O}{\overset{\parallel}{C}}-\overset{H}{\overset{|}{N}}-(CH_2)_6-\overset{H}{\overset{|}{N}}-\overset{O}{\overset{\parallel}{C}}-O-(CH_2)_4-O-\overset{O}{\overset{\parallel}{C}}-\overset{H}{\overset{|}{N}}-(CH_2)_6-\overset{H}{\overset{|}{N}}-\overset{O}{\overset{\parallel}{C}}-O-(CH_2)_4-O-$$

A PORTION OF POLYURETHANE

A polyurethane is structurally similar to a polyamide (nylon). In Europe polyurethanes have applications similar to those of nylon in this country. Polyurethanes have viscosities and melting points that make them useful for foam applications. Foamed polyurethanes are known as "foam rubber" and "foamed hard plastics," depending on the degree of crosslinking.

Polyurethanes are structurally similar to many polyamides.

Table 14–4 lists the trade names and compositions of some of the addition, condensation, and rearrangement polymers on the market today.

POLYMER ADDITIVES — TOWARD AN END USE

Few plastics produced today find end uses without some kind of modification. Polyurethanes are a good example. In order for polyurethane to be useful as insulation in refrigerators and refrigerated trucks and railroad cars and as construction insulation, a **foaming agent** is used. For polyesters to have the strength to compete with metals and certain natural materials such as wood, they are usually modified with a **reinforcing additive.** Inorganic fibers are usually used for this purpose. Tennis racquets, golf clubs, and other sporting equipment have been changed radically by the use of such reinforced plastics in place of the traditional wood and metal (Table 14–5).

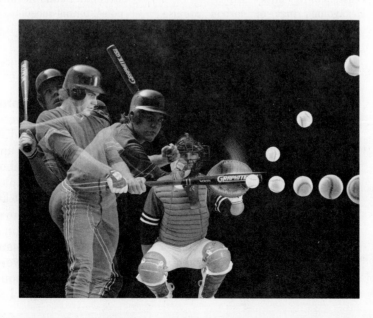

Time-lapse photographic series of a graphite bat in action. (Courtesy of Worth Bat Company.)

TABLE 14-5 Polymer Additives

ADDITIVE	STRUCTURE	USE COMMENTS
Foaming agent Pentane	$CH_3CH_2CH_2CH_2CH_3$	Used to foam polyurethane. Dissolved in liquid polymer under high pressure, then expands in a hot mold.
Plasticizers Dioctyl phthalate (DOP)	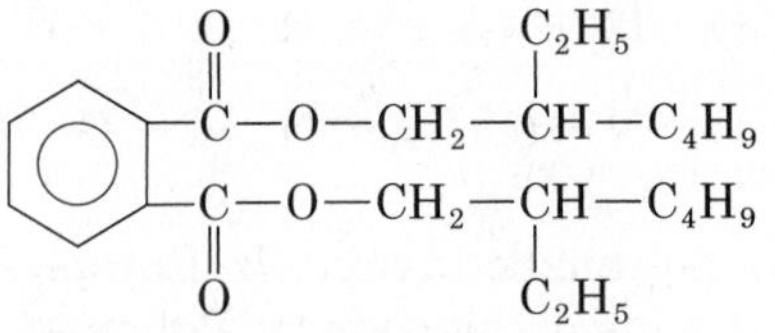 *(structure of dioctyl phthalate)*	Plasticizer in polyvinylchloride to lend flexibility. Gets into the environment.
Dioctyl adipate (DOA)	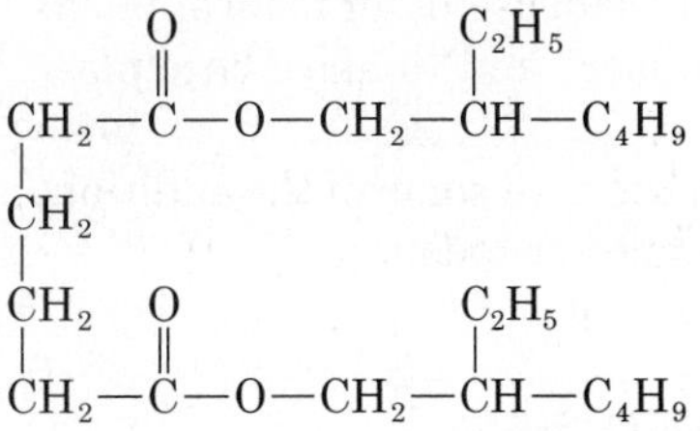 *(structure of dioctyl adipate)*	Used in plastic films to make them flexible. Has Food and Drug Administration approval for food contact.
UV stabilizers Phenylsalicylate Carbon black	*(structure of phenylsalicylate)* Similar to graphite below, but small particles, less structure	*UV – ultraviolet* Absorbs UV light very efficiently. Absorbs UV light and radiates energy as heat. Fine for all-black articles.
Reinforcing agents Glass fibers Boron fibers Graphite fibers	SiO_2 units (see Fig. 11–12) Clusters of B_{12} units Hexagonal rings of carbon atoms, joined on all sides in a layered arrangement	Used in polyesters and other plastics to improve strength. Found in car bodies, boats, fishing poles, tennis racquets, bicycle frames, radio antennas, and so on.

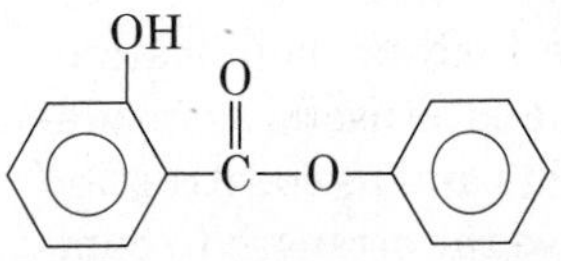

A plastic such as polystyrene, polyethylene, or polypropylene is often too stiff to have an immediate application. For example, polyethylene makes good bread wrap, but only if it is made flexible by an additive called a **plasticizer.** Some plasticizers, such as dioctyl phthalate (DOP), have not been approved for food use but still can be found in samples of living tissue taken from laboratory animals and humans. DOP and other plasticizers also occur in water samples taken from rivers and lakes. The health implications of these findings are not known at this time.

Since most plastics are used where they are exposed to sunlight, almost all contain some form of an additive to absorb the harmful portion of sunlight, ultraviolet (UV) light. These compounds, called **UV stabilizers,** can absorb photons in the 290 to 400 nanometer range of the spectrum, which have sufficient energy to break the chemical bonds found in polymers. If enough bonds are broken, the polymer becomes brittle and will break under stress (Fig. 14–7). Of course, with

FIGURE 14–7 Sunlight (mostly ultraviolet) damages most plastics such as polypropylene webbing. The results are shorter product life and higher costs to the consumer. Plastics containing ultraviolet-absorbing chemicals have a longer outdoor life.

time, even the stabilizers decay, and the polymer article begins to show visible signs of sunlight degradation.

Table 14–5 summarizes several common polymer additives.

THE FUTURE OF POLYMERS

As we have seen in this chapter, the development and use of synthetic polymers is quite recent. Polyethylene, for example, was not discovered until 1933, yet by 1981, its production in the United States amounted to almost 12 billion pounds. Chemists are constantly synthesizing new polymers and finding applications for them. The space age has brought with it the need for new polymers, especially in electronics and as special coatings that can withstand high temperatures without breaking down. Among the newcomers are the polyimides, prepared from the polycondensation of a diacid anhydride and a diamine. Some of these polymers have very high service temperatures (Fig. 14–8).

PYROMELLITIC ANHYDRIDE   1,2-DIAMINOETHANE   A POLYIMIDE

Plastic materials are being improved constantly. Some have been made with the strength and rigidity of steel while having only 15% to 20% of the density of steel. The structural strength of such plastics offers the possibility of self-supporting domes for buildings and automobiles that contain more plastic than metal. New low-temperature polymerizations without the use of a solvent are being developed. An application is to make "spray-on" clothes. Simply spray the monomers onto a mannequin — a little more here, a little less there. Then cut along desired lines, add

Anhydride: an HOH has been removed from the structure.

FIGURE 14–8 An example of how a polyimide film can withstand, for a short period, the flame of a blowtorch.

buttons and zippers, and wear. Some plastics are being developed to replace wood fiber in paper for the printed page. These papers offer smooth surfaces without the graininess of paper, an improvement especially in the quality of microfilming.

Because polymers are used so extensively throughout the world today, the problem of waste disposal is inevitable. Plants are in operation in which solid wastes undergo first a magnetic separation to remove iron and steel objects, then a ballistic separation (Fig. 14–9) based on density, since glass and aluminum objects are denser than plastics. Another method is to have consumers separate waste into garbage, plastics, metals, paper, wood, and glass. Plans are being developed to treat the plastics thus separated in two ways. If suitable separation methods could be developed, thermoplastics could be reprocessed into new items (e.g., if all the nylon could be separated from polystyrene). Thermosetting plastics could not be treated

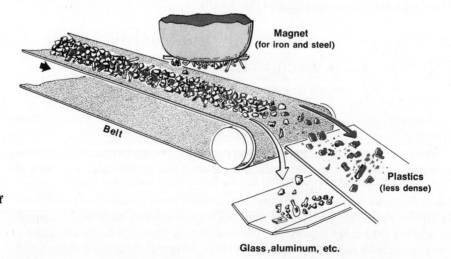

FIGURE 14–9 A method of separation of plastics from other wastes prior to recycling the plastics for reuse.

TABLE 14–6 Energy Requirements to Produce Some Plastics and Metals (Including the Fuel Equivalent of the Monomers)

	MILLION BTU/TON*
Aluminum	244
Copper	112
Low-density polyethylene	106
High-density polyethylene	96
Polystyrene	64
Polyvinyl chloride	49
Steel	24

* Burning 1 ton of coal produces about 25 million Btus of heat energy. The British thermal unit (Btu) is an English unit of energy. 1.00 Btu = 250 cal.

this way, however, because breaking the crosslinking would cause complete molecular degradation. If separation and reuse were not feasible, asphalt could be made from the mixed plastics. In their original composition, combustion units built near cities could actually use plastics as fuels, since they are mostly carbon and hydrogen. There is danger, though, in that some plastics contain elements that could create massive pollution if released into the atmosphere. An example is polyvinylchloride, which on burning yields hydrogen chloride, a very corrosive gas. However, some of the products of incomplete combustion (such as benzene, styrene, and acetylene) could be recycled as raw materials for other chemical syntheses.

The long-range future of plastics looks dismal unless we curtail our ravenous burning of petroleum. Most of the raw materials for plastics come from petroleum and less from coal. Petroleum and coal are the principal sources of energy in this country. Not only are plastics and energy linked through the raw materials – fossil fuels relationship, but also considerable energy is required to purify starting materials and to change them into the desired plastic in the preferred shape. It is the age-old principle that we cannot continue to eat our cake (burn petroleum and coal) and have it, too (use petroleum and coal to make plastics, fibers, and medicines). Table 14–6 illustrates the point that when we consider polymers in the broadest sense in terms of their energy costs of production, we must include their energy value as if they were used as fuel instead of for some object. Considered this way, plastics, fibers, and other items made of polymers derived from petroleum may have only a short history in human existence. Wood, paper, and mineral products such as metals and cement appear either renewable or present in the Earth's crust in far greater abundance than petroleum.

In the United States, 95% of our petroleum is used as fuel; only 5% is used to make products such as medicines, textiles, and plastics.

We hope that these and similar problems will be solved as we begin to understand more fully how to use what we have on this planet and how to live in greater harmony with nature.

SELF-TEST 14–B

1. When $(CH_3)_2SiCl_2$ reacts with water, a representative portion of the structure of the polymer obtained is _____ .

2. Stabilizers protect plastics against the action of _____ .

3. A plastic that is too stiff can be rendered more flexible by the addition of a
   _____ .
4. A silicone polymer contains Si— _____ bonds.
5. The burning of plastics containing chlorine, such as polyvinyl chloride, produces what toxic gas? _____
6. (   ) Ultraviolet or (   ) visible light is more destructive to plastics.
7. Which requires more energy per ton to produce, polyethylene or steel?
   _____

MATCHING SET

_____ **1.** Nylon	**a.** Plastic that forms interlocking bonds when heated
_____ **2.** Block copolymer	**b.** Rubber
_____ **3.** Monomer	**c.** Crosslinking via reaction with sulfur
_____ **4.** Thermoplastic	**d.** —AAAAABBBBBBAAAAA—
_____ **5.** Thermosetting plastic	**e.** Causes room-temperature-vulcanizing silicone to crosslink
_____ **6.** Homopolymer	**f.** Forms polymers of desired structure
_____ **7.** Polymer with a memory	**g.** —AAAAAAAAAAAAAAAA—
_____ **8.** Vulcanize	**h.** Magnetic waste
_____ **9.** Stereochemical control	**i.** A synthetic rubber
_____ **10.** Styrene-butadiene copolymer	**j.** Building unit for a polymer
_____ **11.** Poly-*cis*-isoprene	**k.** Plastic softened by heat
_____ **12.** Polyester	**l.** Graphite
_____ **13.** Moisture	**m.** Formed from a dialcohol and a diacid
_____ **14.** Dioctyl phthalate	**n.** Possibly harmful plasticizer
	o. Magnetite
	p. Natural rubber
	q. A polyamide

QUESTIONS

1. In what ways is a railroad train like polystyrene?
2. Where do you suppose the first chemist who prepared a polymer got the idea for giant molecules?
3. What property does a polymer have when it is extensively crosslinked?
4. Describe on the molecular level the end result of the vulcanization process.
5. What is the origin of the word *polymer?*
6. Is polystyrene a thermoplastic or a thermosetting plastic?
7. What property of the molecular structure of rubber allows it to be stretched?
8. Explain how polymers could be prepared from each of the following compounds. (Other substances may be used.)

a.
$$CH_3-\overset{\overset{\displaystyle H}{|}}{C}=\overset{\overset{\displaystyle H}{|}}{C}-CH_3$$

b.
$$HO-\overset{\overset{\displaystyle O}{\|}}{C}-CH_2-CH_2-\overset{\overset{\displaystyle O}{\|}}{C}-OH$$

c.
$$\underset{\underset{\displaystyle OH}{|}}{CH_2}-\underset{\underset{\displaystyle OH}{|}}{CH}-\underset{\underset{\displaystyle OH}{|}}{CH_2}$$

d.
$$H_2N-CH_2-\langle\bigcirc\rangle-CH_2-NH_2$$

9. What are the monomers used to prepare the following polymers?

a. $-CH_2CH_2CH_2CH_2CH_2CH_2CH_2CH_2CH_2-$
      $\quad CH_3 \quad\quad CH_3 \quad\quad CH_3$

b. $-CHCH_2CHCH_2CHCH_2-$

c. $-CH_2-CCH_2-CCH_2-CCH_2-C-$

10. Write equations showing the formation of polymers by the reaction of the following pairs of molecules:

a. COOH

   and $HOCH_2CH_2OH$

   COOH

b. $HOOCCH_2CH_2COOH$ and $H_2NCH_2CH_2NH_2$

c. CH_2OH
   $HCOH$ and
   CH_2OH

   $C-OH$ (with O)
   $C-OH$ (with O)

11. Is a small molecule eliminated when each monomer unit is added to the chain in addition polymers?

12. Give an example of a copolymer.

13. You are given two specimens of plastic, A and B, to identify. One is known to be nylon and the other polymethyl methacrylate, but you do not know which is which. Analysis of A shows it to contain C, H, and O, while B contains C, H, O, and N. What are A and B?

14. What structural features must a molecule have in order to undergo addition polymerization?

15. What is meant by the term *macromolecule?*

16. Orlon has a polymeric chain structure of

$-CH_2-CH-CH_2-CH-CH_2-CH-$
$\qquad\quad CN \qquad\quad CN \qquad\quad CN$

What is the monomer from which this structure can be made?

17. Which white pigment is banned in interior paints? Explain.

18. What feature do all condensation polymerization reactions have in common?

19. Give an example of the possibilities that exist if a trifunctional acid reacts with a difunctional alcohol.

20. What type of chemical change takes place during the drying of oil paints?

21. What are the starting materials for nylon 66?

22. Suggest a major difference in the bonding of thermosetting and thermoplastic polymers. Which is more likely to have an interlacing (crosslinking) of covalent bonds throughout the structure? Which is more likely to have weak bonds between large molecules?

23. What is a major difference between silicone oils and silicone rubbers?

24. Explain how a plasticizer can make a polymer more flexible.

25. Name one commercial plasticizer found in food wraps.

26. Could an oil-based paint "dry" in a vacuum? Explain.

27. Would a latex paint "dry" in a vacuum? Explain.

28. Describe the properties and structure of Silly Putty.

29. Which is more likely to produce a thermosetting polymer, the monomers of Question 10a or 10c?

30. Draw representative portions of Acrilan, Delrin, Saran, and Plexiglas. Refer to Table 14-4.

31. In what way is the structure of ice like that of a crosslinked polymer? How is it different?

32. What single property must a molecule possess in order to be a monomer?

33. Which do you think is the source of most polymers used today, green plants or petroleum? Do you think this will ever change? Explain.

34. a. Should we stop the burning of petroleum? What are the problems involved?
    b. Should we start to develop research on how to change wood and straw into plastics? What are a few of the problems involved?

35. Would isoprene make a good motor fuel? Explain.

36. What properties of plastics make them superior to metals? What properties of plastics make them inferior to metals?

37. To what would you attribute the superior thermal stability of silicone polymers?

38. A tiny sample of rubber, held in the flame of a match, burns with a small bright flame and gives a *white* flame in contrast to the black smoke of burning tires. Explain.

Chapter 15

BIOCHEMISTRY — CHEMISTRY OF LIVING SYSTEMS

Do you think of yourself as some combination of the chemical elements? Probably not, because it is hard to explain your characteristics in terms of the properties of the elements . . . you are so complex! However, your body *is* chemical in nature, and many relationships are now known to exist between the properties of your body and the chemicals you contain and ingest.

The relationships between chemicals and life forms are in the purview of *biochemistry.* Knowledge in this exciting field of study is expanding currently at an explosive rate. Each year scientific journals publish more articles on biochemical research than on research in any other field of chemistry or chemically related area.

The goal of biochemistry is to develop a chemically based understanding of living cells of all types. This includes the determination of the kinds of atoms present, the investigation of how they are joined together to form the larger structural units present in cells, and the study of the chemical reactions by which living cells obtain the energy required for the life processes of growth, movement, and reproduction.

Some fundamental biochemical substances common to all living systems are fats and oils, carbohydrates, proteins, enzymes, vitamins, hormones, nucleic acids, and compounds for storage and exchange of energy, such as adenosine triphosphate (ATP). Fats and oils were described in Chapter 13; other biochemicals will be described in this chapter along with how components of all of these substances are used in some of life's processes. We shall also note how energy is produced, stored, exchanged, and used in a living cell. In all of this, we believe you will find the study of biochemistry a fascinating and revealing insight into your physical self.

We shall begin with descriptions of carbohydrates and proteins. Both types of biochemicals are polymers, being composed of monomer units. After carbohydrates and proteins, we shall discuss other giant molecules, the nucleic acids, which appear to be in the control center for biochemical change.

CARBOHYDRATES

Carbohydrates contain the elements carbon, hydrogen, and oxygen, with hydrogen atoms and oxygen atoms generally in the ratio of 2 to 1.

Carbohydrates are composed of the three elements carbon, hydrogen, and oxygen. Three structural groups are prevalent in carbohydrates: alcohol (—OH), aldehyde $\left(\begin{smallmatrix} O \\ \| \\ -CH \end{smallmatrix}\right)$, and ketone $\left(\begin{smallmatrix} O \\ \| \\ -C- \end{smallmatrix}\right)$. The carbohydrates can be classified into three main groups: **monosaccharides** (Latin, *saccharum,* meaning "sugar"), **oligosac-charides,** and **polysaccharides.** Monosaccharides are simple sugars that *cannot* be broken down into smaller units by mild acid hydrolysis.

When an oligosaccharide or a polysaccharide undergoes complete hydrolysis, monosaccharides are formed with little change in the monomer's structure. At the point where the break occurs, water furnishes H· and ·OH groups to bond to the broken ends, forming stable structures in the aqueous medium (see Fig. 15–3). The hydrolysis of biochemicals is usually more rapid in acidic or basic media than in a neutral solution. Hydrolysis of a molecule of an oligosaccharide yields two to six molecules of a simple sugar; complete hydrolysis of a polysaccharide produces many monosaccharide units.

Glucose, $C_6H_{12}O_6$, and some of the other simple sugars are quick energy sources for the cell. Large amounts of energy are stored in polysaccharides such as starch. The stored energy is usable by living cells only if polysaccharides are broken down, or hydrolyzed, into monosaccharides.

Some complex carbohydrates are also used by cells of some organisms for structural purposes. Cellulose, for example, partially accounts for the structural properties of wood.

mono—one
oligo—few
poly—many

Hydrolysis is a water-splitting reaction in which H· bonds with one fragment of the attacked molecule and ·OH bonds with the other fragment.

Monosaccharides

Approximately 70 monosaccharides are known; 20 of these simple sugars occur naturally. Unlike many organic compounds, these sugars are very soluble in water owing to the numerous —OH groups present, which can form hydrogen bonds with water.

The most common simple sugar is **D-glucose.** This monosaccharide requires three structures for its adequate representation (Fig. 15–1). Structure (*a*), in which the carbon atoms are numbered for later reference, depicts the "straight-chain" structure with the aldehyde group (—CHO) in position 1. The properties of

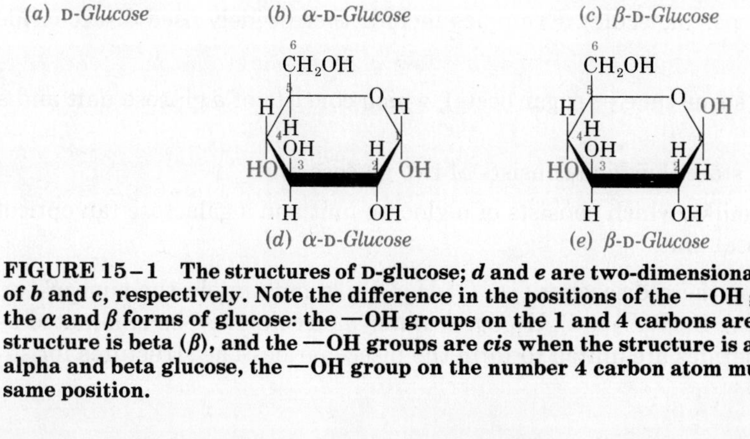

FIGURE 15–1   The structures of D-glucose; *d* and *e* are two-dimensional representations of *b* and *c*, respectively. Note the difference in the positions of the —OH groups *(color)* in the α and β forms of glucose: the —OH groups on the 1 and 4 carbons are *trans* when the structure is beta (β), and the —OH groups are *cis* when the structure is alpha (α), in both alpha and beta glucose, the —OH group on the number 4 carbon atom must be in the same position.

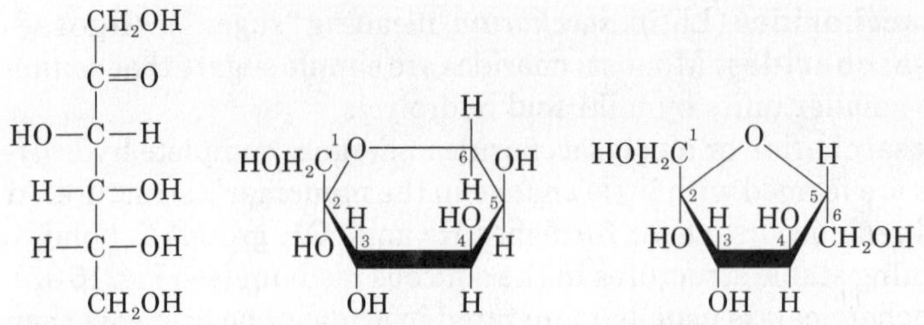

(a) *Ketone structure*

(b) *β-Ring structure (Pyranose structure: 6-membered ring with an oxygen atom in the ring)*

(c) *β-Ring structure (Furanose structure: 5-membered ring with an oxygen atom in the ring)*

FIGURE 15–2   The structures of D-fructose. The α-ring structure (not shown) differs from the β-ring structure in that the CH$_2$OH and OH groups are in reversed positions on carbon 2.

D-glucose, the most prevalent monosaccharide, is found in fruits, blood, and living cells.

a water solution of D-glucose cannot be explained by this structure alone. At any given time, most of the molecules exist in the ring form, structures (b) and (c), which results from a molecular rearrangement in which carbon 1 bonds to carbon 5 through an oxygen atom. Both ring structures are possible since the OH group on carbon 1 may form in such a way to point either along the plane of the molecule or out of the plane. It should be emphasized that a solution of D-glucose contains a mixture of three forms in a dynamic state of change from one form to another. There is more of the ring form and much less of the straight-chain form.

Glucose has a relative sweetness of 74.3, compared with sucrose having an assigned value of 100.0. The value for fructose is 173.3.

Because it is sweet, D-glucose is used in the manufacture of candy and in commercial baking. This simple sugar, also called *dextrose, grape sugar,* and *blood sugar,* is prevalent in fruits, vegetables, blood, and tissue fluids. A solution of D-glucose is fed intravenously when a readily available source of energy is needed to sustain life. As will be discussed later, many polysaccharides, including starch, are composed of glucose units and serve as a source of this important chemical upon hydrolysis of the complex structures.

Another important monosaccharide is D-**fructose.** Its structure, which has a ketone group, is given in Figure 15–2.

Oligosaccharides

The most commonly encountered oligosaccharides are the disaccharides (two simple sugar units per molecule). Examples include these widely used disaccharide sugars:

sucrose (from sugar cane or sugar beets), which consists of a glucose unit and a fructose unit

maltose (from starch), which consists of two glucose units

lactose (from milk), which consists of a glucose unit and a galactose (an optical isomer of glucose) unit.

The formula for these disaccharides, $C_{12}H_{22}O_{11}$, is not simply the sum of two monosaccharides, $C_6H_{12}O_6 + C_6H_{12}O_6$. A water molecule must be eliminated as two monosaccharides are united to form the disaccharide. The structures for su-

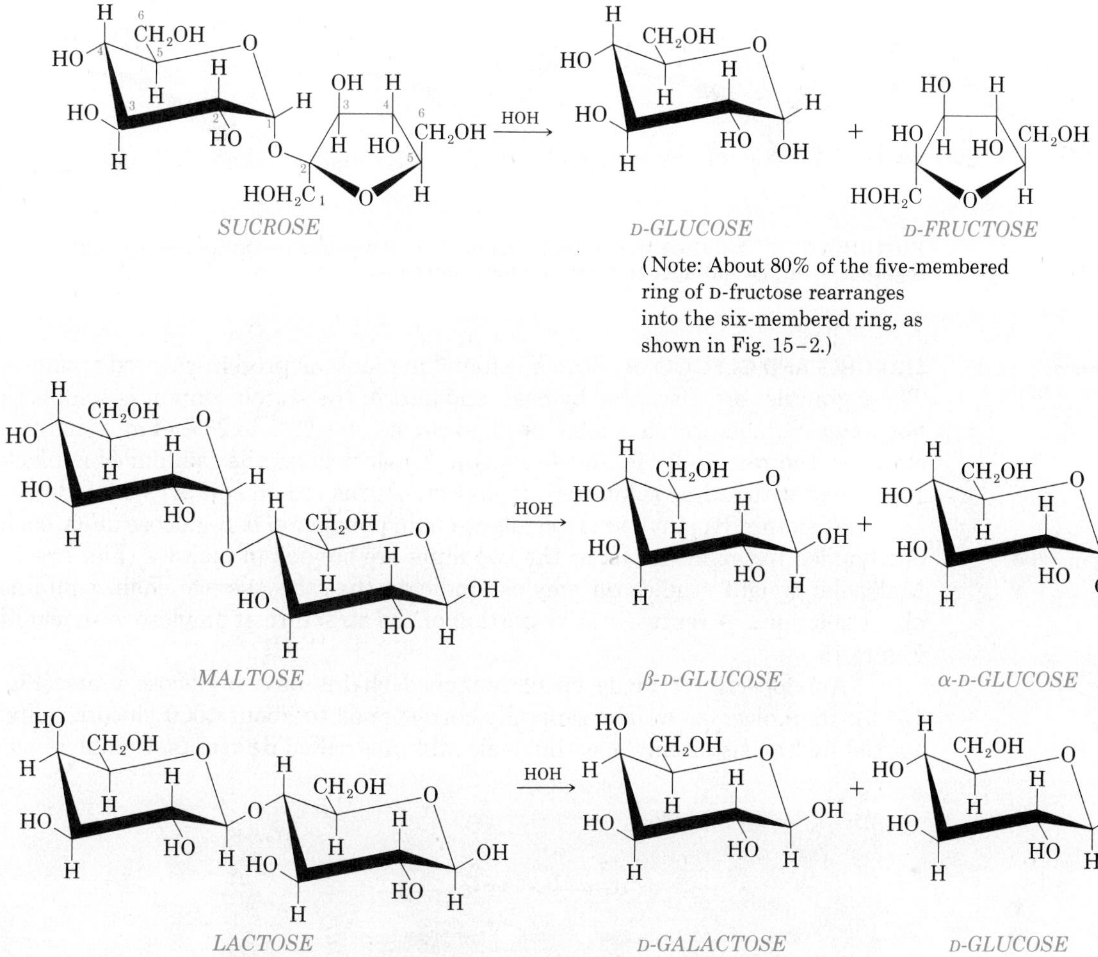

(Note: About 80% of the five-membered ring of D-fructose rearranges into the six-membered ring, as shown in Fig. 15–2.)

FIGURE 15–3 Hydrolysis of disaccharides (sucrose, maltose, and lactose).

crose, maltose, and lactose, along with their hydrolysis reactions, are given in Figure 15–3.

The disaccharides are important as foods. Sucrose is produced in a high state of purity on an enormous scale: the annual production amounts to over 80 million tons per year. Originally produced in India and Persia, sucrose is now used universally as a sweetener. About 40% of the world sucrose production comes from sugar beets and 60% from sugar cane. Sugar provides a high caloric value (1794 kcal per pound); it is also used as a preservative in jams, jellies, and candied fruit.

Disaccharide molecules contain two simple sugars bound together, such as in sucrose, which contains a glucose and a fructose unit in each molecule. A water molecule is eliminated when the bond forms between the two simple sugars.

Polysaccharides

There is an almost limitless number of possible structures in which monosaccharide units (monosaccharide molecules minus one water molecule at each bond between units) can be combined. Molecular weights are known to go above 1 million. Apparently nature has been very selective in that only a few of the many possible monosaccharide units are found in polysaccharides.

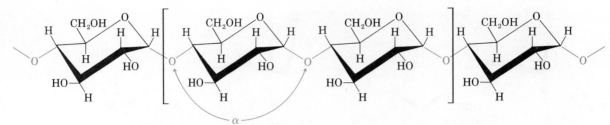

FIGURE 15–4   Amylose structure. From 60 to 300 α-D-glucose units are bonded together by alpha linkages to form amylose molecules.

Starch molecules consist of many glucose units bonded together.

STARCHES AND GLYCOGEN   Starch is found in plants in protein-covered granules. These granules are disrupted by heat, and part of the starch content is soluble in hot water. Soluble starch is **amylose** and constitutes 22% to 26% of most natural starches; the remainder is **amylopectin.** Amylose gives the familiar blue-black starch test with iodine solutions; amylopectin turns red on contact with iodine.

Structurally, amylose is a straight-chain polymer of α-D-glucose units, each one bonded to the next, just as the two units are bonded in maltose (Fig. 15–3). Molecular weight studies on amylose indicate that the average chain contains about 200 units. A representative portion of the structure of amylose is shown in Figure 15–4.

Amylopectin is made up of branched chains of α-D-glucose units (Fig. 15–5). Its molecular weight generally corresponds to about 1000 glucose units. Partial hydrolysis of amylopectin yields mixtures called **dextrins.** Complete hy-

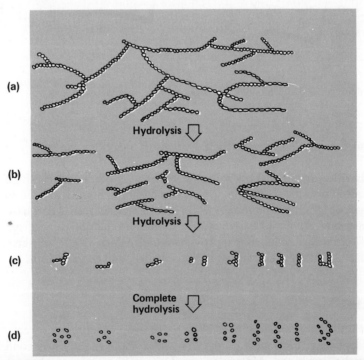

FIGURE 15–5   (*a*) Partial schematic amylopectin structure. (*b*) Dextrins from incomplete hydrolysis of *a*. (*c*) Oligosaccharides from hydrolysis of dextrins. (*d*) Final hydrolysis product: D-glucose. Each circle represents a glucose unit.

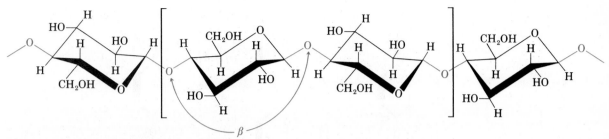

FIGURE 15–6 Cellulose structure. About 2800 β-D-glucose units are bonded together by beta linkages to form a cellulose molecule.

drolysis, of course, yields D-glucose. Dextrins are used as food additives, mucilages, and paste and in finishes for paper and fabrics.

Glycogen serves as an energy reservoir in animals as does starch in plants. Glycogen has a structure similar to that of amylopectin (branched chains of glucose units), except that glycogen is even more highly branched.

CELLULOSE Cellulose is the most abundant polysaccharide in nature. Like amylose, it is composed of D-glucose units. The difference between the structure of cellulose and that of amylose lies in the bonding between the D-glucose units. In cellulose, all of the glucose units are in the β-ring form in contrast to the α-ring form in amylose. (Review the ring forms in Fig. 15–1 and compare the structures in Figs. 15–4 and 15–6.)

The different structures of starch and cellulose account for their difference in digestibility. Human beings and carnivorous animals do not have the necessary enzymes (biochemical catalysts) to break down the cellulose structure as do numerous microorganisms. Cellulose is readily hydrolyzed to D-glucose in the laboratory by heating a suspension of the polysaccharide in the presence of a strong acid. Unfortunately, at present this is not an economically feasible solution to the world's growing need for an adequate food supply.

Paper, rayon, cellophane, and cotton are principally cellulose. A representative portion of the structure of cotton is shown in Figure 15–7. Note the hydrogen bonding between cellulose chains.

Humans do not have an enzyme to split cellulose into its glucose units.

SELF-TEST 15–A

1. Carbohydrates contain the elements _____, _____, and _____.

2. The complete hydrolysis of a polysaccharide yields _____.

3. When a molecule of sucrose is hydrolyzed, the products are one molecule each of the monosaccharides _____ and _____.

4. The sugar referred to as blood sugar, grape sugar, or dextrose is actually the compound _____.

5. Starch is a polymer built up out of _____ units.

6. What type of bonding holds the polysaccharide chains together, side by side, in cellulose? _____

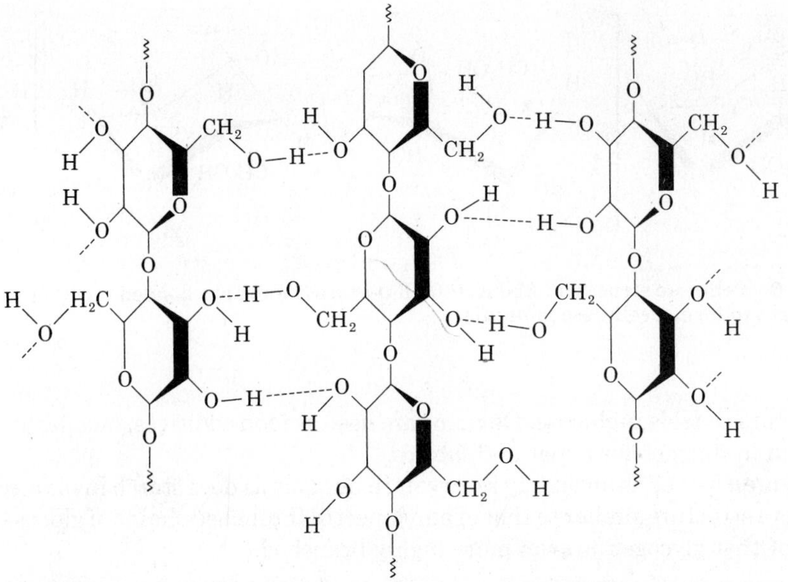

FIGURE 15-7 The properties of cotton, about 98% cellulose, can be explained in terms of this submicroscopic structure. A small group of cellulose molecules, each with 2000 to 9000 units of D-glucose, are held together in an approximately parallel fashion by hydrogen bonding (----). When several of these *chain bundles* cling together in a relatively vast network of hydrogen bonds, a *microfibril* results; the microfibril is the smallest microscopic unit that can be seen. The macroscopic *fibril* is a collection of numerous microfibrils. The absorbent nature of cotton results from the numerous capillaries that exist between the cellulose chains wherein the smaller water molecules are held by hydrogen bonds.

PROTEINS, AMINO ACIDS, AND THE PEPTIDE BOND

Proteins are high molecular weight compounds made up of amino acid units.

Proteins occur in all the major regions of living cells. These compounds serve a wide variety of functions, including motion of the organism, defense mechanism against foreign substances, metabolic regulation of cellular processes, and cell structure. The close relationship between proteins and living organisms was first noted by the German chemist G. T. Mulder in 1835. He named these compounds proteins from the Greek *proteios,* meaning "first," indicating this to be the starting point in the chemical understanding of life.

Proteins are macromolecules with molecular weights ranging from 5000 to several million. Like the polysaccharides, these macrostructures are composed of recurring units of molecular structure. The fundamental units in the case of proteins are **amino acids.** Proteins and amino acids are made primarily from four elements: carbon, oxygen, hydrogen, and nitrogen. Other elements occur in trace amounts; the one most often encountered is sulfur.

Amino acids are compounds that generally have the structure

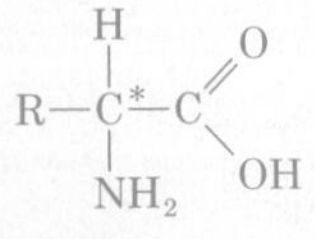

$$R-\overset{\overset{\displaystyle H}{|}}{\underset{\underset{\displaystyle NH_2}{|}}{C^*}}-C\overset{\displaystyle O}{\underset{\displaystyle OH}{\diagup}}$$

C* denotes an asymmetric carbon atom.

Amino Acids

The complete hydrolysis of a typical protein yields a mixture of about 20 different amino acids. Some proteins lack one or more of these acids, others have small amounts of other amino acids characteristic of a given protein, but the 20 given in Table 15-1 are predominant. In a few instances, one amino acid will constitute a

TABLE 15–1   Common Amino Acids

All of the amino acids except proline and hydroxyproline have the general formula

$$R-\overset{\overset{\displaystyle H}{|}}{\underset{\underset{\displaystyle NH_2}{|}}{C^*}}-C\overset{\displaystyle O}{\underset{\displaystyle OH}{\diagup}}$$

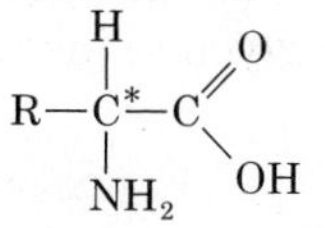

in which R is the characteristic group for each acid. The R groups are as follows.

1. Glycine —H
2. Alanine —CH_3
3. Serine —CH_2OH
4. Cysteine —CH_2SH
5. Cystine —CH_2—S—S—CH_2—
*6. Threonine —$\underset{\underset{\displaystyle OH}{|}}{CH}$—$CH_3$

*7. Valine CH_3—$\underset{|}{CH}$—CH_3

*8. Leucine —CH_2—$\underset{\underset{\displaystyle CH_3}{|}}{CH}$—$CH_3$

*9. Isoleucine —$\underset{\underset{\displaystyle CH_2-CH_3}{\diagdown}}{\overset{\overset{\displaystyle CH_3}{\diagup}}{CH}}$

*10. Methionine —CH_2—CH_2—S—CH_3
11. Aspartic acid —CH_2CO_2H
12. Glutamic acid —CH_2—CH_2—CO_2H
*13. Lysine —CH_2—CH_2—CH_2—CH_2—NH_2

*14. Arginine —CH_2—CH_2—CH_2—$NH\overset{\overset{\displaystyle NH}{||}}{C}NH_2$

*15. Phenylalanine —CH_2—⬡

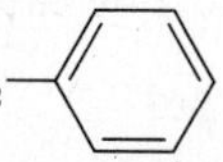

16. Tyrosine —CH_2—⬡—OH

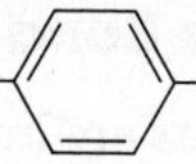

*17. Tryptophan —CH_2— [indole ring]

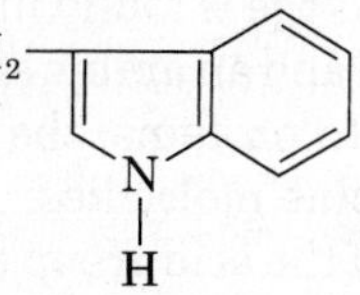

*18. Histidine —CH_2— [imidazole ring N=, N—H]

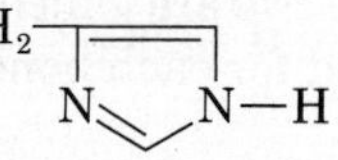

The structures for the other two are:

19. Proline $H_2C\!\!-\!\!\!-\!\!\!-\!\!CH_2$, $H_2C\diagdown\,\underset{\underset{\displaystyle H}{|}}{N}\,\diagup CHCO_2H$

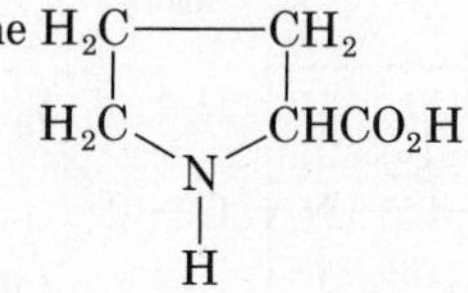

20. Hydroxyproline $HOHC\!\!-\!\!\!-\!\!\!-\!\!CH_2$, $H_2C\diagdown\,\underset{\underset{\displaystyle H}{|}}{N}\,\diagup CHCO_2H$

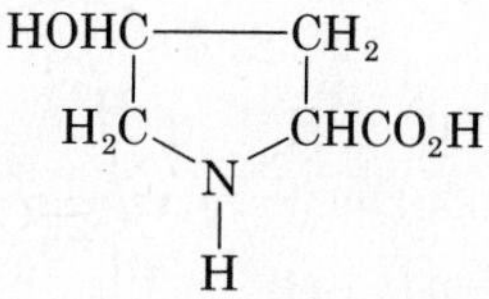

* Essential amino acids; arginine and histidine are essential for children but may not be essential for adults.

major fraction of a protein (the protein in silk, for example, is 44% glycine), but this is not common.

As the name suggests, amino acids contain an amine group ($-NH_2$) and an acid (carboxyl) group ($-COOH$). In all of the amino acids listed in Table 15–1, the amine group and the acid group are bonded to the same carbon atom. Of these acids, 18 have the general formula:

$$R-\overset{\overset{\displaystyle H}{|}}{\underset{\underset{\displaystyle NH_2}{|}}{C^*}}-C\overset{\displaystyle O}{\underset{\displaystyle OH}{\diagup}}$$

There are about 20 common amino acids.

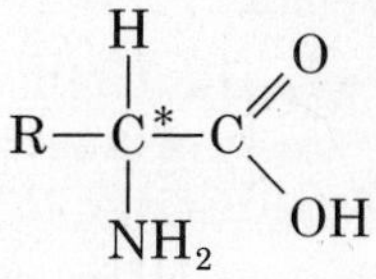

Review the discussion of optically active amino acids in Chapter 12.

where R is a characteristic group for each amino acid, and (*) denotes an asymmetric carbon atom. Recall that all amino acids except glycine can exist as optical isomers, and nature prefers the left-handed amino acids (L-isomers).

The simplest amino acid is glycine, in which R is a hydrogen atom.

Essential amino acids are amino acids that the body needs but cannot make.

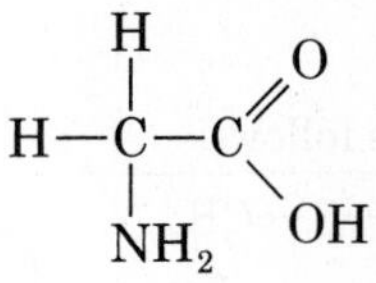

$$
\begin{array}{c}
\text{H} \\
| \\
\text{H}-\text{C}-\text{C} \overset{\displaystyle O}{\underset{\displaystyle OH}{}} \\
| \\
\text{NH}_2
\end{array}
$$

The human body is capable of synthesizing some amino acids needed for protein structures, but it is unable to provide others necessary for normal growth and development. The latter are designated **essential amino acids** and must be ingested in the food supply. The **nonessential amino acids** are just as necessary for life as the essential amino acids but can be made by the body from other compounds. The essential amino acids are indicated in Table 15–1 by asterisks.

For good nutrition we require *all* of the essential amino acids in our daily diet, but the amount required does not exceed 1.5 g per day for any of them.

The Peptide Bond

The peptide bond

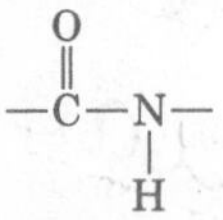

$$
\begin{array}{c}
\text{O} \\
\| \\
-\text{C}-\text{N}- \\
| \\
\text{H}
\end{array}
$$

binds amino acid units together in proteins.

Starch, glycogen, cellulose, and proteins are condensation polymers (Chapter 14).

Amino acid units are linked together in protein structures by peptide bonds. This same linkage is found in polyamides like nylon 66 (Chapter 14), in which a carboxylic acid and an amine are condensed to form the polymer and the amide bond. As it applies to proteins, the peptide bond can be understood by the reaction between two glycine molecules.

If the acid group of one glycine molecule reacts with the basic amine group of another, the two are joined through the peptide linkage, and one molecule of water is eliminated for each bond formed.

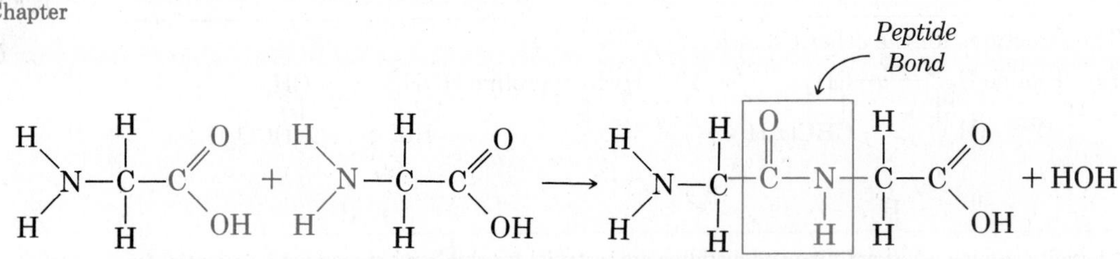

GLYCINE   GLYCINE   GLYCYLGLYCINE

If this hypothetical reaction is carried out with two different amino acids, glycine and alanine, two different **dipeptides** are possible.

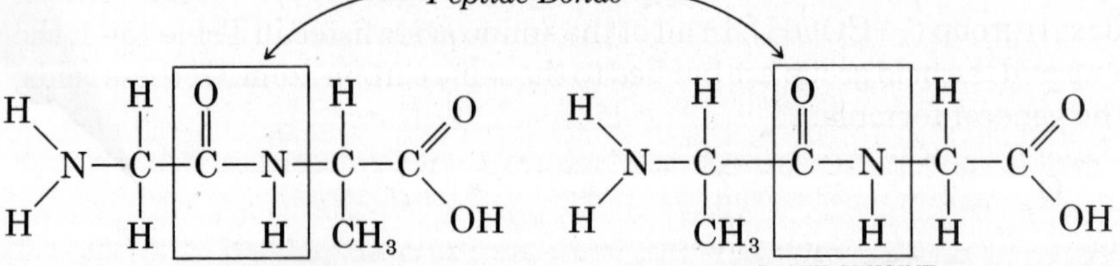

GLYCYLALANINE   ALANYLGLYCINE

Twenty-four **tetra**peptides are possible if four amino acids (for example, glycine, Gly; alanine, Ala; serine, Ser; and cystine, Cys) are linked in all possible combinations. They are:

A very large number of different proteins can be prepared from a small number of different amino acids.

Gly-Ala-Ser-Cys	Ala-Gly-Ser-Cys	Ser-Ala-Gly-Cys	Cys-Ala-Gly-Ser
Gly-Ala-Cys-Ser	Ala-Gly-Cys-Ser	Ser-Ala-Cys-Gly	Cys-Ala-Ser-Gly
Gly-Ser-Ala-Cys	Ala-Ser-Gly-Cys	Ser-Gly-Ala-Cys	Cys-Gly-Ala-Ser
Gly-Ser-Cys-Ala	Ala-Ser-Cys-Gly	Ser-Gly-Cys-Ala	Cys-Gly-Ser-Ala
Gly-Cys-Ser-Ala	Ala-Cys-Gly-Ser	Ser-Cys-Ala-Gly	Cys-Ser-Ala-Gly
Gly-Cys-Ala-Ser	Ala-Cys-Ser-Gly	Ser-Cys-Gly-Ala	Cys-Ser-Gly-Ala

If 17 different amino acids are used, the sequences alone would make 3.56×10^{14} uniquely different 17-unit molecules.* Although there are numerous protein structures in nature, these represent an extremely small fraction of the possible structures. Of all the many different proteins that could possibly be made from a set of amino acids, a living cell will make only a relatively small, select number.

Protein Structures

The **primary structure** of a protein indicates only the sequence of amino acid units in the polypeptide chain. Since the single bonds in the chain allow free rotation around the bond, an almost infinite number of conformations is possible. Because of interactions, such as hydrogen bonds, between atoms in the same chain, certain conformations called **secondary structures** are favored. Linus Pauling, along with R. B. Corey, suggested the two secondary structures for polypeptides discussed subsequently, the sheet structure and the helical structure.

Polyglycine is a synthetic protein made entirely of the amino acid glycine. In polyglycine the hydrogen attached to the nitrogen atom and the oxygen bonded to the carbon are both well suited to engage in hydrogen bonds. In the two stable conformations of polyglycine, maximum advantage is taken of the hydrogen bonds available. In the sheet structure, the hydrogen bonds are between adjacent chains of the polypeptide; in the helical structure, hydrogen bonds occur between atoms within the same chain.

Figure 15–8 illustrates a sheetlike structure in which several chains of the polypeptide are joined by hydrogen bonds. Note that all the oxygen and nitrogen atoms are involved in hydrogen bonds. Most of the properties of silk can be explained in terms of this type of structure for fibroin, the protein of silk.

Hydrogen bonds are possible within a single polypeptide chain if the secondary structure is helical (Fig. 15–9). Bond angles and bond lengths are such that the nitrogen atom forms hydrogen bonds with the oxygen atom in the third amino acid unit down the chain.

Collagen is the principal fibrous protein in mammalian tissue. It has remarkable tensile strength, which makes it important in the structure of bones, tendons, teeth, and cartilage. Three polypeptide chains, each of which is twisted into a left-handed helix, are twisted into a right-handed super-helix to form an

Linus Pauling (1901–   ) is a scientist of great versatility and accomplishment. His interests have included the determination of the molecular structures of crystals by X-ray diffraction and theories of the chemical bond. His work led to the Nobel prize in 1954. For his fight against the nuclear danger confronting the world, he was awarded the 1963 Nobel peace prize.

The amino acids in a protein chain interact with each other via hydrogen bonds.

A coiled spring is helical in structure.

* If the amino acids are all different, the number of arrangements is $n!$ (read n factorial). For five different amino acids, the number of different arrangements is 5! (or $5 \times 4 \times 3 \times 2 \times 1 = 120$).

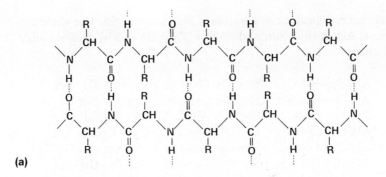

(a)

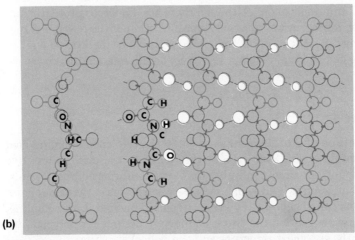

(b)

FIGURE 15-8   Sheet structure for polypeptide. In (a) the two-dimensional drawing emphasizes that all of the oxygen and nitrogen atoms are involved in hydrogen bonds for the most stable structure. (b) Illustrates the bonds in perspective, showing that the sheet is not flat; rather, it is sometimes called a pleated sheet structure.

extremely strong fibril, as shown in Figure 15-10. A bundle of such fibrils forms the macroscopic protein.

The structure of collagen illustrates a third level of protein structure, **tertiary structure.** The primary structure is the sequence of amino acids in the protein, the secondary structure is the helical form of the protein chain, and the tertiary structure is the twisted or folded form of the helix. Another tertiary structure is found in globular proteins. In these structures, the helix chain (secondary structure) is folded and twisted into a definite geometric pattern. This pattern may be held in place by one or more of several different kinds of chemical bonds, such as —S—S—bonds, depending on the particular functional groups in the amino acids involved (Table 15-1). Figure 15-11 illustrates the folded structure of a typical globular protein.

The **quaternary structure** of proteins refers to the degree of aggregation of protein units. Native hemoglobin (molecular weight of 68,000) must have its four polypeptide chains properly aggregated in order to form active hemoglobin. Insulin is also composed of subunits, properly arranged into its quaternary structure.

If hemoglobin, a globular protein, has an abnormal primary, secondary, tertiary, or quaternary structure because of a wrong amino acid in a given position,

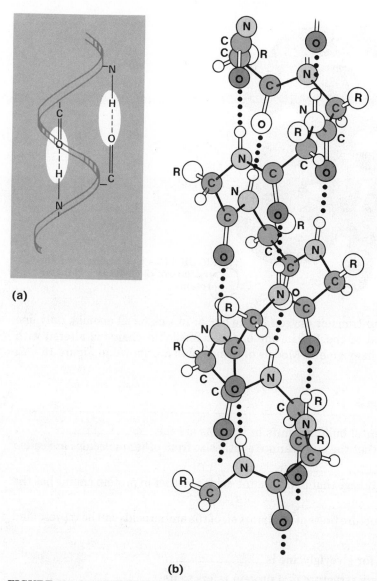

(a)

(b)

FIGURE 15–9 (*a*) Helical structure for a polypeptide in which each oxygen atom can be hydrogen bonded to a nitrogen atom in the third amino acid unit down the chain. (*b*) α-Helix structure of proteins. The sketch represents the actual position of the atoms and shows where intra-chain hydrogen bonds occur.

FIGURE 15–10 The imaginary structure of collagen.

FIGURE 15–11 Imaginary folded structure of the helix in a globular protein.

it may be unable to transfer oxygen in the blood. In sickle cell anemia, only one specific amino acid of the 146 in one of the hemoglobin chains is altered with respect to normal hemoglobin. Models of hemoglobin are shown in Figure 15–12.

SELF-TEST 15–B

1. The fundamental building units in proteins are the _____.
2. Amino acids that the body cannot synthesize from other molecules are called

   _____.
3. The peptide linkage that bonds amino acids together in protein chains has the

   structure _____.
4. The basic structure present in almost all of the amino acids can be represented

   as_____.
5. The formula for glycylglycine is _____.
6. a. The primary structure of a protein refers to its _____;

   b. The secondary structure refers to its _____;

   c. Its tertiary structure refers to _____;

   d. And its quaternary structure refers to _____.
7. a. If we have three different amino acids, we can make a total of

   _____ different tripeptides from them if we can use an amino acid up to three times in any given tripeptide.

   b. If we can use each amino acid only once, there are still _____ possible different tripeptides.
8. Describe how hydrogen bonding is involved in the tertiary structures of pro-

   teins. _____

$$CH_2$$
$$CH$$

(b)

FIGURE 15–12 (*a*) The structure of heme. (*b*) Two views of a model of the hemoglobin structure. Two light-colored chains of protein, two dark-colored chains, and two heme "disks" in proper arrangement compose the quaternary structure of hemoglobin. M. F. Perutz received a Nobel prize for determining this structure. (Courtesy of M. F. Perutz and *Science, 140:*863, 1963.)

ENZYMES

An important group of globular proteins is the **enzymes,** molecules that function as catalysts for reactions in living systems. Like other catalysts, a given enzyme increases the rate of a reaction without requiring an increase in temperature. As an example of a simple type of catalysis, consider the oxidation of glucose, or a cube of sugar, which burns in air with some difficulty and is hard to light with a match. If cigarette ashes or other catalysts are placed on its surface, combustion can be

Enzymes are protein molecules that speed up chemical reactions.

initiated easily with a match. When glucose burns, it liberates a large amount of energy, 688 kcal, or 688,000 calories, per mole.

$$C_6H_{12}O_6 + 6\ O_2 \rightarrow 6\ CO_2 + 6\ H_2O + 688\ \text{kcal}$$
GLUCOSE   OXYGEN   CARBON   WATER
                     DIOXIDE

The energy required to get the reaction started is the **activation energy;** catalysts, in general, work by lowering the activation energy. If an enzyme can lower the activation energy to a point where the average kinetic energy of the molecules in a living cell (or in a laboratory system) is sufficient for reaction, then the reaction can proceed rapidly. Glucose is oxidized rapidly and efficiently at ordinary temperatures in the presence of the proper enzymes. To be sure, the oxidation of glucose in a living cell requires many enzymes and many steps, but enzymatic catalysis produces the same final result as combustion at elevated temperature, namely carbon dioxide, water, and 688 kcal of usable energy per mole of sugar. Figure 15–13 graphically illustrates the concepts of activation energy, the energy available from an energy-producing reaction, and the decrease of the activation energy by an enzyme.

The names of most enzymes end in *-ase.*

Most of the names of enzymes end in **-ase.** There are a few exceptions, such as pepsin and trypsin, which are both digestive enzymes. Hydrolases promote the breakdown of foodstuffs and other substances by hydrolysis. Carbohydrases (such as maltase, lactase, sucrase, ptyalin, and amylase) help to effect hydrolysis of carbohydrates. Proteases (such as pepsin, trypsin, and chymotrypsin) hydrolyze

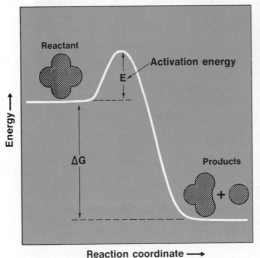 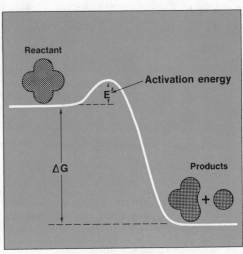

(a) No enzyme present   (b) Enzyme present

FIGURE 15–13 Enzyme effect on activation energy. The vertical coordinate represents increasing energy, and the horizontal one represents time. For energy-producing reactions the reactant molecules are at a higher energy than the product molecules, as illustrated in (a). The difference between these energies is the net useful energy provided by the chemical reaction. The useful energy is tabulated as the free energy change of the reaction (ΔG). However, it is necessary for the reactant molecules to "get over" the energy barrier (acquire the activation energy, E) in going from reactants to products. Note that the activation energy is given back along with the free energy. The enzyme lowers the activation energy to E, as illustrated in (b), while the free energy change remains the same. The net effect is to obtain the free energy of the reaction with a smaller expenditure of activation energy. Free energy is discussed in Chapter 8.

large protein molecules into groups of smaller proteins. Lipases hydrolyze esters such as fats and oils. Examples of the oxidizing enzymes, called oxidases, are catalase, which speeds up the conversion of hydrogen peroxide to water and oxygen, and dehydrogenases, which assist in the removal of hydrogen from molecules. There are other categories of enzymes, but these illustrate the wide variety of biochemical catalysts.

Enzymes are remarkable catalysts because they are highly specific for a given reaction. Maltase, an enzyme, catalyzes the hydrolysis of maltose into two molecules of D-glucose. This is the only known function of maltase, and no other enzyme can substitute for it. The explanation for the specific activity of enzymes can be found in their molecular structures.

Enzymes are specific catalytic molecules with a specific catalytic task.

Enzymes are globular proteins with definite tertiary structures. The highly specific action of maltase can be explained if its globular structure accurately accommodates a maltose molecule at the point where the reaction occurs, the reactive site. When the two units come together, strain is placed on the bonds holding the two simple sugar units together. As a result, water is allowed to enter and hydrolysis occurs. Sucrose cannot be hydrolyzed by maltase because of the different structure involved. Another enzyme, sucrase, hydrolyzes sucrose effectively. Some enzymes, however, are less specific. The digestive enzyme trypsin, for example, acts predominantly on peptide bonds in proteins, but it will also catalyze the hydrolysis of some esters because of somewhat similar structure and polarity at the active site.

How can an enzyme lower the activation energy and be so specific for a given reaction? Just as a key can separate a padlock into two parts and subsequently remain unchanged, ready to unlock other identical locks, so the enzyme makes possible a molecular change (Fig. 15–14). With enough energy the lock could be separated without the key, and with enough energy the molecular alteration could occur without the enzyme. An enzyme cannot make an unnatural chemical reaction occur.

The tremendous speed of enzyme-catalyzed reactions requires more than just random collisions to fit "the key in the lock." For example, a single molecule of β-amylase catalyzes the breaking of 4000 bonds between the α-glucose units in amylose per second. Speed like this requires something to attract the key into the

Enzyme structure is the key to specific catalytic activity.

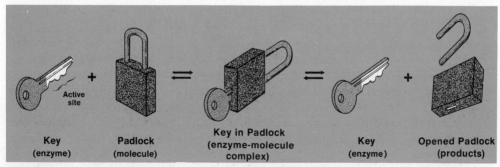

FIGURE 15–14 Lock-and-key theory for enzymatic catalysis. Although it is generally agreed that this analogy is an oversimplification, it does make one very important point: the enzyme makes a difficult job easy by reducing the energy required to get the job started. It also suggests that the enzyme has a particular structure at an active site that will allow it to work only for certain molecules, similar to a key that fits the shape of a particular keyhole and a particular sequence of tumblers.

lock, such as electrically polar regions, partially charged groupings, or ionic sections on the enzyme and the **substrate** (the reactant molecule). These regions attract as well as guide the substrate to the proper position on the enzyme and thereby speed up the reaction. The electrically charged portions of the enzyme are believed to be the chemically **active sites** in the enzyme.

Some enzymes require coenzymes: metal ions or vitamins.

Sometimes an enzyme is more than a globular protein; in such cases, the protein is not a catalyst by itself. In addition to the protein part of the enzyme, there is also another chemical species called a **coenzyme.** The coenzyme, required for catalytic activity, may be an ion (e.g., Co^{3+}, Fe^{3+}, Mg^{2+}, or another of the essential minerals) or may be derived in part from a vitamin. The protein part of such an enzyme is called the **apoenzyme.** The coenzyme alone does not have enzymatic activity; neither does the apoenzyme. Before the enzyme becomes active, the apoenzyme and the coenzyme must combine like the two keys required to open a bank lock-box. Neither your key nor the bank's key will open the box when used alone, but both together will.

Let's use a simple chemical example to demonstrate the relationship between coenzyme, substrate, and apoenzyme. The peptide glycylglycine is hydrolyzed very slowly by water to give two molecules of glycine:

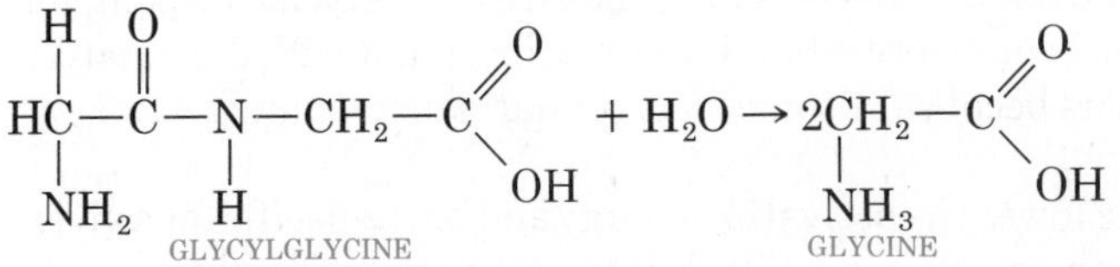

The glycylglycine substrate forms an intermediate compound with an enzyme, and as a result the substrate is activated for further reaction. Activation can result from extensive hydrogen bonds, interaction with a metal ion in the enzyme, or a number of other processes. In this case the coordination of the glycylglycine to the positive charge of a cobalt ion (Co^{2+}), the coenzyme, makes it more susceptible to attack by the negative end of a water molecule (Fig. 15–15).

Many **vitamins** function as coenzymes. The B vitamins are found in every cell as coenzymes in various oxidative processes. For example, niacin (vitamin B_3) becomes part of an enzyme that prevents pellagra, at one time a common vitamin-deficiency disease in the United States. Doctors and biochemists now know that the body suffers from pellagra when it lacks sufficient niacin. Some foods such as yeast, liver, meats, fish, eggs, whole wheat, brown rice, and peanuts contain niacin. The body needs only 0.06 g of niacin each day in order to prevent pellagra. This isn't much, but it is vital. The coenzyme of which niacin is a part is necessary to the energy production in the body. If energy is not provided, the whole process of renewing cells and building needed compounds slows down and eventually stops.

Vitamins' roles in nutrition are discussed in Chapter 20.

The coenzyme involved is a large molecule, **nicotinamide adenine dinucleotide (NAD$^+$).** Riboflavin (vitamin B_2) is a necessary part of another important coenzyme, **flavin adenine dinucleotide (FAD).** FAD and NAD$^+$ help to unlock the energy from energy-rich glucose and glycogen. The process will be described later in this chapter.

Structures of FAD and NAD$^+$ are shown in Chapter 20.

Besides being biochemical middlemen, speeding up and directing all the chemical reactions that go into the continuous breakdown and buildup of our cells (3 million red blood cells are renewed in the human body every second), enzymes may be the answer to future food problems. Scientists have already developed a way to produce sugar (on a limited experimental basis) by bubbling carbon dioxide into

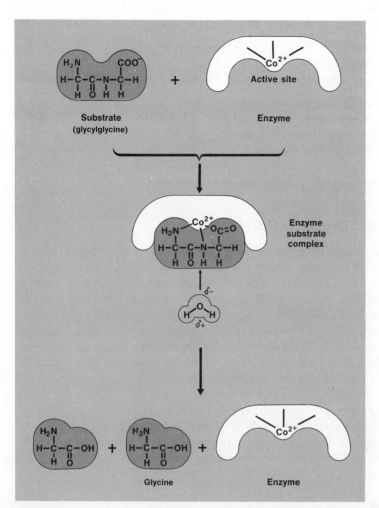

FIGURE 15–15   Action of an enzyme. The substrate molecule is chemically bonded to the enzyme (glycylglycine dipeptidase). The negative oxygen and the nitrogen atoms of the substrate bond to the positive cobalt ion in the enzyme. The bonding of the substrate makes it more susceptible to attack by water. Hydrolysis occurs and the glycine molecules are released by the enzyme, which is then ready to play its catalytic role again.

water containing enzymes. Trash fish can be converted into palatable animal feed by using enzymes. Work is under way to convert oil spills into edible products for sea organisms. A little later in this text we shall encounter the use of enzymes as meat tenderizers.

Enzymes are also important in commercial products.

SELF-TEST 15–C

1. The best term to describe the general function of enzymes is (   ) *catalyst,* (   ) *intermediate,* or (   ) *oxidant.*
2. In the lock-and-key analogy of enzyme activity, the enzyme functions as the _____ while the substrate molecule serves as the _____.
3. Pellagra can be prevented by intake of the vitamin named _____.

4. The activation energy of many biological reactions is decreased if a(n) _____ is present.
5. Apoenzyme + coenzyme → _____ .
6. Riboflavin is a vitamin that is needed because it is part of an essential _____ .
7. The coenzyme nicotinamide adenine dinucleotide (NAD^+) cannot be made by the human body unless it has a supply of _____ .
8. That portion of the enzyme at which the reaction is catalyzed is called the _____ .

ENERGY AND BIOCHEMICAL SYSTEMS

Where do we get the energy to do those things we need to do? The answer is that ultimately all of our energy comes from the Sun. How does the Sun's energy get into us to carry on life's processes? The beginning is photosynthesis, in which green plants absorb radiant energy and then combine carbon dioxide and water to make glucose and oxygen. With its stored energy, glucose is a high-energy substance. When we eat carbohydrates (or proteins or fats), they must be digested (hydrolyzed) into smaller molecules. In subsequent cellular chemistry, some of the stored energy in the foodstuffs is transferred to the bonds in molecules such as ATP. When this energy is needed, the ATP molecules release it to other chemical reactions. These reactions make us tick.

Imagine, for a moment, some light energy travelling from the Sun at 186,000 miles (299,000 kilometers) per second, and the light strikes the leaf of a green plant. How does the light energy ultimately help us to run? The first step in the process is to energize chlorophyll; this is the first step in photosynthesis.

Photosynthesis

Photosynthesis is a very complex process that produces the relatively simple overall reaction in which carbon dioxide and water are converted into energy-rich carbohydrates by solar energy.

$$6\ CO_2 + 6\ H_2O + 688\ \text{kcal} \rightarrow C_6H_{12}O_6 + 6\ O_2$$
CARBON   WATER   ENERGY   GLUCOSE   OXYGEN
DIOXIDE   (SUNLIGHT)

Reduction: gain of electrons or hydrogen. *Oxidation:* loss of electrons or hydrogen.

In photosynthesis, carbon dioxide is **reduced** to form a sugar

$$6\ CO_2 + 24\ H^+ + 24\ e^- \rightarrow C_6H_{12}O_6 + 6\ H_2O$$

and water is **oxidized** to oxygen

$$12\ H_2O \rightarrow 6\ O_2 + 24\ H^+ + 24\ e^-$$

Photosynthesis involves a number of different steps and is a very complex process.

Note that these two reactions, the first reduction and the second oxidation, give the overall reaction when added.

Not all the details of photosynthesis are fully understood. However, photosynthesis is generally considered a series of *light reactions,* which can occur only in the presence of light energy, and a series of *dark reactions,* which can occur in the dark but feed on the high-energy structures produced in the light reactions. The light reactions are unique to green plants, but the dark reactions occur in both plant and some animal cells.

Photosynthesis is initiated by a quantum of light energy. The green plant contains certain pigments that readily absorb light in the visible region of the spectrum. The most important of these are the chlorophylls, **chlorophyll a** and **chlorophyll b.** Note that both chlorophylls are compounds of magnesium and both have complex ring systems. Such ring systems usually absorb light in the visible region of the spectrum; consequently, they are colored. For example, chlorophyll is green because it absorbs light in the violet region (about 400 nanometers) and the red region (about 650 nanometers) and allows the green light between those wavelengths to be reflected or transmitted.

When chlorophyll absorbs photons of light, electrons are raised to higher energy levels. As these electrons move back down to the ground state, very efficient subcellular components of the plant cell known as chloroplasts grab this energy and, through a series of steps that are not all completely known, store the energy as chemical potential energy. As shown in Figure 15–16, one of the chemicals used to store this energy is ATP.

The energy of a photon is captured by chlorophyll by raising an electron to a higher energy state.

Announced in 1985, a blue-green protozoa, *Stentor coeruleus,* contains a light-absorbing substance, stentorin, which allows the microscopic animal to undergo its unique type of photosynthesis.

CHLOROPHYLL A

CHLOROPHYLL B

The energy stored in ATP in the light reactions is transferred to glucose during the dark reactions. Just how do these compounds store energy? The energy is stored in their newly formed chemical bonds.

Each ATP molecule contains two so-called high-energy phosphate bonds. These are marked by wiggle bonds (~) in Figure 15–17.

In the presence of a suitable catalyst, ATP will undergo a three-step hydrolysis. The hydrolysis of ATP to adenosine diphosphate (ADP) and phosphoric acid releases about 12 kcal per mole (Fig. 15–18). The second hydrolysis of ADP to adenosine monophosphate (AMP) also produces about 12 kcal of energy per mole. Finally, the hydrolysis of AMP to adenosine, which involves a low energy bond, releases only about 2.5 kcal per mole.

A visual conception of the release of usable energy by the hydrolysis of ATP, ADP, and AMP is shown in Figure 15–19. Usable energy is also known as "free energy," and it is usable in the sense of doing work (moving matter) and of predict-

Hydrolysis of ATP is exothermic. Synthesis of ATP is endothermic.

Review the discussion of usable energy (free energy) in Chapter 8.

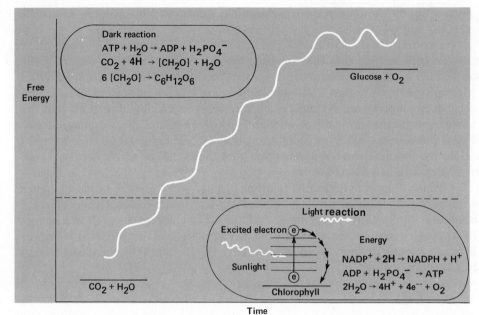

FIGURE 15–16 Usable energy (free energy discussed in Chapter 8) within the chemical system is increased as carbon dioxide and water are converted into glucose and oxygen by photosynthesis. This results in stored, useful energy in glucose.

ing the likelihood that a process will occur. Of interest here is the likelihood that a chemical reaction will occur. When a reaction gives off more usable energy than another reaction, the reaction giving off the greater amount of usable energy is more likely to occur by itself. The ATP hydrolysis gives more usable energy per mole than does the ADP hydrolysis or the AMP hydrolysis. Therefore, of the three reactions, the ATP hydrolysis is most likely to occur. This principle is valuable in trying to decipher a biochemical pathway from a maze of biochemical reactions. But more important to our discussion at this point is the concept that usable energy is given off by all of these reactions, which can run other reactions. So, the hydrolysis of principally ATP and ADP provides the usable energy to run a myriad of

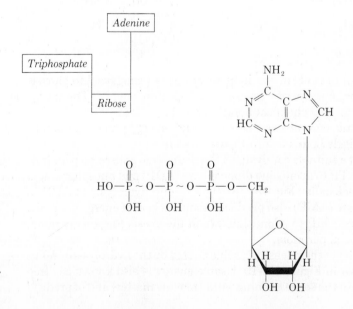

FIGURE 15–17 Molecular structure of ATP.

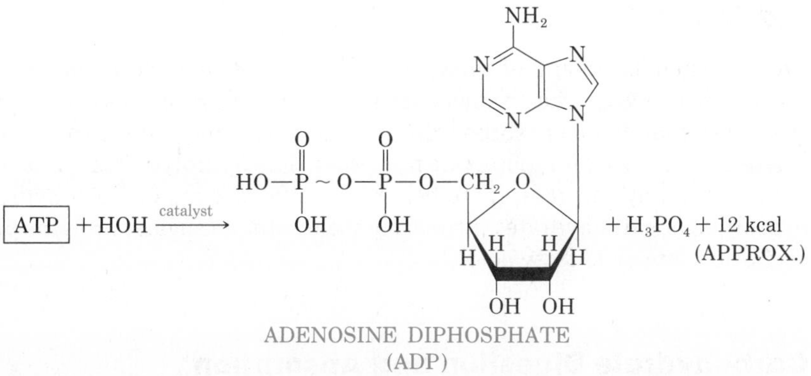

$$\boxed{\text{ATP}} + \text{HOH} \xrightarrow{\text{catalyst}}$$

ADENOSINE DIPHOSPHATE
(ADP)

$+ H_3PO_4 + 12 \text{ kcal}$
(APPROX.)

FIGURE 15–18
Hydrolysis of ATP to ADP.

biochemical reactions in the body. The values of energy given in the preceding paragraph are values of the total energy evolved, which includes the usable energy (the free energy) and the nonuseful energy factor (the entropy, Chapter 8).

As a point of emphasis, the oxygen produced in the light reaction of photosynthesis is the source (and only present source) of all the oxygen in our atmosphere. Only this life-giving gas, given off by trees and grass and greenery, even in the sea, makes possible human life and most animal life on Earth. We are dependent on the plant life of our planet, and we must live in balance with the oxygen output of that plant life, as well as with the food output of the same plants. Hence, the importance of photosynthesis: it is absolutely vital to life on Earth.

After photosynthesis, the living plant may convert the glucose to oligosaccharides, starches, cellulose, proteins, or oils. The end product depends on the type of plant and its degree of development.

Where is the energy now that was absorbed from sunlight? Some of it is stored as potential energy in the bonds of glucose and other high-energy compounds, which are formed subsequently.

The first steps in using the energy stored in high-energy compounds are for them to be eaten as food, digested, and transported to the cells of the body.

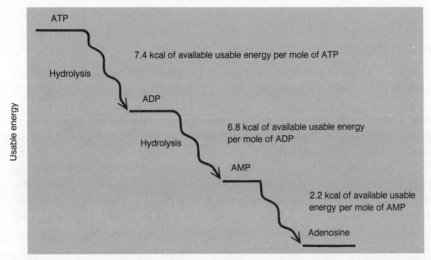

FIGURE 15–19   Usable energy furnished by successive hydrolyses of phosphate groups from ATP. This energy is available to bring about other useful chemical reactions.

DIGESTION

From a chemical point of view, digestion is the breakdown of ingested foods through hydrolysis. The products of these hydrolytic reactions are relatively small molecules that can be absorbed through the intestinal walls into the body fluids, where they are used for metabolic processes. The hydrolytic reactions of digestion are catalyzed by enzymes, there being a specific enzyme for each hydrolysis. The hydrolysis of carbohydrates ultimately yields simple sugars, proteins yield amino acids, and fats yield fatty acids.

Carbohydrate Digestion and Absorption

The principal forms of carbohydrates in our food are (1) high molecular weight polymers such as starch and glycogen, (2) disaccharides such as sucrose and lactose, and (3) simple sugars such as glucose and fructose. The first enzyme capable of acting on ingested food is furnished by the saliva and is named salivary amylase, or ptyalin. Its action on starch or glycogen produces limited amounts of the disaccharide maltose. Ptyalin is inactivated by the high acidity in the stomach. The stomach furnishes no enzymes that can catalyze the splitting of carbohydrate polymers. Indeed, the high acidity of the gastric juice would destroy or inactivate all such complex structures.

When food passes from the stomach, the acidity is neutralized by a secretion of the pancreas. The enzymes secreted by the pancreas can split some of the polysaccharides to maltose and maltose to glucose and can catalyze other hydrolytic reactions. The final result is a mixture of simple sugars such as glucose, fructose, and galactose. These simple sugars are then absorbed into the bloodstream, where the concentration of blood sugar is regulated by the hormone insulin. If the sugar level is too high, the simple sugars are converted into the polysaccharide glycogen in the liver; if the blood sugar level is too low, the stored glycogen is hydrolyzed to raise the level. Malfunctions of our biological systems can lead to too much blood sugar (hyperglycemia) or too little blood sugar (hypoglycemia); either condition, if sustained, indicates one type or another of diabetes.

Fat and Oil Digestion and Absorption

The term *lipid* denotes a group of compounds that includes fats, oils, and other substances whose solubility characteristics are similar to those of fats and oils. These compounds are not all structurally related, and we shall consider only the triglycerides (triesters of fatty acids and glycerol) in this chapter. A typical triglyceride is palmitooleostearin. Its structure and its hydrolytic products are shown in the equation at the top of page 375.

Fats and oils are digested primarily in the intestinal tract, where bile salts secreted by the liver aid in the process. The enzyme that aids in the hydrolysis of the fatty acid esters is water soluble, whereas the fats and oils are insoluble in water. Bile salts emulsify the oil, that is, they break up the oil into very tiny drops and prevent the drops from recombining readily. The tiny drops provide more surface area for the enzyme to attack so digestion can occur. The bile salts form an interface between the nonpolar oil and the polar water and make it possible for the oil to

$$\begin{array}{l}
\underset{\substack{\\}}{H_2C}-O-\overset{\overset{\displaystyle O}{\|}}{C}-(CH_2)_{14}CH_3 \\[2mm]
HC-O-\overset{\overset{\displaystyle O}{\|}}{C}-(CH_2)_7CH{=}CH(CH_2)_7CH_3 \; + \; 3\,HOH \\[2mm]
H_2C-O-\overset{\overset{\displaystyle O}{\|}}{C}-(CH_2)_{16}CH_3
\end{array}$$

PALMITOOLEOSTEARIN
(A TRIGLYCERIDE)   WATER

$$CH_3(CH_2)_{14}COOH$$
PALMITIC ACID

$$CH_3(CH_2)_7CH{=}CH(CH_2)_7COOH$$
OLEIC ACID

$$CH_3(CH_2)_{16}COOH$$
STEARIC ACID

$$\begin{array}{l} HO-CH_2 \\ HO-CH \\ HO-CH_2 \end{array}$$
GLYCEROL

"dissolve" in water. For a molecule to be an emulsifier between polar and nonpolar molecules, the emulsifier must have characteristics of both. One of the principal bile salts is derived from glycocholic acid.

SODIUM SALT OF GLYCOCHOLIC ACID

Notice that the bulky hydrocarbon groups of this bile salt are compatible with oil or fat and that the —OH and ionic groups anchor to water molecules. The bile salts emulsify oil in a manner similar to the action of a soap or detergent during the cleaning process (Chapter 22).

Protein Digestion and Absorption

The hydrolysis of proteins begins in the stomach and continues in the small intestine. Several different types of enzymes are known to be involved. These enzymatic systems must be controlled very carefully, for they have the potential of digesting the walls of the stomach and intestines. A number of these enzymes are secreted in an inactive form. For example, pepsin, which is secreted in the stomach, is first present in a form called pepsinogen. The molecular weight of pepsinogen is 42,600. In the presence of the acid of the stomach, pepsinogen is broken down to **pepsin**. The molecular weight of pepsin is 34,500. It is reasonable to believe that pepsinogen, pepsin, other enzymes, and the stomach acid normally have no effect on the stomach wall. However, pepsin would have considerable action on the stomach protein if it were formed under the mucous lining, a lining that is constantly sloughing off like the outer skin.

The stomach is protected from protein-splitting enzymes by a mucous lining.

Pepsin facilitates the breakdown of only about 10% of the bonds in a typical protein, leaving polypeptides with molecular weights from 600 up to 3000. In the small intestine, hydrolysis is completed to amino acids, which are absorbed through the intestinal wall.

Some protein enzymes are sold commercially. Meat tenderizers are proteases, materials that speed up partial digestion of meat. Enzymes are used as stain removers in detergents, although they may irritate the skin of some individuals. Related enzymes are also used to free the lens of the eye prior to cataract surgery.

THE CENTRAL BANK OF THE BODY — THE LIVER

After digestion, most food nutrients pass directly to the liver for distribution to the body. Fats are broken down to fatty acids and glycerol. Glucose is used for energy in the liver, to prepare glucose phosphate as the first step in preparation of the storage carbohydrate, glycogen, and about one third goes on in the bloodstream to nourish the cells. From the liver, a fraction of the amino acids is sent to the cells for protein building. In the liver, amino acids are used in the formation of enzymes, and some are oxidized to obtain energy. In these respects, and others, the liver is the central nutrient bank of the body, in that the liver stores, converts, and classifies nutrients. In addition to these activities, the liver detoxifies pollutants invading the body.

ENZYMES AND HEREDITY

Genetic effects are often observed in the pattern of enzymes produced by individuals or races. An example of this is found in "lactose intolerance," common in certain peoples of Asia (e.g., Chinese and Japanese) and Africa (many black tribes), whose diets have traditionally contained little milk after the age of weaning. While infants, such people manufacture the enzyme **lactase** that is necessary to digest lactose, a sugar occurring in all mammals' milk. As they grow older, their bodies stop producing this enzyme because their diets normally contain no milk, and the ingestion of milk products containing lactose can lead to considerable discomfort in the form of stomach aches and diarrhea. People whose ancestral adult diets contained substantial amounts of milk or milk products (African tribes such as the Masai), continue to produce lactase as adults and can eat such foods and digest the lactose they contain. It is quite possible that this is only one of several similar cases in which a traditional tribal diet has altered the pattern of production of digestive enzymes.

GLUCOSE METABOLISM

The absorbed energy of sunlight is now stored in the bonds of small molecules such as glucose, amino acids, or fatty acids. This stored energy is then transferred from the bonds of these small and energy-rich molecules to the high-energy bonds in ATP. In the process, the small, energy-rich molecules are changed into energy-poor compounds such as carbon dioxide, water, and urea.

Although several Nobel prizes have been awarded for discoveries in glucose metabolism, we shall center on two Nobel prize – winning discoveries. After studying a few of the details of how glucose is converted to carbon dioxide and water with the accompanying transfer of energy, perhaps you will agree the awards were deserved. The series of chemical reactions had to be discovered in a maze of complexity fraught with many variables. Rather than say the work and results are either complex or simple, we will simply show you the results of the work and let you make your own evaluation.

The initial sequence of reactions by which energy is obtained from glucose and similar compounds can follow two courses; one sequence does not use elemental oxygen **(anaerobic),** but the other sequence does **(aerobic).**

The anaerobic process was discovered by the German chemist Otto Fritz Meyerhof (1884–1951). More details were discovered by Gustav Embden (1874–1933). In 1918, Meyerhof showed that animal cells broke down sugar in much the same way yeast, a plant, did. For the first time, this work made clear that, with only minor differences, metabolism follows the same sequences in all creatures. Details of the individual steps were discovered between 1932 and 1933 by Embden and between 1937 and 1941 by Carl Ferdinand Cori (1896–   ) and his wife Gerty Theresa (1896–1957) (they shared a Nobel prize in 1947). Meyerhof shared the Nobel prize in physiology and medicine in 1922 with Archibald Vivian Hill (1886–1977), who had investigated muscle from the viewpoint of its heat production. The anaerobic sequence is known as the *Embden-Meyerhof pathway.*

The aerobic sequence of reactions was discovered by the German biochemist Sir Hans Adolf Krebs (1900–   ). In 1933, Krebs fled from Hitler's Germany to England, where he studied at Cambridge, later joined the faculty at Oxford, and was knighted in 1958. In 1953, he shared the Nobel prize in physiology and medicine with Fritz Albert Lipmann (1899–   ), who discovered the roles of ATP and other such compounds in the storing of energy in phosphorus-oxygen bonds. The aerobic sequence of reactions is known as the *Krebs cycle.*

When a muscle is used, glucose is anaerobically converted by a series of steps in the **Embden-Meyerhof pathway** to lactic acid (Fig. 15–20). We are concerned here only with the starting material, the energy flow, and the final products. The overall reaction can be represented by the equation:

$$C_6H_{12}O_6 + 2\ ADP + 2\ H_3PO_4 \rightarrow 2\ CH_3 - \underset{\underset{\text{LACTIC ACID}}{|}}{\overset{\overset{\displaystyle H\quad O}{|\quad \|}}{C}} - C - OH + 2\ ATP + 2\ H_2O$$

GLUCOSE

If the muscle is used strenuously for a sufficiently long period of time, the lactic acid buildup will produce tiredness and a painful sensation in the muscles. The bloodstream eventually carries away the lactic acid, but time and oxygen are needed to convert the lactic acid to carbon dioxide and water, which are excreted. This is accomplished by the slower, aerobic (with air) process of the **Krebs cycle.** As you look at Figure 15–21, imagine how a molecule of citric acid reacts with the enzyme aconitase. The citric acid has —OH and —H exchange sites, which exchange and produce isocitric acid. The isocitric acid dissociates from aconitase and goes to the next enzyme, isocitric dehydrogenase, where two hydrogen atoms are removed by $NADP^+$, and so on around the cyclic pathway. Hydrogen atoms are removed, carboxyl groups are destroyed as carbon dioxide is formed, and hydrolysis occurs. Since two carbon atoms are fed into the cycle at oxaloacetic acid by acetyl coenzyme A, two carbon atoms in the form of two molecules of carbon dioxide (CO_2) must be eliminated before the cycle returns to oxaloacetic acid. This is exactly what happens. See whether you can find the two molecules of CO_2 formed in the Krebs cycle.

If the Krebs cycle is aerobic, where does the oxygen enter the cycle? Oxygen is required to regenerate the coenzyme $NADP^+$ from NADPH. It is also required to remove the hydrogen atoms from NADH and $FADH_2$. There are about seven

Aerobic: use elemental oxygen.
Anaerobic: use no elemental oxygen.

Gerty Cori was the third woman to receive a Nobel prize. The first two were Marie Curie and her daughter, Irene Joliet.

Muscular activity converts glucose to lactic acid, which produces fatigue in muscles as the lactic acid accumulates.

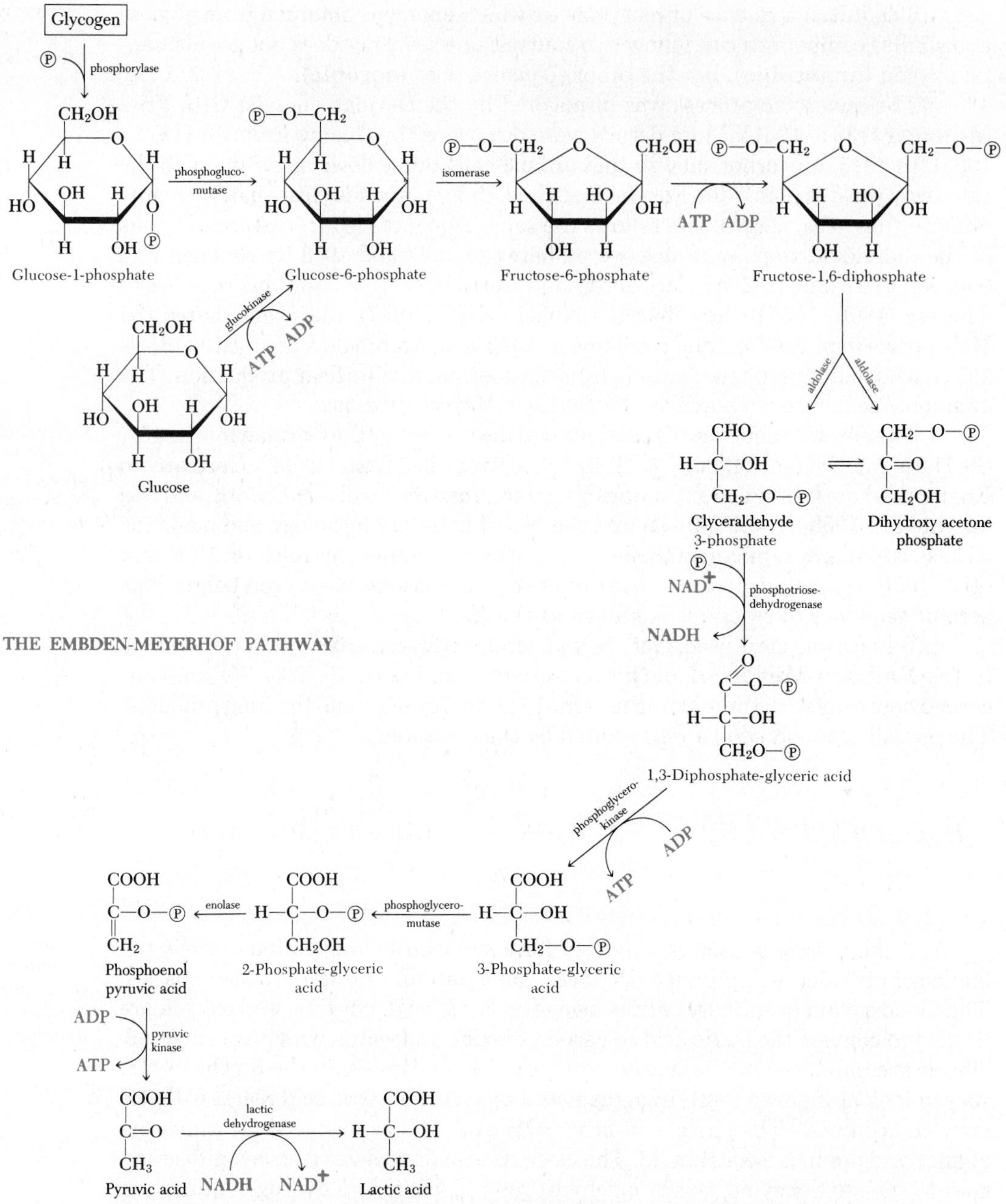

THE EMBDEN-MEYERHOF PATHWAY

FIGURE 15–20   The Embden-Meyerhof pathway—anaerobic oxidation of glucose and glycogen. The ℗ indicates inorganic phosphate, PO_4^{3-}. (Adapted from J. I. Routh: *Introduction to Biochemistry*, p. 100. Philadelphia. Saunders College Publishing, 1971.)

known steps involved in the removal of hydrogen from NADH and NADPH and six steps for $FADH_2$. The steps can be summarized by the following equations:

$$NADPH + 3\ ADP + 3\ H_3PO_4 + H^+ + \tfrac{1}{2}\ O_2 \rightarrow NADP^+ + 3\ ATP + 4\ H_2O$$

$$NADH + 3\ ADP + 3\ H_3PO_4 + H^+ + \tfrac{1}{2}\ O_2 \rightarrow NAD^+ + 3\ ATP + 4\ H_2O$$

$$FADH_2 + 2\ ADP + 2\ H_3PO_4 + \tfrac{1}{2}\ O_2 \rightarrow FAD + 2\ ATP + 3\ H_2O$$

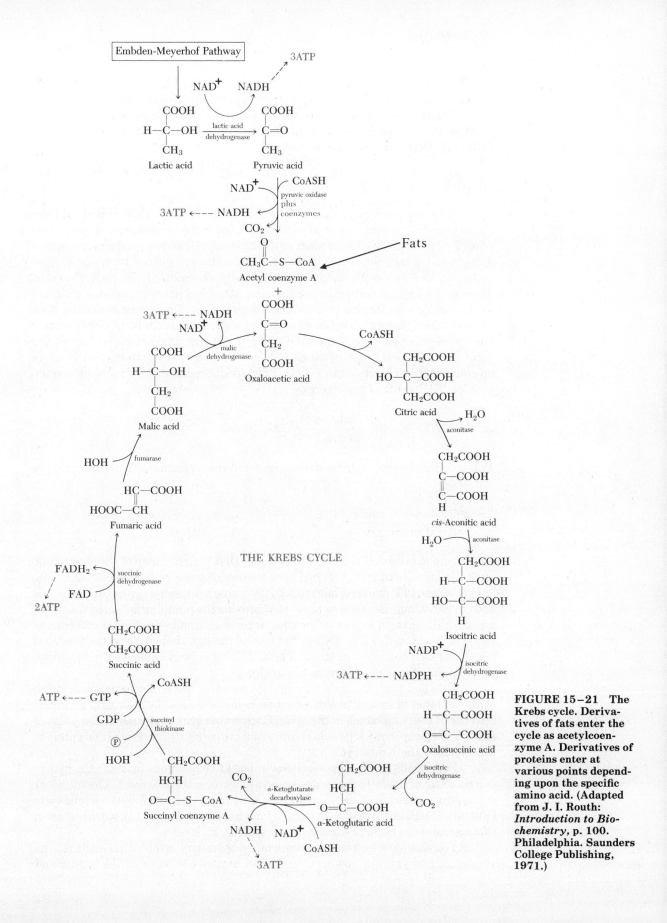

FIGURE 15–21 The Krebs cycle. Derivatives of fats enter the cycle as acetylcoenzyme A. Derivatives of proteins enter at various points depending upon the specific amino acid. (Adapted from J. I. Routh: *Introduction to Biochemistry*, p. 100. Philadelphia. Saunders College Publishing, 1971.)

The NADP$^+$, NAD$^+$, and FAD are now ready to be fed back into the Krebs cycle, where the process is continued by extracting other hydrogen atoms.

When these equations are combined with the principal sequence of reactions for the Krebs cycle, we have an equation for the conversion of lactic acid into carbon dioxide and water aerobically.

$$\underset{\text{LACTIC ACID}}{C_3H_6O_3} + 18\ ADP + 18\ H_3PO_4 + 3\ O_2 \rightarrow 3\ CO_2 + 21\ H_2O + 18\ ATP$$

The key products of glucose metabolism are ATP (energy), CO_2, and H_2O.

The key product of this oxidation reaction is ATP. Energy derived from glucose (and lactic acid) is stored and transported in the phosphorus-oxygen bonds of ATP. This is the usable product of the Krebs cycle. Waste products are carbon dioxide and water. Since two lactic acid molecules are formed from one glucose molecule, 36 ATPs are formed per glucose molecule oxidized via the Krebs cycle. Two more ATPs are formed in the Embden-Meyerhof pathway, making a total of 38 ATP molecules formed in the complete oxidation of a glucose molecule. If we burn a mole of glucose to carbon dioxide and water outside the body, 688,000 cal of energy are released. This is the total energy available. Earlier in this chapter, the point was made that breaking a mole of P—O—P bonds and forming $H_2PO_4^-$ gives up about 7400 cal of free energy (ΔG). Thirty-eight moles of ATP molecules would provide 281,200 cal of free energy.

$$38\ \cancel{\text{moles}}\ \text{ATP} \times 7400\ \frac{\text{calories}}{\cancel{\text{mole}}\ \text{ATP}} = 281{,}200\ \text{cal}$$

This is an efficiency of 41% for obtaining stored energy from the total usable energy available.

$$\frac{281{,}200\ \text{calories} \times 100\%}{688{,}000\ \text{calories}} = 41\%$$

This figure is remarkable when you consider that the efficiency of the automobile engine is only about 20%, and real heat engines of any size seldom go above 35%.

When ATP converts back to ADP by losing a phosphate group, the energy is used to move muscles, such as those that provide the pumping action of the heart, move the diaphragm so we can breathe, or produce hundreds of other movements required for everyday life. This is the fate of the light energy from the Sun that began our discussion of photosynthesis. This energy is conserved in quantity, although all of it is not conserved in quality.

Fats and proteins (as amino acids) can enter the Krebs cycle also. Amino acids can enter at several places after first being converted by a series of reactions into one of the compounds in the cycle. Each amino acid has its particular point of entry. For example, aspartic acid is converted into α-ketoglutaric acid and enters at that point in the Krebs cycle.

Fatty acids go through a series of at least five reactions in which the hydrocarbon chain of the fatty acid is decreased by a two-carbon fragment. The fragment completes a molecule of acetyl coenzyme A, which is important in the Krebs cycle (Fig. 15–21). Palmitic acid, a 16-carbon fatty acid (see Chapter 13), would do seven turns around the cycle to form eight acetyl CoA molecules.

You now have seen some of the detailed chemistry involved in simply raising your arm, and you are now aware of what happens to some of the sugars and starches that disappear down the hatch. Of course, there is much more known than is presented here, and there appears to be no end to what is left to be discovered.

SELF-TEST 15 – D

1. The source of energy for photosynthesis is _____.
2. Most of the energy obtained by food oxidation is used immediately to synthesize the molecule _____.
3. The hydrolysis of ATP results in the molecules _____ and _____. The other "product" is _____.
4. Energy available to do work is called _____ energy.
5. The reactants in the photosynthesis process are _____ and _____; _____ must also be supplied.
6. Energy absorbed by chloroplasts in the green cells of a plant is transferred by means of the molecule _____ to ATP.
7. Digestion is the breakdown of foodstuffs by _____.
8. Substances whose solubility characteristics are similar to those of fats and oils are termed _____.
9. Bile salts act as ( ) catalysts, ( ) emulsifying agents, ( ) enzymes.
10. The two products of the Embden-Meyerhof pathway are _____ and _____.
11. The end products of the Krebs energy cycle are _____ and _____.

NUCLEIC ACIDS, HEREDITY, AND PROTEIN SYNTHESIS
Nucleic Acids

Like the polysaccharides and the polypeptides, the **nucleic acids** are polymeric substances with molecular weights up to several million. Nucleic acids are found in all living cells, with the exception of the red blood cells of mammals. The almost infinite variety of possible structures for nucleic acids allows information in coded form to be recorded in molecular structures in a somewhat similar fashion to the way a few language symbols can be used to convey the many ideas in this book. Such stored information is believed to control the inherited characteristics of the next generation as well as many of the ongoing life processes of the organism.

The coded information of nucleic acids tells cells which molecules to synthesize.

Hydrolysis of nucleic acids yields one of two simple sugars, phosphoric acid (H_3PO_4), and a group of nitrogen compounds that have basic (alkaline) properties. The structures of the two sugars in nucleic acids are shown in Figure 15–22. The names and formulas of the basic nitrogen compounds are given in Figure 15–23.

The nucleic acids can be classified as those containing the sugar **α-2-deoxy-D-ribose** and those containing **α-D-ribose.** The former are called **deoxyribonucleic acids (DNA)** and the latter **ribonucleic acids (RNA).** DNA is found primarily in the nucleus of the cell, whereas RNA is found mainly in the cytoplasm outside the nucleus (Fig. 15–24).

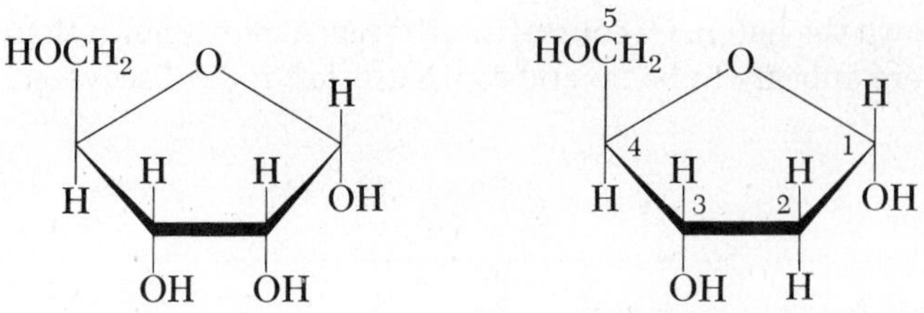

FIGURE 15–22   The structure of α-D-ribose and α-2-deoxy-D-ribose. In the IUPAC names given, α indicates the one of two ring forms possible; D distinguishes the isomers that rotate plane polarized light in opposite directions, and the 2 indicates the carbon to which no oxygen is attached in the second sugar.

Nucleotides

One nucleotide is joined to another by an ester-forming reaction.

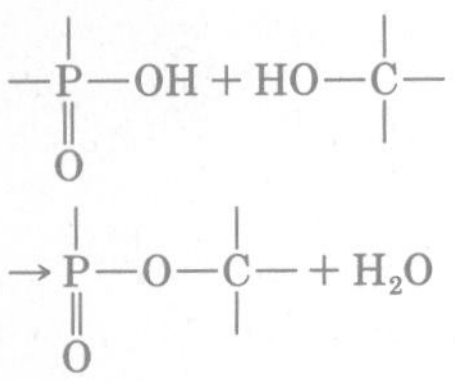

$$-\overset{|}{\underset{\overset{\|}{O}}{P}}-OH + HO-\overset{|}{\underset{|}{C}}-$$

$$\rightarrow \overset{|}{\underset{\overset{\|}{O}}{P}}-O-\overset{|}{\underset{|}{C}}- + H_2O$$

The repeating units of DNA and RNA are **nucleotides.** These substances contain a simple sugar unit, one of the nitrogenous bases, and one or two units of phosphoric acid. An example of a nucleotide structure is illustrated in Figure 15–25.

Polynucleotides

In addition to the mononucleotides, partial hydrolysis of DNA or RNA yields oligonucleotides that have a few nucleotide units bonded, as shown in Figure 15–26. Obviously, a large number of oligonucleotides is possible when one considers the choice of base structures and the different sequence possibilities for the chain of nucleotides.

DNA and RNA are polynucleotides. The number of possible structures for these molecules, which have molecular weights as high as a few million, appears to be almost limitless. Since DNA is a major part of the chromosome material in the nucleus of a cell, it seems reasonable to assume that the organism's characteristics are coded in the DNA structure. Even if we assume that there are over 2 million different species of organisms and that each individual of each species requires a different DNA structure, there are ample combinations of nucleotides for each individual to be unique. It is now known that some kinds of RNA transfer the information coded in the DNA structure to the cytoplasmic region of the cell, where they control the thousands of reactions that occur.

The inherited traits of an organism are controlled by DNA molecules.

Three major types of RNA have been identified. They are messenger RNA (mRNA), transfer RNA (tRNA), and ribosomal RNA (rRNA). Each has a characteristic molecular weight and base composition. Messenger RNAs are generally the

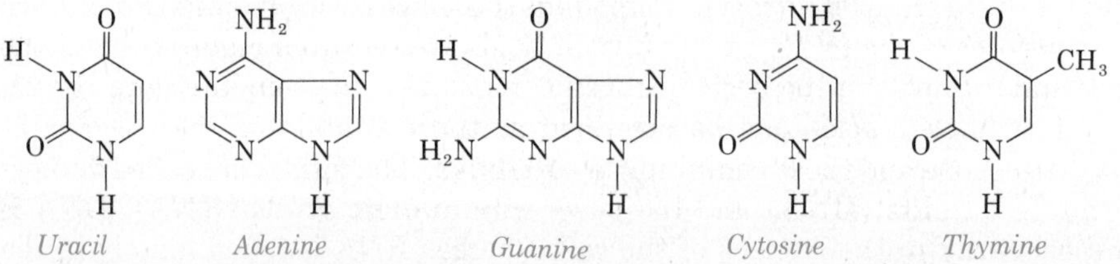

Uracil   *Adenine*   *Guanine*   *Cytosine*   *Thymine*

FIGURE 15–23   Some nitrogenous bases obtained from the hydrolysis of nucleic acids.

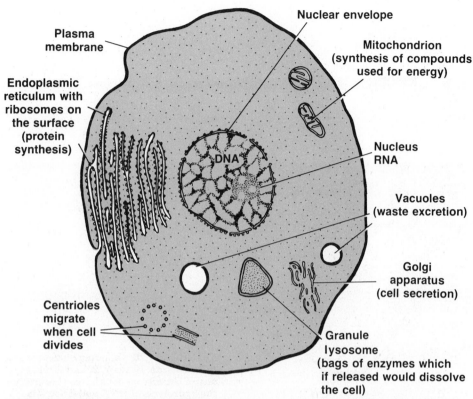

FIGURE 15–24 Diagrammatic generalized cell to show the relationships between the various components of the cell. Cytoplasm is the material of the cell, exclusive of the nucleus. Many of the components shown are not visible through an ordinary optical microscope.

largest, with molecular weights between 25,000 and 1 million. They contain from 75 to 3000 mononucleotide units. Transfer RNAs have molecular weights in the range of 23,000 to 30,000 and contain 75 to 90 mononucleotide units. Ribosomal RNAs have molecular weights between those of mRNAs and tRNAs and make up as much as 80% of the total cell RNA.

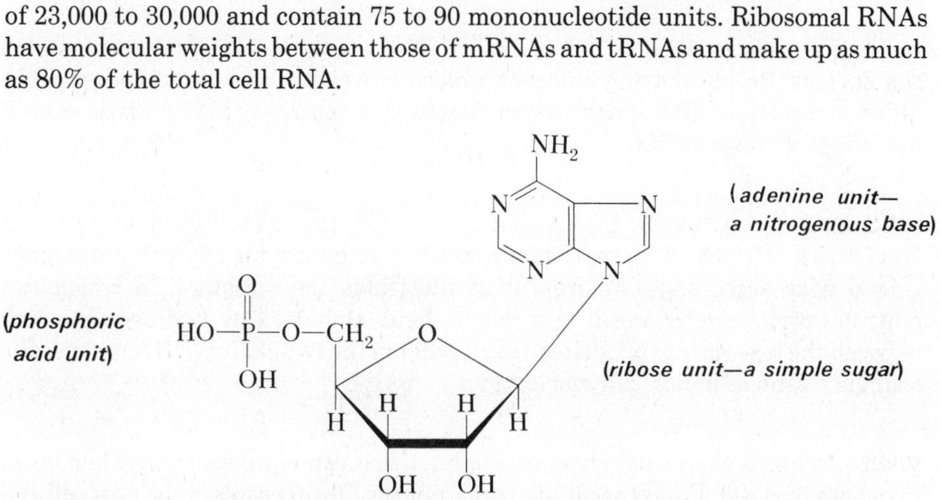

FIGURE 15–25 A nucleotide. If other bases are substituted for adenine, a number of nucleotides are possible for each of the two sugars shown in Figure 15–22. There is ample evidence that the nucleotides found in both DNA and RNA have the general structure indicated for this nucleotide.

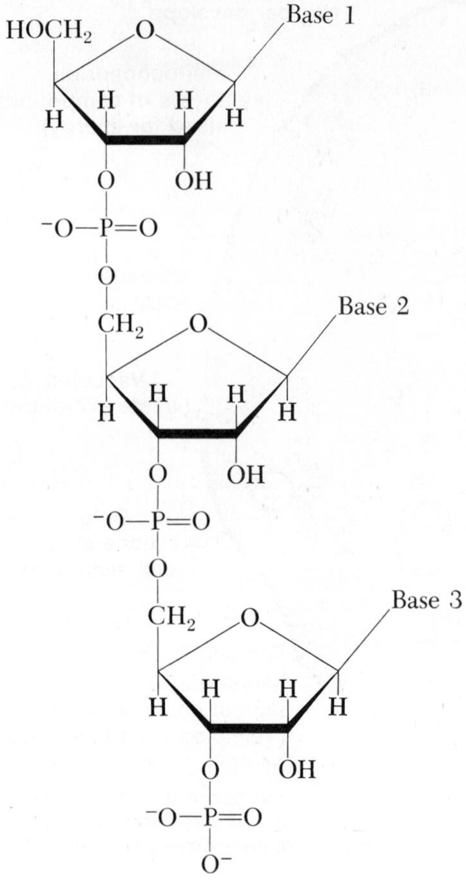

FIGURE 15–26 Bonding structure of a trinucleotide. Bases 1, 2, and 3 represent any of the nitrogenous bases obtained in the hydrolysis of DNA and RNA. The primary structure of both DNA and RNA is an extension of this structure to produce molecular weights as high as a few million.

Most RNA is found in the cytoplasm and ribosomes of the cell (Fig. 15–24), but in liver cells as much as 11% (largely mRNA) of the total cell RNA is found in the nucleus. Besides having different molecular weights, the three types of RNA differ in function. One difference in function is described in the discussion of natural protein synthesis.

SECONDARY STRUCTURE OF DNA AND RNA In 1953, James D. Watson and Francis H. C. Crick (Fig. 15–27) proposed a secondary structure for DNA that has since gained wide acceptance. Figure 15–28 illustrates the structure, in which two polynucleotides are arranged in a double helix stabilized by hydrogen bonding between the base groups opposite to each other in the two chains. RNA is generally a single strand of helical polynucleotide.

VIRUS A virus is a parasitic chemical complex that can reproduce only when it has invaded a host cell. It has the ability to disrupt the life processes of the host cell and order its cell contents to reproduce the virus structure. The isolated virus unit has neither the enzymes nor the smaller molecules necessary to reproduce itself alone.

A virus is a polynucleotide surrounded by a layer of protein.

FIGURE 15–27 Francis H. C. Crick (1916–) *(right)* and James D. Watson (1928–) *(left)*, working in the Cavendish Laboratory at Cambridge, built scale models of the double helical structure of DNA based on the X-ray data of Maurice H. F. Wilkins. Knowing distances and angles between atoms, they compared the task to the working of a three-dimensional jigsaw puzzle. Watson, Crick, and Wilkins received the Nobel prize in 1962 for their work relating to the structure of DNA.

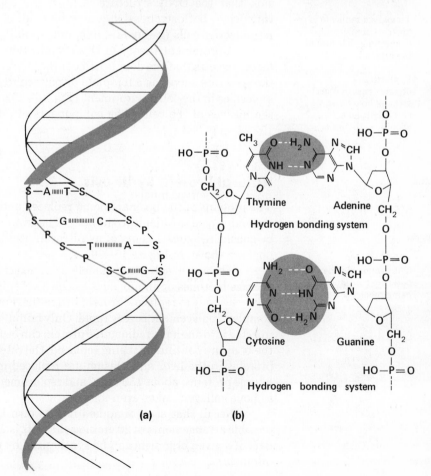

FIGURE 15–28 (*a*) Double helix structure proposed by Watson and Crick for DNA. S-sugar, P-phosphate, A-adenine, T-thymine, G-guanine, C-cytosine. (*b*) Hydrogen bonds in the thymine-adenine and cytosine-guanine pairs stabilize the double helix. Adenine will also pair with uracil in the mRNA since there is no thymine in mRNA.

SYNTHESIS OF LIVING SYSTEMS

Our current knowledge is far from sufficient to allow the synthesis of a living cell.

In the quest for a molecular understanding of living systems, theories must finally be put to the ultimate test of synthesis. If some of the complex biochemical substances can be synthesized and can successfully participate in the life processes, we can be reassured that we are on the right track. It should be emphasized that the biochemist is currently working at the molecular level, and as yet only a relatively few of the giant molecules have been characterized. Most of the syntheses in a living cell are far too complex for our present methods to duplicate. However, despite the enormity of this undertaking, remarkable strides have been made recently, and interest in current research in this area is intense.

Replication of DNA — Heredity

The DNA molecule is capable of causing the synthesis of its duplicate.

To make a replicate is to make a complement (something that fits) of the original.

Replicates:
object — cast of object
screw — hole for screw
bolt — bolt hole
adenine — thymine

Almost all of the cells of one organism contain the same chromosome structure in their nuclei. This structure remains constant regardless of whether the cell is starving or has an ample supply of food materials. Each organism begins life as a single cell with this same chromosome structure; in sexual reproduction one half of this structure comes from each parent. These well-known biological facts, along with recent discoveries concerning polynucleotide structures, lead to the conclusion that the DNA structure is faithfully copied during normal cell division (**mitosis**—both strands), and only one half is copied in cell division producing reproductive cells (**meiosis**—only one strand).

A prominent theory of DNA replication, based on verifiable experimental facts, suggests that the double helix of the DNA structure unwinds and each half of the structure serves as a template or pattern to reproduce the other half from the molecules in the cell environment (Fig. 15–29). Replication of the DNA occurs in the nucleus of the cell. (The components of a typical cell are illustrated in Fig. 15–24 on p. 383.)

Natural Protein Synthesis

The proteins in the human body are continuously being replaced.

The proteins of the body are being replaced and resynthesized continuously from the amino acids available to the body. The amino acids and proteins in the body can be considered constituents of a "nitrogen pool"; additions to and losses from the pool are shown in Figure 15–30.

The use of isotopically labelled amino acids has made possible studies of the average lifetimes of amino acids as constitutents in proteins — that is, the time it takes the body to replace a protein in a tissue. For a process that must be extremely complex, replacement is very rapid. Only minutes after radioactive amino acids are injected into animals, radioactive protein can be found. Although all the proteins in the body are continually being replaced, the rates of replacement vary. Half of the proteins in the liver and plasma are replaced in *six days*. The time is longer for muscle proteins, about 180 days, and replacement of protein in other tissues, such as bone collagen, takes even longer.

Recall that each organism has its own kinds of proteins. The number of possible arrangements of 20 amino acid units is 2.43×10^{18}, yet proteins characteristic of a given organism can be synthesized by the organism in a matter of a few minutes.

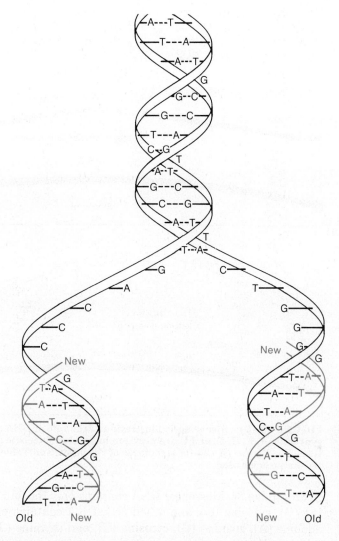

FIGURE 15–29 Replication of DNA structure. When the double helix of DNA *(black)* unwinds, each half serves as a template on which to assemble subunits *(color)* from the cell environment.

The DNA in the cell nucleus holds the code for protein synthesis. Messenger RNA, like all forms of RNA, is synthesized in the cell nucleus. The sequence of bases in one strand of the chromosomal DNA serves as the template for monoribonucleotides to order themselves into a single strand of mRNA (Fig. 15–31). The bases of the mRNA strand complement those of the DNA strand. A pair of complementary bases is so structured that each one fits the other and forms one or more

The DNA molecule tells the cell what kind of protein to synthesize.

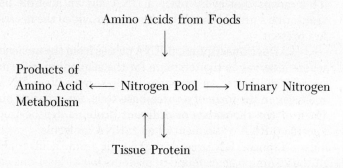

FIGURE 15–30 The nitrogen pool.

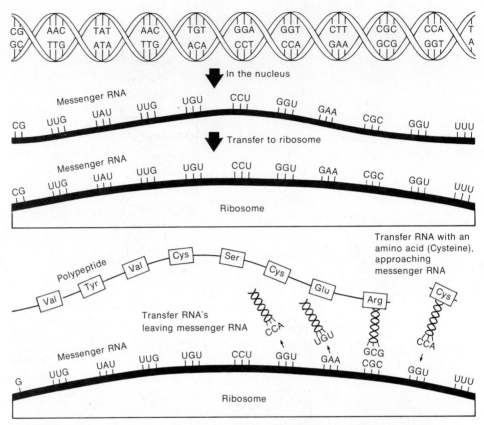

FIGURE 15–31   A schematic illustration of the role of DNA and RNA in protein synthesis. A, C, G, T, and U are nitrogen bases characteristic of the individual nucleotides. See Figure 15–23 for structures of the bases, and Table 15–2 for abbreviations of the amino acids used.

hydrogen bonds. Messenger RNA contains only the four bases adenine (A), guanine (G), cytosine (C), and uracil (U). DNA contains principally the four bases adenine (A), guanine (G), cytosine (C), and thymine (T). The base pairs are as follows:

DNA	mRNA
A	U
G	C
C	G
T	A

This means that every place a DNA has an adenine base (A), the mRNA will transcribe a uracil base (U), and so on, provided the necessary enzymes and energy are present.

After transcription, mRNA passes from the nucleus of the cell to a ribosome, where it serves as the template for the sequential ordering of amino acids during protein synthesis. As its name implies, messenger RNA contains the sequence message, in the form of a three-base code, for ordering amino acids into proteins. Each of the thousands of different proteins synthesized by cells is coded by a specific mRNA or segment of an mRNA molecule.

Transfer RNAs carry the specific amino acids to the messenger RNA. Each of the 20 amino acids found in proteins has at least one corresponding tRNA, and

Transfer RNA carries specific amino acids to the messenger RNA.

some have multiple tRNAs (Table 15–2). For example, there are five distinctly different tRNA molecules specifically for the transfer of the amino acid leucine in cells of the bacterium *Escherichia coli*. At one end of a tRNA molecule is a trinucleotide base sequence (called the **anticodon**) that fits a trinucleotide base sequence on mRNA (the **codon**). At the other end of a tRNA molecule is a specific base sequence of three terminal nucleotides — CCA — with a hydroxyl group on the sugar exposed on the terminal adenine nucleotide group. This hydroxyl group reacts with a specific amino acid by an esterification reaction with the aid of enzymes.

The bases in groups of three are called codons.

$$(\text{MONONUCLEOTIDES})_{75-90}\underset{\text{tRNA}}{\text{CCA}}-\text{OH} + \underset{\text{AMINO ACID}}{\text{HO}\overset{\displaystyle O}{\overset{\|}{\text{C}}}\text{CH(NH}_2)\text{R}} \rightarrow$$

$$(\text{MONONUCLEOTIDES}]_{75-90}\underset{\text{tRNA-AMINO ACID}}{\text{CCA}-\text{O}\overset{\displaystyle O}{\overset{\|}{\text{C}}}\text{CH(NH}_2)\text{R}} + \text{H}_2\text{O}$$

The P — O bonds of ATP provide energy for the reaction between an amino acid and its transfer RNA. A molecule of ATP first activates an amino acid.

ATP + Amino acid → ATP — amino acid-activated species

The activated complex then reacts with a specific transfer RNA and forms the products shown in Figure 15–32.

TABLE 15–2   Messenger RNA Codes for Amino Acids*

AMINO ACID	SHORTENED NOTATION USED FOR AMINO ACIDS IN FIG. 15–31	BASE CODE ON mRNA
Alanine	Ala	GCA, GCC, GCG, GCU
Arginine	Arg	AGA, AGG, CGA, CGG, CGC, CGU
Asparagine	Asp-NH$_2$	AAC, AAU
Aspartic acid	Asp	GAC, GAU
Cysteine	Cys	UGC, UGU
Glutamic acid	Glu	GAA, GAG
Glutamine	Glu-NH$_2$	CAG, CAA
Glycine	Gly	GGA, GGC, GGG, GGU
Histidine	His	CAC, CAU
Isoleucine	Ileu	AUA, AUC, AUU
Leucine	Leu	CUA, CUC, CUG, CUU, UUA, UUG
Lysine	Lys	AAA, AAG
Methionine	Met	AUG
Phenylalanine	Phe	UUU, UUC
Proline	Pro	CCA, CCC, CCG, CCU
Serine	Ser	AGC, AGU, UCA, UCG, UCC, UCU
Threonine	Thr	ACA, ACG, ACC, ACU
Tryptophan	Try	UGG
Tyrosine	Tyr	UAC, UAU
Valine	Val	GUA, GUG, GUC, GUU

* In groups of three (called codons), bases of mRNA code the order of amino acids in a polypeptide chain. A, C, G, and U represent adenine, cytosine, guanine, and uracil, respectively. Some amino acids have more than one codon, and hence more than one transfer RNA can bring the amino acid to messenger RNA. The research on this coding was initiated by Nirenberg. (Adapted from J. I. Routh, D. P. Eyman, and D. J. Burton: *Essentials of General, Organic, and Biochemistry.* 3rd ed. Philadelphia, Saunders College Publishing, 1977.

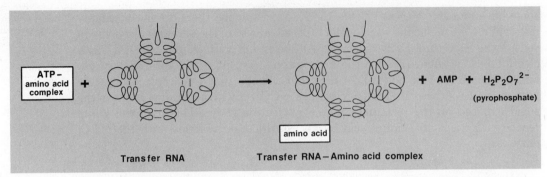

FIGURE 15–32   Bonding of activated amino acid to transfer RNA. AMP is adenosine monophosphate.

The transfer RNA and its amino acid migrate to the ribosome, where the amino acid is used in the synthesis of a protein. The transfer RNA is then free to migrate back to the cell cytoplasm and repeat the process.

Messenger RNA is used only once, or at most a few times, before it is depolymerized. While this may seem to be a terrible waste, it allows the cell to produce different proteins on very short notice. As conditions change, a different type of messenger RNA comes from the nucleus, a different protein is made, and the cell responds adequately to a changing environment.

The ribosome is the part of the cell in which protein synthesis takes place.

Synthetic Nucleic Acids

Progress in the synthesis of polynucleotides has been difficult, principally because of the difficulties involved in determining the proper blocking groups. Progress, although slow, is being made.

In 1959, Arthur Kornberg synthesized a DNA-type polynucleotide, for which he received a Nobel prize. He used natural enzymes to arrange the nucleotides in the order of a desired polynucleotide. His product was not biologically active. In 1965, Sol Spiegelman synthesized the polynucleotide portion of an RNA virus. This polynucleotide was biologically active and reproduced itself readily when introduced into living cells. In 1967, Mehran Goulian and Kornberg synthesized a fully infectious virus of the more complicated DNA type.

In 1970, Gobind Khorana synthesized a complete, double-stranded, 77-nucleotide gene. He, too, used natural enzymes to join previously synthesized, short, single-stranded polynucleotides into the double-stranded gene.

Biogenetic engineering— alteration of genes for a desired purpose.

If scientists can construct DNA, can they then control the genetic code? Genes are the submicroscopic, theoretical bodies proposed by early geneticists to explain the transmission of characteristics from parents to progeny. It was thought that genes composed the chromosomes, which are large enough to be observed through the microscope as the central figures in cell division. It is now generally believed that DNA structures carry the message of the genes; hence, DNA contains the **genetic code.** If scientists can construct DNA, they could very well alter its structure and thereby control the genetic code. The ability to alter genes has led to a new field of science known as **biogenetic engineering.** This is a very active field of research that has developed (among other accomplishments) bacteria that can clean up an oil spill. In this case a patent was granted to the General Electric Co. for the production of life — a unique patent. Other bacteria have been produced that can synthesize protein, human growth hormone, and insulin. The method of producing bacteria for a particular function involves removing a gene from the bacterium, splicing in part of a gene from a human or other organism (that part that produces human insulin, for example), placing the spliced gene back into the bacterium, and letting the bacterium make millions of other insulin-producing bacteria. The process of splicing and recombining genes is referred to as **recombinant DNA technology.** The implications of gene splicing are tremendous — for both good and bad — and will demand responsible human decision-making for guidance toward the common good.

> The process for forming and cloning recombinant DNA is discussed in Chapter 1. See Figure 1–9 for an outline of this process.

A **mutation** occurs whenever an individual characteristic appears that has not been inherited but is duly passed along as an inherited factor to the next generation. A mutation can readily be accounted for in terms of an alteration in the DNA genetic code; that is, some force alters the nucleotide structure in a reproductive cell. Some sources of energy, such as gamma radiation, are known to produce mutations. This is entirely reasonable because certain kinds of energy can disrupt some bonds, which can re-form in another sequence.

> A mutation results when there has been an alteration of the genetic code contained within the DNA molecule.

If scientists can control the genetic code, can they control hereditary diseases such as sickle cell anemia, gout, some forms of diabetes, or mental retardation? If the understanding of detailed DNA structure and the enzymatic activity in building these structures continues to grow, it is reasonable to believe that some detailed relationships between structure and gross properties will emerge. If this happens, it may be possible to build compounds that, when introduced into living cells, can combat or block inherited characteristics.

SELF-TEST 15–E

1. a. The basic code for the synthesis of protein is contained in the _____ molecule.

    b. The synthesis of a protein is carried out when _____ molecules bring up the required amino acids to messenger RNA.

2. When DNA replicates itself, each nitrogenous base in the chain is matched to another one via _____ bonds.

3. The energy for DNA replication and natural protein synthesis is supplied by substances such as _____.

4. What nitrogenous base complements (matches through hydrogen bonding) adenine (A)? _____ cytosine (C)? _____ guanine (G)? _____ thymine (T)? _____ uracil (U)? _____

5. At this time, a gene has been synthesized from individual nucleotides in the laboratory without the aid of natural enzymes. ( ) True or ( ) False

6. Replication means the same as duplication. ( ) True or ( ) False

7. The sugar in RNA is _____, whereas the one in DNA is

    _____.

8. A nucleotide contains _____, _____, and _____.

9. The secondary structure of DNA is in the shape of a(n) _____.

10. A virus is a chemical compound that can reproduce itself. ( ) True or ( ) False

MATCHING SET I

_____ 1. Energy "cash" in the living cell   a. Ptyalin
_____ 2. Mutation
_____ 3. Enzyme that splits polysaccharides in the mouth
_____ 4. Natural protein
_____ 5. High-energy compound formed during photosynthesis
_____ 6. Occurs under aerobic (with air) conditions
_____ 7. Product of ATP hydrolysis
_____ 8. Molecules that absorb light energy
_____ 9. Due to inadequate supply of lactase

a. Ptyalin
b. ADP + energy
c. Krebs cycle
d. Chlorophylls
e. Structure determined by DNA and RNA
f. Altered DNA
g. ATP
h. Glucose
i. Lactose intolerance

MATCHING SET II

_____ 1. D-glucose
_____ 2. Methionine
_____ 3. Enzymes
_____ 4. Carbohydrate storage in animals
_____ 5. Starch
_____ 6. Polypeptides
_____ 7. DNA
_____ 8. Fibrous protein
_____ 9. Cellulose
_____ 10. Vitamins
_____ 11. Enzyme that splits sucrose into fructose and glucose

a. Polymer consisting of α-D-glucose units
b. Proteins
c. Sugar present in blood
d. An essential amino acid
e. Sucrase
f. A polynucleotide
g. Biochemical catalysts
h. Coenzymes
i. Glycogen
j. Collagen
k. Polymer consisting of β-D-glucose units

QUESTIONS

1. Show the structure of the product that would be obtained if two alanine molecules (Table 15–1) react to form a dipeptide.
2. Biochemistry is a special field within what branch of chemistry?
3. What is an essential amino acid?
4. The ketone structure of D-fructose has three asymmetric carbon atoms per molecule. How many isomers result from the asymmetric centers?
5. Name a polysaccharide that yields only D-glucose upon complete hydrolysis. Name a disaccharide that yields the same hydrolysis product.
6. What is the difference between the starch, amylopectin, and the "animal starch," glycogen?
7. What is the chief function of glycogen in animal tissue?
8. Explain the basic difference between starch, amylose, and cellulose.
9. Why does cotton, a cellulose material, absorb moisture so well in contrast to nylon 66?

10. What functional groups are always present in each molecule of an amino acid?

11. Give the name and formula for the simplest amino acid. What natural product has a high percentage of this amino acid?

12. The amino acid of silk is glycine. Name the peptide bond in silk.

13. If six different amino acids formed all the possible different tripeptides, how many would there be?

14. a. What element is necessarily present in proteins that is not present in either carbohydrates or fats?
   b. Name another element that is probably present in proteins but not present in either carbohydrates or fats.

15. What is the meaning of the terms *primary, secondary,* and *tertiary structures of proteins?*

16. In a protein, what type of bond holds the helix structure in place?

17. Enzymes are what type of proteins?

18. How many kilocalories of heat energy are liberated when 90 g of glucose, $C_6H_{12}O_6$, are burned to carbon dioxide and water?

19. What is another name for niacin?

20. a. Which of the following biochemicals are polymers: starch, cellulose, glucose, fats, proteins, DNA, RNA?
   b. What are the monomer units for those that are polymers?

21. What is a chemical function of vitamins? Give some examples.

22. Why are carbohydrates considered "energy rich"?

23. Why is it that humans cannot digest cellulose?

24. The molecular structures of enzymes (particularly apoenzymes) are most closely related to which structures: proteins, fats, carbohydrates, or polynucleic acids?

25. Write an equation for the digestion of:
   a. Starch to a disaccharide
   b. A disaccharide to a simple sugar
   c. A protein to amino acids
   d. A triglyceride to fatty acids

26. What is the metal in chlorophyll?

27. What are the two major divisions of reactions in photosynthesis? Express in words what is accomplished in each.

28. What is the source of oxygen in photosynthesis?

29. What part of photosynthesis could take place in an animal cell?

30. Since chlorophyll loses electrons because of light, it must subsequently gain electrons from somewhere. Where do they come from?

31. What is the basic nature of the digestion processes for large molecules?

32. The chemical changes in the Krebs cycle can be classified as dehydrogenation (removal of hydrogens, a type of oxidation), dehydration (removal of water), hydrolysis (reaction with water in which water loses its molecular identity), decarboxylation (removal of —COOH group and formation of CO_2), and phosphorylation (adding a phosphate group, such as $H_2PO_4^-$). Beginning with citric acid and progressing around the cycle to oxaloacetic acid, determine the total number of each kind of chemical change.

33. What compound produces soreness in the muscles after a period of vigorous exercise?

34. Which compounds in the CO_2 fixation scheme in photosynthesis could enter directly the reactions of the Embden-Meyerhof pathway or the Krebs cycle?

35. If protein digestion is facilitated by enzymes, and these enzymes are produced in body organs made of proteins, explain why the enzymes do not cause rapid digestion of the organs themselves.

36. What is pyruvic acid? Why is it so important in extracting energy from sugars?

37. Give the structure of ATP and point out the region of the molecule that contains bonds that are hydrolyzed in reactions with water and energy.

38. What is a storehouse chemical for biochemical energy?

39. What are the end products in the digestion of carbohydrates? of fats? of proteins?

40. How do living beings store and transfer energy?

41. When water reacts with ATP, is this an energy-releasing or an energy-requiring process?

42. Triphosphoric acid has the formula $H_5P_3O_{10}$. After drawing a structure for triphosphoric acid, look at the structure of ATP. Do you see any similarities?

43. What is the role of enzymes in digestion?

44. What is a purpose of ATP?

45. The importance of water in living systems is emphasized by incidences of hydrolysis. Cite examples.

46. In the structure of ATP, indicate the particular bond that yields the most energy when hydrolyzed. Why is this bond more energy rich than other P—O bonds?

47. What important type of chemical can function as a coenzyme?

48. What three molecular units are found in nucleotides?

49. Based on the structures in Figure 15–22, explain the meaning of the prefix *deoxy-* in deoxyribonucleic acid.

50. What is the acid of a nucleotide?

51. How many trinucleotides with the structure indicated in Figure 15–26 could be made with the nitrogenous bases listed in Figure 15–23?

52. What are the basic differences between DNA and RNA structures?

53. What stabilizing forces hold the double helix together in the secondary DNA structure proposed by Watson and Crick?

54. Calculate the molecular weight of the nucleotide shown in Figure 15–25. Recall that the representation of the adenine unit omits two hydrogen atoms.

55. What is recombinant DNA?

56. What happens if an amino acid is needed for protein synthesis and the amino acid can neither be made by the body nor obtained from the diet? Does the modern theory of protein synthesis include an explanation of the role of essential amino acids? (These are amino acids that cannot be manufactured by the human body; they must be obtained in the diet.)

57. Does a strand of DNA actually duplicate itself base for base in the formation of a strand of messenger RNA? Explain.

58. a. Describe the general method of synthesizing DNA in vitro (in a test tube) at present.
    b. Why would the synthesis of a polynucleotide from the individual phosphoric acid, sugar, and nitrogenous bases be a breakthrough in controlling the genetic code?

59. A mutation can be explained in terms of a change in which chemical in the cell?

60. The replication of DNA occurs in which part of the cell?

61. What is meant by a base pair in protein synthesis? What type of bonds holds base pairs together?

62. What two molecular structures are present in viruses?

63. Discuss some aspect of the feasibility of solving sociological and psychological problems via religion, DNA alteration, chemical suppressants, political pressures, and/or persuasive dialogue.

64. Check the recent issues of *Science* or other scientific news publications to find out about more recent work done on synthesis of polynucleotides.

65. What happens chemically to nutrients, fats, glucose, and amino acids in the liver?

66. Why is the liver called the central nutrient bank of the body?

67. Using paper, pen, scissors, and the technique of making paper dolls, make a molecular model of cellulose.

68. Using the same materials and technique, make a molecular model of a polypeptide in the sheet structure. What advantage does your model have over the diagram in Figure 15–8?

Chapter 16

TOXIC SUBSTANCES — INTERFERENCES IN BIOCHEMICAL SYSTEMS

Toxic substances upset the incredibly complex system of chemical reactions occurring in the human body. Sometimes toxic substances cause mere discomfort; sometimes they cause illness, disability, or even death. Toxic symptoms can be caused by very small amounts of extremely toxic materials (an example is sodium cyanide) or larger amounts of a less toxic substance. The term *toxic substances* usually is limited to materials that are dangerous in small amounts. However, as most of us know, ill effects can be caused by excessive intake of substances normally considered harmless (eating too much candy, for example). Fortunately, in most cases the human body is capable of recognizing "foreign" chemicals and ridding itself of them. In this chapter, we shall focus on the chemical mechanisms by which toxic substances act.

A large enough dose of any compound can result in poisoning.

DOSE

Lethal doses of toxic substances are customarily expressed in milligrams (mg) of substance per kilogram (kg) of body weight of the subject. For example, the cyanide ion (CN^-) is generally fatal to human beings in a dose of 1 mg of CN^- per kg of body weight. For a 200-pound (90.7 kg) person, about one tenth of a gram of cyanide is a lethal dose. Examples of somewhat less toxic substances and the range of lethal doses for human beings follow:

"Dosis sola facit venenum" — the dose makes the poison.

Morphine	1 to 50 mg per kg
Aspirin	50 to 500 mg per kg
Methyl alcohol	500 to 5000 mg per kg
Ethyl alcohol	5000 to 15,000 mg per kg

A quantitative measure of toxicity is obtained by introducing into laboratory animals (such as rats) various dosages of substances to be tested. That dosage that is found to be lethal in 50% of a large number of the animals under controlled conditions is called the LD_{50} (lethal dosage — 50%) and is reported in milligrams of poison per kilogram of body weight. Thus, if a statistical analysis of data on a large population of rats showed that a dosage of 1 mg per kg was lethal to 50% of the population tested, the LD_{50} for this poison would be 1 mg per kg. Obviously, meta-

Metabolism (from the Greek, *metaballein*, meaning "to change or alter") is the sum of all the physical and chemical changes by which living organisms are produced and maintained.

TABLE 16–1 Approximate Comparison of LD$_{50}$ Values with Lethal Doses for Human Adults

ORAL LD$_{50}$ FOR ANY ANIMAL (mg/kg)	PROBABLE LETHAL ORAL DOSE FOR HUMAN ADULT
Less than 5	A few drops
5 to 50	"A pinch" to 1 teaspoonful
50 to 500	1 teaspoonful to 2 tablespoonfuls
500 to 5000	1 ounce to 1 pint (1 pound)
5,000 to 15,000	1 pint to 1 quart (2 pounds)

bolic variations and other differences between species will produce different LD$_{50}$ values for a given poison in different kinds of animals. For this reason such data cannot be extrapolated to human beings with any assurance, but it is safe to assume that a substance with a low LD$_{50}$ value for several animal species will also be quite toxic to humans (Table 16–1).

Toxic substances can be classified according to the way in which they disrupt the chemistry of the body. Some modes of action of toxic substances can be described as **corrosive, metabolic, neurotoxic, mutagenic, teratogenic,** and **carcinogenic,** and these will serve as the bases of our discussion.

CORROSIVE POISONS

Toxic substances that actually destroy tissues are corrosive poisons. Examples include strong acids and alkalies and many oxidants such as those found in laundry products, which can destroy tissues. Sulfuric acid (found in auto batteries) and hydrochloric acid (also called muriatic acid, used for cleaning purposes) are very dangerous corrosive poisons. So is sodium hydroxide, used in clearing clogged drains. Death has resulted from the swallowing of 1 ounce of concentrated (98%) sulfuric acid, and much smaller amounts can cause extensive damage and severe pain.

Concentrated mineral acids such as sulfuric acid act by first dehydrating cellular structures. The cell dies because its protein structures are destroyed by the acid-catalyzed hydrolysis of the peptide bonds.

PEPTIDE LINK (IN PROTEIN) — CARBOXYL END OF SMALLER PEPTIDE OR AMINO ACID — AMINE END OF SMALLER PEPTIDE OR AMINO ACID

Strong acids and bases destroy cell protoplasm.

In the early stages of this process there will be a large proportion of larger fragments present. Subsequently, as more bonds are broken, smaller and smaller fragments result, leading to the ultimate disintegration of the tissue.

Chemical "warfare gases," such as phosgene, were outlawed by an international conference in 1925.

Some poisons act by undergoing chemical reaction in the body to produce corrosive poisons. Phosgene, the deadly gas used during World War I, is an example. When inhaled, it is hydrolyzed in the lungs to hydrochloric acid, which causes pulmonary edema (a collection of fluid in the lungs) due to the dehydrating effect of the strong acid on tissues. The victim dies of suffocation because oxygen cannot be absorbed effectively by the flooded and damaged tissues.

$$\underset{\substack{\text{PHOSGENE}}}{\underset{\text{Cl}\qquad\text{Cl}}{\overset{\overset{\text{O}}{\|}}{\text{C}}}} + H_2O \rightarrow \quad \underset{\substack{\text{HYDROCHLORIC} \\ \text{ACID}}}{2\ HCl} \quad + \underset{\substack{\text{CARBON} \\ \text{DIOXIDE}}}{CO_2}$$

Sodium hydroxide, NaOH (caustic soda — a component of drain cleaners), is a very strongly alkaline, or basic, substance that can be just as corrosive to tissue as strong acids. The hydroxide ion also catalyzes the splitting of peptide linkages:

$$\underset{}{R-\overset{\overset{\text{O}}{\|}}{C}-\overset{\overset{\text{H}}{|}}{N}-R'} + H_2O \xrightarrow[\text{base}]{OH^-} R-\overset{\overset{\text{O}}{\|}}{C}-OH + H-\overset{\overset{\text{H}}{|}}{N}-R'$$

Both acids and bases, as well as other types of corrosive poisons, continue their action until they are consumed in chemical reactions.

Some corrosive poisons destroy tissue by oxidizing it. This is characteristic of substances such as ozone, nitrogen dioxide, and possibly iodine, which destroy enzymes by oxidizing their functional groups. Specific groups, such as the —SH and —S—S— groups in the enzyme, are believed to be converted by oxidation to nonfunctioning groups; alternatively, the oxidizing agents may break chemical bonds in the enzyme, leading to its inactivation.

A summary of some common corrosive poisons is presented in Table 16–2.

TABLE 16–2   Some Corrosive Poisons

SUBSTANCE	FORMULA	TOXIC ACTION	POSSIBLE CONTACT
Hydrochloric acid	HCl	Acid hydrolysis	Tile and concrete floor cleaner; concentrated acid used to adjust acidity of swimming pools
Sulfuric acid	H_2SO_4	Acid hydrolysis, dehydrates tissue — oxidizes tissue	Auto batteries
Phosgene	ClCOCl	Acid hydrolysis	Combustion of chlorine-containing plastics (PVC or Saran)
Sodium hydroxide	NaOH	Base hydrolysis	Caustic soda, drain cleaners
Trisodium phosphate	Na_3PO_4	Base hydrolysis	Detergents, household cleaners
Sodium perborate	$NaBO_3 \cdot 4\,H_2O$	Base hydrolysis — oxidizing agent	Laundry detergents, denture cleaners
Ozone	O_3	Oxidizing agent	Air, electric motors
Nitrogen dioxide	NO_2	Oxidizing agent	Polluted air, automobile exhaust
Iodine	I_2	Oxidizing agent	Antiseptic
Hypochlorite ion	OCl^-	Oxidizing agent	Bleach
Peroxide ion	O_2^{2-}	Oxidizing agent	Bleach, antiseptic
Oxalic acid	$H_2C_2O_4$	Reducing agent, precipitates Ca^{2+}	Bleach, ink eradicator, leather tanning, rhubarb, spinach, tea
Sulfite ion	SO_3^{2-}	Reducing agent	Bleach
Chloramine	NH_2Cl	Oxidizing agent	Produced when household ammonia and chlorinated bleach are mixed
Nitrosyl chloride	NOCl	Oxidizing agent	Mixing household ammonia and bleach

METABOLIC POISONS

Metabolic poisons are more subtle than the tissue-destroying corrosive poisons. In fact, many of them do their work without actually indicating their presence until it is too late. Metabolic poisons can cause illness or death by interfering with a vital biochemical mechanism to such an extent that it ceases to function or is prevented from functioning efficiently.

Carbon Monoxide

The interference of carbon monoxide with extracellular oxygen transport is one of the best understood processes of metabolic poisoning. As early as 1895, it was noted that carbon monoxide deprives body cells of oxygen (asphyxiation), but it was much later before it was known that carbon monoxide, like oxygen, combines with hemoglobin:

O_2 + hemoglobin $\rightarrow$ oxyhemoglobin

CO + hemoglobin $\rightarrow$ carboxyhemoglobin

Laboratory tests show that carbon monoxide reacts with hemoglobin to give a compound (carboxyhemoglobin) that is 140 times more stable than the compound of hemoglobin and oxygen (oxyhemoglobin) (Fig. 16–1). Since hemoglobin is so effectively tied up by carbon monoxide, it cannot perform its vital function of transporting oxygen.

An organic material that undergoes incomplete combustion will always liberate carbon monoxide. Sources include auto exhausts, smoldering leaves, lighted cigars or cigarettes, and charcoal burners. In the United States alone, combustion sources of all types dump about 200 million tons of carbon monoxide per year into the atmosphere.

ppm — parts per million — a measure expressing concentration. 50 ppm CO means 50 ml CO for every million ml of air.

While the best estimates of the maximum global background level of carbon monoxide are of the order of 0.1 parts per million (ppm), the background concen-

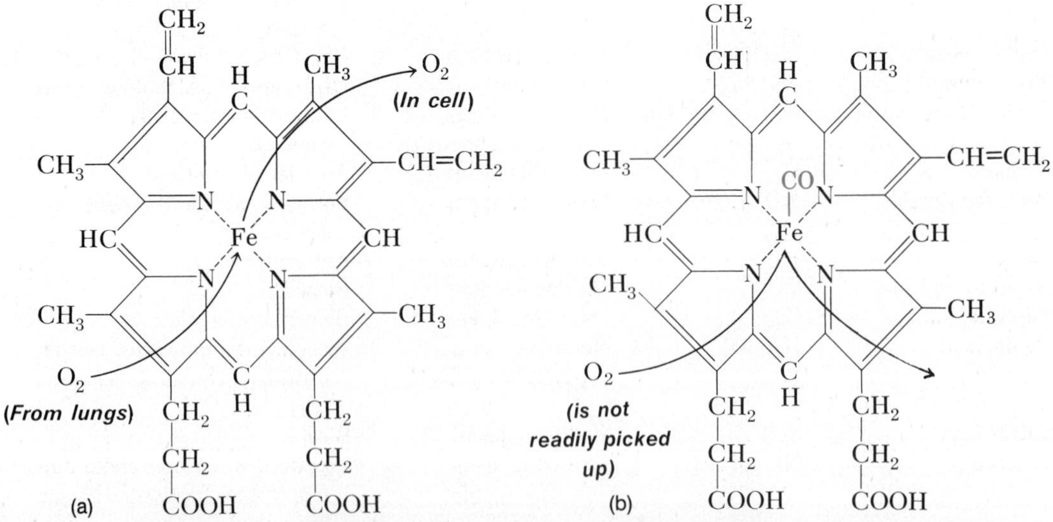

FIGURE 16–1   Structure of the heme portion of hemoglobin. (*a*) Normal acceptance and release of oxygen. (*b*) Oxygen blocked by carbon monoxide.

TABLE 16–3 Concentration of CO in Atmosphere Versus Percentage of Hemoglobin (Hb) Saturated*

CO concentration in air	0.01% (100 ppm)	0.02% (200 ppm)	0.10% (1,000 ppm)	1.0% (10,000 ppm)
Percentage of hemoglobin molecules saturated with CO†	17	20	60	90

* A few hours of breathing time is assumed.
† Normal human blood contains up to 5% of the hemoglobin as carboxyhemoglobin (HbCO).

tration in cities is higher. In heavy traffic, sustained levels of 100 or more ppm are common; for offstreet sites an average of about 7 ppm is typical for large cities. A concentration of 30 ppm for 8 hrs is sufficient to cause headache and nausea. Breathing an atmosphere that is 0.1% (1000 ppm) carbon monoxide for 4 hrs converts approximately 60% of the hemoglobin of an average adult to carboxyhemoglobin (Table 16–3), and death is likely to result (Fig. 16–2).

Since both the carbon monoxide and oxygen reactions with hemoglobin involve easily reversed reactions, the concentrations, as well as relative strengths of bonds, affect the direction of the reaction. In air that contains 0.1% CO, oxygen molecules outnumber CO molecules 200 to 1. The larger concentration of oxygen helps to counteract the greater combining power of CO with hemoglobin by shifting the reaction equilibrium to the right. Consequently, if a carbon monoxide victim is exposed to fresh air or, still better, pure oxygen (provided he or she is still breathing), the carboxyhemoglobin (HbCO) is gradually decomposed, owing to the greater concentration of oxygen:

$$HbCO + O_2 \rightleftarrows HbO_2 + CO$$
EQUILIBRIUM SHIFTED TO RIGHT BECAUSE OF GREATER CONCENTRATION OF OXYGEN

Although carbon monoxide is not a cumulative poison, permanent damage can occur if certain vital cells (e.g., brain cells) are deprived of oxygen for more than a few minutes.

Individuals differ in their tolerance of carbon monoxide, but generally those with anemia or an otherwise low reserve of hemoglobin (e.g., children) are more

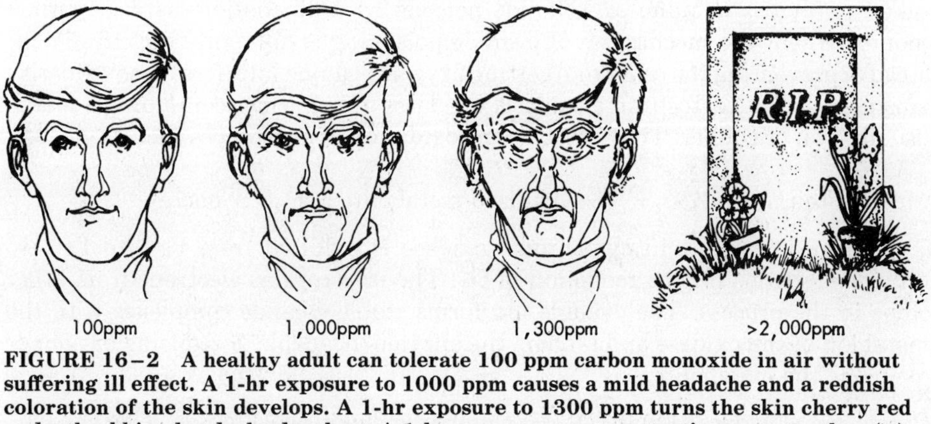

To convert ppm to percent, divide by 10,000.

Air is 21% O_2 by volume; in 1 million "air molecules" there would be 210,000 O_2 molecules.

The equilibrium concept was discussed in Chapter 9.

100ppm   1,000ppm   1,300ppm   >2,000ppm

FIGURE 16–2 A healthy adult can tolerate 100 ppm carbon monoxide in air without suffering ill effect. A 1-hr exposure to 1000 ppm causes a mild headache and a reddish coloration of the skin develops. A 1-hr exposure to 1300 ppm turns the skin cherry red and a throbbing headache develops. A 1-hr exposure to concentrations greater than (>) 2000 ppm will likely cause death.

susceptible. No one is helped by carbon monoxide, and smokers suffer chronically from its effects. It is a subtle poison, since it is odorless and tasteless.

Cyanide

The cyanide ion, CN^- is the toxic agent in cyanide salts such as sodium cyanide used in electroplating. Since the cyanide is a relatively strong base, it reacts easily with many acids (weak and strong) to form volatile hydrogen cyanide, HCN:

$$CH_3COOH + Na^+CN^- \rightleftharpoons HCN + Na^+CH_3COO^-$$

ACETIC ACID   SODIUM CYANIDE   HYDROGEN CYANIDE   SODIUM ACETATE

Since HCN boils at a relatively low temperature ($26°C$), it is a gas at temperatures slightly above room temperature. It is often used as a fumigant in storage bins and holds of ships because it is toxic to most forms of life and, in gaseous form, can penetrate into tiny openings, even into insect eggs.

Natural sources of cyanide ions include the seeds of the cherry, plum, peach, apple, and apricot fruits. Hydrogen cyanide is produced by hydrolysis of certain compounds, such as amygdalin, contained in the seeds:

AMYGDALIN   HYDROGEN CYANIDE   GLUCOSE   BENZALDEHYDE

The cyanide is not toxic as long as it is tied up in the amygdalin, but presumably if enough apple or peach seeds were hydrolyzed in warm acid, sufficient HCN would result to cause considerable danger. There are a few recorded instances of humans poisoned by eating large numbers of apple seeds. Amygdalin is not confined to the seeds; amounts as high as 66 mg per 100 g have been reported in peach leaves.

The cyanide ion is one of the most rapidly working poisons. Lethal doses taken orally act in minutes. Cyanide poisons by asphyxiation, as does carbon monoxide, but the mechanism of cyanide poisoning is different (Fig. 16–3). Instead of preventing the cells from getting oxygen, cyanide interferes with oxidative enzymes, such as cytochrome oxidase. Oxidases are enzymes containing a metal, usually iron or copper. They catalyze the oxidation of substances such as glucose:

A metabolite is any substance in a metabolic process.

$Fe^{2+} \rightarrow Fe^{3+} + e^-$
OXIDATION

$$\text{Metabolite } (H)_2 + \tfrac{1}{2} O_2 \xrightarrow{\text{oxidase}} \text{Oxidized metabolite} + H_2O + \text{energy}$$

The iron atom in cytochrome oxidase is oxidized from Fe^{2+} to Fe^{3+} to provide electrons for the reduction of O_2. The iron regains electrons from other steps in the process. The cyanide ion forms stable cyanide complexes with the metal ion of the oxidase and renders the enzyme incapable of reducing oxygen or oxidizing the metabolite.

$$\text{Cytochrome oxidase (Fe)} + CN^- \rightarrow \text{Cytochrome oxidase} \underbrace{\text{(Fe)} \cdots CN^-}_{\text{complex}}$$

FIGURE 16–3 The mechanism of cyanide (CN^-) poisoning. Cyanide binds tightly to the enzyme cytochrome C, an iron compound, thus blocking the vital ADP-ATP reaction in cells.

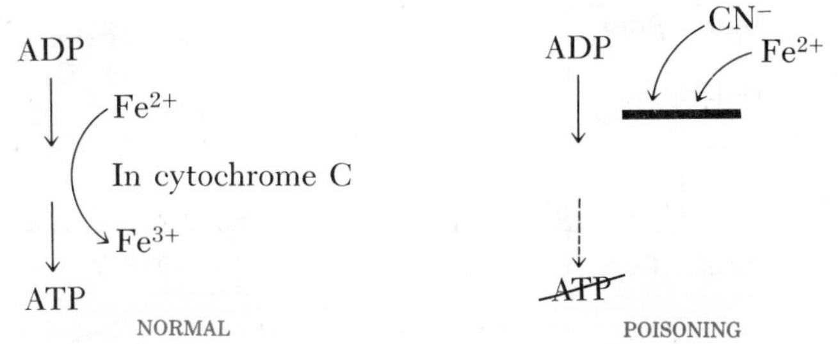

In essence, the electrons of the iron ion are "frozen" — they cannot participate in the oxidation-reduction processes. Plenty of oxygen gets to the cells, but the mechanism by which the oxygen is used in the support of life is stopped. Hence the cell dies, and if this occurs fast enough in the vital centers, the victim dies.

The body has a mechanism for ridding itself slowly of cyanide ions. The cyanide-oxidative enzyme reaction is reversible, and other enzymes such as rhodanase, found in almost all cells, can convert cyanide ions to relatively harmless thiocyanate ions. For example,

$$CN^- + \underset{\text{THIOSULFATE}}{S_2O_3^{2-}} \xrightarrow{\text{rhodanase}} \underset{\text{THIOCYANATE}}{SCN^-} + SO_3^{2-}$$

This mechanism is not as effective in protecting a cyanide-poisoning victim as it might appear, since only a limited amount of thiosulfate is available in the body at a given time.

The body can rid itself of many toxic substances if the dose is small enough and sufficient time is allowed.

Fluoroacetic Acid

Nature has used the synthesis of fluoroacetic acid as part of a defense mechanism for certain plants. Native to South Africa, the *gilbaar* plant contains lethal quantities of fluoroacetic acid. Cattle that eat these leaves usually sicken and die.

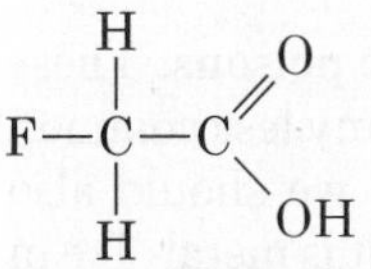

FLUOROACETIC ACID

Sodium fluoroacetate, the sodium salt of this acid (Compound 1080), is a potent rodenticide (rat poison). Because it is odorless and tasteless it is especially dangerous, and its sale in this country is strictly regulated by law.

Fluoroacetate is toxic because it enters the Krebs cycle, producing fluorocitric acid, which in turn blocks the Krebs cycle (Fig. 16–4). The C—F linkage apparently ties up the enzyme aconitase, thus preventing it from converting citrate to isocitrate.

In this instance, the poison is similar enough to the normal substrate to compete effectively for the active sites on the enzyme. If a poison has sufficient affinity for the active site on the enzyme, it blocks the normal function of the enzyme. The blocking of the Krebs cycle by fluorocitrate is a typical example of this affinity. If fluoroacetate is not present in excessive amounts, its action can be reversed simply by increasing the concentration of available citrate.

Some poisons can act by mimicking other compounds.

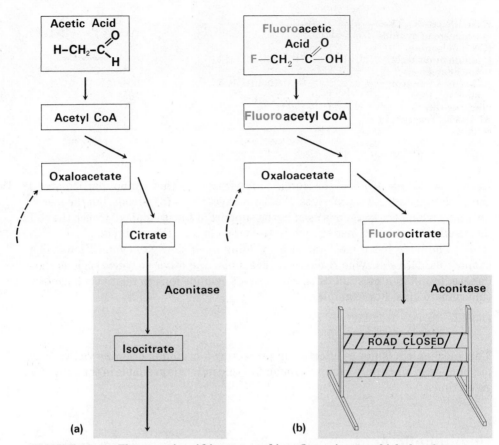

FIGURE 16–4 Fluoroacetic acid is converted into fluorocitrate, which then forms a stable bond with the enzyme aconitase. This blocks the normal Krebs cycle, a portion of which is shown.

Heavy Metals

Heavy metals are perhaps the most common of all the metabolic poisons. These include such common elements as lead and mercury, as well as many less common ones such as cadmium, chromium, and thallium. In this group we should also include the infamous poison, arsenic, which is really not a metal but is metal-like in many of its properties, including its toxic action.

Arsenic, a classic homicidal poison, occurs naturally in small amounts in many foods. Shrimp, for example, contain about 19 ppm arsenic, while corn may contain 0.4 ppm arsenic. Some agricultural insecticides contain arsenic (Table 16–4), and so some arsenic is observed in very small amounts on some fruits and vegetables. The Federal Food and Drug Administration (FDA) has set a limit of 0.15 mg of arsenic per pound of food, and this amount apparently causes no harm. Several drugs, such as arsphenamine, which has found some use in treating syphilis, contain covalently bonded arsenic. In its ionic forms, arsenic is much more toxic.

Arsenic and heavy metals owe their toxicity primarily to their ability to react with and inhibit sulfhydryl (—SH) enzyme systems, such as those involved in the production of cellular energy. For example, glutathione (a tripeptide of glutamic acid, cysteine, and glycine) occurs in most tissues; its behavior with metals illustrates the interaction of a metal with sulfhydryl groups. The metal

The effects of cumulative poisons add up.

Most heavy metals are cumulative poisons.

TABLE 16–4 Some Arsenic-Containing Insecticides

NAME	FORMULA
Lead arsenate	$Pb_3(AsO_4)_2$
Monosodium methanearsenate	$CH_3-\overset{\overset{\displaystyle O}{\|}}{\underset{\underset{\displaystyle OH}{\|}}{As}}-O^-Na^+$
Paris green (copper acetoar-senite)	$3\ CuO \cdot 3\ As_2O_3 \cdot Cu(C_2H_3O_2)_2$

replaces the hydrogen on two sulfhydryl groups on adjacent molecules (Fig. 16–5), and the strong bond that results effectively eliminates the two glutathione molecules from further reaction. Glutathione is involved in maintaining healthy red blood cells.

The typical forms of toxic arsenic compounds are inorganic ions such as arsenate (AsO_4^{3-}) and arsenite (AsO_3^{3-}). The reaction of an arsenite ion with sulfhydryl groups results in a complex in which the arsenic unites with two sulfhydryl groups, which may be on two different molecules of protein or on the same molecule:

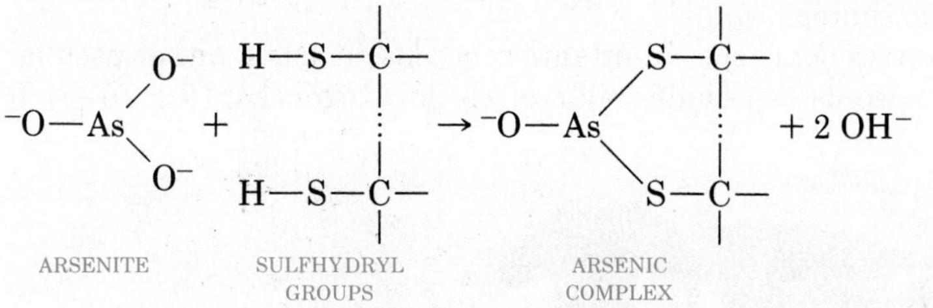

ARSENITE   SULFHYDRYL GROUPS   ARSENIC COMPLEX

The problem of developing a compound to counteract *Lewisite,* an arsenic-containing poison gas used in World War I, led to an understanding of how arsenic acts as a poison and subsequently to the development of an antidote. Once it was

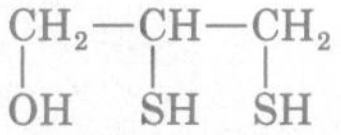

$$CH_2-CH-CH_2$$
$$\ \ |\quad\ \ |\quad\ \ |$$
$$OH\quad SH\quad SH$$

BAL
BRITISH ANTI-LEWISITE

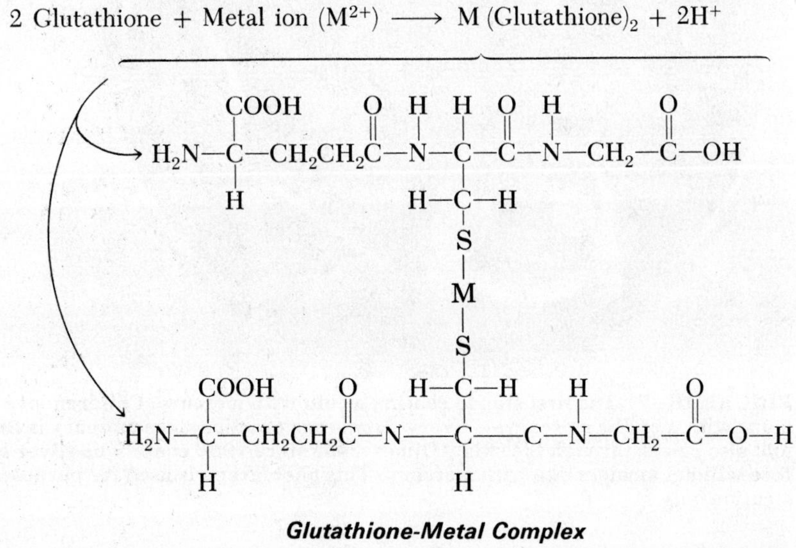

$$2\ \text{Glutathione} + \text{Metal ion}\ (M^{2+}) \longrightarrow M\ (\text{Glutathione})_2 + 2H^+$$

Glutathione-Metal Complex

FIGURE 16–5 Glutathione reaction with a metal (M).

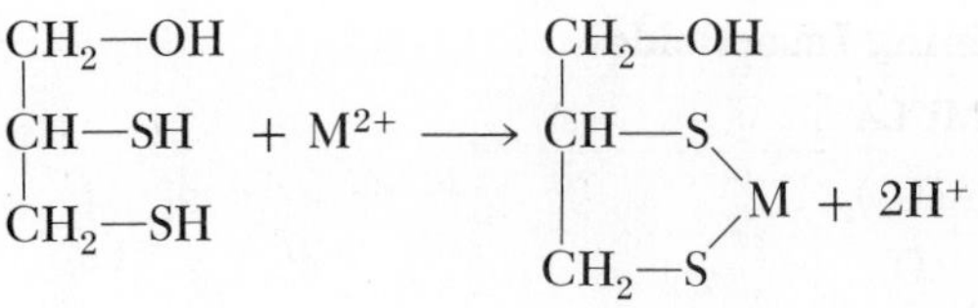

$$
\begin{array}{c}
\text{CH}_2\text{—OH} \\
| \\
\text{CH—SH} \\
| \\
\text{CH}_2\text{—SH}
\end{array}
\;+\; M^{2+} \longrightarrow \;
\begin{array}{c}
\text{CH}_2\text{—OH} \\
| \\
\text{CH—S} \\
\qquad\qquad \searrow \\
\qquad\qquad\quad M \;+\; 2H^+ \\
\qquad\qquad \nearrow \\
\text{CH}_2\text{—S}
\end{array}
$$

BAL   *Heavy metal ion*   *Chelated metal ion*

FIGURE 16–6 BAL chelation of arsenic or a heavy metal ion such as lead.

A chelating agent encases an atom or ion like a crab or an octopus surrounds a bit of food.

understood that Lewisite poisoned people by the reaction of arsenic with protein sulfhydryl groups, British scientists set out to find a suitable compound that contained highly reactive sulfhydryl groups that could compete with sulfhydryl groups in the natural substrate for the arsenic, and thus render the poison ineffective. Out of this research came a compound now known as British Anti-Lewisite (BAL).

The BAL, which bonds to the metal at several sites, is called a **chelating agent** (Greek, *chela,* meaning "claw"), a term applied to a reacting agent that envelops a species such as a metal ion. BAL is one of many compounds that can act as chelating agents for metals (Fig. 16–6).

With the arsenic or heavy metal ion tied up, the sulfhydryl groups in vital enzymes are freed and can resume their normal functions. BAL is a standard therapeutic item in a hospital's poison emergency center and is used routinely to treat heavy metal poisoning.

Mercury deserves some special attention because it has a rather peculiar fascination for some people, especially children, who love to touch it (Fig. 16–7). It

FIGURE 16–7 The first step to coating a coin with mercury. Children love to coat coins with metallic mercury—a very dangerous practice since mercury is easily inhaled and also passes through the skin. (Dimes made since 1966 contain no silver and therefore will not amalgamate with mercury. This photograph is used for purposes of demonstration only.)

is poisonous and, to make matters worse, mercury and its salts accumulate in the body. This means the body has no quick means of ridding itself of this element and there tends to be a buildup of the toxic effects leading to **chronic** poisoning.

Although mercury is rather unreactive compared with other metals, it is quite volatile and easily absorbed through the skin. In the body, the metal atoms are oxidized to Hg_2^{2+} [mercury (I) ion] and Hg^{2+} [mercury (II) ion]. Compounds of both Hg_2^{2+} and Hg^{2+} are known to be toxic.

Today mercury poisoning is a potential hazard to those working with or near this metal or its salts, such as dentists (who use it in making amalgams for fillings), various medical and scientific laboratory personnel (who routinely use mercury compounds or mercury pressure gauges), and some agricultural workers (who employ mercury salts as fungicides).

Mercury can also be a hazard when it is present in food. It is generally believed that mercury enters the food chain through small organisms that feed at the bottom of bodies of water that contain mercury from industrial waste or mercury minerals in the sediment. These in turn are food for bottom-feeding fish. Game fish in turn eat these fish and accumulate the largest concentration of mercury, the accumulation of poison building up as the food chain progresses.

Lead is another widely encountered heavy-metal poison. The body's method of handling lead provides an interesting example of a "metal equilibrium" (Fig. 16–8). Lead often occurs in foods (100–300 μg per kg), beverages (20–30 μg per liter), public water supplies (100 μg per liter, from lead-sealed pipes), and even air

A vivid description of the psychic changes produced by mercury poisoning can be found in the Mad Hatter, a character in Lewis Carroll's *Alice in Wonderland*. The fur felt industry once used mercury (II) nitrate, $Hg(NO_3)_2$, to stiffen the felt. Chronic mercury poisoning accounted for the Mad Hatter's odd behavior; it also gave the workers in hat factories symptoms known as "hatter's shakes."

Amalgam: Any mixture or alloy of metals of which mercury is a constitutent.

$1 \mu g$ (microgram) $= 10^{-6}$ g

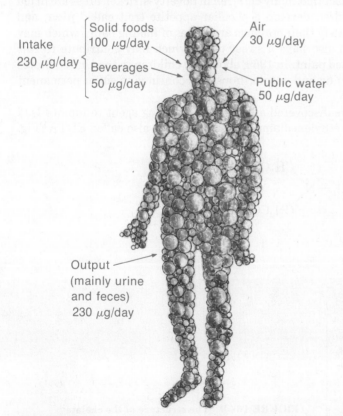

Intake
230 μg/day

Solid foods
100 μg/day

Beverages
50 μg/day

Air
30 μg/day

Public water
50 μg/day

Output
(mainly urine
and feces)
230 μg/day

FIGURE 16–8 Lead equilibrium in humans. Figures chosen for intake are probable upper limits.

1 mg (milligram) =
10^{-3} g = 1000 μg

(2.5 μg per cubic meter, from lead compounds in auto exhausts). With this many sources and contacts per day, it is obvious that the body must be able to rid itself of this poison; otherwise everyone would have died long ago of lead poisoning! The average person can excrete about 2 mg of lead a day through the kidneys and intestinal tract; the daily intake is normally less than this. However, if intake exceeds this amount, accumulation and storage result. In the body lead not only resides in soft tissues but also is deposited in bone. In the bones lead acts on the bone marrow, while in tissues it behaves like other heavy-metal poisons, such as mercury and arsenic. Lead, like mercury and arsenic, can also affect the central nervous system.

Unless they are very insoluble, lead salts are always toxic, and their toxicity is directly related to the salt's solubility. One common covalent lead compound, tetraethyllead, $Pb(C_2H_5)_4$, until recently a component of most gasolines, is different from most other metal compounds in that it is readily absorbed through the skin. Even metallic lead can be absorbed through the skin; cases of lead poisoning have resulted from repeated handling of lead foil, bullets, and other lead objects.

One of the truly tragic aspects of lead poisoning is that even though lead-pigmented paints have not been used for interior painting in this country during the past 30 years, children are still poisoned by lead from old paint. Health experts estimate that up to 225,000 children become ill from lead poisoning each year, with many experiencing mental retardation or other neurological problems. The reason for this is twofold. Lead-based paints still cover the walls of many older dwellings. Coupled with this is the fact that many children in poverty-stricken areas are ill fed and anemic. These children develop a peculiar appetite trait called **pica,** and among the items that satisfy their cravings are pieces of flaking paint, which may contain lead. Lead salts also have a sweet taste, which may contribute to this consumption of lead-based paint. In 1969, about 200 children in the United States alone died of lead poisoning and untold thousands continue to suffer permanent neurological damage.

Toxicologists have discovered an effective chelating agent to remove lead from the human body — ethylenediaminetetraacetic acid, also called EDTA (Fig. 16–9).

EDTA
(ETHYLENEDIAMINETETRAACETIC ACID)

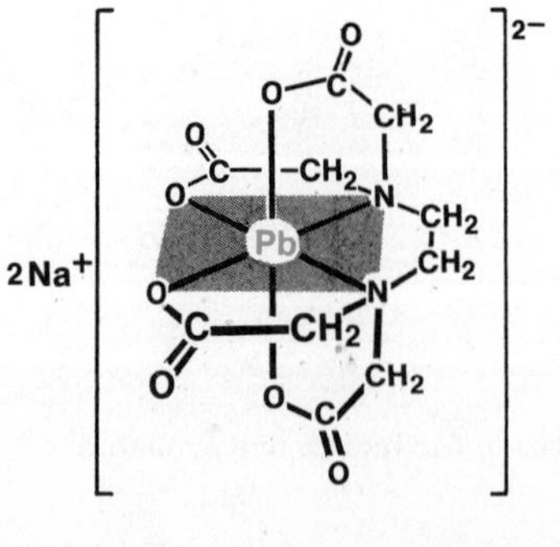

FIGURE 16–9 The structure of the chelate formed when the anion of EDTA envelops a lead (II) ion.

The calcium disodium salt of EDTA is used in the treatment of lead poisoning because EDTA by itself would remove too much of the blood serum's calcium. In solution, EDTA has a greater tendency to complex with lead (Pb^{2+}) than with calcium (Ca^{2+}). As a result, the calcium is released and the lead is tied up in the complex:

$$[CaEDTA]^{2-} + Pb^{2+} \rightarrow [PbEDTA]^{2-} + Ca^{2+}$$

The lead chelate is then excreted in the urine.

SELF-TEST 16–A

1. Corrosive poisons such as sulfuric acid destroy tissue by _____ followed by _____ of proteins.
2. Corrosive poisons, such as ozone, nitrogen dioxide, and iodine, destroy tissue by _____ it.
3. Carbon monoxide poisons by forming a strong bond with iron in _____ and thus preventing the transport of _____ from the lungs to the cells throughout the body.
4. CO is a cumulative poison. ( ) True or ( ) False
5. The cyanide ion has the formula _____. It poisons by complexing with iron in the enzyme _____ _____, thus preventing the use of _____ in the oxidative processes in the cells.
6. Give an example of a metabolic poison that is toxic because its structure is so similar to a useful substance that it can mimic the useful substance.

   _____

7. BAL is an antidote for _____. BAL is effective because its sulfhydryl (—SH) groups _____ arsenic and heavy metals and render them ineffective toward enzymes.
8. Mercury is a cumulative poison. ( ) True or ( ) False

NEUROTOXINS

Some metabolic poisons are known to limit their action to the nervous system. These include poisons such as strychnine and curare (a South American Indian dart poison), as well as the dreaded nerve gases developed for chemical warfare. The exact modes of action of most neurotoxins are not known for certain, but investigations have discovered the action of a few.

A nerve impulse or stimulus is transmitted along a nerve fiber by electrical impulses. The nerve fiber connects either with another nerve fiber or with some other cell (such as a gland or cardiac, smooth, or skeletal muscle) capable of being stimulated by the nerve impulse (Fig. 16–10). Neurotoxins often act at the point where two nerve fibers come together, called a **synapse.** When the impulse reaches the end of certain nerves, a small quantity of **acetylcholine** is liberated. This activates a receptor on an adjacent nerve or organ. The acetylcholine is thought to activate a nerve ending by changing the permeability of the nerve cell membrane. The method of increasing membrane permeability is not clear, but it may be related

Investigations of the actions of neurotoxins have provided insight into how the nervous system works.

$$\underset{\text{ACETYLCHOLINE}}{CH_3\overset{\displaystyle O}{\overset{\|}{C}}OCH_2CH_2\overset{\displaystyle CH_3}{\underset{\displaystyle CH_3}{\overset{|}{\underset{|}{N^+}}}}CH_3, OH^-}$$

Permeability: The ability of a membrane to let chemicals pass through it.

10^{-6} of a mole of acetylcholine is 6×10^{17} molecules.

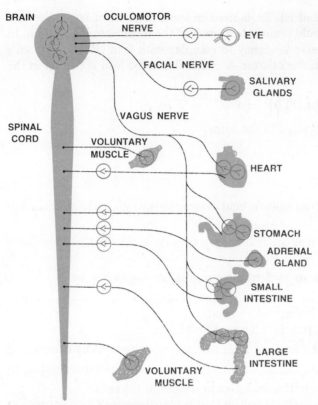

FIGURE 16–10 "Cholinergic" nerves, which transmit impulses by means of acetylcholine, include nerves controlling both voluntary and involuntary activities. Exceptions are parts of the "sympathetic" nervous system that utilize norepinephrine instead of acetylcholine. Sites of acetylcholine secretion are circled in color; poisons that disrupt the acetylcholine cycle can interrupt the body's communications at any of these points. The role of acetylcholine in the brain is uncertain, as is indicated by the broken circles.

to an ability to dissociate fat-protein complexes or to penetrate the surface films of fats. Such effects can be brought about by as little as 10^{-6} mole of acetylcholine, which could alter the permeability of a cell so ions can cross the cell membrane more freely.

To enable the receptor to receive further electrical impulses, the enzyme **cholinesterase** breaks down acetylcholine into acetic acid and choline (Fig. 16–11):

$$\underset{\substack{\text{ACETYLCHOLINE}}}{CH_3COCH_2CH_2\overset{\displaystyle CH_3}{\underset{\displaystyle CH_3}{N^+}}-CH_3,\ OH^-} + \underset{\text{WATER}}{H_2O} \xrightarrow{\text{cholinesterase}} \underset{\text{ACETIC ACID}}{CH_3\overset{\displaystyle O}{C}OH} + \underset{\substack{\text{CHOLINE}}}{HOCH_2CH_2\overset{\displaystyle CH_3}{\underset{\displaystyle CH_3}{N^+}}-CH_3,\ OH^-}$$

In the presence of potassium and magnesium ions, other enzymes such as acetylase resynthesize new acetylcholine from the acetic acid and the choline within the incoming nerve ending:

Acetic acid + Choline $\xrightarrow{\text{acetylase}}$ Acetylcholine + H_2O

The new acetylcholine is available for transmitting another impulse across the gap.

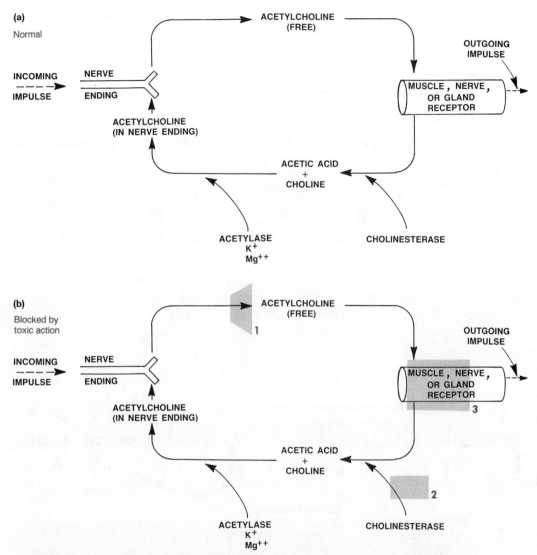

FIGURE 16-11 The acetylcholine cycle, a fundamental mechanism in nerve impulse transmission, is affected by many poisons. An impulse reaching a nerve ending in the normal cycle (*a*) liberates acetylcholine, which then stimulates a receptor. To enable the receptor to receive further impulses, the enzyme cholinesterase breaks down acetylcholine into acetic acid and choline; other enzymes resynthesize these into more acetylcholine. (*b*) Botulinus and dinoflagellate toxins inhibit the synthesis, or the release, of acetylcholine (*1*). The "anticholinesterase" poisons inactivate cholinesterase and therefore prevent the breakdown of acetylcholine (*2*). Curare and atropine desensitize the receptor to the chemical stimulus (*3*).

Neurotoxins can affect the transmission of nerve impulses at nerve endings in a variety of ways. The **anticholinesterase poisons** prevent the breakdown of acetylcholine by deactivating cholinesterase. These poisons are usually structurally analogous to acetylcholine, so they bond to the enzyme cholinesterase and deactivate it (Fig. 16-12). The cholinesterase molecules bound by the poison are held so effectively that the restoration of proper nerve function must await the manufacture of new cholinesterase. In the meantime, the excess acetylcholine overstimulates nerves, glands, and muscles, producing irregular heart rhythms,

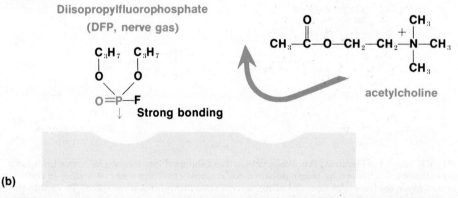

FIGURE 16–12   (*a*) **The mechanism of cholinesterase breakdown of acetylcholine.** (*b*) **The tie-up of cholinesterase by an anticholinesterase poison like the nerve gas DFP blocks the normal hydrolysis of acetylcholine since the acetylcholine cannot bind to the enzyme.**

Curare, used by South American Indians in poison darts, was brought to Europe by Sir Walter Raleigh in 1595. It was purified in 1865, and its structure was determined in 1935.

convulsions, and death. Many of the organic phosphates that are widely used as insecticides are metabolized in the body to produce anticholinesterase poisons. For this reason, they should be treated with extreme care. Some poisonous mushrooms also contain an anticholinesterase poison. Figure 16–13 contains the structures of some anticholinesterase poisons.

Neurotoxins such as **atropine** and **curare** are able to occupy the receptor sites on nerve endings of organs that are normally occupied by the impulse-carrying acetylcholine. When atropine or curare occupies the receptor site, no stimulus is transmitted to the organ. Acetylcholine in excess causes a slowing of the heartbeat,

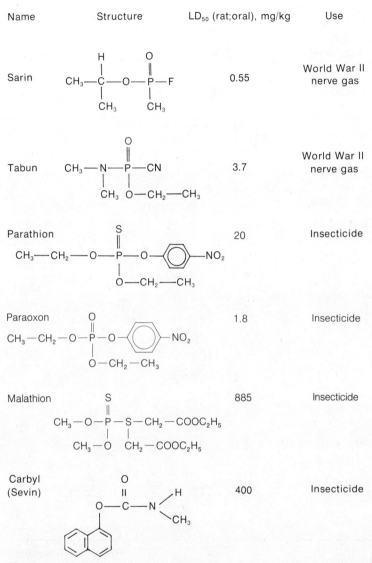

Name	Structure	LD_{50} (rat;oral), mg/kg	Use
Sarin		0.55	World War II nerve gas
Tabun		3.7	World War II nerve gas
Parathion		20	Insecticide
Paraoxon		1.8	Insecticide
Malathion		885	Insecticide
Carbyl (Sevin)		400	Insecticide

FIGURE 16–13 Some anticholinesterase poisons. In animals, parathion is converted into paraoxon in the liver. Carbyl and malathion do not bind to cholinesterase as strongly. Malathion was the insecticide used in California in July 1981 to eradicate Medflies.

a decrease in blood pressure, and excessive saliva, whereas atropine and curare produce excessive thirst and dryness of the mouth and throat, a rapid heartbeat, and an increase in blood pressure. The normal responses to acetylcholine activation are absent, and the opposite responses occur when there is sufficient atropine present to block the receptor sites.

Neurotoxins of this kind can be extremely useful in medicine. For example, atropine is used to dilate the pupil of the eye to facilitate examination of its interior. Applied to the skin, atropine sulfate and other atropine salts relieve pain by deactivating sensory nerve endings on the skin. Atropine is also used as an antidote for anticholinesterase poisons. Curare has long been used as a muscle relaxant.

TABLE 16–5   Alkaloid Neurotoxins That Compete with Acetylcholine for the Receptor Site*

NAME	NORMAL CONTACT	LETHAL DOSE (FOR A 70-kg HUMAN)	FORMULA
Atropine	Dilation of pupil of the eye	0.1 g	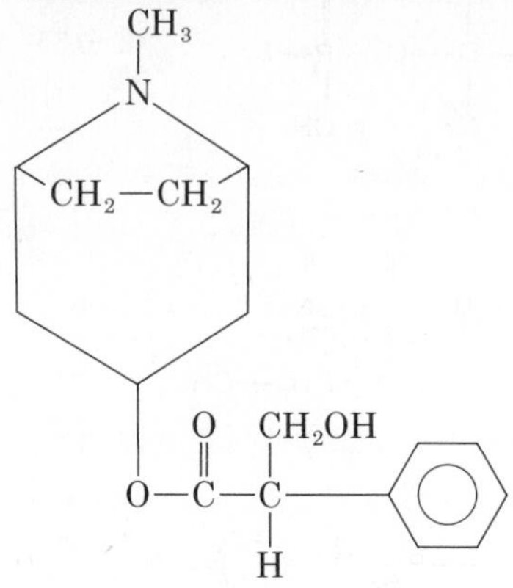
Curare	Muscle relaxant	20 mg	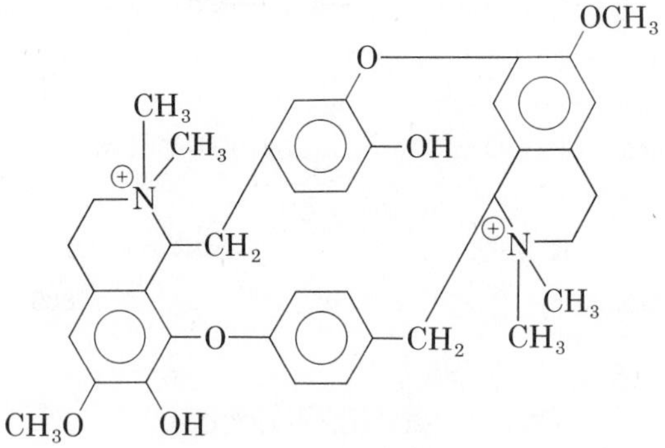
Nicotine	Tobacco, insecticide	75 mg	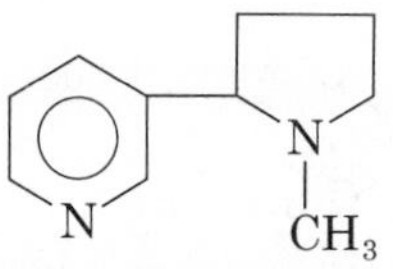

(continued on next page)

Morphine is the most effective pain killer known.

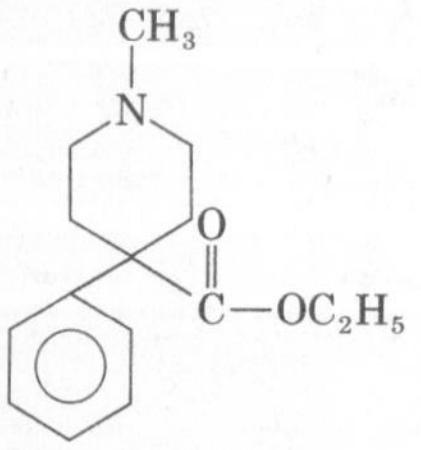

MEPERIDINE

A well-known, natural organic compound that blocks receptor sites in a manner similar to that of curare and atropine is **nicotine.** This powerful poison causes stimulation and then depression of the central nervous system. The probable lethal dose for a 70-kg person is less than 0.3 g. It is interesting to note that pure nicotine was first extracted from tobacco and its toxic action observed *after* tobacco use was established as a habit.

Natural or synthetic **morphine** is the most effective pain reliever known. It is widely used to relieve short-term acute pain resulting from surgery, fractures, burns, and so on, as well as to reduce suffering in the later stages of terminal illnesses such as cancer. The manufacture and distribution of narcotic drugs are stringently controlled by the federal government through laws designed to keep these products available for legitimate medical use. Under federal law, some preparations containing small amounts of narcotic drugs may be sold without a prescription (for example, cough mixtures containing codeine), but not many.

TABLE 16–5 (continued)

NAME	NORMAL CONTACT	LETHAL DOSE (FOR A 70-kg HUMAN)	FORMULA
Caffeine	Coffee, tea, cola drinks	13.4 g (one cup of coffee contains about 40 mg caffeine)	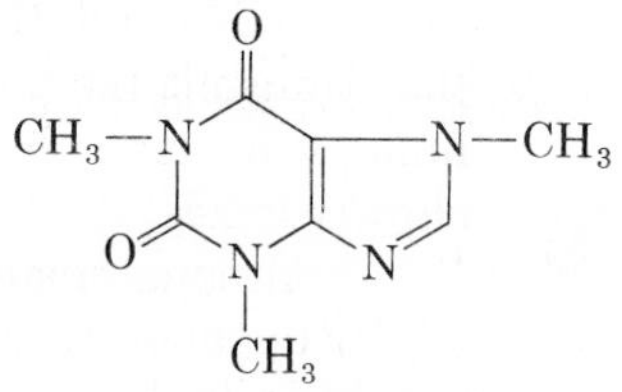
Morphine	Opium — pain killer	100 mg	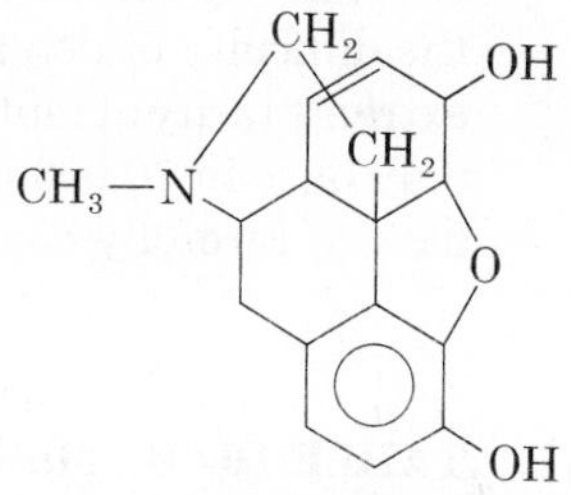
Codeine	Opium — pain killer	0.3 g	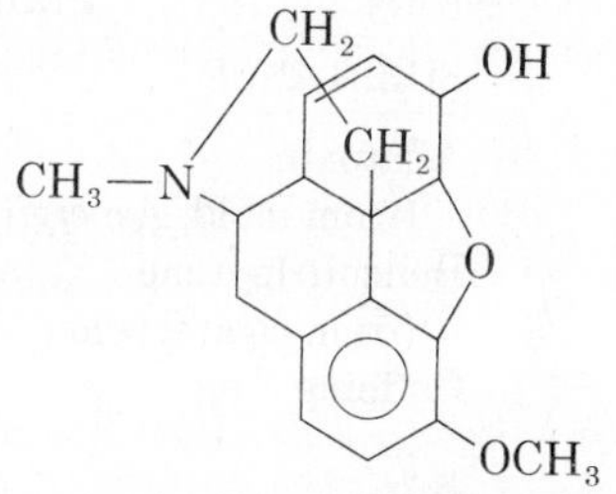
Cocaine	Leaves of *Erythroxylon coca* in South America	1 g	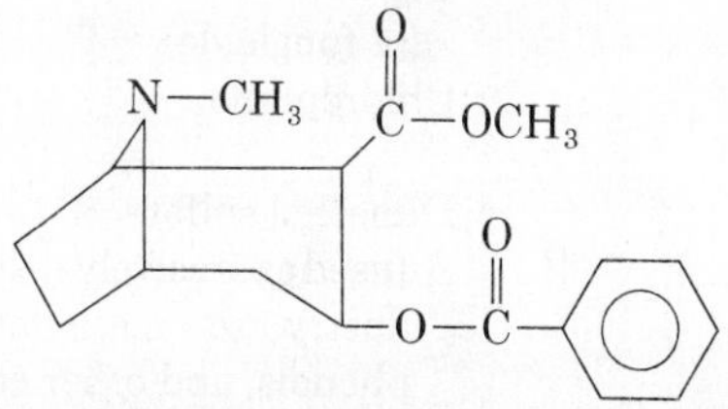

* **Alkaloid** is broadly defined as a physiologically active compound found in plants and containing amino nitrogen atoms (consequently it has basic properties). The nitrogen atom, or atoms, are frequently found as part of rings.

In spite of stringent controls, drugs like **morphine, heroin, meperidine,** and **methadone** are abused and illicitly used. Heroin is prepared from morphine, which is derived from sap in the opium poppy. It takes about 10 pounds of opium to prepare one pound of morphine. Morphine reacts with acetic anhydride in a one-to-one reaction to form heroin. Street-grade heroin is only 9% to 10% pure.

Meperidine and **methadone** are products of chemical laboratories rather than of poppy fields. Meperidine was claimed to be nonaddictive when first produced. Experience, however, proved otherwise (as it did with morphine and heroin). A major difference between methadone and morphine and heroin is that when methadone is taken orally, under medical supervision, it prevents withdrawal symptoms for approximately 24 hours.

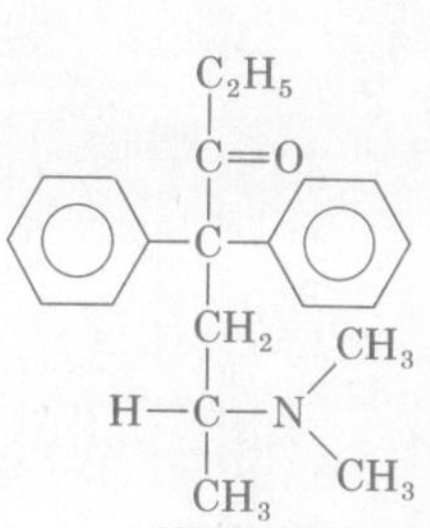

Some other natural products that affect the central nervous system and can be neurotoxic in comparatively small amounts are listed in Table 16–5.

METHADONE

MUTAGENS

A mutagen is a chemical that can change the hereditary pattern of a cell.

Mutagens are chemicals capable of altering the genes and chromosomes sufficiently to cause abnormalities in offspring. Chemically, mutagens alter the structures of DNA and RNA, which compose the genes (and, in turn, the chromosomes) that transmit the traits of parent to offspring. Mature sex, or germinal, cells of humans normally have 23 chromosomes; body, or somatic, cells have 23 *pairs* of chromosomes.

Although many chemicals are under suspicion because of their mutagenic effects on laboratory animals, it should be emphasized that no one has yet shown conclusively that any chemical induces mutations in human germinal cells. Part of the difficulty of determining the effects of mutagenic chemicals in humans is the extreme rarity of mutation. A specific genetic disorder may occur as infrequently as only once in 10,000 to 100,000 births. Therefore, to obtain meaningful statistical data, a carefully controlled study of the entire population of the United States

TABLE 16–6   Mutagenic Substances as Indicated by Experimental Studies on Plants and Animals

SUBSTANCE	EXPERIMENTAL RESULTS
Aflatoxin (from mold, *Aspergillus flavus*)	Mutations in bacteria, viruses, fungi, parasitic wasps, human cell cultures, mice
Benzo(α)pyrene (from cigarette and coal smoke)	Mutations in mice
Caffeine	Chromosome changes in bacteria, fungi, onion root tips, fruit flies, human tissue cultures
Captan (a fungicide)	Mutagenic in bacteria and molds; chromosome breaks in rats and human tissue cultures
Chloroprene	Mutagenic in male sex cells; results in spontaneous abortions
Dimethyl sulfate (used extensively in chemical industry to methylate amines, phenols, and other compounds)	Methylates DNA base guanine; potent mutagen in bacteria, viruses, fungi, higher plants, fruit flies
LSD (lysergic acid diethylamide)	Chromosome breaks in somatic cells of rats, mice, hamsters, white blood cells of humans and monkeys
Maleic hydrazide (plant growth inhibitor; trade names Slo-Gro, MH-30)	Chromosome breaks in many plants and in cultured mouse cells
Mustard gas (dichlorodiethyl sulfide)	Mutations in fruit flies
Nitrous acid (HNO_2)	Mutations in bacteria, viruses, fungi
Ozone (O_3)	Chromosome breaks in root cells of broadleaf plants
Solvents in glue (glue sniffing) (toluene, acetone, hexane, cyclohexane, ethyl acetate)	4% more human white blood cells showed breaks and abnormalities (6% versus 2% normal)
TEM (triethylenemelamine) anticancer drug, insect chemosterilants)	Mutagenic in fruit flies, mice

would be required. In addition, the very long time between generations presents great difficulties, and there is also the problem of tracing a medical disorder to a single specific chemical out of the tens of thousands of chemicals with which we come in contact.

If there is no direct evidence for specific mutagenic effects in human beings, why, then, the interest in the subject? The possibility of a deranged, deformed human race is frightening; the chance for an improved human body is hopeful; and the evidence for chemical mutation in plants and lower animals is established. A wide variety of chemicals is known to alter chromosomes and to produce mutations in rats, worms, bacteria, fruit flies, and other plants and animals. Some of these are listed in Table 16–6.

Experimental work on the chemical basis of the mutagenic effects of nitrous acid (HNO_2) has been very revealing. Repeated studies have shown that nitrous acid is a potent mutagen in bacteria, viruses, molds, and other organisms. In 1953, at Columbia University, Dr. Stephen Zamenhof demonstrated experimentally that nitrous acid attacks DNA. Specifically, nitrous acid reacts with the adenine, guanine, and cytosine bases of DNA by removing the amino group of each of these compounds. The eliminated group is replaced by an oxygen atom (Fig. 16–14). The changed bases may garble a part of DNA's genetic message, and in the next replication of DNA, the new base may not form a base pair with the proper nucleotide base.

For example, adenine (A) typically forms a base pair with thymine (T) (Fig. 16–15). However, when adenine is changed to hypoxanthine, the new compound forms a base pair with cytosine (C). In the second replication, the cytosine forms its usual base pair with guanine (G). Thus, where an adenine-thymine (A-T) base pair

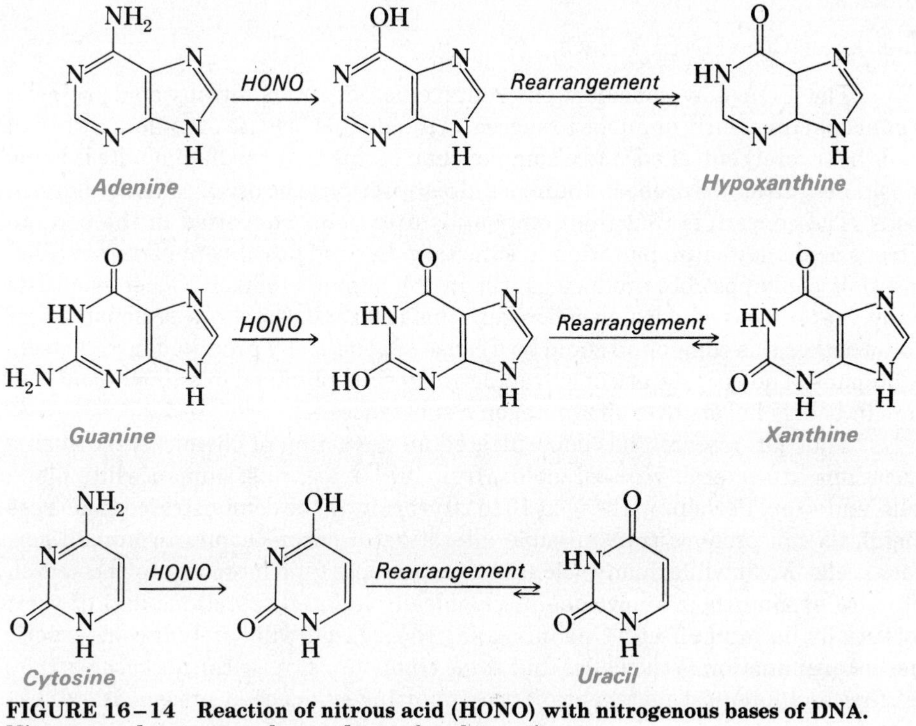

FIGURE 16–14 Reaction of nitrous acid (HONO) with nitrogenous bases of DNA. Nitrogen and water are also products of each reaction.

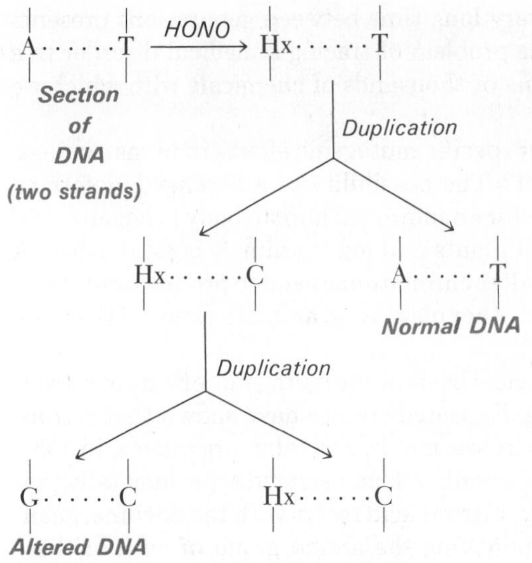

FIGURE 16-15 Alteration of DNA genetic code by base-pairing nitrous acid–converted nitrogenous bases. The bases are adenine (A), cytosine (C), guanine (G), hypoxanthine (Hx), and thymine (T).

existed originally, a guanine-cytosine (G-C) pair now exists. The result is an alteration in the DNA's genetic coding, so that a different protein is formed later.

Do all of these findings mean that nitrous acid is mutagenic in humans? Not necessarily. We do know that **sodium nitrite** has been widely used as a preservative, color enhancer, or color fixative in meat and fish products for at least the past 30 years. It is currently used in such foods as frankfurters, bacon, smoked ham, deviled ham, bologna, Vienna sausage, smoked salmon, and smoked shad. The sodium nitrite is converted to nitrous acid by hydrochloric acid in the human stomach:

Sodium nitrite produces nitrous acid in the stomach.

$$NaNO_2 + HCl \rightarrow HNO_2 + NaCl$$

The FDA now considers the mutagenic effects of nitrous acid in lower organisms sufficiently ominous to suggest strongly that the use of sodium nitrite in foods be severely curtailed, and a complete ban of this use of sodium nitrite is being considered. Several European countries already restrict the use of sodium nitrite in foods. The concern is that this compound, after being converted in the body to nitrous acid, may cause mutation in somatic cells (and possibly in germinal cells) and thus could possibly produce cancer in the human stomach. Other scientists doubt that nitrous acid is present in germinal cells and, therefore, seriously question whether this compound could be a cause of genetically produced birth defects in humans. The uncertainty of extrapolating results obtained in animal studies to human beings hovers over the mutagenic substances.

Thus far, research has concentrated on the action of chemicals in causing mutations in bacterial viruses, molds, fruit flies, mice, rats, human white blood cells, and so on. Perhaps in the next 10 to 20 years it will be demonstrated that these chemicals can produce transmissible alteration of chromosomes in human germinal cells. Meanwhile, many scientists are pressing for a more vigorous research effort to expand their knowledge of chemically induced mutations and of their potentially harmful effects. One intriguing theory that will surely invoke experimental examination is the belief that some compounds cause cancer because they are first and foremost mutagenic. The supporting evidence at present is still extremely inconclusive.

TERATOGENS

The effects of chemicals on human reproduction are a frightening aspect of toxicity. The study of birth defects produced by chemical agents is the discipline of **teratology.** The word root *terat* comes from the French word meaning "monster." There are three known classes of teratogens: radiation, viral agents, and chemical substances.

Birth defects occur in 2% to 3% of all births. About 25% of these occur from genetic causes, some possibly due to contact with mutagens, and 5% to 10% are the result of teratogens. The remaining 60% or so result from unknown causes.

In the development of the newborn, there are three basic periods during which the fetus is at risk. For a period of about 17 days between conception and implantation in the uterine wall, a chemical "insult" will result in cell death. The rapidly multiplying cells often recover, but if a lethal dose is administered, death of the organism occurs followed by abortion or reabsorption. During the critical embryonic stage (18 to 55 days) organogenesis, or development of the organs, occurs. At this time the fetus is extremely sensitive to teratogens. During the fetal period (56 days to term), the fetus is less sensitive. Contact with teratogens results in reduction of cell size and number. This is manifested in growth retardation and failure of vital organs to reach maturity.

The horrible thalidomide disaster in 1961 focused worldwide attention on chemically induced birth defects. Thalidomide, a tranquilizer and sleeping pill, caused gross deformities (flipperlike arms, shortened arms, no arms or legs, and other defects) in children whose mothers used this drug during the first two months of pregnancy. The use of this drug resulted in more than 4000 surviving malformed babies in West Germany, more than 1000 in Great Britain, and about 20 in the United States. With shattering impact, this incident demonstrated that a compound can appear to be remarkably safe on the basis of animal studies (so safe, in fact, that thalidomide was sold in West Germany without prescription) and yet cause catastrophic effects in humans. While the tragedy focused attention on chemical mutagens, thalidomide presumably does not cause genetic damage in the germinal cells and is really not mutagenic. Rather, thalidomide, when taken by a woman during early pregnancy, causes direct injury to the developing embryo.

THALIDOMIDE
(A TERATOGEN)

Any chemical substance that can cross the placenta is a potential teratogen, and any activity resulting in the uptake of these chemicals into the mother's blood might prove a dangerous act for the health and well-being of the fetus. Smoking a cigarette results in higher-than-normal blood levels of such substances as carbon monoxide, hydrogen cyanide, cadmium, nicotine, and benzo(α)pyrene. Of course, many of these substances are present in polluted air as well. Table 16-7 lists a number of chemical substances known to be teratogenic in humans and laboratory animals.

CARCINOGENS

Carcinogens are chemicals that cause cancer. **Cancer** is an abnormal growth condition in an organism that manifests itself in at least three ways. The rate of cell growth (that is, the rate of cellular multiplication) in cancerous tissue differs from the rate in normal tissue. Cancerous cells may divide more rapidly or more slowly than normal cells. Cancerous cells spread to other tissues; they know no bounds. Normal liver cells divide and remain a part of the liver. Cancerous liver cells may

Cancer of the epithelial tissue — *carcinoma.*

Cancer of the connective tissue — *sarcoma.*

TABLE 16–7   Teratogenic Substances

SUBSTANCES	SPECIES	EFFECTS ON FETUS
METALS		
Arsenic	Mice	Increase in males born with eye defects, renal
	Hamsters	damage
Cadmium	Mice	Abortions
	Rats	Abortions
Cobalt	Chickens	Eye, lower limb defects
Gallium	Hamsters	Spinal defects
Lead	Humans	Low birth weights, brain damage, stillbirth,
	Rats	early and late deaths
	Chickens	
Lithium	Primates	Heart defects
Mercury	Humans	Minamata disease (Japan)
	Mice	Fetal death, cleft palate
	Rats	Brain damage
Thallium	Chickens	Growth retardation, abortions
Zinc	Hamsters	Abortions
ORGANIC COMPOUNDS		
DES (diethylstilbestrol)	Humans	Uterine anomalies
Caffeine (15 cups per day equivalent)	Rats	Skeletal defects, growth retardation
PCBs (polychlorinated biphenyls)	Chickens	Central nervous system and eye defects
	Humans	Growth retardations, stillbirths

leave the liver and be found, for example, in the lung. Most cancer cells show partial or complete loss of specialized functions. Although located in the liver, cancer cells no longer perform the functions of the liver.

Attempts to determine the cause of cancer have evolved from early studies in which the disease was linked to a person's occupation. It was first noticed in 1775 that persons employed as chimney sweeps in England had a higher rate of skin cancer than the general population. It was not until 1933 that **benzo(α)pyrene,** $C_{20}H_{12}$ (a 5-ringed aromatic hydrocarbon), was isolated from coal dust and shown to be metabolized in the body to produce one or more carcinogens. In 1895, the German physician Rehn noted three cases of bladder cancer, not in a random population, but in employees of a factory that manufactured dye intermediates in the Rhine Valley. Rehn attributed these cancers to his patients' occupation. These and other cases confirmed that at times as many as 30 years passed between the time of the initial employment and the occurrence of bladder cancer. The principal product of these factories was aniline. Although aniline was first thought to be the carcinogenic agent, it was later shown to be noncarcinogenic. It was not until 1937 that continuous long-term treatment with **2-naphthylamine,** one of the suspected dye intermediates, in dosages of up to 0.5 g per day produced bladder cancer in dogs. Since then other dye intermediates have been shown to be carcinogenic.

A vast amount of research has verified the carcinogenic behavior of a large number of diverse chemicals. Some of these are listed in Table 16–8. This research has led to the formulation of a few generalizations concerning the relationship between chemicals and cancer. For example, carcinogenic effects on lower animals are commonly extrapolated to humans. The mouse has come to be the classic

BENZO(α)PYRENE

ANILINE   2-NAPHTHYLAMINE

animal for studies of carcinogenicity. Strains of inbred mice and rats have been developed that are genetically uniform and show a standard response.

Some carcinogens are relatively nontoxic in a single, large dose, but may be quite toxic, often increasingly so, when administered continuously. Thus, much patience, time, and money must be expended in carcinogenic studies. The development of a sarcoma in humans, from the activation of the first cell to the clinical manifestation of the cancer, takes from 20 to 30 years. With life expectancy of an average person in the United States now set at about 70 years, it is not surprising that the number of deaths due to cancer is increasing.

Cancer does not occur with the same frequency in all parts of the world. Breast cancer occurs less frequently in Japan than in the United States or Europe. Cancer of the stomach, especially in males, is more common in Japan than in the United States. Cancer of the liver is not widespread in the western hemisphere but accounts for a high proportion of the cancers among the Bantu in Africa and in certain populations in the Far East. The widely publicized incidence of lung cancer is higher in the industrialized world and is increasing at an appreciable rate.

Some compounds cause cancer at the point of contact. Other compounds cause cancer in an area remote from the point of contact. The liver, the site at which most toxic chemicals are removed from the blood, is particularly susceptible to such compounds. Since the original compound does not cause cancer on contact, some other compound made from it must be the cause of cancer. For example, it appears that the substitution of an $\diagup$NOH group for an $\diagup$NH group in an aromatic amine derivative produces at least one of the active intermediates for carcinogenic amines. If R denotes a two- or three-ring aromatic system, then the process can be represented as follows:

$$
\begin{array}{ccccc}
\text{H} & & \text{OH} \\
| & & | \\
\text{RNCOCH}_3 & \rightarrow & \text{RNCOCH}_3 & \rightarrow & \text{RX?} \rightarrow \text{RY?} \xrightarrow{\text{tissue}} \text{Tumor cell} \\
\text{INACTIVE} & & \text{ACTIVE ON} & & \text{OTHER UNKNOWN} \\
\text{ON CONTACT} & & \text{CONTACT} & & \text{INTERMEDIATES}
\end{array}
$$

As indicated by the variety of chemicals in Table 16–8, many molecular structures produce cancer, whereas closely related ones do not. The 2-naphthylamine mentioned earlier is carcinogenic, but repeated testing gives negative results for 1-naphthylamine.

1-NAPHTHYLAMINE
(NONCARCINOGENIC)

2-NAPHTHYLAMINE
(CARCINOGENIC)

For some types of cancer there are distinct stages that ultimately result in cancer. These may be identified as the *initiation period,* the *development* or *promotion period,* and the *progression period.* A single, minute dose of a carcinogenic polynuclear aromatic hydrocarbon, such as benzo(α)pyrene, applied to the skin of mice produces the permanent change of a normal cell to a tumor cell. This is the initiation step. No noticeable reaction occurs unless further treatment is made. If the area is painted repeatedly with noncarcinogenic croton oil, even up to one year later, carcinomas appear (Fig. 16–16). This is the development period. Additional

Cancer spreads from one tissue to another via *metastases.*

An abnormal growth is classified as cancerous or malignant when examination shows it is invading neighboring tissue. A growth is benign if it is localized at its original site.

Smoking is thought to play both an initiation and promotion role in cancer causation.

(Text continued on p. 422)

TABLE 16–8 Chemicals Carcinogenic for Humans

COMPOUND	FORMULA	USE OR SOURCE	SITE AFFECTED	CONFIRMING ANIMAL TESTS*
Inorganic Compounds				
Arsenic (and compounds)	As	Insecticides, alloys	Skin, lung	−
Asbestos	$Mg_6(Si_4O_{11})(OH)_6$	Brake linings, insulation	Liver Respiratory tract	+
Beryllium	Be	Alloy with copper	Bone, lung	+
Cadmium	Cd	Metal plating	Kidney, lung	+
Chromium	Cr	Metal plating	Lung	+
Organic Compounds				
Benzene	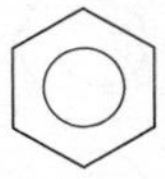	Solvent, chemical intermediate in syntheses	Blood (leukemia)	+
Acrylonitrile	$CH_2{=}CH(CN)$	Monomer	Colon, lung	+
Aflatoxins B$_1$ (shown)	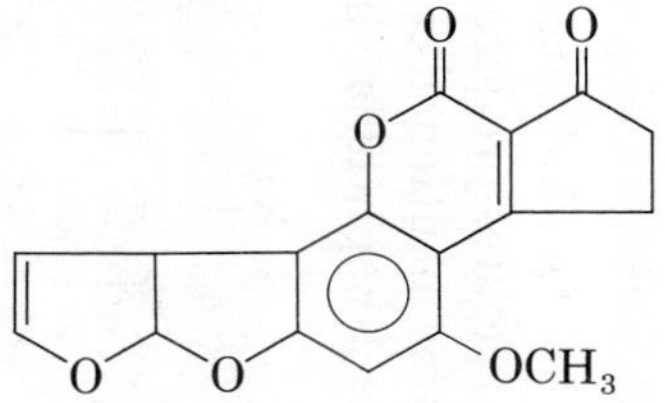	Mold or peanuts	Liver	+
Carbon tetrachloride	CCl_4	Solvent	Liver	+

				+
Diethylstilbestrol	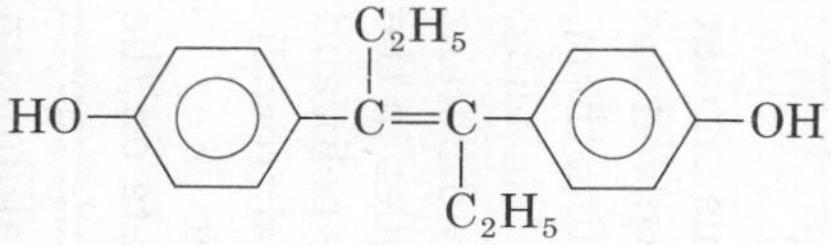	Hormone	Female genital tract	+
Benzo(α)pyrene	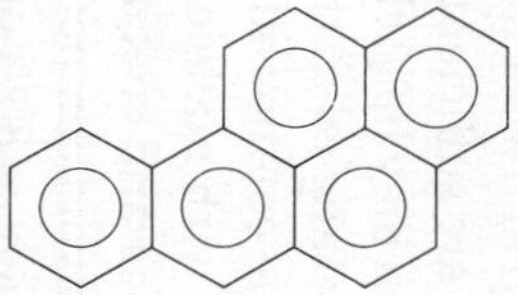	Cigarette and other smoke	Skin, lung	+
Benzidine	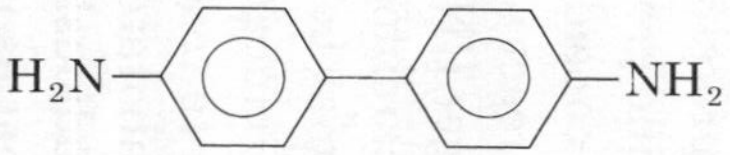	Dye manufacture, rubber compounding	Bladder	+
Ethylene oxide	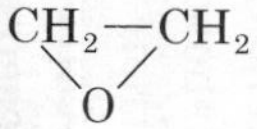 $CH_2{-}CH_2$ with O	Chemical intermediate used to make ethylene glycol, surfactants	Gastrointestinal tract	±
Soots, tar, and mineral oils		Roofing tar, chimney soot, oils of hydrocarbon nature	Skin, lung, bladder	+
Vinyl chloride	$CH_2{=}CHCl$	Monomer for making PVC	Liver, brain, lung, lymphatic system	+

* For animal tests, (+) means positive supporting data, (−) means a lack of supporting data, (±) means conflicting data.

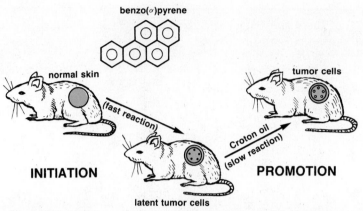

FIGURE 16–16   A second chemical can promote tumor growth in mice after an initiation period. Treatment of mice with croton oil produces no tumor nor does treatment with small quantities of benzo (α) pyrene alone. Croton oil is an irritant oil similar to castor oil. Both are derived from plants.

fundamental alterations in the nature of the cells occur during the progression period. If there is no initiator, there are no tumors. If there is initiator but no promoter, there are no tumors. If the initiator is followed by repeated doses of promoter, tumors appear. This seems to indicate that cancer cannot be contracted from chemicals unless repeated doses are administered or applied.

Just how do these toxic substances work? Cancer might be caused if the carcinogen combines with growth control proteins, rendering them inactive. During the normal growth process the cells divide and the organism grows to a point and stops. Cancer is abnormal in that cells continue to divide and portions of the organism continue to grow. One or more proteins are thought to be present in each cell with the specific duty of preventing replication of DNA and cell division. Virtually all of the carcinogens bind firmly to proteins, but so do some similar compounds that are noncarcinogenic. The specific growth proteins involved are not yet known for any of the carcinogens, despite considerable efforts to find them.

Another theory suggests that carcinogens react with and alter nucleic acids, so the proteins ultimately formed on the messenger RNA are sufficiently different to alter the cell's function and growth rate. The carcinogen may be included in the DNA or RNA strands by covalent bonding, or it may be entangled in the helix and held by weak van der Waals attractions. The carcinogenic compounds nitrosodimethylamine and mustard gas have been shown to react with nucleic acids.

While researchers collect data in their laboratories and speculate on the theoretical structural causes of cancer, we can studiously avoid compounds known to cause cancer in man. It has been proposed that as much as 80% of all human cancer has its origin in carcinogenic chemicals.

HALLUCINOGENS

Hallucinogens can produce temporary changes in perception, thought, and mood. These substances include **mescaline, lysergic acid diethylamide (LSD), tetrahydrocannabinol** (the active component of marihuana), and a broadening field of more than 50 other substances. They are included in this chapter as a separate section because they can be toxic in comparatively small doses.

Almost all chemical carcinogens have an induction period.

Smoking is associated with over 20% of all cancers; asbestos with between 3% and 18%.

All of the hallucinogenic drugs are toxic.

Several characteristics of some of the more famous hallucinogens are given in Table 16–9. Each of these substances is capable of disturbing the mind and producing bizarre and even colored interpretations of visual and other external stimuli. Mescaline is one of the oldest known hallucinogens, having been isolated from the peyote plant in 1896 by Heffter. Indeed, as early as 1560 the Mexican Indians who ate or drank the peyote were described by Spanish explorers as experiencing "terrible or ludicrous visions; the inebriation lasting for two or three days and then disappearing."

Although known for nearly 5000 years, marihuana is one of the least understood of all natural drugs. Very early in China's history it was used as a medicine, and in the United States it had early use as an analgesic and a poultice for corns.

TABLE 16–9   Some Hallucinogenic Chemicals

CHEMICAL	ALIASES	NOTES	ACTIVE COMPOUND	FORMULA
LSD	Acid Crackers The chief The hawk	Very powerful hallucinogen	D-lysergic acid diethylamide	
Mescaline	Peyote	Extracted from mescal buttons from peyote cactus in South and Central America	3,4,5-trimethoxy-phenethylamine	
Marihuana	Grass Pot Reefers Locoweed Hash Mary Jane	Leaves of the 5-leaf-per-frond marihuana plant *(Cannabis sativa)*	3,4,-*trans*-tetra-hydrocannabinol	
Phencyclidine	PCP Hog Elephant Peace pills Angel dust	Powerful analgesic-anesthetic, used in veterinary medicine	Phencyclidine	
Methaqualone	Quaalude Ludes	Hypnotic	Methaqualone	

Marihuana has little acceptable medical use in this country at present, although it is being studied experimentally for the treatment of glaucoma.

The body does not become dependent on the continuing use of marihuana as it does with heroin or other narcotics. However, reliable scientific data are not available with regard to chronic toxicity resulting from long-term use of the drug. It is known to cause dryness of the mouth, leading to dental problems.

Today there are many substances that produce hallucinogenic experiences; the most powerful one known is LSD. Our brief discussion will be restricted to this compound.

LSD

Literally hundreds of scientists are doing research on the effects of hallucinogens, including LSD. They are investigating the influence of these drugs on nerve and brain function, the possibility of chromosome alteration, tissue damage, psychological changes, and a host of other physiological and psychological effects. Scientists are not near a consensus on the toxic effects that LSD can cause. Collecting valid data is very difficult; many users of LSD overestimate the quantity of the drug they have used, its purity is often in question, and some take other drugs in addition. Even if the purity of the LSD is known, the results are difficult to interpret. For example, a study in which mice were given 0.05 to 1.0 μg of LSD on the seventh day of pregnancy showed a 5% incidence of badly deformed mouse embryos. Another study, probably equally valid and meticulous, showed no unusual fetal damage to rat embryos when 1.5 to 300 μg of LSD were administered during the fourth or fifth day of pregnancy. Babies with depressed skulls were aborted from two women who had taken LSD during the early weeks of pregnancy. Two other women, also LSD users, delivered full-term, apparently normal babies.

LSD has been linked with birth defects.

There are, however, some dangers that are well documented. These drugs destroy one's sense of judgment. Such things as height, heat, or even a moving truck may seem to hold no danger for the person under the influence of a hallucinogen. A dose of 50 to 200 μg will take the user on a "trip" for approximately 8 to 16 hours. Excessive or prolonged use of LSD can cause a person to "freak out" and possibly to sustain permanent brain damage. After one "trip," a user can experience another "trip" some time later, unexpectedly, without taking any more of the drug. The debate over LSD's dangers is continuing and is not likely to be resolved soon.

CH$_2$CH$_2$NH$_2$

NH

OH
SEROTONIN

OH

OH

HOCHCH$_2$NH$_2$
NOREPINEPHRINE

Out of the darkness of LSD-induced suicides and brain damage has come some new understanding of how the brain works. It is an interesting coincidence that soon after LSD was found localized in areas of the brain responsible for eliciting a human being's deep-seated emotional reactions, the compounds were discovered that are probably responsible for transmitting the impulse across synapses in the brain. Serotonin and norepinephrine are thought to act in a manner similar to acetylcholine, which was discussed earlier. These compounds carry the message from the end of one neuron to the end of another across the synapse. Somewhat later, Dilworth Wooley discovered that LSD blocks serotonin action (Fig. 16–17). This has led to an interesting theory to explain the LSD trip.

The theory of the hallucinogenic mechanism is, roughly, as follows. The LSD releases in some chemical way those emotional experiences that are generally hidden away, or chemically stored in the lower midbrain and the brain stem.

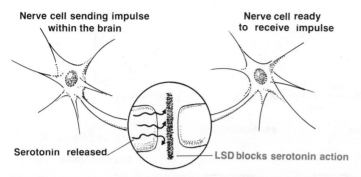

FIGURE 16–17   LSD blocks the flow of serotonin from one brain nerve cell to another.

Serotonin or norepinephrine, or both, inhibit the escape of these experiences into consciousness, as they turn off certain excitations. LSD interacts with serotonin or norepinephrine at the synapse and nullifies its blocking effect. Thus, these stored, previously rejected thoughts are allowed to enter the conscious part of the brain where the imaginary trip occurs.

This theory is built on the experimental fact that LSD blocks the action of serotonin, but there are other theories as well as some unexplained phenomena involved. Psychiatry is fraught with theories of why LSD causes an uplifting experience in some people and a frightening one in others. There are theories about dosage, purity of the compound, psychiatric state of the tripper, and conditions of administration. If there is anything close to a consensus, it is that the more controlled and relaxed the surroundings, the "better" the trip will be; even this, however, is debatable. Many trips, even under medical supervision, have not been satisfactory.

ALCOHOLS

Alcohols have some well-known toxic effects, but a complete chemical explanation has eluded scientists so far.

Methyl alcohol (methanol or wood alcohol) is highly poisonous, and unlike the other simple alcohols, it is, in effect, a cumulative poison in human beings. It has a specific toxic effect on the optic nerve, causing blindness with large doses. After its rapid uptake by the body, oxidation occurs in which the alcohol is first converted into formaldehyde and then to formic acid, which is eliminated in the urine.

$$CH_3OH \xrightarrow{\text{oxidation}} H{-}\overset{\overset{\textstyle O}{\|}}{C}{-}H \xrightarrow{\text{oxidation}} HCOH$$

METHYL ALCOHOL   FORMALDEHYDE   FORMIC ACID

This is a slow process and, for this reason, daily exposure to methyl alcohol can cause an extremely dangerous buildup of the alcohol in the body. The toxic effect on the optic nerve is thought to be caused by the oxidative products.

Ethyl alcohol (ethanol, grain alcohol) is found in alcoholic beverages, yet it is toxic like the other simple alcohols and is quantitatively absorbed by the gastrointestinal tract. About 58% of a dose is absorbed in 30 min, 88% in 1 hr, and 93% in 90 min. Over 90% of the ethyl alcohol is then slowly oxidized to carbon dioxide and water, mainly in the liver.

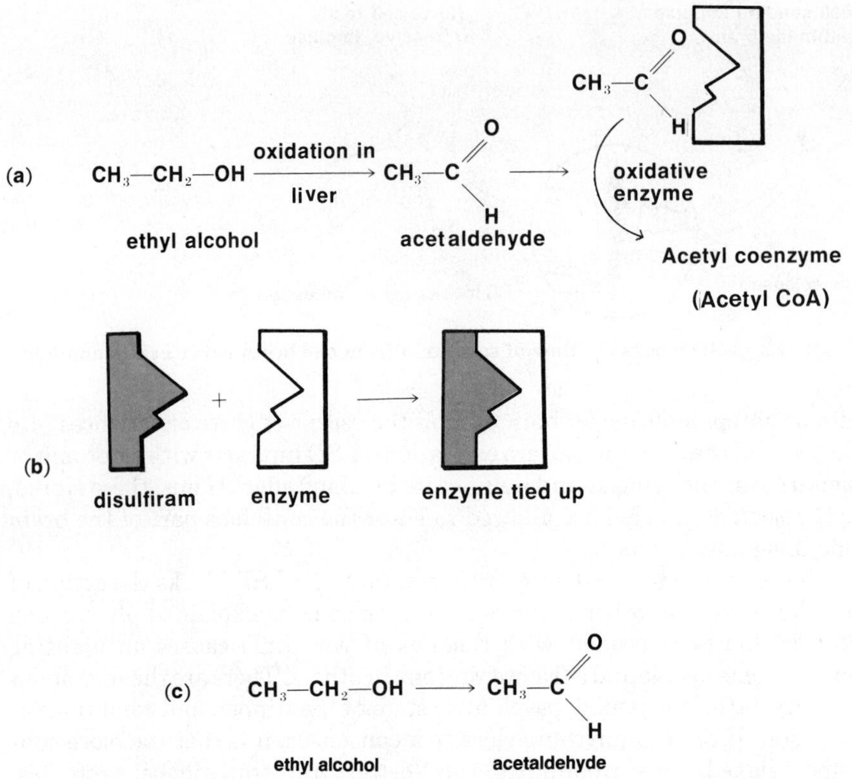

(a) $CH_3—CH_2—OH$ $\xrightarrow[\text{liver}]{\text{oxidation in}}$ $CH_3—C\overset{O}{\underset{H}{\lesssim}}$ $\longrightarrow$ oxidative enzyme

ethyl alcohol   acetaldehyde

$CH_3—C\overset{O}{\underset{H}{\lesssim}}$

Acetyl coenzyme A

(Acetyl CoA)

(b)

disulfiram   enzyme   enzyme tied up

(c) $CH_3—CH_2—OH$ $\longrightarrow$ $CH_3—C\overset{O}{\underset{H}{\lesssim}}$

ethyl alcohol   acetaldehyde

FIGURE 16–18 The blocking of ethyl alcohol oxidation by disulfiram (Antabuse). When the normal oxidative process (*a*) is stopped by blockage of the oxidative enzyme (*b*), acetaldehyde builds up in the body (*c*).

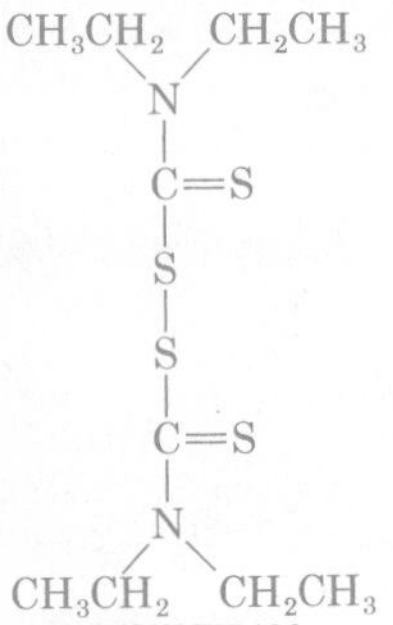

CH_3CH_2 CH_2CH_3
N
$C=S$
S
S
$C=S$
N
CH_3CH_2 CH_2CH_3
DISULFIRAM

The intoxicated person's staggering gait, stupor, and nausea are caused by the presence of acetaldehyde, but the chemical reactions involved are not fully understood. The compound disulfiram (Antabuse) is sometimes given as a treatment for chronic alcoholism because it blocks the oxidative steps beyond acetaldehyde (Fig. 16–18). The accumulation of acetaldehyde causes nausea, vomiting, blurred vision, and confusion. This is supposed to encourage the partaker to avoid this severe sickness by avoiding alcohol. Interestingly, this drug was discovered by two researchers who took a dose for another purpose and got violently ill that evening at a cocktail party.

SELF-TEST 16–B

1. Substrates that poison the nervous system are called _____.
2. Most neurotoxins affect chemical reactions that occur in the opening between two nerve cells. These openings are called _____.
3. The electrical impulse is carried across a synapse by the chemical

   _____.

4. Mutagens alter the structures of _____ or _____.
5. If a substance is mutagenic in test animals, particularly dogs, it must necessarily be mutagenic in human beings. ( ) True or ( ) False
6. The first occupation definitely linked to cancer was _____.

7. An active component of marihuana is _____.

8. The nausea and stupor of drunkenness from ethyl alcohol are caused by _____ and not by the alcohol itself.

9. Two dangers associated with smoking are _____ and _____.

10. A chemical that can cross the placenta and harm the fetus is called a _____.

MATCHING SET

_____ 1. Metabolic poison
_____ 2. Metabolism
_____ 3. Corrosive poison
_____ 4. Neurotoxin
_____ 5. Mutagen
_____ 6. Carcinogen
_____ 7. Carcinoma
_____ 8. Metastases
_____ 9. Sarcoma
_____ 10. Chelating agent
_____ 11. Hallucinogen

a. Lysergic acid diethylamide
b. Cyanide ion
c. Cancer in connective tissue
d. Benzo(α)pyrene metabolite
e. Cancerous growths of lung tissue located in liver
f. Use of chemicals in the body
g. Cancer of skin
h. Sodium hydroxide (caustic soda)
i. Atropine
j. Alters DNA
k. EDTA
l. Acetylcholine
m. Pica

QUESTIONS

1. Give an example of a toxic substance that is toxic as a result of:
    a. Binding to an oxygen-carrying molecule
    b. Disguising itself as another compound
    c. Attack on an enzyme
    d. Hydrolysis

2. True or False. Explain each answer concisely.
    a. Lead is a corrosive poison.
    b. Carbon monoxide and the cyanide ion poison in the same way.
    c. There are no known chemical compounds that cannot be toxic under some circumstances.

3. The application of a single, minute dose of a fused-ring hydrocarbon such as benzo(α)pyrene fails to produce a tumor in mice. Does this mean this compound is definitely noncarcinogenic? Give a reason for your answer.

4. Describe the chemical mechanism by which the following substances show their toxic effects.
    a. Fluoroacetic acid
    b. Phosgene
    c. Curare

5. Should any laws and regulations be placed on the use of any of the following? Justify your answers.
    a. LSD
    b. Marihuana
    c. Ethanol

6. Discuss some of the pros and cons of testing toxic substances on animals.

7. Give chemical reactions in words for
    a. Action of NaOH on tissue
    b. Action of carbon monoxide in blood
    c. Reaction of EDTA and lead ion

8. What questions do you think need to be answered before the action of ethyl alcohol is understood?

9. Assume a normal diet has the quantity of lead in a given quantity of food stated in the text. What would a person's total food intake of lead be per day?

10. What is the meaning of the symbolism LD_{50}?

11. Write chemical equations for:
    a. The hydrolysis of acetylcholine
    b. Acid hydrolysis of a protein having a glycine-glycine primary structure

12. Describe how the corrosive poisons lye (NaOH), NO_2, and the hypochlorite ion (OCl^-) destroy tissue.

13. What are some common sources of carbon monoxide?

14. If a relatively small amount of carbon monoxide is inhaled, are the chemical reactions reversible, or is carbon monoxide a cumulative poison?

15. What poisons can be rendered ineffective by wrapping a large molecule around them?

16. What is the cause of pica?

17. What is a common structural feature of alkaloids?

18. Give two examples of each: (a) corrosive poison, (b) metabolic poison, (c) neurotoxic poison, (d) mutagenic poison, (e) carcinogenic poison.

19. Phosgene hydrolyzes in the lungs to produce what acid?

20. What concentration of carbon monoxide in the air is likely to cause death in 1 hr?

21. Is it possible that one molecule of a mutagen could cause a problem to human life?

22. Two new chemicals are prepared in the lab. One is a relatively simple acid with corrosive properties. The other chemical has carcinogenic properties. Which chemical's toxic property is more likely to be discovered?

23. An old laboratory chemical is discovered to have a new property. It reacts with amine groups to produce —OH groups. Could this chemical be a possible mutagen?

24. If a poisoning victim has pinpoint pupils and is salivating excessively, what type of poison was the probable cause of these symptoms?

25. If your drinking water contains numerous toxic substances, why are you not normally harmed by drinking it?

Chapter 17

WATER — PLENTY OF IT, BUT OF WHAT QUALITY?

UNIQUENESS OF WATER

There would be no life on Earth without water *and* its unique properties. What are these unique properties, and what is their effect on life as we know it?

1. The density of solid water (ice) is less than that of liquid water. Put another way, water expands when it freezes. If ice were a normal solid, it would be more dense than liquid water, and lakes would freeze from the bottom up. This would have disastrous consequences for marine life, which could not survive in areas with winter seasons.

2. Water is a liquid at room temperature. However, the hydrogen compounds of all nonmetals around oxygen in the periodic table are toxic, corrosive gases such as NH_3, H_2S, and HF.

3. Water has a high heat capacity per unit of weight. This means it can absorb relatively large quantities of heat without large changes in temperature. For comparison, the heat capacity of water is about ten times that of copper or iron for equal weights. This property accounts for the moderating influence of lakes and oceans on the climate. Huge bodies of water absorb heat from the Sun and release it at night or in cooler seasons. The Earth would have extreme temperature variations if it weren't for this unique property of water.

   > Heat capacity is defined as the amount of heat required to raise the temperature of a sample of matter 1°C (Celsius).

4. Water has the highest heat of vaporization of all known substances. The heat needed to vaporize 1 g of water at 100°C is 540 cal. A consequence of this is the cooling effect that occurs with perspiration because of the heat absorbed from the skin when water evaporates.

5. Water has a large surface tension. The large surface tension of water and its ability to wet surfaces are the bases for capillary action, which carries water to leaves in plants and trees.

   > Surface molecules of a liquid are pulled inward by the intermolecular interactions with molecules below the surface. Surface tension is a measure of this force.

6. Water is an excellent solvent, often referred to as the universal solvent. As a result, natural water is not pure water but a solution of substances dissolved by contact with water.

What causes these unique properties of water? Hydrogen bonding between water molecules and the polarity of the water molecule are the causes of the unique properties of water. The extensive hydrogen bonding between water molecules in liquid water and in ice was described in Chapter 5. Recall that each water molecule

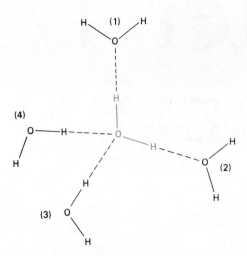

FIGURE 17–1 Tetrahedral cluster of four water molecules around a fully coordinated water molecule in the center.

Review the discussion of hydrogen bonding in Chapter 5.

can be hydrogen bonded to four other water molecules as shown in Figure 17–1. Liquid water is made up of tetrahedral clusters of hydrogen-bonded water molecules, and in ice the tetrahedral clusters are connected in three dimensions to give the structure shown in Figure 17–2.

Ice floats on water because the open structure of ice requires more volume for a given number of molecules than the volume required for the same number of

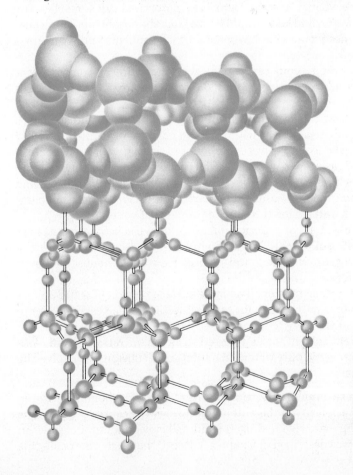

FIGURE 17–2 The crystal structure of ice. Note the open structure that gives ice its low density. (From *College Chemistry*, 3rd ed., by Pauling, Linus. W. H. Freeman and Company, Copyright © 1964.)

molecules in the liquid state. Since density equals mass per volume, ice has a lower density than liquid water. Other familiar consequences of this property are the bursting of water pipes in cold weather and the need for antifreeze to prevent cracks in engine blocks.

The melting of ice breaks about 15% of the hydrogen bonds holding the ice structure together. This collapses the structure shown in Figure 17–2 to give the more dense liquid. The structure of liquid water can be viewed as large clusters of hydrogen-bonded molecules, in which the hydrogen bonds are continually breaking and reforming, and the extent of hydrogen bonding per cluster is a function of temperature.

The high boiling point, high heat of vaporization, and large heat capacity of water are a result of the energy needed to break the hydrogen bonds in liquid water as it is heated or vaporized.

Hydrogen bonding also accounts for the large surface tension of liquid water. The water molecules on the surface are pulled inward by hydrogen bonding to water molecules below the surface. This unbalanced force at the surface causes the surface layer to contract, and energy is required to break this surface. This is why insects can walk on water even though they are more dense than water.

Floating a needle on water illustrates the phenomenon of surface tension.

WATER — THE MOST ABUNDANT COMPOUND

Water is the most abundant substance on the Earth's surface. Oceans cover about 72% of the Earth, and the average depth is 2.5 miles. The oceans are the source of 97.2% of all water. The rest consists of 2.1% in glaciers, 0.6% fresh water in lakes and rivers, 0.6% groundwater, and 0.1% in brine wells and brackish waters.

Brackish water contains dissolved salts but at a lower level than sea water.

Water is the major component of all living things. For example, the water content of human adults is 70%—the same proportion as the Earth's surface (Table 17–1).

It is estimated that an average of 4350 billion gallons of rain and snow fall on the contiguous United States each day. Of this amount, 3100 billion gallons return to the atmosphere by evaporation and transpiration. The discharge to the sea and

Transpiration is the release of water by leaves of plants. An acre of corn is estimated to release 3000 gallons per day, while a large oak tree releases 110 gallons per day.

TABLE 17–1 Water Content

Marine invertebrates	97%
Human fetus (1 month)	93%
Adult human	70%
Body fluids	95%
Nerve tissue	84%
Muscle	77%
Skin	71%
Connective tissue	60%
Vegetables	89%
Milk	88%
Fish	82%
Fruit	80%
Lean meat	76%
Potatoes	75%
Cheese	35%

to underground reserves amounts to 800 billion gallons daily, leaving 450 billion gallons of surface water each day for domestic and commercial use. The 48 contiguous states withdrew from natural sources 40 billion gallons per day in 1900 and 430 billion gallons in 1980, and it is estimated that the demand will be at least 900 billion gallons per day by the year 2000. The demand for water by our growing population is already greater than the resupply by natural resources in many parts of the country.

ARE WE FACING A WATER CRISIS?

Yes, we are entering a water crisis era that may be as serious as the highly publicized energy crisis. We are not running out of water because the 430 billion gallons per day is only one tenth of the daily supply. However, we need to examine the sources of usable water before we can understand the seriousness of the water crisis.

The two sources of usable water are surface water (lakes, rivers) and groundwater. Groundwater is that part of underground water that is below the water table. Figure 17–3 shows the various parts of the water cycle and the flow of groundwater. About 90 billion gallons of the 430 billion gallons per day of water usage come from groundwater supplies. These groundwater supplies are from wells drilled into the aquifers. The supply and demand for surface water and groundwater are uneven across the country, and in many areas the quantity and quality of the withdrawn water are not being resupplied to the lakes, rivers, and aquifers at the rate needed.

An aquifer is a water-bearing stratum of permeable rock, sand, or gravel, as illustrated in Figure 17–3.

In the arid West, wells used to pump water for irrigation either are going dry or are requiring drilling so deep that irrigation is no longer economically feasible. The huge Ogallala aquifer that stretches from South Dakota to Texas has 150,000 wells tapping it for irrigation of 10 million acres. As a result, the Ogallala aquifer is being drawn down at a rate that has reduced the average thickness of the aquifer from 58 feet in 1930 to 8 feet today. At the current rate, the effectiveness of the Ogallala aquifer as a source of groundwater will be used up in 20 to 30 years!

The depletion of the Ocala aquifer along the eastern seaboard has caused large sinkholes in Georgia and Florida when the limestone rock strata of the aquifer collapse as the water is withdrawn. Many coastal cities are also experiencing problems with brackish drinking water that comes from aquifers where drawing off fresh water causes sea water to flow into the depleted aquifer.

Depletion of underground sources has also caused sinkholes in Texas. Houston has sunk several feet as the result of extensive use of the underground water sources in that area! Figure 17–4 is a graphic illustration of the change in surface level in California's San Joaquin Valley as a result of groundwater depletion. So much groundwater has been pumped out for irrigation that the land has sunk 29 feet between 1925 and 1977.

The dispute between Arizona and California about Colorado River water is another example of the water crisis. The U.S. Supreme Court has ruled in Arizona's favor, setting California's allotment of the Colorado River at 4.4 million acre-feet per year, which is less than the current intake of 5 million acre-feet per year.

An acre-foot is the volume that would cover 1 acre to the depth of 1 foot.

These water shortages together with the problems of polluted water require careful study of possible solutions and widespread cooperation if we are to solve the water crisis. There are no quick fixes. A reliable long-term solution will require conservation, reuse, and less reliance on fresh water sources.

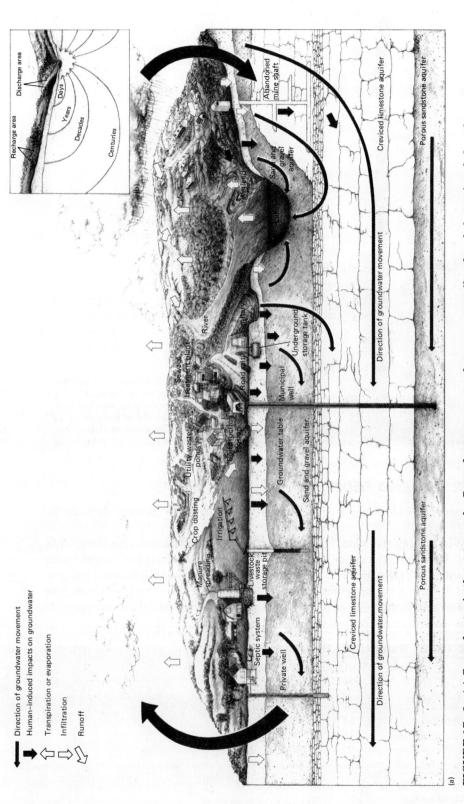

Direction of groundwater movement

Human-induced impacts on groundwater

Transpiration or evaporation

Infiltration

Runoff

FIGURE 17–3   (*a*) Groundwater in the water cycle. Groundwater—water that saturates soil and rock formations below the water table—plays an integral role in the hydrologic cycle and supplies the drinking water for half of this country's population. Bodies of groundwater stored in underground geologic formations, known as aquifers, may range in thickness from a few feet to several hundred feet and extend in area for many square miles. Shallow aquifers close to the land surface may be recharged by rainfall and surface runoff that percolates through pores and cracks in overlying soil and rock. Deep aquifers may contain large quantities of water, though they typically receive no surface recharge. All groundwater tends to move toward an area of discharge in lowlands where the water table intersects the land surface in streams, lakes, or wetlands. But rates of movement vary greatly, depending on the local geology. Groundwater moves more quickly in and through sandy aquifers but is slowed by clay soils or lightly fractured limestone. Threats to groundwater quality can come from the full spectrum of human activities on or near the surface, including septic tanks, agricultural fertilizers and pesticides, waste landfills and ponds, underground storage tanks, and other sources of soluble material. (*b*) Flow within groundwater systems. The time required for groundwater (and contaminants that may be in the groundwater) to move through aquifers and reach surface streams or drinking-water wells can range from days and years to decades and centuries. Many factors affect the travel time, including distance, hydraulic gradient, the nature of local geologic media, and the chemical nature of the contaminants themselves. Groundwater in sandy or coarse rock formations, for example, may move as much as several feet a day, while less permeable geology may limit the rate of movement to a few inches a year. Water in some very deep aquifers has remained virtually in situ for thousands of years. (Courtesy of Electric Power Research Institute; reprinted from *EPRI Journal*, October 1985, p. 6. Original source: Wisconsin Bureau of Water Resources Management.)

FIGURE 17–4   Markers on a utility pole in California's San Joaquin Valley indicate the large drop in surface level caused by withdrawal of groundwater for irrigation. (Courtesy of U.S. Geological Survey.)

WATER USE AND REUSE

Water reuse will be a major consideration in meeting the growing demand for water.

Who's using the water? Table 17–2 shows the breakdown for 1950, 1965, and 1980. Industry is the major user and also accounts for most of the increase since 1950. Table 17–3 gives an idea of how much water is needed for different products.

However, it is important to note that much of the industrial usage now involves recycled water. For example, the largest single industrial use of water is in

TABLE 17–2   Water Use in the United States

USE	BILLIONS OF GALLONS PER DAY		
	1950	1965	1980
Public supplies	14	26	36
Agricultural irrigation	110	115	140
Industry	77	165	255
Total	201	306	431

From W. W. Hales: Use and reuse of water. *Chemtech*, Vol. 12, pp. 532–537, 1982.

TABLE 17–3   Sample of Water Usage by Industry in 1980

| | WATER USED | |
INDUSTRY	Per Unit Production	Per Finished Product
Paper	20,000 gallon/ton	1 gallon/8 sheets typing paper
Oil refinery	20,000 gallon/ barrel crude oil	80 gallon/gallon gasoline
Steel	50,000 gallon/ton	25 gallon/1-lb. box nails
Power	360 gallon/min/Mw	51 gallon/100-W bulb burning for 24 hours

From W. W. Hales: Use and reuse of water. *Chemtech*, Vol. 12, pp. 532–537, 1982.

plant cooling systems. Recirculating cooling water systems, which often use cooling towers (Fig. 17–5), are an important means of water reuse. Such reuse also helps to reduce thermal pollution of the river or lake where the used water is discharged. In addition, the high heat capacity of water enables the industrial user to recycle this important source of heat energy during the reuse water cycle.

Thermal pollution is discussed on p. 442.

An indication of how the paper industry has reduced the amount of water by reuse is shown in Figure 17–6. The steel industry, another large water user, uses 50,000 gallons per ton of steel produced. However, 31,200 gallons of this are recycled water, 1200 gallons are vaporized, and the rest is discharged as wastewater.

These examples illustrate the difference between water use and water consumption. Much of the industrial use is temporary. The water is used to cool equipment or to provide steam, and then some fraction of it is discharged (treated if necessary) back to the river or lake source.

FIGURE 17–5   Industrial cooling towers. (Courtesy of Betz Laboratories, Inc.)

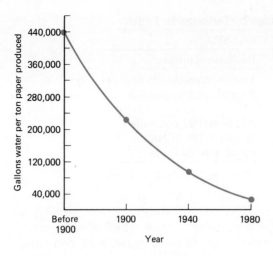

FIGURE 17–6 Drop in water usage by paper industry.

PUBLIC USE OF WATER

In the United States, the average use of potable (drinkable) water per day is 60 gallons per person. Table 17–4 gives the average amounts for various personal uses. One of the contributing factors to the water crisis is the fact that only **one half gallon** of water per person per day must be of drinkable quality. Although urban water delivery systems are not set up to deliver two types of water — drinkable water and water for other purposes — dual water supply systems are feasible. Only a small fraction of municipal water would have to be of drinking water quality. The largest portion of nonpotable water supply would be disinfected and bacteriologically safe, but would not meet drinking water regulations. This second source would be suitable for irrigation of parks and golf courses, air conditioning, industrial cooling, and toilet flushing.

Table 17–4 illustrates the inefficiency of residential water systems. We use 40% of residential water for flushing toilets and 30% for bathing. Conventional showers use up to 10 gallons per minute, which can be reduced by as much as 70% by the installation of inexpensive water-saving shower heads.

Residential water conservation is a way to cut demand for fresh water supplies. It has been effective during drought conditions. This was illustrated during the California drought of 1976–1977, when individual usage was reduced by

TABLE 17–4 Average Water Usage (by Gallons) per Person per Day

Flushing toilets	24
Bathing	18
Laundering	9
Dishwashing	3.6
Drinking and cooking	3.0
Miscellaneous	2.4
Total	60.0

63% in Marin County. Although Table 17–2 shows that residential use is a small part of the total, home water conservation is an important step in cutting demand for fresh water supplies in large urban areas. Home conservation will also help focus attention on the need for a total and continuing analysis of the problems relating to water supply.

Direct reuse of water is even possible for drinkable water. The idea of obtaining potable water from sewage is psychologically difficult for many persons to accept, but the technology has been developed. The rate of depletion of aquifers in southwestern United States has led to use of water recycling plants in several cities. For example, El Paso, Texas, uses a water recycling plant to obtain 10 million gallons of pure water per day from sewage effluent. The recycled water is pumped into the underground aquifer that is the main source of water for El Paso.

Where do we get the water we drink? Groundwater and surface water sources each provide half of the more than 35 billion gallons of potable water per day used in the United States. The water listed for public supplies in Table 17–2 is potable water. How pure is this water? The U.S. Public Health Service has set standards for potable water, and these are listed in Table 17–5. Wastewater from sewage treatment plants, industries, and agriculture runoff goes into lakes and rivers, so water purification is necessary.

The Clean Water Act of 1977 shifted the burden of producing water suitable for reuse from the user to the wastewater discharger. This action was a crucial step in improving the quality of our rivers and lakes, since it is easier to clean the wastewater prior to dumping than to clean the river water after the untreated waste has been discharged. In addition, the quality of the wastewater effluent is often high enough to be used as a resource of water for other purposes, such as irrigation or cooling towers.

TABLE 17–5   U.S. Public Health Service Standards for Potable Water

CONTAMINATING ION(S)	MAXIMUM CONCENTRATION, mg/liter
Arsenic	0.05
Barium	1.00
Cadmium	0.01
Chloride	250.00
Chromium	0.05
Copper	1.00
Cyanide	0.20
Fluoride	2.00
Iron	0.30
Lead	0.05
Manganese	0.05
Nitrate	45.00
Organics	0.20
Selenium	0.01
Silver	0.05
Sulfate	250.00
Zinc	5.00
Total dissolved solids	500

WATER PURIFICATION IN NATURE

Water is a natural resource that, within limitations, is continuously renewed. The familiar water cycle (Fig. 17–3) offers a number of opportunities for nature to purify its water. The worldwide **distillation** process results in rain water containing only traces of nonvolatile impurities, along with gases dissolved from the air. **Crystallization** of ice from ocean saltwater results in relatively pure water in the form of icebergs. **Aeration** of groundwater as it trickles over rock surfaces, as in a rapidly running brook, allows volatile impurities to be released to the air. **Sedimentation** of solid particles occurs in slow-moving streams and lakes. **Filtration** of water through sand rids the water of suspended matter such as silt and algae. Next, and of very great importance, are the **oxidation processes.** Practically all naturally occurring organic materials—plant and animal tissue, as well as their waste materials—are changed through a complicated series of oxidation steps in surface waters to simple substances common to the environment. Finally, another process used by nature is **dilution.** Most, if not all, pollutants found in nature are rendered harmless if reduced below certain levels of concentration by dilution with pure water.

Before the advent of the exploding human population and the industrial revolution, natural purification processes were quite adequate to provide ample water of very high purity in all but desert regions. Nature's purification processes can be thought of as massive but somewhat delicate. In many instances the activities of humans push the natural purification processes beyond their limit, and polluted water accumulates.

A simple example of nature's inability to handle increased pollution comes from dragging gravel from stream beds. This excavation leaves large amounts of suspended matter in the water. For miles downstream from a source of this pollutant, aquatic life is destroyed. Eventually, the solid matter settles, and normal life can be found again in the stream.

A more complex example, and one for which there is not nearly so much reason to hope for the eventual solution by natural purification, is the degradability of organic materials. A **biodegradable** substance is composed of molecules that are broken down to simpler ones in the natural environment by microorganisms. For example, cellulose suspended in water will eventually be converted to carbon dioxide and water. Some organic compounds, notably some of those synthetically produced, are not easily biodegradable; these substances simply stay in the natural waters or are absorbed by life forms and remain intact for long periods of time. An example is DDT.

Even nature's pure rain water is in jeopardy. If the acidic air pollutants, such as the oxides of sulfur, are concentrated enough, the absorbing rain water will become acidic enough to harm life forms and mar metal and stone structures. The government of Canada has complained to Washington because of acid rains arising from the industrial Northeast. Acid rain is discussed in more detail in Chapter 18. In areas in which heavy concentrations of automobile fumes collect, poisonous lead compounds have been found in rain water in concentrations many times higher than the 0.01 ppm generally allowed in drinking water. The concentration of the lead can be correlated with the concentration of exhaust fumes in the air. Fortunately, lead does not long remain in water, since it generally forms insoluble compounds.

Rain water in clean air is very pure.

Volatile: goes easily into the gaseous state.

Pure water:
Chemist: "Pure H_2O—no other substance."
Parent: "Nothing harmful to human beings."
Game and Fish Commission: "Nothing harmful to animals."
Sunday boater: "Pleasing to the eye and nose, no debris."
Ecologist: "Natural mixture containing necessary nutrients."

Only about 1% of groundwater supplies in the United States are now considered unsafe.

Some synthetic compounds are not biodegradable and therefore are very persistent in natural waters.

Acid rain in Pasadena, California (1976–1977), contained:
1. Acid (pH = 4.06)
2. NH_4^+
3. K^+
4. Ca^{2+} — Concentrations in 10^{-5} to 10^{-6} molar range.
5. Mg^{2+}
6. Cl^-
7. NO_3^-
8. SO_4^{2-}

TABLE 17–6   Classes of Water Pollutants, with Some Examples

1. Oxygen-demanding wastes	Plant and animal material
2. Infectious agents	Bacteria and viruses
3. Plant nutrients	Fertilizers, such as nitrates and phosphates
4. Organic chemicals	Pesticides, such as DDT, detergent molecules
5. Other minerals and chemicals	Acids from coal mine drainage, inorganic chemicals such as iron from steel plants
6. Sediment from land erosion	Clay silt on stream bed may reduce or even destroy life forms living at the solid-liquid interface
7. Radioactive substances	Waste products from mining and processing of radioactive material, radioactive isotopes after use
8. Heat from industry	Cooling water used in steam generation of electricity

THE SCOPE OF WATER POLLUTANTS

There was a time when polluted water could be thought of in terms of dissolved minerals, natural silt, and contaminants associated with the natural wastes of animals and humans. As our use of water has increased, the pollution has become more diversified. The U.S. Public Health Service now classifies water pollutants into the eight broad categories listed in Table 17–6.

When natural purification processes cannot cope with materials added to water, pollution results.

BIOCHEMICAL OXYGEN DEMAND

The way in which organic materials are oxidized in the natural purification of water deserves special attention. The process opposes **eutrophication.** Even in the natural state, living organisms found in natural waters are constantly discharging organic debris into the water. To change this organic material into simple inorganic substances (such as CO_2 and H_2O) requires oxygen. The amount of oxygen required to oxidize a given amount of organic material is called the **biochemical oxygen demand (BOD).** The oxygen is required by microorganisms, such as many forms of bacteria, to metabolize the organic matter that constitutes their food. Ultimately, given near normal conditions and enough time, the microorganisms will convert huge quantities of organic matter into the following end products:

A quantitative relationship exists between oxygen needs and organic pollutants to be destroyed. This is BOD.

Organic carbon $\rightarrow CO_2$

Organic hydrogen $\rightarrow H_2O$

Organic oxygen $\rightarrow H_2O$

Organic nitrogen $\rightarrow NO_3^-$

A standardized solution is one of known concentration.

One way to determine the amount of organic pollution is to determine how much oxygen a given sample of polluted water will require for complete oxidation. For example, a known volume of the polluted water is diluted with a known volume of standardized sodium chloride solution of known oxygen content. This mixture is then held at $20°C$ for five days in a closed bottle. At the end of this time the amount of oxygen that has been consumed is taken to be the biochemical oxygen demand.

Highly polluted water often has a high concentration of organic material, with resultant large biochemical oxygen demand (Fig. 17–7). In extreme cases,

Fish cannot live in water that has less than 0.004 g O_2 per liter (4 ppm).

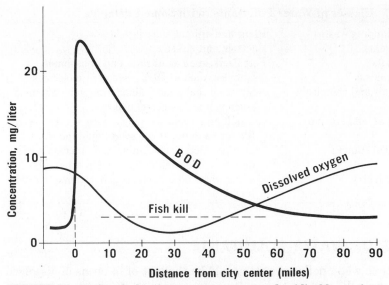

FIGURE 17–7 Graph showing oxygen content and oxidizable nutrients (BOD) as a result of sewage introduced by a city. The results are approximated on the basis of a river flow of 750 gallons per second. Note that it takes 70 miles for the stream to recover from a BOD of 0.023 g oxygen per liter. (From A. Turk, et al.: *Environmental Science,* 2nd ed., Philadelphia, Saunders College Publishing, 1978.)

Temperature °C	Solubility of O_2 g_{O_2}/liter H_2O
0	0.0141
10	0.0109
20	0.0092
25	0.0083
30	0.0077
35	0.0070
40	0.0065

These data are for water in contact with air at 760 mm mercury pressure.

Characteristic BOD Levels	g_{O_2}/liter
Untreated municipal sewage	0.1–0.4
Runoff from barnyards and feed lots	0.1–10
Food processing wastes	0.1–10

more oxygen is required than is available from the environment, and putrefaction results. Fish and other freshwater aquatic life can no longer survive. The aerobic bacteria (those that require oxygen for the decomposition process) die. As a result of the death of these organisms, even more lifeless organic matter results and the BOD soars. Nature, however, has a backup system for such conditions. A whole new set of microorganisms (anaerobic bacteria) takes over; these organisms take oxygen from oxygen-containing compounds to convert organic matter to CO_2 and water. Organic nitrogen is converted to elemental nitrogen by these bacteria. Given enough time, enough oxygen may become available, and aerobic oxidation will then return.

A stream containing 10 parts per million (ppm) by weight (just 0.001%) of an organic material, the formula of which can be represented by $C_6H_{10}O_5$, will contain 0.01 g of this material per liter. The calculation used to obtain this is:

?g = 1 liter of water

$$?g = 1 \text{ liter} \times \frac{1000 \text{ ml}}{1 \text{ liter}} \times \frac{1 \text{ g}}{\text{ml}} = 1000 \text{ g}$$

0.001% of this is the pollutant:

0.001% of 1000 g = (0.00001)(1000 g) = 0.010 g

To transform this pollutant to CO_2 and H_2O, the bacteria present use oxygen as described by the equation:

$$\underset{\substack{\text{RELATIVE} \\ \text{WEIGHT} \\ 162}}{C_6H_{10}O_5} + \underset{\substack{\text{RELATIVE} \\ \text{WEIGHT} \\ 192}}{6\,O_2} \rightarrow 6\,CO_2 + 5\,H_2O$$

The 0.010 g of pollutant requires 0.012 g of dissolved oxygen.

$$\frac{?\text{g oxygen}}{\text{liter}} = \frac{0.010\text{ g pollutant}}{\text{liter}} \times \frac{192\text{ g oxygen}}{162\text{ g pollutant}} = \frac{0.012\text{ g oxygen}}{\text{liter}}$$

At 68°F (20°C), the solubility of oxygen in water under normal atmospheric conditions is 0.0092 g of oxygen per liter.

    Since the BOD (0.012 g per liter) is greater than the equilibrium concentration of dissolved oxygen (0.0092 g per liter), as the bacteria utilize the dissolved oxygen in this stream, the oxygen concentration of the water will soon drop too low to sustain any form of fish life (Fig. 17–8). Life forms can survive in water where the BOD exceeds the dissolved oxygen if the water is flowing vigorously in a shallow stream (this facilitates the absorption of more oxygen from the air via aeration).

    BOD values can be greatly reduced by treating industrial wastes and sewage with oxygen and/or ozone. Numerous commercial cleanup operations now being developed and used employ this type of "burning" of the organic wastes. Another benefit of treating waste water with oxygen is that some of the nonbiodegradable material becomes biodegradable as a result of partial oxidation.

High concentration of organic pollutants
↓
Low oxygen concentration
↓
Dead organisms
↓
Higher concentration of organic pollutants
↓
Lower oxygen concentration
↓
Anaerobic conditions

FIGURE 17–8   Fish kills can be caused by the lack of a substance necessary for life, such as oxygen, or by the presence of toxic materials that interfere with the life processes. A heavy concentration of organic matter in a stream may depress the oxygen concentration below that required to support fish life.

THERMAL POLLUTION

Thermal pollution results when water is used for cooling purposes and in the process has its own temperature raised. Recall that water has a high heat capacity per unit weight and a high heat of vaporization. This combination makes water an ideal cooling fluid for thermal power stations, nuclear energy generators, and industrial plants. If the warm water from these sources is discharged into a lake or river, the temperature of that natural body of water will be raised. This problem is not as serious as it once was since reuse of water by industry reduces the amount of warm water discharged into natural bodies of water. In addition, higher energy costs have led to industrial extraction of the heat energy from the warm water as part of the recycling process.

If the warm water is returned to the lake or river before it has cooled, it can affect the aquatic life.

The solution of oxygen in water is facilitated by:
1. Exposed surface area of water;
2. Low temperature;
3. Low concentration of oxygen in the water.

The oxygen content of water in contact with air is dependent on the temperature of the water, since more oxygen can dissolve in a quantity of cold water than in the same quantity of warm water. Also, the *rate* at which water dissolves oxygen is directly proportional to the difference between the actual concentration of oxygen present and the equilibrium value. This is extremely fortunate, since it means the rate of solution of oxygen increases sharply as the oxygen is consumed. Since a larger surface area will allow quicker absorption of oxygen, the rate of absorption of oxygen from the air is much greater in a shallow, cold mountain stream than in a deep lake behind a dam in a warm river.

Aquatic life is very sensitive to temperature. Lethal temperatures for various species of fish in Wisconsin and Minnesota:

Trout	77°F
White sucker	84–85°F
Walleye	86°F
Yellow perch	84–88°F
Fathead minnow	93°F

By 1950 the increased temperature of the Thames River decreased the oxygen content by 4% over what it otherwise would have been. However, the biochemical results of this factor alone could not be determined, since the river was anaerobic as a result of other pollutants. Even though more is to be learned about thermal pollution, two conclusions seem obvious: (1) thermal pollution aggravates the problem of oxygen supply; and (2) a significant rise in the temperature of a stream can drastically change or even destroy entire biological populations.

Thermal pollution is sometimes beneficial. Alligators are no longer an endangered species because they have thrived in the warm water from the Savannah River reactors. Manatees, still an endangered species, are seen in the warm water near the power station in Fort Myers, Florida.

SELF-TEST 17–A

1. The property of water that accounts for its moderating influence on climate is _____.

2. Which process does nature not use in purifying natural water: distillation, crystallization, sedimentation, chlorination, or aeration? _____

3. BOD stands for _____ _____ _____.

4. _____ causes the unique properties of water.

5. The average person in the United States uses _____ gallons of water per day.

6. The actual amount of drinkable water a person needs is _____ per day.

7. What fraction of the water that you use each day must be pure enough to drink? _____

8. Three common water pollutants are _____ , _____ , and

_____ .

9. Name three ways water is purified by natural processes: _____ ,

_____ , and _____ .

10. A stratum of porous rock that holds water is called a(n) _____ .

11. Approximately what percentage of the human body is water? _____

IMPACT OF HAZARDOUS WASTES ON WATER QUALITY

Industrial wastes can be an especially vexing sort of pollution problem because often they either are not removed or are removed very slowly by naturally occurring purification processes, and are generally not removed at all by a typical municipal water treatment plant. Table 17–7 lists some of the industrial pollutants that result from products important to us.

Disposal of hazardous wastes in landfills has been the principal method of disposal for industries, agriculture, and municipalities for decades. Incidents such as the Love Canal disaster drew attention to the serious contamination of groundwater by hazardous wastes. Action on local, state, and federal levels began in the 1970s to solve problems caused by past disposal and to develop workable methods for future disposal of hazardous wastes. In 1980, Congress established the "Superfund," a $1.6 billion program designed to clean up hazardous waste sites that were threatening to contaminate the nation's underground water supplies. By October 1985, the Environmental Protection Agency (EPA) had placed 850 hazardous waste sites on its National Priorities List for cleanup under the Superfund law. The distribution of these sites is shown in Figure 17–9. In 1985 the Office of Technology Assessment estimated that the number of hazardous waste sites requiring cleanup will increase, perhaps to as high as 10,000, and the cost of cleanup may reach $100 billion.

TABLE 17–7   Important Industrial Products and Consequent Hazardous Wastes

THE PRODUCTS WE USE	THE POTENTIALLY HAZARDOUS WASTE THEY GENERATE
Plastics	Organic chlorine compounds
Pesticides	Organic chlorine compounds, organic phosphate compounds
Medicines	Organic solvents and residues, heavy metals like mercury and zinc, for example
Paints	Heavy metals, pigments, solvents, organic residues
Oil, gasoline, and other petroleum products	Oils, phenols, and other organic compounds, lead, salts, acids, alkalies
Metals	Heavy metals, fluorides, cyanides, acids, and alkaline cleaners, solvents, pigments, abrasives, plating salts, oils, phenols
Leather	Chromium, zinc
Textiles	Heavy metals, dyes, organic chlorine compounds, organic solvents

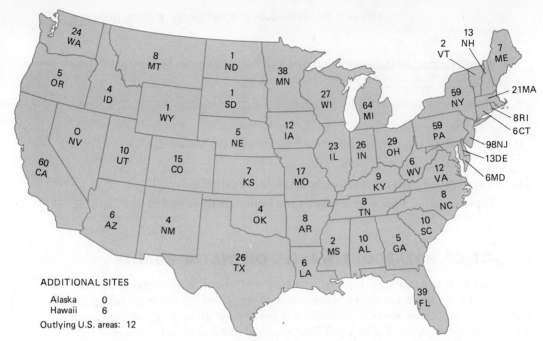

FIGURE 17–9 Distribution of 850 Superfund priority sites as of October 1985.

Although only 1% of the aquifers have been polluted by hazardous wastes, many of these aquifers are near large population centers, so the problem is a serious one. The basic problem with land disposal of hazardous wastes is the contamination of groundwater as it moves through the disposal area (Fig. 17–3). Water pollution from these sites generally occurs as seepage into an underlying aquifer.

The EPA has designated 34 industrial categories that produce polluted wastewater and 65 classes of pollutants (Table 17–8). Under the law it is now illegal for these pollutants to be discharged into the U.S. waterways without a permit.

In 1976 the federal government passed the Resource Conservation and Recovery Act (RCRA). This law is designed to give "cradle-to-grave" (origin to disposal) responsibility to generators of hazardous wastes. The RCRA regulations cover generation, transportation, storage treatment, and disposal of hazardous wastes.

Considerable attention has been given to safe disposal of hazardous wastes, monitoring groundwater near hazardous waste sites and reducing the quantity of hazardous wastes by recycling chemicals. The technology for safe disposal exists, but the costs are high. Data reported by the EPA in 1980 (Table 17–9) indicate that 90% of hazardous wastes were being disposed of by environmentally unsound methods. The effect of the present government regulations and the greater public awareness are making current disposal methods safer, but the cleanup of Superfund sites and other landfills that are contaminating groundwater will take time.

The following are examples of groundwater contamination by seepage from hazardous waste sites.

In 1978, the area around an old chemical dumpsite in Love Canal, a community in southeastern Niagara Falls, New York, was declared a disaster area by President Carter. Record rainfall caused leaching of chemicals from corroding waste-disposal barrels buried in the old chemical dumpsite. Over 230 families were relocated, and

TABLE 17–8   Categories of Industrial Materials Producing Pollutants and Related Toxic Pollutant Classes — EPA

34 INDUSTRIAL CATEGORIES

Adhesives	Plastics processing	Pesticides
Leather tanning and finishing	Porcelain enamel	Pharmaceuticals
Soaps and detergents	Gum and wood chemicals	Plastic and synthetic materials
Aluminum forming	Paint and ink	Rubber
Battery manufacturing	Printing and publishing	Auto and other laundries
Coil coating	Pulp and paper	Mechanical products
Copper forming	Textile mills	Electric and electronic components
Electroplating	Timber	Explosives manufacturing
Foundries	Coal mining	Inorganic chemicals
Iron and steel	Ore mining	
Nonferrous metals	Petroleum refining	
Photographic supplies	Steam electric	
	Organic chemicals	

65 TOXIC POLLUTANT CLASSES

Acenapthene	DDT and metabolites	Nitrobenzene
Acrolein	Dichlorobenzenes	Nitrophenols
Acrylonitrile	Dichlorobenzidine	Nitrosamines
Aldrin/dieldrin	Dichloroethylenes	Pentachlorophenol
Antimony and compounds	2,4-dimethylphenol	Phenol
Arsenic and compounds	Dinitrotoluene	Phthalate esters
Asbestos	Diphenylhydrazine	Polychlorinated biphenyls (PCBs)
Benzene	Endosulfan and metabolites	Polynuclear aromatic hydrocarbons
Benzidine	Endrin and metabolites	
Beryllium and compounds	Ethylbenzene	Selenium and compounds
Cadmium and compounds	Fluoranthene	Silver and compounds
Carbon tetrachloride	Haloethers	2,3,7,8-tetrachlorodibenzo-p-dioxin (TCDD)
Chlordane	Halomethanes	
Chlorinated benzenes	Heptachlor and metabolites	Tetrachloroethylene
Chlorinated ethanes	Hexachlorobutadiene	Thallium and compounds
Chloralkyl ethers	Hexachlorocyclopentadiene	Toluene
Chlorinated phenols	Hexachlorocyclohexane	Toxaphene
Chloroform	Isophorone	Trichloroethylene
2-chlorophenol	Lead and compounds	Vinyl chloride
Chromium and compounds	Mercury and compounds	Zinc and compounds
Copper and compounds	Napthalene	
Cyanides	Nickel and compounds	

TABLE 17–9   Hazardous Waste Disposal Methods in 1980

METHOD	PERCENTAGE OF TOTAL
UNACCEPTABLE	
Unlined surface impoundment	48
Land disposal	30
Uncontrolled incineration	10
Other	2
ACCEPTABLE	
Controlled incineration	6
Secure landfills	2
Recovered	2

the area was fenced off. In 1980, new boundaries that affected an additional 800 families were established. The emergency declarations by President Carter in 1978 and 1980 provided federal funds to assist the state in relocating families. This was the first use of federal emergency funds for something other than a "natural" disaster.

Groundwater supplies in Toone and Teague, Tennessee, were contaminated by organic wastes from a nearby landfill in 1978. The landfill, closed six years earlier, held 350,000 drums, and pesticide wastes were leaking from many of them. The towns must pump water from other locations.

Groundwater in a 30-mile-square area near Denver was contaminated from disposal of pesticide waste between 1943 and 1957 in unlined disposal ponds.

At least 1500 drums containing wastes from a metal-finishing operation were buried near Bryon, Illinois, until 1972. Surface water, soil, and groundwater were contaminated with cyanide, heavy metals, and organic toxic compounds.

About 17,000 waste drums littered a 7-acre site in Kentucky that became known as the "Valley of the Drums" (Fig. 17–10). Many drums have been leaking their contents onto the ground. In 1979, an EPA survey identified about 200 toxic organic chemicals and 30 heavy metals in the soil and in water samples near the dump.

Chemical contamination has forced the closing of more than 600 groundwater wells in the New York City area since 1980.

The Love Canal containment project illustrates the type of action necessary to reduce or eliminate the problem of groundwater contamination from improperly disposed hazardous wastes. A 3-foot-thick clay cap has been placed over the old canal. The clay has been compacted to maximize its resistance to seepage of rain water to the underground wastes. The cap is graded to divert storm water into surface drains. Beneath the surface, the chemical landfill has been surrounded with a 3-foot-wide barrier drain filled with crushed stone and sand and extending 12 to 19 feet below ground level. A pipe at the bottom of this trench carries drain water

FIGURE 17–10
"Valley of the Drums" site in Kentucky. (Courtesy of Environmental Protection Agency.)

FIGURE 17–11 Type of protective gear worn by workers at hazardous waste sites. EPA workers remove a soil sample from the Times Beach, Missouri, site in November 1982. (Courtesy of Environmental Protection Agency.)

from the dump site to a treatment plant that removes the chemicals. This barrier drain isolates the canal from the surrounding environment.

WATER PURIFICATION: CLASSICAL AND MODERN PROCESSES

The outhouses of some rural dwellers had their counterparts in city cesspools. The terrible job of cleaning led to the development of cesspools that could be flush-cleaned with water, followed by a connecting series of such pools that could be flushed from time to time. City sewer systems with no holding of the wastes were the next step.

Since there were not enough pure wells and springs to serve the growing population, water purification techniques were developed. The classical method, which is now termed **primary water treatment,** involved settling and filtration (Fig. 17–12).

Cesspools were an early and crude form of the modern activated sludge process.

Sewage is still 99.9% water!

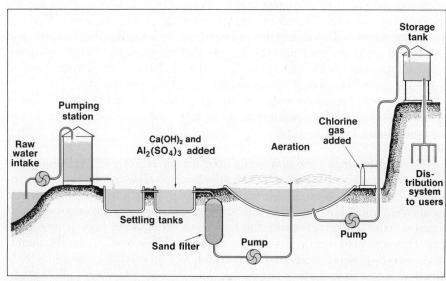

FIGURE 17–12 Primary water purification removes particles that will settle or can be filtered, usually by sandbed filters. Aeration adds oxygen and gets rid of foul gases, and chlorination kills microbes. These are more recent additions. The system outlined here is common in communities throughout the civilized world.

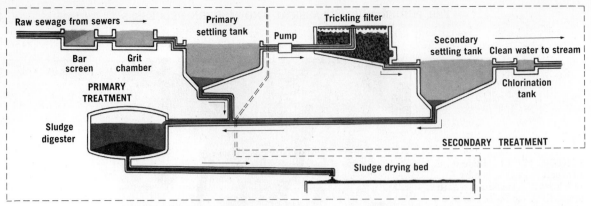

FIGURE 17–13 Sewage plant schematic, showing facilities for primary and secondary treatment. (From *The Living Waters.* U.S. Public Health Service Publication No. 382.)

In the settling stage, calcium hydroxide and aluminum sulfate are added to produce aluminum hydroxide. Aluminum hydroxide is a sticky, gelatinous precipitate that settles out slowly, carrying suspended dirt particles and bacteria with it.

$$3\ Ca(OH)_2 + Al_2(SO_4)_3 \rightarrow 2\ Al(OH)_3 + 3\ CaSO_4$$

If the intake water is polluted enough with biological wastes, the primary treatment, even with chlorination, cannot render the water safe. To be sure, enough chlorine or other oxidizing agents could be added to kill all life forms, but the result would be water loaded with a wide variety of noxious chemicals, especially chlorinated organics, many of which are suspected carcinogens. Some way had to be found to coagulate and separate out the organic material that passed through the primary filters.

Secondary water treatment revives the old cesspool idea under a more controlled set of conditions and acts only on the material that will not settle or cannot be filtered (Fig. 17–13). Modern secondary treatment operates in an oxygen-rich environment (aerobic), whereas the cesspool operates in an oxygen-poor environment (anaerobic). The results are the same: the organic molecules that will not settle are consumed by organisms; the resulting sludge will settle. Bacteria and even protozoa are introduced into the oxygen-rich environment for this purpose. Two techniques, the trickle filter (Fig. 17–14) and the activated sludge method (Fig. 17–15), have been widely used in secondary water treatment.

Primary and secondary water treatment systems will not remove dissolved inorganic materials such as poisonous metal ions or even residual amounts of organic materials. These materials are removed by a variety of **tertiary water treatments.**

Three technologies are now being used for the removal of toxic materials from wastewater; these are carbon adsorption, activated sludge, and steam stripping. Carbon black has been used for many years for adsorbing vapors and solute materials from liquid streams. Many toxic organic materials can be removed with activated or baked carbon granules that have been activated by high-temperature baking. This activated carbon has a high surface area that readily adsorbs chemicals from the wastewater. Activated sludge is a hurry-up version of natural stream purification. Bacteria and other microorganisms degrade the water pollutants in the sludge medium. Steam stripping (Fig. 17–16), involves the removal of volatile organic pollutants from wastewater through steam distillation.

FIGURE 17–14 Picture of a trickle filter with a section removed to show construction details. Rotating pipes discharge the water over a bed of stones. As a result, the organic molecules are "eaten" by microorganisms.

FRESH WATER FROM THE SEA

Since sea water covers 72% of the Earth, it is not surprising that this source would be a major consideration for areas where fresh water supplies aren't sufficient to meet the demand. The oceans contain an average 3.5% dissolved salts by weight, a concentration too high for most uses. The solvent properties of water are illustrated by the average composition of sea water in Table 17–10. If you add these up in terms of the number 0.001 g/kg, you have over 35,000 parts per million (ppm) of dissolved ions. The total must be reduced to below 500 ppm before the water is suitable for human consumption.

The technology has been developed for the conversion of sea water to fresh water. The extent to which this technology is actually put to use depends on the availability of fresh water and the cost of the energy for the conversion. Over 2200 desalination plants were in operation throughout the world in 1981. Although 70%

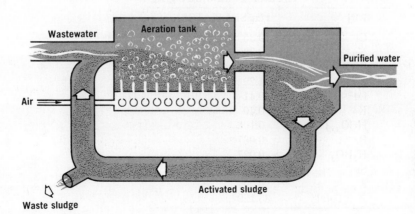

FIGURE 17–15 The activated sludge process provides a closed-system environment in which organic compounds can be consumed by organisms that will settle readily.

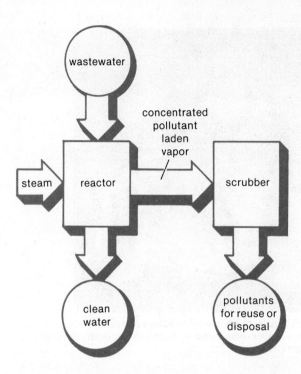

FIGURE 17–16   Steam stripping. Volatile pollutants are extracted from wastewater by steam.

of the capacity is from plants that use **multistage flash distillation** techniques, **reverse osmosis** has gained ground in recent years because of lower energy requirements. The top desalination processes in 1981 are listed in Table 17–11. The number of plants using reverse osmosis is about the same as those using distillation, although the output is only one third as large. Other techniques include **electrodialysis, freezing, ion exchange,** and **solar distillation.**

Multistage Flash Distillation

Water can be separated from dissolved solids by distillation (Fig. 2–4). However, the normal distillation process is not suitable for large-scale separation of salts

TABLE 17–10   Ions Present in Sea Water at Concentrations Greater Than 0.001 g/kg

ION	g/kg SEA WATER
Cl^-	19.35
Na^+	10.76
SO_4^{2-}	2.71
Mg^{2+}	1.29
Ca^{2+}	0.41
K^+	0.40
HCO_3^-, CO_3^{2-}	0.106
Br^-	0.067
$H_2BO_3^-$	0.027
Sr^{2+}	0.008
F^-	0.001
Total	35.129

TABLE 17–11 Top Desalination Processes in 1981

PROCESS	NUMBER OF PLANTS	CAPACITY, mgd*
Distillation	965	1459
Reverse osmosis	929	391
Electrodialysis	310	73

* mgd = million gallons per day

from water since the salts will begin to deposit in the distillation vessel as the distillation proceeds. The salt deposits reduce the efficiency of the distillation. This problem can be avoided by heating the sea water in a series of coiled pipes before releasing the warm sea water into a partially evacuated chamber. A schematic illustration of the multistage flash distillation process is shown in Figure 17–17. This technique capitalizes on the fact that water boils at lower temperature as the pressure is decreased.

When the water enters the chamber, some vaporizes instantly. The water vapor is then condensed by contact with cold pipes carrying unheated sea water. In this way the heat released when the water condenses (540 cal/g) can be used to preheat the sea water. The whole process can be repeated by having a series of chambers, each of which is at a slightly lower pressure than the previous one.

Reverse Osmosis

When a membrane is permeable to water molecules but not to ions or molecules larger than water, it is called a **semipermeable membrane.** If a semipermeable membrane is placed between sea water and pure water, the pure water will pass through the membrane to dilute the sea water. This is **osmosis.** The liquid level on the sea water side rises as more water molecules enter than leave, and pressure is exerted on the membrane until the rates of diffusion of water molecules in both

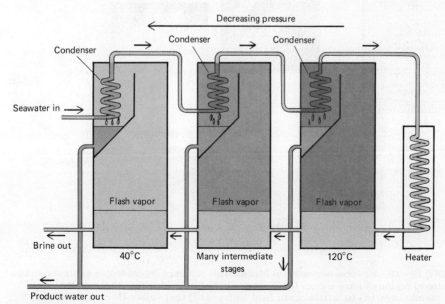

FIGURE 17–17 Schematic diagram of a multistage flash distillation unit. A plant at Key West, Florida, produces 2.6 million gallons per day by this method.

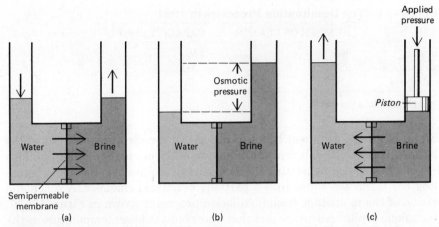

FIGURE 17-18 Normal osmosis is represented by *a* and *b*. Water molecules create osmotic pressure by passing through the semipermeable membrane to dilute the brine solution. Reverse osmosis, represented in *c*, is the application of an external pressure in excess of osmotic pressure to force water molecules to the pure water side.

directions are equal. **Osmotic pressure** is defined as the external pressure required to prevent osmosis. Figure 17-18 illustrates the concept of osmosis and osmotic pressure.

      Reverse osmosis is the application of pressure to cause water to pass through the membrane to the pure water side (Figs. 17-18*c* and 17-19). The osmotic pressure of normal sea water is 24.8 atmospheres. As a result, pressures

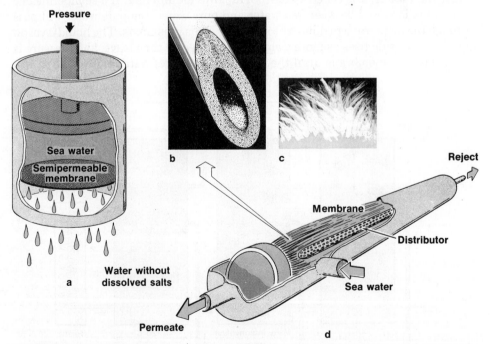

FIGURE 17-19 Reverse osmosis. (*a*) Mechanical pressure forces water against osmotic pressure to region of pure water. (*b*) Enlargement of individual membrane. (*c*) Mass of many membranes. (*d*) Industrial unit; feed water (salt) that passes through membranes collects at the left end (permeate). The more concentrated salt solution flows out to the right as the reject.

greater than 24.8 atmospheres must be applied to cause reverse osmosis. Pressures up to 100 atmospheres are used to provide a reasonable rate of filtration and to account for the increase in salt concentration that occurs as the process proceeds.

The most common semipermeable membrane used in reverse osmosis is a modified cellulose acetate polymer, although several polyamide polymers also have been used. The largest reverse osmosis plant in operation today is the Yuma Desalting Plant in Arizona. This plant, which began operation in 1982, can produce 100 million gallons per day. The plant was built to reduce the salt concentration of irrigation wastewater in the Colorado River from 3200 ppm to 283 ppm. This project is part of a U.S. commitment to supply Mexico with a sufficient quantity of water suitable for irrigation.

Irrigation water of desert fields dissolves about 2 tons of salt per acre per year. Irrigation wastewater carries the salt back to the Colorado River.

Electrodialysis

We have learned that in the electrolysis of salt water, positive ions migrate toward the negative electrode and negative ions move toward the positive electrode. If an electrolysis cell is divided into three compartments by semipermeable membranes, one permeable to positive ions and the other permeable to negative ions (Fig. 17–20), the resulting ionic separation is called **electrodialysis.** Dialysis is the passage of selected species in solution through membranes while other species are excluded. Electrodialysis is the special case in which the passage of ions is influenced by an electrical field.

In Figure 17–20, note that positive ions can move to the left out of the center compartment but cannot move through the negative ion membrane from the right compartment to the center one. In a similar way, negative ions move out of the

In electrodialysis, ions are attracted to oppositely charged electrodes on the other side of a membrane.

Home treatment devices on the market:
1. Disinfection: chlorine, ultraviolet light, ozone, iodine, bromine, silver salts
2. Filtration
3. Adsorption
4. Deionization
5. Reverse osmosis

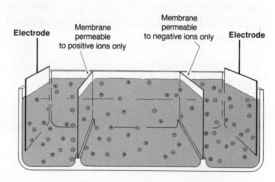

All three compartments filled with brackish water

FIGURE 17–20 The essential features of the electrodialysis process. Each compartment of the cell contains brackish water. Application of an electrical potential across the cell causes the positive ions to move from the center into the left compartment and the negative ions to move into the right compartment. The salt content of the water in the center compartment is thus reduced and raised in the end compartments.

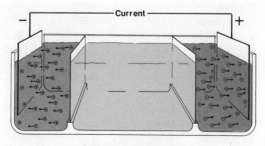

Salts removed from water in center compartment

center compartment to the right. Thus, the ionic concentration of the water in the central compartment is reduced. If the process is continued long enough, the water in the central compartment loses most of its salt content.

Freezing

When cold sea water is sprayed into a vacuum chamber, the evaporation of some of the water cools the remainder, and ice crystals form in the brine. When the crystals of ice form, they tend to exclude the salt ions. Any solid separating from a liquid will tend to take only the molecules or ions that fit into the particular solid pattern. Recall that this generalization is the basis for purification by recrystallization. Even though the separation of salt and water is not complete in one step, the ice has less salt than the same weight of liquid solution. The ice crystals can be collected on a filter, washed with a small amount of fresh water, and then melted to obtain purified water. The process is repeated until the desired degree of purity is achieved. Plants have produced as much as 250,000 gallons of pure water per day by this purification technique.

Solar Distillation

Since heating sea water in the multistage flash distillation process is expensive, solar evaporation units are a possible alternative in areas that receive a lot of sunlight. The main disadvantage of solar units is the amount of land required to produce appreciable amounts of fresh water. Table 17–12 lists the location and size of large solar stills. The output of these units is about 3 liters per square meter per day, or 7000 gallons per day for the larger units. Figure 17–21 shows a basic design used for solar stills. Smaller units can be constructed that will provide enough fresh water for homes.

Ion Exchange

In the process called ion exchange, brackish water or sea water is first passed through an ion exchange resin to replace the metal ions with H^+, and then through

TABLE 17–12   Large Solar Stills

LOCATION	COUNTRY	AREA (METER^2)	YEAR BUILT	FEED TYPE
Gwadar	Pakistan	9072	1972	Sea water
Patmos	Greece	8667	1967	Sea water
Las Salinas	Chile	4757	1872	Sea water
Coober Pedy	Australia	3160	1966	Brackish
Megisti	Greece	2528	1973	Sea water
Kimolos	Greece	2508	1968	Sea water
Klonlon	Greece	2400	1971	Sea water
Fiskardo	Greece	2200	1971	Sea water
Nisiros	Greece	2005	1969	Sea water
Awania	India	1867	1978	Brackish
St. Vincent	West Indies	1710	1967	Brackish
Mahdia	Tunisia	1300	1968	Brackish

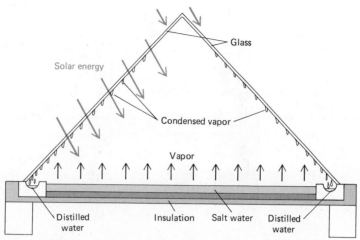

FIGURE 17–21   Principle of solar still. Radiation heats salt water in black trough. Vapor condenses on sloping glass surfaces and runs off into distilled water troughs.

another exchange resin to replace the negative ions with OH^-; the H^+ ions and the OH^- ions then neutralize each other and form water.

Modern ion exchange resins are high molecular weight polymers containing firmly bonded functional groups that can exchange one ion for another as an ionic solution is passed over the giant molecules. **Positive-ion exchange** resins swap their hydrogen ions for the positive ions present in solution. They are usually organic derivatives of sulfuric acid, and their action can be depicted as:

$$\underset{\text{EXCHANGE RESIN}}{\text{Polymer-}SO_3^-H^+} + \underset{\text{BRACKISH WATER}}{Na^+ + Cl^-} \rightarrow \text{Polymer-}SO_3^-Na^+ + H^+ + Cl^-$$

When the resin is saturated with metal ions, the resin can be regenerated by treatment with strong acid, which reverses the above reaction. Because they generate weakly acidic solutions, such exchange resins themselves do not do a complete job. When they are followed by **negative ion exchange resins,** they provde a good route to very pure water (Fig. 17–22). A negative ion exchange resin can replace the negative ions in solution by hydroxide ions. These resins are again high molecular weight polymers, but now the polymer contains a nitrogen atom bonded to the polymer and three other groups. The reaction is:

The chloride ion is shown on both sides of this reaction to indicate that it is not removed.

Ion exchange resins can be regenerated.

$$\underset{\substack{\text{NEGATIVE ION} \\ \text{EXCHANGE RESIN}}}{\text{Polymer}-\overset{|}{\underset{|}{N}}{}^+OH^-} + \underset{\substack{\text{FROM POSITIVE ION} \\ \text{EXCHANGE RESIN}}}{H^+ + Cl^-} \rightarrow \text{Polymer}-\overset{|}{\underset{|}{N}}{}^+Cl^- + \underset{\text{PURE WATER}}{H_2O}$$

When the negative ion exchange resin has exchanged all its hydroxide, the resin can be regenerated by treatment with a strong solution of sodium hydroxide:

$$\text{Polymer}-\overset{|}{\underset{|}{N}}{}^+Cl^- + Na^+ + OH^- \rightarrow \text{Polymer}-\overset{|}{\underset{|}{N}}{}^+OH^- + Na^+ + Cl^-$$

Unfortunately, the amount of sea water that can be purified by a given amount of ion exchange resin is quite small, and the resulting water is relatively costly.

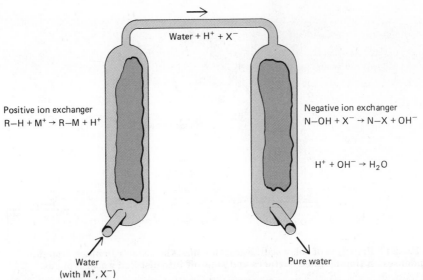

FIGURE 17–22 Ion exchange purification. The tank on the left contains an exchange resin that replaces metal ions in solution with hydrogen ions. The tank on the right contains an exchange resin that replaces nonmetal ions in solution with hydroxide ions. One might well ask why this kind of process cannot be used to turn sea-water into pure water. The answer is that the process can be used, but its cost is much *greater* than that of distillation or reverse osmosis. It is best suited for the removal of small amounts of salts from water that is already quite pure.

SOFTENING OF HARD WATER

Hard water contains metal ions that react with soaps and give precipitates.

The presence of Ca^{2+}, Mg^{2+}, Fe^{3+}, or Mn^{2+} will impart "hardness" to waters. Hardness in water is objectionable because (1) it causes precipitates (scale) to form in boilers and hot water systems, (2) it causes soaps to form insoluble curds (this reaction does not occur with some synthetic detergents), and (3) it can impart a disagreeable taste to the water.

Hardness due to calcium or magnesium, present as their bicarbonates, is produced when water containing carbon dioxide trickles through limestone or dolomite:

$$\underset{\text{LIMESTONE}}{CaCO_3} + CO_2 + H_2O \rightarrow Ca^{2+} + 2\,HCO_3^-$$

$$\underset{\text{DOLOMITE}}{CaCO_3 \cdot MgCO_3} + 2\,CO_2 + 2\,H_2O \rightarrow Ca^{2+} + Mg^{2+} + 4\,HCO_3^-$$

Water softeners that act like ion exchange resins are used to make soft water. They remove the hard water ions, Ca^{2+}, Mg^{2+}, and Fe^{3+}, and put Na^+ ions in the water in exchange.

Such "hard water" can be softened by removing these compounds. The principal methods for softening water are (1) the lime-soda process and (2) ion exchange processes.

The lime-soda process is based on the fact that calcium carbonate ($CaCO_3$) is much less soluble than calcium bicarbonate [$Ca(HCO_3)_2$] and that magnesium hydroxide is much less soluble than magnesium bicarbonate. The raw materials added to the water in this process are hydrated lime [$Ca(OH)_2$] and soda (Na_2CO_3). In the system, several reactions take place, which can be summarized:

$$HCO_3^- + OH^- \rightarrow CO_3^{2-} + H_2O$$
$$Ca^{2+} + CO_3^{2-} \rightarrow CaCO_3\downarrow$$
$$Mg^{2+} + 2\,OH^- \rightarrow Mg(OH)_2\downarrow$$

(The downward arrow denotes a precipitate.)

The overall result of the lime-soda process is to precipitate almost all the calcium and magnesium ions and to leave sodium ions as replacements.

Iron present as Fe^{2+} and manganese present as Mn^{2+} can be removed from water by oxidation with air (aeration) to higher oxidation states. If the pH of the water is 7 or above (either naturally or by adding lime), the insoluble compounds $Fe(OH)_3$ and $MnO_2(H_2O)_x$ are produced and precipitate from solution.

The desire for and achievement of soft water for domestic use has sparked a rather heated health debate during the past two decades. Soft water is usually acidic and contains Na^+ ions in the place of di- and trivalent metal ions. An increased intake of Na^+ is known to be related to heart disease. Also, the acidic soft water is more likely to attack metallic pipes, resulting in the solution of dangerous ions such as Pb^{2+}. One way to avoid sodium ions in drinking water and to use less soap when washing would be to drink naturally hard water and wash in soft water.

Soft water—less than 65 mg of metal ion per gallon

Slightly hard—65–228 mg

Moderately hard—228–455 mg

Hard—455–682 mg

Very hard—above 682 mg

CHLORINATION

With the advent of chlorination of water supplies in the early 1900s, the number of deaths in the United States that were caused by typhoid and other water-borne diseases dropped from 35 per 100,000 population in 1900 to 3 per 100,000 population in 1930.

Chlorine is introduced into water as the gaseous free element (Cl_2), and it acts as a powerful oxidizing agent for the purpose of killing bacteria that remain in water after preliminary purification. The principal water-borne diseases spread by bacteria include cholera, typhoid, paratyphoid, and dysentery.

FIGURE 17–23   This apparatus adds chlorine in sufficient amounts to meet health standards (1 ppm residual) for a 60-million-gallon per day water treatment plant. (Courtesy of Robert L. Lawrence Jr. Filtration Plant, Nashville, Tennessee.)

Most city water supplies are not bacteria-free. Surviving bacteria will usually produce counts numbering in the tens of thousands but only rarely do these surviving bacteria cause disease. The most common bacterial disease borne by water today is giardiasis, a gastrointestinal disorder. Most often this disease comes from surface water but, on occasion, it can be traced to city water systems.

Chlorination of industrial wastes and city water supplies presents a potential threat because of the reaction of chlorine with residual concentrations of organic compounds. Traditional purification methods do not remove chlorinated hydrocarbons, or for that matter, most organics. The chlorinated hydrocarbons, which may be present at levels of a few parts per million or less, include dichloromethane, chloroform, trichloroethylene, and chlorobenzene, all suspected carcinogens.

A number of these chemicals in the same concentration range have been shown to be mutagenic to salmonella bacteria. Studies show an increased risk of 50% to 100% in rectal, colon, and bladder cancers in individuals who drink chlorinated water. According to the EPA, mutagenic or carcinogenic chemicals have been found in 14 major river basins in the United States. It is estimated that more than 500 water systems in the United States exceed EPA's maximum of 0.1 ppm for chlorinated hydrocarbons. The presence of these chlorinated hydrocarbons can be prevented either by using another disinfectant or by removing the low-level organic compounds before chlorination.

Ozone, a disinfectant widely used in Europe, does not produce harmful compounds when it reacts with low-level organic compounds in water. Over 1000 water treatment plants around the world were reported to use ozone in 1977. Table 17–13 indicates that only five of these are in the United States. Ozone is produced on site by passing oxygen or air through an electrical discharge. This normally gives about a 20% ozone/oxygen mixture.

TABLE 17–13 Number of Water Treatment Plants Using Ozone in 1977

COUNTRY	NUMBER OF PLANTS
France	593
Switzerland	150
Germany	136
Austria	42
Canada	23
England	18
The Netherlands	12
Belgium	9
Poland	6
Spain	6
United States	5
Italy	5
Japan	4
Denmark	4
Russia	4
Other countries	22
Total	1039

An efficient process for reducing the level of organic compounds in water is to pass the water through biologically activated carbon. (See p. 448 for a description of the use of activated carbon.)

WHAT ABOUT THE FUTURE?

The good news is that water quality in the United States is improving. A 1985 EPA report on water quality indicates a large improvement over water quality in 1972. The bad news is that additional improvements are necessary. Thirty-seven states reported elevated levels of toxic pollutants in some of their waters, evidenced by elevated levels of toxic substances in fish tissue. Metals were the most frequently reported, followed by pesticides and other organic chemicals. Thirty-five states reported some problems with groundwater contamination from industrial and municipal landfills, underground storage tanks, pesticide applications, septic tanks, and chemical and oil spills. The contaminants included chlorinated solvents, pesticides, gasoline, salts, and radionuclides.

The loss of massive supplies of cheap irrigation water will require changes to more efficient irrigation techniques, such as drip and trickle irrigation, in which water is applied slowly and uniformly to a crop at or below soil level. Industrial reuse is already a major factor in water and energy conservation and will continue to be a part of the long-term solution to the water crisis. Residential conservation such as was practiced in the California drought of 1977 should be evaluated in every community. Communities should consider dual water supply systems for delivery of drinkable water and water for other purposes. Expansion of the capacity for desalination of sea water, particularly for industrial and agricultural uses, will help ease the demand for groundwater. Large cities such as Boston and New York City are considering replacement of leaky plumbing, which accounts for up to one third of their water use. Pumping sludge from sewage plants directly onto fields offers a way to recycle both water and nutrients to the soil.

The discussion in this chapter has focused on water quality in the United States. A combined program of water conservation, protection of water quality, and water recycling will help to alleviate the water crisis in the United States and other industrialized nations. However, contaminated water is still a serious problem for 75% of the world's population. It has been estimated that 80% of the sickness in the world is caused by contaminated water. For years, many countries and international organizations have provided financial and technical aid to help improve the water quality in developing countries. However, much work remains to be done to reduce sickness caused by contaminated water.

SELF-TEST 17-B

1. The four metal ions that are present in sea water at concentrations of 400 ppm or higher are _____, _____, _____, and _____.

2. Heavy metal ions in more than trace concentrations are usually _____ to life forms.

3. Select the ions that may cause water to be hard: sodium, calcium, magnesium, potassium. _____

4. The element _____ is added to water to kill microbes.

5. Electrodialysis involves ion migration in a(n) _____ field.

6. Ion exchange systems for removing salts from water involve _____ (how many) types of resins.

7. Ice formed from impure water will probably be ( ) more pure than, ( ) as pure as, ( ) less pure than the original water. _____

8. Primary water treatment involves _____ and _____ of particles.

9. Tertiary water treatment removes _____ ions and trace amounts of _____ .

MATCHING SET

_____	1.	Sedimentation
_____	2.	Biodegradable
_____	3.	Detergent
_____	4.	BOD
_____	5.	Thermal pollution
_____	6.	Phosphate
_____	7.	Activated sludge process
_____	8.	Reverse osmosis
_____	9.	Water hardness
_____	10.	Ozone

a. A measure of organic material in water
b. Widely used as a detergent builder
c. Results mostly from water used to cool a process
d. Caused by metal ions such as Ca^{2+} and Mg^{2+} in solution
e. Primary purification process
f. Disinfectant used in water treatment plants in Europe
g. Secondary purification process
h. Naturally reducible to simpler compounds
i. Soap substitute
j. Requires high pressure

QUESTIONS

1. If four fifths of the Earth are covered with water, why is there a problem with water supply for humans?

2. Which can dissolve more oxygen to support marine or aquatic life, cold or warm water?

3. Find out what industrial wastes are produced in your community. Are you satisfied with the way these wastes are handled? Explain.

4. Name three ways to obtain fresh water from sea water. Name one way that does not involve a change in state.

5. What are some processes that *decrease* the amount of dissolved oxygen in a stream? What are some processes that *increase* the amount of dissolved oxygen in a stream? Which ones are most readily subject to human control?

6. Explain why each of the following introduces a pollution problem when its wastes are emptied into a stream:
   a. A chlorine-producing plant
   b. A steel mill
   c. An electricity-generating plant burning oil or coal
   d. An agricultural area that is intensively cultivated

7. An old rule of thumb is, "Water purifies itself by running 2 miles from the source of incoming waste." What processes are active in purifying the water? Is this adage foolproof? Explain.

8. Explain how hydrogen bonding causes the unique properties of water.

9. Obtain some distilled water and evaluate its taste. What can you conclude about drinking pure water?

10. Describe the molecular structure of ice.
11. Debate the topic: Since water pollution is a national problem, the federal government should license water districts to supply the water for U.S. citizens.
12. Describe how these processes purify water.
    a. Multistage flash distillation
    b. Electrodialysis
    c. Solar distillation
    d. Freezing
    e. Ion exchange
13. What is the chemical cause of hard water? Describe how hard water can be made soft.
14. What are some ecological consequences (both good and bad) of thermal pollution?
15. From your experience, add one additional example for each of the classes of pollutants listed in Table 17–6.

16. What pertinent facts would you try to gather if it were your responsibility to vote on a bill to regulate water pollution?
17. At what point should pollutants be removed from used water? Who should be responsible for this removal? Would you distinguish between industrial wastes and household wastes?
18. The most abundant elements in organic compounds are carbon, hydrogen, oxygen, and nitrogen. What are the oxidation products for these elements in the decomposition that occurs in nature?
19. Classify water pollutants into as few major groups as you can. Describe some effects of each group and a removal process.
20. In your judgment, what are the most serious pollution problems? Be ready to defend your points in class discussion.
21. What is natural osmosis? Explain the significance of the word *reverse* in reverse osmosis.

Chapter 18

CLEAN AIR — SHOULD IT BE TAKEN FOR GRANTED?

Planet Earth is enveloped by a few vertical miles of chemicals, which compose the gaseous medium in which we exist — the atmosphere. Close to the Earth's surface and near sea level, the atmosphere is mostly nitrogen (80%) and life-sustaining oxygen (20%). It is the few little fractions of a percentage point of other chemicals that make a difference in the quality of life in various spots on Earth. Extra water in the atmosphere can mean a rain forest; a little less water produces a balanced rainfall; and practically no water results in a desert.

An unhealthful, unpleasant medium for the existence of human life was created with the advent of urbanization, through a vast number of transportation vehicles, and with the advent of industrialization, through an unwanted (and for a while ignored) increase in some of the pesky, naturally produced "minor" chemicals in the atmosphere (nitrogen oxides, sulfur dioxide, carbon monoxide, carbon dioxide, and ozone). Early in the 1960s, air pollution became generally recognized as a problem and caused widespread concern, although devastating air pollution was prevalent earlier in certain geographical areas such as London, England, and where volcanic eruptions and burning of large areas occurred.

In some parts of the world, smog and air pollution control devices are a way of life. In most areas of the United States, the "blanket in the sky" is cleaner now than it was five years ago, and the trend seems to be toward even cleaner air. Although improvements have been made in maintaining air quality standards, air pollution is still a problem.

Air pollution knows no political boundaries. Progress in cleaner air is dependent on efforts at the local, regional, national, and international levels (since molecules and weather do not know where the borders are located). Air pollution has caused international controversy and concern over acid rain that is largely produced in the United States and falls in Canada.

Now, let's put air pollution in context, understand it a bit better, and focus on the sources, reactions, and removal of the polluting chemicals.

Prior to 1960 there was little concern about air pollution. Most smoke, carbon monoxide, sulfur dioxide, nitrogen oxides, and organic vapors were emitted into the air with little apparent thought of their harmful nature as long as they were scattered into the atmosphere and away from human smell and sight (Fig. 18–1).

A few decades ago, we operated on the principle "Dilution is the solution to pollution."

FIGURE 18–1 Coppertown Basin (Ducktown), Tennessee, as photographed in 1943. Copper ore (principally copper sulfide, Cu_2S) had been mined and smelted in this area since 1847. In the early years, large quantities of sulfur dioxide, a by-product, were discharged directly into the atmosphere and killed all vegetation for miles around the smelter. Today the sulfur is reclaimed in the exhaust stacks to make sulfuric acid, but the denuded soil remains a monument to the misuse of the atmosphere.

Earth inhabitants acted as though the atmosphere were infinite. But not so (Fig. 18–2). Ninety-nine percent of the estimated 5500 trillion tons of gases that compose the atmosphere is below an altitude of 19 miles. Sufficient oxygen to sustain life extends upward to only about 4 miles above sea level, and most of our weather takes place within an average altitude of just 7 miles. In this limited region, pollutants collect and react; they do not escape Earth and venture into outer space.

By the early 1960s, air pollution achieved notoriety as air pollution disasters began to occur more often. Some of the more devastating disasters were:

October 27–31, 1948, Donora, Pennsylvania, 18 excess deaths (Fig. 18–3)

November 26–December 1, 1948, London, 700 to 800 excess deaths

December 5–9, 1952, London, 4000 excess deaths

January 3–6, 1956, London, 1000 excess deaths

December 5–10, 1962, London, 700 excess deaths

December 7–10, 1962, Osaka, Japan, 60 excess deaths

January 29–February 12, 1963, New York, 200 to 400 excess deaths

February 27–March 10, 1964, New York, 168 excess deaths

Increased pollution of the air was brought on by increased urbanization, industrialization, and transportation via the automobile and airplane. More public and governmental concern about air pollution led to the federal Clean Air Act of 1970 and additional state regulations. As a result, controls were placed on emissions from automobiles, industries, and power companies. The effects of the controls can be seen in Figure 18–4.

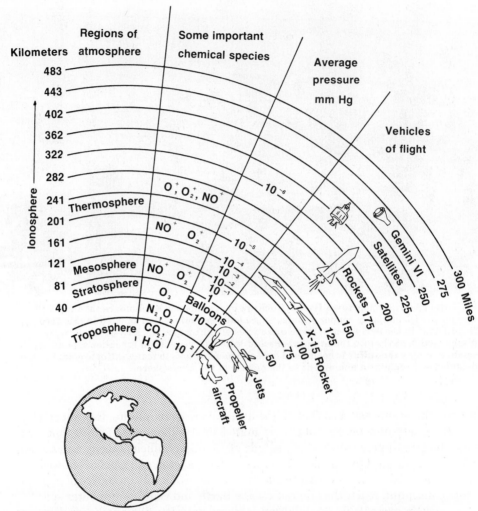

FIGURE 18-2   Some facts about our limited atmosphere. The troposphere was named by British meteorologist Sir Napier Shaw from the Greek word _tropos,_ meaning "turning." The stratosphere was discovered by the French meteorologist Leon Philippe Teisserenc de Bort, who believed that this region consisted of an orderly arrangement of layers with no turbulence or mixing. The word _stratosphere_ comes from the Latin word _stratum,_ meaning "layer."

By 1982, actual air pollutant emissions from the major human-controlled sources were at the levels shown in Table 18-1. Volcanic action, forest fires, dust storms, and even growing plants are natural sources of air pollutants, and sometimes in some places they contribute significantly to the composition of a breath of air.

Even energy crises, which have caused relaxation of some of the air quality standards (current standards shown in Table 18-2), did not reverse the downward trends in the emissions of pollutants.

Although pollutants comprise relatively small proportions of the atmosphere compared with oxygen, nitrogen, and carbon dioxide, still a resident of Los

FIGURE 18–3   Region of Donora, Pennsylvania. In late October 1948, Donora had a five-day siege of extreme air pollution. Before rain cleansed the air, more than 800 domestic animals died and 43% of the population, or 5910 people, became ill. Eighteen died; the normal rate was two deaths every five days. (From A. Turk, et al.: *Environmental Science*, 2nd ed. Philadelphia. Saunders College Publishing, 1978.)

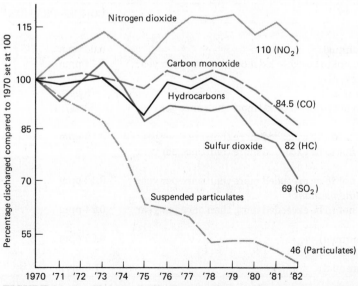

FIGURE 18–4   Changes in discharges of air pollutants since 1970, the birth date of the Clean Air Act. (Courtesy of *Fortune* Magazine. May 4, 1981, updated 1985.)

TABLE 18–1 Estimated Air Pollutant Emissions in the United States in 1982* (Millions of Tons per Year)

	TOTALS	% OF TOTALS	CARBON MONOXIDE	SULFUR OXIDES	HYDRO-CARBONS	NITROGEN OXIDES	PARTIC-ULATES
Transportation	77.1	51	57.9	0.8	6.6	10.5	1.3
Fuel combustion in stationary sources	32.4	21	†	19.3	†	10.0	3.1
Industry	19.6	13	6.1	3.3	8.0	†	2.2
Solid waste disposal and miscellaneous	22.2	15	16.5	0	3.3	1.1	1.3
Totals	151.3	100	80.5	23.4	17.9	21.6	7.9

* The total pollutant emissions decreased 20% from 1979 to 1982.
† Included in solid waste disposal and miscellaneous.

A quadrillion is 10^{15}.

Angeles inhales about 200 quadrillion *pollutant* molecules per breath on a *clear* day in Los Angeles. An average breath would contain:

Carbon monoxide	175 quadrillion molecules
Hydrocarbons	10 quadrillion
Peroxides	5 quadrillion
Nitrogen oxides	4 quadrillion
Lower aldehydes	3.5 quadrillion
Ozone	3 quadrillion
Sulfur dioxide	2.5 quadrillion

along with 1.0×10^{22} nitrogen molecules and 2.6×10^{21} oxygen molecules in a usual breath of half a liter.

TABLE 18–2 Federal Air Quality Standards as of August 1981

	CONCENTRATION*
Sulfur Dioxide	
Arithmetic mean (annual)	0.03 ppm
24-hr concentration not to be exceeded more than once per year	0.14 ppm
Suspended Particulates	
Geometric mean (annual)	75 $\mu g/m^3$
24-hr concentration not to be exceeded more than once per year	260 $\mu g/m^3$
Carbon Monoxide	
8-hr concentration not to be exceeded more than once per year	9 ppm
1-hr concentration not to be exceeded more than once per year	35 ppm
Ozone	
1-hr concentration not to be exceeded more than once per year	0.12 ppm
Hydrocarbons (excluding methane)	
3-hr concentration not to be exceeded more than once per year	0.24 ppm
Nitrogen Dioxide	
Arithmetic mean (annual)	0.05 ppm
Lead	
Arithmetic mean (annual)	1.5 $\mu g/m^3$

* μg is microgram ($1\ \mu g = 10^{-6}$ g); ppm is parts per million (1 ppm is one molecule per million molecules).

On a smoggy day, the pollutants increase by a factor of five or more. The air you are now inhaling could contain pollutant molecules in comparable amounts, give or take a few quadrillion.

The long-range effects of air pollution on materials and the health of plants, animals, and human beings are beginning to emerge. The lung cancer rate in large metropolitan areas is twice as great as the rate in rural areas, even after full allowance is made for differences in cigarette smoking habits. The incidence of the serious pulmonary disease emphysema shot up eightfold during the 1960s. However, only after we comprehend the short- and long-range effects of air pollution can we evaluate wisely its relative importance.

A serious study of the material in this chapter, combined with a study of the chapter on energy (Chapter 8), provides a sound basis for the decisions society must make concerning abundant energy, low pollution, and good health, and its willingness to pay for them.

DO AIR POLLUTANTS SOLO OR AGGREGATE?

Pollutants may exist and react as single, isolated molecules, ions, or atoms. More often, because of the polar nature of pollutants such as SO_2 and NO_2, the pollutants are attracted into water droplets and form **aerosols,** or onto larger particles called **particulates.**

Aerosols range upward from a diameter of 1 nanometer to about 10,000 nanometers and may contain as many as a trillion atoms, ions, or small molecules per particle. They are small enough to remain suspended in the atmosphere for long periods of time. Smoke, dust, clouds, fog, mist, and sprays are typical aerosols. Since they are small, many aerosol particles can exist in a small volume of gas. Because of their vast combined surface area, aerosol particulates have enormous capacities to *adsorb* and concentrate gases on the surfaces of the particles. At other times, liquid aerosols *absorb* air pollutants, thereby concentrating them and providing a water medium in which reactions can occur readily. Thanks to the concentration and reaction effects, aerosols can be more devastating than isolated air pollutant molecules.

Air pollutants tend to attack the site where they first enter the body, that is, the lungs.

Aerosol particles are intermediate in size between small molecules and easily visible small particles.

1 nanometer = 10^{-9} meter.

Adsorption is the attachment of particles to a surface.

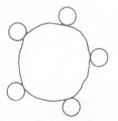

Absorption is pulling particles inside.

1 micron = 10^{-6} meter.

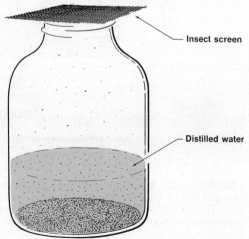

FIGURE 18–5 A dust sample collector for use in the backyard. The open jar containing water is exposed to the air for a known interval. Fifteen days is sufficient exposure time. The water is then evaporated and the residue weighed. Balances sensitive to 0.001 g should be used; the area (cm²) of the opening must be known.

Insect screen

Distilled water

Particulates are generally large enough to be seen as individual particles. They range in size from 1 to 10 microns in diameter. Millions of tons of soot, dust, and smoke particulates are deposited into the atmosphere of the United States each year. Average suspended particulate concentrations in the United States range from 0.00001 g per cubic meter (g/m^3) of air in remote rural areas to about six times that value in urban locations. In heavily polluted areas, concentrations up to 0.002 g/m^3, or 200 times the usual value, have been measured.

Particulates can cause damage in several ways. As small solid particles, particulates may cause damage by abrasive action, fouling and shorting electrical contacts and switches, and by blocking breathing passages.

Some particulates are intrinsically toxic to human beings and animals. The toxic effects of lead and arsenic were described in Chapter 16. Lead components are emitted from automobiles that use leaded gasoline. Arsenic compounds are used as insecticides to dust growing plants. Particulates containing fluorides, commonly emitted from aluminum-producing and fertilizer factories, have caused weakening of bones and loss of mobility in animals that have eaten plants covered with the dust. Asbestos particulates have been shown to be carcinogenic.

Like aerosols, particulates can adsorb and concentrate air pollutants. Sulfur dioxide, nitrogen oxides, hydrocarbons, and carbon monoxide do their greatest damage when concentrated on the surface of particulates or aerosols.

Particulates in the atmosphere can cool the Earth by partially shielding the Earth from the Sun. Large volcanic eruptions such as that from Mt. St. Helens in 1980 have a cooling effect on the Earth.

Particulates and aerosols are removed naturally from the atmosphere by gravitational settling and by rain and snow. They can be prevented from entering the atmosphere by treating industrial emissions by one or more of a variety of physical methods such as filtration, centrifugal separation, spraying, and electro-

Particulate effects depend heavily on the chemical nature of the particle.

Major contributors to the amount of atmospheric particulates are volcanic eruptions by: Krakatoa, Indonesia, 1883; Mt. Katmai, Alaska, 1912; Hekla, Iceland, 1947; Mt. Spurr, Alaska, 1953; Bezymyannaya, U.S.S.R., 1956; Mt. St. Helens, Washington, 1980.

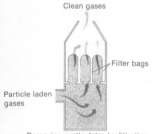

Clean gases

Filter bags

Particle laden gases

Removing particulates by filtration

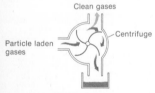

Clean gases

Centrifuge

Particle laden gases

Removing particulates by centrifugal separation

FIGURE 18–6 The General Electric CF6-6D engines that power this McDonnel Douglas DC-10 were designed to eliminate exhaust smoke and reduce by one half the noise level at takeoff. American Airlines advertises that it spent in a 10-year period an amount equal to 43% of its profits on noise and air pollution control systems. (Courtesy of American Airlines.)

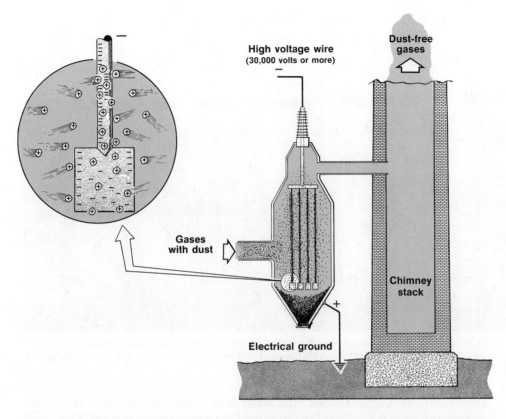

FIGURE 18-7 The Cottrell electrostatic precipitator.

static precipitation. A method often used is electrostatic precipitation, which is better than 98% effective in removing aerosols and dust particulates even smaller than 1 micron from exhaust gases of industrial plants. A diagram of a Cottrell electrostatic precipitator is shown in Figure 18-7. The central wire is connected to a source of direct current at high voltage (about 50,000 volts). As dust or aerosols pass through the strong electrical field, the particles attract ions that have been formed in the field, become strongly charged, and are attracted to the electrodes. The collected solid grows larger and heavier and falls to the bottom, where it is collected.

SMOG — INFAMOUS AIR POLLUTION

The poisonous mixture of smoke, fog, air, and other chemicals was first called **smog** in 1911 by Dr. Harold de Voeux in his report on a London air pollution disaster that caused the deaths of 1150 people. Through the years, smog has been a technological plague in many communities and industrial regions (Figs. 18-8 and 18-9).

Two general kinds of smog have been identified. One is the chemically reducing type that is derived largely from the combustion of coal and oil and contains sulfur dioxide mixed with soot, fly ash, smoke, and partially oxidized organic compounds. This is the **London type,** which is diminishing in intensity and frequency as less coal is burned and more controls are installed. A second type of smog is the chemically oxidizing type, typical of Los Angeles and other cities where exhausts from internal combustion engines are highly concentrated in the

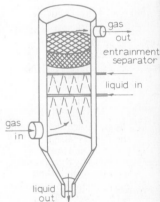

Removing particulates and aerosols by scrubbing. Schematic drawing of a spray collector, or scrubber.

Industrial or London-type smog:
fog + SO_2
Photochemical smog:
fog + NO_x + hydrocarbons.

FIGURE 18–8   Smog over New York City. A heavy haze hangs over Manhattan Island, viewed from the roof of the RCA Building. The Empire State Building is barely visible in the background. (Wide World Photos, Inc.)

atmosphere. This type is called **photochemical** smog because light—in this instance sunlight—is important in initiating the photochemical process. This smog is practically free of sulfur dioxide but contains substantial amounts of nitrogen oxides, ozone, ozonated olefins, and organic peroxide compounds, together with hydrocarbons of varying complexity.

What general conditions are necessary to produce smog? Although the chemical ingredients of smogs often vary, depending on the unique sources of the pollutants, certain geographical and meterological conditions exist in nearly every instance of smog.

There must be a period of windlessness so that pollutants can collect without being dispersed vertically or horizontally. This lack of movement in the ground air can occur when a layer of warm air rests on top of a layer of cooler air. This sets the conditions for a **thermal inversion,** which is an abnormal temperature arrangement for air masses (Fig. 18–10). Normally the warmer air is on the bottom nearer the warm Earth, and this warmer, less dense air rises and transports most of the pollutants to the upper troposphere where they are dispersed. In a thermal inversion the warmer air is on top, and the cooler, more dense air retains its position nearer the Earth. The air becomes stagnated. If the land is bowl shaped (surrounded by mountains, cliffs, or the like), this stagnant air mass can remain in place for quite some time.

When these natural conditions exist, humans supply the pollutants by combustion and evaporation in automobiles, electrical power plants, space heating, and industrial plants. The chief pollutants are sulfur dioxide (from burning coal and some oils), nitrogen oxides, carbon monoxide, and hydrocarbons (chiefly from the automobile). Add to these ingredients the radiation from the Sun, and a massive smog is in the offing.

"Olefin" is another name for an unsaturated hydrocarbon.

Organic peroxides contain the R—O—O—R' structure and are produced by ozone reacting with organic molecules. Hydrogen peroxide is H—O—O—H.

Thermal inversion: mass of warmer air over a mass of cooler air.

FIGURE 18–9 Photochemical smog (a brown haze) enveloping the city of Los Angeles. (Los Angeles Times photo.)

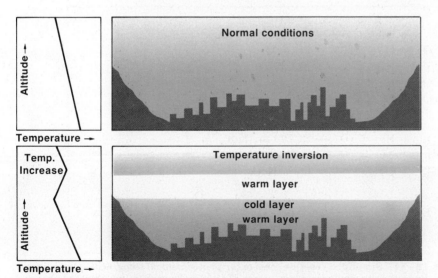

FIGURE 18–10   A diagram of a temperature inversion over a city. Warm air over a polluted air mass effectively acts as a lid, holding the polluted air over the city until the atmospheric conditions change. The line on the left of the diagram indicates the relative air temperature.

Photochemical Smog

Primary pollutants: pollutants emitted into the air.

Secondary pollutants: pollutants formed in the air by chemical reaction.

A city's atmosphere is an enormous mixing bowl of frenzied chemical reactions. Ferreting out the exact chemical reactions that produce smog has been a tedious job, but in 1951, insight into the formation process was gained when smog was first duplicated in the laboratory. Detailed studies have subsequently revealed that the chemical reactions involved in the smog-making process are photochemical and that aerosols serve to keep the reactants together long enough to form secondary pollutants. Ultraviolet radiation from the sun is the energy source for the formation of this **photochemical smog.**

A chemical radical is a species with an unpaired valence electron. They are usually very reactive and short lived.

The exact reaction scheme by which primary pollutants are converted into the secondary pollutants found in smog is still not completely understood, but the reactions shown in Figure 18–11 account for the major secondary pollutants. The process is thought to begin with the absorption of a quantum of light by nitrogen dioxide, which causes its breakdown into nitrogen oxide and atomic oxygen, a chemical radical. The very reactive atomic oxygen reacts with molecular oxygen to form ozone (O_3), which is then consumed by reacting with nitrogen oxide to form the original reactants—nitrogen dioxide and molecular oxygen. Atomic oxygen, however, also reacts with reactive hydrocarbons—olefins and aromatics—to form other chemical radicals. These radicals, in turn, react to form other radicals and secondary pollutants such as aldehydes (e.g., formaldehyde). About 0.2 ppm of nitrogen oxides and 1 ppm of reactive hydrocarbons are sufficient to initiate these reactions. The hydrocarbons involved come mostly from unburned petroleum products like gasoline.

Industrial Smog

The type of smog formed in London and around some industrial and power plants is thought to be caused by sulfur dioxide. Laboratory experiments have shown that

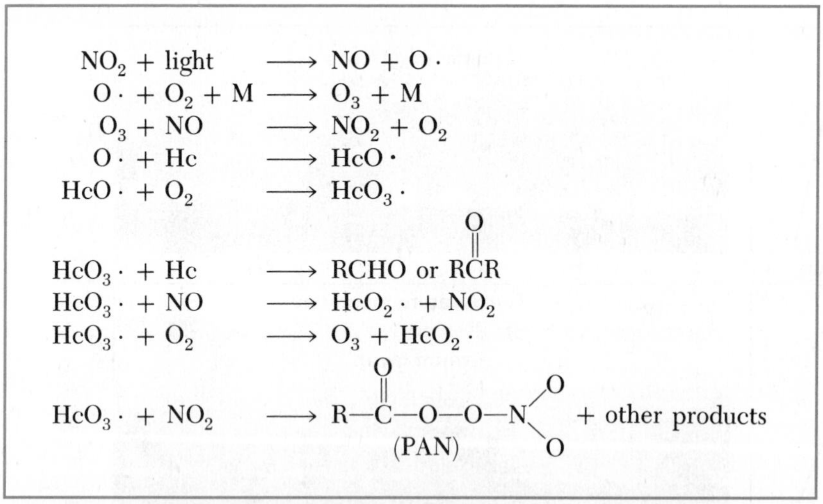
NO2 + light ⟶ NO + O·
O· + O2 + M ⟶ O3 + M
O3 + NO ⟶ NO2 + O2
O· + Hc ⟶ HcO·
HcO· + O2 ⟶ HcO3·
HcO3· + Hc ⟶ RCHO or RCOR
HcO3· + NO ⟶ HcO2· + NO2
HcO3· + O2 ⟶ O3 + HcO2·
HcO3· + NO2 ⟶ R—C(=O)—O—O—NO2 + other products
(PAN)

FIGURE 18–11 Simplified reaction scheme for photochemical smog. Ultraviolet light initiates the process to produce oxygen atoms. Hc is a hydrocarbon (unsaturated or aromatic); M is a third body to absorb the energy released from forming the ozone. Among many possibilities, M could be a N_2 molecule, an O_2 molecule, or a solid particle. A species with a dot, as HcO ·, is a chemical radical. R is a saturated hydrocarbon group; PAN is peroxyacyl nitrate, a very reactive secondary pollutant that causes eyes to water and lungs to hurt.

sulfur dioxide increases aerosol formation, particularly in the presence of mixtures of hydrocarbons, nitrogen oxides, and air energized by sunlight. For example, mixtures of 3 ppm olefin, 1 ppm NO_2, and 0.5 ppm SO_2 at 50% relative humidity form aerosols that have sulfuric acid as a major product. Even with 10 to 20% relative humidity, sulfuric acid is a major product. Sulfuric acid, which is formed in this kind of smog, is very harmful to people suffering from respiratory diseases such as asthma or emphysema. At a concentration of 5 ppm for one hour, SO_2 can cause constriction of bronchial tubes. A level of 10 ppm for 1 hour can cause severe distress. In the 1962 London smog, readings as high as 1.98 ppm of SO_2 were recorded. The sulfur dioxide and sulfuric acid are thought to be the primary causes of deaths in the London smogs.

We shall next examine some of the principal components of air pollution and their chemistry. Emphasis will be placed on how the compounds are produced and how they can be eliminated from our atmosphere.

A MAJOR AIR POLLUTANT — SULFUR DIOXIDE

Sulfur dioxide is produced by burning sulfur or sulfur-containing substances in air:

$$S + O_2 \rightarrow SO_2 \text{ (gas)}$$

Most of the atmospheric SO_2 comes from electrical power plants, smelting plants (which treat sulfide ores), sulfuric acid plants, and burning coal or oil for home heating. Almost 50% of the nation's total SO_2 output is isolated in the seven industrialized states of New York, Pennsylvania, Michigan, Illinois, Indiana, Ohio, and Kentucky.

When coal or petroleum is burned as in electrical power plants, the sources of sulfur are elemental sulfur (S), iron pyrite (FeS_2), and organic compounds such as mercaptans (compounds containing —SH groups). The average sulfur content of all coal mined in the United States is about 2.0%. Much of the petroleum used in the eastern United States is Caribbean residual fuel oil, which has an average sulfur content of 2.6%. If coal and petroleum containing up to 5% sulfur are the fuel for a 1000-megawatt electrical power plant, about 600 tons of SO_2 are produced each day. More than 23 million tons of SO_2 were put into the air in 1982.

Once in the air, what happens to the primary pollutant, SO_2? It can be oxidized to a secondary pollutant, SO_3, in the presence of oxygen, sunlight, and water vapor:

$$2\,SO_2 + O_2 \rightarrow 2\,SO_3$$

When SO_3 dissolves in water, sulfuric acid is formed.

$$SO_3 + H_2O \rightarrow H_2SO_4$$

You will recall that H_2SO_4 is a strong acid.

$$\underset{\substack{\text{STRONG} \\ \text{ACID}}}{H_2SO_4} \rightarrow H^+ + HSO_4^-$$

Sulfurous acid, a weak acid, can be formed in the presence of SO_2 and H_2O.

$$H_2O + SO_2 \rightleftharpoons \underset{\substack{\text{WEAK} \\ \text{ACID}}}{H_2SO_3} \rightleftharpoons H^+ + HSO_3^-$$

Once SO_2 (or as SO_3, H_2SO_3, or H_2SO_4) is on Earth, then what?

Relative humidity is a measure of the amount of water vapor air contains compared to the maximum amount it can contain.

ppm = number of molecules of pollutant per million molecules of air.

To change from percent to ppm, multiply by 10,000.

1 ppm is the same as 1 inch in 16 miles, 1 minute in 2 years, 1 cent in $10,000.

UNPAIRED VALENCE ELECTRONS

$:N::\overset{..}{O}\quad\overset{..}{\underset{..}{O}}:$

The nation's power plants accounted for 80% of the atmospheric SO_2 in 1976.

CH_3—SH
METHYL MERCAPTAN

A large modern power station (e.g., a station of 2000 megawatts capacity) annually produces about the same amount of SO_2 as an industrial city of a million inhabitants.

Sulfurous acid, H_2SO_3, is a weak acid; sulfuric acid, H_2SO_4, is a strong acid. Both can make the water in streams and lakes too acidic for fish.

When SO_2 is dissolved in rivers, lakes, and streams, the acidity can increase considerably. If the pH varies much, aquatic life suffers. Salmon, for example, cannot survive if the pH is as low as 5.5. The lower limit of tolerance for most organisms is a pH of 4.0. Several years ago, certain sections of the Netherlands had precipitation with a pH less than 4.

Sulfur dioxide and its attendant forms can damage vegetation, affect breathing, corrode metals, and decay building stones, in particular marble and limestone. Both marble and limestone are forms of calcium carbonate ($CaCO_3$), which reacts readily with acid (H^+) and with SO_2 and H_2O.

Sulfur dioxide in the air is harmful to people, animals, plants, and buildings.

$$CaCO_3 + 2\,H^+ \rightarrow Ca^{2+} + H_2O + CO_2$$
$$CaCO_3 + SO_2 + 2\,H_2O \rightarrow \underset{\text{(soluble)}}{CaSO_3 \cdot 2\,H_2O} + CO_2$$

An alarming example is the disintegration of marble statues and buildings on the Acropolis in Athens, Greece. As all coatings have failed to protect the marble adequately, the only known solution is to bring the prized objects into air-conditioned museums protected from SO_2 and other corroding chemicals.

How can we have less SO_2 in the usable air and still maintain industrial productivity? Three methods are now in use to accomplish the goal: use low-sulfur fuels, trap the SO_2 before it gets to the atmosphere, and (or) dilute the SO_2 high in the sky.

If low-sulfur fuels are used, there should be less SO_2 formed. This obvious conclusion was confirmed by a six-year study in Bayonne, New Jersey. The sulfur in the fuel was decreased from 1% to an average of 0.25% over the six-year span. The fourfold reduction of sulfur in the fuel correlated well with a fourfold reduction in the average SO_2 concentration in the air (200 micrograms per cubic meter to 50 micrograms per cubic meter).

What, then, is the problem? Why not use only low-sulfur fuels?

TABLE 18–3   Physiological and Corrosive Effects of SO_2

SO_2 EXPOSURE (ppm)	DURATION	EFFECT	COMMENT
0.03–0.12	Annual average	Corrosion	Moist temperate climate with particulate pollution
0.3	8 hr	Vegetation damage (bleached spots, suppression of growth, leaf drop, and low yield)	Laboratory experiment; other environmental factors optimal. Field studies are consistent but dose is difficult to estimate.
0.47	<1 hr	Odor threshold (50% of subjects detect)	May be higher for many persons or when other methods are used
0.2	Daily average	Respiratory symptoms	Community exposure exceeding 0.2 ppm more than 3% of the time
>0.05	Long-term average	Respiratory symptoms	With particulates $>100\ \mu g/m^3$
0.2	Daily average	Respiratory symptoms	With particulates
0.9	Hourly average	Respiratory symptoms	With particulates
>0.05	Monthly average	Respiratory symptoms, including impairment of lung function in children	With particulates

Most low-sulfur coals are mined far from the major metropolitan areas where they are most needed for power generation. The cleansing of sulfur from closer, high-sulfur coal is costly and incomplete. One method is to pulverize the coal to the consistency of talcum powder and remove the pyrite (FeS_2) by magnetic separation. Technology is available to decrease the sulfur content of fuel oil to 0.5%, but this process, too, is costly. It involves the formation of hydrogen sulfide (H_2S) by bubbling hydrogen through the oil in the presence of metallic catalysts, such as a platinum-palladium catalyst.

Several efficient methods are available to trap SO_2. In one method, limestone is heated to produce lime. The lime reacts with SO_2 to form calcium sulfite, a solid particulate, which can be removed from an exhaust stack by an electrostatic precipitator.

$$\underset{\text{LIMESTONE}}{CaCO_3} \xrightarrow{\text{heat}} \underset{\text{LIME}}{CaO} + CO_2$$

$$CaO + SO_2 \rightarrow \underset{\text{CALCIUM SULFITE}}{CaSO_3} \text{ (solid)}$$

Another trapping method involves the passage of SO_2 through molten sodium carbonate. Solid sodium sulfite is formed.

$$SO_2 + \underset{\text{SODIUM CARBONATE}}{Na_2CO_3} \xrightarrow{800°C} \underset{\text{SODIUM SULFITE}}{Na_2SO_3} + CO_2$$

The less desirable method of dissipating SO_2 is by tall stacks. Although tall stacks emit SO_2 into the upper atmosphere away from the immediate vicinity and give SO_2 a chance to dilute itself on the way down, the fact remains that SO_2 will come down, and the longer it stays up the greater chance it has to become sulfuric acid. A nickel smelter in Sudbury, Ontario, Canada, has a 1250-foot smokestack that emits about 2500 tons of SO_2 a day (Fig. 18–12). A 10-year study in Great Britain showed that although SO_2 emissions from power plants increased by 35%, the construction of tall stacks decreased the ground level concentrations of SO_2 by as much as 30%. The question is, who got the SO_2? In this case, Britain's solution was others' pollution. In the United States, the EPA may have added to a pollution problem unwittingly with rules in 1970 that caused plants to increase the height of smokestacks and caused pollutants to be carried longer distances by winds. There are about 179 stacks in the United States that are 500 feet or higher, and 20 stacks are 1000 or more feet tall.

MAJOR AIR POLLUTANTS — NITROGEN OXIDES

There are eight known oxides of nitrogen, two of which are recognized as important components of the atmosphere: dinitrogen oxide (N_2O) and nitrogen dioxide (NO_2).

Most of the nitrogen oxides emitted are in the form of NO, a colorless reactive gas. In a combustion process involving air, some of the atmospheric nitrogen reacts with oxygen to produce NO:

$$N_2 + O_2 + \text{heat} \rightarrow 2 NO$$

Nitrogen oxide is formed in this manner during electrical storms. Since the formation of nitrogen oxide requires heat, it follows that a higher combustion temperature would produce relatively more NO. This will be an important point to consider

About 97% of the nitrogen oxides in the atmosphere are naturally produced and only 3% result from human activity.

FIGURE 18–12
World's largest chimney (as of 1972) standing 1250 feet high (as tall as the Empire State Building) was built at a cost of $5.5 million for the Copper Cliff smelter in the Sudbury District of Ontario, Canada.

later in Chapter 23 (automobile), since one way to achieve greater burning efficiency of fuels in automobile engines is to operate them at higher temperatures.

In the atmosphere NO reacts rapidly with atmospheric oxygen to produce NO_2:

$$2\,NO + O_2 \rightarrow \underset{\substack{\text{NITROGEN}\\\text{DIOXIDE}}}{2\,NO_2}$$

Normally the atmospheric concentration of NO_2 is a few parts per billion (ppb) or less.

Fixed nitrogen (nitrogen oxides are one type) is necessary to perpetuate nature's cycle (Fig. 18–13). In this respect, nitrogen oxides are useful. However, too large a quantity of nitrogen oxides in the air can lead to photochemical smog and bronchial problems for those who breathe this air. In these respects and many others, nitrogen oxides are harmful.

Fixed nitrogen is nitrogen chemically bonded to another element.

If NO_2 does not react photochemically, it can react with water vapor in the air to form nitric and nitrous acids:

$$2\,NO_2 + H_2O \rightarrow \underset{\substack{\text{NITRIC}\\\text{ACID}}}{HNO_3} + \underset{\substack{\text{NITROUS}\\\text{ACID}}}{HNO_2}$$

In addition, nitrogen dioxide and oxygen yield nitric acid:

$$4\,NO_2 + 2\,H_2O + O_2 \rightarrow 4\,HNO_3$$

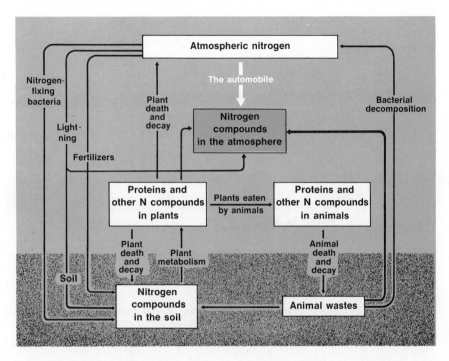

FIGURE 18–13   The nitrogen cycle.

These acids in turn can react with ammonia or metallic particles in the atmosphere to produce nitrate or nitrite salts. For example,

$$\underset{\text{AMMONIA}}{NH_3} + HNO_3 \rightarrow \underset{\substack{\text{AMMONIUM NITRATE}\\\text{(A SALT)}}}{NH_4NO_3}$$

Nitrates are important components of fertilizers.

The acids or the salts, or both, ultimately form aerosols, which eventually settle from the air or dissolve in raindrops. Nitrogen dioxide, then, is a primary cause of haze in urban or industrial atmospheres because of its participation in the process of aerosol formation. Normally nitrogen dioxide has a lifetime of about three days in the atmosphere.

At present, the emission of nitrogen oxides by human activities is significant in some urban areas but globally is minor compared with that emitted by natural processes such as lightning fixation of nitrogen and oxygen and the natural decay of plants and animals (Fig. 18–13).

In laboratory studies, nitrogen dioxide in concentrations of 25 to 250 ppm inhibits plant growth and causes defoliation. The growth of tomato and bean seedlings is inhibited by 0.3 to 0.5 ppm NO_2 applied continuously for 10 to 20 days.

In 1979, haze near Abbeville, Louisiana, had two times more ozone and particulates and four times more $(NH_4)_2SO_4$ than normal.

In a concentration of 3 ppm for 1 hour, nitrogen dioxide causes bronchio-constriction in humans, and short exposures at high levels (150–220 ppm) produce changes in the lungs that produce fatal results. A seemingly harmless exposure one day can cause death a few days later.

ACID RAIN

Acids formed from sulfur oxides (sulfuric, H_2SO_4, and sulfurous, H_2SO_3) and acids formed from nitrogen oxides (nitric, HNO_3, and nitrous, HNO_2) can be leached from the air by rain or snow and produce precipitation with a pH below 7 (Fig.

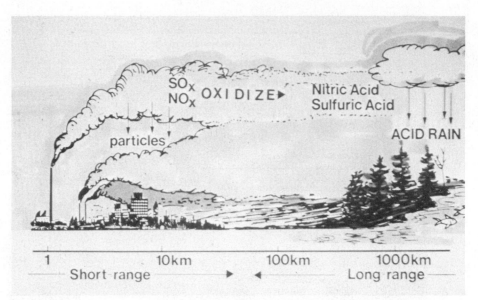

FIGURE 18–14   Major sources and components of acid rain. (From J. Turk and A. Turk: *Environmental Science* 3rd ed. Philadelphia, Saunders College Publishing, 1984.)

Tall stacks contribute to the formation of acid rain.

18–14). The average annual pH of precipitation in much of northeastern Europe and large areas of northeastern United States is between 4 and 4.5. Specific storms have dumped pH 2.7 precipitation on Kane, Pennsylvania, and pH 1.5 precipitation on Wheeling, West Virginia. "Clean" rain is slightly acidic because of dissolved carbon dioxide, which forms carbonic acid in water and produces a pH of 5.6. Any precipitation with a pH *below* 5.6 is considered *acid rain*. For a review of the pH scale and a view of where acid rain fits into the scale, see Figure 18–15.

Acid rain destroys lakes.

The extent of the problems with acid rain can be seen in dead (fishless) ponds and lakes, dying or dead forests, and crumbling buildings. Because of wind patterns, Norway and Sweden have received the brunt of western Europe's emission of sulfur oxides and nitrogen oxides as acid rain. As a result, of the 100,000 lakes in Sweden, 4000 have become fishless, and 14,000 other lakes have been acidified to some degree. In the United States, 6% of all ponds and lakes in the Adirondack Mountains of New York are now fishless, and 200 lakes in Michigan are dead. For the most part, these "dead" lakes are still picturesque, but no fish can live in the acidified water. Lake trout and yellow perch die at pH below 5.0, and smallmouth bass die at pH below 6.0. Mussels die at pH below 6.5.

Acid rain kills forests.

Trees have been affected by acid rain along the eastern coast of the United States and especially in West Germany's Black Forest. As the second highest recipient of SO_2 in Europe, West Germany receives about 2560 tons of sulfur dioxide per month in its precipitation. (Czechoslovakia is highest, with 2870 tons of SO_2 received as fallout per month.) Fifty percent of the trees in the Black Forest show damage such as yellowing of needle tips on conifers (fir, pine, spruce), premature defoliation on deciduous trees, and slowing of growth rate and dieback on all types of trees. Trees at higher elevations are most vulnerable. At the rate going in 1985, the fir and spruce in the Black Forest will all be dead before the end of this century.

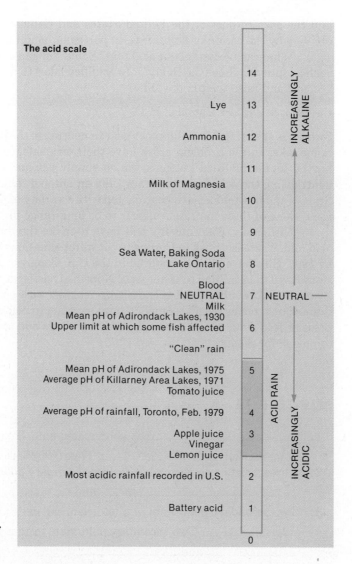

The acid scale

	14	INCREASINGLY ALKALINE
Lye	13	
Ammonia	12	
	11	
Milk of Magnesia	10	
	9	
Sea Water, Baking Soda Lake Ontario	8	
Blood NEUTRAL Milk	7	NEUTRAL
Mean pH of Adirondack Lakes, 1930 Upper limit at which some fish affected	6	
"Clean" rain		
Mean pH of Adirondack Lakes, 1975 Average pH of Killarney Area Lakes, 1971 Tomato juice	5	ACID RAIN
Average pH of rainfall, Toronto, Feb. 1979	4	INCREASINGLY ACIDIC
Apple juice Vinegar Lemon juice	3	
Most acidic rainfall recorded in U.S.	2	
Battery acid	1	
	0	

FIGURE 18–15 The pH of acid rain is compared with the pH of other mixtures.

Acid rain damages trees in several ways. It disturbs the stomata (openings) in tree leaves and causes increased transpiration and a water deficit in the tree. Acid rain can acidify the soil, damaging fine root hairs, and thus diminish nutrient and water uptake. The acid can leach out needed minerals in the soil, and the minerals are carried off in the groundwater. The surface structures of the bark and the leaves can be destroyed by the acid in the rain.

The effects of acid rain and other pollution on stone and metal structures are more subtle. These effects are especially devastating because of their irreversibility. By damaging stone buildings in Europe, acid rain is slowly but surely dissolving the continent's historic heritage. The bas-reliefs on the Cologne (West Germany) cathedral are barely recognizable. The London Tower, St. Paul's Cathedral, and the Lincoln Cathedral in London (England) have suffered the same fate. Other beautifully carved statues and bas-reliefs on buildings throughout Europe

Acid rain crumbles buildings.

and the eastern part of the United States and Canada are slowly passing into oblivion by the action of pollutants, in particular acid rain.

What can be done about acid rain? Some stopgap measures are being taken, such as spraying lime ($Ca(OH)_2$) into acidified lakes to neutralize at least some of the acid and raise the pH toward 7.

$$Ca(OH)_2 + 2\,H^+ \rightarrow Ca^{2+} + 2\,H_2O$$

Sweden is spending $40 million per year to neutralize the acid in some of its lakes. Some lakes in the problem areas have their own safeguard against acid rain by having limestone-lined bottoms, which supply calcium carbonate ($CaCO_3$) for neutralizing the acid from acid rain (like an antacid tablet relieves indigestion). Statues and bas-reliefs have been coated with a variety of plastics and other materials. None of these materials appear to be long-range protectors.

Ultimately, governments will have to make the difference in diminishing acid rain. There is a push in Europe to cut sulfur emissions 30% from the 1980 base by 1993. Bills have been introduced in the U.S. Congress to reduce the SO_2 emissions by 10 to 12 million tons per year. None had passed at the time of this writing. Even if no more sulfur dioxide and nitrogen oxides were emitted, lakes not fed and drained by streams would require 30 or more years to recover naturally from their present dead state. Some ornate, historic buildings and statues have deteriorated too much for repair.

SELF-TEST 18–A

1. _____ is abbreviated ppm. To change from percent to ppm, multiply by _____. Thus 0.092% is _____ ppm.

2. Three substances normally considered primary pollutants are _____, _____, and _____.

3. Two secondary pollutants in photochemical smog are _____ and _____. Two secondary pollutants in industrial smog are _____ and _____.

4. Particles that can remain suspended in air for long periods of time and are intermediate in size between individual molecules and particulates are called _____.

5. Much of the effect of aerosols is due to their large _____.

6. A device that imparts changes to dust and aerosol particles so they can be attracted out of a gaseous effluent stream is known as a(n) _____.

7. The initial step in the formation of photochemical smog occurs when ultraviolet light decomposes a molecule of _____ _____.

8. During a thermal inversion, a ( ) warm or ( ) cool mass of air is above a mass of ( ) warm or ( ) cool air.

9. For initiation, industrial smog requires the substance _____, and photochemical smog requires the substance _____.

10. When sulfur is burned in air, the major product is _____.

11. The major source of sulfur dioxide is the burning of _____.

12. How does the production of electrical power rank as a producer of sulfur dioxide? _____

13. Fixed nitrogen is nitrogen combined with _____ .

14. The reaction of nitrogen with oxygen is ( ) endothermic or ( ) exothermic.

15. In all combustion processes in air, some nitrogen _____ are formed.

16. The artificial emission of nitrogen oxides has greatly disrupted the nitrogen cycle. ( ) True or ( ) False

17. Nitrogen oxides remain in the atmosphere for an indefinite length of time. ( ) True or ( ) False

18. Acid rain is formed by dissolving _____ and(or) _____ in rain water.

A MAJOR AIR POLLUTANT — CARBON MONOXIDE

Carbon monoxide (CO) is the most abundant and widely distributed air pollutant found in the atmosphere. It is produced in combustion processes when carbon or some carbon-containing compound is burned in an insufficient amount of oxygen:

$$2\,C + \underset{\substack{\text{LIMITED} \\ \text{SUPPLY}}}{O_2} \rightarrow 2\,CO$$

The toxic effects of carbon monoxide are discussed in Chapter 16.

For every 1000 gallons of gasoline burned, 2300 pounds of CO are emitted if no emission controls are placed on the engine.

The highest concentration of CO is around heavy traffic areas where levels of 50 ppm or more are encountered, although 7 ppm or less is normal. In the countryside, the CO level is close to the global level of about 0.1 ppm. A bit of a mystery is that the low global level does not seem to be changing in spite of huge amounts of CO being dumped into the atmosphere (about 10^{14} g per year in the United States, or about 1000 pounds per person per year). Although polar CO will dissolve readily in water droplets, and CO will react slowly with oxygen to form CO_2, the bulk of its disappearing act is probably accounted for in the natural carbon cycle.

At least ten times more CO enters the atmosphere from natural sources than from all industrial and automotive sources combined. Of the about 3.8 billion tons of CO emitted per year, about 0.3 billion come from human sources and about 3 billion from the oxidation of methane produced by decaying organic matter.

MAJOR AIR POLLUTANTS — CERTAIN HYDROCARBONS

As we have seen in Chapter 12, hydrocarbons come in all shapes and sizes, beginning with methane, CH_4, and continuing to molecules containing hundreds of carbon atoms. Some have all single bonds, some have double bonds, and a few have triple bonds. Some are aromatic hydrocarbons, and some are nonaromatic. Literally hundreds of these hydrocarbons and their oxygen, sulfur, nitrogen, and halogen derivatives find their way into the atmosphere.

Trees and other plants silently release turpentine, pine oil, and thousands of other hydrocarbons into the air. We can smell some of them. Bacterial decomposition of organic matter emits very large amounts of marsh gas, principally methane. We contribute our share of 15% (of the total global emissions; a greater quantity in urban areas) through incomplete incineration, leakage of industrial solvents, un-

A commercial synthetic fuel plant, if not controlled, could emit as much as 10,000 kg of polynuclear aromatic hydrocarbons per day.

Benzo(α)pyrene, a carcinogenic polynuclear aromatic hydrocarbon found in smoke.

In situ air pollution: cigarette smoking.

TOLUENE

Alkyl: hydrocarbon group such as ethyl, $-C_2H_5$, and octyl, $-C_8H_{17}$.

burned fuel from the automobile, evaporation of gasoline from tanks, incomplete combustion of coal and wood, and petroleum processing, transfer, and use. An estimated 163,000 pounds of polynuclear aromatic hydrocarbons pass from the air into Lake Superior each year.

In Chapter 16 we saw that polynuclear aromatic hydrocarbons such as benzo(α)pyrene are capable of causing cancer in mice and in humans. In the late 1950s, the U.S. Public Health Service, Division of Air Pollution, surveyed 103 urban and 28 nonurban areas of the United States and found that the air in all of the 103 urban areas contained benzo(α)pyrene (BaP). Concentrations ranged from 0.11 to 61 micrograms per 1000 cubic meters of air, with the average concentration being 6.6 micrograms. In 1967, the estimated annual emission of BaP in the United States was 422 tons from burning coal, oil, and gas, 20 tons from refuse burning, 19 tons from industries (petroleum catalytic cracking, asphalt road mix, and the like), and 21 tons from motor vehicles. British researchers report that lung cancer in nonsmokers closely parallels the ten-times-greater amount of BaP in city air than in rural air; there is nine times more lung cancer in cities than in rural areas. A resident of a large town may inhale 0.20 g BaP a year. If he or she is a heavy smoker (two packs a day without filters), add another 0.15 g for a total of 0.35 g. This is about 40,000 times the amount of BaP necessary to produce cancer in a mouse. Coal smoke contains about 300 ppm BaP. Every million tons of coal burned in England in 1958 produced smoke laden with 750 tons of BaP. Many authorities attribute England's high lung cancer rate today to this enormous production of BaP.

Other polynuclear aromatics have also shown carcinogenic activity. Particulates from the atmosphere around Los Angeles, London, Newcastle, Liverpool, and eight other urban sites were extracted with organic solvents. The extracts produced cancer in mice.

In the section on smog, we discussed the role of unsaturated and aromatic hydrocarbons in photochemical smog formation. In a study made in Los Angeles in 1970, an average of 0.106 ppm (maximum of 0.33 ppm) aromatics was found in that city's atmosphere. (About 38% of the total was toluene and 40% the more reactive dialkyl- and trialkylbenzenes.) These compounds are about as reactive as propylene and higher-molecular-weight unsaturated hydrocarbons in causing smog formation. The automobile is responsible for emitting most of these hydrocarbons to the atmosphere. In fact, the automobile without pollution control emits more than 200 different hydrocarbons and hydrocarbon derivatives.

OZONE—A SECONDARY POLLUTANT AND A SUNSCREEN

Ozone is a pungent-smelling gas that can be detected by the human nose at concentrations as low as 0.02 ppm. Sparking electrical appliances, lightning, and even silent electrical discharges convert oxygen into ozone, which is a more reactive form of oxygen.

$$\text{Energy} + 3\,O_2 \rightarrow 2\,\underset{\text{OZONE}}{O_3}$$

Pure oxygen can be breathed for weeks by humans and animals without apparent injurious effects. Several studies have shown that concentrations of 0.3 to 1.0 ppm ozone, well within the recorded range of photochemical oxidant levels, after 15 minutes to 2 hours cause marked respiratory irritation accompanied by choking, coughing, and severe fatigue. For these reasons, outdoor recreation classes in Los Angeles public schools are cancelled on days when the ozone level reaches 0.35 ppm. Ozone at these levels for 1 hour depresses the body temperature, perhaps by an impairment of the brain center that regulates body temperature or by opening the pores of the skin. These levels (0.2–0.5 ppm) cause a considerable decrease in night vision in addition to other effects on vision.

Ozone attacks mercury and silver, which are not affected by molecular oxygen at room temperature. A typical reaction might be

$$6 \, Ag + O_3 \rightarrow 3 \, Ag_2O$$

but

$$Ag + O_2 \rightarrow No \ reaction$$

Even with all the electrical sparks from lightning, electrical motors, and such, very little ozone is emitted into the air. It decomposes into molecular oxygen or reacts with other molecules too quickly to leave its source.

Ozone is found in the lower troposphere only as a secondary pollutant; that is, it is formed from other substances, as in photochemical smog (Fig. 18–11). When sunlight impinges on automobile exhaust fumes, a considerable amount of ozone is produced. The stratosphere contains about 10 ppm of ozone in a layer that has the important function of filtering out some of the Sun's ultraviolet light and providing an effective shield against radiation damage to living things.

Atmosphere: troposphere — sea level to about 7 miles up (N_2, O_2, H_2O, CO_2); stratosphere — 7 to about 50 miles up (N_2, O_2, ozone).

HALOGENATED HYDROCARBONS AND THE OZONE LAYER

Most pollutants are adsorbed on surfaces or react with other pollutants in the troposphere and eventually wash out in the rain. One group of reluctantly reactive pollutants, the halogenated hydrocarbons, does not react quickly enough to be consumed in the troposphere, so these may eventually mix with air in the stratosphere. The common halogenated hydrocarbon pollutants are listed in Table 18–4.

A relatively large amount of Freon-12 escapes into the atmosphere from air conditioners, which are often recharged yearly without a thought of what happened to all that Freon.

In the stratosphere, where the ultraviolet radiation from the Sun is most intense, bonds of the halogenated hydrocarbons are broken and species with unpaired electrons are formed. The most common reactive species produced is the chlorine atom ($Cl\cdot$), as the breakdown of Freon-11 illustrates.

$$
\begin{array}{ccc}
 & Cl & & Cl \\
 & | & & | \\
F - & C - Cl + light \rightarrow F - & C\cdot + Cl\cdot \\
 & | & & | \\
 & Cl & & Cl
\end{array}
$$

TABLE 18–4   The Major Halogenated Hydrocarbons

NAME	FORMULA	USES
Freon-11	CCl_3F	Aerosol propellant
Freon-12	CCl_2F_2	Propellant, refrigerant
Carbon tetrachloride	CCl_4	Making fluorocarbons
Methyl chloroform	CH_3CCl_3	Metal cleaning
Perchloroethylene	C_2Cl_4	Dry cleaning

The United States accounts for about half the emissions of each of these chemicals. Released, they become atmospheric pollutants.

The reactive chlorine atom then combines with ozone (O_3), which is present in high concentration in the stratosphere.

$$O_3 + Cl\cdot \rightarrow ClO\cdot + O_2$$

Since many oxygen atoms are available in the upper atmosphere as participants in the production of ozone,

$$O_2 + light \rightarrow \dot{O}\cdot + \dot{O}\cdot$$
$$\dot{O}\cdot + O_2 \rightarrow O_3$$

the $ClO\cdot$ species can react with an oxygen atom to release the $Cl\cdot$ radical

$$ClO\cdot + \dot{O}\cdot \rightarrow O_2 + Cl\cdot$$

to react with another ozone molecule.

The net effect of these reactions is the destruction of ozone. The total number of ozone molecules destroyed by a single chlorine atom can run into the thousands. Eventually the chlorine atom reacts with a water molecule to form HCl, which mixes into the troposphere and washes out in rain.

In 1974, M. J. Molina and F. S. Roland of the University of California predicted that the increasing use of halogenated hydrocarbons could seriously deplete the stratospheric ozone, with a corresponding increase in ultraviolet rays reaching the Earth's surface. Since living things are sensitive to ultraviolet rays (for instance, ultraviolet light increases skin cancer in humans), the depletion theory was judged to suggest a serious threat to human safety and health. Later research has shown that this theory is probably correct in its predictions if halogenated hydrocarbon emissions go unchecked. The EPA has not limited the emissions of these compounds.

INDOOR POLLUTION

A five-year study by the EPA ending in 1985 concluded that pollution is greater indoors than outdoors for a number of toxic organic chemicals. According to the study, the indoor pollution levels are about the same for houses in industrialized areas as for houses in rural areas.

What are the sources of indoor pollution? Tobacco smoke, if present, is an obvious source. Benzene, a known carcinogen, was 30% to 50% higher in the air of the homes of smokers in the study than in the homes of nonsmokers. Building materials and other consumer products are also sources for indoor pollution. As buildings have become more tight and energy efficient, more emissions from materials in the building are trapped inside.

The National Aeronautics and Space Administration has tested 10,000 materials for gaseous emissions because astronauts were experiencing "sick-building syndrome." Xylenes, for example, were found in 800 of the materials tested.

According to the EPA study, toxic chemicals found to be prevalent indoors were 1,1,1-trichloroethane (source: drycleaning solvent), tetrachloroethylene (solvent), benzene (paint, tobacco, gasoline), o-xylene (paint, gasoline, marking pens), m,p-xylene (paint, gasoline), ethylbenzene (paint, gasoline), carbon tetrachloride (solvent), trichloroethylene (solvent), chloroform (tap water), styrene (insulation, plastics), and p-dichlorobenzene (moth crystals, deodorants).

Air conditioners, dehumidifiers, smoke removers, vacuum cleaners, and dish and bath cleaning water down the drain help remove some of the indoor pollution.

CARBON DIOXIDE — AN AIR POLLUTANT . . . OR IS IT?

How can carbon dioxide be considered a pollutant when it is a natural product of respiration and a required reactant for photosynthesis? CO_2 is not a pollutant per se; its increasing amount is. Between 1900 and 1970, the global concentration of CO_2 increased from 296 ppm to 318 ppm, an increase of 7.4%. The level in 1985 was 340 ppm, and if the present trend continues, the concentration will be 640 ppm by the year 2020. Look at Figure 18–16 for a pretty curve of an ugly trend in industrial emissions of CO_2. CO_2 is the only atmospheric substance whose global concentration is known to be rising.

CO_2 is a necessary ingredient for photosynthesis.

The increased amount of CO_2 comes primarily from electrical power plants, internal combustion engines, and the manufacture of cement. But there are numerous sources: home heating, trash burning, forest fires, and bacterial oxidation of soil humus, to mention a few. CO_2 has been entering the atmosphere faster than oceans and growing plants can remove it.

A substantial increase of CO_2 in the air can cause two detrimental effects: CO_2 can increase the temperature of the atmosphere, and it can increase the acidity of the oceans.

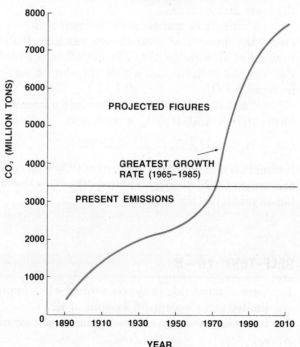

FIGURE 18–16 U.S. industrial emissions of CO_2 into the atmosphere in millions of tons per year.

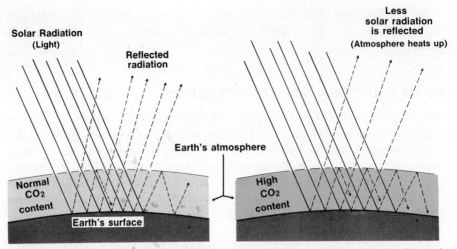

FIGURE 18–17 The greenhouse effect. Owing to a balance of incoming and outgoing energy in the Earth's atmosphere, the mean temperature of the Earth is 14.4° C (58° F). Carbon dioxide permits the passage of visible radiation from the Sun to the Earth but traps some of the heat radiation attempting to leave the Earth.

Carbon dioxide in the atmosphere produces a greenhouse effect.

Another chance for "acid rain."

Pollution emitted mainly by automobiles and trucks is discussed in Chapter 23.

Carbon dioxide increases the temperature of the atmosphere by the **greenhouse effect** (Fig. 18–17). Carbon dioxide, like the glass or plastic of a greenhouse, lets the shorter wavelengths of light through but absorbs the longer wavelengths as they are emitted from the surface of the Earth. The carbon dioxide (and the glass), in turn, emits the absorbed energy as heat radiation; some returns to Earth and some energizes other molecules. The net effect is an increase in the temperature of the lower atmosphere.

Particulates and aerosols counteract the warming effect of CO_2. They decrease the amount of solar energy reaching the Earth by scattering incoming sunlight of all wavelengths. The net effect is a cooling of the lower atmosphere. Calculations indicate that a 25% increase in aerosols would counteract a 100% increase in CO_2.

Carbon dioxide can make surface waters acidic by reacting with water to form carbonic acid, H_2CO_3, a weak acid.

$$CO_2 + H_2O \rightleftharpoons H_2CO_3 \rightleftharpoons H^+ + HCO_3^-$$

It would take an enormous amount of CO_2 in an ocean to affect its acidity. A major consumer of CO_2 (through photosynthesis) is the huge amount of phytoplankton (small plants) in the oceans.

SELF-TEST 18–B

1. Carbon monoxide is always formed when hydrocarbons are burned in a (   ) limited or (   ) plentiful amount of air.
2. In the human body, carbon monoxide is a formidable competitor with oxygen for _____.

3. The amount of carbon monoxide is definitely increasing in the atmosphere. ( ) True or ( ) False

4. A polynuclear aromatic hydrocarbon commonly found in the air and in cigarette smoke that is known to cause cancer is _____ .

5. A secondary pollutant around human beings but an upper atmosphere screen for ultraviolet energy is the substance _____ .

6. Carbon monoxide, CO, is very soluble in water because both substances are ( ) polar or ( ) nonpolar.

7. Two destructive effects of CO_2 on the environment are that it _____ the temperature of the atmosphere and _____ the acidity of the oceans.

WHAT DOES THE FUTURE HOLD?

There will undoubtedly be an abatement of air pollution in the future; the sheer pressures of population increase will demand it. But life also will undoubtedly have to be different. Perhaps the first major change will be the disappearance of the automobile from the city, followed by a gradual modification of the power plant of the automobile until it is relatively nonpolluting. One interesting effect of the lower legal speed limit on our nation's highways was the reduction of NO_x emissions due to the lower operating temperatures of the auto engines.

Most pollution exists because we demand the benefits of a technology that, for the most part, has given little consideration to the long-range effects of its products. When industry, automobile manufacturers, or power plant operators add equipment to stop noxious waste products from getting into the air, the costs are added to the already considerable manufacturing expense without adding one cent to the market value of the product being made. The cost to the consumer, however, will go up and will be reflected in the increased price of consumer goods, but what is the value of clean air?

At this time, it is uncertain whether the knowledge of the harmful, long-range effects of air pollution will bring people to the point where they are willing to give up the immediate activities that give rise to the pollution. Not many seem willing to use less energy or to give up the automobile.

In addition to existing studies and regulations, other problems in air pollution will be addressed, such as indoor air pollution, regulation of all hazardous chemicals, a standard for the transient fine particles, and a significant reduction in SO_2 to control acid rain.

There will be increased litigation to bring industry into conformance with the Clean Air Act. In 1981, of the estimated 6500 major sources of industrial pollution, 2400 had never complied or agreed to comply with the Clean Air Act.

In the final analysis, we all pollute the atmosphere. Much of the pollution is due to the misapplication of chemical techniques, yet the eradication of most forms of air pollution is within the capabilities of chemical technology. At present, there is some awareness of the problems, but the trade-offs among pollution, energy, and inflation are far from being settled. The process will be very slow — it is up to us.

In 1981, the estimated cost to bring 147 plants in the Chicago area under the level of 250 micrograms of NO_2 per liter was $130 million.

The cost of air pollution control in the United States is about $19.3 billion annually.

MATCHING SET

_____	1.	Smog
_____	2.	Primary pollutant
_____	3.	Secondary pollutant
_____	4.	Aerosol
_____	5.	Micron
_____	6.	ppm
_____	7.	Unsaturated hydrocarbon
_____	8.	Photochemical
_____	9.	Abatement
_____	10.	Emphysema
_____	11.	Brown gas
_____	12.	Endothermic
_____	13.	Polynuclear hydrocarbon
_____	14.	Chlorofluorocarbon

a. Heat absorbing
b. Mixture of fog, SO_2, and (or) hydrocarbons and NO_2
c. Parts per million
d. Disease of lungs
e. Hydrocarbon with double bond
f. Eradication
g. CO
h. Destroys stratospheric ozone
i. Process of decreasing
j. Intermediate in size between individual molecules and particulates
k. Benzo(α)pyrene
l. Peroxyacyl nitrate
m. Chemical reaction energized by light
n. 10^{-6} meter
o. NO_2

QUESTIONS

1. The formation of photochemical smog involves very reactive chemical species. What structural feature makes these species reactive?
2. Write a balanced chemical equation for the burning of iron pyrite (FeS_2) in coal to sulfur dioxide and Fe_2O_3, iron (III) oxide.
3. What conditions are necessary for thermal inversion?
4. What are the basic chemical differences between industrial smog and photochemical smog?
5. What are the major sources of the following pollutants?
   a. Carbon monoxide
   b. Sulfur dioxide
   c. Nitrogen oxides
   d. Ozone
6. What is a photochemical reaction? Give an example.
7. If air pollutants rise from the Earth into the atmosphere, why do they not continue on into space?
8. If nature emits more than 90% of the particulates (volcanic eruptions), nitrogen oxides (lightning), and carbon monoxide (decaying organic matter), why the concern about air pollution caused by people?
9. What effects does weather have on local air pollution problems? On regional air pollution problems?
10. Knowing the chemistry of photochemical smog formation, list some ways to prevent its occurrence.
11. What part do aerosols play in the formation of smogs?
12. Which substance, naturally found in the atmosphere, seems to be increasing in concentration over the years?
13. Describe an air pollution problem in your community. How can this problem be solved?
14. Discuss the merits of abatement versus eradication of air pollution.
15. Of the following air pollutants — particulates, sulfur dioxide, carbon monoxide, ozone, and nitrogen dioxide —
    a. Which is normally a secondary pollutant?
    b. Which is emitted almost exclusively by human-controlled sources?
    c. Which can be removed from emissions by centrifugal separators?
    d. Which two react with water to form acids?
16. Define:
    a. Micron
    b. ppm
17. Should human beings make a major effort to alter their activities in order to stop the increase of carbon dioxide in the atmosphere? What would be some of the costs to achieve this goal?
18. Why are small particulates in the atmosphere dangerous to human beings?

19. Which are more effective in producing smog, hydrocarbons with all single bonds or hydrocarbons with some double bonds?

20. Why is natural gas a good substitute for oil and coal as far as air pollution is concerned? What problem is related to its substitution?

21. What is the approximate concentration of air pollutants in the atmosphere during smoggy conditions?

22. What brown oxide of nitrogen is a necessary component of photochemical smog?

23. What is the chief source of air pollution in the United States?

24. Why should CO_2, SO_2, and CO be more soluble in water than N_2 and O_2?

25. What is the molecular weight of BaP? See p. 482.

26. Sketch these generalized (no actual data required) graphs:

a. Y axis: SO_2 emissions; x axis: concentration of S in coal

b. Y axis: CO emissions; x axis: amount of oxygen present

c. Y axis: CO emissions; x axis: gasoline consumed

d. Y axis: CO emissions; x axis: controls applied

27. Why are our desires for cheap energy and a clean environment in conflict?

28. The Cottrell precipitator makes use of what property of particulate particles?

29. Why does a temperature inversion tend to act as a lid over polluted air?

30. Does sulfur dioxide have a long or short atmospheric life?

31. What are some sources of indoor pollutants not necessarily found outdoors?

32. How does acid affect trees, buildings, and lakes?

Chapter 19

AGRICULTURAL CHEMISTRY

Human hunters have successfully established civilizations on six of the seven continents, and even Antarctica has prolific mammalian and bird populations that exist on the basis of being able to find natural food. The extent of all such populations, human and animal, is dependent on the abundance of the food supply. Figure 19–1 traces the world human population and the population explosion that began in about 1600 A.D. Among the several factors that have contributed to the population explosion, such as the control of disease, a major factor has been the developments in modern agriculture. Although agriculture can be traced through 8000 to 10,000 years of human history, there was little application of scientific information to food production prior to 1800. The approximately 5 billion people who are alive now can be fed using modern scientific agriculture; however, mass starvation threatens if climate or political-economic conditions cause a slowdown in the flow of food.

The British Isles recorded 201 famines between 10 A.D. and 1846, with none since. In China, there were 1846 famines between 108 B.C. and 1828 A.D.

The relationship between population and food supply is complex, with an oversupply of food production capacity in the industrialized West and the population pressure constantly testing the food supply in Asia and Africa. However, the human race has moved to the point that the only hope for an adequate food supply on a worldwide basis is through scientific agriculture, which is based on a chemical understanding of food production.

A number of the ancient civilizations apparently developed good farming practices that are not directly recorded in history. In Roman times, Cato the Elder described seed selection, green manuring with legumes, testing the soil for acidity, the use of marl, the value of alfalfa and clover, composting, the preservation and use of animal manures, pasture management, early-cut hay, and the importance of livestock in any general farming operation. Such progress based on practical experience was apparently gained and lost a number of times in the course of history.

In 18th-century England, Arthur Young set the stage for modern scientific agriculture: as the first noted agricultural extension worker, he set in motion a system of dissemination of agricultural information that has now been developed on a worldwide basis. In 1840, Justus von Liebig published his *Organic Chemistry in Its Applications to Agriculture and Physiology*. Liebig began the description of the chemicals required by plants, and his work set the stage for the chemical fertilizer industry. As a result of his work, Liebig has been called the Father of Modern Soil Science. The Industrial Revolution provided better tools and power that make the

A STATEMENT OF PHILOSOPHY

"Perfect agriculture is the true foundation of all trade and industry — it is the foundation of the riches of nations. But a rational system of agriculture cannot be formed without the application of scientific principles, for such a system must be based on an exact acquaintance with the means of vegetable nutrition. This knowledge we must seek through chemistry."

Justus von Liebig (1803–1878)

AN AGRICULTURAL LAW, THE "LAW OF THE MINIMUM:"

If one of the nutritive elements is deficient or lacking, plant growth will be poor even when all other elements are abundant. If the deficient element is supplied, growth will be increased up to the point where the supply of that element is no longer the limiting factor.

Liebig

physical tasks in agriculture relatively easy. Even now, however, the best estimate is that two thirds of the world's agriculture is "backward." How far will we be able to go in the application of the molecular understanding of foods to food production, and what will be the consequence in the human population? These are interesting questions, and they relate to the quality of life for all of us as 10 billion hands reach for "our daily bread."

In this chapter we will see how a knowledge of agricultural chemistry will enable us to understand how natural soils support life, how natural soils can be selectively supplemented when deficient in a required nutrient, how fertilizers can be produced from inorganic sources, how chemicals can be used to destroy unwanted plants and pests, and how chemicals can be used to cause growth, inhibit growth, or produce selective growth for a desired agricultural product. A nagging question will persist at the end of this study: Is adequate consideration being given to the control of unwanted side effects from the use of agricultural chemicals?

Annual U.S. crop exports: $40 billion, including 38% of our corn production, 43% of the soybeans, 27% of the wheat, and 64% of the rice.

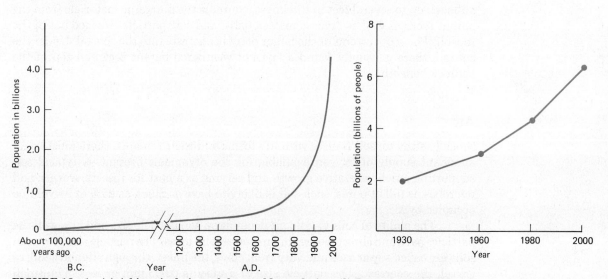

FIGURE 19–1 (*a*) A historical view of the world population. (*b*) Present trend in the world population and a projection for the year 2000. (Figures published by the United Nations.)

SOILS AND SOIL CONTENTS

Cold astronomical bodies develop soils as a result of natural forces that tend to break up the solidified crust materials. Interstellar radiation provides a portion of the energy for these processes. The surface of the moon illustrates the pulverizing effect of meteoric materials ranging from subatomic particles to massive meteorites. The soil-forming processes are accelerated on planets and moons that have fluids on their surfaces; weather and ocean currents, resulting from convection in the fluids, exert the powerful forces of freezing and thawing along with the grinding action of moving matter to pulverize the surface structures. Also, volcanic activity, resulting from the heat energy stored within, contributes significantly to the particles available for soil formation.

Soil on Earth has a special character, as yet unknown on other heavenly bodies, owing to the life forms that have thrived within and on it for millions of years. These living organisms have contributed both to the chemical composition of the soil and to the physical makeup of the gross material.

There are literally thousands of soil types. The qualities of a virgin soil on Earth depend on five factors:

1. Living matter — The association of trees, grasses, shrubs, microorganisms, and animals
2. Climate — Heat, cold, rain, snow, and their distribution
3. Parent rock — Kinds of minerals and fineness of particles
4. Slope — Degree, length, and shape of surface
5. Time — The age of the formation

In addition to this list, human activities have altered the composition and physical nature of the Earth's soils.

Soil scientists refer to the layers within the soil as **horizons.** The **topsoil** contains most of the presently living material and humus from dead organisms. It is not uncommon to find as much as 5% organic matter in topsoil. The topsoil is usually several inches thick, though in places more than 3 feet can be found. The **subsoil,** up to several feet in thickness, contains the inorganic materials from the parent rocks as well as organic matter, salts, and clay particles washed out of the topsoil. The root systems of the larger plants penetrate into the subsoil. Under the subsoil, there is usually found a layer of weathered parent rocks on top of the bedrock beneath.

Air

Since healthy topsoil is alive with life forms and their remains, there must be an abundant supply of oxygen available for the organisms present. A typical soil supporting a rich vegetative growth and serving as a host for insects, worms, and microbes, is full of pores; such soil is likely to have as much as 25% of its volume occupied by air.

The ability of soils to hold air depends on soil particle size and how well the particles pack and cling together in forming a solid mass. The particle size groups in soils are called **separates** and vary from clays, the finest, through silt and sand, to gravel, the coarsest. The particle size of a **clay** is 0.005 mm or less. The small particles in a clay deposit can pack close together, eliminating essentially all of the air and thus supporting little or no life. A typical soil horizon will be composed of several separates. A **loam,** for example, is a soil consisting of a friable mixture of varying proportions of clay, sand, and organic matter; a loam will be rich in air content. The porosity of soil also depends on the natural shape of the soil particles, which, in turn, is dependent on the crystalline structure of the parent rock and the physical action involved in breaking it apart. On close examination, soil particles are likely to be plates, blocks, or granules.

Air in soil has a different composition from the air above. Normal dry air at sea level contains about 21% oxygen and 0.03% carbon dioxide. In soil, the percentage of oxygen may drop to as low as 15%, and the percentage of carbon dioxide may rise above 5%. This results from the organic matter being partially oxidized in the closed space. The carbon in the organic material uses oxygen to form carbon dioxide. This increased concentration of carbon dioxide tends to cause the ground water flowing through to become acidic; acidic soils are described as **sour soils** because of the sour taste of aqueous acids.

$$CO_2 + H_2O \rightarrow H^+ + HCO_3^-$$

Crushed limestone, $CaCO_3$, applied to the soil furnishes carbonate ions to combine with the hydrogen ions to form bicarbonate ions, and it produces a pH that is slightly basic. A slightly basic soil is a **sweet soil.**

$$H^+ + CO_3^{2-} \rightarrow HCO_3^-$$

Water

Soils can hold water in three ways: water can be absorbed into the structure of the particulate material, it can be adsorbed onto the surface of the soil particles, and it can occupy the pores ordinarily filled with air. **Sorption** is the general term covering absorption and adsorption; water held by either method of sorption is not readily released by soils. For example, if you take the driest soil you can find (perhaps from under an old house) and heat it in a test tube, you will find it will release considerable moisture. "Dry soil" is not really dry in the chemical sense.

The sorption of water in soils can be explained in terms of the chemical composition of a typical soil. Nine elements make up the great bulk of the Earth's soils; they are oxygen, silicon, aluminum, iron, calcium, magnesium, sodium, potassium, and hydrogen. Note the relationship between this list and the composition of the crust of the Earth (Fig. 2–6). Notice that in this group of elements, oxygen is the only strongly electronegative element. As a result, oxygen forms strongly polar bonds or ionic bonds with each of these elements. Since water is a polar molecule, there is a strong interaction between water molecules and most of the bonded atoms in soil, especially oxygen. To illustrate with something that you have observed, recall the small enclosed envelope of desiccant that you have unpacked with optical instruments or electronic equipment. The water-absorbing desiccant is usually silica gel. Silica gel is silicon dioxide with the same chemical composition as white sand. If silica gel is baked at 2000 °C for a few hours, it becomes essentially dry and is ready to take on all of that sorbed water through hydrogen bonding to the numerous oxygen locations in the silicon dioxide structure (Fig. 11–13).

Water is removed from soil in four ways: plants transpire water in carrying on the life processes, soil surfaces evaporate water, water is carried away in plant products, and gravity pulls water to the subsoil and rock formations below. **Percolation** is the ability of a solid material to drain a liquid from the spaces between the

Friable material easily crumbles under slight pressure.

George Washington had marl, an earthy mixture of limestone and clay, dug from the Potomac River bed for application to his fields.

solid particles. Soils with good percolation will drain water from all but the small pores in the natural flow of the water. The flow of water through soils is a necessity, as it will take several hundred pounds of water for the typical food crop to make one pound of food. A negative aspect of the massive flow of water through soil is the **leaching effect.** Water, known as the universal solvent because of its ability to dissolve so many different materials, dissolves away, or leaches, many of the chemicals needed for a productive soil. If a method is not maintained for replacing the leached material, the soil cannot be as productive as before.

The percolation of a soil depends on the soil particle size and on the chemical composition of the soil material. The soil particle size is determined by either the degree of breakup of the parent rock or the fineness of the soil aggregates after it has formed. Depending on moisture conditions, the soil may be in fairly large aggregates surrounded by water, or it may be broken up into exceedingly fine particles, each surrounded by adsorbed water. Clays, and silts to a lesser degree, because of the small particle sizes involved, tend to pack together in an impervious mass with little or no percolation. Of course, sand, gravel, and rock pass water readily. Water-logged soils that will not percolate will support few crops because of their lack of air and oxygen. Rice is an exception.

Soils become acidic, or sour, not only because of the oxidation of organic matter but also because of selective leaching by the passing groundwater. Salts of the alkali and alkaline earth metals are more soluble than the salts of the Group III and transition metals. For example, a soil containing calcium, magnesium, iron, and aluminum is likely to be slightly alkaline, or sweet, prior to leaching with water. If calcium and magnesium are removed in excess over the iron and aluminum, the soil becomes acidic because of the acidic nature of these trivalent ions. Each of these ions will tie up hydroxide ions from water and release an excess of hydrogen ions:

$$Fe^{3+} + H_2O \rightarrow FeOH^{2+} + H^+$$

$$Al^{3+} + H_2O \rightarrow AlOH^{2+} + H^+$$

Actually, these trivalent ions may release more than one hydrogen ion each, depending on the overall soil pH that is determined, as we have seen, by more than one factor. In some arid regions, calcium tends to collect as calcium carbonate just under the **solum** or true soil.

Leaching is not altogether bad because unwanted elements are removed by this natural process. In dry regions selenium is present in relatively high concentrations because of its concentration in the parent rocks. The selenium would be removed by leaching if this soil had been formed in a moist climate. Plants that grow in soil rich with selenium often take up large amounts of this element and are, as a result, poisonous to animals and humans. This is the case with several plants in the western United States and Mexico.

Humus

Organic matter varies in soil from the relatively fresh remains of leaves, twigs, and other plant and animal parts to peat, the precursor of coal and oil. Humus is not far removed in time from the living debris. However, it is well decomposed, dark colored, and rather resistant to further decomposition. As a source of nutrients for plants, humus is almost like a time-release capsule, taking considerable time to release its contents while holding them in an insoluble form. Finally, the humus is decomposed into minerals and inorganic oxides. In soils with poor drainage, the decomposition of humus is stopped, and peat is formed. Such mucky soils are worthless for crop production unless thay are drained and fertilized with phosphorus and potash; then they become highly productive.

In addition to being a source of plant nutrients, humus is important in maintaining good soil structure, often keeping the soil friable in a soil rich in clay. Soil, rich in humus, may contain as much as 5% organic matter. Soils in the grasslands of North America are rich in humus to a considerable depth in contrast to only a thin film on the ground surface in a humid forest region.

Maintaining humus in the soil is of major concern to the agriculturist. Humus such as peat moss or organic fertilizer can be added. However, there is no real substitute for natural plant growth that is returned to the ground for humus formation. Clover is often grown for this purpose and plowed under at the point of its maximum growth. The compost pile of the gardener is another effort to maintain humus for a productive soil.

If large amounts of organic matter such as leaves or sawdust are added to soil to promote humus formation, it should be remembered that the oxidation of this material produces acid that would have to be counterbalanced by lime if the soil is to remain sweet.

Chemical Composition

The chemical compositions of soils reflect the Earth crust composition, the composition of the parent rocks, and the chemical and physical activity during and subsequent to the formation of the particular soil. Even though it is second to oxygen in percent composition, silicon is the central element in explaining the soil chemicals. Sand is silicon dioxide, clays are mixtures of silicates, and the different kinds of silicate rock fragments are numerous. The bulk of most soil horizons is composed of silicate materials. Trace elements in micro amounts vary considerably from soil to soil, and a wide variety of the elements is present in a given sample.

Black soils are usually rich in organic matter and consequently contain the elements required for plant life. However, a word of caution is needed here—the same amount of organic matter that will make a soil black in a temperate region will only make it brown in a tropical region. Red soils are likely to be rich in iron, and soils that are nearly white have been heavily leached and are likely to be poor in quality. Poorly drained soils are likely to be of uneven chemical texture, with several colors such as gray, brown, and yellow appearing in a spadeful of the material.

Nutrients

At least 18 known elemental nutrients are required for normal green plant growth (Table 19–1). Three of these are **nonmineral nutrients** and are obtained from air and water. The mineral nutrients must be absorbed through the plant root system as water solutes. The 15 mineral nutrients fall into three groups: **primary, secondary,** and **micronutrients,** depending on the amounts necessary for healthy plant growth. Also, it should be noted that there is the possibility that other

Sir Humphrey Davies argued the Humus Theory, "Carbon for plants came from humus." A Swiss, de Saussure, showed the carbon to come from carbon dioxide.

TABLE 19–1   Essential Plant Nutrients

NONMINERAL	PRIMARY	SECONDARY	MICRONUTRIENTS
Carbon	Nitrogen	Calcium	Boron
Hydrogen	Phosphorus	Magnesium	Chlorine
Oxygen	Potassium	Sulfur	Copper
			Iron
			Manganese
			Molybdenum
			Sodium
			Vanadium
			Zinc

elements are essential in micro amounts, but the chemistry involved is not presently known.

NONMINERAL   Carbon, hydrogen, and oxygen are readily available from the air and water. Carbon is taken up by the plant in the form of carbon dioxide, hydrogen is taken up in the form of water, and oxygen can be absorbed in water or as the element. Through the process of photosynthesis, green plants produce an excess of oxygen during the light period that is released through the leaves and other green tissue.

PRIMARY   The primary nutrients are nitrogen, phosphorus, and potassium. Although bathed in an atmosphere of nitrogen, most plants are unable to use this source as a supply for this vital element. **Nitrogen fixation** is the process of changing atmospheric nitrogen into the compounds of this element that can be dissolved in water, absorbed through the plant roots, and assimilated by the plant (Fig. 19–2). Most plants thrive on soils rich in nitrates, but many plants that grow in swamps, where there is a lack of oxidized materials, can use reduced forms of nitrogen such as the ammonium ion. An element can be in several oxidation states, depending on how many electrons it has lost or has tended to lose in combinations

The air above each acre of Earth's surface contains 36,000 tons of nitrogen.

Chlorophyll requires nitrogen and magnesium from the soil.

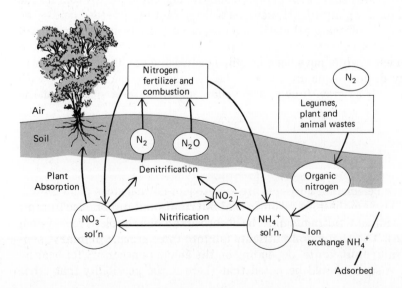

FIGURE 19–2   Nitrogen pathways through soil.

with other atoms. The nitrate ion is the most highly oxidized form of combined nitrogen, and the ammonium ion is the most reduced form of nitrogen.

Nature fixes nitrogen on a massive scale in two ways. Nitrogen is oxidized under highly energetic conditions such as in a discharge of lightning or, to a lesser extent, in a fire. The reaction is:

$$N_2 + O_2 \rightarrow 2\, NO$$

The nitric oxide, NO, is easily oxidized in air to nitrogen dioxide, which dissolves in water to form nitric acid, HNO_3, and nitrous acid, HNO_2.

$$2\, NO + O_2 \rightarrow 2\, NO_2$$
$$H_2O + 2\, NO_2 \rightarrow HNO_3 + HNO_2$$

The nitric acid is readily soluble in the rain, clouds, or ground moisture, and thus the nitrate concentration in the soil is increased. A legume plant such as beans or clover lives in a symbiotic relationship with bacteria, which live in nodules on the root system of the host plant. The bacteria are able to fix atmospheric nitrogen in far greater amounts than their own requirements, or even those of the host plant. As a result of legume fixation, more than 100 pounds of nitrogen can be added to an acre of soil in one growing season. In fact, if a legume is grown often enough on a particular plot of ground, there will be no need to supplement this soil with nitrogen fertilizer.

Another major source of nitrogen replenishment for the soil is the return of dead organisms and animal wastes to the soil. Even in the absence of legumes, this can be an adequate source of nitrogen in the absence of wholesale plant or plant-product removal from the plot.

Like nitrogen, phosphorus has to be in mineral or inorganic form before it can be used by plants. Unlike nitrogen, phosphorus has a natural content that comes totally from the mineral content of the soil. Orthophosphoric acid, H_3PO_4, loses hydrogen ions to form the dihydrogen and monohydrogen ions, $H_2PO_4^-$, HPO_4^{2-}, which are the dominate forms in soils of normal pH (Fig. 19–3). Because of the greater concentration of electrical charge associated with the trivalent phosphate ion, phosphates tend to be held to positive centers in the soil structure and are not as easily leached by groundwater as are the nitrate compounds. The nitrate ion is a rather large ion with only one negative charge.

Potassium is a key element in the enzymatic control of the interchange of sugars, starches, and cellulose. Although potassium is the seventh most abundant element in the Earth's crust, soil heavily used in crop production can be depleted of

Nitrogen is the distinguishing element in amino acids and proteins.

A German, Hellriegel, showed in 1886 that leguminous plants "fix" nitrogen.

The protein content of corn and other grains is directly related to the amount of nitrogen in the soil.

Phosphorus is concentrated in fast-growing tissue. A plant can translocate phosphorus from old to growing tissue.

Potassium is absorbed as the free ion, $K^+(aq)$.

Potassium, if available, is constantly released by ion exchange reactions.

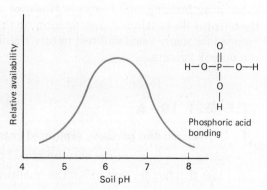

FIGURE 19–3 The availability of phosphate in the soil is a function of the pH. The dominant species present for phosphoric acid at pH 5–8 are $H_2PO_4^-$ and HPO_4^{2-}. At very low pH values, the phosphorus would be in the form of the acid H_3PO_4 (all three protons on the acid structure). At very high pH, all three protons would be removed, and the phosphorus would be in the form of the PO_4^{3-} ion. Low soil temperatures in temperate regions significantly reduce phosphorus uptake by plants.

this important metabolic element, especially soil that is regularly fertilized with nitrate with no regard to the potassium content. Some of the fungus plants in the soil produce chemicals that cause the bound potassium to be released into a soluble form that can be taken in through the plant root system in excessive amounts or simply leached out by the flow of soil water.

SECONDARY Calcium and magnesium are available in Ca^{2+} and Mg^{2+} forms in small amounts as well as in complex ions and crystalline formations. These abundant elements are bound tightly enough so they are not readily leached and loosely enough to be readily available to plants. Sulfur, in the form of inorganic sulfates from the minerals and organic sulfates from the organic material, is readily available if present in the soil.

> Magnesium deficiencies, like nitrogen deficiencies, cause chlorosis—a low chlorophyll content.

MICRONUTRIENTS Because of the very small amounts of micronutrients required by plants, probability is high that micronutrients will be present in adequate amounts in most soils deemed suitable for farming operations prior to extensive cropping on that soil. Only in recent years has there been extensive research at the molecular level on the flow of the micronutrients from the soil structures through their vital roles in plant metabolism. It is becoming increasingly clear that deficiencies in the micronutrients can be identified in some virgin soils and that extensive farming and the associated leaching of soil contents result in unproductive soils owing to these deficiencies.

> Lime, as CaO or $Ca(OH)_2$, is the number four commercial chemical. See inside the front cover.
>
> The addition of lime, a basic substance, raises the pH of the soil.

Iron is an essential component of the catalyst involved in the formation of chlorophyll, the green plant pigment. When the soil is iron deficient or when too much "lime," $Ca(OH)_2$, is present in the soil, iron availability will decrease. This condition is usually present when plant leaves lighten in color or even turn yellow. Often a gardener or lawn worker will apply phosphate and lime to adjust soil acidity, only to see the green plants turn yellow. What is happening is that both phosphate and the hydroxide from the lime tie up the iron and make it unavailable to the plants.

$$\underset{\text{PHOSPHATE}}{Fe^{3+} + 2\,PO_4^{3-}} \rightarrow \underset{\text{TIGHTLY BOUND COMPLEX}}{Fe(PO_4)_2^{3-}}$$

$$Fe^{3+} + 3\,OH^- \rightarrow \underset{\text{INSOLUBLE HYDROXIDE}}{Fe(OH)_3}$$

Boron is absolutely necessary in trace amounts, but there is a relatively narrow concentration range above which the boron is toxic to most plants.

Currently, there is a considerable debate in the agricultural community about how much investment should be made in soil testing for micronutrients and consequent farming modifications. However, two things are sure: scientific curiosity will provoke continued investigations in this area, and soils that are extensively cropped for many years will test nature's ability to provide sufficient amounts of the micronutrients.

SELF-TEST 19-A

1. Pulverized and partially separated crust materials constitute the

_____ of the astronomical bodies.

2. List five factors that affect the qualities of a virgin soil: _____

_____

3. Which of the following types of soils will have, on average, the smallest soil particles: silts, sandy soils, loams, clays? _____

4. Acidic soils are described as _____ because of the common taste of aqueous acids.

5. Carbon dioxide causes soils to be ( ) acidic or ( ) basic. Limestone, $CaCO_3$, causes soils to be ( ) acidic or ( ) basic.

6. In soil sorption, water that fills pores is referred to as adsorbed or absorbed water? _____

7. The two factors that determine the percolation of a soil are _____ and _____.

8. Which is more acidic, a monovalent ion like Na^+ or a trivalent ion like Fe^{3+}? _____

9. A well-decomposed, dark-colored plant residue that is rather resistant to further decomposition is known as _____.

10. The bulk of most soil horizons is composed of _____ materials.

11. The secondary elemental plant nutrients are _____, _____, and _____.

12. Nitrogen fixation involves the oxidation or reduction of nitrogen? _____

13. Some micronutrients can poison plants. ( ) True or ( ) False

CHEMICAL FERTILIZERS — KEYS TO THE WORLD'S FOOD PROBLEMS

Primitive peoples learned to raise crops on a cultivated plot until it lost its fertility and then to move on to a virgin piece of ground. In many cases, the slash-burn-cultivate cycle was no more than a year in length, and few have found a piece of ground anywhere that could support successful cropping for more than five years without fertilization. The farming villages, developed in ancient times and prevalent through the Middle Ages, demanded innovation in fertilization, as the same land had to be used through the years. With the use of legumes in crop rotations, manures, dead fish, or almost any organic matter available, the land was kept in production even if to very disappointing yields. This situation continues in two thirds of the world today.

With approximately 5 billion people in the world, it is estimated that less than 4 billion acres are used worldwide in the cultivation of crops for food, less than 0.8 acre per person. This ratio would likely be acceptable if modern farming practices with good chemical fertilization were employed. Estimates are that if $40 per acre were spent on fertilizer for all cultivated acreage, the world crop production would increase by 50%, which is equivalent to 1.7 billion more acres under cultivation. Of course, the cost to produce this additional food would approach $160 trillion and appears to be prohibitive, but it is obvious that there is a lot of room between status quo and this rich abundance of food.

Fertilizers that contain only one nutrient are called **straight** fertilizers, and those containing a mixture of the three primary nutrients are **complete** or **mixed** fertilizers. Urea for nitrogen and potassium chloride for potassium are examples of straight products. The macronutrients are absorbed by plant roots as simple inor-

Crop yield explosions: (1) U.S. corn—25 bushels per acre in 1800, 110 bushels per acre in the 1980s; (2) English wheat—below 10 bushels per acre from 800 A.D. to 1600, above 75 bushels per acre in the 1980s; (3) Rice in Japan, Korea, and Taiwan—fourfold increase in the last 40 years.

Chinese farmers added calcined bones to their soil 2000 years ago.

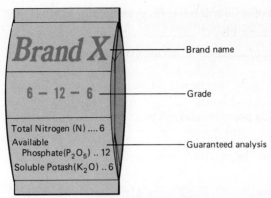

Brand X — Brand name

6 — 12 — 6 — Grade

Total Nitrogen (N) 6
Available
    Phosphate(P_2O_5) .. 12 — Guaranteed analysis
Soluble Potash(K_2O) .. 6

FIGURE 19–4 **The numbers on the fertilizer bag. Samual William Johnson, an American student of Liebig and author of** *How Crops Grow,* **following the lead of Liebig, burned plants and analyzed their ashes. He expressed the nutrient concentrations in the oxide form present in the ashes as P_2O_5, K_2O, and so on, a practice that has continued to this day. The numbers for some of the chemicals and common commercial fertilizers are: ammonia 82-0-0; urea, 46-0-0; ammonium nitrate, 33.5-0-0; ammonium sulfate 21-0-0; normal superphosphate, 0-18-0 to 0-20-0; phosphoric acid, 0-52-0 to 0-54-0; superphosphoric acid, 0-68-0 to 0-72-0; phosphous pentoxide, 0-100-0; potash, 0-0-100; and leading commercial fertilizers, 6-24-24, 10-10-10, and 13-13-13.**

ganic ions: nitrogen in the form of nitrates, NO_3^-, phosphorus as $H_2PO_4^-$ or HPO_4^{2-}, and potassium as the K^+ ion. Organic fertilizers can supply these ions, but it takes a lot of fertilizer and a lot of time. For example, a manure might be a 0.5-0.24-0.5 fertilizer in contrast to a typical chemical fertilizer that would carry the numbers

TABLE 19–2   Some Chemical Sources for Plant Nutrients

ELEMENT	SOURCE COMPOUND(S)
	NONMINERAL NUTRIENTS
C	CO_2, carbon dioxide
H	H_2O, water
O	H_2O, water
	PRIMARY NUTRIENTS
N	NH_3, ammonia; NH_4NO_3, ammonium nitrate; H_2NCONH_2, urea
P	$Ca(H_2PO_4)_2$, calcium dihydrogen phosphate
K	KCl, potassium chloride
	SECONDARY NUTRIENTS
Ca	$Ca(OH)_2$, calcium hydroxide (slaked lime); $CaCO_3$, calcium carbonate (limestone), $CaSO_4$, calcium sulfate (gypsum)
Mg	$MgCO_3$, magnesium carbonate, $MgSO_4$, magnesium sulfate (epsom salts)
S	Elemental sulfur, metallic sulfates
	MICRONUTRIENTS
B	$Na_2B_4O_7 \cdot 10\ H_2O$, borax
Cl	KCl, potassium chloride
Cu	$CuSO_4 \cdot 5\ H_2O$ copper sulfate pentahydrate
Fe	$FeSO_4$, iron (II) sulfate, iron chelates
Mn	$MnSO_4$, manganese (II) sulfate, manganese chelates
Mo	$(NH_4)_2MoO_4$, ammonium molybdate
Na	NaCl, sodium chloride
V	V_2O_5, VO_2, vanadium oxides
Zn	$ZnSO_4$, zinc sulfate, zinc chelates

6-12-6. These numbers indicate the **grade** or **analysis,** in order, of the percentage of nitrogen, phosphorus as P_2O_5, and potassium as K_2O (Fig. 19–4). In addition to having many times the concentration of the desired ion, the chemical fertilizer places the ion in the soil in a form that can be directly absorbed by the plant. The problem is that these inorganic ions are relatively easily leached from the soil and may pose pollution problems if not contained. The much slower organic fertilizer tends to stay put. **Quick-release** fertilizers are water soluble as opposed to **slow-release** products, which require days to weeks for the material to dissolve completely. Table 19–2 lists the nutrients and suitable chemical sources for each. The list is extended from time to time as more is learned about trace elements in biochemistry.

Table 19–3 presents the nutrients known to be necessary to produce 150 bushels of corn.

Manure releases about one half of its total nitrogen in the first growing season.

Nitrogen

Commercial fertilizers contain nitrogen fixed by the Haber process, the direct reaction of nitrogen with hydrogen to produce ammonia (Fig. 19–5):

$$N_2 + 3\ H_2 \rightleftharpoons \underset{\text{AMMONIA}}{2\ NH_3}$$

Pure nitrogen is obtained by distilling oxygen and other gases from liquid air. Hydrogen is more difficult to obtain. At present, petroleum products such as propane, $CH_3CH_2CH_3$, are made to react with steam in the presence of catalysts to produce hydrogen:

$$\underset{\text{STEAM}}{CH_3CH_2CH_3 + 6\ H_2O} \xrightarrow{\text{catalysts}} 3\ CO_2 + 10\ H_2$$

This is one of the principal reasons why ammonia fertilizer costs are so closely tied to petroleum prices. Hydrogen can also be prepared by the electrolysis of water

$$2\ H_2O \xrightarrow[\text{KOH}]{\text{electricity}} 2\ H_2 + O_2$$

Fixed nitrogen refers to nitrogen present in chemical compounds, that is, in combinations other than N_2.

Major energy problems in the fertilizer industry began in the 1970s.

TABLE 19–3   Approximate Amounts of Nutrients Required to Produce 150 Bushels of Corn

NUTRIENT	APPROXIMATE POUND PER ACRE	SOURCE
Oxygen	10,200	Air
Carbon	7800	Air
Water	3225–4175 tons	29–36 inches of rain
Nitrogen	310	1200 lbs. of high-grade fertilizer
Phosphorus	120 (as phosphate)	1200 lbs. of high-grade fertilizer
Potassium	245 (as K_2O)	1200 lbs. of high-grade fertilizer
Calcium	58	150 lbs. of agricultural limestone
Magnesium	50	275 lbs. of magnesium sulfate (epsom salt)
Sulfur	33	33 lbs. of powdered sulfur
Iron	3	15 lbs. of iron sulfate
Manganese	0.45	1.3 lbs. of manganese sulfate
Boron	0.05	1 lb. of borax
Zinc	Trace	Small amount of zinc sulfate
Copper	Trace	Small amount of copper sulfate
Molybdenum	Trace	Trace of ammonium molybdate

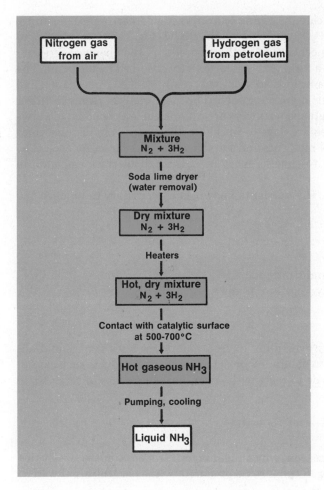

FIGURE 19–5 The Haber process for synthesizing ammonia.

and by several other methods, but all of these require great quantities of energy. So, as energy costs continue to rise, food costs will necessarily rise as a result of the added costs of fertilizers.

Ammonia can be injected in liquid form, **anhydrous ammonia,** directly into the soil because of its very high affinity for moisture (Fig. 19–6). There is some danger involved, since liquid ammonia will immediately evaporate if exposed to the air above ground at ordinary temperatures when released from a pressure tank, thus creating a poisonous atmosphere. Water solutions of ammonia, containing as much as 30% nitrogen by weight, can also be applied successfully to the soil. Again, special equipment and training are required to prevent the loss of ammonia through volatilization.

Research shows solid and liquid fertilizers to be equally effective.

AMMONIA TO NITRATES Nitrogen, in the form of ammonia, is readily available for plant growth. Under usual soil conditions, the ammonia molecules pick up a hydrogen ion to form the ammonium ion.

$$NH_3 + H^+ \rightarrow NH_4^+$$

The soil, rich in oxygen, is an oxidizing medium, and the ammonium ion, through a process called **nitrification,** is oxidized to the nitrate ion, NO_3^-, and it is the nitrate ion that is taken up and used by the plants.

FIGURE 19-6 Liquid ammonia is being injected directly into the ground to provide nitrogen for the growing plants.

Solid nitrogen fertilizers can be prepared readily from ammonia. Ammonia, from the Haber reaction, can be oxidized and converted to nitric acid. The ammonia is burned over a platinum catalyst in oxygen to obtain nitric oxide, NO. From there, the reactions are similar to the natural fixation of nitrogen by lightning described earlier.

$$4\ NH_3 + 5\ O_2 \rightarrow 4\ NO + 6\ H_2O$$

The NO reacts readily with O_2 from the air to form NO_2.

$$2\ NO + O_2 \rightarrow 2\ NO_2$$

This, in turn, reacts with water to give nitric acid and nitrous acid. The nitrous acid, being unstable, is separated with heat, and the NO can be volatilized and recycled.

$$H_2O + 2\ NO_2 \rightarrow HNO_3 + HNO_2$$
$$2\ HNO_2 \rightarrow H_2O + 2\ NO$$

Additional ammonia then reacts with the nitric acid to produce ammonium nitrate, NH_4NO_3. Solid ammonium nitrate contains 35% nitrogen and as a solid is easily handled. It should be noted that ammonium nitrate is an oxidizing agent and can explode when mixed in a large volume with reducing materials. However, this poses no problem to the agriculturist when handling the pure chemical.

Nitrogen fertilizers were not "cheap" until after World War II. In 1943, nitrate production was greater than war demands, and the nitrogen went into fertilizers.

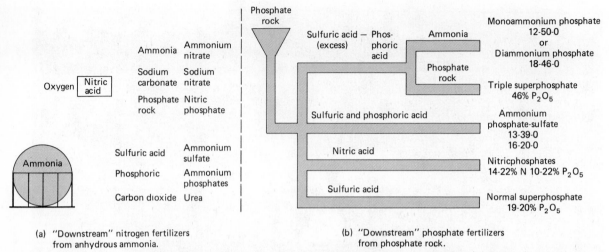

(a) "Downstream" nitrogen fertilizers from anhydrous ammonia.

(b) "Downstream" phosphate fertilizers from phosphate rock.

FIGURE 19–7 "Downstream" nitrogen fertilizers, fertilizers produced from anhydrous ammonia. (From Ronald D. Young and Frank J. Johnson: *Fertilizer Products.* Muscle Shoals, Alabama, National Fertilizer Development Center, TVA.)

A flooded soil quickly becomes a reducing medium as the air supply of oxygen is cut off. **Denitrification** occurs when conditions are such that nitrate is reduced to elemental nitrogen, which escapes into the atmosphere. It is estimated that soil fertilized with a soluble nitrate and then flooded for three to five days will lose 15% to 30% of the nitrogen as a result of denitrification.

Nitric acid is the number 11 commercial chemical. See inside the front cover.

UREA Urea (NH_2CONH_2) is probably one of the world's most important chemicals because of its wide use as a fertilizer and as a feed supplement for cattle. Ammonia and carbon dioxide react under high pressure near 200°C to produce first ammonium carbamate, which then decomposes into urea and water:

$$2\,NH_3 + CO_2 \rightarrow H_2N-\overset{\displaystyle O}{\underset{\displaystyle O^-NH_4^+}{C}} \rightarrow H_2N-\overset{\displaystyle O}{\underset{\displaystyle \|}{C}}-NH_2 + H_2O$$

AMMONIUM CARBAMATE   UREA

Sulfur-coated urea (SCU) slows the release of nitrogen from urea.

Urea synthesis also produces the compound biuret as a byproduct. Biuret must be removed from fertilizer-grade urea because it retards seed germination.

$$H_2N-\overset{\displaystyle O}{\underset{\displaystyle \|}{C}}-NH-\overset{\displaystyle O}{\underset{\displaystyle \|}{C}}-NH_2$$

BIURET

A slurry of water, urea, and ammonium nitrate is often applied to crops under the name of "liquid nitrogen." Such a solution can contain up to 30% nitrogen and is easy to store and apply.

"Acidulated" bones (bones in dilute H_2SO_4) were sold commercially by Sir James Murray (Ireland) in 1808.

Urea, applied to the surface of the ground around plants, is subject to considerable nitrogen loss unless it is washed into the soil by rain or irrigation. When urea hydrolyzes (decomposed by water), ammonia is formed and is lost to the air unless surrounded by moist soil particles. Such nitrogen losses could be as much as half of the nitrogen applied.

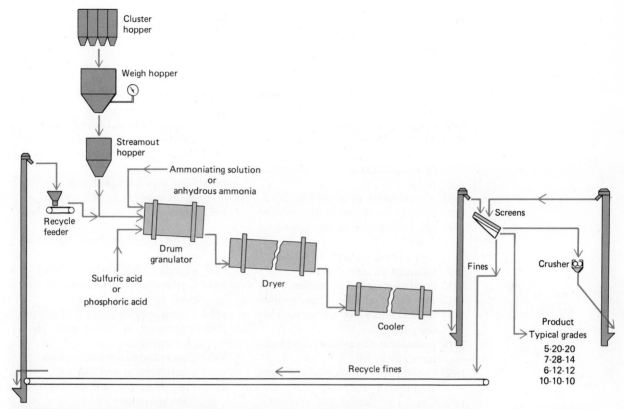

FIGURE 19-8 A typical granulation fertilizer process. Superphosphate could be made by adding phosphate rock to the weigh hopper and treating it with an aqueous solution of sulfuric acid. Micronutruents could be added to either the solid or liquid feed stock if there were no interfering side reactions.

Phosphorus and Potassium (Phosphate Rock and Potash)

Phosphate rock and potash are two minerals that can be mined, pulverized, and dusted directly onto deficient soil. Often they are specially treated to produce desirable mixing properties. Phosphorus, for example, is found scattered throughout the world in deposits of **phosphate rock,** which when treated with sulfuric acid becomes more soluble and hence produces a product of greater phosphorus availability called "superphosphate."

$$\underset{\substack{\text{PHOSPHATE} \\ \text{ROCK}}}{Ca_3(PO_4)_2} + 2\ H_2SO_4 \rightarrow \underset{\text{"SUPERPHOSPHATE"}}{\underline{Ca(H_2PO_4)_2 + 2\ CaSO_4}}$$

Phosphate rock itself is not as useful to a growing plant because of its very low solubility. Recently, it has become evident that phosphate rock demand will eventually exceed supply unless large new deposits are discovered.

Potassium in the form of **potash,** K_2CO_3, exists in enormous quantities throughout the world. A soluble form of potassium is its chloride (KCl), called muriate of potash or simply potash. Because this compound often occurs with sodium chloride, which is toxic to plants, the potash ores must be treated by some process such as recrystallization to separate the two compounds.

Phosphoric acid, H_3PO_4, is the number eight commercial chemical. It is prepared from phosphate rock. See inside the front cover.

The world has limited deposits of phosphate rock, which is essential to the manufacture of fertilizers.

Sir John Lawes (England) set up a superphosphate fertilizer factory in 1842. By 1853, there were 14 superphosphate factories in England.

Two major problems in fertilizer mineral supply: (1) higher impurity concentrations in available phosphate rock; (2) requirement for much greater depth in the deep mining of potash.

CHRONOLOGICAL EVENTS IN THE U.S. FOR THE "BIG THREE":

NITROGEN

1824 Peruvian guano first imported into U.S.
1830 Chilean nitrate of soda first imported
1893 Ammonium sulfate from coke ovens
1910 Cyanamid produced
1913 Ammonia first synthesized in Germany
1921 First commercial U.S. ammonia plant
1928 Ammoniation of mixed fertilizers
1929 Synthetic sodium nitrate produced
1932 Ammoniating solutions placed on market
      Ten ammonia plants on stream
1935 Granulation introduced
1935 Solid urea placed on market
1943 Physical condition of ammonium nitrate improved, rendering wide use possible
1944 Anhydrous ammonia applied commercially in field
1950 Continuous ammoniation and granulation
1956 Ureaform introduced commercially
1958 Fifty-eight ammonia plants on stream
1959 Ammonium polyphosphate introduced
1965 First vast (100 tons/day) ammonia plant on stream.

POTASH

1608 Potash salts from wood ashes shipped to England for glass manufacture in lieu of taxes
1790 George Washington signed patent for potash manufacture
1839 Discovery of potash in Germany
1870 First potash imported from Germany
1915 First potash produced from Searles Lake, California
1925 Potash discovered in New Mexico while drilling for oil
1931 Potash mining at Carlsbad, N.M.
1962 Canadian potash mining
1963 Commercial potassium nitrate produced in U.S.

PHOSPHATE

1825 U.S. farmers began using ground bones
1837 Phosphate ore discovered in South Carolina
1842 Sir John B. Lawes got patent for making superphosphate from sulfuric acid and phosphate rock (England)
1850 Production of mixed fertilizers in Baltimore
1852 First superphosphates (made from spent bone black) sold in U.S.
1867 Acidulation of rock phosphate began in U.S.
1867 South Carolina rock phosphate mining
1887 Florida rock phosphate mining
1890 Concentrated superphosphate first made
1894 Tennessee rock phosphate mining
1906 Idaho and Wyoming rock phosphate mining
1907 Concentrated superphosphate commercial production
1917 First U.S. production of ammonium phosphate
1929 Montana rock phosphate mining
1930 Superphosphate supplied over 90 percent of U.S. phosphate
1935 Granulated superphosphate produced
1942 Enriched superphosphate commercial production
1945 Cone mixer and continuous processing developed
1946 Start of rapid rise in concentrated superphosphate production
1954 Commercial production of diammonium phosphate
1956 Start of rapid rise in ammonium phosphate fertilizer
1957 High-analysis superphosphate from superphosphoric acid

From: John D. Hardesty, Plant Food Review, 1966.

Calcium, Magnesium, and Sulfur

Calcium is rarely in short supply as a plant nutrient. In addition to its widespread occurrence in clays and minerals, lime is used to adjust soil pH in the alkaline direction and at the same time serve as a rich supply of calcium. Calcium is insoluble enough in neutral and alkaline media that there is no danger of excessive uptake by the plants.

Magnesium deficiencies occur in acid sandy soils and in other soils in which the magnesium complexes are so strong that an insufficient amount of the element is released in solution form. Magnesium sulfate or a double salt of magnesium and potassium sulfate is effective in supplying the element. Liming with dolomitic limestone has the advantage of adding magnesium from the dolomite, while adjusting the pH and adding calcium.

A dolomite is a soil mixture containing both calcium and magnesium carbonate.

A sulfur deficiency usually occurs in sandy and well-drained soils and has been most often noted in the U.S. Southeast and Northwest. However, positive crop responses to added sulfate have been documented in 29 of the states. Several sulfur chemicals are effective, including ammonium sulfate, ammonium nitrate sulfate, ammonium thiosulfate, potassium sulfate, potassium magnesium sulfate, and ordinary superphosphate.

Micronutrients

The biggest difficulty in fertilizing for the micronutrients is the inability of the agriculturalist to find out simply what element is needed and in what amount. Precise analytical methods are available that will tell the analyst the concentration of each element in the soil, but these tests will not indicate how much of these elements are available for plant absorption. However, the analytical tests can be calibrated in terms of crop response in greenhouse and field studies. Obviously, this involves a costly and laborious process, and the results would likely vary from soil type to soil type and from crop to crop.

It is a problem to keep micronutrients uniform when mixed with granulated fertilizers.

Micronutrients are sometimes applied alone to the soil, the plant, or to the seed prior to planting. Iron is often used in a foliar application, and molybdenum compounds are dusted onto the seeds. The micronutrients are mixed with the macronutrients in many applications, but care must be taken to allow for the possible chemical reactions between the ingredients in the mixture. For example, manganese sulfate is effective as a source of the metal, but it is not as effective in the presence of polyphosphate fertilizers as in insoluble precipitate form. Another complication of micronutrient fertilization is the need for uniform application. In the case of boron, in which it is so easy to get into the toxic range, nonuniform application of the correct amount per acre would result in deficiencies and poisonous amounts in the same plot.

Waxes and oils have been used to bind micronutrients to the surface of fertilizer particles.

Considerable chemical innovation has been required in addressing some of the micronutrient deficiencies that have surfaced in crop production. We shall illustrate with iron. In the 1950s, severe iron deficiencies appeared on thousands of acres of both acid and alkaline soils in the Florida citrus groves. Applications of iron sulfate to the extent of 100 pounds per acre were ineffective in correcting the problem as the iron was bound by the soil chemicals and was therefore unavailable to the plants. It occurred to researchers working on this problem that they might introduce into the soil another complexing agent, which would hold the iron with just the right tenacity. The new complexes would keep the iron from the "gripping"

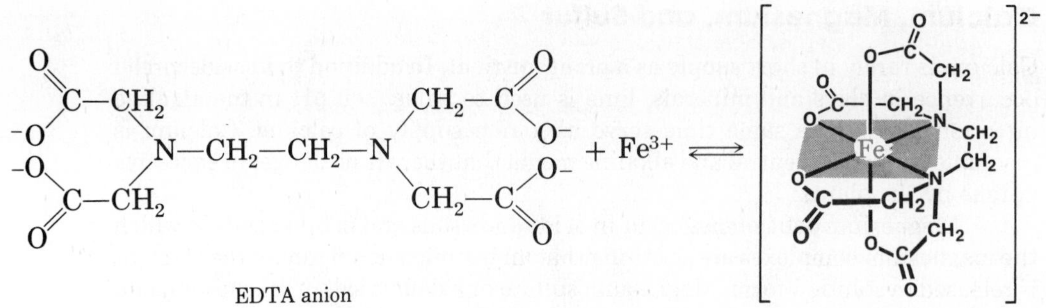

FIGURE 19-9   The EDTA anion formula and the structure of the iron-EDTA complex ion.

complexing agents already in the soil and at the same time would release enough of the iron to the roots. Ethylenediaminetetraacetic acid (EDTA) was tried, and it worked! The iron is complexed or **chelated** by EDTA in a complex structure in which the organic structure folds around the iron ion in six different directions (Figure 19-9). The word *chelate* comes from the Greek *chela,* meaning "claw." (The iron is "pinched" at six different bonding sites.) Since the work with the iron-EDTA complex, it has been learned that a number of the micronutrients can be introduced into the plants using this technique.

Table 19-4 lists the sensitivities of various crops to low levels of five micronutrients. This semiquantitative information shows considerable differences between the plant species in micronutrient requirements. An interesting exercise is to apply Liebig's Law of the Minimum to the information presented in this table.

TABLE 19-4   **Sensitivities of Various Crops to Low Levels of Five Micronutrients in Soil**

SENSITIVE	MODERATELY TOLERANT	TOLERANT
Zinc		
Apples	Barley	Alfalfa
Citrus	Clover	Carrots
Corn	Cotton	Fescue
Field beans	Potatoes	Grapes
Flax	Sorghum	Oats
Peaches	Soy beans	Peas
Pears	Sugar beets	Rye
Rice	Table beets	
Sudangrass	Tomatoes	
Sweet corn	Wheat	
*Molybdenum**		
Alfalfa	Cauliflower	Corn
Clover	Field beans	Cotton
Cowpeas	Grasses	Flax
Field beans	Soybeans	Fruit trees
Lespedeza	Sugar beets	Grasses
Peas	Tomatoes	Oats
Soybeans		Rice
		Wheat

TABLE 19–4 *(Continued)*

SENSITIVE	MODERATELY TOLERANT	TOLERANT
Manganese*		
Alfalfa	Barley	Barley
Citrus	Corn	Corn
Fruit trees	Cotton	Cotton
Oats	Field beans	Field beans
Onions	Fruit trees	Fruit trees
Potatoes	Oats	Rice
Soybeans	Potatoes	Rye
Sugar beets	Rice	Soybeans
Wheat	Rye	Vegetables
	Soybeans	Wheat
	Vegetables	
	Wheat	
Copper		
Alfalfa	Apples	Asparagus
Barley	Broccoli	Fescue
Carrots	Cabbage	Field beans
Citrus	Cauliflower	Peas
Lettuce	Celery	Potatoes
Oats	Clover	Rice
Spinach	Corn	Rye
Sudangrass	Cotton	Soybeans
Table beets	Cucumber	
Wheat	Peaches	
	Pears	
	Sorghum	
	Sugar beets	
	Sweet corn	
	Tomatoes	
Iron*		
Berries	Alfalfa	Alfalfa
Citrus	Barley	Barley
Field beans	Corn	Corn
Flax	Cotton	Cotton
Forage sorghum	Field beans	Flax
Fruit trees	Field peas	Grasses
Grain sorghum	Flax	Millet
Grapes	Forage legumes	Oats
Mint	Fruit trees	Potatoes
Ornamentals	Grain sorghum	Rice
Peanuts	Grasses	Soybeans
Soybeans	Oats	Sugar beets
Sudangrass	Orchard grass	Vegetables
Vegetables	Ornamentals	Wheat
Walnuts	Rice	
	Soybeans	
	Vegetables	
	Wheat	

* Some crops are listed under two or three categories because of variations in soil, growing conditions, and differential response of varieties of a given crop. (From J. J. Mortvedt: Know your fertilizers. *Farm Chemicals,* Nov. and Dec., 1980.)

PESTICIDES

A wide variety of organisms compete with humans for the agricultural product. Chemical control has been applied to each type of pest. Although all pesticides will not be treated in this brief account, Table 19–5 lists the classes. The agricultural market consumes about three fourths of the nearly $4.5 billion worth of pesticides produced in the United States each year. The Environmental Protection Agency has registered more than 2600 active ingredients for pesticides.

Plants are the ultimate source of all food. The natural enemies of plants include over 80,000 diseases due to viruses, bacteria, fungi, algae, and similar organisms, 30,000 species of weeds, 3000 species of nematodes, and about 10,000 species of plant-eating insects. One third of the food crops in the world are lost to pests each year, with the figure going above 40% for some of the emerging countries. Crop losses to pests amount to over $20 billion per year. Pesticides make a big difference as indicated in Table 19–6. **Organic farming** has come to mean farming without the aid of chemical fertilizers and pesticides. The production and use of these chemicals require less than 1% of the energy consumed in the United States. Without these chemicals, crop production, on average, would be 20% lower than at present. More land, and less productive land, would have to be used to produce the same amount of products; the increased crop land likely would have to be in excess of 30% more than is used today.

Insecticides

Before World War II, the list of insecticides included only a few arsenicals, petroleum oils, nicotine, pyrethrum, rotenone, sulfur, hydrogen cyanide gas, and cryo-

TABLE 19–5   Classes of Pesticides

CLASS	FUNCTION: KILLS	WORD ORIGIN (LATIN [L] OR GREEK [G])
Acaricide	Mites	(G) *akari* for mite or tick
Algicide	Algae	(L) *alga* for seaweed
Avicide	Birds	(L) *avis* for bird
Bactericide	Bacteria	(L) *bacterium*
Fungicide	Fungi	(L) *fungus*
Herbicides	Plants	(L) *herbum* for grass or plant
Insecticide	Insects	(L) *insectum*
Larvicide	Larvae	(L) *lar* for mask
Molluscicide	Snails, slugs, etc.	(L) *molluscus* for soft or thin shell
Nematicide	Round worms	(G) *nema* for thread
Ovicide	Eggs	(L) *ovum* for egg
Pediculicide	Lice	(L) *pedis* for louse
Piscicide	Fish	(L) *piscis* for fish
Predicide	Predators (coyotes, wolves, etc.)	(L) *praeda* for prey
Rodenticides	Rodents	(L) *rodere* for gnawing
Silvicide	Trees and brush	(L) *silva* for forest
Slimicide	Slimes	English
Termiticide	Termites	(L) *termes* for wood-boring worm

TABLE 19–6 Increased Yields (Test Plots Versus Controls) Due to Insecticide Use

CROP	INSECT	INCREASED YIELD (%)
Corn	Southwestern corn borer	24.4
Corn	Leafhopper or silage corn	38.4
Corn	Corn rootworm	10.7
Cotton	Boll weevil	11.9
Cotton	Bollworm	78.7
Cotton	Pink bollworm	25.5
Cotton	Thrips	40.3
Potatoes	Colorado potato beetle	45.6
Potatoes	European corn borer	52.8
Potatoes	Potato leafhopper	42.8
Soybeans	Mexican bean beetle	25.6
Soybeans	Stink bugs	6.5
Soybeans	Velvet bean caterpillar	14.2
Soybeans	Looper caterpillar	15.0
Wheat	Brown wheat mite	79.0
Wheat	Cutworms	29.7
Wheat	White grubs	27.7

Source: Washington Farmletter, 1979.

lite. DDT, the first of the chlorinated organic insecticides, was first prepared in 1873, but it was not until the beginning of the war that it was recognized as an insecticide.

The use of synthetic insecticides increased enormously on a worldwide basis after World War II (Table 19–7). As a result, insecticides such as DDT have found their way into lakes and rivers. There is a great variety of pesticides, and their use frequently leads to severe damage to other forms of animal life, such as fish and birds. The toxic reactions and peculiar biological side effects of many of the pesticides were not thoroughly studied or understood prior to their widespread use.

A good case in point is **DDT.** This insecticide, which had not been shown to be toxic to humans in doses up to those received by factory workers involved in its manufacture (400 times the average exposure), does have peculiar biological consequences. The structure of DDT is such that it is *not* metabolized (broken down) very rapidly by animals; it is deposited and stored in the fatty tissues. The biological half-life of DDT is about eight years; that is, it takes about eight years for an animal to metabolize one half of an amount it assimilates. The enzymes are just not present for rapid breakup of this molecule. If ingestion continues at a steady rate, it is evident that DDT will build up within the animal over a period of time. For many animals this is not a problem, but for some predators, such as eagles and ospreys, which feed on other animals and fish, the consequences are disastrous. The DDT in the fish eaten by such birds is concentrated in the bird's body, which attempts to metabolize the large amounts by an alteration in its normal metabolic pattern. This alteration involves the use of compounds that normally regulate the calcium metabolism of the bird and are vital to its ability to lay eggs with thick shells. When these compounds are diverted to their new use, they are chemically modified and are no longer available for the egg-making process. As a consequence, the eggs the bird does lay are easily damaged, and the survival rate decreases drastically. This

DDT

Half-life is the time required for half of the substance to disappear (see Chapter 7).

Do we really have to decide between pests and eagles?

TABLE 19–7   Insecticide Types and Examples

TYPE	EXAMPLE(S)	COMMENT(S)
Organochlorines	DDT	Insecticide of greatest impact (banned in U.S. in 1973)
	Chlordane Aldrin Dieldrin Endrin	Persistent, choice for termites (banned by EPA for agriculture, 1975–1980)
Polychloroterpenes	Toxaphene	Persistent, use peaked in 1976
Organophosphates	Malathion	Effective for plant and animal use
	Diazinon	Very versatile
Organosulfurs	Aramite	Choice for mite control
Carbamates	Carbaryl (Sevin)	Lawn and garden choice, low toxicity
Formamidines	Chlordimeform	Promising new group for resistant insects
Thiocyanates	Lethane	Limited use in aerosols
Dinitrophenols	Dinoseb	Choice for mildew fungi
Organotins	Cyhexatin	Most selective acaricide
Botanicals	Pyrethrum	Extracted from chrysanthemum
Synergists	Sesamex	Increases insecticide activity
Inorganics	Sulfur	Oldest insecticide
	Arsenicals	Stomach poisons
Fumigants	Methyl bromide	Kills insects in stored grain
Microbials	Heliothis virus	Specific for corn earworm and cotton ballworm
Insect growth regulators	Ecdysone	Environmentally sound
Insect repellents	Delphene	Superior to other repellents

process has led to the nearly complete extinction of eagles and ospreys in some parts of the United States where formerly they were numerous.

Since DDT is not readily biodegradable, there is a resultant buildup of this substance in natural waters. However, it is not an irreversible process; the Environmental Protection Agency reported a 90% reduction of DDT in Lake Michigan fish by 1978 as a result of the ban on the use of this insecticide. DDT and other insecticides such as **dieldrin** and **heptachlor** are referred to as **persistent pesticides.** * Substitutions of other substances with biodegradable structures are now made where possible. The compound **chlordan** is an example of just such a substitution. It is interesting to note that the structural differences between heptachlor (persistent) and chlordan (short-lived) are relatively slight (look at the chlorine atom on the lower five-membered ring).

City water treatment generally does not remove DDT.

Dieldrin

and aldrin

were banned by the courts from their major uses (such as on corn) in 1974.

HEPTACHLOR   CHLORDAN

* The use of DDT was banned in the United States in 1973, although it is still in use in some other parts of the world. Over 1.8 billion kg of DDT have been produced and used.

There are many other insecticides that are actually much more toxic to humans than is DDT. These include inorganic materials based on arsenic compounds, as well as a wide variety of phosphorus derivatives based on structures of the type

$$\begin{array}{c} R \\ \diagdown \\ \\ R' \diagup \end{array} \overset{\overset{\displaystyle Z}{\parallel}}{P} - X$$

Heptachlor and chlordan were banned for most garden and home use in December 1975.

where Z is oxygen or sulfur, R and R′ are alkyl, alkoxy, alkylthio, or amide groups, and X is a group that can be split easily from the phosphorus. Insecticides of this type include **parathion,** which is effective against a large number of insects but is also *very* poisonous to human beings. These compounds are anticholinesterase poisons. One of their most important properties, however, is that they are readily hydrolyzed to less toxic substances that are not residual poisons.

Perhaps the most publicized case of a poorly handled insecticide was the manufacture of **kepone** ($C_{10}Cl_{10}O$, a complex molecule containing several fused rings) in a converted filling station. The resulting contamination of the James River was extensive, as was the personal suffering of the workers. The dredging costs to clean the river are estimated by the EPA to be $1 billion, taking one dredge 120 years to complete the job. Total cleanup costs are estimated at $7.2 billion. Obviously, there are no immediate plans to clean up this chemical spill.

The choice of solutions to our problems with pesticides is not an easy one. The use of insecticides introduces them into our environment and our water supplies. A refusal to use insecticides means that we must tolerate malaria, plague, sleeping sickness, and consumption of a large part of our food supply by insects. It is obvious that continuing research is needed on new methods and materials for the control of insect populations.

PARATHION

MALATHION

The goal of the insecticide quest; a selectively toxic chemical that is quickly biodegradable.

Herbicides

Herbicides kill plants. They may be selective and kill only a particular group of plants, such as the broad-leaved plants or the grasses, or they may be nonselective, making the ground barren of plant life. The use of a particular herbicide may be in a granular form, which is worked into the soil prior to planting the crop in a preemergence application, or the herbicide may be best applied as a liquid spray at various stages after planting. The liquid spraying may be either a preemergence or a postemergence application; the choice depends on the particular chemical, weed, soil type, and crop involved. Also, different species of plants in the same class respond differently; some require only one application of the herbicide and others require as many as three applications before being controlled. The use of herbicides is a complicated matter, and several factors must be considered for each type of application.

Nonselective herbicides usually interfere with photosynthesis and thereby starve the plant to death. On application, the plant quickly loses its green color, withers in the lack of energy to carry on the life processes, and dies. The selective herbicide acts like a hormone, a very selective biochemical catalyst that controls a particular chemical change in a particular type of organism at a particular state of its development. Most selective herbicides in use today are growth hormones; in causing abnormal growth in a plant, the symptoms would be swelling cells, such as a leaf becoming so thick that chemicals cannot be transported through the leaf, and thick roots that are unable to absorb needed water and nutrients. In theory, selective herbicides can become far more specific as the particular biochemistry of each plant species is understood, and it is expected that many new products will follow the intense research in this field.

The traditional method for the control of weeds in agriculture was tillage. Only in the early 1900s was it recognized that some fertilizers were also weed killers. For example, while calcium cyanamide, CaNCN, was used as a source of nitrogen, it was observed to retard the growth of weeds. Arsenites, arsenates, sulfates, sulfuric acid, chlorates, and borates have found use as weed killers. A typical product still in commercial use contains 40% sodium chlorate, $NaClO_3$; 50% sodium metaborate, $NaBO_2$; and 10% inert filler. These herbicides are nonselective and must be used with considerable care in protecting the desired plants.

Nitrophenol was used in 1935 as the first selective, organic herbicide, and it was also in the 1930s that work began on the auxins or hormone-type weed killers. The most widely used herbicide today came out of this work and has the common name 2,4-D; the chemical name and formula are:

Nitrophenols:

ORTHO   META   PARA

2,4-D
2,4-DICHLOROPHENOXYACETIC ACID

The corresponding trichloro- compound (common name: 2,4,5-T) has also been produced and shown to be highly effective. The only difference is the additional chlorine atom on the benzene ring in the fifth position. Agent Orange, widely used as a defoliant in the Vietnam War, is a mixture of these two compounds. The second compound, 2,4,5-T, has been removed by law from many markets because of a number of health problems associated with its use. Probably most of the problems were caused by an impurity, dioxin, present in the product. Dioxin, described in Chapter 16, is a severe poison with a toxicity equaled by few compounds. It will be interesting to see if 2,4,5-T, which can now be commercially produced free of the dioxin, will be reestablished as a herbicide. Both of these compounds cause an abnormally high level of RNA in the cells of the affected plants, and the plants literally grow themselves to death.

Several different triazines have been effective as herbicides, the most famous one being atrazine. Atrazine is widely used in no-till corn production or as a weed control in minimum tillage.

1,3,5-TRIAZINE

ATRAZINE
2-CHLORO-4-ETHYLAMINO-6-ISOPROPYLTRIAZINE

Atrazine is a poison to any green plant if it is not quickly changed into another compound. Corn and certain other crops have the ability to render the atrazine harmless, while the weeds cannot. Hence, the weeds die, and the corn shows no ill effect.

A typical fact sheet by a state agricultural extension service lists six formulations using atrazine for no-till corn and nine mixtures for preplant and preemergence applications on tilled ground. For example, one formulation for no-till calls for a mixture of 2 to 3 pounds of atrazine, 0.25 to 0.50 pounds of paraquat, and a surfactant. The surfactant lowers the surface tension of the liquid spray and makes it easier for the liquid to wet and penetrate the plant surface.

Paraquat is also used as a contact herbicide. When applied directly to affected plants, they quickly develop a frostbitten appearance and die. Paraquat has received considerable attention because of its use in spraying poppy and marijuana fields by government aircraft in Mexico and other places where these crops are grown illegally. Like atrazine, paraquat has a nitrogen atom in each aromatic ring of the two-ring system.

Soap is a surfactant in water, allowing it to wet glass or greasy surfaces.

PARAQUAT
1,1'-DIMETHYL-4,4'-BIPYRIDINIUM DICHLORIDE

To further illustrate the complexity of herbicide selection, the soybean grower might be presented a fact sheet telling about 20 different herbicides in over 30 different formulations. Prominently used are trifluralin (α,α,α-trifluoro-2,6-dinitro-N,N-dipropyl-p-toluidine), the sodium salt of acifluorfen (sodium 5-[2-chloro-4-(trifluoromethyl)phenoxyl-2-nitrobenzoate], and bentazon (3-isopropyl-1-H-2,1,3-benzothiadiazion-4[3 H]-one-2,3-dioxide).

The amount of energy saved in no-till farming is enormous, a savings made possible by herbicides. Also the saving of topsoil is considerable, as the cover from the last crop holds the soil against wind and water runoff. However, agriculturists who use herbicides are highly dependent on a close relationship with the agricultural research institutions for the selection of the herbicide that will do the job desired without harmful side effects. Such selections are dependent on considerable and ongoing research, much of which is on a trial-and-error basis on test plots. The recommended procedure today may be outdated in the next growing season.

Rodenticides

Rodenticides include coumarins, which are anticoagulants, botanicals such as strychnine, zinc phosphide (Zn_3P_2), and organophosphates illustrated by phorazetim. The rodenticide most often used is the coumarin **warfarin.** Warfarin is relatively safe to use around children and pets because, as an anticoagulant, it requires repeated dosages to be lethal. Rodents, unlike children and pets, will feed on warfarin bait several times. The poison inhibits the formation of prothrombin, a blood-clotting agent, and also causes capillary damage; the animal bleeds to death internally. Two new anticoagulants were introduced in the 1980s, talon and maki; they are more effective and also more dangerous because one dose will kill.

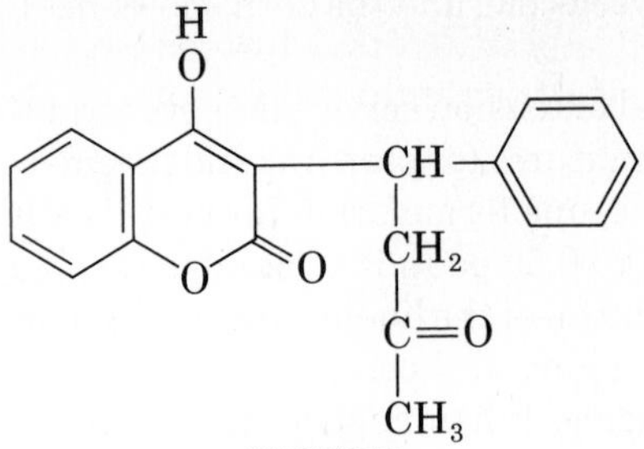

WARFARIN

Avicides

When birds become pests, they can be effectively fought with repellents such as avitrol (4-aminopyridine), perch treatment with insecticides such as endrin and entrex, which are absorbed through the feet, toxic baits like strychnine, and a newer approach, using chemical chemosterilants. Ornitrol, a derivative of cholesterol, produces temporary sterility in pigeons but has little or no effect on mammals. Detergents, or other surfactants, are sometimes sprayed on roosts of starlings with limited success. The detergent destroys the oil protection in the feathers, and the skin of the bird becomes exposed to the cold water. If the weather is cold enough, the birds die of exposure.

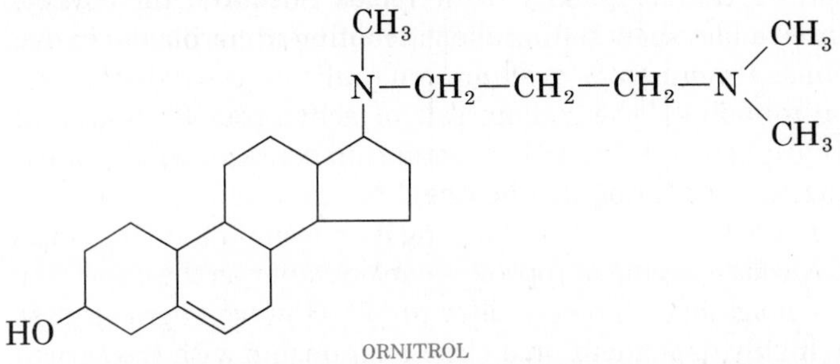

ORNITROL

Apparently because of the societal sensitivity that we have for the beauty and importance of bird life, essentially all of the formulations prepared as avicides are controlled by law to be dispensed by those licensed for this purpose.

Nematicides

The nematodes are smooth round worms, and some species of this group infest plant root systems, where they damage the roots and encourage other organisms such as the fungi. The most effective treatment for the offensive worms is fumigation with nematicides such as 1,3-dichloropropene or ethylene dibromide.

The use of ethylene dibromide has now been banned in the U.S. because of its link to cancer.

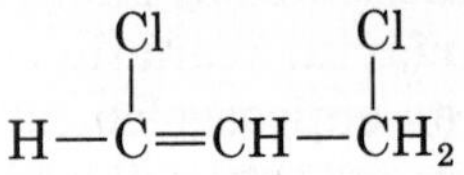

1,3-DICHLOROPROPENE

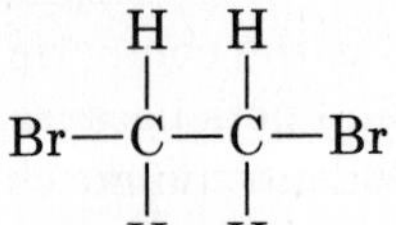

ETHYLENEDIBROMIDE
or
1,2-DIBROMOETHANE

Molluscides

Several different species of mollusks, especially of the snail group, inhabit soils. The primary problem associated with these organisms is that they host numerous parasites that afflict farm animals and humans. **Metaldehyde** was introduced in 1936 and has been effective in the control of mollusks.

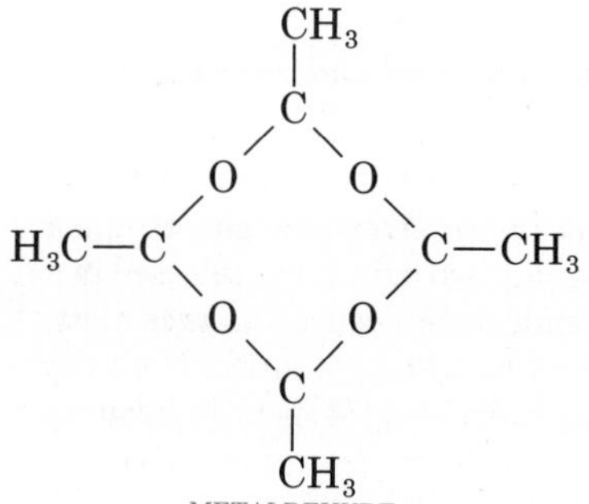

METALDEHYDE

PLANT GROWTH REGULATORS

Every phase of plant growth and development can be altered and controlled to some degree by chemicals known as plant growth regulators. In 1932, it was noticed that ethylene or acetylene promotes flowering in pineapples. Two years later, it was discovered that a group of compounds that came to be known as auxins induce elongation in shoot cells. Auxins are used to thin apples and pears, to increase yields in beans, sugar cane, and potatoes, to assist the rooting of cut plants, and to increase flower formation. Perhaps the most remarkable plant growth regulator is a group of compounds known as the gibberellins. Figure 19–10 presents the basic structure for gibberellic acid, and the following quotation from the *Farm Chemical Handbook,* published by the Meister Publishing Company of Willoughby, Ohio, illustrates its many uses:

> The gibberellins, the acid and its derivatives, are used to elongate cluster, increase berry size, and reduce bunch rot of grapes; to maintain color, delay yellowing, and reduce percentage of small tree-ripe fruit in lemons; reduce rind staining water spots, and tacky rind in Navel oranges; to overcome certain symptoms of cherry yellow virus diseases in sour cherries; to produce taller, thicker stalks of celery harvested in cooler seasons; to prevent head formation and induce production of the seed stalk in lettuce; to increase fruit set and yields of the Orlando tangelo; to accelerate maturity of artichokes and to shift the harvest to an earlier date; to stimulate uniform sprouting

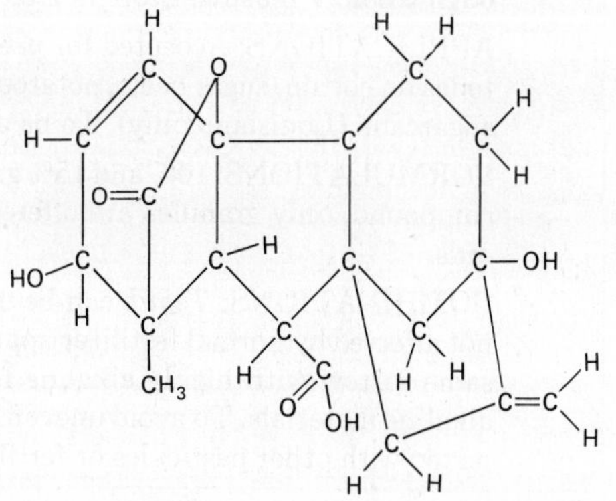

FIGURE 19–10 Gibberellic acid.

FEATURE—DANGERS IN PREPARATION AND USE OF A PESTICIDE

Two thousand people were poisoned to death in 1984 in Bhopal, India, by methyl isocyanate, $CH_3—N=C=O$. The gas was released into the atmosphere by accident in a heavily populated area of the city. Methyl isocyanate is used as an intermediate in the preparation of aldicarb oxime, a systemic insecticide, acaricide, and nematicide. The pesticide is a solid material designed for soil use.

$$CH_3—S—\underset{\underset{CH_3}{|}}{\overset{\overset{CH_3}{|}}{C}}—CH=N—O—\overset{\overset{O}{||}}{C}—\underset{\underset{H}{|}}{N}—CH_3$$

Aldicarb oxime, sold under the trade name Temik

CHEMICAL NAME: 2-METHYL-2(METHYLTHIO) PROPIONALDEHYDE 0-(METHYLCARBAMOYL) OXIME

The pesticide, though considerably less dangerous than the intermediate gas, still is a chemical to be treated with respect. On August 11, 1985, aldicarb oxime was released in a cloud from a chemical plant in West Virginia. This chemical spill caused an area alert, sent 125 people to the hospital, but apparently produced no lasting injuries. Note the following technical information from the *Farm Chemical Handbook* (Meister Publishing Company) presented for the user of this pesticide.

TEMIK

CHEMICAL NAME: 2-Methyl-2(methylthio) propionaldehyde *O*-(methylcarbamoyl) oxime.

COMMON NAME: *aldicarb* (ANSI, BSI, ISO).

OTHER NAMES: *UC21149, OMS771* (WHO assigned no.).

ACTION: Systemic insecticide, acaricide and nematicide for soil use.

CHEMICAL PROPERTIES: Technical *aldicarb;* white, crystalline solid; specific gravity, 1.1950 at 25/20°C; vapor pressure, 1×10^{-4} mm Hg at 25°C; solubility in water 0.6% at 25°C; non-corrosive to common metals and plastics.

TOXICITY: Technical *aldicarb;* acute oral LD_{50} (rat), 0.9 mg/kg; acute dermal LD_{50} (rabbit), >5.0 mg/kg. *Temik* 10G; approximate oral LD_{50} (rat), 7.0 mg/kg; acute dermal LD_{50} (rat), dry: 2100–3970 mg/kg.

SIGNAL WORD: DANGER.

ANTIDOTE: Atropine sulfate. Do not use 2-PAM, opiates and cholinesterase inhibiting drugs.

HANDLING AND STORAGE CAUTIONS: Do not mix *Temik* granules with water or the resultant solution may be seriously hazardous. Do not use applicators that will grind the granules. Wear long-sleeved clothing and protective gloves when handling. Wash hands and face before eating or smoking. Bathe at the end of work day, washing entire body and hair with soap and water. Change contaminated clothing daily and wash in strong washing soda solution and rinse thoroughly before reusing. Store in a clean, dry and well-ventilated area.

APPLICATIONS: Accepted for use to control certain insects, mites and nematodes on cotton, sugar beets, potatoes, peanuts, ornamentals, sweet potatoes, and sugarcane (Louisiana only). To be used only as a soil application.

FORMULATIONS: 10% and 15% granules. Because of high toxicity of the parent compound, only granules are offered, thus significantly reducing handling hazards.

COMBINATIONS: *Temik* can be used concurrently with many pesticides. It is not affected by normal fertilizer applications; however, it should not be applied in same furrow with highly alkaline forms. Do not use with lime or other highly alkaline materials. To avoid uneven distribution in the field, *Temik* should not be mixed with other pesticides or fertilizers prior to application.

The tank in the West Virginia spill contained 65% methylene chloride and 35% aldicarb oxime.

of seed potatoes that have not had a full rest period; to delay harvesting, to produce a brighter colored, firmer fruit, and to increase size of sweet cherries; for reduction of internal browning and watery pits of the Italian prune and to increase yields of marketable forced rhubarb and to break dormancy on plants receiving insufficient chilling; to increase yield and pickability of Fluggle hops; to improve fruit set in blueberries where natural honeybee pollination may be insufficient, and to overcome cold stress effects on certain grasses.

There are six classes of plant growth regulators, according to the American Society for Horticultural Science: auxins, cytokinins, ethylene generators, gibberellins, growth retardants, and inhibitors. The herbicide 2,4-D is also an auxin, which is used to prevent preharvest on older fruit trees, to prevent leaf and fruit drop following insecticide use, and to delay fruit maturity in order to increase fruit size. The cytokinins are derivatives of the nitrogenous base adenine (Chapter 15), and as a group induces cell division. This results in the prolongation of the storage life of green vegetables, cut flowers, and mushrooms. Benzoic acid is an effective inhibitor and can be used to retard seed germination.

SELF-TEST 19–B

1. Most virgin soils can support crop production for a decade or more before fertilization is needed. (   ) True or (   ) False
2. Would the numbers, such as 6-12-6, for a manure fertilizer be higher or lower than for a typical chemical fertilizer? _____
3. Which would be properly termed a quick-release fertilizer, potassium nitrate for potassium or manure for nitrogen? _____
4. Would the nitrogen in ammonia be considered "fixed" nitrogen? _____
5. Pure ammonia under ordinary conditions is a solid, liquid, or gas? _____
6. If ammonia is oxidized to NO_2 and dissolved in water, two acids are produced. They are _____ and _____.
7. Flooded soils cause nitrification or denitrification? _____
8. The chemical formula for potash is _____.
9. Which is more likely to be in short supply as a plant nutrient in soils — calcium or magnesium? _____
10. Approximately what percentage of the food crops of the world is lost to pests each year? _____
11. The first chlorinated organic insecticide was _____.
12. Which of the following is not a persistent insecticide: DDT, dieldrin, heptachlor, or chlordan? _____
13. Which is more likely to be a hormone, a selective or a nonselective herbicide? _____
14. The most widely used herbicide today is _____.
15. The rodenticide most often used is _____.
16. Which is more likely to be a compound similar to soap, an insecticide, a nematicide, or an avicide? _____
17. Gibberellic acid is a _____ _____ _____.

MATCHING SET

_____	1.	Humus
_____	2.	Cato
_____	3.	Liebig
_____	4.	Young
_____	5.	Subsoil
_____	6.	Loam
_____	7.	Sweet soil
_____	8.	Percolation
_____	9.	Leaching
_____	10.	Selenium
_____	11.	Compost pile
_____	12.	Silicon
_____	13.	Zinc
_____	14.	Phosphate
_____	15.	Calcium
_____	16.	Lime
_____	17.	Ammonia
_____	18.	Haber process
_____	19.	Urea
_____	20.	Sulfur deficiency
_____	21.	Manganese
_____	22.	EDTA
_____	23.	DDT
_____	24.	2,4-D
_____	25.	Ornitrol

a. Solid nitrogen compound
b. Friable material
c. Readily available secondary nutrient
d. Result of dissolution
e. Herbicide
f. Source of humus
g. Insecticide
h. Provides nitrogen to soil
i. Central element in soil
j. Likely in sandy soils
k. Applied organic chemistry to agriculture
l. A calcium compound
m. Alkaline
n. A micronutrient
o. Described seed selection
p. Fortunately leached
q. Avicide
r. Extension worker
s. An alkaline anion
t. Several feet in thickness
u. A gas
v. Passes water
w. Ammonia production
x. Trace nutrient
y. Complexing agent for micronutrient
z. Rodenticide

QUESTIONS

1. What are the major factors that determine the qualities of a virgin soil?
2. Describe the horizons that would be found in a typical soil.
3. Why is the air in soil of a different chemical composition than the air around us?
4. Which will contain more air per soil volume, a clay or a loam? Give a reason for your answer.
5. What causes a soil to be sour? Sweet?
6. If crushed limestone is spread on soil, will it raise or lower the pH of the soil? Explain.
7. What would be a typical ratio for the number of pounds of water required to produce one pound of food?
8. What are two types of sorption for water in soils? Distinguish between the two types.
9. Explain the relationship between the description of water as the universal solvent and the leaching effect in the removal of soil nutrients.
10. What are the principal elements in the soils of the Earth? Contrast the elements in this list to the elemental composition of the crust of the Earth.
11. Which groups of elements are first leached from soils, the alkali and alkaline earth metals or the

Group III and transition metals? What is the effect of this selective leaching on soil pH?

12. What is the effect of selenium in the soil on plant and animal life? How is the concentration of selenium reduced through natural processes?

13. What are two important roles of humus in the soil? Leaves turned into the soil to produce humus will raise or lower the soil pH?

14. Relate the colors of soils to the plant nutrients that they likely contain.

15. The oxide of what element predominates in the soils of the Earth?

16. What three elements are obtained from the air and water as nonmineral plant nutrients?

17. What are the three primary mineral plant nutrients that are considered first in fertilizer formulations?

18. Explain the necessity of nitrogen fixation for plant growth. Give a physical, a biological, and a chemical method for nitrogen fixation.

19. Are nitrates or phosphates more easily leached from the soil? Explain.

20. Soil phosphates will be in different ionic forms depending on the pH. What are the predominant forms of ionic phosphate in sweet soils?

21. Which is more likely to be a problem in farming, a soil shortage of N, P, and K or a shortage of Ca, Mg, and S? Give a reason for your answer.

22. Explain the numbers 6-12-6, which you might find on a fertilizer bag.

23. What is the danger in the use of anhydrous ammonia as a chemical fertilizer?

24. What is superphosphate? How is it made?

25. What two herbicides were formulated to produce Agent Orange? Which of these herbicides is presently banned in the United States for agricultural use?

26. Why are the rodenticides talon and maki much more dangerous to use around pets and children than the older rodenticide warfarin?

27. List some uses of plant growth regulators that are of interest to you. Is there a relationship between the use of these chemicals and the feeding of antibiotics to cattle to increase beef production? Explain.

28. Investigate a no-till farming operation. What herbicides are used? How is energy saved and, at the same time, how is additional energy required in no-till farming? What is the effect of no-till farming on the conservation of topsoil?

29. Trace the history of DDT in the rise and fall of its use in agriculture. Debate the question of whether it has been more good than bad for the human race.

30. Debate the proposition that herbicides, such as paraquat, should be sprayed from airplanes to destroy crops grown to produce illegal drugs.

31. Contact the U.S. Department of Agriculture through the Soil Conservation Service in your area. Find out if there is documentation of a micronutrient problem in the agriculture in your state. Define the problem if one is found, and outline a chemical solution.

32. How is it possible that your great-grandfather may have been happy with 25 bushels of corn per acre, and farmers today, who cannot average over 100 bushels per acre, cannot adequately compete in the economics of corn production?

33. In the period after World War II, most farmers fertilized "enough to be sure." It is likely now that the farmer will have the soil analyzed and have a fertilizer formulated on prescription. What is the cause for this change? Can you see an effect that this change in farming practice might have on water pollution?

34. Explain how Liebig's Law of the Minimum was a consideration of macro and trace plant nutrients.

35. Based on your present level of knowledge and understanding, how safe do you think a synthetic food diet would be?

36. Argue for or against the proposition: Early agriculture suffered not so much from the lack of available scientific information as it did from the lack of the dissemination of the available knowledge. Which, in your judgment, is the greater of the two problems today?

37. Give the approximate dates for the beginning of the world population explosion and the beginning of the application of scientific information to agriculture. In your opinion, is there a direct cause-and-effect relationship between these phenomena?

38. Guano was used as a fertilizer in colonial America. If you do not know the meaning of this word, look it up in the dictionary. What nutrients would guano add to the soil?

Chapter 20

NUTRITION — THE BASIS OF HEALTHY LIVING

Nutrition is the science of seeking the proper diet to maintain health, supply energy, maintain proper temperature, promote growth, and replace worn or injured tissues, especially in human beings. The old saying " we are what we eat" is true in the sense that we must replace parts of the body continually, and the material to make the replacements comes from our food. The skin that covers us now is not the same skin that covered us seven years ago; our skin now is made entirely from new cells. The fat beneath our skin is not the same fat that was there just a year ago. Our oldest red blood cell is 120 days old. The entire lining of our digestive tract is renewed every three days. Many chemical reactions are required to replace these tissues, and all of these reactions are supplied ultimately by what we eat.

Nutrition, then, is concerned with the chemical requirements of the body — the nutrients. These are the raw materials of our chemical, living factory. The six classes of nutrients are carbohydrates, fats, proteins, vitamins, minerals, and water. The preparation, molecular structures, and fundamental properties of these nutrients were discussed in Chapters 13 and 15 (fats). In this chapter we shall focus on the health effects of too much or too little of these nutrients, why we need the nutrients (their physiological functions), general ways to assess nutritional status, and the recommended requirements of nutrients.

Chapter 15 and part of Chapter 13 (fats) are basic to material in this chapter.

In our society, a discussion of needed food would be incomplete without mention of food additives. While the vitamins, minerals, and sugars added to natural foods often have nutritional value, chemicals added to preserve food, to make the food taste and look better, or to give the food a desired consistency often do not have nutritional value. It is difficult to find food that has not been doctored up with palate-pleasing food additives. We shall discuss how these additives act chemically and why they are added to food.

THE SETTING OF MODERN NUTRITION

The concern for nutrition differs in different parts of the world. In the United States, we no longer have the problem of diseases such as beriberi, scurvy, and rickets, which are due to a lack of a particular nutrient, but we do confront diseases such as cancer, heart and circulatory disorders, and so on, some forms of which are perhaps caused by an excess of certain nutrients.

One of the amazing observations of life is the ability of each form of life to select and eat the right foods at the correct time in the proper amount. Except for famines and similar disasters, it seems that even primitive people with no knowledge of modern nutrition eat well and exhibit generally good health. In a modern society with the large variety of prepared foods, the instinctive nature of human beings is set aside, and training and education must step in to provide proper nutrition.

Recognition of human nutritional needs began with the determination of the gross chemical composition of food. By the beginning of the 20th century, the existence of carbohydrates, fats, and proteins was well recognized, and the heat-producing values of the various foodstuffs had been determined. The role of iron and several other minerals had been recognized in human nutrition.

At the turn of the present century, an increasing amount of food was obtained from stores rather than from fresh farm products. In the store-bought food, too often there were appreciable amounts of dirt and other filth. False or misleading labels described the food. As a result of these problems, a federal food and drug law was passed in 1906. This law was replaced in 1938 by a more comprehensive act, known as the Pure Food and Drug Law. While these laws cleaned up the food and made labelling more reliable, in one sense they caused the food to be too clean chemically. In the process of food purification, nutrients such as vitamins and minerals were removed or destroyed. The replacement of needed nutrients began the food additive process.

As early as 1915, E. V. McCollum had discovered the fat-soluble group of closely related organic compounds known as vitamin A and a water-soluble collection of organic compounds known as B vitamins. Following this, other investigators discovered other vitamins, essential minerals, essential fats, and essential amino acids.

With the chemical raw materials identified — the nutrients — diversified investigative research led to the molecular structures, chemical reactions, and effects (excesses and deficiencies) of the nutrients. Much has been learned, but there is more to be discovered about the nutrients in an ongoing process.

The study of the effects of nutrients on the human body is interesting and has some general precautions. The obvious method to test the effect of a particular nutrient on the human body is to give an excess of the nutrient and observe its effect or to withhold the nutrient and observe the effect of its absence. This method of testing would be relatively simple except for two major obstacles. One hindrance is the ethics and morality involved in using human subjects for testing when health may be harmed. Even volunteers for the study do not offset the ethics and morality issues. A second obstacle to using human beings is the individuality of each of us. What would be learned by one would not necessarily be carried over to the other of us. For example, in healthy adults, the variation in weight of the human liver is about fourfold, and the free hydrochloric acid in the stomach varies from 0.03 to 0.13 M (molar). Because of individual anatomical and physiological differences in human beings, what is learned from one person does not necessarily apply per se to another person. Individual variations with respect to the need for many of the essential nutrients are quite large, as we shall see.

When animals (such as rats, dogs, cats, guinea pigs, monkeys) are used for nutritional studies, anatomical and physiological differences occur within each species as such differences occur among human beings. When animals are used for nutritional studies, the question "Are the systems sufficiently alike to apply the results in animal studies to humans?" always arises. One hindrance in extrapolating the results is that many animals do not have the same biochemical pathways as humans. As just one example, human beings require the intake of vitamin C in their diets. Rats and dogs, however, do not require the intake of vitamin C in their diets because they can produce vitamin C in their bodies. The guinea pig requires vitamin C but does not make the vitamin in its body. Even when an animal such as the guinea pig is used in vitamin C studies, we are not always sure what chemicals the animal is manufacturing that the human would have to ingest to obtain the same results. The scaling up or scaling down of amounts is not always proportional for the same effect.

While valuable information can be learned from animal studies, the human body has a thinking and psychologically active mind that also affects nutrition. Emotional stress and upset can alter eating habits and digestive functions. We respond to TV and other forms of advertisement and eat what is suggested. Nervousness can lead to overeating.

GENERAL NUTRITIONAL NEEDS

Basically we need the elements that compose the tissues, organs, and systems of the body. The major elements in the human body are oxygen (65%), carbon (18%), hydrogen (10%), and nitrogen (3%) on a weight percentage basis. Practically all of these elements are in the form of water or organic compounds. Other elements present (and required) are calcium (2%), phosphorus (1%), potassium (0.35%), sulfur (0.25%), sodium (0.15%), chlorine (0.15%), and magnesium (0.05%). The total thus far equals over 99.9% of the total body weight. The rest of the body is composed of trace amounts of other elements. Table 20–1 lists all of the elements found in the human body, the total amount of each element in a 70-kg man, and the daily intake. Either a lack of proper nutrients or an excess of improper nutrients can produce *malnutrition*.

How is nutritional status assessed?

General nutritional needs are determined by one's nutritional status. There are many ways to assess nutritional status of a human being: (1) size and weight — neither too thin nor too fat; (2) effect of stress — if a person is well nourished, stress is more bearable; (3) intelligence — undernourished and malnourished individuals are more dull and unresponsive; (4) ability to reproduce — undernourished individuals are sometimes not able to reproduce; and (5) biochemical and clinical analysis — analysis of urine, blood, appearance, weight change, posture, and other chemical and medical tests and observations. Items 1 and 5 are used more often to determine nutritional status because they are more reliable (not as many variables) and are determined more quickly.

Elements in the body are combined into many chemical compounds in a precise fashion in order to build the various tissues and organs of the body. Knowledge of the composition and chemistry of these organs and tissues is needed to understand the nutritional requirements of humans. For example, liver, fat tissue, and muscle tissue are criteria for nutritional status of an individual. If one consumes an excess of alcohol and has a poor diet, the liver grows and the cells become fatty. Fat tissue in humans can vary from a low of 13% to a high of 70% of body weight. The amount of fat tissue depends on the amount and type of food consumed, the age of the person, and certain inherited traits. Muscle tissue ranges from 25% to 45% by weight of the body. A weightlifter may possess a large percentage of muscle tissue. As we grow older, the amount of muscle tissue decreases in

TABLE 20–1   Chemical Elements in the Adult Human Body*

ELEMENT	TOTAL AMT. IN BODY (mg)	DAILY INTAKE (mg)	ELEMENT	TOTAL AMT. IN BODY (mg)	DAILY INTAKE (mg)
Aluminum	61	34	Manganese	12	3.7
Antimony	ca.8	ca.50	Mercury	15	0.015
Arsenic	18	1	Molybdenum	9	0.3
Barium	22	ca.0.8	Nickel	10	0.4
Beryllium	0.036	0.012	Niobium	ca.120	0.6
Bismuth	ca.0.2	0.02	Nitrogen	1,800,000	16,000
Boron	20	1.3	Oxygen	43,000,000	3,500,000
Bromine	200	7.5	Phosphorus	780,000	1,400
Cadmium	50	0.15	Potassium	140,000	3,300
Calcium	1,000,000	1,000	Radium	3×10^{-8}	$(1-7) \times 10^{-15}$
Carbon	16,000,000	300,000	Rubidium	680	2.2
Cesium	1.5	0.01	Selenium	15	0.15
Chlorine	95,000	5,000	Silver	0.8	0.07
Chromium	ca.6	0.15	Sodium	100,000	ca.5,000
Cobalt	ca.1.5	0.30	Strontium	32	1.9
Copper	72	3.5	Sulfur	140,000	850
Fluorine	2,600	1.8	Tellurium	9	0.6
Gold	ca.9	—	Tin	ca.16	4
Hydrogen	7,000,000	3,500,000	Titanium	9	0.9
Iodine	13	0.20	Uranium	0.09	0.002
Iron	4,200	15	Vanadium	ca.10	2
Lead	120	0.44	Zinc	2,300	13
Lithium	80	2	Zirconium	ca.450	4.2
Magnesium	19,000	340			

* The approximate values given here refer to a 70-kg man. Adapted from W.S. Synder, M. J. Cook, E. S. Nasset, L. R. Karhausen, G. P. Howells, and I. H. Tipton: *Report of the Task Group on Reference Man,* ICRP Pub. 23, Oxford, England: Pergamon Press, 1975.

most of us. Despite common belief, the amount of muscle tissue is not determined by the amount of protein eaten but rather by the amount of exercise.

NUTRIENT REQUIREMENTS — MDR, RDA, USRDA, FAO, AND WHO

Early studies in nutrition searched for levels of nutrients that would prevent serious disease. When sufficient data had been collected from animal and biochemical studies, a value called the MDR, or minimum daily requirement, was established for each nutrient. Although used extensively throughout the 1960s in dietary surveys, the MDR is no longer used. As it was set up, the MDR, in most cases, did not allow for differences in the nutrient requirements for age, sex, and such factors as stress.

Minimum daily requirement (MDR).

A more specific list of recommendations for nutrient intake is the Recommended Dietary Allowance (RDA), published by the Food and Nutrition Board of the National Academy of Sciences and the National Research Council. Sample data are shown in Table 20–2, RDAs are given for age groups, sex differences, energy requirements, and special categories of pregnant and lactating (milk-producing) women. The RDA is the intake level of a nutrient that should insure adequate nutrition for good health for as large a percentage of the population as possible, taking into account the effects of some stress and biochemical differences

Recommended Dietary Allowance (RDA).

TABLE 20–2 A Selection of Recommended Dietary Allowances (RDA), 1980

	AGE (YEARS)	WEIGHT (POUNDS)	HEIGHT (INCHES)	PROTEIN (g)	VITAMINS				CALCIUM (mg)
					D (μg)	C (mg)	E (mg)	B_6 (mg)	
Infants	0.5–1.0	20	28	18	10	4	35	0.6	540
Children	4–6	44	44	30	10	6	45	1.3	800
Males	19–22	154	70	56	7.5	10	60	2.2	800
Females	19–22	120	64	44	7.5	8	60	2.0	800
Pregnant women				+30	+5	+2	+20	+0.6	+400
Lactating women				+20	+5	+3	+40	+0.5	+400

of humans. The listing was first published in 1968 and was revised in 1974 and 1980. For most nutrients, each revision produced slightly lower values.

United States recommended dietary allowances (USRDA).

The set of recommendations used for labelling products is the United States Recommended Dietary Allowances (USRDA) published by the Food and Drug Administration (FDA). The USRDA is based on the 1968 version of the RDA, but the USRDA lists slightly higher values than the RDA and does not give ranges of recommendations (Table 20–3). On packages of products, such as cereal boxes, the datum for each nutrient is usually the percentage of the USRDA recommendation contained in a serving of the product.

The RDA and USRDA are recommendations for the population of the United States. Other countries such as Canada have a separate set of recommendations. Two world organizations have set recommendations for nutrient intake. The

Other countries' nutrient recommendations; FAO and WHO.

Food and Agriculture Organization (FAO) and the World Health Organization

TABLE 20–3 United States Recommended Daily Allowances (USRDA) for Adults and Children over 4 Years Old

NUTRIENTS	AMOUNTS
Protein	45 or 65 g*
Vitamin A	5,000 International Units
Vitamin C (ascorbic acid)	60 mg
Thiamine (vitamin B_1)	1.5 mg
Riboflavin (vitamin B_2)	1.7 mg
Niacin	20 mg
Calcium	1.0 mg
Iron	18 mg
Vitamin D	400 International Units
Vitamin E	30 International Units
Vitamin B_6	2.0 mg
Folic acid (folacin)	0.4 mg
Vitamin B_{12}	6 μg
Phosphorus	1.0 g
Iodine	150 μg
Magnesium	400 mg
Zinc	15 mg
Copper	2 mg
Biotin	0.3 mg
Panthothenic acid	10 mg

* 45 g if protein quality is equal to or greater than milk protein, 65 g if protein quality is less than milk protein.

(WHO) have sought to set acceptable levels of nutrient intake in a world where poverty is prevalent. There is some concern that the FAO and WHO recommendations are too low to promote good health, and there is continuing study on their proposals.

Some perspective on the values of RDA, USRDA, and the world listings can be gained when the following points are considered.

1. These are recommendations — not commandments or requirements. Do not expect ill health if each recommendation is not met daily. They are thought to include a margin of safety for most people, being about 30% higher than the average requirement in the case of RDA.
2. They are published by government agencies, but the recommendations come from nutritionists and other scientists, not politicians.
3. They are updated as more scientific evidence is established.
4. They are for healthy persons only because medical problems and certain kinds of stress are known to alter nutrient needs.

Limitations of RDA, USRDA recommendations.

General dietary recommendations proposed by the Department of Health and Human Services and the United States Department of Agriculture in 1980 are:

1. Eat a variety of foods.
2. Maintain ideal body weight.
3. Avoid too much saturated fat and cholesterol.
4. Eat foods with adequate starch and fiber.
5. Avoid excess sugar.
6. Avoid too much sodium.
7. Drink alcoholic beverages in moderation.

Dietary guidelines for Americans.

CALORIC NEED

Heat is needed by humans to maintain body temperature at about 37°C (98.6°F, under the tongue) and to energize endothermic chemical reactions. The principal source of this heat is the oxidation of certain nutrients, namely fats and carbohydrates. The oxidation of proteins and various other exothermic reactions provide the rest of the heat for the body.

Since fat contains less oxygen per gram than do carbohydrates, the exothermic reaction of a fat with oxygen to form carbon dioxide and water produces more heat per gram of fat. Some specific oxidations representative of the three major sources of energy from food are the oxidation of glucose (a sugar),

$$\underset{\text{GLUCOSE}}{C_6H_{12}O_6} + \underset{\text{OXYGEN}}{6\ O_2} \rightarrow \underset{\substack{\text{CARBON}\\\text{DIOXIDE}}}{6\ CO_2} + \underset{\substack{\text{WATER}\\\text{(LIQUID)}}}{6\ H_2O} + 670 \text{ kcal (3.7 kcal/g glucose)}$$

A food Calorie is a kilocalorie (kcal).

the oxidation of a fatty acid (representing a fat),

$$\underset{\text{PALMITIC ACID}}{C_{16}H_{32}O_2} + 23\ O_2 \rightarrow 16\ CO_2 + 16\ H_2O + 2385 \text{ kcal (9.3 kcal/g palmitic acid)}$$

and the oxidation of an amino acid (representing a protein).

$$2\ \underset{\text{ALANINE}}{C_3H_7O_2N} + 6\ O_2 \rightarrow \underset{\text{UREA}}{CO(NH_2)_2} + 5\ CO_2 + 5\ H_2O + 416 \text{ kcal (2.3 kcal/g alanine)}$$

For comparison purposes, the oxidation of ethanol is given.

$$\underset{\text{ETHANOL}}{C_2H_5OH} + 3\ O_2 \rightarrow 2\ CO_2 + 3\ H_2O + 327 \text{ kcal (7.1 kcal/g ethanol)}$$

Calorie values for other foods are listed in Table 20–6.

Some calorie outputs for the oxidation of a few other specific foodstuffs are starch (4.12 kcal/g), table sugar (sucrose, 3.95 kcal/g), butter (9.12 kcal/g), olive oil (9.38 kcal/g), lard (9.37 kcal/g), meat protein (5.35 kcal/g), and egg protein (5.58 kcal/g). Average calorie outputs for the oxidation of the three main foodstuff sources are listed in Table 20–4. Although proteins have less oxygen per gram than carbohydrates, the heat output per gram of protein is lower because of the diminishing effect of the endothermic formation of some nitrogen-containing compounds when proteins are oxidized.

The values from Table 20–4 can be used to calculate the caloric value of foods, if the composition is known. For example, if a steak is 49% water, 15% protein, 0% carbohydrate, 36% fat, and 0.7% minerals, a 3.5-ounce steak (about 100 g) would produce about 384 kcal, or 384 food Calories.

Nutrient	Weight kcal/g	Total
Water	49 g × 0 kcal/g =	0 kcal
Protein	15 g × 4 kcal/g =	60
Carbohydrate	0 g × 4 kcal/g =	0
Fat	36 g × 9 kcal/g =	324
Minerals	0.7 g × 0 kcal/g =	0
	Total	384 kcal

By this method, calorie values of most foods are calculated, and these are the values listed in diet books.

When calories are counted in diet plans, the concern is not the extra heat produced by an excess of food. The concern is the gain in weight produced by certain foods. Calorie intake is a convenient way to put numbers on what we eat or should not eat. The exact amount of calories necessary for an adult ranges from 2000 to 3000 calories per day. The higher number is needed by a person under stress, involved in much physical activity, or exposed to a colder climate. The lower value is required by a person with minimal physical activity, such as at a desk job. Calorie counting is not the only criterion for proper food selection. For example, if weight reduction is desired, fewer fats should be eaten, not necessarily because of the high calorie production but because fats once digested into fatty acids and glycerol can be reformed by the body into and stored as fats. Sugars and simple starches in excess may contribute to an existing diabetes problem and can be synthesized into and stored as fats.

Physical activity is one way to consume the foods that would be stored as fat (use them to produce heat and energy). Some of the average calorie values for various activities are listed in Table 20–5.

TABLE 20–4   Calorie Data for Fats, Carbohydrates, and Proteins

FOODSTUFF	Kcal/g	RDA	ACTUALLY CONSUMED IN U.S. DAILY DIET	Kcal PRODUCED BY DAILY INTAKE	PERCENTAGE OF DAILY CALORIE OUTPUT
Fat	9	—	100–150 g	900–1350	30–50
Carbohydrate	4	—	300–400 g	1200–1600	35–45
Protein	4	46–56 g (10 oz meat)	80–120 g	320–480	10–15

TABLE 20–5   Approximate Energy Expenditure by a 150-Pound Person in Various Activities

ACTIVITY	ENERGY (kcal/hr)	ACTIVITY	ENERGY (kcal/hr)
Bicycling, 5.5 mph	210	Roller skating	350
13 mph	660	Running, 10 mph	900
Bowling	270	Skiing, 10 mph	600
Domestic work	180	Square dancing	350
Driving an automobile	120	Squash and handball	600
Eating	150	Standing	140
Football, touch	530	Swimming, 0.25 mph	300
tackle	720	Tennis	420
Gardening	220	Volleyball	350
Golf, walking	250	Walking, 2.5 mph	210
Lawn mowing, power mower	250	3.75 mph	300
Lying down or sleeping	80	Wood chopping or sawing	400

Energy spent for normal maintenance activities of the body is the basal metabolic rate (BMR). These maintenance activities include the beating of the heart, breathing, the ongoing maintenance of life in each cell, the maintenance of body temperature, and the sending of nerve impulses from the brain to direct these automatic activities. Energy for these activities must be supplied before energy can be taken for digesting food, running, walking, talking, and our other activities. For people keeping normal hours, the rate of energy expenditure is lowest around 4 A.M. in bed, corresponding to about 0.9 kcal/minute or 54 kcal/hour, depending on the individual. This would correspond very closely with the BMR, because basal metabolism, usually expressed as kcal per hour, is defined as the energy spent by a body at rest after a 12-hour fast.

Basal metabolic rate (BMR).

A rough estimate of your daily BMR is to multiply your weight (in pounds) by 10.

The BMR can be affected by many factors. Increased BMR can come from anxiety, stress, lack of sleep, food intake, congestive heart failure, fever, increased heart activity, and drugs, including caffeine, amphetamine, and epinephrine. Decreased BMR can result from malnutrition, menopause, inactive tissue due to obesity, and low-functioning adrenal glands.

SELF-TEST 20–A

1. The six classes of nutrients are _____, _____, _____, _____, _____, and _____.
2. The most abundant element in the human body is _____, and the most abundant mineral is _____.
3. Which set of recommendations for nutrient intake is printed on packaged products? _____
4. All humans are sufficiently alike for nutrition studies to be extrapolated from one to the other. ( ) Definitely ( ) Only partially ( ) Absolutely not
5. The nutrient that produces the most heat per gram is _____.

6. A food Calorie is the same value as _____ kilocalorie(s), and it will raise the temperature of 1000 g of water _____ degree(s) centigrade.

7. The amount of heat required to operate the body at rest is the _____, which in kcal is about ten times your _____.

INDIVIDUAL NUTRIENTS — WHY WE NEED THEM IN A BALANCED AMOUNT

In this section, we shall focus on how the nutrients are used in the human body, what effects nutrient excesses or deficiencies have on the human body, and what foods supply each nutrient.

Proteins

Properties and structures of proteins are given in Chapter 15.

Histidine is required for wound healing.

See Table 15–1, and relate the essential amino acids to the letters TV Till PM HA.

Of the some 22 amino acids identified in human protein, 10 are considered essential in that the human body cannot synthesize these amino acids and, therefore, must obtain these amino acids from ingested food. Infants require arginine because they cannot make it fast enough to have a supply for both protein synthesis and urea synthesis. The lack of an essential amino acid is not supplied by an excess of the particular amino acid in another meal since excess amino acids are not stored very long except in functioning proteins. If proteins are eaten at only one meal per day, the liver must store a full day's supply from that one meal.

Hormone (Greek *hormaein,* "to set in motion, spur on"); a chemical substance, produced by the body that has a specific effect on the activity of a certain organ.

FUNCTIONS IN THE HUMAN BODY Humans must have proteins because proteins provide the structural tissue for muscles and most organs. Proteins are part (the apoenzyme) of the some 80,000 known enzymes. Some hormones, transport molecules (such as hemoglobin and transferrin), antibodies, and fibrinogin (for blood clotting) contain proteins.

DAILY NEEDS Proteins are nearly the only source of nitrogen in the diet. An adult male has about 10 kg of protein, and about 300 g of this protein is replaced daily. Part of the 300 g is recycled, and part comes from intake. Various studies indicate that on the average 25 to 38 g of high-quality protein (such as in meat, chicken eggs, cow's milk) or 32 to 42 g of lower-quality proteins (such as in corn and wheat) are required in the diets of healthy adult humans in order to maintain nitrogen equilibrium in the body. The average daily intake has remained near 100 g of protein per person since 1910, although there was a small drop in the intake during the Depression of the 1930s. Methionine is the essential amino acid required in the greatest amount (2 g of the total of 7.1 g of all of the essential amino acids). Protein losses occur in the urine (as urea, a byproduct of protein metabolism), fecal material, sweat, hair and nail cutting, and sloughed skin.

The USRDA requirement for protein is 46 g for young female adults and 56 g for adult males.

FOOD SOURCES Table 20–6 lists some foods relatively high in protein content. Generally speaking, persons who are reasonably well fed and eat meat, fish, eggs, or dairy products every day have no worry about their protein intake.

PROTEIN-RELATED PROBLEMS If the diet does not contain the proper balance of the essential amino acids, protein synthesis is curtailed. The excess amino acids are

TABLE 20–6 **The Approximate Percentages of Carbohydrates, Fats, Proteins, and Water in Some Whole Foods as Normally Eaten***

FOOD	WATER	PROTEIN	FAT	CARBOHYDRATES	Kcal/100 g	FOOD	WATER	PROTEIN	FAT	CARBOHYDRATES	Kcal/100 g
Vegetables						*Grains and Grain Products*					
Spinach, raw	90.7	3.2	0.3	4.3	26	Wheat grain, hard	13.0	14.0	2.2	69.1	330
Collard greens, cooked	89.6	3.6	0.7	5.1	33	Brown rice, dry	12.0	7.5	1.9	77.4	360
Lettuce, Boston, raw	91.1	2.4	0.3	4.6	25	Brown rice, cooked	70.3	2.5	0.6	25.5	119
Cabbage, cooked	93.9	1.1	0.2	4.3	20	Whole-wheat bread	36.4	10.5	3.0	47.7	243
Potatoes, cooked	75.1	2.6	0.1	21.1	93	White bread	35.8	8.7	3.2	50.4	269
Turnips, cooked	93.6	0.8	0.2	4.9	23	Whole-wheat flour	12.0	14.1	2.5	78.0	361
Carrots, raw	88.2	1.1	0.2	19.7	42	White cake flour	12.0	7.5	0.8	79.4	364
Squash, raw summer	94.0	1.1	0.1	4.2	19	*Dairy Products and Eggs*					
Tomatoes, raw	93.5	1.1	0.2	4.7	22	Milk, whole	87.4	3.5	3.5	4.9	65
Corn kernels, cooked on cob	74.1	3.3	1.0	21.0	91	Yogurt, whole-milk	89.0	3.4	1.7	5.2	50
Snap beans, cooked	92.4	1.6	0.2	5.4	25	Ice cream	62.1	4.0	12.5	20.6	207
Green peas, cooked	81.5	5.4	0.4	12.1	71	Cottage cheese	79.0	17.0	0.3	2.7	86
Lima beans, cooked	70.1	7.6	0.5	21.1	111	Cheddar cheese	37.0	25.0	32.2	2.1	398
Red kidney beans, cooked	69.0	7.8	0.5	21.4	118	Eggs	73.7	12.9	11.5	0.9	163
Soybeans, cooked	73.8	9.8	5.1	10.1	118	*Fruits, Berries, and Nuts*					
Meats and Fish						Apples, raw	84.4	0.2	0.6	14.5	58
Lean beef, broiled	61.6	31.7	5.3	0	183	Pears, raw	83.2	0.7	0.4	15.3	61
Beef fat, raw	14.4	5.5	79.9	0	744	Oranges, raw	86.0	1.0	0.2	12.2	49
Lean lamb chops, broiled	61.3	28.0	8.6	0	197	Cherries, sweet	80.4	1.3	0.3	17.4	70
Lean pork chops, broiled	69.3	17.8	10.5	0	171	Bananas, raw	75.7	1.1	0.2	22.2	85
Lard, rendered	0	0	100.0	0	902	Blueberries, raw	83.2	0.7	0.5	15.3	62
Calf's liver, cooked	51.4	29.5	13.2	4.0	261	Red raspberries, raw	84.2	1.2	0.5	13.6	57
Beef heart, cooked	61.3	31.3	5.7	0.7	188	Strawberries, raw	89.9	0.7	0.5	8.4	37
Brains	78.9	10.4	8.6	0.8	125	Almonds	4.7	18.6	54.2	19.5	598
Chicken, whole broiled	71.0	23.8	3.8	0	136	Pecans	3.4	9.2	71.2	14.6	689
Cod, raw	81.2	17.6	0.3	0	78	Walnuts	3.5	14.8	64.0	15.8	651
Salmon, broiled	63.4	27.0	7.4	0	182						
Fresh-water perch, raw	79.2	19.5	0.9	0	91						
Oysters, raw	84.6	8.4	1.8	3.4	66						

* The caloric value per 100 g of food is listed for each food.

consumed for energy or converted into glucose and then glycogen in the liver. *Kwashiorkor* (pronounced: kwash-ee-OR-core) is a protein-deficiency disease. To Ghanaians, who named the disease, kwashiorkor originally meant "the evil spirit which infects the first child when the second child is born." If a child is nursing when a second child is born, the first child is weaned from the mother's protein-rich milk to a starchy-protein-poor sustenance of gruel. The first child begins to sicken and dies within a few years. If kwashiorkor sets in around the age of two, by the time the child is four, growth is stunted, hair has lost its color, the skin is patchy and scaly with sores, the belly, limbs, and face are swollen by the collection of fluid in the intercellular spaces (edema), and the child sickens easily (lowered supply of antibodies) and is weak, fretful, and apathetic. If the child with kwashiorkor is given nutritional therapy before the disease has progressed to its last stages, he or she has a good chance to recover.

If proteins occupy too large a proportion of the intake (carbohydrates too low), **uremia** can occur. Uremia is marked by nausea, vomiting, headache, vertigo, dimness of vision, coma or convulsions, and a urinous odor of the breath and perspiration. Protein metabolized completely forms ammonia. The liver converts the ammonia to urea, some of which is used to make "nonessential" amino acids. Excess urea is excreted in the urine. If insufficient carbohydrates are available to oxidize for energy, too much urea from the oxidation of proteins is sent to the kidneys, which may become overworked, and uremia sets in.

Athletes who are building muscles sometimes eat 250 to 300 g of protein per day. According to a study by Dr. Ralph Nelson (*The Physician and Sports Medicine,* November 1975), the excess consumption of protein is useless and may be dangerous. Any extra protein consumed goes into calorie requirements or into fat. Muscles increase in size by exercise and not by extra, excess proteins. Enough nutrients (carbohydrates, fats, or proteins) must be present to supply the calorie needs for the exercise performed. In fact, studies verify that intakes of significant amounts of carbohydrates result in muscle-building chemical reactions. The insulin released to escort glucose into cells also accelerates the uptake of amino acids by muscles and reduces the breakdown of muscle proteins by depressing the levels of free amino acids in the blood.

Fats

Properties and structures of fats are given in Chapter 13.

Triacylglycerol is used by some sources for the older term, *triglyceride.*

Structures of glycerol and triglycerides are in Chapter 13, and the structure of cholesterol is in Chapter 21.

Fats and lipids are often considered synonymous, but lipid is the larger category of compounds. A lipid is an organic substance that has a greasy feel and is insoluble in water but soluble in alcohol, ether, chloroform, and other fat solvents. As used in the United States, lipid includes neutral fats and oils, waxes, steroids, phospholipids, and similar compounds. When we refer to fats, we are usually referring to triglycerides, composed of one glycerol molecule esterified by three fatty acid molecules. Almost all (95%) of the lipids in the diet are triglycerides. The other 5% are composed of phospholipids (lecithin is an example) and steroids (cholesterol is the major one in food).

The only truly essential fatty acid is linoleic acid (must be eaten in the diet; cannot be synthesized in the body). Arachidonic and linolenic fatty acids were thought to be essential until it was discovered that the two acids can be synthesized in the body from linoleic acid.

2 fatty acids

From choline
Cl⁻

From glycerol

Lecithin

(The plus charge on the N is balanced by a negative ion—usually chloride—that stays nearby)

FUNCTIONS IN THE HUMAN BODY Fats are essential structural parts of cell membranes. They provide the highest energy per gram of the nutrients and serve as energy storage reservoirs in the body. They insulate thermally, pad the body, and are packing material for various organs. Fat is transported in the body by the blood and the lymph system. Fatty or adipose tissue is composed mainly of specialized cells, each consisting mainly of a large globule of triglycerides.

Fatty acids are precursors of prostaglandins; the oxidation of arachidonic acid gives several possible prostaglandins. The prostaglandins function as bioregulators to influence the action of certain hormones and nerve transmitters. They inhibit high blood pressure, ulcer formation, and inflammation.

Structures of prostaglandins are in Chapter 13.

DAILY NEEDS The daily consumption of fat in the United States has risen on a continuing trend from 125 g per person in 1910 to about 155 g per person today. In the present day diet, about 40% is saturated fat, 40% monounsaturated fat, and 20% polyunsaturated fats. There are no RDA or USRDA recommendations for fats. Dietary Guidelines for Americans recommend that we avoid too much fat, saturated fat, and cholesterol. "Choose low-fat protein sources such as lean meats, fish, poultry, dry peas and beans; use eggs and organ meats in moderation; limit intake of fats on and in foods; trim fats from meats; broil, bake, or boil — don't fry; read labels for fat contents."

Dietary Guidelines for Americans were stated earlier in this chapter.

FOOD SOURCES Table 20–6 lists the fat content in some foods. The fat in the edible portions of the food, as normally eaten, is difficult to ascertain since some people trim more fat than others before the food is eaten. Animal fats (oils) are higher in saturated and monounsaturated fatty acids. Vegetable fats (oils) have a higher percentage of polyunsaturated fatty acids. Nearly all diets supply enough linoleic acid to meet the needs of the human body. Pork (lard) and chicken fat are nearly all in the form of linoleic acid. Even in a totally fat-free diet, only one teaspoon (5 g) of corn oil would supply the daily need of linoleic acid. Two of the highest sources of cholesterol are brains (2.5 g/100 g) and egg yolk (1.15 g/100 g).

PROBLEMS ASSOCIATED WITH EATING FATS Too much fat in the diet can lead to obesity. After digestion of a fat, if the components glycerol and fatty acids are not used otherwise, they are resynthesized into fats and stored as such. Problems associated with obesity will be discussed later.

If there are too much fat in the diet and too little carbohydrate, **ketosis** can occur. Ketosis is the combination of high blood ketones **(ketonemia)** and ketones in the urine **(ketonuria).** Ketones are formed when fats are broken down to form glucose when no glucose is readily available to the body. Glycerol derived from fat destruction forms pyruvate, then glucose. The fatty acids form ketones (the simplest being acetone, CH_3COCH_3) and keto-acids (such as acetoacetic acid, CH_3COCH_2COOH, and the substance in largest amount, 3-hydroxybutanoic acid, $CH_3CH(OH)CH_2COOH$). One noticeable characteristic of ketosis is "acetone" breath. Although some cells in the body can use ketones for fuel, other cells must have glucose. There is a small amount of ketones in the blood normally. Excess ketones and keto-acids lead to **ketoacidosis,** a potentially fatal condition.

If too little fats are eaten, especially the essential fatty acid, problems ensue with coarsened, sparse hair and eczema (skin disease characterized by lesions, watery discharge, and the development of crusts and scales).

Cholesterol and saturated fats may have received some bad publicity relative to their being the primary cause of heart attacks. There are about 145 g of cholesterol in the body of a normally healthy man. One gram of cholesterol is synthesized by the body per day, and moderate diets ingest half a gram per day. Cholesterol is vital to cell membranes. It is a fact that saturated fats do increase blood cholesterol more than do dietary unsaturated fats. **Atherosclerotic plaque** formed in coronary arteries is composed of lipids, mainly triglycerides and cholesterol. As more plaque forms, the vessel is constricted and loses its elasticity. Both effects increase blood pressure, which can lead to ballooning of a weak section of the vessel (aneurysm). If enough plaque forms to shut off the flow of blood in an artery, the tissues that depend on the oxygen and nutrients delivered by the artery die.

There are many documented studies of high intake of dietary cholesterol and saturated fats without any more coronary artery blockage (plaque) than in the general population. In some cases, the excess fat intake causes plumpness, but coronary heart disease was practically unknown. Some of the cases studied were the community of Roseto, Pennsylvania, the Samburu tribe in Kenya, Africa, and the Masai people of Tanganyika. In these studies and others, there was high fat intake without increased coronary disease.

A different approach to the fat-heart problem is the large amount of literature on the subject of polyunsaturated fats and the lowering of cholesterol. Studies

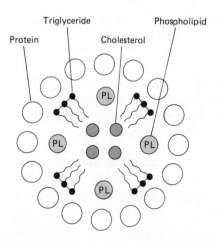

have been done on both animals and humans. Since the mid-1960s, there has been a pronounced increase in the general consumption of unsaturated fats (and less saturated fats). During this time, there has been no decrease in heart disease or in deaths due to coronary disease. At the time of this writing, there is no practical conclusion that definitely links fat intake as the primary cause of coronary disease.

If cholesterol and saturated fats are not the primary cause of heart problems, what is? Maybe there is a combination of contributing factors such as excessive fatness that overworks the heart, sedentary living, diabetes mellitus, other food effects such as those of refined foods and some food additives, and social or emotional stress. One of the more accepted theories proposes that plaque begins with some type of injury to the lining of the arterial wall. The proliferation of new smooth muscle cells in some way causes free cholesterol and triglycerides and mushy cell debris to accumulate at the place of injury, and plaque is formed. As discussed in the next chapter on medicine, if an individual's system has enough **high-density lipoprotein,** cholesterol and plaque are decreased in the bloodstream.

There is a problem caused by the hydrogenation of polyunsaturated fats (oils). The process converts oils into solid fats to make margarine, cooking fats, and similar products. Hydrogenation of vegetable oils (liquid fats) decreases some of the double bonds, forms unnatural trans-fatty acids from natural cis-fatty acids, and moves double bonds around to form conjugated structures. The trans-fatty acids are not metabolized in the human system, but they can be stored for the life of the individual. The conjugated structures are too active chemically for the human system. The problem is solved by eating fewer of the processed vegetable fats.

When saturated or unsaturated fats are used in cooking, they should not be heated to temperatures at which they smoke. Under these conditions, the fats produce toxic peroxides, and the unsaturated fats can polymerize. An undesirable oxidation of unsaturated fatty acids can lead to "runaway" free radical reactions, which are believed to be associated with premature aging. Vitamin E prevents free radical formation. All fats and oily foods should be smelled for rancidity when the package or bottle is opened and should be returned to the store for credit if there is any evidence of rancidity.

Carbohydrates

Carbohydrates in foods include digestible simple sugars (glucose, fructose, galactose), disaccharides (sucrose, maltose, and lactose), and polysaccharides (amylose, amylopectin, glycogen). Indigestible carbohydrates consumed include cellulose, insulin, hemicellulose, lignin, plant gums, sulfated polysaccharides, carrageenan, and cutin.

Properties and structures of carbohydrates are given in Chapter 15.

FUNCTIONS IN THE HUMAN BODY The only beneficial function of digestible carbohydrates is to provide energy at the rate of approximately 4 kcal per gram of glucose oxidized. Excess digestible carbohydrates are stored first as glycogen principally in the liver. Further excesses are converted into fats and stored as such. The indigestible carbohydrates serve as roughage in the diet along with bran and fruit pulp.

DAILY NEEDS In order to provide 2000 kcal per day, about 500 g of glucose (or equivalent) would have to be eaten. Fats and proteins are also oxidized for energy, so less digestible carbohydrate is required. The daily consumption of digestible

Table 20–6 lists the carbohydrate content of some foods.

carbohydrate has declined from 500 g per person in 1910 to 380 g per person in the 1980s. The decline in total amount of carbohydrates is really a rise in the amounts of refined sugars: 150 g in 1910 to 200 g in the 1980s. Problems associated with refined sugar are discussed later. There are no RDA or USRDA recommended intake amounts for carbohydrates. The 1979 Dietary Guidelines for Americans recommend that we decrease eating concentrated sweets (candy, soft drinks, cookies, etc.) and substitute with starches (complex carbohydrates), fresh fruits, and fiber.

PROBLEMS ASSOCIATED WITH EATING CARBOHYDRATES Problems can be encountered when the glucose level is too low in the blood and when there is too little roughage in the diet. The general medical term for too low a concentration of glucose in the blood is **hypoglycemia.** Persons with hypoglycemia need a regular intake of sugar to avoid the lows characterized by inability to think clearly and being emotionally disturbed with the feeling of being indisposed.

Diseases associated with lack of dietary fiber (roughage) are appendicitis, diverticular disease (herniation of mucous membrane lining of a tubular organ), and benign or malignant tumors of colon and rectum. In the gastrointestinal tract, fiber absorbs water, swells, and facilitates regular bowel movements and prevents stagnation of foods, particularly refined foods, in the intestines. Fiber also absorbs cholesterol (this action may prevent coronary disease) and inhibits the use of glucose in the body. Bran is a good source of dietary fiber. The pulp of fruit is another good source if, of course, the whole fruit is eaten. Some bakers incorporate wood or cotton cellulose into their high-fiber foods.

There are some problems with specific carbohydrates such as lactose, galactose, and oligosaccharides involving galactose and sucrose. **Lactose intolerance** caused by hereditary lack of some enzymes is discussed in Chapter 15. **Galactosemia** affects only 10 to 20 persons per million and is caused by the absence of one of the two enzymes required to convert galactose into glucose in the liver. A derivative of galactose deposits in the lens of the eye and causes cataracts.

Why do beans produce gas?

About 4% to 9% by weight of beans are oligosaccharides composed of one or more galactose molecules bonded in a row to the glucose end of the sucrose molecule. The simplest of these oligosaccharides is raffinose (galactose-glucose-fructose). The human body has no enzymes to break up these oligosaccharides, so they pass into the intestines undigested. Intestinal microorganisms consume these oligosaccharides and produce some hydrogen, carbon dioxide, and less methane, CH_4. As these gases are expelled, the odor is due to protein residues in the lower intestine.

Clinical problems associated with a diet rich in carbohydrates, particularly refined sugar, are obesity, dyspepsia, atherosclerosis, and diabetes mellitus.

Dyspepsia is the name applied to chronic indigestion caused by consuming large amounts of sugar by adults.

An international statistical study involving 23 countries reports that the death rate from heart disease about equally parallels the consumption of sucrose and fat, but it showed no correlation with the intake of complex carbohydrates. Several studies have shown that a decreased intake of sucrose decreases high levels of triglycerides, but low levels of sucrose have no changing effect on the levels of cholesterol or on changing low levels of triglycerides. Experiments on rats and volunteer humans reveal that diets with high sugar cause increased fat content of the blood, decreased insulin activity of blood serum, and damaging effects for those people with diabetes mellitus.

Insulin escorts glucose to the fatty cell membrane.

Diabetes mellitus is characterized by elevated blood glucose, multiple hormonal and metabolic disturbances in the secretion of insulin and the growth hormone, thirst, hunger, weakness, lowered resistance to infection, slowness to heal, and, in later stages, blindness and coma. About 15% of people in higher age groups have diabetes, but the disease is generally rare in parts of the world where the people eat no refined or processed food. While the several types of diabetes relate to decreased glucose metabolism, diabetes mellitus is caused by too little insulin due to defective *islets of Langerhans* in the pancreas. The pancreas does not produce sufficient insulin for one or more reasons. There is less insulin (or no insulin) because of hereditary capacity, diseases or injuries that destroyed parts or all of the pancreas, interruption of the mechanism of producing insulin by lack of proper reactants or the presence of toxicants, or too much demand for insulin invoked by too high glucose intake. If the pancreas produces even a small amount of insulin, diabetes can be held at bay by a diet low or absent in sugar. If the pancreas produces no insulin, daily injections of insulin are required.

Insulin, a protein, is hydrolyzed (digested) in the GI tract if taken orally.

The yearly sugar consumption in the United States was 2 pounds per person in 1750. Today, each person consumes 110 to 135 pounds of sugar annually. Sixty percent of the sugar comes from sugar cane, and the other 40% comes from sugar beets. The average American drinks 500 eight-fluid-ounce soft drinks per year, each containing about 10% refined sugar by weight.

Some problems with refined sugar are due to the removal of required nutrients during the refining process and dumping too much refined sugar into the bloodstream too quickly. The production of white sugar (almost pure sucrose) removes all other nutrients such as B vitamins, manganese, and chromium, which generally coexist in natural foods with sucrose in the appropriate amounts for proper metabolism in the human body. Therefore, a large refined sugar intake requires that the B vitamins and certain minerals be obtained from another food source. With respect to minerals, brown sugar supplies more minerals than white sugar because brown sugar is darkened with molasses, the residue from sugar manufacturing that is rich in essential minerals.

Unrefined and unprocessed sugar is often contained in cellular structures, which are not easily digestible. Sugar such as from sugar cane or from apples goes to the bloodstream much more slowly than refined sugar. The slower transfer allows the body to metabolize the sugar for energy more efficiently and avoids both the buildup of glucose in the bloodstream and the consequent storage of excess glucose as fat. If we obtain our sugar from the unrefined, unprocessed source, we would likely eat less sugar. For example, it takes four apples to supply the same amount of

TABLE 20–7   Refined Sugar Added to Some Commercially Processed Foods*

FOOD	SUGAR (PERCENT)
Cherry Jello	82.6
Coffeemate	65.4
Shake'N Bake, Barbecue Style	50.9
Wishbone Russian Dressing	30.2
Heinz Ketchup	28.9
Sealtest Chocolate Ice Cream	21.4
Libby's Peaches (in Heavy Syrup)	17.9
Skippy Peanut Butter	9.2
Coca Cola	8.8

* According to *Consumer Reports* (1978); percents by weight.

Unrefined carbohydrates
contain factors that
destroy the bacteria that
consume carbohydrates
and produce acid and
storage carbohydrates
(tooth plaque).

sugar as in one 12-ounce cola drink. Or, it would take 3 pounds of apples to supply our normal intake of sugar per day. Studies on animals indicate that high-molecular-weight, digestible (complex) carbohydrates, such as in potatoes, are less harmful (and more helpful) to the body than refined sugars.

About two thirds of all of the grains consumed by humans in the United States are milled, white flour. Milled flour has less lysine and fat. The fiber content is reduced to 10% of that in wheat grains. Vitamins are reduced to between 10% and 50% of the original content, mostly by removal of the wheat germ. Enriched flour has some of the removed vitamins and minerals added back. Some millers add white paper pulp to replace the roughage provided by the removed bran. Some people supplement white flour with wheat germ, though this practice is inadvisable since wheat germ separated from the wheat degrades nutritionally and becomes rancid too easily.

Refined flour, like refined sugar, digests more quickly and more completely than unrefined (whole wheat) flour. The stomach empties more slowly (less content); the small intestine has less bulk; and the rate of passage through the intestine is decreased considerably. Wastes in the intestines have more time to effect any toxicity they might have. Studies have shown that it takes 40 to 140 hours for refined carbohydrates to pass through the human body and only 15 to 45 hours for a traditional diet (not supplemented by Western foods) to pass through the body.

The process of bleaching to make whiter flour also destroys vitamins. Commonly used bleaching agents are chlorine dioxide and benzoyl peroxide. This problem can be averted by purchasing unbleached flour.

A large proportion of the caloric intake of our times is purified sugar and refined white flour, and they do not satisfy the appetite. Not satisfying, refined sugars, in particular, lead to wanting more . . . and more . . . and more . . . and winding up as a sugar addict. An effective way to reduce or eliminate sugar addiction is to reduce sugar consumption gradually without resorting to artificial sweeteners because the real taste sensations of foods will continue to be smothered by the sugary-sweet taste, and the desire for sugars will continue. According to several studies, those addicted to sugar, sugarholics, readily turn to alcoholics. Alcoholic drinks provide no nutrients, only 1.8 more calories per gram than for a gram of sugar. The concentration of alcohol in the blood is caused by a rapid absorption but a limited oxidation rate.

Interest in a normal diet is
lost by the high caloric
supply from drinking
alcohol.

SELF-TEST 20–B

1. Proteins compose the _____ part of enzymes.
2. Kwashiokor is a _____ deficient disease.
3. Muscles are built primarily by eating excess proteins. ( ) True or ( ) False
4. Uremia is caused by excess protein intake and the excretion of _____.
5. A common phospholipid in food is _____, and a common steroid in food is _____.
6. The major proportion of the lipids in the diet is the _____.
7. The one essential fatty acid is _____.
8. Bioregulators of hormones and nerve transmitters are called _____, and they are made from _____.

9. Natural sources rich in linoleic acid are _____ and _____.

10. Ketosis is caused by too much _____ and too little _____ in the diet.

11. Eczema is caused by too little _____ in the diet.

12. Polyunsaturated fats lower cholesterol in the bloodstream. ( ) True or ( ) False

13. High-density lipoprotein (HDL) ( ) decreases or ( ) increases cholesterol and arterial plaque.

14. Heating fats until they smoke can produce _____.

15. All ingested carbohydrates are digestible. ( ) True or ( ) False

16. Refined _____ and refined _____ are absorbed more quickly than unrefined carbohydrates.

Minerals

As nutrients, minerals are substances that contain elements other than C, H, O, and N that are needed for good health. On vitamin and mineral supplement labels and elsewhere, the nutrient element is called a mineral since the two terms are used interchangeably in nutrition. Most of the elements needed for nutrition are obtained from soluble inorganic salts either in foods or in supplements. Magnesium is an exception in that Mg is obtained primarily from organic chlorophyll.

The required inorganic nutrients can be grouped into two classes. Calcium, phosphorus, and magnesium are required in amounts of a gram or more per day. Trace elements such as iron, iodine, zinc, copper, and many others are needed in milligram or microgram quantities each day.

The nutrient minerals find varied uses, including components of enzymes, structural components (calcium and phosphorus in bones and teeth), electrolyte balance in body fluids, and transport vehicles (iron in hemoglobin for transport of oxygen and iron and cobalt in electron transport cycles). Not only does the human body need minerals for its functions, the minerals must be maintained in balanced amounts with no deficiencies and no excesses. Since many of the body's minerals are excreted daily in the feces, urine, and sweat, there is the need to replenish excreted minerals. The amount excreted is very nearly the amount ingested each day for most of the elements.

The specific needs we have for minerals are borne out by the mineral's functions, excess effects, and deficiency effects listed in Table 20–8. Once the need for minerals is realized, we naturally want to know some food sources, which are given in the last column of Table 20–8. One way to assure an ample supply of the mineral nutrients, particularly the ones required in trace amounts, is to eat a variety of whole foodstuffs grown in different places. For convenience, the USRDA values are also listed in this table (and in Table 20–3). The amounts of elements in the body are listed in Table 20–1.

The **glucose tolerance factor (GTF)** improves the tolerance of glucose for some persons having mild diabetes. A molecule of GTF is composed of a **chromium** atom in the center surrounded by a niacin molecule and three amino acids (glutamic acid, glycine, and cysteine). Inorganic chromium as $CrCl_3$ improves the tolerance of glucose, too. Studies indicate that chromium may be a co-factor with

TABLE 20–8 Some Functions, Deficiency Effects, Excess Effects, Food Sources, USRDA Values, and Amount Ingested per Day of Elements Required by the Human Body

ELEMENT	USRDA*	AMOUNT INGESTED PER DAY	FUNCTION	DEFICIENCY EFFECTS	EXCESS EFFECTS	FOOD SOURCES
Calcium (Ca)	1.0 g	1.0 g	Component of bone as hydroxy apatite, bone collagen; involved in fertilization, cell division, synthesis of ATP, muscle contraction, blood clotting, and transmission of nerve impulses	Bone dissolution	Slows heart beat, kidney stones	Milk, yogurt, cheeses, dark leafy vegetables, soybeans, almonds, whole wheat breads
Chromium (Cr)	—	0.15 mg	Crosslinks collagen, component of glucose tolerance factor	Cholesterol level increases	Growth depression, damage to kidneys and liver, lung cancer by breathing chromate	Honey, maple syrup, grape juice
Chlorine (Cl)	—	5 g	Counter ion to cations, as the chloride ion		Infant deaths from salt overloading from cow's milk	Salt (NaCl), clams, oysters, meats, eggs, milk, molasses
Cobalt (Co)	—	0.30 mg	Component of vitamin B_{12}	Wasting disease, pernicious anemia (see vitamin B_{12})	Death when consumed in beer (added to maintain foamy head): now forbidden by FDA	Meat, clams, oysters, egg yolks
Copper (Cu)	2 mg	3.5 mg	Component of enzymes (tyrosinase, lactase, ascorbic acid oxidase, cytochrome oxidase)	Anemia, scurvy-like bone abnormalities, uncoordination in newborn, reproduction failure, cardiac disease, diarrhea in infants, infertility, affects growth and appearance of hair	Leukemia, anemia, blockage of coronary arteries in copper miners; restricts iron flow and red blood synthesis, psychoses, histapenia,† schizophrenia	Oysters, crabs, lobsters, liver, wheat grain, spinach, molasses, bitter chocolate (from copper water pipes)

Element			Function	Deficiency	Excess / Toxicity	Sources
Fluorine (F)	—	1.8 mg	Fluorapatite of bones and teeth		Elemental F_2 is extremely toxic; F^- in dust: loss of appetite and body weight, digestive upsets, muscular weakness, congestion of lungs, convulsions, respiratory or heart failure	From toothpaste and drinking water
Iodine (I)	0.150 mg	0.20 mg	Thyroxin hormone (regulates rate at which body uses energy)	Goiter, cretin children from I-deficient mother	Goiter	Kelp, cod liver oil, seafood, egg yolk, added to some vegetable oils, iodized salt
Iron (Fe)	18 mg	15 mg	Component of hemoglobin, myoglobin,‡ ferritin,§ transferrin‖	Anemia	Fatigue, dizziness, loss of weight, gray hue of skin	Meats, egg yolk, prunes, raisins
Magnesium (Mg)	400 mg	340 mg	Component of bone, dentine of teeth, enzymes, such as activate ATP reactions	Circulatory problems, mental depression, irritability, muscle cramps, liver and kidney damage, red bulbous nose of alcoholics	Inhibits central nervous system, decreases blood pressure, cathartic, laxative, sedative, anticonvulsant	Foods with chlorophyll (Mg is a component), green leafy vegetables, nuts, whole-grain breads
Manganese (Mn)	—	3.7 mg	Component of melanin (skin pigment) and enzymes involved in using several vitamins (B and C) and choline	Impaired growth, skeletal abnormalities, reproduction problems, atasia¶ in newborn, less glucose metabolism	Elevates blood pressure in older people, may cause permanent crippling, schizophrenia	Spinach, green beans, white beans, wheat, wheat germ, rice, oats, pecans, cloves, tea leaves
Molybdenum (Mo)	—	0.3 mg	Component of enzymes, xanthine oxidase and sulfite oxidase	Cancer of esophagus, sexual impotency in older men	"Teart" in cattle (decreasing Cu by formation of copper thiomolybdate)	Legumes, cereal grains, leafy vegetables, liver
Nickel (Ni)	—	0.4 mg	Activates several enzymes	Harms liver	$Ni(CO)_4$ is a toxic gas and a suspected carcinogen.	Oysters, cheeses, dark green vegetables, tea, bitter chocolate

TABLE 20–8 *(Continued)*

ELEMENT	USRDA*	AMOUNT INGESTED PER DAY	FUNCTION	DEFICIENCY EFFECTS	EXCESS EFFECTS	FOOD SOURCES
Phosphorus (P)	1.0 g	1.4 g	Component of phospholipids (in fatty and nerve tissues), nucleic acids, ATP and ATP-like compounds, bone and teeth structure	Weakened bones, impairment of action of ATP and ATP-like compounds	Convulsions in infants, decreased calcium level	Lima beans, peas, corn on cob, soybeans, wheat grain, oatmeal, most meats, cheese, eggs, oysters, nuts, milk
Potassium (K)	—	3.3 g	Involved in protein synthesis, component of enzyme, pyruvate kinase, transmits nerve impulses, electrolyte balance	Weakness in the legs, heart attacks, rheumatoid arthritis	More water retention, decreases sodium chloride intake	Green vegetables (especially spinach), salmon, wheat grains, peanuts, bananas, raisins, milk
Selenium	—	0.15 mg	Component of several enzymes and selenoproteins, growth stimulator (with vitamin E)	Degeneration of skeletal muscles, low with kwashiorkor (see proteins)	Disgusting H_2Se in the breath, nervousness, diarrhea, headaches, breathing problems, dizziness, clouded senses, mental depression, emotional instability	Wheat, brewer's yeast
Sodium (Na)	—	7 g	Electrolyte and water balance, component of bones, nerve impulse transmission		Increased water retention, heart attacks, yearning for salty foods, increased blood pressure in some individuals	Most foods (except fruit) beets, shrimp, drinking water, table salt

Element			Function	Deficiency Symptoms	Excess Symptoms	Sources
Sulfur (S)	—	0.85 g	Component of amino acids (methionine, cystine, and cysteine), vitamins (thiamin and biotin), heparins, chondroitin sulfates (bones), keratin sulfates (connective tissue), muciotin sulfates (lubricate GI tract)	Same as deficiency of essential amino acids since methionine is essential	Excess of methionine produces feelings of unreality	Eggs, meat, garlic, onions, mustard, horseradish
Vanadium (V)	—	2.0 mg	Inhibits biosynthesis of cholesterol; involved in growth, bone development, reproduction			Parsley, lettuce, liver, lobster, sardines
Zinc (Zn)	15 mg	13 mg	Component of the choroid, insulin, some cell membranes, and enzymes involved in synthesis of ribonucleic acids and proteins. Involved in nerve transmission, utilization of vitamin A, and in healing wounds	Dwarfism, underdeveloped sex organs in young men, stretch marks on skin, painful knee and hip joints, cold extremities, white spots in finger nails, subject to shock, finicky eaters, impaired acuity of taste	Decreases metabolism of copper and iron, increases cholesterol synthesis	Oysters, meats, whole cereals, nuts, clams, shrimp, perch, cocoa (from tapwater running through galvanized pipes)

* USRDA values for adults and children over 4 years old.
† Histapenia is a type of schizophrenia.
‡ Myoglobin stores oxygen in the muscles.
§ Ferritin stores iron in the liver.
‖ Transferrin transports iron in the bloodstream.
¶ Atasia is poor coordination in newborns.
The amount of each element in the body is listed in Table 20–1. USRDA values are listed in Table 20–3 and repeated here for convenience.

insulin, enhance the activity of insulin, and thus give more tolerance to glucose. GTF may become ranked as a vitamin.

Calcium slows down the heart beat by increasing the electrical resistance across nerve membranes. The movement of potassium and sodium ions across the membrane is constrained, and the nerve impulse rate is thus decreased.

Calcium is metabolized in the body by a hormone synthesized from calciferol (vitamin D). The calciferol also brings about synthesis of a substance called a calcium-binding protein (CBP), which acts to carry calcium through the small intestine wall. Fat slows down the transfer, and lactose speeds up the calcium absorption.

> The role of calcium in regulating the heartbeat is discussed in Chapter 21.

Calcium chloride is used to "firm" canned tomatoes so the tomatoes will retain their shape, and the packer puts fewer tomatoes in a can for a full appearance. Two unlikely sources of calcium are vinegar "sweetened" with egg shells and thousand-year eggs popular in Chinese culture. First, the egg shells are coated with a paste made of slaked lime $(Ca(OH)_2)$ and wood ashes (active ingredient: KOH or K_2CO_3), and then the eggs are buried for several months (not 1000 years). Results are a green yolk, a brown, gelatinous white, and increased calcium in the egg.

Excess calcium may lead to the formation of kidney stones, but the body has a protein, calmodulin, which collects excess calcium and then binds to a number of enzymes to mediate their activity. By the use of calmodulin, the body monitors the amount of calcium in the bloodstream. A benefit of excess calcium is that it makes people taller. This is a benefit only if you can argue that "taller is better."

A deficiency in calcium can occur in women beyond menopause, who at that stage in life produce less estrogen. The estrogen suppresses bone dissolution. Since continued medication with estrogen has produced some toxic side effects in some women, continued medication seems unwise. Eating more turnip greens and collards and (or) taking a calcium supplement might help.

> Copper is carried in the blood by proteins such as cerulaplasmin.

Copper water pipes in modern homes may lead to high intake of copper. The condition, **dementia dialytica,** occurs when there is too much copper in the tapwater for kidney dialysis patients.

The treatment of teeth with **fluoride** has been established as a deterrent to tooth decay in children. For this reason, toothpaste contains stannous fluoride, SnF_2, and some municipal water supplies are fluoridated (0.8 to 1.2 ppm F). Fluoride has essentially no effect on adult teeth since the fluoride incorporates into apatite during the time of tooth development (under 20 years of age). An excess of fluoride (6 g F per kg of tooth weight) causes mottled or brown teeth due to chalkiness.

Iodine in the human body has a principal use in the proper operation of the thyroid glands located at the base of the neck. Two of the thyroid hormones are T_3 and T_4.

3,5,3'-TRIIODOTHYRONINE T_3 THYROXINE (T_4)

These hormones and other similar ones are collectively known as thyroxin. They go into every cell and regulate the rate at which the cell uses oxygen. This is the same as saying thyroxin regulates the BMR. Iodine is absolutely necessary to produce thyroxin. If there is a deficiency of iodine, the thyroid glands sometimes swell to as large as a person's head. The swelling is called a goiter. In 1960, it was estimated that 7% of the world's population (200 million) had goiters. Treatment with iodized salt (0.1% KI) and (or) with the hormone thyroxine decreases or eliminates the small goiters. Larger goiters may require surgery. Since an excess amount of iodine also causes goiters, balance is the key to health.

Anemia can be caused by a deficiency of **iron,** but the deficiency of iron is not necessarily the only cause. Heredity, improper level of vitamin B_6, lack of folic acid (vitamin B_9), and lack of vitamin B_{12} (pernicious anemia, described later in the discussion of this vitamin) are other causes of anemia. Iron-deficient anemia is not necessarily fatal because a person with only 20% of the normal amount of hemoglobin still has the energy and strength to walk.

The lack of either vitamin B_9 or vitamin B_{12} can cause pernicious anemia.

For some nutrient elements, good health depends on having the element in proper amount and having the element in *proper ratio* to one or more other elements. Two ratios that must be maintained are the calcium/phosphorus ratio and the potassium/sodium ratio.

The **Ca/P weight ratio** from traditional diets appears to be 2.5, although the USRDA levels give a ratio of 1.0. The Melvin Page method uses the Ca/P ratio and the glucose levels in the blood to diagnose and treat various diseases. With diet changes (mainly from processed to unprocessed foods and omitting all sweets) and the administration of hormones, the Ca/P ratio is brought to 2.5. When the ratio is reached, symptoms of a variety of diseases are lessened or eliminated, and the patient is considered cured. The investigation, begun years ago, continues in trying to find the optimum Ca/P ratio in the human body and in the human diet. However, the evidence is sufficient at present to realize the need to keep essential nutrients balanced in the body.

Typical values of the **K/Na ratio** are greater than one. Some K/Na ratios for specific tissues are for muscle (4), liver (2.5), heart (1.8), brain (1.7), and kidney (1.0). For individual cells, potassium ions, K^+, concentrate inside the cell, and sodium ions, Na^+, concentrate outside in the fluid that bathes the cell. Natural, unprocessed food has high K/Na weight ratios. Fresh, leafy vegetables average a K/Na ratio of 35. Fresh, nonleafy vegetables and fruits average a ratio of 360, with extreme values of 3 for beets and 840 for bananas. K/Na ratios in meats range from 2 to 12. So, what is the problem? The body has K/Na ratios greater than 1, and fresh, natural foods provide a K/Na ratio greater than 1. Problems occur with processed and cooked foods. Potassium and sodium compounds are quite soluble in water. During processing (and cooking, if boiled), both potassium and sodium compounds are dissolved by water and discarded. The sodium is replenished by "salting" the food (adding sodium chloride). Potassium is usually not added to the food. One solution to the problem is to eat unprocessed, natural food, which "naturally" has the proper K/Na ratios. Another solution is to "salt" food with one of the commercial products that contain both potassium and sodium, such as Morton's Lite Salt. In summary, do not add much NaCl, if any, to food, and eat fresh vegetables and fruits high in potassium.

The normal urinary excretion of **sodium** per day is in the range of 1.4 to 7.8 g for adults. If excess sodium is not eliminated, more water is retained, which may lead to edema (swollen legs and ankles). Different clinical studies have shown that increased levels of sodium raise the blood pressure in some individuals but have no effect on the blood pressure of others. The high-salt diets of 70 g NaCl per day in certain areas of Japan have traditionally produced an unusually high frequency of heart attacks among these Japanese. Sodium levels in the bloodstream are regulated by **aldosterone,** which is secreted from the adrenal gland. Aldosterone works in the kidney to reabsorb sodium from the urine. The secretion of aldosterone is controlled by receptors, which measure the salt concentration in the blood. If the sodium concentration is too high in the blood, less aldosterone is excreted, and less sodium is reabsorbed from the urine.

Sodium is also excreted in sweat as sodium chloride. The salt concentration in sweat is dependent on dietary sodium intake, environmental temperature, the amount of sweating, and the degree of acclimation to the environment. Abrupt overheating is a very stressful problem for the body and involves an increase in skin and rectal temperatures, more rapid beating of the heart, and greatly increased sweat rate. After about a week of working at high temperatures, the body adapts by lowering the pulse rate and body temperature to normal levels. A high rate of sweating continues, but the concentration of NaCl in the sweat is decreased. As the person becomes acclimated to the heat, more water is needed, but no more salt is needed than during normal temperatures. Salt tablets may be needed and beneficial during the acclimation period but are not advisable after the body becomes acclimated. It should be noted that some never adapt to the heat and should avoid overheating.

Some nutrient elements affect the biochemical action of other elements. Examples already seen are the K/Na and Ca/P ratios and the effect of **molybdenum** on **copper** (Table 20–8). Another example is the enhancement effect the trace element **arsenic** has on the action of **zinc.** Deficiency symptoms are severe when both are absent. There is really no concern about deficiency, however, since arsenic is widely distributed in foods (especially seafood, such as shrimp) and is needed in such tiny amounts. Arsenic in larger amounts is poisonous.

Silicon, a trace element, is ingested at the rate of about 4 mg per day. More comes to the body from dust in the lungs. This dust can lead to **silicosis** and **asbestosis,** which are common in persons mining silica, silicates, and asbestos. There is considerable concern that breathing this dust can lead to lung cancer. Silicon stimulates the synthesis of collagen and is associated with calcium in bone calcification.

Tin, a trace element, has a low toxicity, which is probably good since we have several accesses to tin from nonfood sources: bronze plumbing fixtures, tin cans, tinning on household pots and pans, and stannous fluoride, SnF_2, in toothpaste. There is no known function for the trace amount of tin in the human body. Rats given diets supplemented with tin showed an increased growth rate.

Salary derives from the Latin *sal,* for "salt." Roman soldiers were given an allowance for salt.

SELF-TEST 20–C

1. Our mineral needs can be obtained by eating a _____ of _____ foodstuffs grown in _____.

2. Anemia can be caused by a deficiency of the mineral _____ because _____ is used to make _____.

3. The glucose tolerance factor (GTF) contains the mineral _____. The mineral and GTF enhance the activity of _____ so more glucose can be used by the body.

4. Which mineral is required in the mitochondria, melanin, and enzymes for using vitamins? _____

5. Excesses of the mineral _____ cause mental disorders (psychoses).

6. The mineral _____ aids in healing wounds because it functions in the synthesis of proteins.

7. A deficiency of the mineral _____ causes stretch marks in the skin.

8. Finicky eaters may have a deficiency of the mineral _____ .

9. Minerals involved in transmitting a nerve impulse are _____ and _____ .

10. For proper balance, the K/Na weight ratio in the body should be slightly ( ) greater than or ( ) less than 1.

11. The mineral obtained from chlorophyll in green vegetables is _____ .

12. The red, bulbous, swollen nose of a chronic alcoholic is attributed to a deficiency of the mineral _____ .

13. The mineral with the largest weight in the body is _____ , which is used mostly in the _____ .

14. Calmodulin, a protein, collects excess calcium from the bloodstream and helps to prevent _____ stones.

15. Goiter is caused by a deficiency of _____ .

16. Excess fluoride can cause _____ teeth.

17. How many amino acids contain sulfur? _____

Vitamins

A vitamin is an organic constituent of food that is consumed in relatively small amounts (less than 0.1 g/kg of body weight per day) and is essential to maintain life although vitamins are not synthesized by human beings. Vitamins are synthesized by plants, our principal natural source of vitamins. Of the some million organic compounds eaten in a normal diet, only about 100 are of proper size and stability to be absorbed from the digestive tract into the bloodstream without digestion or breakdown. Vitamins have these characteristics.

The structures of vitamins divide them into the two classes of oil-soluble vitamins and water-soluble vitamins. The oil-soluble vitamins—vitamins A, D, E, F, and K—tend to be stored in the fatty tissues of the body (especially the liver). For good health and nutrition, it is important to store enough of these vitamins, but not too much . . . again, a balance. The water-soluble vitamins tend to pass through the body and are not stored readily. Water-soluble vitamins are the B group (called vitamin B complex) and C. Fewer problems are caused by excesses of water-soluble vitamins.

In the testing for the effects, especially of the deficiencies, of vitamins in a diet, it is experimentally very difficult to identify and to remove selectively trace amounts of vitamins in the presence of large amounts of other organic compounds. Trace amounts of minerals are much more readily identified and removed. Because of the analytical and clinical problems associated with the discovery of vitamins, "new" vitamins are not likely to surface soon, although there is evidence that the liver contains health-giving factors that may not have been identified chemically as yet.

Listed in Table 20–9 with the vitamins, their USRDA advisements, and molecular structures is a summary of some of the biochemical functions and some effects associated with deficiencies of vitamins. The **IU** (**international unit,**

TABLE 20–9 What Vitamins Are and Why We Need Them

VITAMIN	USRDA*	STRUCTURE AND NAME	BIOCHEMICAL FUNCTIONS	DEFICIENCY EFFECTS	FOOD SOURCES (OTHER SOURCES)
Oil Soluble					
A	5000 IU	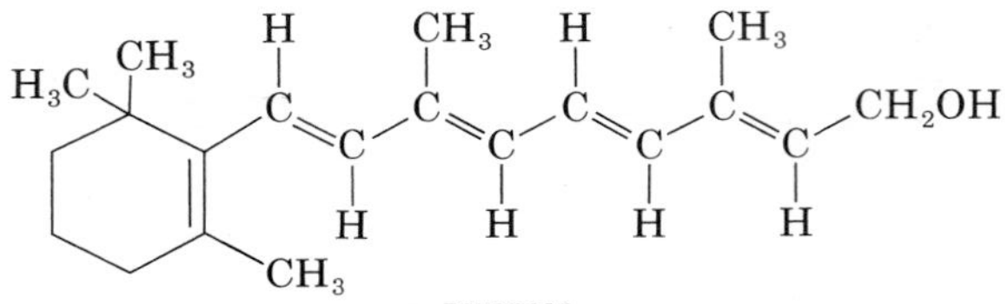 RETINOL	Prevents night blindness by regeneration of rhodopsin (visual purple), protects against infection, membrane activator	Excessive light sensitivity, night blindness, increased susceptibility to infection, xerophthalmia,† nerve degeneration in spinal column	Cod liver oil, halibut oil, egg yolk, apricots, peaches; for β-carotene: green vegetables, carrots
D	400 IU	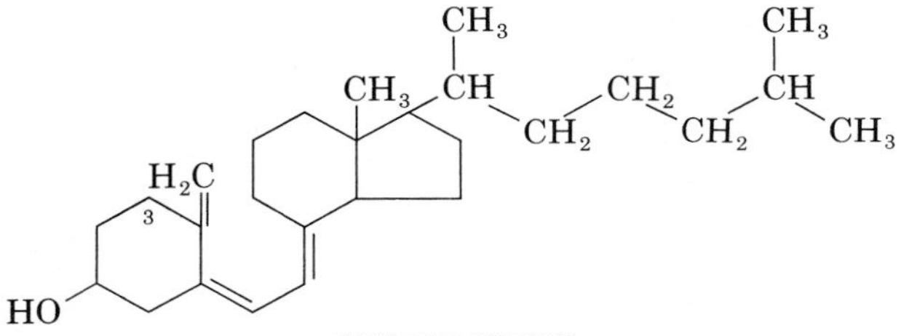 CHOLECALCIFEROL 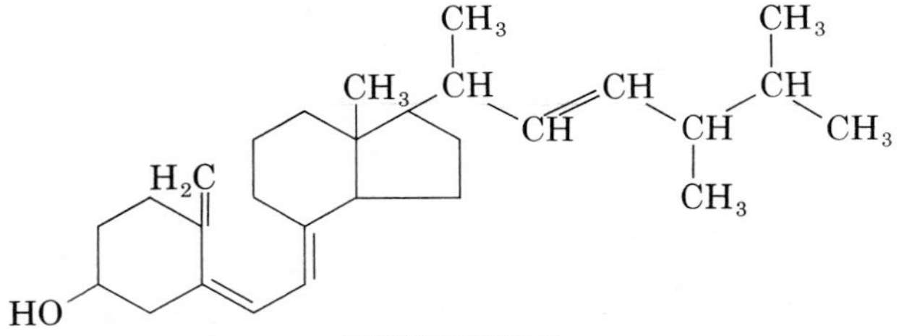 ERGOCALCIFEROL	Ca and P metabolism, transport of Ca in bloodstream, important in building strong bones	Abnormal development of bones and teeth, rickets‡	Ocean fish such as tuna, halibut, herring, cod (sunlight), milk fortified with vitamin D
E	30 IU	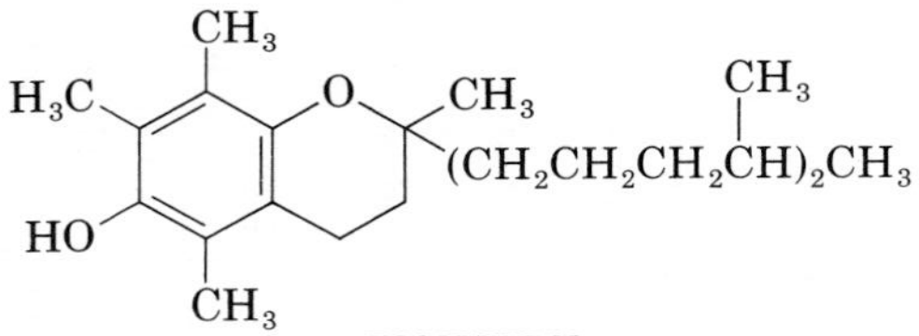 α-TOCOPHEROL	Antioxidant (e.g., of fats), maintains red blood cell membranes	Edema§ and anemia in infants	Yeast, lettuce, wheat grains, meat, butterfat, salmon, eggs, nuts

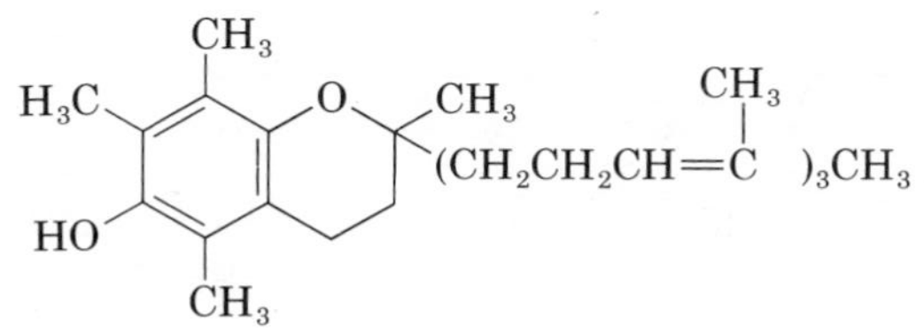

F	—	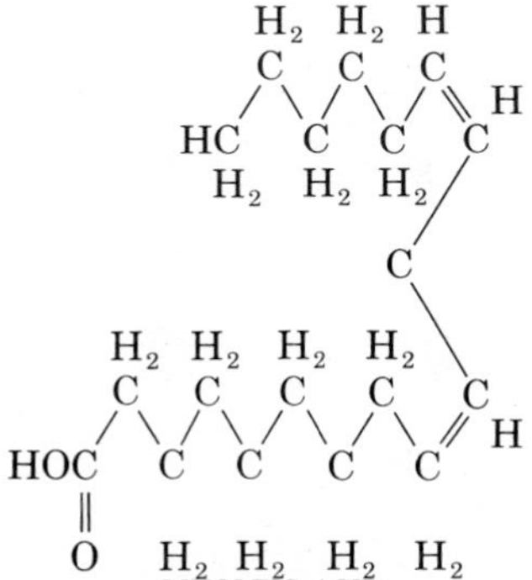 α-TOCOTRIENOL LINOLEIC ACID	Structural parts of cell membranes	Coarsened, sparse hair, eczema||	Fatty foods, especially pork lard
K	est¶	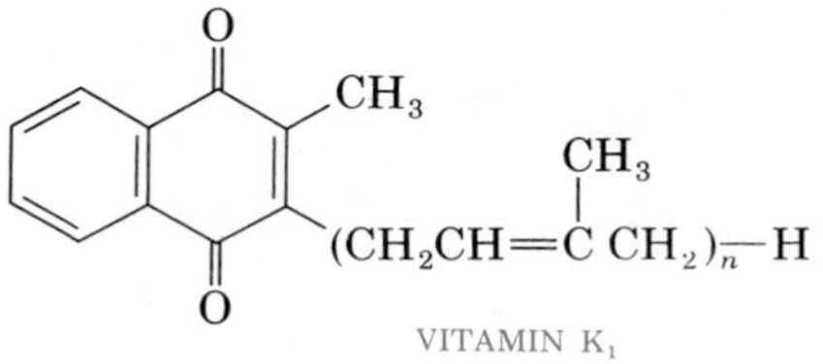 VITAMIN K₁ VITAMIN K₂ (n = 5 to 13)	Synthesis of prothrombin (blood coagulant)	Hemorrhages, slow clotting of blood	Leafy vegetables, cauliflower

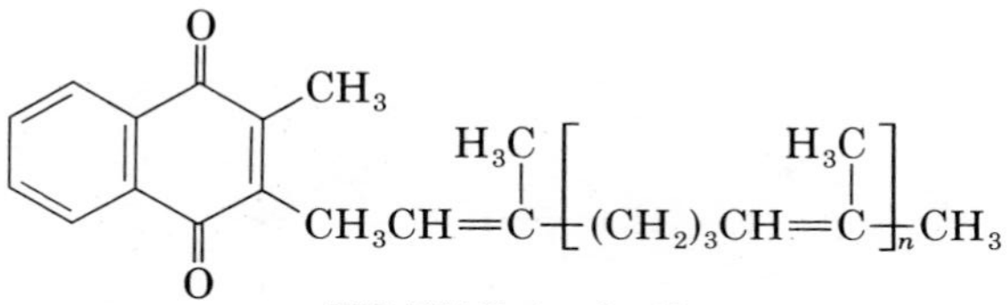

α-TOCOTRIENOL

$(CH_2CH_2CH=C \underset{CH_3}{\quad})_3 CH_3$

LINOLEIC ACID

VITAMIN K$_1$

$(CH_2CH=C\,CH_2)_n$—H

VITAMIN K$_2$ (n = 5 to 13)

$CH_3CH=C$—$[(CH_2)_3CH=C]_n$—CH_3

TABLE 20–9 (*Continued*)

VITAMIN	USRDA*	STRUCTURE AND NAME	BIOCHEMICAL FUNCTIONS	DEFICIENCY EFFECTS	FOOD SOURCES (OTHER SOURCES)
Water Soluble B₁	1.5 mg	 PYRIMIDINE THIAZOLE (THIAMIN)	Carbohydrate metabolism, proper functioning of heart and circulatory system, provides adequate ATP, maintains low levels of pyruvate and lactate ions, energy production, necessary for growth	Beriberi;** serious nervous disorders, muscular atrophy, serious circulatory changes, absence of knee jerk on tapping the knee	Soybeans, bran, corn flakes, hominy grits, pork loin, raw peanuts, brewer's yeast, wild rice, liver, eggs
B₂	1.7 mg	 ribitol purine flavin RIBOFLAVIN	Coenzyme in electron transfer reactions (such as in metabolism of fats, proteins, and pyruvate (Krebs cycle), necessary for healthy eyes and skin	Cessation of tissue growth, pale urine, oily skin at lower corners of nose, inflamed eyelids, bloodshot eyes, sore cracks around mouth, excessive fatigue, pellagrasine-pellagra	Most natural foods, especially liver, milk, eggs, fresh, dark-green, leafy vegetables
B₃	2.0 mg	 NICOTINIC ACID NICOTINAMIDE (ACTIVE FORM) NIACIN	Metabolism of fats, proteins, carbohydrates, synthesis of proteins, fats, necessary for healthy skin and nervous system	Pellagra,†† canker sores, anemia, trench mouth, stunted growth	Yeasts, liver, peanuts, soybeans, whole wheat breads, potatoes

Vitamin	Amount	Structure	Function	Deficiency symptoms	Sources
B$_5$	10 mg	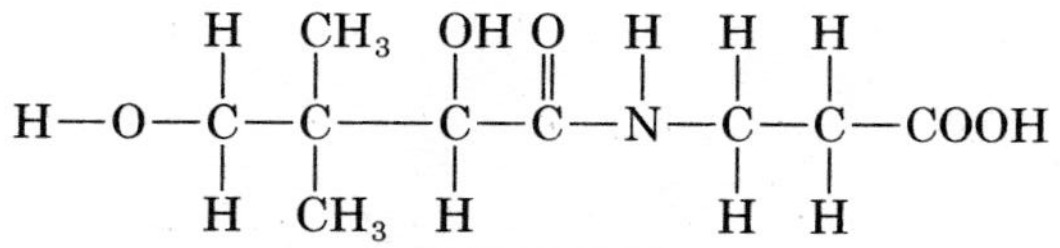 PANTOTHENIC ACID	Holds and transfers two carbon fragments in metabolism of carbohydrates and fats. Fragments are used in synthesis of sex hormones, heme, cholesterol, fatty acids, amino acids	Retarded growth	Yeasts, broccoli, cauliflower, mushrooms, bran, egg yolk, nuts, liver, salmon
B$_6$	2 mg	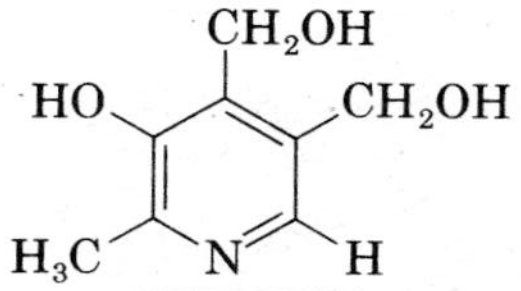 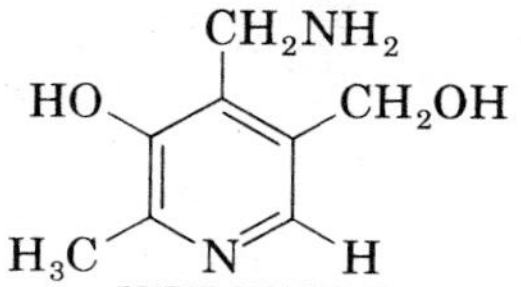 PYRIDOXINE, 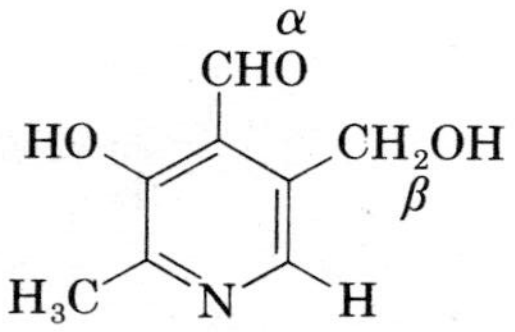 PYRIDOXAMINE, PYRIDOXAL	60 known enzymatic reactions mostly in metabolism and synthesis of proteins, necessary for growth and healthy nervous system (more, see the text)	Oily dermatitis, nausea, vomiting, weakness, dizziness, tingling in hands, impaired finger flexibility	Yeast, bran, raw carrots, avocados, bananas, liver, egg yolks, most meats if not overcooked
B$_7$	0.3 mg	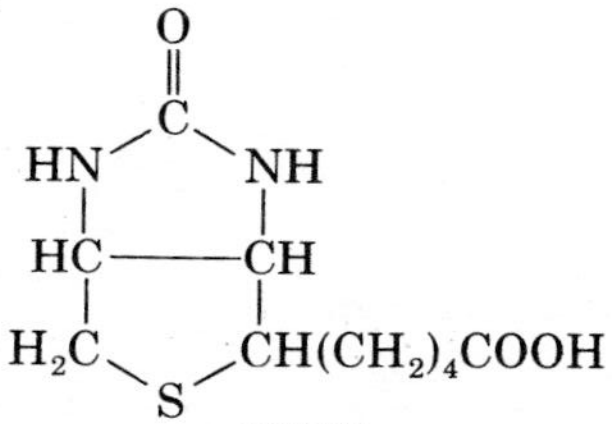 BIOTIN, 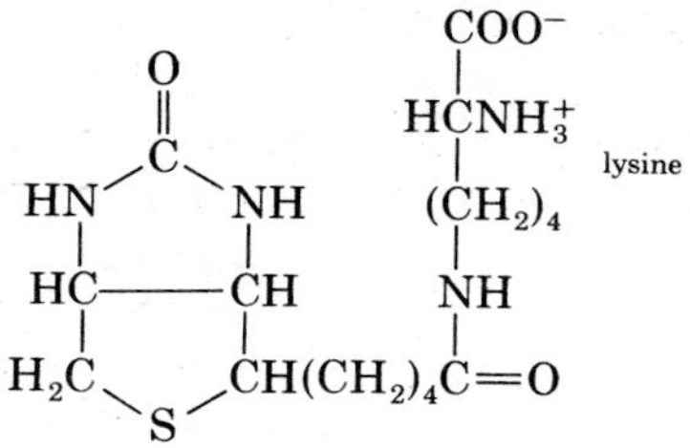 BIOCYTIN‡‡ (lysine)	Coenzyme of carboxylases (transfers CO_2 in food metabolism)	Sleepiness, muscle pains, hypersensitivity, aversion to food, large increase in blood cholesterol, grayish, dry, scaly skin	Most vegetables, brewer's yeast, beef, peanuts, liver, milk (intestinal bacteria produce the vitamin)
B$_9$	0.4 mg	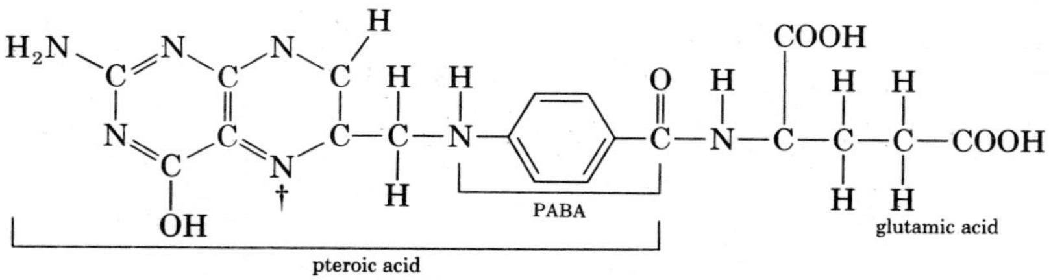 FOLACIN (OR FOLIC ACID)§§ (pteroic acid, PABA, glutamic acid)	Coenzymes related to growth, citrovorum factor,‖ important in proper development of red blood cells	Anemia, fatigue, poor appetite, sleeplessness, forgetfulness	Liver, whole grains, bran, yeast, asparagus, spinach

TABLE 20–9 *(Continued)*

VITAMIN	USRDA*	STRUCTURE AND NAME	BIOCHEMICAL FUNCTIONS	DEFICIENCY EFFECTS	FOOD SOURCES (OTHER SOURCES)
B_{12}	6 μg	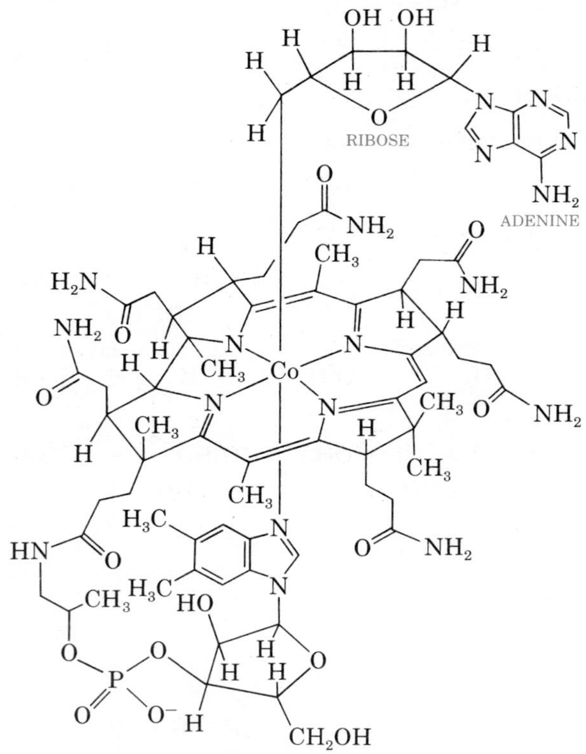 ADENOSYLCOBALAMIN (Cobalamin has a CN on cobalt in place of ribose and adenine.)	Coenzyme in reduction of ribonucleotides to deoxyribonucleotides, promotes growth, catalyzes the synthesis of nucleic acids, red and white blood cells, maintains reproductive systems and myelin coating of nerve fibers	Poor cell division, pernicious anemia,*** pallor and jaundice of skin, poor coordination, psychoses	Meat, clams, oysters, egg yolks

C | 60 mg

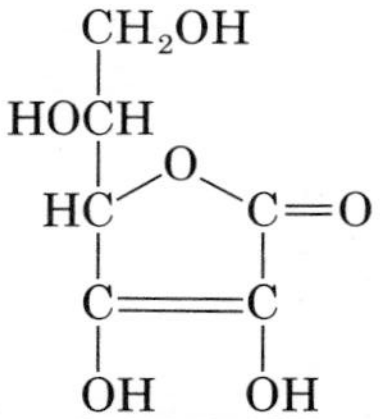

CH$_2$OH
HOCH
HC—O—C=O
C   C
OH OH
ASCORBIC ACID

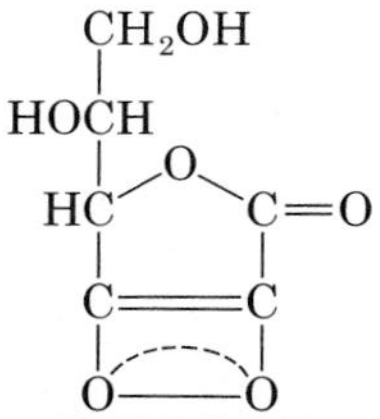

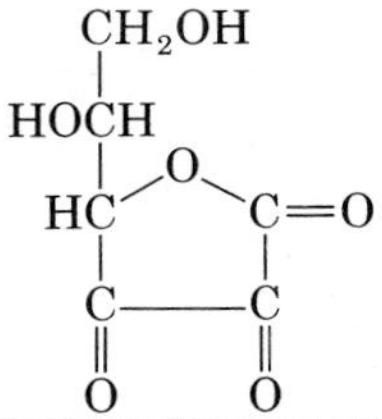

CH$_2$OH
HOCH
HC—O—C=O
C   C
O   O
DEHYDROASCORBIC ACID

CH$_2$OH
HOCH
HC—O—C=O
C   C
O   O
FREE RADICAL

Reducing agent, free radical former and scavenger, complexes metal ions, resistor of disease, aids in healing wounds

Scurvy†††

Fresh fruit and vegetables

* USRDA values are for adults and children over 4 years old.
† Xerophthalmia is a dry, lusterless condition of the eyeball.
‡ Rickets causes soft and poorly mineralized bones because without sufficient calcium, the synthesis of the organic part of the bone surpasses the hard apatite synthesis.
§ Edema is the condition of excessive fluid in tissue beneath the skin.
‖ Eczema is an inflammatory skin disease characterized by lesions, watery discharge, scales, and crusts.
¶ There is no USRDA value for vitamin K, but an estimated need is for 0.1 mg per day.
** Symptoms of beriberi are labored breathing, enlarged heart, extreme pain, mental confusion, and, with some types, paralysis of the arms and legs. Beriberi is Singhalese for "I cannot," meaning I am too ill to do anything.
†† Pellagra's symptoms are diarrhea, dermatitis, and dementia (loss of memory, mental confusion).
‡‡ Active form of biotin.
§§ PABA is para-aminobenzoic acid.
‖‖ The citrovorum factor, a coenzyme and derivative of folacin, is involved in synthesis of purines and pyrimidines, nitrogenous bases which hydrogen bond strands of nucleic acids to form the double helix structure of DNA.
¶¶ Adenosylcobalamin is the coenzyme form of cobalamin (vitamin B$_{12}$).
*** Pernicious anemia is characterized by abnormally large red blood cells.
††† Scurvy produces swollen legs, spotted skin with spots of blood, stinking mouth with rotten gums, black lungs, and damaged liver.

really units) listed for vitamins A, D, and E is commonly seen on packaged, enriched foods. A remnant from the past, IUs were employed before chemical analyses of the specific vitamins were possible. In modern units, one IU of vitamin A is 0.344 microgram of crystalline vitamin A acetate. Therefore, the USRDA advisement of 5000 IU for vitamin A is the same as 1.72 mg of vitamin A acetate. Another unit that expresses the efficiency of use of vitamin A is the *retinal equivalent* (RE), where one RE equals 5 IU. For vitamin D, 1 IU is 0.025 microgram of cholecalciferol; the USRDA advisement of 400 IU is the same as 0.01 mg. For vitamin E, 1 IU is 1 mg; the USRDA advisement of 30 IU is the same as 30 mg of vitamin E. The more important datum on the cereal box is the percentage of the USRDA provided by a serving of the cereal.

As with the major nutrients and minerals, now with the vitamins, the purpose of presenting biochemical functions and malnutrition effects of the nutrient is to submit proof for the essentiality of the nutrient in a balanced intake. The reader is then left to apply this evidence to his or her health goals and needs, life style, and diet availability. In the discussion of the individual vitamins, some biochemical functions and malnutrition problems will be elaborated on, along with other scientific findings related to nutrition.

See Chapter 24 for the structure of β-carotene and for the role of vitamin A in vision.

Contrary to popular belief, carrots provide the **provitamin,** β-carotene, and not retinol **(vitamin A).** The body converts β-carotene into retinol during the transfer of the provitamin through the intestinal wall.

Vitamin A aids in the prevention of infection by barring bacteria from entering and passing through cell membranes. The vitamin performs its sentinel duty by producing and maintaining mucus-secreting cells. Bacteria stick to the mucus and are trapped.

Not that you likely would, but if you took a bite (30 g) of polar bear liver, you would ingest 450,000 IU of retinol. Continued eating of polar bear liver has caused peeling of the skin from head to foot.

Vitamin D may be a hormone and not a vitamin. Vitamins are not produced by the body, but hormones are produced and have enhancing action on specific organs. With the aid of ultraviolet light in the 290 to 300 nanometer (nm) wavelength range, calciferol is synthesized by the human body. The series of enzymatically controlled reactions begins with acetate being converted into squalene, which is further changed into 7-dehydrocholesterol. The greatest concentration of 7-dehydrocholesterol is in the skin. The considerable energy required for the next step is supplied by ultraviolet light, which converts 7-dehydrocholesterol into cholecalciferol. In the liver first and then in the kidney, cholecalciferol is hydroxylated (substitute — OH groups for — H) to form the active 1α, **25-dihydroxycholecalciferol.** Since the body can, with ultraviolet light, synthesize vitamin D, some contend calciferol is not a vitamin but a hormone. Humans have no need for dietary calciferol if their bodies are exposed to sufficient sunlight or a sunlamp. Window glass blocks out the 290 to 300 nm ultraviolet light. Suntanning seems to be a protective measure to prevent the formation of too much calciferol. Any exposure to the sun *may* cause skin cancer. Again, a balance is needed.

Vitamin D is more toxic than the other (or true) vitamins. Intakes of 0.25 mg per day for four months or 5 mg per day for two weeks cause toxicity in children and, if prolonged, in adults. Symptoms include nausea, vomiting, diarrhea, headache, brown pigmentation of the skin, and kidney stones. Calcification occurs in the soft tissues and joints.

Although **vitamin E** deficiency is suspected of causing sterility in male human beings, no clinical evidence to support this hypothesis exists at this time.

The function of **vitamin E** as an antioxidant has been well established. Vitamin E is particularly effective in preventing the oxidation of polyunsaturated fatty acids, which readily form peroxides. This is perhaps the reason vitamin E is always found distributed among fats in nature. The fatty acid peroxides are particularly damaging because they can lead to runaway oxidation in the cells. Since cell membranes contain considerable fat, vitamin E protects the integrity of the membranes. In addition, vitamin E helps to maintain the integrity of the circulatory and central nervous systems, and it is involved in the proper functioning of the kidneys, lungs, liver, and genital structures. Vitamin E also detoxifies poisonous materials absorbed into the body.

Since some theories view aging as the cumulative effects of the action of free radicals running wild, **vitamin E** with its antioxidant properties, is a continued speculative candidate as an agent to inhibit aging or at least help to avoid premature aging.

Vitamin E appears to be the only vitamin destroyed in the freezing of foods.

Vitamin K is involved in the first three of four phases in the complicated process of blood coagulation. The four phases of blood coagulation are the generation of thromboplastin, its activation, the interaction of thromboplastin with prothrombin to form thrombin, and the interaction of thrombin, fibrinogen, and another factor to form the fibrin making the clot. One of the several antagonists to vitamin K is dicoumarol, which is found in clover and alfalfa. Dicoumarol and other anticoagulants are used in medicine to *decrease* the coagulability of blood. As rodenticides, they cause rodents (particularly rats) to bleed to death.

Vitamin K_2 is synthesized by intestinal bacteria.

The B group of vitamins (the **B-complex vitamins**) work together as a group to make life better. They are primarily involved as coenzymes in biochemical reactions leading to growth and to energy production. Their place of action is in the mitochondria of the cells. Being water soluble, the B vitamins are easily eliminated during the processing and cooking of food. The effectiveness of vitamins B_3 and B_6 is diminished in the presence of light, especially if the food is hot.

Before many of the B vitamins can serve as coenzymes, they must be converted into other substances, most of which are familiar names from the chapter (15) on biochemistry. For example, **riboflavin (B_2)** becomes part of the coenzyme flavin adenine dinucleotide (FAD).

FAD

Nicotinamide **(niacin, B_3)** appears predominantly as part of two coenzymes, nicotinamide adenine dinucleotide (NAD) and its phosphate (NADP). In the structure of the following coenzymes shown, R = H for NAD and R = P(O)(OH)$_2$ for NADP.

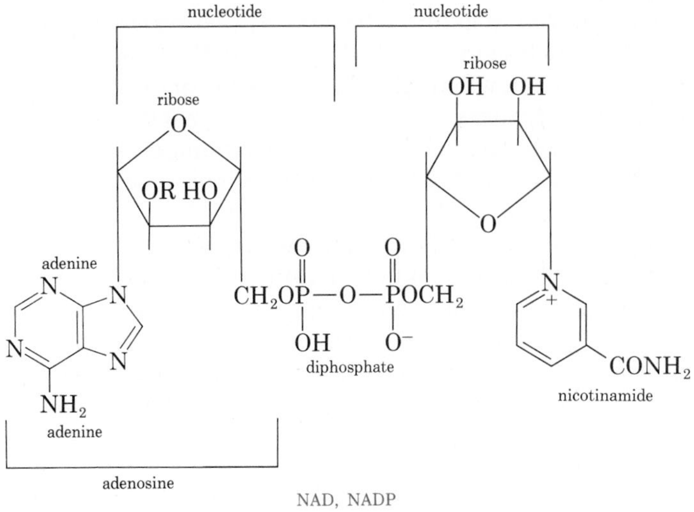

NAD, NADP

Pantothenic acid (B$_5$) (from the Greek word meaning "coming from every side") is incorporated into coenzyme A.

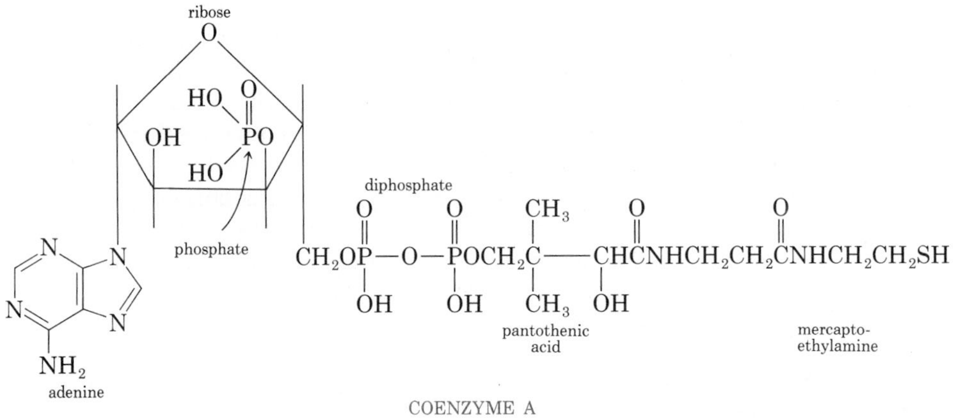

COENZYME A

Niacin is synthesized in the human body from **l-tryptophan,** an essential amino acid, assisted by vitamin B$_6$. The efficiency of the series of reactions is low since it takes 60 mg of tryptophan to make 1 mg of nicotinamide. Tryptophan is effective in treating niacin-deficient diseases if tryptophan is administered in the 60 to 1 ratio.

Pyridoxine (B$_6$) is considered the "master vitamin." Without sufficient vitamin B$_6$, there is impairment to the endocrine system (hormones), the metabolism of calcium and phosphorus, nerve activity, and absorption of vitamin B$_{12}$. The vitamin is also involved in the formation of red blood cells, control of blood sugar level, absorption of iron and zinc, maintenance of skin health (with biotin), and the metabolism of unsaturated fats (with vitamin E). An increased intake of vitamin B$_6$ has been reported to decrease or eliminate diabetic symptoms, certain anemias, diarrhea, edema, nervousness, numbness, muscular weakness, sore joints, halitosis, insomnia, hair loss, headaches, and excess infections. Along with magnesium, vitamin B_6 is needed for the synthesis of lecithin in the liver. With all of this and more, one wonders what vitamin B_6 cannot do.

Several substances interfere with the biochemical action of **vitamin B_6**, and, in effect, make it appear there is a deficiency of the vitamin. Several naturally occurring components of beans and mushrooms interfere with the action of the vitamin. When *hydralazine,* a commonly used drug for hypertension, is used, more vitamin B_6 is required. When oral contraceptives are used, the body requires more of the vitamin than normally supplied by the diet due to abnormal metabolism of tryptophan and other proteins. A supplement of 30 mg of pyridoxine hydrochloride daily has been suggested for oral contraceptive users. During pregnancy, there is an increase of tryptophan metabolites, and the need for B_6 supplementation is increased. Vitamin B_6 is effective in treating nausea and vomiting (morning sickness) during pregnancy.

The eating of raw egg whites ingests a **biotin (B_7)** antagonist, **avidin,** a protein-like substance. Avidin renders biotin ineffective, and, in effect, creates a deficient-like situation of biotin in the body. A classic case of the effects of avidin is a 66-year-old man who since youth had consumed raw eggs in wine. During the six years prior to his hospitalization for cancer, his diet was 1 to 4 quarts of wine and 4 to 10 eggs daily and very little else. In the hospital, vitamin supplementation caused his scaly skin and other vitamin-deficient symptoms to improve rapidly. Since avidin is inactivated by heat and (or) light, cooked eggs have no avidin antagonism to biotin.

Folacin (B_9) may be the most common vitamin deficiency among the world's population, and its area of influence is so vital — cell multiplication through a derivative of folacin, the **citrovorum** factor. The citrovorum factor is a coenzyme in the synthesis of purines and pyrimidines, which are nitrogenous bases. These bases match each other and hold together strands of nucleic acids by hydrogen bonds, thus forming the double-helix structure of DNA. The vitamin is also involved in the synthesis of choline and methionine, in the metabolism of amino acids, and in the formation of red blood cells. The heaviest activity of the vitamin is in the liver, bone marrow, lymph, and kidneys.

Cobalamin (B_{12}) is unique among the B-complex vitamins because it is the largest molecule, and it contains a metal, cobalt. The name *cobalamin* derives from the inclusion of cobalt in its structure. The larger size of the molecule requires some special arrangements to move through the various membranes on its way to the liver or kidneys.

A special protein, called the **intrinsic factor,** is exuded at the exit region of the stomach. The intrinsic factor and cobalamin bind together and travel to the lower end of the small intestine where cobalamin is absorbed into the wall, leaving the protein behind. In the bloodstream, one of three other proteins (transcobalamine I, II, or III, by name) bind with cobalamin and travel in concert to the vitamin's destination.

Since **vitamin B_{12}** is synthesized naturally only by bacteria, there is none to very little in plants. Pure vegetarians who consume no milk, eggs, or other animal products could be deficient in this vitamin. Although vegetarians may ingest some vitamin B_{12} from the manure accumulated on plants, fermented foods, and edible seaweeds, they are advised to supplement their diets with vitamin B_{12}.

When the body has less than 0.3 g of **vitamin C,** the symptoms of **scurvy** begin to appear. Scurvy has been known for several centuries, particularly on seagoing vessels. Descriptions from a ship's log in 1536 describe the horrors of the disease: swollen legs, spotted skin with spots of blood, stinking mouths, rotten gums, white rotten heart with a quart of red water about it, black lungs, and "an indifferent" liver. From this kind of wretchedness came the discovery of vitamin C. Captain James Cook found in his Pacific voyages (1768–1780) that if he collected and consumed fresh fruits, vegetables, berries, and green plants everywhere he had the opportunity, his men had no scurvy. By 1795, the British admiralty ordered that the sailors have a daily ration of lime juice. Scurvy was prevented, and English sailors came to be known as Limeys.

Vitamin C does sentinel duty in being consumed in consortium with the destruction of invading bacteria, in being involved in the synthesis and activity of interferon, which prevents the entry of viruses into cells, and in decreasing the ill effects of toxic substances, including drugs and pollutants.

The question of whether or not **vitamin C** will decrease the incidence of the common cold has been studied for many years and is now a discussion of interest. Results of the studies show an average decrease of about 30% in illness (particularly upper respiratory infection) as a result of taking vitamin C supplements. Not as well publicized or studied, **vitamin A** in large doses also decreases colds and the effects due to colds. In avoiding or breaking colds, some persons respond better to vitamin A than to vitamin C. Others respond better to vitamin C than to vitamin A, and some respond to neither vitamin as far as reducing the number of colds or symptoms of colds. In any case, response to either vitamin requires taking it preferably before but no later than at the early onset of a cold. It is recommended not to take excesses of both vitamins to break colds since this seems to prolong the cold symptoms.

SELF-TEST 20-D

1. The only vitamin containing a metal is vitamin _____, which has the name _____.

2. Vitamins are synthesized by the body. (   ) True or (   ) False

3. Vitamins A, D, E, F, and K are _____-soluble, whereas vitamins B-complex and C are _____-soluble.

4. The relationship between β-carotene and vitamin A is that β-carotene is a _____ of vitamin A.

5. A good vegetable source of vitamin A is _____, and a good animal source of the vitamin is _____.

6. Polar bear liver is exceptionally rich in vitamin _____.

7. Vitamin E is effective as a(n) _____, particularly in preventing the deterioration of polyunsaturated fatty acids.

8. What vitamin is destroyed in the freezing of foods? _____

9. The vitamin involved in coagulation of blood is vitamin _____.

10. What is the relationship of dicoumarol (found in clover and alfalfa) to vitamin K? _____

11. Which vitamin can be synthesized by the body in the presence of sunlight (or ultraviolet light)? _____

12. Rickets is caused by a deficiency of vitamin _____.

13. The role of vitamin D is to absorb the mineral _____ and transport it in the body.

14. The most toxic vitamin is _____.

15. Origin of the nickname "Limeys" for English sailors came from giving the sailors limes to prevent _____.

16. B-complex vitamins function generally as _____ involved generally in the process of _____.

17. A deficiency of thiamin, vitamin _____, causes the disease _____.

18. Pellagra is caused by a deficiency of the vitamin called _____.

19. What are the symptoms of pellagra? _____, _____, and _____.

20. Which vitamin holds two carbon fragments and thus is involved in the metabolism of carbohydrates and fats? _____

21. The master vitamin is _____ because it is involved in so many biochemical reactions.

22. Oral contraceptive users may need a supplement containing vitamin _____.

23. The eating of raw eggs can provide avidin, which renders _____ ineffective.

24. The vitamin so involved in cell multiplication is also the most commonly deficient among the vitamins. This is vitamin _____.

25. Which vitamin has the largest molecular structure? _____

26. Which B-complex vitamin is not found in vegetables? _____

OBESITY

Since so much money, time, and effort are spent nowadays on dieting (not being obese), the question necessarily posed revolves around "Is dieting really worth it?" or "Is it bad to be fat?" Primarily from a nutritional viewpoint, we shall examine the question(s) under the topics of causes of obesity, problems with obesity, ways to avoid or decrease obesity, and dieting taken too far — **anorexia.**

The fundamental cause of obesity is an unbalanced energy budget. More food energy (kcal) is consumed than expended, and the surplus is stored in fat cells. The energy balance is experimentally verified, and we cannot pretend it does not exist or that it will not affect us. Perhaps more fundamental questions are "What is obesity?" and "Why do people overeat?"

According to standards set by insurance companies, anyone over the ideal weight for his or her age group is considered obese. According to these standards, 10% to 25% of teenagers and 25% to 50% of adults in the United States are overweight. Another test for obesity is the skinfold test. By measuring the thickness of a big pinch of skin (skinfold) on the back of the upper arm, the back, or the waist, fatness is defined as a skinfold thicker than one inch. The problem with both of these tests and other such tests is that each individual has his or her own "set point," ideal weight for optimum health, which may appear fat (or, for that matter, skinny) to others. However, for the person's heredity, bone density, muscle conformations, occupation, and other such factors, his or her set-point weight is ideal for body functions, and this weight may differ from other general recommendations.

As to why people overeat, answers are probably more personal and individual than one's set-point weight. Do people overeat because of hunger (physiological), appetite (mostly psychological), previous consumption, habit, or satiety (degree of fullness)? Or do people overeat because of their genetic makeup? Some use more blood lipid, whereas others use more blood glucose for energy, and the latter are more hungry. Is the drive to overeat metabolic? Is there an insensitive monitor in the brain to check on blood glucose levels? Are there too many fat cells sending signals to fill me up? Is the set-point too high? Is the overeating motivated by an insulin problem? Enlarged fat cells resist insulin intake, which leaves glucose in the bloodstream. More insulin is supplied, and more fat storage occurs. Is there no nerve or hormonal switch to turn off when the body's physiological need has been met? Is the motivation to overeat environmental or psychological? Are there too many external cues (temptations such as being a cook, a candy store operator, or a soda jerk or having vending machines too handy)? Do people overeat because of stress or to relieve tension? Hormones released in response to physical or emotional stress favor rapid metabolism of energy nutrients, which can be used to fuel muscular activity quickly. If the fuel is not used during these stressful moments, the fuel is stored as fat. In contrast, overly thin people reject food during stress, while overweight people tend to eat during stress. Does overeating stem from a fear of starvation, or a response to fundamental sensations of yearning, craving, addiction, or compulsion? Does one eat too much just to relieve boredom or to avoid depression? The causes of overeating that lead to obesity are complex and individually assorted. However, if one knows the cause (or causes) of one's overeating, he or she will know where to start in correcting the problem.

Some factors have been tested experimentally and sufficiently to generalize about their effects on obesity and overeating. For example, on the basis of studies with identical twins, overeating is more of an inherited trait than an environmental response, but both do have an effect.

Those people who eat small but frequent meals tend to store less fat than those who eat large meals at irregular intervals.

Children who are obese develop sturdy muscles and bones to support the excess weight and, as a result, have more lean body mass. It is more difficult for these children to lose weight, and even after losing some weight, they will still look stocky. Conversely, people who gain weight later in life and were not obese as a child do not have the bone and muscle structure to support the newly gained weight.

The number of fat cells becomes fixed by adulthood. The number of these cells is proportional to one's diet and appetite at an early age. The more fat cells one has, the harder it is to lose weight.

According to insurance statistics, fat people die younger from a host of causes, including heart attacks, strokes, and diabetes because they have high levels of blood fat, more hypertension, strains on the skeletal system (causing arthritis, gout, and abdominal hernias), and more varicose veins (the leg muscles are too fat to contract sufficiently to pump blood back to the heart).

In our society, fat is not usually considered attractive. This perhaps unfair viewpoint leads to fewer marriages, high life insurance premiums, fewer job opportunities, less sports participation, more difficulty in finding clothes that fit, and, for children, there is more ridicule and the fat child is often chosen last for teams.

According to the American Medical Association, "the only way to lose weight is to eat less and exercise more." Despite this advice, a multimoney set of enterprises has flourished from the promotion of multidiet plans to lose weight. Many of these diet plans are nutritionally unsound, and some are even dangerous. Before participating in a diet plan, it may be wise to consult a source that examines the plans to see if they are based on sound, scientific studies or if they are simply wild statements. One such source is the Committee on Nutritional Misinformation of the Food and Nutrition Board, National Academy of Sciences, National Research Council.

When dieting, there is one immutable law: fat loss always must experience an expenditure of 3500 kcal for each pound lost. There is no magic escape or circumvention of this principle. The kilocalories can be used by the body for the BMR or for additional activities, but activity is required to use the energy. A person or a machine moving your muscles does not expend nearly as much energy; you must move your muscles to expend the energy.

The priority of weight control is, of course, on prevention and control, and, if at all possible, before adulthood so fewer fat cells will be formed. Where prevention has failed, the treatment must involve a simultaneous attack on three fronts. Eat fewer calories, do more exercise, and change behavior. Changed behavior can mean not eating when upset or tense or eating slowly because the satiety signal indicating you are full is sent after a 20-minute lag.

From what has been learned in the previous discussion of nutrition concerning the functions and deficiency effects of each nutrient, the problems with refined foods, and energy requirements from food, changing the lifestyle to eat less would mean:

1. To include all nutrients in the diet in appropriately needed amounts, including having the Ca/P and K/Na ratios correct
2. To choose from low-calorie foods rich in nutrients such as tasty vegetables, fruits, whole-grain breads and cereals and a limited amount of lean protein-rich foods like poultry, fish, eggs, cottage cheese, and skim milk (find the ones you like and use them more often)
3. To include some fats (about one third of the kilocalories) because fats make the meals more satisfying
4. To omit refined sugar, pure fat and oil, and alcohol (use complex carbohydrates instead; the damage done by alcohol on the liver is like tinkering with the major and only transportation terminal of the world)
5. To include roughage such as bran or fruit pulp so the food track runs well

Compare these conclusions with the Dietary Guidelines for Americans stated earlier in this chapter.

Dieting taken too far eventually leads to the **anorexia nervosa** stage. The self-imposed diet becomes an obsession. The cycle of eating followed by self-induced vomiting is common among anorexics. The person, usually a young female, becomes severely undernourished, reaching a body weight of 70 pounds or less. At this point, self-imposed starvation rules her life, and she is on the verge of incurring permanent brain damage and chronic invalidism or death. Although she has symptoms of starvation, she sees them as desirable and prides herself on holding out against her extreme hunger. Among the physical symptoms are arresting of sexual development, stopping of menstruation, drying and yellowing of the skin, loss of texture of hair, pain on touch, lowered blood pressure and metabolic rate, anemia, and severe sleep disturbances.

Dr. Hilde Bruch, an authority on anorexics, describes the treatment of the anorexic as first, restoring normal nutrition by tube feeding directly into the stomach if necessary. As the anorexic begins to return to normal weight, family interactions have to be dealt with. In the final stage of therapy, she can begin to be taught some new principles and clear up some of her misconceptions about nutrition and acceptable appearance.

FOOD ADDITIVES

Many chemicals with little or no nutritive value are added to food for a variety of reasons. The chemicals are added during the processing and preparation of food for purposes of preserving the food from oxidation, microbes, and the effects of metals on food. Food additives add and enhance flavor. They color the food, control pH, prevent caking, stabilize, thicken, emulsify, sweeten, leaven, and tenderize foods, among other effects.

The GRAS List

The GRAS list is a noble effort—but at present it is not foolproof.

The Food and Drug Administration lists about 600 chemical substances **"generally recognized as safe" (GRAS)** for their intended use. A small portion of this list is given in Table 20–10. It must be emphasized that an additive on the GRAS list is safe *only if it is used in the amounts and in the foods specified.* The GRAS list was published in several installments in 1959 and 1960. It was compiled by asking experts in nutrition, toxicology, and related fields to give their opinions about the safety of using various materials in foods. Since its publication, few substances have been added to the GRAS list and some, such as the cyclamates, carbon black, safrole, and Red Dye no. 2, have been removed.

It is evident, in view of the more than 2500 known food additives that many more chemicals than those that appear on the GRAS list are approved (or at least, not banned) for use as food additives by the FDA. It is quite expensive to introduce a new food additive with the approval of the FDA. Allied Chemical Corporation began research in 1964 on a new synthetic food color, Allura Red AC. It was approved by the FDA and went on the market in 1972. The cost for introducing this product was $500,000, and about half of this amount was spent on safety testing.

Preservation of Foods

Dry foods tend to be stable.

Foods generally lose their usefulness and appeal a short time after harvest. Bacterial decomposition and oxidation are the prime reasons steps must be taken to lengthen the time that a foodstuff remains edible. Any process that prevents the growth of microorganisms and/or retards oxidation is generally an effective preservative process for food. Perhaps the oldest technique is the drying of grains, fruits, fish, and meat. Water is necessary for the growth and metabolism of microorganisms, and it is also important in oxidation. Dryness thus thwarts both the oxidation of food and the microorganisms that feed on it.

A hypertonic solution is more concentrated than solutions in its immediate environment.

Osmosis is the flow of water from a more dilute solution through a membrane into a more concentrated solution.

Chemicals may also be added as preservatives. Salted meat, and fruit preserved in a concentrated sugar solution, are protected from microorganisms. The abundance of sodium chloride or sucrose in the immediate environment of the microorganisms forms a **hypertonic** condition in which water flows by **osmosis** from the microorganism to its environment. The salt and sucrose have the same effect on the microorganism as does dryness. Both dehydrate the microorganism.

The canning process for preserving food was developed around 1810, and involves first heating the food to kill all bacteria and then sealing it in bottles or

TABLE 20–10   A Partial List of Food Additives Generally Recognized as Safe*

Anticaking Agents
  Calcium silicate
  Iron ammonium citrate
  Silicon dioxide
  Yellow prussiate of soda

Acids, Alkalies, and Buffers
  Acetates: Ca, K, Na
  Acetic acid
  Calcium lactate
  Citrates: Ca, K, Na
  Citric acid
  Fumaric acid
  Lactic acid
  Phosphates, CaH, Ca_3, Na_2, Na_3, NaAl
  Potassium acid tartrate
  Sorbic acid
  Tartaric acid

Surface Active Agents (Emulsifying Agents)
  Glycerides: mono- and diglycerides of fatty acids
  Polyoxyethylene (20) sorbitan monopalmitate
  Sorbitan monostearate

Polyhydric Alcohols
  Glycerol
  Sorbitol
  Mannitol
  Propylene glycol

Preservatives
  Benzoic acid
    Na benzoate
  Methylparaben
  Propylparaben
  Propionic acid
    Ca propionate
    Na propionate

  Sorbic acid
    Ca sorbate
    K sorbate
    Na sorbate
  Sulfites, Na^+, K^+

Antioxidants
  Ascorbic acid
    Ca ascorbate
    Na ascorbate
  Butylated hydroxyanisole (BHA)
  Butylated hydroxytoluene (BHT)
  Lecithin
  Propyl gallate
  Sulfur dioxide and sulfites
  Trihydroxybutyrophenone (THBP)

Flavor Enhancers
  Monosodium glutamate (MSG)
  5′-Nucleotides
  Maltol

Sweeteners
  Aspartame
  Mannitol
  Saccharin
  Sorbitol

Sequestrants
  Citrate esters: isopropyl, stearyl
  Citric acid
  EDTA, Ca^{2+} and Na^+ salts
  Pyrophosphate, Na^+
  Sorbitol
  Tartaric acid
    NaK tartrate

Stabilizers and Thickeners
  Agar-agar
  Algins: NH_4^+, Ca^{2+}, K^+, Na^+
  Carrageenin
  Gum acacia
  Gum tragacanth
  Sodium carboxymethyl cellulose

Flavorings (1,700)
  Acetanisole (slight haylike)
  Amyl butyrate (pearlike)
  Bornyl acetate (piney, camphor)
  Carvone (spearmint)
  Cinnamaldehyde (cinnamon)
  Citral (lemon)
  Ethyl cinnamate (spicy)
  Ethyl formate (rum)
  Ethyl propionate (fruity)
  Ethyl vanillin (vanilla)
  Eucalyptus oil (bittersweet)
  Geraniol (rose)
  Geranyl acetate (geranium)
  Ginger oil (ginger)
  Linalool (light floral)
  Menthol (peppermint)
  Methyl anthranilate (grape)
  Methyl salicylate (wintergreen)
  Orange oil (orange)
  Peppermint oil (peppermint) (menthol)
  Pimenta leaf oil (allspice) (eugenol, cineole)
  Vanillin (vanilla)
  Wintergreen oil (wintergreen) (methyl salicylate)

If past history is a guide, at least some of these compounds will be taken off the GRAS list in the future.

* For precise and authoritative information on levels of use permitted in specific applications, the regulations of the U.S. Food and Drug Administration and the Meat Inspection Division of the U.S. Department of Agriculture should be consulted.

cans to prevent access of other microorganisms and oxygen. Some canned meat has been successfully preserved for over a century. Newer techniques for the preservation of food include vacuum freezing, pasteurization, cold storage, irradiation, and chemical preservation.

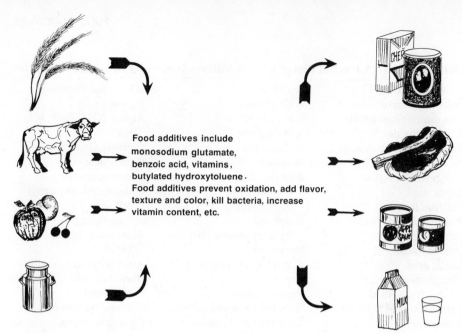

FIGURE 20–1 Between the harvested and the consumer-ready food, one often finds the addition of a large variety of food additives.

ANTIMICROBIAL PRESERVATIVES Food spoilage caused by microorganisms is a result of excretion of toxins. A preservative is effective if it prevents multiplication of the microbes during the shelf-life of the product. Sterilization by heat or radiation, or inactivation by freezing, is often undesirable since the quality of the food is impaired. Chemical agents seldom achieve sterile conditions but can preserve foods for considerable lengths of time.

A preservative must interfere with microbes but be harmless to the human system — a delicate balance.

Antimicrobial preservatives are widely used in a large variety of foods. For example, in the United States sodium benzoate is permitted in nonalcoholic beverages and in some fruit juices, fountain syrups, margarines, pickles, relishes, olives, salads, pie fillings, jams, jellies, and preserves. Sodium propionate is legal in bread, chocolate products, cheese, pie crust, and fillings. Depending on the food, the weight of the preservative permitted ranges up to a maximum of 0.1% for sodium benzoate and 0.3% for sodium propionate.

SODIUM BENZOATE   SODIUM PROPIONATE

Postulated mechanisms for the action of food preservatives may be grouped into three categories: (1) interference with the permeability of cell membranes of the microbes in foodstuffs, so the bacteria die of starvation; (2) interference with bacterial genetic mechanisms so the reproduction processes are hindered; and (3) interference with intracellular enzyme activity so that metabolic processes such as the Krebs cycle cease.

ATMOSPHERIC OXIDATION Microbial activity results in oxidative decay of food, but it is not the only means of oxidizing food. The direct action of oxygen in the air, **atmospheric oxidation,** is the chief factor in destroying fats and and fatty portions of foods. Chemically, oxygen reacts with the fat to form a hydroperoxide (R—OOH).

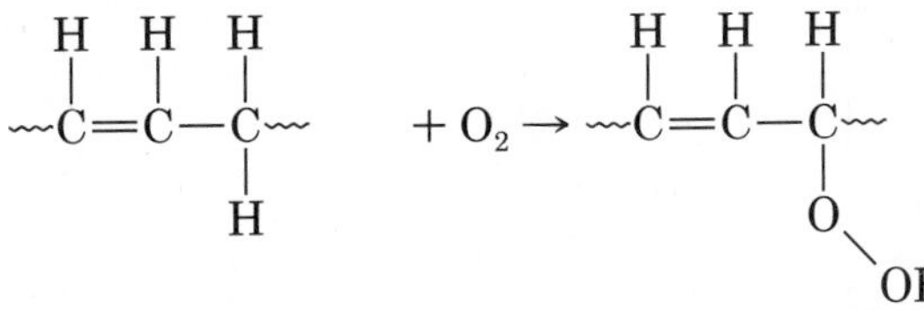

PORTION OF AN UNSATURATED
FAT MOLECULE
HYDROPEROXIDE

The mechanism involves the formation of reactive free radicals (species with one or more unpaired valence electrons) in a chain reaction process. For example:

$$RH + O_2 \rightarrow R\cdot + HO_2\cdot$$

FAT   FREE RADICALS

Free radicals usually have short lives; in rare cases they have a stable structure.

$$R\cdot + O_2 \rightarrow ROO\cdot$$

$$ROO\cdot + RH \rightarrow ROOH + R\cdot$$

HYDROPEROXIDE

Foods kept wrapped, cold, and dry are relatively free of oxidation. An antioxidant added to the food can also hinder oxidation. Antioxidants most commonly used in edible products contain various combinations of butylated hydroxyanisole (BHA), butylated hydroxytoluene (BHT), or propyl gallate:

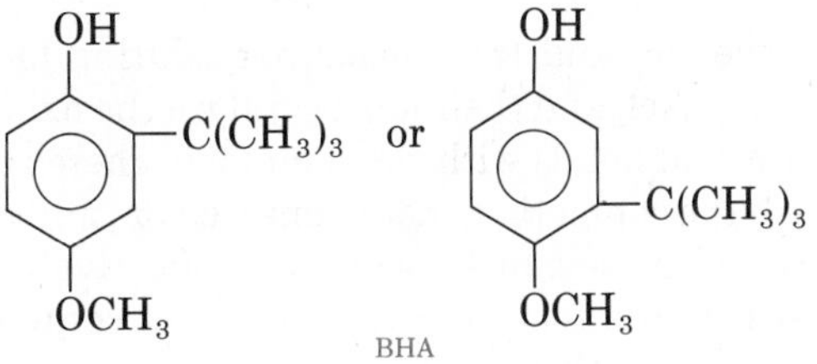

BHA

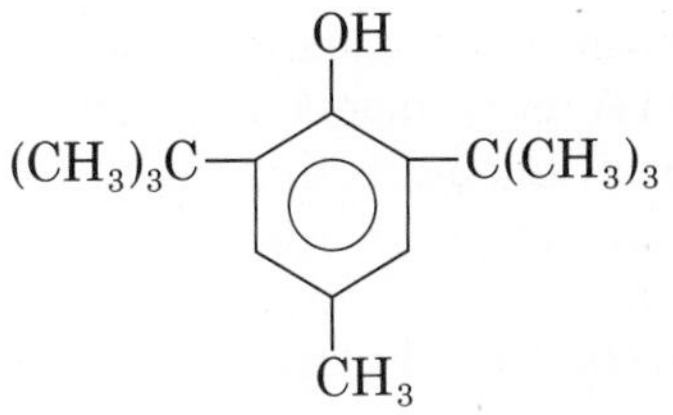

BHT 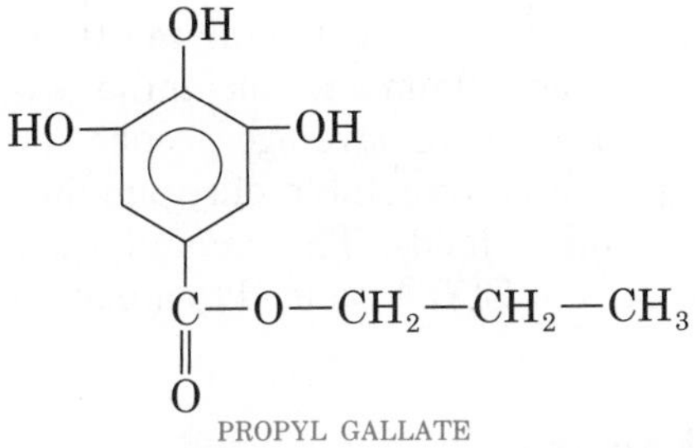   PROPYL GALLATE

To prevent the oxidation of fats, the antioxidant can donate the hydrogen atom in the —OH group to the free radicals and stop the chain reactions. The bulky aromatic radicals formed are relatively stable and unreactive; they add the unpaired electrons to their supply of delocalized electrons:

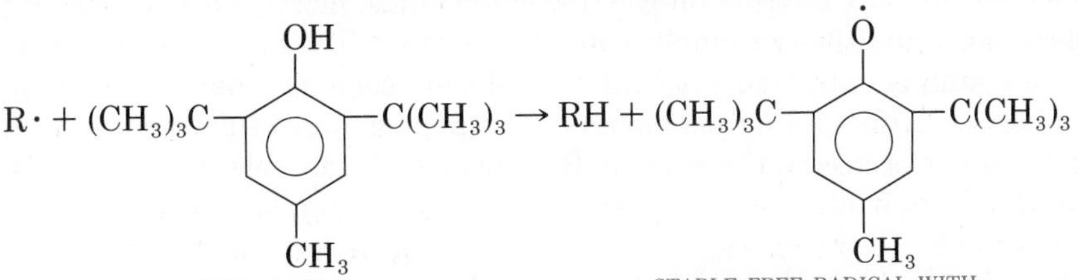

ANTIOXIDANT

STABLE FREE RADICAL WITH
LITTLE TENDENCY TO REACT

If antioxidants are not present, the hydroperoxy group will attack a double bond. This reaction leads to a complex mixture of volatile aldehydes, ketones, and acids which cause the odor and taste of rancid fat.

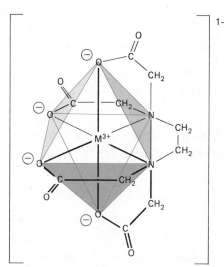

FIGURE 20–2 The structural formula for the metal chelate of ethylenediaminetetraacetic acid (EDTA).

Sequestrants

Metals get into food from the soil and from machinery during harvesting and processing. Copper, iron, and nickel, and their ions, catalyze the oxidation of fats. However, a molecule of citric acid bonds with the metal ion, thereby rendering it ineffective as a catalyst. With the competitor metal ions tied up, antioxidants such as BHA and BHT can accomplish their task much more effectively.

To sequester means "to withdraw from use." The sequestering ability of EDTA accounts for its use in treating heavy metal poisoning (Chapter 16).

Citric acid belongs to a class of food additives known as **sequestrants.** For the most part sequestrants react with trace metals in foods, tying them up in complexes so the metals will not catalyze the decomposition or oxidation of food. Sequestrants such as sodium and calcium salts of EDTA (ethylenediaminetetraacetic acid) are permitted in beverages, cooked crab meat, salad dressing, shortening, lard, soup, cheese, vegetable oils, pudding mixes, vinegar, confectioneries, margarine, and other foods. The amount ranges from 0.0025% to 0.15%. The structural formula of EDTA bonded to a metal ion is shown in Figure 20–2.

Flavor in Foods

Flavors result from a complex mixture of volatile chemicals. Since we have only four tastes (sweet, sour, salt, bitter), much of the sensation of taste in food is smell. For example, the flavor of coffee is determined largely by its aroma, and this in turn is due to a very complex mixture of over 100 compounds, mostly volatile oils.

Some 1700 natural and synthetic substances are used to flavor foods, making flavors the largest category of food additives.

Most flavor additives originally came from plants. The plants are crushed, and the compound is extracted with various solvents such as ethanol or carbon tetrachloride. Sometimes a single compound is extracted; more often, a mixture of several compounds occurs in the residue. By repeated efforts, relatively pure oils are obtained. Oils of wintergreen, peppermint, orange, lemon, and ginger, among others, are still obtained this way. These oils, alone or in combination, are then added to foods to obtain the desired flavor. Gradually, analyses of the oils and flavor components of plants have revealed the active compounds responsible for the flavor. Today, the synthetic preparation of the same flavors actively competes with natural sources.

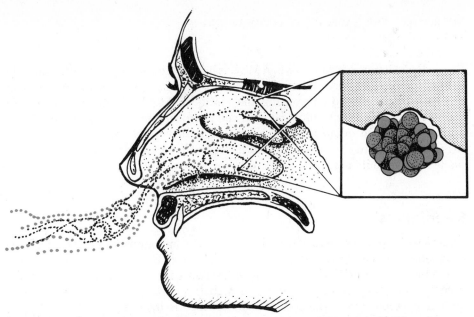

FIGURE 20–3 A stereochemical interpretation of the sensation of smell. The substance fits a cavity in the back of the oral cavity. If the atoms are properly spaced, they sensitize nerve endings that transmit impulses to the brain. The brain identifies these sensations as a particular smell. A complete explanation of smell is certainly more involved than the simple idea presented here.

The FDA has banned some of the naturally occurring flavoring agents that were formerly used, including safrole, the primary root beer flavor, found in the root of the sassafras tree.

SAFROLE

Flavor Enhancers

Flavor enhancers have little or no taste of their own but amplify the flavors of other substances. They exert synergistic and potentiation effects. Synergism is the cooperative action of discrete agents such that the total effect is greater than the sum of the effects of each used alone. Potentiators do not have a particular effect themselves but exaggerate the effect of other chemicals. The 5′-nucleotides, for example, have no taste, but they enhance the flavor of meat and the effectiveness of salt. Potentiators were first used in meats and fish but now are also used to intensify the flavor and cover unwanted flavors in vegetables, bread, cakes, fruits, nuts, and beverages. Three commonly used flavor enhancers are *monosodium glutamate (MSG), 5′-nucleotides* (similar to inosinic acid; Chapter 15), and *maltol.*

In some people, MSG causes the so-called "Chinese restaurant syndrome," an unpleasant reaction that includes headaches, sweating, and other symptoms usually occurring after an MSG-rich Chinese meal. Tomatoes and strawberries affect some individuals in the same way.

MALTOL
(FROM PINE NEEDLES)

MONOSODIUM GLUTAMATE

INOSINIC ACID
(a 5′-nucleotide)

MSG is a natural
constituent of many foods,
such as tomatoes and
mushrooms.

Brain damage is caused when MSG is injected in very high dosages under the skin of ten-day-old mice. When these laboratory results were reported, considerable discussion ensued concerning the merits of MSG. National investigative councils have suggested that it be removed from baby foods since infants do not seem to appreciate enhanced flavor. However, in the absence of hard evidence that MSG is harmful in the amounts used in regular food, no recommendations were made relative to its use.

Sweeteners

Insulin is a hormone that
regulates glucose
metabolism.

Sweetness is characteristic of a wide range of compounds, many of which are completely unrelated to sugars. Lead acetate, $Pb(CH_3COO)_2$, is sweet but poisonous. A number of **artificial sweeteners** are allowed in foods. These are primarily used for special diets such as those of diabetics. Artificial sweeteners have no known metabolic use in the body and do not require insulin.

SACCHARIN   The most common artificial sweetener is saccharin.

Saccharin is a synthetic
chemical.

SACCHARIN

Saccharin is about 300 times sweeter than ordinary sugar (sucrose). When ingested, saccharin passes through the body unchanged. It therefore has no food value other than to render an otherwise bland mixture more tasty. Saccharin has a somewhat bitter aftertaste, which renders it unpleasant to some users. Glycine, the simplest amino acid, which is also sweet tasting, is often added to counteract this bitter taste.

Laboratory studies have shown that high doses of saccharin cause cancer in mice. After months of consideration, the Institute of Medicine of the National Academy of Science joined the FDA in its 1978 statement that saccharin should be banned in U.S. foods. The U.S. Congress passed the ban and then suspended it pending the results of some ongoing studies in Canada. As of 1986, saccharin was still being sold, but with a warning label, "Use of this product may be hazardous to your health. This product contains saccharin which has been determined to cause cancer in laboratory animals."

ASPARTAME   A new entry into the sweetener market, aspartame is chemically an ester of a two amino acid peptide with the name N-L-α-aspartyl-L-phenylalanine methyl ester and trade names of NutraSweet, Equal, and Tri-Sweet. The sweetener was approved by the FDA in 1974 and subsequently withdrawn by its maker, G. D. Searle Co., when toxicity questions were raised. With these questions resolved, aspartame again received FDA approval in 1981.

Aspartame is about 180 times sweeter than table sugar (sucrose). The caloric value of aspartame is similar to that of proteins. The caloric intake of consumers using this product is reduced, since much smaller amounts are needed to produce the same sweetening effect. Aspartame does not have the bitter aftertaste asso-

$$HO-\overset{\overset{\displaystyle O}{\|}}{C}-\overset{\overset{\displaystyle H}{|}}{\underset{\underset{\displaystyle H}{|}}{C}}-\overset{\overset{\displaystyle H}{|}}{\underset{\underset{\displaystyle NH_2}{|}}{C}}-\overset{\overset{\displaystyle O}{\|}}{C}-\overset{\overset{\displaystyle H}{|}}{N}-\overset{\overset{\displaystyle H}{|}}{\underset{\underset{\displaystyle H-C-H}{|}}{C}}-\overset{\overset{\displaystyle O}{\|}}{C}-O-CH_3$$

ASPARTIC ACID

PHENYLALANINE ESTER

ASPARTAME

ciated with other artificial sweeteners. It is normally metabolized in the body as a peptide.

Mannitol and sorbitol, polyhydric alcohols, are sweeteners used in such products as sugarless gum. These sweeteners have some caloric value in the body.

Food and Esthetic Appeal

FOOD COLORS There are about 30 (33 in 1982) chemical substances used to color food. All are under investigation by the FDA, and some food colors may be prohibited as the investigations progress. About half of the food colors are laboratory synthesized, and half are extracted from natural materials. Most food colors are large organic molecules with several double bonds and aromatic rings. The electrons of these conjugated structures can absorb certain wavelengths of light and pass the rest; the wavelengths passed give the substance its characteristic color. β-carotene, an orange-red substance in a variety of plants that gives carrots their characteristic color, has a conjugated system of electrons and is used as a food color. β-carotene is a precursor (provitamin) of vitamin A.

> Colored organic substances often are *conjugated* molecules, having alternating double and single bonds in the carbon chain or ring.
>
> The structure of β-carotene is in Chapter 24.

Because one of the food colors, Yellow No. 5, causes allergic reactions (mainly rashes and sniffles) in an estimated 50,000 to 90,000 Americans, the FDA has required maufacturers to list Yellow No. 5 on the labels of any food products containing it.

pH CONTROL IN FOODS Weak organic acids are added to such foods as cheese, beverages, and dressings to give a mild acidic taste. They often mask undesirable aftertastes. Weak acids and acid salts, such as tartaric acid and potassium acid tartrate, react with bicarbonate to form CO_2 in the baking process.

Some acid additives control the pH of food during the various stages of processing as well as in the finished product. In addition to single substances, there are several combinations of substances that will adjust and then maintain a desired pH. These mixtures are called **buffers**. An example of one type of buffer is potassium acid tartrate, $KHC_4H_4O_6$.

> Buffer solutions resist change in acidity and basicity; pH remains constant.

Adjustment of the pH of a fruit juice is allowed by the FDA. If the pH of the fruit is too high, it is permissible to add acid (called an **acidulant**). Citric acid and lactic acid are the most common acidulants used since they are believed to impart good flavor; but phosphoric, tartaric, and malic acids are also used. These acids are often added at the end of the cooking time to prevent extensive hydrolysis of the sugar. In the making of jelly they are sometimes mixed with the hot product immediately after pouring. To raise the pH of a fruit that is unusually acid, buffer salts such as sodium citrate or sodium potassium tartrate are used.

> Small amounts of certain acids are allowed to be added to some foods.

The versatile acidulants also function as preservatives to prevent growth of microorganisms, as synergists and antioxidants to prevent rancidity and browning, as viscosity modifiers in dough, and as melting point modifiers in such food products as cheese spreads and hard candy.

ANTICAKING AGENTS   Anticaking agents are added to hygroscopic foods—in amounts of 1% or less—to prevent caking in humid weather. Table salt (sodium chloride) is particularly subject to caking unless an anticaking agent is present. The additive (magnesium silicate, for example) incorporates water into its structure as water of hydration and does not appear wet as sodium chloride does when it absorbs water physically on the surface of its crystals. As a result, the anticaking agent keeps the surface of sodium chloride crystals dry and prevents crystal surfaces from codissolving, which would join the crystals together.

Hygroscopic substances absorb moisture from the air.

STABILIZERS AND THICKENERS   Stabilizers and thickeners improve the texture and blends of foods. The action of carrageenan (a polymer from edible seaweed) is shown in Figure 20–4. Most of this group of food additives are polysaccharides (Chapter 15) having numerous hydroxyl groups as a part of their structure. The hydroxyl groups form hydrogen bonds with water to prevent the segregation of water from the less polar fats in the food, and to provide a more even blend of the water and oils throughout the food. Stabilizers and thickeners are particularly effective in icings, frozen desserts, salad dressing, whipped cream, confectioneries, and cheeses.

Stabilizers and thickeners are types of emulsifying agents.

SURFACE ACTIVE AGENTS   Surface active agents are similar to stabilizers, thickeners, and detergents in their chemical action. They cause two or more normally incompatible (nonpolar and polar) chemicals to disperse in each other. If the chemicals are liquids, the surface active agent is called an **emulsifier.** If the surface active agent has a sufficient supply of hydroxyl groups, such as cholic acid has, the groups form hydrogen bonds to water. Cholic acid and its associated group

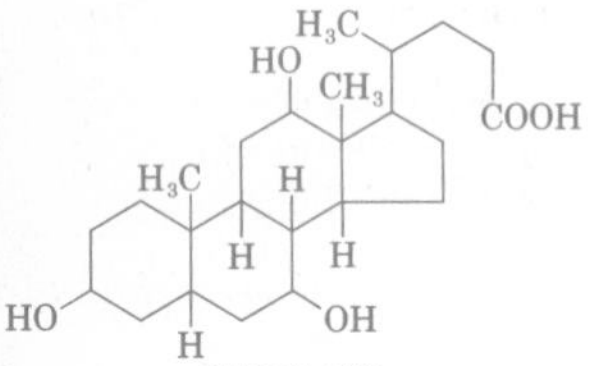

CHOLIC ACID

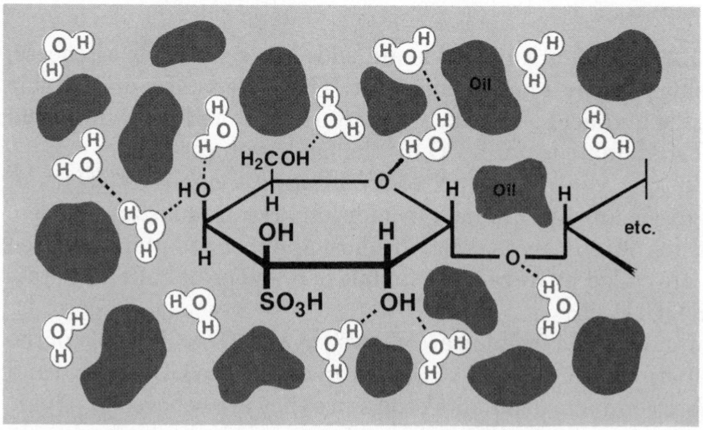

FIGURE 20–4   The action of carrageenan to stabilize an emulsion of water and oil in salad dressing. An active part of carrageenan is a polysaccharide, a portion of which is shown above. The carrageenan hydrogen-bonds to the water, which keeps it dispersed. The oil, not being very cohesive, disperses throughout the structure of the polysaccharide. Gelatin (a protein) undergoes similar action in absorbing and distributing water to prevent ice crystals in ice cream.

of water molecules are distributed throughout dried egg yolk in a manner quite similar to that of carrageenan and water in salad dressing.

Some surface active agents have both hydroxyl groups and a relatively long nonpolar hydrocarbon end. Examples are diglycerides of fatty acids, polysorbate 80, and sorbitan monostearate. The hydroxyl groups on one end of the molecule are anchored via hydrogen bonds in the water, and the nonpolar end is held by the nonpolar oils or other substances in the food. This provides tiny islands of water held to oil. These islands are distributed evenly throughout the food.

Hydrogen bonding plays a major role in stabilizers, thickeners, surface active agents, and humectants.

POLYHYDRIC ALCOHOLS Polyhydric alcohols are allowed in foods as humectants, sweetness controllers, dietary agents, and softening agents. Their chemical action is based on their multiplicity of hydroxyl groups that hydrogen-bond to water. This holds water in the food, softens it, and keeps it from drying out. Tobacco is also kept moist by the addition of polyhydric alcohols such as glycerol. An added feature of polyhydric alcohols is their sweetness. The two polyhydric alcohols mentioned earlier for their sweetness are mannitol and sorbitol. The structures of these alcohols are strikingly similar to the structure of glucose (Chapter 15), and all three have a sweet taste.

$$
\begin{array}{cc}
\text{CH}_2\text{OH} & \text{CH}_2\text{OH} \\
| & | \\
\text{H}-\text{C}-\text{OH} & \text{HO}-\text{C}-\text{H} \\
| & | \\
\text{HO}-\text{C}-\text{H} & \text{HO}-\text{C}-\text{H} \\
| & | \\
\text{H}-\text{C}-\text{OH} & \text{H}-\text{C}-\text{OH} \\
| & | \\
\text{H}-\text{C}-\text{OH} & \text{H}-\text{C}-\text{OH} \\
| & | \\
\text{CH}_2\text{OH} & \text{CH}_2\text{OH} \\
\text{D-SORBITOL} & \text{D-MANNITOL}
\end{array}
$$

Kitchen Chemistry

LEAVENED BREAD Sometimes cooking causes a chemical reaction that releases carbon dioxide gas, and the trapped carbon dioxide causes breads and pastries to rise. Yeast has been used since ancient times to make bread rise, and remains of bread made with yeast have been found in Egyptian tombs and the ruins of Pompeii. The metabolic processes of the yeast furnish gaseous carbon dioxide, which creates bubbles in the bread and makes it rise:

Leavened bread is as old as recorded history.

$$
\underset{\text{GLUCOSE}}{\text{C}_6\text{H}_{12}\text{O}_6} \xrightarrow[\text{from yeast}]{\text{zymase}} 2 \underset{\text{(GAS)}}{\text{CO}_2} + 2 \underset{\text{ETHANOL}}{\text{C}_2\text{H}_5\text{OH}}
$$

When the bread is baked, the CO_2 expands even more to produce a light, airy loaf.

Carbon dioxide can be generated in cooking by other processes. For example, baking soda (which is simply sodium bicarbonate, $NaHCO_3$, a base) can react with acidic ingredients in a batter to produce CO_2.

About 350 million pounds of phosphates are added to foods in the U.S. each year. This is 20% to 25% of our phosphorus intake. Some phosphates are used as leavening agents. Sodium phosphate thickens puddings, retains juices and makes hams tender, and prevents canned milk from thickening on standing.

$$\text{NaHCO}_3 + \text{H}^+ \rightarrow \text{Na}^+ + \text{H}_2\text{O} + \text{CO}_2 \text{ (gas)}$$

Baking powders contain sodium bicarbonate and an added acid salt or a salt that hydrolyzes to produce an acid. Some of the compounds used for this purpose are potassium hydrogen tartrate, $KHC_4H_4O_6$, calcium dihydrogen phosphate monohydrate, $Ca(H_2PO_4)_2 \cdot H_2O$, and sodium acid pyrophosphate, $Na_2H_2P_2O_7$. The reactions of these white, powdery salts with sodium bicarbonate are similar, although the compounds all have somewhat different appearances. For example:

$$\text{KHC}_4\text{H}_4\text{O}_6 + \text{NaHCO}_3 \xrightarrow{\text{water}} \text{KNaC}_4\text{H}_4\text{O}_6 + \text{H}_2\text{O} + \text{CO}_2 \text{ (gas)}$$

COOKING AND PRECOOKING — "PRELIMINARY DIGESTION" The cooking process involves the partial breakdown of proteins or carbohydrates by means of heat and hydrolysis (Fig. 20–5). The polymers that must be degraded if cooking is to be effective are the carbohydrate cellular wall materials in vegetables and the collagen

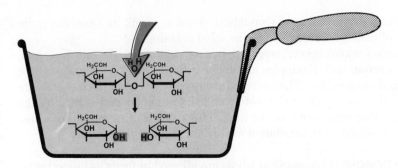

FIGURE 20–5 The hydrolysis of starch during the cooking of foods such as potatoes and rice.

Cooking starts the digestive process, although it is rarely needed for this purpose. Since it may destroy nutrients, it is done mostly for esthetic reasons.

or connective tissue in meats. Both types of polymers are subject to hydrolysis in hot water or moist heat. In either case, only partial depolymerization is required.

In recent years several precooking additives have become popular; the **meat tenderizers** are a good example. These are simply enzymes that catalyze the breaking of peptide bonds in proteins via hydrolysis at room temperature. As a consequence, the same degree of "cooking" can be obtained in a much shorter heating time. Meat tenderizers are usually plant products such as papain, a proteolytic (protein-splitting) enzyme from the unripe fruit of the papaw tree. Papain has considerable effect on connective tissue, mainly collagen and elastin, and shows some action on muscle fiber proteins. On the other hand, microbial protease enzymes (from bacteria, fungi, or both) have considerable action on muscle fibers. A typical formulation for the surface treatment of cuts of beef contains 2% commercial papain or 5% fungal protease, 15% dextrose, 2% monosodium glutamate (MSG), and salt.

SELF-TEST 20–E

1. Flavor enhancers exert a(n) _____ or _____ effect on the flavors of foods.
2. Name two of the oldest means for preserving food. _____
3. Antimicrobial preservatives make foods sterile. ( ) True or ( ) False
4. Flavors result from ( ) volatile or ( ) nonvolatile compounds.
5. Antioxidants are ( ) more or ( ) less easily oxidized than the food into which they are placed.
6. Citric acid is an example of a(n) _____. Such compounds tie up metals in stable complexes.
7. A flavor in a food can usually be traced to a single compound. ( ) True or ( ) False
8. Monosodium glutamate is a(n) _____ _____.
9. Salt is effective in preserving foods because it kills microorganisms by _____ them.
10. BHT serves as an antioxidant by destroying _____.
11. Which has the sweetest taste when an equal weight of each is tasted: table sugar, aspartame, or saccharin? _____
12. What molecular characteristic do most food colors have? _____
13. What acids are added to foods to lower the pH? _____ and _____
14. The gas released by leavening agents is _____

15. Cooking _____ some chemical bonds.
16. The GRAS list is the FDA list of food additives that are _____.
17. Hydrogen bonding generally plays a very important role in the action of surface _____ agents.

MATCHING SET I

Match the nutrient with the disease caused by a *deficiency* in the nutrient. Use all of the diseases.

Nutrient		Disease	
_____	1. Cobalamin	a.	Atasia in newborns
_____	2. Iodine	b.	Kwashiokor
_____	3. Vitamin A	c.	Goiter
_____	4. Vitamin D	d.	Scurvy
_____	5. Vitamin C	e.	Pernicious anemia
_____	6. Thiamin	f.	Rickets
_____	7. Riboflavin	g.	Anemia
_____	8. Niacin	h.	Finicky eaters, stretch marks
_____	9. Protein	i.	Pellagra
_____	10. Fat	j.	Cretinism
_____	11. Sugar	k.	Eczema
_____	12. Iron	l.	Night blindness
_____	13. Manganese	m.	Beriberi
_____	14. Zinc	n.	Xerophthalmia
		o.	Hypoglycemia
		p.	Pellagra-sine-pellagra

MATCHING SET II

_____	1. β-carotene	a.	Nutrient supplement in food
_____	2. Monosodium glutamate	b.	Food color
_____	3. Copper, nickel, and iron	c.	Flavor enhancer
_____	4. Sodium benzoate	d.	Catalyze oxidation of fats
_____	5. Potentiator	e.	Antimicrobial perservative
_____	6. Mineral	f.	Exaggerates some chemical effects
_____	7. Mannitol	g.	Sequestering agent
		h.	Sweetener
		i.	pH adjuster

QUESTIONS

1. What two factors influenced the United States government to pass laws on food? What problem was caused by the government regulations?
2. How moral is it to use animals and humans for testing of nutrient effects? What is an alternative? Do we need to know nutrient effects in the first place?
3. If all of our blood cells are renewed each 120 days, must new nutrients be ingested to make the new cells? Explain.

4. Other than moral issues, what are some problems with doing nutrition studies on humans? On animals?

5. Distinguish among RDA, USRDA, WHO (and FAO) recommendations on the bases of:
   a. Which are recommended for the United States, which for foreign countries
   b. Which have the higher recommendations, which the lower
   c. Which (between RDA and USRDA) have the greater breakdown with respect to age, sex, etc.
   d. Which are used on labelling packages

6. Why would WHO and FAO be lower recommendations than those in the United States? Are not all people alike?

7. If the USRDA is ingested each day, is good health assured? Explain.

8. Give three examples in which balance is especially important in the area of nutrition.

9. a. What activities go on during a determination of the basal metabolic rate (BMR)?
   b. What three factors affect BMR, other than weight, height, and age?

10. How would you determine the energy value of your daily intake of food?

11. What are two sources of dietary fiber?

12. Why is dietary fiber important in the diet?

13. What is the name of the essential fatty acid? What vitamin is designated as this fatty acid?

14. What complications could arise if one were to go on a starvation of a very low carbohydrate diet?

15. Name the ten essential amino acids.

16. What is the cause of kwashiorkor?

17. What foods are good sources of thiamin, vitamin B_{12}, niacin, ascorbic acid, vitamin A, vitamin D, and vitamin E?

18. What diseases or symptoms are caused by a deficiency of niacin, thiamin, proteins, calciferol, and ascorbic acid?

19. What group of vitamins is most easily destroyed or removed during food processing and cooking?

20. What functions do sodium, potassium, calcium, iodine, and phosphorus have in the human body?

21. What is the relationship between iodine and goiter?

22. Based on their structures and solubilities, what are the two classes of vitamins? Which class causes fewer problems because of an excess of the vitamin?

23. Which two ratios of minerals must be controlled for good health?

24. Discuss the basic causes of obesity?

25. Devise a good weight reduction program. What guidelines should be used for good nutrition?

26. Why do people overeat?

27. What are the functions of fat, protein, and carbohydrates in the body?

28. a. Discuss the relationship (if any) between cholesterol, saturated fats, and atherosclerosis.
    b. What causes arterial plaque?
    c. What effect does high-density lipoprotein have on arterial plaque?

29. What problem is caused by hydrogenation of polyunsaturated fats (oils)?

30. What problem is caused by heating fats (especially unsaturated fats) to temperatures at which they smoke?

31. Distinguish between refined sugar and complex carbohydrates as to how they are assimilated by the body.

32. What are some good foodstuff sources of complex carbohydrates?

33. What problems may arise from consuming too much
    a. Refined sugar?
    b. Refined grains (white flour)?

34. Why is vitamin B_6 called the "master vitamin"?

35. Why should one eat the whole fruit (where delectable) rather than, for example, suck the juice out of an orange and throw the rest away?

36. Does sugar, particularly refined sugar, cause arterial plaque? Discuss the pros and cons.

37. Discuss three possible causes of diabetes mellitus.

38. Will artificial sweeteners kill the desire for sugar? Explain. Discuss another alternative to artificial sweeteners.

39. Rate the factors that most influence proper nutrition, such as knowledge, tradition, economics, advertisement, pride, and other factors that you may like to add.

40. Use Table 20–4 to calculate the calories in some fast-food hamburgers, French fries, and milkshakes.

	Burgers	Shakes	Fries
Chain A			
Protein (g)	11	10	2
Carbohydrates (g)	29	72	20
Fat (g)	9	9	19
Chain B			
Protein (g)	12	10	3
Carbohydrates (g)	30	66	26
Fat (g)	10	9	12
Chain C			
Protein (g)	13	11	2
Carbohydrates (g)	29	55	25
Fat (g)	11	7	10

Which chain of fast-food restaurants offers the *lowest* calorie total for a hamburger, milkshake, and an order of French fries? What is the total number of kcal?

41. List some functions, and hence need, for chromium, copper, potassium, sodium, magnesium, phosphorus, iron, calmodulin, and manganese.

42. Which mineral is in the glucose tolerance factor (GTF)? How might the GTF function?

43. How are iron and copper ions carried in the blood?

44. Should salt tablets be taken after the body is acclimated to the heat? Why should or why should they not be taken?

45. a. What is the typical K/Na weight ratio in the body?
    b. What is a problem in maintaining this ratio?
    c. What can be done nutritionwise to maintain the proper K/Na ratio?

46. What substance is used to "firm" canned tomatoes?

47. How are Chinese thousand-year eggs made from chicken eggs?

48. Why may a calcium deficiency occur in some women after menopause?

49. What voyager discovered the effects of Vitamin C years before the vitamin was discovered?

50. What is the usual Ca/P weight ratio in the human body? What are some symptoms if the Ca/P ratio deviates from the normal stated in answer to the first question?

51. If niacin can be synthesized in the human body from tryptophan, an amino acid, is niacin a vitamin, or is tryptophan the vitamin? Explain.

52. Why do we need each specific mineral and each specific vitamin? List one function or one deficiency effect.

53. Which vitamin has PABA (para-aminobenzoic acid) in its structure? What else is PABA used for (consult the index)?

54. Describe the stages in blood coagulation. What vitamin is involved?

55. Which vitamins are produced by bacteria in the intestine?

56. Is calciferol a vitamin or a hormone? Explain why both are considered.

57. Which vitamin is most toxic?

58. What is the role of the intrinsic factor relative to vitamin B_{12}?

59. Does vitamin C decrease the symptoms of the common cold? Does vitamin A?

60. Describe pernicious anemia. What vitamins cause this disease by being deficient in the body? Is the disease usually fatal?

61. List the vitamins that
    a. Do not have aromatic rings as part of their structures
    b. Have a chain-conjugated structure system
    c. Are water soluble
    d. Are fat soluble

62. What are the primary reasons that food spoils?

63. How does salt preserve food?

64. Which of the following food additives should be avoided? Give your reasons.
    a. Butter yellow   c. Glycerin
    b. Propionic acid   d. Sodium cyclamate

65. Why does it take less time to cook food in a pressure cooker than in an open pot of boiling water?

66. Why does cooking aid digestion?

67. What would happen to your ability to digest protein if you kept the acid in your stomach neutralized all the time? Is acid bad for your stomach?

68. Why is saccharin preferable to chloroform as a sweetener?

69. A label on a brand of breakfast pastries lists the following additives: dextrose, glycerin, citric acid, potassium sorbate, vitamin C, sodium iron pyrophosphate, and BHA. What is the purpose of each substance?

70. What is a common flavor enhancer? How do flavor enhancers work?

71. How do BHA and BHT prevent potato chips from becoming rancid?

72. Choose a label from a food item, and try to identify the purpose of each additive.

73. Describe some of the chemical changes that occur during the cooking of
    a. A carbohydrate   c. A fat
    b. A protein

74. What causes fat in foods to become rancid? How can this be avoided?

75. What causes bread to rise?

76. What do the letters FDA represent, and what does this governmental agency do?

77. What are the pros and cons of eating "natural" foods as opposed to foods containing chemical additives?

78. Why were cyclamates taken off the market?

79. Do you think it is wise to use animals in safety tests for drugs and food additives? Should mental patients and prisoners be used for this purpose?

80. Many consumer products are almost identical in chemical composition but are sold at widely different prices under different trade names. Do you think the products should be identified by their chemical names or their trade names? Why?

81. What is the GRAS list?

82. What foods have you eaten during the past week that did not have chemicals added or applied to them?

83. See what you can find out about correlation between taste and smell. Are they the same sensation? Are they independent of each other?

84. Do you think it would be possible to live on a diet of entirely synthetic foods?

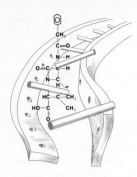

Chapter 21

MEDICINES AND DRUGS — SAVING LIVES WITH CALCULATED RISKS

The average life expectancy for men in the United States has risen from 53.6 years in 1920 to 71.4 years in 1983, a rise of 32.5%. During this same period, the life expectancy for women has risen from 54.6 years to 78.3 years, a rise of 43.4% (see Table 21–1).

The reasons for these differences by sex are not altogether clear, but an overall major contributing factor to the longer life span for both sexes has been the widespread use of a large assortment of new medicinal compounds.

Medicines rise and fall in popularity.

The contents of the medicine cabinet have changed drastically in the past few decades. A survey of physicians shortly before World War I revealed the ten most essential drugs (or drug groups) to be ether, opium and its derivatives, digitalis, diphtheria antitoxin, smallpox vaccine, mercury, alcohol, iodine, quinine, and iron. When another survey was made at the end of World War II, at the top of the list were sulfonamides, aspirin, antibiotics, blood plasma and its substitutes, anesthetics and opium derivatives, digitalis, antitoxins and vaccines, hormones, vitamins, and liver extract. Today there is an even wider array of medicinal chemicals, but drugs for reducing fever, relieving pain, and fighting infection still head the list in all areas of medical practice (Table 21–2).

Many new drugs are tested each year, but few ever reach the marketplace.

TABLE 21–1 Life Expectancy in the United States for Men and Women (1920–1983)

YEAR	MEN	WOMEN
1920	53.6	54.6
1930	58.1	61.6
1940	60.8	65.2
1950	65.6	71.1
1960	66.6	73.1
1970	67.1	74.8
1980	70.0	77.5
1982	70.8	78.2
1983	71.0	78.3

TABLE 21–2 Some Widely Prescribed Drugs Chosen to Illustrate the Variety of Uses

GENERIC NAME*	MEDICAL USE
Tetracycline HCl	Antibiotic
Ampicillin	Antimicrobial (kills bacteria but is not derived from a plant, as is an antibiotic)
Phenobarbital	Sedative, hypnotic, anticonvulsant
Thyroid	Increases rates of metabolism
Prednisone	Antiinflammatory, antiallergic agent similar to cortisone
Digoxin	Decreases the rate of the heartbeat but increases the force of the heartbeat, similar to digitalis
Meprobamate	Tranquilizer
Erythromycin	Antimicrobial
Penicillin G potassium	Antibiotic
Nitroglycerin	Dilates the blood vessels of the heart
Penicillin VK	Antibiotic
Quinidine sulfate	Slows the heartbeat, also used for malaria and hiccups
Paregoric	A tincture (alcoholic solution) of camphorated opium used as an analgesic
Reserpine	Tranquilizer
Nicotinic acid (niacin)	Dilates blood vessels; also, essential B vitamin with antipellagra activity

The generic name for a drug is its widely accepted chemical name.

* The generic name for a drug is its generally accepted chemical name rather than a specific brand name. For example, the generic drug tetracycline HCl is sold under brand names such as Acromycin V, Ambracyn, Artomycin, Diacycline, Quatrex, and others. Medical doctors can prescribe either the generic name or a brand name. If the generic name is used, the prescription is often cheaper, particularly if the drug is not protected by patents and can be manufactured and marketed competitively by several companies.

Americans spent $18.8 billion on medicines in 1983. This amounts to about $77 per person. Today, the top ten prescription drugs based on the number of prescriptions written (Table 21–3) show an interesting cross section of medicinal uses.

In this chapter we will look at several classes of medicines and how they work. Those drugs that are of interest because of their widespread application, their unique chemistry, or some other readily understood factor will be emphasized. We will begin with a few drugs that help us with discomfort, aches, and pains.

TABLE 21–3 Ten Most Prescribed Drugs in the United States in 1984, Ranked by Total Prescriptions Dispensed

TRADE NAME	GENERIC NAME	USE
1. Inderal	Propranolol	Heart diseases
2. Dyazide	Triamterine	Diuretic/antihypertensive
3. Lanoxin	Digitalis glycoside	Heart disease
4. Valium	Diazepam	Tranquilizer
5. Tylenol with codeine	Acetaminophen	Pain relief
6. Orthonovum	Norethindrone	Birth control
7. Tagamet	Cimetidine	Stomach ulcer control
8. Lasix	Furosemide	Diuretic
9. Amoxil	Amoxicillin	Antibiotic
10. Motrin	Ibuprofen	Antiinflammatory

(Data from *Drug Topics*, March 1985)

TABLE 21–4   The Chemistry of Some Antacids

COMPOUND	REACTION IN STOMACH	COMMENTS
Magnesium oxide MgO	$MgO + 2\,H^+ \rightarrow Mg^{2+} + H_2O$	MgO is white and tasteless.
Milk of magnesia $Mg(OH)_2$ in water	$Mg(OH)_2 + 2\,H^+ \rightarrow Mg^{2+} + 2\,H_2O$	The water suspension has an unpleasant chalky consistency.
Calcium carbonate $CaCO_3$	$CaCO_3 + 2\,H^+ \rightarrow Ca^{2+} + H_2O + CO_2$	Calcium carbonate is purified limestone.
Sodium bicarbonate $NaHCO_3$	$NaHCO_3 + H^+ \rightarrow Na^+ + H_2O + CO_2$	Baking soda, like $CaCO_3$, produces CO_2 gas in the stomach.
Aluminum hydroxide $Al(OH)_3$	$Al(OH)_3 + 3\,H^+ \rightarrow Al^{3+} + 3\,H_2O$	$Al(OH)_3$ is a clear gel.
Dihydroxyaluminum sodium carbonate $NaAl(OH)_2CO_3$	$NaAl(OH)_2CO_3 + 4\,H^+ \rightarrow Na^+ + Al^{3+} + 3\,H_2O + CO_2$	Sold as Rolaids, will not ordinarily cause pH to go above 5
Sodium citrate $Na_3C_6H_5O_7 \cdot 2\,H_2O$	$Na_3C_6H_5O_7 \cdot 2\,H_2O + 3\,H^+ \rightarrow 3\,Na^+ + H_3C_6H_5O_7 + 2\,H_2O$	Mild

ANTACIDS

The contents of the stomach are highly acidic.

The walls of a human stomach contain thousands of cells that secrete hydrochloric acid, the main purposes of which are to suppress growth of bacteria and to aid in the hydrolysis (digestion) of certain foodstuffs. Normally, the stomach's inner lining is not harmed by the presence of this hydrochloric acid, since the mucosa, the inner lining of the stomach, is replaced at the rate of about a half million cells per minute. When too much food is eaten the stomach often responds with an outpouring of acid, which lowers the pH to a point where discomfort is felt.

If the reduction of acidity is too great, the stomach responds by secreting an excess of acid. This is "acid rebound."

Antacids are compounds used to decrease the amount of hydrochloric acid in the stomach. The normal pH of the stomach ranges from 0.9 to 1.5. Some alkaline compounds used for antacid purposes and their modes of action are given in Table 21–4.

ANALGESICS

Analgesics relieve pain, but they are harmful in large doses.

Analgesics are pain killers. Most people need these compounds at one time or another. When we have a headache we take aspirin. When we have a tooth filled or extracted, the dentist uses Novocain. Intense suffering requires a strong pain killer, such as codeine or morphine. While these compounds are immensely useful, they are nevertheless dangerous if taken or used improperly. They can even become killers if taken in overdose.

Early societies may well have used opium. Although not all opium derivatives have therapeutic value, most of them are efficient pain killers. Their chief disadvantage lies in their addictive properties.

Opium is obtained from the opium poppy by scratching the seed pod with a sharp instrument. From this scratch flows a sticky mass that contains about 20 different compounds called **alkaloids** (organic nitrogenous bases containing basic nitrogen atoms). About 10% of this mass is the alkaloid **morphine,** which is primarily responsible for opium's effects.

Morphine and its derivatives are addictive drugs.

MORPHINE

Two derivatives of morphine are of interest. One of these is **codeine,** a methyl ether of morphine, which is less addictive than morphine and is about as powerful an analgesic. The other compound is **heroin,** the diacetate ester of morphine. Heroin is much more addictive than morphine and for that reason finds no medical uses in the United States.

Analgesics may or may not be habit forming.

CODEINE   (METHYL ETHER)

HEROIN   (ACETATE ESTER)

Two acetate groups

MEPERIDINE (DEMEROL)

One of the most effective substitutes for morphine is **meperidine,** first reported in 1931, and now sold as Demerol. It is less addictive than morphine. Two other relatively strong pain relievers used today are **pentazocine** (Talwin) and **propoxyphene** (Darvon). Talwin is slightly addictive, while Darvon has not been shown to be. However, Darvon has been much abused; there were approximately 600 Darvon-related deaths in 1977. Critics argue Darvon to be more dangerous and less effective in killing pain than other available opiates. Note that in the structures of these compounds there is a strong resemblance to the morphine structure.

Considerable progress has been made in understanding the drug action of the opiates. Solomon Snyder, along with co-workers, discovered in 1973 that the brain and spinal cord contain specific bonding sites into which the opiate molecules fit as a key fits a lock. John Hughes and Hans Kosterlitz followed in 1975 with the discovery that vertebrates produce their own opiates, which they named **enkephalins** (from Greek, meaning "in the head"). Individuals with a high tolerance for pain produce more enkephalins and consequently tie up more receptor sites; hence there is less pain. A dose of heroin would temporarily bond to a high percentage (or all) of the sites, and there would be little or no pain. Continued use of heroin would cause the body to reduce or cease its production of enkephalins. If the use of the narcotic is stopped, the receptor sites are empty and withdrawal symptoms appear.

Many of the **local analgesics,** or local anesthetics, are nitrogen compounds, like the alkaloids (Table 21–5). Local analgesics include the naturally

PENTAZOCINE (TALWIN)

PROPOXYPHENE (DARVON)

TABLE 21-5　Some Local Analgesics

Cocaine		Probably the first local analgesic used
Procaine (Novocain)		Often used in dental work
Lidocaine (Xylocaine)		More potent than procaine, can be applied to the skin

occurring **cocaine,** derived from the leaves of the coca plant of South America, and the familiar **Novocain** (procaine). All of these drugs act by some blockage of the nerves that transmit pain. Acetylcholine appears to be the "opener of the gate" for sodium (Na^+) and potassium (K^+) ions to flow into a nerve cell (Fig. 21-1).

There are times when milder, general analgesics are required, and few compounds work as well for as many people as **aspirin.** Not only is aspirin an analgesic, but it is also an antipyretic, a fever reducer.

The synthesis of aspirin was outlined in Chapter 13. Each year, about 40 million pounds are manufactured in the United States. Aspirin is thought to inhibit cyclooxygenase, the enzyme that catalyzes the reaction of oxygen with polyunsatu-

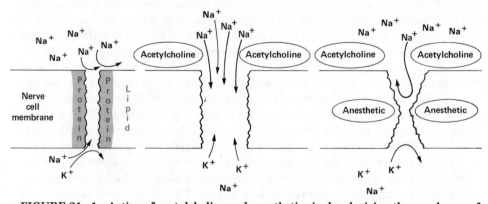

FIGURE 21-1　Action of acetylcholine and anesthetics in depolarizing the membrane of a nerve cell. Acetylcholine makes it possible for sodium and potassium ions to neutralize the negative charge associated with a nerve impulse so another impulse can be transmitted. Anesthetics block the action of acetylcholine and do not allow repetitive impulses to travel along the nerve.

rated fatty acids to produce prostaglandins. Excessive prostaglandin production causes fever, pain, and inflammation—just the symptoms aspirin relieves.

A great danger presented by aspirin is stomach bleeding, caused when an undissolved aspirin tablet lies on the stomach wall. As the aspirin molecules pass through the fatty layer of the mucosa, they appear to injure the cells, causing small hemorrhages. The blood loss for most individuals taking two 5-grain tablets is between 0.5 mL and 2 mL. Some people are more susceptible. Early aspirin tablets were not particularly fast dissolving, which aggravated this problem greatly. Today, aspirin tablets are formulated to disintegrate quickly, although crushing the tablet in a little water might not be a bad idea.

Another potential danger of aspirin is its possible link to **Reye's syndrome,** a brain disease that also causes fatty degeneration in organs such as the liver. Reye's syndrome can occur when children are recovering from the flu or chicken pox. Vomiting, lethargy, confusion, and irritability are the symptoms of the disease. Studies have shown a high degree of association between aspirin dosages and the onset of Reye's syndrome. About one quarter of the 200 to 600 cases per year have proved fatal! Beginning in 1982, aspirin products were required to contain a warning regarding the possible link between aspirin and Reye's syndrome. As of now, there is no explanation for the relationship between aspirin and this disease.

An alternative for pain sufferers who have trouble with aspirin may be acetaminophen (Tylenol). Although known for about as long as aspirin (a century), acetaminophen is a relative newcomer to the market. Like aspirin, acetaminophen is both an analgesic and an antipyretic. Of course, no drug should be taken without proper caution. Acetaminophen is toxic to the liver when taken in large doses.

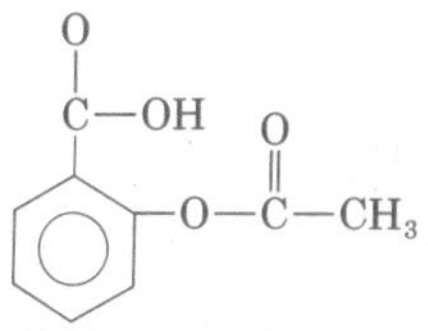

ACETYLSALICYLIC ACID
(ASPIRIN)

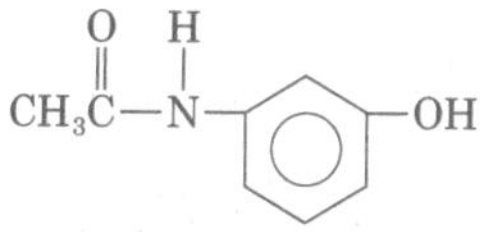

ACETAMINOPHEN
(TYLENOL)

Many aspirin tablets contain starch to hasten their disintegration in the stomach.

DESIGNER DRUGS

Designer drugs are chemical substances structurally similar to legal drugs. Because of their action, they are potential drugs of abuse. For example, fentanyl (Fig. 21–2) is a powerful narcotic marketed under the trade name Sublimaze. Fentanyl is about

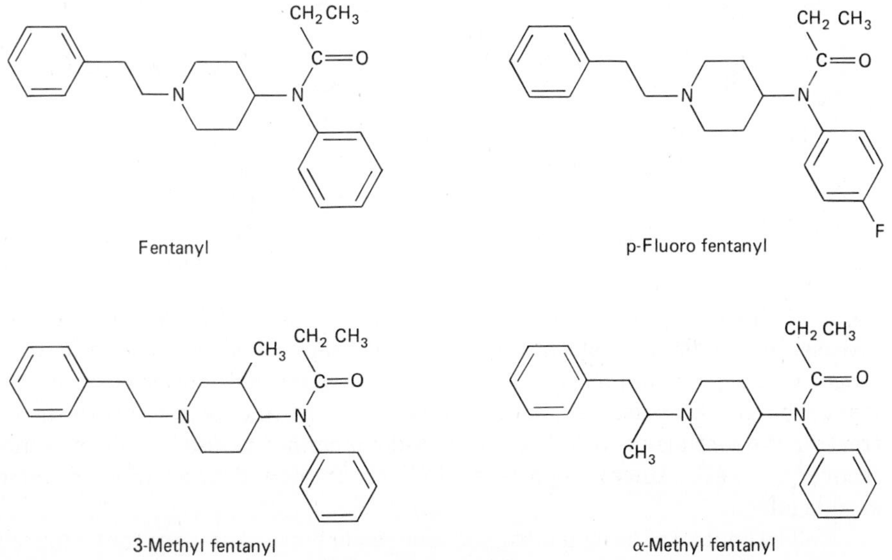

Fentanyl

p-Fluoro fentanyl

3-Methyl fentanyl

α-Methyl fentanyl

FIGURE 21–2 Fentanyl and several of its derivatives.

150 times more potent than morphine and just as addictive, but fentanyl is very short acting. Fentanyl is used in up to 70% of all surgical procedures in the United States. The derivatives of the fentanyl molecule are also potent narcotics. These drugs were called **designer drugs** when they first appeared on the streets because they had been obviously designed by some unscrupulous chemists for consumption by drug addicts. These fentanyl derivatives were every bit as potent as heroin, but because they were not listed on the U.S. Drug Enforcement Administration (DEA) list of controlled substances, they could be sold legally. Until a compound is recognized as being abused and is classified as a dangerous drug — a process known as *scheduling* — no laws apply to it.

In the past few years several fentanyl derivatives (Fig. 21-2) have appeared in California. Samples ranged from pure white powder, sold as China White, to a brown material. First came α-methyl fentanyl, then p-fluoro fentanyl, then α-methyl acetyl fentanyl, and in early 1984 3-methyl fentanyl, a compound 3000 times more potent than morphine. Because of this potency and because heroin addicts can use the fentanyl derivatives interchangeably with heroin, fentanyl derivatives have been responsible for over 100 overdose deaths in California. Most of these have been due to 3-methyl fentanyl.

Another group of designer drugs are those derived from meperidine (Demerol). A designer drug based on meperidine is MPPP, which is short for 1-methyl-4-phenyl-4-propionoxy-piperidine. This compound was first synthesized in 1947, never used commercially, and never scheduled as a controlled substance. MPPP is about 3 times more potent than morphine and 25 times more potent than meperidine. MPPP is structurally so close to meperidine that you have to look closely at the structures to see the difference (hint: look at the ester linkage). If the synthesis of MPPP is carried out at too high a temperature or at too low a pH, the product is MPTP, 1-methyl-4-phenyl-1,2,5,6-tetrahydropyridine. In 1982 a batch of MPTP-tainted MPPP was sold in San Jose, California, as "synthetic heroin." The MPTP produced terrible side effects. It seems MPTP causes the symptoms of Parkinson's disease, which include stiffness, impaired speech, rigidity, and tremors.

MPTP   MPPP   MEPERIDINE

Users of this batch of synthetic heroin became victims of advanced Parkinson's disease. In this disease, cells in the area of the brain called the *substantia nigra* no longer produce dopamine, which is necessary for normal muscle control. A substance that had been used to treat Parkinson's disease, L-dopa, also proved useful in treating the victims of MPTP toxicity. L-dopa cannot be used to effect complete recovery, however, since L-dopa causes hallucinations and exaggerated movements as side effects.

The control of designer drugs is now easier since passage of the Comprehensive Crime Control Act of 1984, which gave the DEA emergency scheduling author-

Doonesbury

BY GARRY TRUDEAU

ity. Now any drug can be designated a controlled substance within 30 days. This scheduling lasts for one year while additional data are gathered to determine final scheduling authority.

SELF-TEST 21–A

1. The accepted chemical name for a drug is called its _____ name.

2. The most widely prescribed drug in the U.S. is used for _____ disease.

3. Which antacids will not produce a gaseous byproduct? (a) magnesium oxide, MgO (b) calcium carbonate, $CaCO_3$ (c) sodium bicarbonate, $NaHCO_3$ (d) sodium citrate, $Na_3C_6H_5O_7$

4. Another name for a pain-killing drug is a(n) _____.

5. Morphine comes from the _____ plant.

6. Codeine is less addictive than morphine. (   ) True or (   ) False

7. Another name for a fever reducer is _____.

8. Acetaminophen is toxic to what organ? _____

9. Fentanyl derivatives that have been sold on the streets tend to act like what other drug of abuse? _____

ALLERGENS AND ANTIHISTAMINES

A person may have an unpleasant physiological response to poison ivy, pollen, mold, food, cosmetics, penicillin, aspirin, and even cold, heat, and ultraviolet light. In the United States about 5000 people die yearly from bronchial asthma, at least 30 from the stings of bees, wasps, hornets, and other insects, and about 300 from ordinary doses of penicillin. The reason: **allergy.** About one person in ten suffers from some form of allergy; more than 16 million Americans suffer from hay fever.

An allergy is an adverse response to a foreign substance or to a physical condition that produces no obvious ill effects in *most* other organisms, including humans. An **allergen** (the substance that initiates the allergic reaction) is, in most

An allergy is a physiological response such as sneezing, runny nose, coughing, or dermatitis to the introduction of a foreign substance. This foreign substance is called an *allergen*.

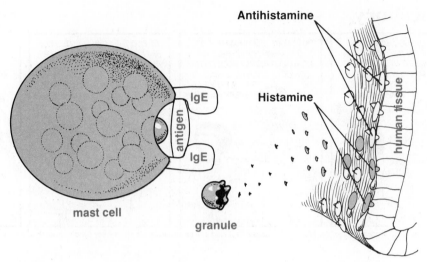

FIGURE 21–3   A postulated mechanism for the cause of and relief from hay fever. The details are described in the text.

Most allergens are high molecular weight substances.

cases, a highly complex substance — usually a protein. Some are polysaccharides or compounds formed by combining a protein and a polysaccharide. Usually allergens have a molecular weight of 10,000 or more.

The principal allergen of ragweed pollen, a major allergy producer, has been isolated and is named ragweed antigen E. It is a protein with a molecular weight of about 38,000; it represents only about 0.5% of the solids in ragweed pollen but contributes about 90% of the pollen's allergenic activity. A mere 1×10^{-12} g of antigen E injected into an allergic person is enough to induce a response.

10^{-12} is 0.000000000001

The allergens come in contact with special cells in the nose and breathing passages to which a particular type of antibody is attached, the IgE antibody, which has a molecular weight of about 196,000. Allergic individuals have 6 to 14 times more IgE in their blood serum than nonallergic people. The IgE is formed in the nose, bronchial tubes, and gastrointestinal tract and binds firmly to specific cells, called **mast cells,** in these regions.

Antibodies are high molecular weight proteins, called immunoglobulins, that attack foreign proteins such as allergens.

Histamine causes runny noses, red eyes, and other hay fever symptoms.

Antigen E from ragweed reacts with the IgE antibody attached to the mast cells, forming antigen-antibody complexes. The formation of these antigen-antibody complexes leads to the release of so-called "allergy mediators" from special granules in the mast cells (Fig. 21–3). The most potent of these mediators found so far is **histamine.** Although it is widely distributed in the body, it is especially concentrated in the 250 to 300 granules of the mast cells. Histamine accounts for many, if not most, of the symptoms of hay fever, bronchial asthma, and other allergies.

$$H_2NCH_2CH_2 - \underset{\underset{\displaystyle H}{\overset{\displaystyle |}{N}}}{\boxed{}} N$$

HISTAMINE

The chemical mediators such as histamine must be released from the cell to cause the symptoms of allergy. The release mechanism is an energy-requiring process in which the granules may move to the outer edge of the living cell and, without leaving the cell, discharge their contents of histamine through a temporary gap in the cell membrane. This sends histamine on its way to produce the toxic effects of hay fever.

Treatment consists of three procedures: avoidance (Fig. 21–4), desensitization, and drug therapy. Desensitization therapy is costly and inconvenient, since 20 or more injections are required to achieve what is usually a partial cure. One chemical idea of desensitization is to inject a blocking antibody that preferentially reacts with the allergen so that it cannot react with the IgE allergy-sensitizing antibody. This breaks the chain of events leading to the release of histamine or other allergy-producing mediators. Many small injections, spaced in time, are required to build up a sufficient level of the blocking antibody.

Epinephrine (adrenalin), steroids, and antihistamines are effective drugs in treating allergies. The first two are particularly effective in treating bronchial asthma, whereas the **antihistamines,** introduced commercially in the United States in 1945, are the most widely used drugs for treating allergies. More than 50 antihistamines are offered commercially in the United States. Many of these contain, as does histamine, an ethylamine group ($-CH_2CH_2N{<}$):

PYRIBENZAMINE
(AN IMPORTANT ANTIHISTAMINE)

These drugs act competitively by occupying the receptor sites normally occupied by histamine on cells. This, in effect, blocks the action of histamine.

FIGURE 21–4   In this Abbott Laboratories map, the size of each circle represents the amount of all late-summer and fall pollens found in the air in each city. Dark portions show amount of ragweed pollen. Shaded areas are regions of low pollen count.

TABLE 21–6   Some Common Antiseptics

Iodine	Phenols
Sodium hypochlorite	Mercurochrome
Potassium permanganate	Metaphen
Hydrogen peroxide	Merthiolate
Iodophors	Pine oil
Ethanol	Soap
Quaternary ammonium compounds	Hexylresorcinol
Chloramine-T	Mercuric chloride

ANTISEPTICS AND DISINFECTANTS

An antiseptic is a compound that prevents the growth of microorganisms. It now has the legal meaning **"germicide,"** or a compound that *kills* microorganisms. A disinfectant is a compound that destroys pathogenic bacteria or microorganisms, but usually not bacterial spores. Disinfectants are generally poisonous and therefore suitable only for external use as on the skin or a wound.

Some common germicides are listed in Table 21–6. Some of these, such as the halogens, sodium hypochlorite, hydrogen peroxide, and potassium permanganate, are effective because of their oxidizing properties. This is a general property and allows them to oxidize any kind of cell, including human cells. For this reason they are used mostly as disinfectants in destroying the germs on nonliving objects. Phenol (carbolic acid) is readily absorbed by cells and is a general poison. The quaternary ammonium compounds are surface active agents, and their bactericidal effect seems to be related to their ability to weaken the cell wall so the cell contents cannot be contained.

One of the newer developments solves the problem of applying antiseptics to children. You may remember the sting of "iodine" when applied to a scratch or a wound. Old-fashioned iodine is a solution of iodine (I_2) in alcohol with a little potassium iodide (KI) to increase the solubility of the iodine. The alcohol causes most of the pain. Now there are polymers, such as polyvinylpyrrolidone, that complex iodine molecules (Fig. 21–5); the products (iodophors) are soluble in water. The resultant solution is a very efficient and painless disinfectant. The iodophors are active ingredients in a popular mouthwash and in restaurant glassware disinfectants.

Because such compounds are generally toxic to living matter, it is necessary to utilize only dilute solutions and then only on the skin. Although they help to prevent the spread of disease, they are practically useless in its treatment because they act nonspecifically against all cells with which they come in contact. They are to be distinguished from antibiotics, which act more selectively against infecting bacteria than against "host" cells within the human body.

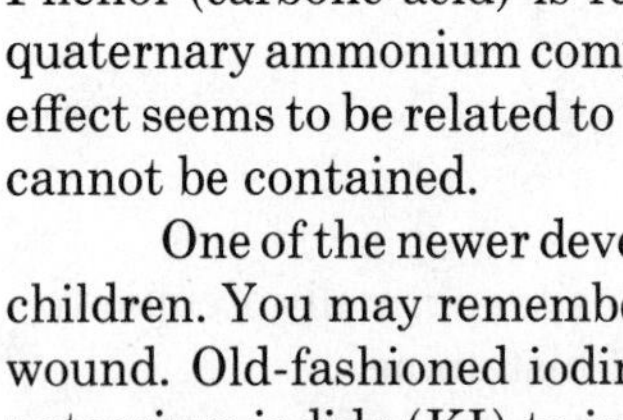

Pathogenic bacteria cause many illnesses.

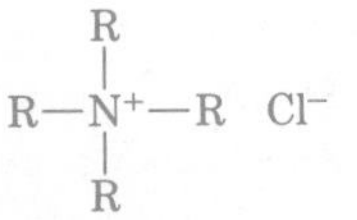

$$R-\overset{\displaystyle R}{\underset{\displaystyle R}{N^+}}-R \quad Cl^-$$

A QUATERNARY AMMONIUM CHLORIDE

A tincture is an alcoholic solution. Tincture of iodine is a solution of water, iodine, and potassium iodide in ethanol.

FIGURE 21–5   An iodophor: complex of polyvinylpyrrolidone (PVP) and iodine.

TABLE 21–7 Deaths per 100,000 Americans Due to Different Causes

	1900	1977
Infectious diseases	500	40
Influenza and pneumonia	210	23
Diphtheria	40	Fewer than 1
Typhoid and paratyphoid	30	Fewer than 1
Whooping cough	10	Fewer than 1
Gastrointestinal problems	150	Fewer than 10

ANTIMICROBIAL MEDICINES

In our time, the quest for drugs to wipe out disease due to microorganisms has been virtually fulfilled by the **antibiotics** (Table 21–7). In the original sense, an antibiotic is a substance such as penicillin, produced by a microorganism, that inhibits the growth of another organism. It has become common practice to include synthetic chemicals such as the sulfa drugs in a discussion of antibiotics.

Since the antibiotics are so efficient, they were the first of what came to be called "miracle" drugs. Their job generally is to aid the white blood cells by stopping bacteria from multiplying. When a person falls victim to or is killed by a disease, it means that the invading bacteria have multiplied faster than the white blood cells could devour them and that the bacterial toxins increased more rapidly than the antibodies could neutralize them. The action of the white blood cells and antibodies plus an antibiotic is generally enough to repulse an attack of the disease germs.

An antibody is a specific protein produced to protect the organism from harmful invading molecules.

The Sulfa Drugs

Sulfa drugs represent a group of compounds discovered in a conscious search for antibiotics. In 1904, the German chemist Paul Ehrlich (1854–1915; Nobel prize in 1908) realized that infectious diseases could be conquered if toxic chemicals could be found that attacked parasitic organisms within the body to a greater extent than they did host cells. Ehrlich achieved some success toward his goal; he found that certain dyes that were used to stain bacteria for microscopic examination could also kill the bacteria. This led to the use of dyes against organisms causing African sleeping sickness and arsenic compounds against those causing syphilis.

In 1935, after experimenting with several drugs, Gerhard Domagk, a pathologist in the I. G. Farbenindustrie Laboratories in Germany, found that Prontosil, a dye, was somewhat effective against bacterial infection in mice. Prontosil can be changed to **sulfanilamide,** which is very effective.

Large doses of sulfa drugs are required compared with the doses of "true antibiotics."

$$H_2N-\bigcirc-N{=}N-\bigcirc-SO_2NH_2 \xrightarrow{H_2O} H_2N-\bigcirc-SO_2-NH_2$$

with NH_2 group below

PRONTOSIL   SULFANILAMIDE

This discovery led to the synthesis and testing of many related compounds in the search for drugs that are more effective or less toxic to the infected experimental animal. By 1964, more than 5000 sulfa drugs had been prepared and tested.

A sulfa drug mimics an
essential compound.

Sulfa drugs inhibit bacteria by preventing the synthesis of folic acid, a vitamin essential to their growth. The drugs' ability to do this apparently lies in their structural similarity to a key ingredient in the folic acid synthesis, *para*-aminobenzoic acid.

SULFANILAMIDE,
A TYPICAL SULFA DRUG

p-AMINOBENZOIC ACID

The close structural similarity of sulfanilamide and *p*-aminobenzoic acid permits sulfanilamide to be incorporated into the enzymatic reaction sequence instead of *p*-aminobenzoic acid. By bonding tightly, sulfanilamide shuts off the production of the essential folic acid, and the bacteria die of vitamin deficiency. In humans and the higher animals, *p*-aminobenzoic acid is not necessary for folic acid synthesis, so sulfa drugs have no effect on this mechanism.

The Penicillins

Penicillin was discovered in 1928 by Alexander Fleming (Fig. 21–6), a bacteriologist at the University of London, who was working with cultures of *Staphylococcus aureus,* a germ that causes boils and some other types of infections. In order to examine the cultures with a microscope, he had to remove the cover of the culture plate for a while. One day as he started work he noticed that the culture was contaminated by a blue-green mold. For some distance around the mold growth, the bacterial colonies were being destroyed. Upon further investigation, Fleming found that the broth in which this mold had grown also had an inhibitory or lethal effect

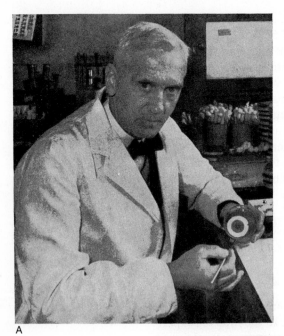

A

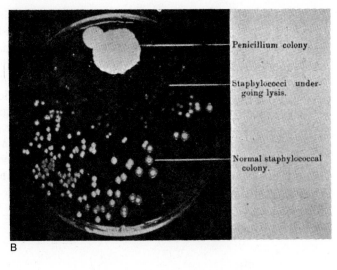

Penicillium colony.

Staphylococci undergoing lysis.

Normal staphylococcal colony.

B

FIGURE 21–6 (*a*) Sir Alexander Fleming. (*b*) A culture-plate showing the dissolution of staphylococcal colonies in the neighborhood of a colony of penicillium.

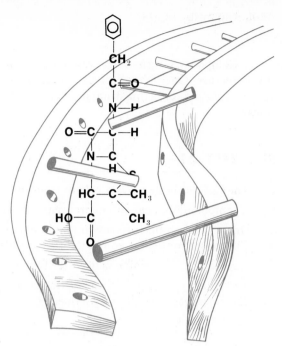

FIGURE 21–7 Penicillin kills bacteria by interfering with the formation of cross links in the cell wall.

against many pathogenic (disease-causing) bacteria. The mold was later identified as *Penicillin notatum* (the spores sprout and branch out in pencil shapes, hence the name).

  Penicillin, the name given to the antibacterial substance produced by the mold, apparently has no toxic effect on animal cells, and its activity is selective. The structure of penicillin (see margin) has now been determined. Many different penicillins exist, differing in the structure of the R group. Penicillin G is the most widely used in medicine. Several antibiotics, such as penicillin and bacitracin, are known to prevent cell-wall synthesis in bacteria, as shown in Figure 21–7.

Streptomycin and the Tetracyclines

In 1937, following collaboration with René Dubos, Selman Waksman isolated a compound from a soil organism, *Streptomyces griseus,* which came to be known as **streptomycin** and was released to physicians in 1947. This compound was quite successful in controlling certain types of bacteria but later had to be withdrawn because of its adverse side effects.

  In 1945, B. M. Duggan discovered that a gold-colored fungus, *Streptomyces aureofaciens,* produced a new type of antibiotic, **Aureomycin,** the first of the **tetracyclines.** Research then stepped up to a fever pitch. Pfizer Laboratories tested 116,000 different soil samples before they discovered the next antibiotic, which they named **Terramycin.**

PENICILLIN G

Streptomycin has many undesirable side effects.

Tetracyclines get their names from their four-ring structures.

AUREOMYCIN

TERRAMYCIN

Compounds of the tetracycline family are so named because of their four-ring structure. One side effect of taking these drugs is diarrhea, caused by the killing of the patient's intestinal flora (the bacteria normally residing in the intestines).

SELF-TEST 21-B

1. The IgE antibody plays a role in what disorder? _____
2. A compound that can kill microorganisms can be called a(n) _____.
3. Drugs that inhibit the growth of microorganisms are called _____ ____.
4. The drug Prontosil is converted to sulfanilamide, which resembles _____, which is used for folic acid synthesis.
5. Penicillin is produced from cultures of what kind of organism? _____
6. The tetracycline drugs are produced from cultures of what kind of organism? _____

HEART DISEASE DRUGS

Heart disease is the number one killer of Americans, claiming approximately 750,000 lives each year, or 38% of the total yearly deaths. Although more than 100 years ago nitroglycerin was found to be effective in relieving pain associated with an insufficient flow of blood to the heart itself, newer drugs have been used in recent years to improve dramatically the quality of life of those suffering heart disease.

The effects of substances on the buildup of plaque is discussed in Chapter 20.

Heart disease is actually an assortment of diseases, but the basic cause of many of them is **atherosclerosis,** which is the buildup of fatty deposits called **plaque** on the inner walls of arteries. Cholesterol, a lipid, is a major component of atherosclerotic plaque. Many scientists believe that a high level of cholesterol in the blood, along with high blood levels of triglycerides, which are also lipids, seems to contribute to the buildup of this plaque. The plaque buildups reduce the flow of blood to the heart. If a coronary artery is blocked by plaque, a heart attack occurs as a result of the reduced blood flow carrying oxygen to the heart. This attack often causes part of the heart muscle to be destroyed. Of about 1.5 million persons experiencing heart attacks each year, one third die from the attack. About 98% of all heart attack victims have atherosclerosis. The treatment of heart disease with drugs has developed along the line of easing the flow of blood to the heart. This lowers the forces the heart must extend to pump the blood throughout the circulatory system. Other heart disease drugs lower the buildup of plaque by regulating the amount of lipids in the blood.

In 1948, Raymond P. Ahlquist, at the Medical College of Georgia, discovered that heart muscle contains receptors, which he called **beta receptors.** Stimulation of these receptors by epinephrine and norepinephrine results in an increase in the number of heart beats. In 1967, Alonzo M. Lands, a pharmacologist in Rensselaer, New York, discovered two different beta receptors, $beta_1$ and $beta_2$. $Beta_1$ sites are located primarily in the heart but also in the kidneys. $Beta_2$ receptors are involved in relaxation of the peripheral blood vessels and the bronchial tube.

With the knowledge about beta receptors gained by Ahlquist, chemists began to explore the action of chemicals that would compete with epinephrine and norepinephrine at the beta receptor sites. If these sites could be blocked, then the heart rate would decrease. For a heart already overworked from the buildup of plaque in the arteries supplying it with blood (and oxygen), this might just relax the heart enough to allow it to recover from an impending attack. In addition to this desirable property, these drugs might be able to relieve high blood pressure (**hypertension**) and **migraine** headache by their relaxation of the blood vessels.

The first drugs of this type, called **beta blockers** because of their action of blocking beta receptor sites, came into use in the late 1950s and early 1960s but were withdrawn because of undesirable characteristics and side effects. In 1967, the beta blocker propranolol (Table 21–8), trade name Inderal, was first prescribed. Its first use was for cardiac arrhythmias, but now it has been approved by the FDA for treating angina, hypertension, and migraine headache. Propranolol has high lipid solubility, so it passes through the blood-brain barrier and builds up in the central nervous system, where it is more slowly metabolized. This buildup of the chemical in the central nervous system causes the side effects of fatigue, lethargy, depression, and confusion. In spite of these side effects, propranolol is the most widely prescribed drug of any type in the United States (see Table 21–3). Sales increased from almost $12 million in 1972 to over $225 million in the late 1980s.

A second beta blocker, metoprolol (trade name Lopressor) was introduced in the United States in 1978 for treatment of hypertension. Metoprolol is selective in its beta blocking effects, blocking only $beta_1$ sites and not the $beta_2$ sites of the peripheral blood vessels or the bronchial tube. This means metoprolol is safer for asthma sufferers and for those patients with severe blood vessel disorders. Metoprolol is not as soluble in lipids as propranolol and does not accumulate in the central nervous system. It is also more slowly metabolized by the liver, which allows more widely spaced doses, usually twice a day. A third beta blocker, nadolol (trade name Corgard), was introduced in the United States in 1979. This chemical is not metabolized to a great extent by the liver and passes mostly unchanged through the kidneys to the urine. Because it is largely unmetabolized, nadolol may be taken only once a day. Nadolol is also less soluble in lipids than either propranolol or metoprolol and thus produces very few side effects associated with accumulation in the central nervous system. The three agents have quite similar structures, as may be seen in Table 21–8.

TABLE 21–8 Beta-Blocking Drugs Used in Treating Heart Disease

TRADE NAME	GENERIC NAME	STRUCTURE
Inderal	Propranolol	
Lopressor	Metoprolol	
Corgard	Nadolol	

Another class of heart disease drugs are the **calcium channel blockers.** Research has found that calcium ions move into the heart muscle by means of holes or channels in the phospholipid membrane surrounding the muscle. In the muscle cells, the calcium ions cause an interaction between the parallel protein filaments myosin and actin, which causes the cell to contract (Fig. 21–8). In addition, the double positive charge on the calcium ion neutralizes some of the negative charge of the muscle cell. This, too, causes the muscle to contract. Movement of the calcium ions out of the cell restores the negative charge, and the cell relaxes. Blocking the flow of calcium ions into the cell causes the muscles to relax. When the smooth muscles in the walls of the coronary arteries are relaxed, these arteries expand and increase the supply of blood to the heart. Some calcium blockers also decrease the force of contraction and thus decrease the oxygen requirements of the heart.

Calcium blockers, unlike beta blockers, can prevent spasms of the coronary arteries. These spasms cause a blockage of blood flow to the heart and the intense pain of the angina attack. The causes of these spasms are poorly understood, but the calcium blockers do dilate the arteries and lessen the possibilities of angina attacks.

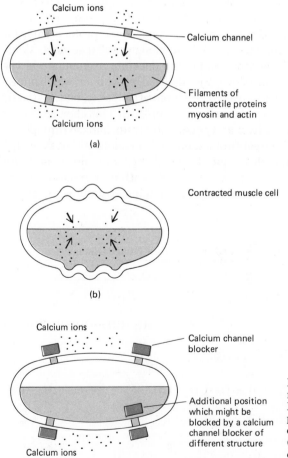

(a)

Calcium ions

Calcium channel

Filaments of contractile proteins myosin and actin

Calcium ions

(b)

Contracted muscle cell

(c)

Calcium ions

Calcium channel blocker

Additional position which might be blocked by a calcium channel blocker of different structure

Calcium ions

FIGURE 21–8 (a), Calcium ions flowing into a muscle cell. (b), Muscle cell contracted caused by presence of calcium ions in contractile protein fiber bundles. (c), Relaxed muscle cell with calcium channels blocked by drug molecules.

GENERIC NAME	TRADE NAME	STRUCTURE
Verapamil	Isotin, Calan	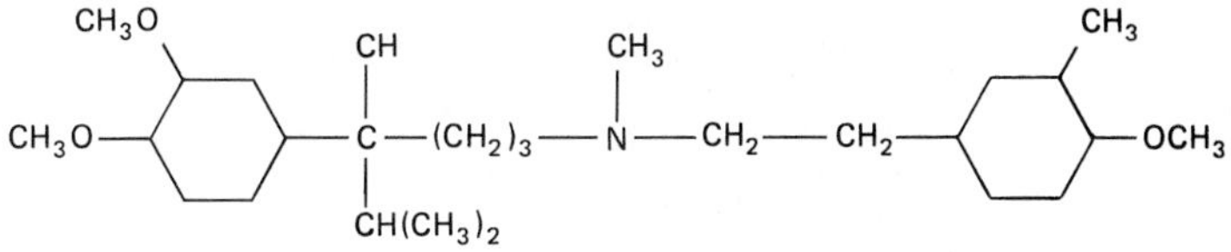
Nifedipine	Procardia	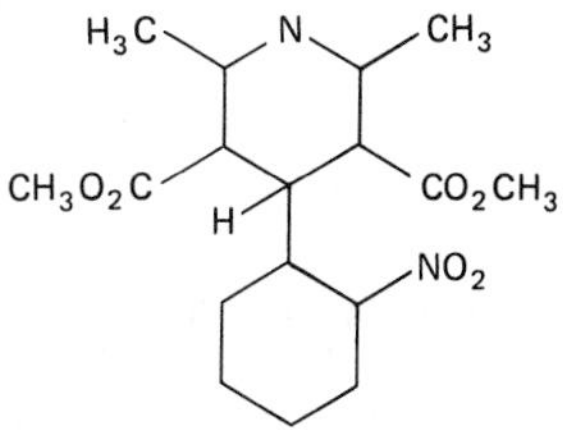

FIGURE 21-9 Some calcium channel blocking drugs used in treating heart disease. Note the major differences in structure.

Calcium blockers are unlike beta blockers in that they are not structurally similar (Fig. 21-9). The effectiveness of such structurally dissimilar chemicals on the same process may be due to different types of calcium channels in muscle cell walls, or the differently shaped molecules may stop up the calcium ion channels at different places (Fig. 21-8).

Verapamil (trade names Isoptin and Calan) became available in the United States in 1981 after 19 years of use in Europe. In 1982 verapamil was first used to treat angina, and in the same year, a second calcium channel blocker, nifedipine (trade name Procardia), was introduced. Nifedipine is a powerful dilator of coronary arteries. It has an immediate effect on patients suffering from angina. The calcium blockers are not without their problems, however. Most produce the side effects of headaches, dizziness, flushing of the skin, and light-headedness.

The ultimate solution to preventing most forms of heart disease would be to prevent the buildup of the atherosclerotic plaque in the arteries. In 1954, John W. Gofman, at the University of California at Berkeley, discovered four separate categories of lipoproteins in the blood, based on their densities. About 65% of the cholesterol in the blood is carried by the low-density lipoproteins, whereas only 25% is carried by the high-density lipoproteins. In 1968, John A. Glomset at the University of Washington showed that the high-density lipoproteins are effective in removing cholesterol from arterial walls and transporting it to the liver, where it is metabolized. This type of discovery means that it may be possible to dissolve atherosclerotic plaque already formed! For patients whose blood levels of lipids cannot be controlled by proper diet, scientists have been attempting to develop drugs that either raise the level of high-density lipoproteins or lower the levels of low-density lipoproteins.

Niacin (nicotinic acid) lowers blood lipid concentrations by interfering with the synthesis of cholesterol in the liver. Several drugs have been introduced that cause the liver to convert more cholesterol into bile acids by lowering the concentration of bile acids in the intestines. One of these, cholestyramine resin (Fig. 21-10), binds bile acids.

GENERIC NAME	TRADE NAME	STRUCTURE
Nicotinic acid (niacin)	Nicolar	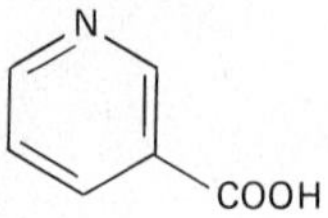
Cholestyramine resin	Questran	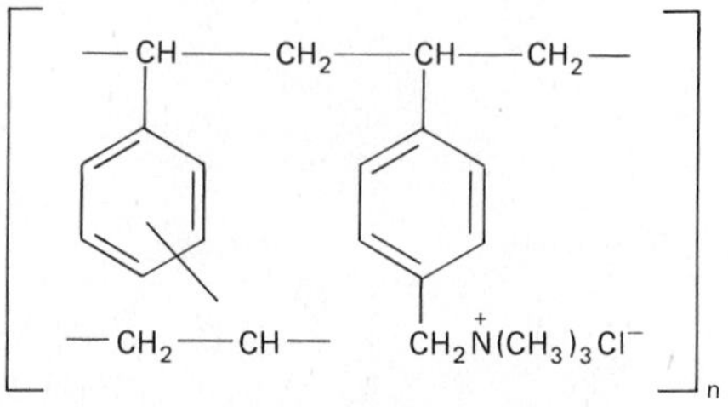

FIGURE 21-10   The structures of two drugs used to lower cholesterol in the blood.

SELF-TEST 21-C

1. What are two ingredients in atherosclerotic plaque? _____ and _____

2. When beta sites in the heart are stimulated by epinephrine, the heart beats (   ) faster or (   ) slower.

3. When propranolol passes through the blood-brain barrier, it causes the side effects of _____ and _____.

4. A drug that could block calcium ions from flowing into heart muscles would have the effect of (   ) exciting or (   ) relaxing the heart.

5. Cholesterol is carried in the blood by both high- and low-density lipoproteins. Which carries the greater percentage of cholesterol? _____

6. Would a drug that removed cholesterol from the bloodstream be useful in treating some forms of heart disease? (   ) Yes or (   ) No

THE STEROID DRUGS

A large and important class of naturally occurring compounds is derived from the following tetracyclic structure.

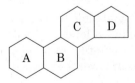

These compounds are known as **steroids,** and they occur in all plants and animals. The most abundant animal steroid is cholesterol, $C_{27}H_{46}O$. The human body synthesizes cholesterol and also readily absorbs dietary cholesterol through the intestinal wall. It is associated with gallstones and atherosclerosis.

Biochemical alteration or degradation of cholesterol leads to many steroids of great importance in human biochemistry. When **cortisone,** one of the adrenal

Cortisone is a "powerful drug" having a major effect on biological systems.

cortex hormones, is applied topically or injected into a diseased joint, it acts as an antiinflammatory agent and is of great use in treating arthritis.

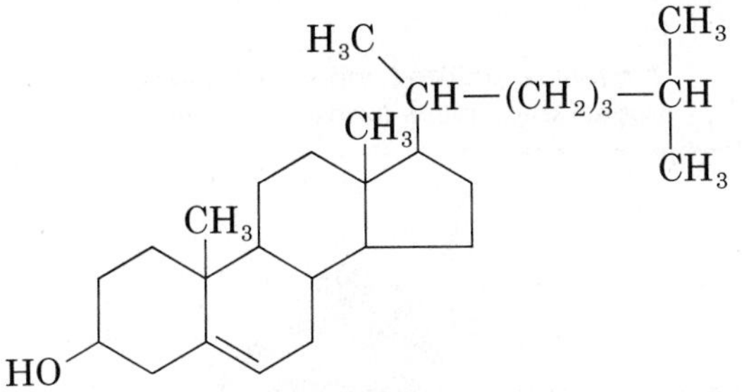

CHOLESTEROL

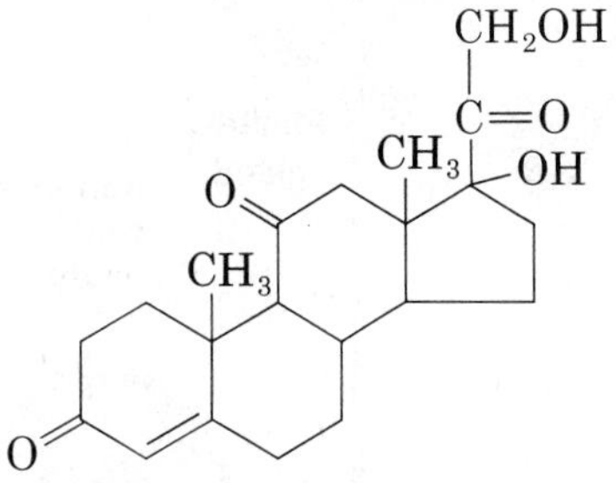

CORTISONE

Structurally related to cholesterol and cortisone are the sex hormones. One female sex hormone, **progesterone,** differs only slightly from an important male hormone, **testosterone.**

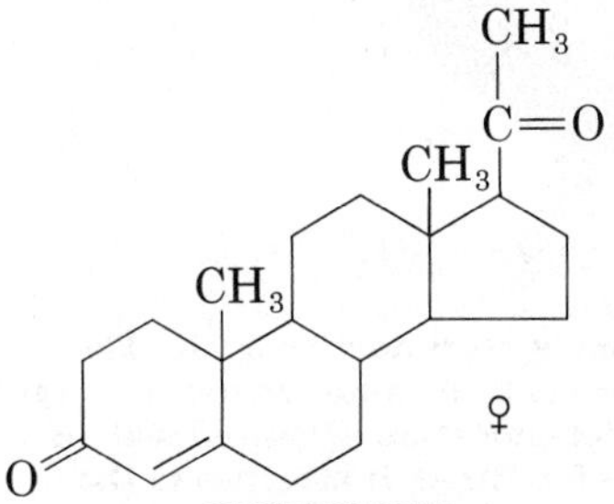

PROGESTERONE

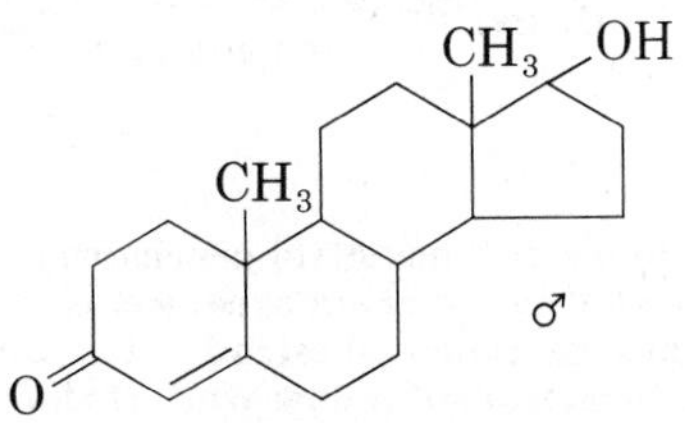

TESTOSTERONE

Other female hormones are estradiol and estrone, called **estrogens.** The estrogens differ from the other steroids discussed earlier in that they contain an aromatic A ring (in color).

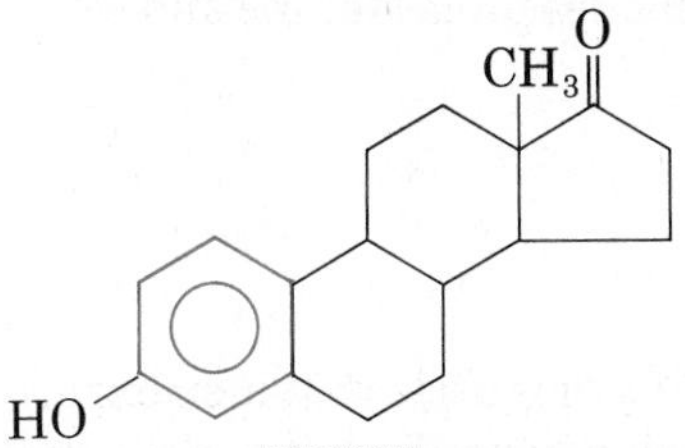

ESTRONE

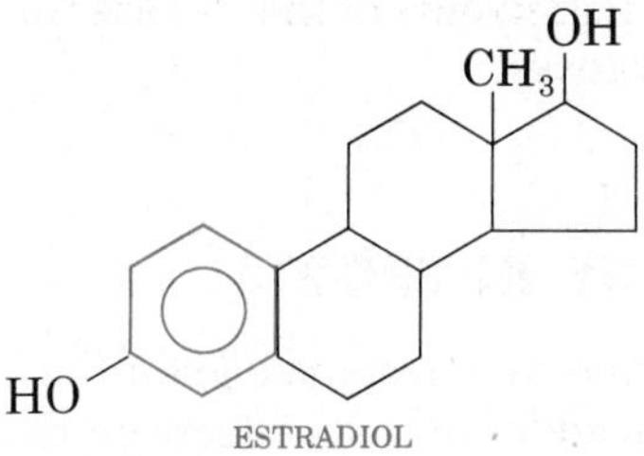

ESTRADIOL

BIRTH CONTROL PILLS

One of the most revolutionary medical developments of the 1960s was the world-wide introduction and use of "The Pill." More than 10 million women in the United States use birth control pills. The basic feature of oral contraceptives for women is their chemical ability to simulate the hormonal processes resulting from pregnancy and, in so doing, prevent ovulation. Ovulation, the production of eggs by the ovary, ceases at the onset of pregnancy because of hormonal changes (Fig. 21–11). This same result can be produced by the administration of a variety of steroids, some of which are effective when taken orally, although the mechanism of their action and their long-term effects are not known in detail.

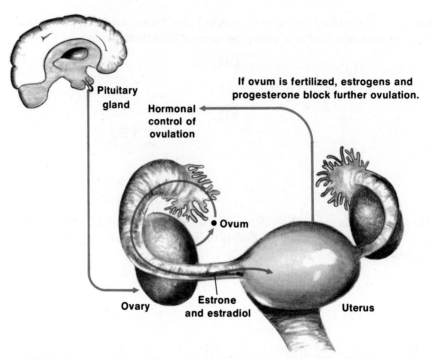

Pituitary gland

If ovum is fertilized, estrogens and progesterone block further ovulation.

Hormonal control of ovulation

Ovum

Ovary

Estrone and estradiol

Uterus

FIGURE 21–11   Some of the control processes in the female reproductive cycle. The pituitary gland, located in the brain area, sends out hormones that cause the ovary to release other hormones. Estrone and estradiol inititate the deterioration of the old wall of the uterus and the formation of a new wall. If the ovum is fertilized, it attaches to the uterine wall, and estrogens and progesterone are released to prevent further ovulation. The Pill serves the same function as the estrogens and progesterone in that it prevents the ovulation process.

The active ingredients of the Pill are the hormones progesterone and estrogen, or their derivatives.

STEROID DRUGS IN SPORTS

The steroid testosterone is responsible for the building of muscle that men experience at puberty, in addition to its effects on the development of adult male sexual characteristics. Synthetic steroids have been developed in part to separate the masculinizing (androgenic) effects and muscle-building (anabolic) effects of testosterone. These steroids have been prescribed by physicians to correct hormonal imbalances or to prevent the withering of muscle in persons who are recovering from surgery or starvation.

Healthy athletes discovered that these steroids appeared to have an anabolic effect on them as well. Initially these anabolic steroids were used by weight lifters and by athletes in track-and-field events like the shot-put and hammer throw. Later, some inconclusive evidence surfaced suggesting that anabolic steroids increased endurance, and this caused runners, swimmers, and cyclists to begin using them.

Such sports organizations as the International Olympic Committee (see box) have banned the use of anabolic steroids and other drugs among athletes to

---

DRUGS BANNED BY INTERNATIONAL OLYMPIC COMMITTEE 1982

PSYCHOMOTOR STIMULANTS

Amphetamine
Benzphetamine
Chlorphentermine
Cocaine
Diethylpropion
Dimethylamphetamine
Ethylamphetamine
Fencamfamin
Meclofenoxate
Methylamphetamine
Methylphenidate
Norpseudoephedrine
Pemoline
Phendimetrazine
Phenmetrazine
Phentermine
Pipradol
Prolintane
Related compounds

ANABOLIC STEROIDS

Clostebol
Dehydrochlormethyltes-
tosterone
Fluoxymesterone
Mesterolone
Methenolone
Methandienone
Methyltestosterone
Nandrolone
Norethandrolone
Oxymesterone
Stanozolol
Testosterone*
Related compounds

NARCOTIC ANALGESICS

Anileridine
Codeine
Dextromoramide
Dihydrocodeine
Dipipanone
Ethylmorphine
Heroin
Hydrocodone
Hydromorphone
Levorphanol
Methadone
Morphine
Oxycodone
Oxymorphone
Pentazocine
Pethidine
Phenazocine
Piminodine
Thebacon
Trimeperidine
Related compounds

MISCELLANEOUS CENTRAL NERVOUS SYSTEM STIMULANTS

Amiphenazole
Bemegride
Caffeine†
Cropropamide
Crotethamide
Doxapram
Ethamivan
Leptazol
Nikethamide
Picrotoxin
Strychnine
Related compounds

SYMPATHOMIMETIC AMINE STIMULANTS‡

Clorprenaline
Ephedrine
Etafedrine
Isoetharine
Isoprenaline
Methoxyphenamine
Methylephedrine
Related compounds

* Ratio of total concentration of testosterone to that of epitestosterone in the urine must not exceed six.
† Concentration in urine must not exceed 15 μg/mL.
‡ Mimicking the effects in the sympathetic nervous system.
(Source: U.S. Olympic Committee)

---

achieve advantage over opponents for several reasons. First, there have been few human studies made concerning their use by healthy individuals. Most physicians believe these steroids are ineffective, and any perceived effects are due to nutritional, psychological, and environmental effects. If these steroids are effective at all, they may work by allowing an athlete to get off a plateau when further exercise fails to offer improvement.

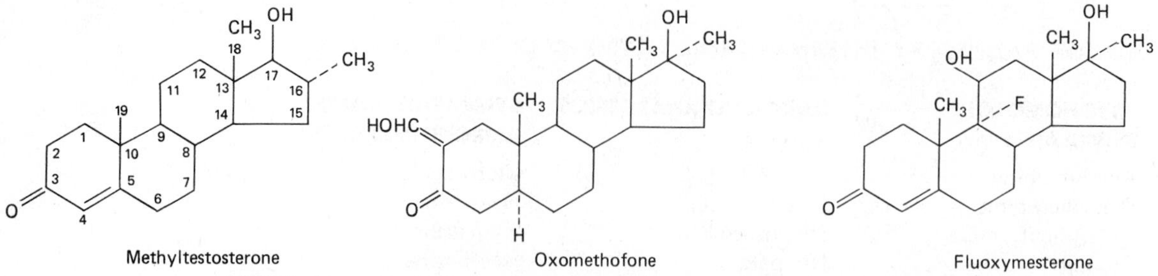

FIGURE 21-12 Structure of some anabolic steroids. The carbon-17 position in some oral dose steroids is occupied by an alkyl group.

The side effects of anabolic steroid use include acne, baldness, and changes in sexual desire. Some men experience enlargement of the breasts. Accompanying these noticeable changes are testicular atrophy and decreased sperm production. This is caused by an imbalance between the testes, pituitary, and hypothalamus due to the increased concentration of these male sex hormones in the bloodstream. High levels of male sex hormones cause the hypothalamus to signal the pituitary gland to lower production of two other hormones, luteinizing hormone and follicle-stimulating hormone, which stimulate sperm production in the testes. While these changes appear to be reversible, additional testing is needed.

In women, the use of anabolic steroids produces facial hair, male-pattern baldness, deepening of the voice, and changes in the menstrual cycle. Most of these changes are not reversible.

In addition to these problems, oral dose anabolic steroids are toxic to the liver. Testosterone taken orally is not very effective since most of it is rapidly metabolized by the liver before it reaches the bloodstream. Several of the common anabolic steroids are active when taken orally. This activity is due in part to an alkyl group in addition to the hydroxyl group at the carbon-17 position of the steroid nucleus (Fig. 21-12). This alkyl structure slows the metabolism in the liver and thus allows more of the dose to reach the bloodstream, but this structural change also increases liver toxicity. Some liver cancer has been reported in anabolic steroid users.

ANTICANCER DRUGS

Cancer is not one disease but perhaps 100 different diseases. Cancers are caused by a number of factors. A cancer begins when a cell in the body begins to multiply without restraint and produces descendants that invade the tissues in the vicinity. It seems reasonable, then, that some drugs might exist that would be able to either stop this undesirable spreading of cancer cells or prevent cancer from happening at all. Cancers are treated by (1) surgery to remove whole areas affected by cancers as well as the cancerous growths themselves, (2) irradiation to kill cancer cells, and (3) chemicals (**chemotherapy**) that also kill cancer cells. These treatment methods have resulted in some dramatic improvements in survival of patients with certain cancers. A group of cancer patients can be considered cured if, after their treatment, they die at about the same rate as the general population. Another way of judging success in cancer therapy is by the number of patients surviving five years after the treatment. As Table 21-9 shows, some cancers have shown marked increases in survival rates over the past two decades.

TABLE 21–9   Five-Year Survival Rates for Various Cancers

TYPE	1960–1963 (%)	1973–1980 (%)
Breast	63	74
Bladder	53	73
Hodgkin's disease	40	70
Colon	43	51
Leukemia	14	32
Lung	8	12
Pancreas	1	3

(Source: National Cancer Institute, 1984)

In World War I the toxic effects of a class of the chemical warfare gases called mustard gases were recognized. They were found to cause damage to the bone marrow and to be mutagenic. In these ways the mustard gases were acting like X rays, which were also toxic to cells and caused mutations.

$$Cl-CH_2CH_2-S-CH_2CH_2-Cl$$
MUSTARD GAS

Beginning around 1935 other mustards of the nitrogen family were synthesized. They, too, caused mutations in some laboratory animals. In addition, they caused cancers in some animals.

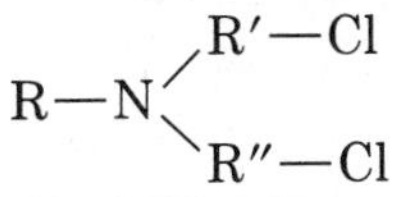

$$R-N\begin{cases} R'-Cl \\ R''-Cl \end{cases} \qquad CH_3CH_2-N\begin{cases} CH_2CH_2-Cl \\ CH_2CH_2-Cl \end{cases}$$

NITROGEN MUSTARD   A NITROGEN MUSTARD
GENERAL FORMULA

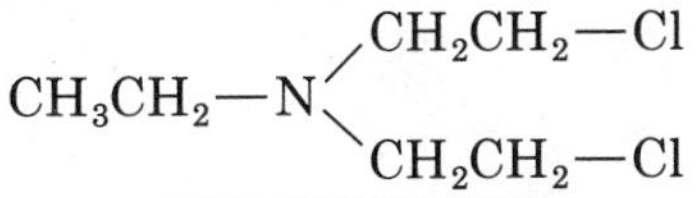

After World War II the secrecy surrounding the mutagenic nature of these chemicals was lifted, and it occurred to cancer researchers that cancers might be treated with chemicals that selectively destroy unwanted cells.

One of the most widely used anticancer drugs is cyclophosphamide, a compound that contains the nitrogen mustard group (shown in color).

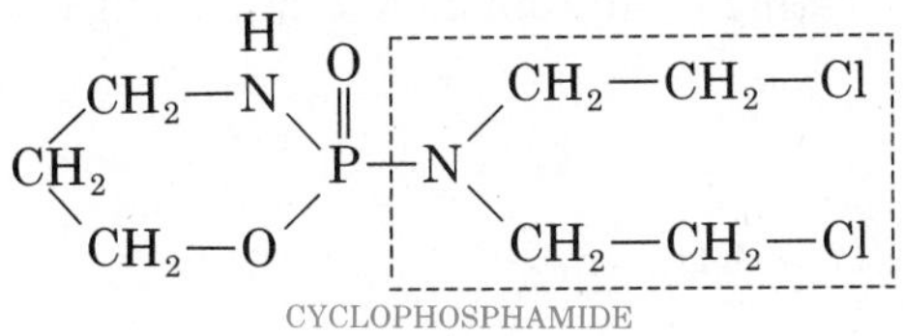

$$\begin{array}{c} CH_2-N\overset{H}{} \quad \overset{O}{\underset{\|}{}} \\ CH_2 \qquad \quad P-N \\ CH_2-O \end{array} \begin{cases} CH_2-CH_2-Cl \\ CH_2-CH_2-Cl \end{cases}$$

CYCLOPHOSPHAMIDE

Compounds such as cyclophosphamide belong to the **alkylating class** of anticancer drugs. Alkylating agents are reactive organic compounds that transfer alkylating groups in chemical reactions. Their effectiveness as anticancer agents is due to the transfer of alkyl groups to the nitrogen bases in DNA, particularly guanine (Fig. 21–13). The presence of the alkyl group in the guanine molecule blocks base pairing and prevents DNA replication. As a result, cell division is stopped. Although alkylating agents attack both normal cells and cancer cells, the effect is greater for cancer cells because they divide rapidly.

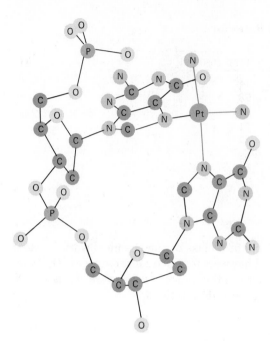

FIGURE 21–13   The structural arrangement of atoms when a cis[Pt(NH₃)₂] fragment (shown in color) binds to two nitrogen atoms in guanine rings of a dinucleotide.

Another widely used anticancer drug, **cisplatin,** blocks DNA replication by a similar mechanism.

Review Chapter 1 for a historical account of cisplatin as an anticancer agent.

CISPLATIN

Inside the cell, the Cl⁻ ions are displaced, and the $>$Pt$<^{NH_3}_{NH_3}$ unit binds to the nitrogen sites on the guanine bases in DNA. As with alkylating agents, this prevents base pairing and DNA replication.

One alkylating agent with the ability to cross the blood-brain barrier is BCNU (carmustine). BCNU is used for treating brain tumors and other cancers located in the brain.

$$Cl-CH_2-CH_2-N-\overset{\overset{O}{\|}}{C}-N-CH_2-CH_2-Cl$$

BCNU
1,3-BIS-(2-CHLOROETHYL)-1-NITROSOUREA

Another group of chemotherapeutic agents, called **antimetabolites,** interferes with DNA synthesis. One of these chemicals, 5-fluorouracil, gets involved in the synthesis of a nucleotide, which inhibits the formation of a thymine-containing nucleotide necessary for DNA synthesis. Lack of proper DNA synthesis slows cell division. 5-Fluorouracil has proved useful in the treatment of cancers of the breast.

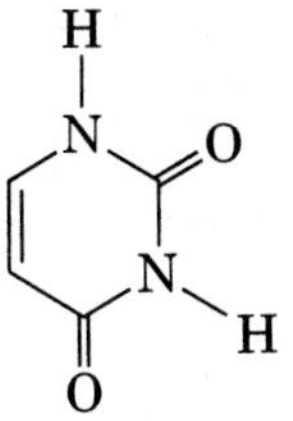

URACIL 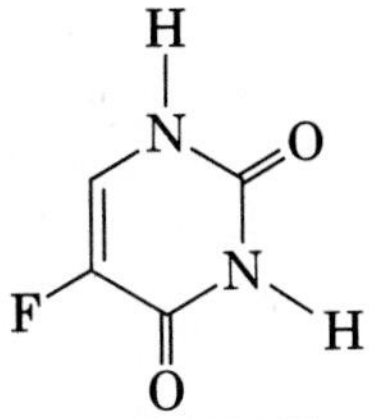   5-FLUOROURACIL

Another antimetabolite is methotrexate, which has a structure similar to folic acid. In a manner similar to how the drug sulfanilamide inhibits

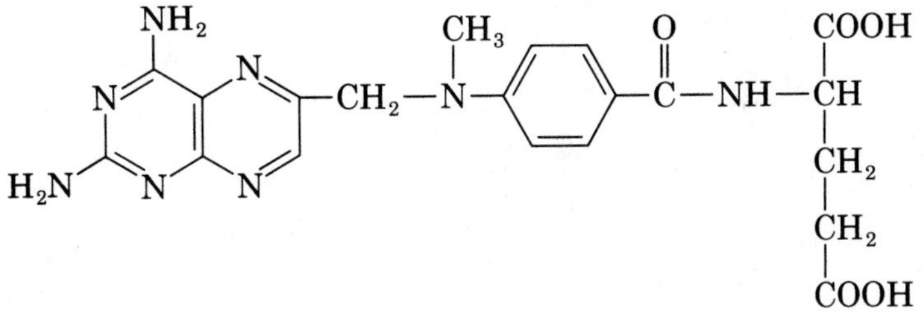

METHOTREXATE

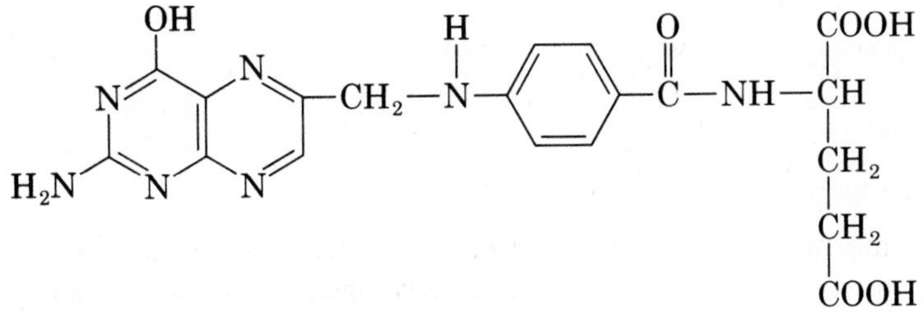

FOLIC ACID

bacterial cell wall synthesis (see page 588), methotrexate takes the place of folic acid on an enzyme that promotes cell growth. This results in a slowdown of cell growth. Leukemia is treated with methotrexate.

Other classes of anticancer drugs include **plant alkaloids** such as vincristine from the periwinkle plant, **antibiotics** such as the actinomycins, and the **sex hormones** such as the estrogens.

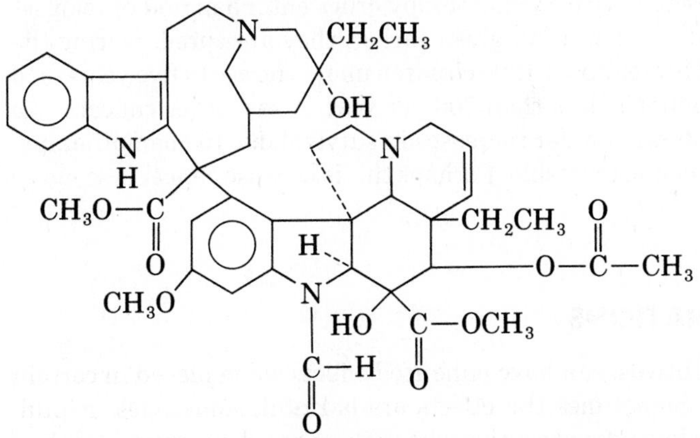

VINCRISTINE

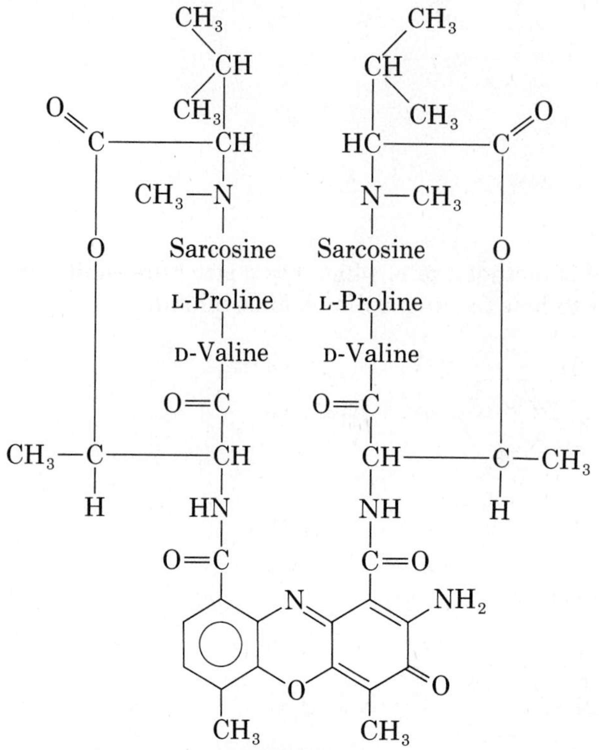

DACTINOMYCIN
(ACTINOMYCIN D)

All cancer chemotherapy is tedious and has its own risks. In addition to being highly toxic, most of the useful chemotherapy agents are carcinogenic (cancer-causing) themselves. Often very high doses are necessary to effect treatment. As a result, single-agent chemotherapy has largely given way to combination chemotherapy because of the success of finding additive or even **synergistic effects** when two or more anticancer drugs are used. For example, a combination that is used in the treatment of several cancers is cisplatin and cyclophosphamide. When these are given together, lower doses of each compound can be used because of the synergistic action, and this reduces the harmful side effects of chemotherapy.

Childhood cancers respond most favorably to chemotherapy. Most children with leukemia who are treated with chemotherapy drugs enter a period of relapse-free survival. In terms of the definition given earlier, they are cured. During the early 1950s in the United States about 1900 children under the age of five years died of cancer. Today the amount is less than 700 per year. A few other cancers like Hodgkin's disease have shown similar increases in survival due to chemotherapy, but not all cancers have been so treatable. Perhaps this is because cancer is so many different diseases.

Synergism is the working together by two things to produce an effect greater than the sum of the individual effects.

DRUGS IN COMBINATIONS

Drugs, like some food additives, can have enhanced effects when placed in certain chemical environments. Sometimes the effects are harmful, sometimes helpful. Take the case of an aging business executive who took an antidepressant and then ate a meal that included aged cheese and wine. The antidepressant is an inhibitor of monoamine oxidase, an enzyme that helps to control blood pressure. Both the aged cheese he ate and the wine he drank contained pressor amines, which raise blood pressure. Without the controlling effect of the enzyme, these amines skyrocketed his blood pressure and caused a stroke. Neither the amines nor the antidepressant alone would have been likely to cause the stroke, but the combination did.

Likewise, people who take digitalis for heart trouble and for reducing the sodium level in the blood should take aspirin only under medical supervision. Aspirin can cause a 50% reduction in salt excretion for 3 or 4 hours after it is taken.

Pressor amines tend to increase blood pressure.

Alcohol increases the action of many antihistamines, tranquilizers, and drugs such as reserpine (for lowering blood pressure) and scopolamine (contained in many over-the-counter nerve and sleeping preparations), making such combinations extremely dangerous. Staying away from dangerous alcohol-drug combinations is not as easy as it may seem. Many people fail to realize that a large number of over-the-counter preparations — such as liquid cough syrup and tonics — contain appreciable amounts of alcohol.

Not all drug combinations are bad. The synergistic effect observed in the treatment of cancer by drugs in combination has already been mentioned. Doctors have been highly successful in prolonging the lives of leukemia and other cancer victims with combinations of drugs that individually could not do the job. Resistant kidney disease has also responded to drug combinations in cases in which single drugs were ineffective.

Perhaps the best advice is to take medicine only when you are seriously ill, making sure that a physician knows what you are taking.

THE ROLE OF THE FDA

Since 1940, new drugs introduced into the U.S. market have numbered 1119. In this number are almost all of the drugs you probably recognize on your medicine shelf at home. Drug companies must petition the U.S. Food and Drug Administration (FDA) with data showing a new drug is safe and effective for its intended use. Usually this involves extensive animal tests. If the FDA gives its approval, then the drug undergoes limited controlled testing on healthy human test subjects. A second phase involves testing on research subjects who have the disease that the drug is intended to treat. Further tests are then carried out on larger groups to gauge the drug's effectiveness and safety. After all of this testing, only about one in ten drugs passes. The FDA has in the past been accused of being overly cautious, but the thalidomide incident (see box) showed that caution can often prevent tremendous human anguish and needless suffering.

Thalidomide, a tranquilizer prescribed in Europe in the late 1950s, was being sponsored for introduction into the U.S. market. Dr. Frances O. Kelsey, a new drug investigator at the FDA in Washington, refused to allow thalidomide to be listed as safe and effective in light of some evidence she read in the data supplied concerning the drug and its effects on laboratory animals. Only after her refusal to certify thalidomide did the mutagenic effects of the drug become known. Truly thousands of young people born in the early 1960s owe Dr. Kelsey a debt of gratitude for their good health.

Countless other stories could be told of how experimental drugs are withheld from the market by the FDA. Perhaps none of these will ever again be as dramatic as the thalidomide story. Often, the FDA gets embroiled in controversy regarding

The drug thalidomide was prescribed for controlling morning sickness during pregnancy. Its mutagenic effects were discovered only later when children of mothers who had taken the drug were born with deformities such as flipper-like arms and legs. Later it was discovered that of the two stereoisomers of thalidomide, D-thalidomide was active and safe. The L-isomer turned out to be the active mutagen.

THALIDOMIDE

* Indicates the asymmetric carbon atom

drugs that have claimed effectiveness in other countries. Because the drug isn't tested properly, the FDA has no recourse but to withhold approval. Two drugs of this type are Laetrile, a cyanide-containing compound found in the seeds of apples and peaches. This compound has been reputed to cure certain cancers. Yet, the FDA refuses to approve its use as a chemotherapy agent because sufficient evidence is lacking. Laetrile clinics are run in a number of countries where U.S. citizens travel and receive treatments without FDA approval.

Another chemical involved in controversy is **cyclosporin A,** a cyclic peptide consisting of 11 amino acid residues with a molecular weight of 1202. Cyclosporin was discovered in 1970 by J. F. Borel in Switzerland as a metabolite in a culture broth of the fungus *Tolypocladium inflatum* Gams. It was soon discovered that this peptide had powerful immunosuppressive effects, and by 1977 it was being used in organ transplants in animals. A year later cyclosporin A was being used to suppress rejection in human liver transplants and in marrow transplants for patients with leukemia. Because cyclosporin A so effectively involves itself with the immune system, it is not surprising to see it tried for treatment of diseases in which the immune system is out of control (after all, the most effective anticancer drugs are themselves carcinogens). Cyclosporin A has recently been tried in the treatment of acquired immune deficiency syndrome (AIDS), but not in the United States because it has not been approved for testing for that purpose. Whether cyclosporin A will ever be used in the treatment of AIDS in the United States depends on the data available to the FDA regarding its intended use.

If the FDA continues to do its job, we can be reasonably certain the drugs we take have been thoroughly tested before being approved for widespread human use.

SELF-TEST 21–D

1.   The male sex hormone is called _____.

2.   The female sex hormone is called _____.

3.   Anabolic as used in the term *anabolic steroid* means _____.

4.   Name three undesirable side effects in male athletes who use anabolic steroids.

_____ , _____ , and _____

5. Most anticancer drugs can also cause cancer. ( ) True or ( ) False
6. The nitrogen mustards act on cancer cells by blocking _____ replication.
7. Chemicals that interfere with DNA synthesis are called _____ .
8. Which cancer has shown the greatest response to chemotherapeutic drugs?
   _____

MATCHING SET

_____ 1. Histamine
_____ 2. Cortisone
_____ 3. Tetracycline
_____ 4. Penicillin
_____ 5. Sulfa drug
_____ 6. Iodine
_____ 7. Propranolol
_____ 8. Procaine (Novocain)
_____ 9. Opium poppy
_____ 10. Dihydroxyaluminum sodium carbonate
_____ 11. 3-methyl fentanyl
_____ 12. Anabolic
_____ 13. Alkylating drug
_____ 14. Antimetabolite drug
_____ 15. Immune suppressant
_____ 16. Calcium ion
_____ 17. Nicotinic acid

a. 3000 times more powerful than heroin
b. Source of morphine
c. Interferes with cell wall synthesis
d. Analgesic used in dentistry
e. Related to Reye's syndrome
f. Polymer that lowers blood cholesterol
g. Causes symptoms of hay fever
h. Causes heart muscles to contract
i. Sulfanilamide
j. Muscle producing
k. Cyclosporin A
l. Methotrexate
m. Female sex hormone
n. Cyclophosphamide
o. Relaxes heart muscle
p. Antibiotic containing a four-ring structure
q. Antacid
r. Antiseptic
s. A steroid

QUESTIONS

1. Discuss some reasons why women on the average live longer than men.
2. Why does the stomach not dissolve itself?
3. Name two alkaloid narcotics that are derived from morphine.
4. What is Reye's syndrome? What common drug has it been associated with?
5. Name three ions that neutralize negative charges associated with nerve and muscle cells.
6. Acetaminophen is toxic to what organ?
7. What designer drug is toxic to the part of the brain called the substantia nigra? What disease does this drug cause?
8. Look up the drugs scheduled by the Drug Enforcement Agency. How many are Schedule 1 drugs?
9. Describe the action of antihistamines.
10. How is iodine toxic to a living cell?
11. Describe how the sulfa drug sulfanilamide works.
12. How was penicillin discovered?
13. What happens when beta receptor sites in heart muscles are stimulated?
14. What can happen to a patient when a drug passes through the blood-brain barrier?
15. What is atherosclerotic plaque? What are its two major ingredients?
16. Explain how a beta blocker drug can lower blood pressure.
17. (a) What effect does the calcium ion have on the heart? (b) How can different calcium channel blockers be structurally so dissimilar and affect

the flow of calcium ions into muscle cells in such a similar way?

18. What organ uses cholesterol to produce bile acids?

19. How do the estrogens differ structurally from the other female hormone progesterone?

20. Name an antiinflammatory steroid produced in the body.

21. Anabolic steroids that can be taken orally are unlike testosterone in what structural feature?

22. Nitrogen mustards are alkylating agents. These interfere with DNA replication. Explain.

23. Explain how the antimetabolite methotrexate works to kill cells.

24. If the chemotherapeutic agents kill living cells, why do they have a preferential effect on cancer cells?

25. What is one of the dangers of chemotherapy using an alkylating agent or an antimetabolite?

Chapter 22

BEAUTY AND CLEANSING AGENTS — CHEMICAL FORMULATIONS FOR LOOKS, HEALTH, AND COMFORT

The use of chemical preparations, which are applied to the skin to cleanse, beautify, disinfect, or alter appearance or smell, is older than recorded history. Such preparations are known as **cosmetics.** Although the distinction is only clearly made in the legal definitions, cosmetics are conceptually distinguished from drugs in that drugs are applied to alter body functions. Thus, in concept, an antiperspirant should be a drug even though it is generally regulated as a cosmetic. Soaps and detergents have been given special status in legal definitions, setting them apart from cosmetics. In this chapter we will consider the basic chemistry of the major classes of preparations which you will find at the cosmetic counter, along with the chemistry of important cleansing agents.

There is relatively little legal control in the use of cosmetics. While a drug has to be proved safe prior to its general use, a cosmetic product must be removed from use if it is found to be unsafe. This trial-and-error approach to cosmetic materials has resulted in many abuses. Skin and scalp irritations have been many, and there are extreme cases such as blindness from ill-conceived eye preparations. However, good business has dictated safer cosmetics through the years, and the user today can approach the cosmetic counter with a fairly high level of confidence.

A knowledge of the chemistry of cosmetics can be of considerable value to you as you use these materials. First, an understanding of the chemical effect that accompanies the cosmetic use will help you decide whether the cosmetic should be used. Second, the active ingredient may be sold in one brand (in a pretty package with a lot of hype) at several times the cost of the same ingredient in another brand. As a third consideration, our present level of understanding of allergenic reactions dictates a need for a personal evaluation of the safety, or danger, in the use of cosmetic chemicals.

A word of caution — do not conclude, based on this brief study, that you will be prepared to formulate your own cosmetics from the basic ingredients. Although you could undoubtedly make a number of safe preparations, there are just too many dangers from impure chemicals and improper mixtures. This level of chemical knowledge can and should be very useful to the user of cosmetics, but in general, much more knowledge is needed for the safe manufacture of cosmetics.

Cosmetic products require careful testing before they are used on humans.

SKIN, HAIR, NAILS, AND TEETH — A CHEMICAL VIEW

The skin, hair, and nails are protein structures. Skin (Figs. 22–1 and 22–2), like other organs of the body, is not composed of uniform tissue and has several functions made possible by its structure; they are protection, sensation, excretion, and body temperature control. The exterior of the epidermis is called the **stratum, corneum,** or **corneal** layer and is where most cosmetic preparations for the skin act. The corneal layer is composed principally of dead cells with a moisture content of about 10%. The principal protein of the corneal layer is **keratin,** which is composed of about 22 different amino acids. Its structure renders it insoluble in, but slightly permeable to, water. Dry skin is uncomfortable, and excessively moist skin is a good host for fungus organisms. An oily secretion, **sebum,** is secreted by the sebaceous glands to protect from excessive moisture loss, which results in dry skin. In order to control the moisture content of the corneal layer so that it does not dry out and slough off too quickly, moisturizers may be added to the skin. Normal skin is slightly acidic with a pH of about 4.

Hair is composed principally of keratin (Fig. 22–3). An important difference between hair keratin and other proteins is its high content of the amino acid **cystine.** About 16% to 18% of hair protein is cystine, but only 2.3% to 3.8% of the keratin in corneal cells is cystine. This amino acid plays an important role in the structure of hair.

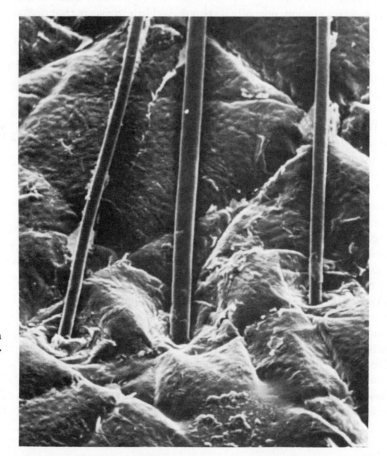

FIGURE 22–1 Replica of the surface of human forearm skin, showing three hairs emerging from the skin (×225). (Courtesy of E. Bernstein and C. B. Jones: *Science,* Vol. 166, pp 252–253, 1969. Copyright 1969 by the American Association for the Advancement of Science.)

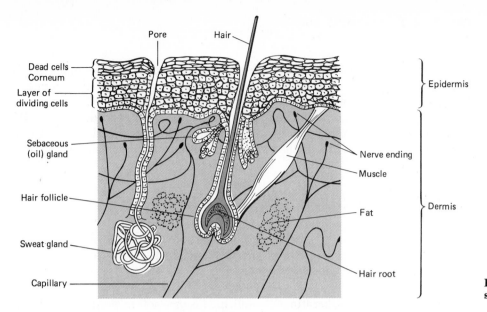

Pore • Hair

Dead cells — Corneum
Layer of — dividing cells

Epidermis

Sebaceous — (oil) gland

Nerve ending

Muscle

Hair follicle —

Dermis

Fat

Sweat gland —

Capillary —

Hair root

FIGURE 22–2 Cross section of the skin.

The toughness of both skin and hair is due to the bridges between different protein chains, such as hydrogen bonds and —S—S— linking bonds, called **disulfide bonds**:

$$
\begin{array}{cccc}
O{=}C & & & N{-}H \\
| & & & | \\
H{-}CCH_2{-}S{-}S{-}CH_2C{-}H \\
| & & & | \\
H{-}N & & & C{=}O \\
| & & & |
\end{array}
$$

DISULFIDE BONDS (CROSSLINKS)

$$
\begin{array}{c}
NH_2 \\
| \\
HOOC{-}CH{-}CH_2{-}S \\
| \\
HOOC{-}CH{-}CH_2{-}S \\
| \\
NH_2
\end{array}
$$

CYSTINE

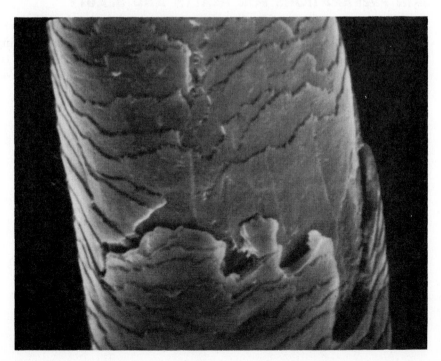

FIGURE 22–3 Electron micrograph of human hair. Note the layers of keratinized cells.

Another type of bridge between two protein chains, which is important in keratin as well as in all proteins, is the **ionic bond.** Consider the interaction between a lysine amine group —NH_2 and a carboxylic group —COOH of glutamic acid on a neighboring protein chain. At pH 4.1, protons are added to the —NH_2 groups and removed from the —COOH groups, resulting in —NH_3^+ and —COO^- groups on adjacent chains. If the two charged groups approach closely, an ionic bond is formed.

The structures of protein tissues are due in part to disulfide crosslinks and to ionic bonds between "molecules."

$$\overset{|}{\underset{|}{H}C}CH_2CH_2CH_2CH_2NH_2 + HOOCCH_2CH_2\overset{|}{\underset{|}{C}H} \xrightarrow{\text{at pH 4.1}}$$

LYSINE   GLUTAMIC ACID

$$\overset{|}{\underset{|}{H}C}CH_2CH_2CH_2CH_2NH_3^+ \; {}^-OOCCH_2CH_2\overset{|}{\underset{|}{C}H}$$

IONIC BOND

As the pH rises above 4, keratin will swell and become soft as these crosslinks are broken. This is an important aspect of hair chemistry.

Finger- and toenails are composed of **hard keratin,** a very dense type of this protein. These epidermal cells grow from epithelial cells lying under the white crescent at the growing end of the nail. Like hair, the nail tissue beyond the growing cells is dead.

The mineral content, or the hard part, of bones and teeth consists of two compounds of calcium. Calcium carbonate ($CaCO_3$) is present in bones and teeth in the crystalline form, known to mineralogists as aragonite. The second calcium compound found in teeth is calcium hydroxyphosphate, $[Ca_5(OH)(PO_4)_3]$, or apatite (Fig. 22–4).

SKIN PREPARATIONS FOR HEALTH AND BEAUTY

To remain healthy, the moisture content of skin must stay near 10%. If it is higher, microorganisms grow too easily; if lower, the corneal layer flakes off. Washing skin removes fats that help retain the right amount of moisture. If dry skin is treated with a fat after washing, it will be protected until enough natural fats have been regenerated.

Skin with a low fat content tends to be dry.

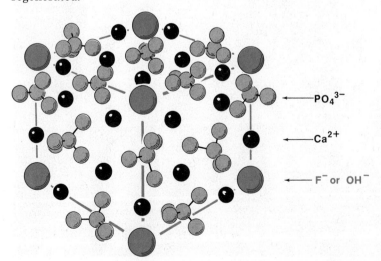

$\longleftarrow PO_4{}^{3-}$

$\longleftarrow Ca^{2+}$

$\longleftarrow F^-$ or OH^-

FIGURE 22–4 Structure of apatite and fluoroapatite. The small dark circles denote Ca^{2+} ions, the groups of four circles tied together by lines represent the PO_4^{3-} groups, and the largest circles represent OH^- groups in apatite.

Lanolin is an excellent skin softener *(emollient)* and is a component of many cosmetics. It is a complex mixture of esters from hydrated wool fat. The esters are derived from 33 different alcohols of high molecular weight and 37 fatty acids. Cholesterol, a common alcohol in lanolin, is found both free and in esters. Cholesterol appears to give fat mixtures the property of absorbing water. This is one factor that makes lanolin an excellent emollient. With its high proportion of free alcohols, particularly cholesterol, and hydroxyacid esters, lanolin has the structural groups (—OH) to hydrogen-bond water (to keep the skin moist) and to anchor within the skin (the fatty acid and ester hydrocarbon structures; see Fig. 22–5).

> Lanolin is grease from wool.

Creams

Creams are generally emulsions of either an oil-in-water type or a water-in-oil type. An *emulsion* is simply a colloidal suspension of one liquid in another. The oil-in-water emulsion has tiny droplets of an oily or waxy nature dispersed throughout a water solution (homogenized milk is an example). The water-in-oil emulsion has tiny droplets of a water solution dispersed throughout an oil (natural petroleum and melted butter are examples). An oil-in-water emulsion can be washed off the hands with tap water, while a water-in-oil emulsion gives the hands a greasy, water-repellent surface.

> Colloids are particles intermediate in size between small molecules and clumps of molecules sufficiently large to precipitate.

Cold cream originally was an emulsion of rose water in a mixture of almond oil and beeswax. Subsequently, other ingredients were added to get a more stable emulsion. An example of a modified cold cream composition is: almond oil, 35%; beeswax, 12%; lanolin, 15%; spermaceti (from whale oil), 8%; and strong rose water, 30%. Other oils can be substituted for some or all of the almond oil. Lanolin stabilizes the emulsion. Any oil preparation that holds moisture in the skin is a *moisturizer*.

> Creams add oil or fat content to surface skin.

Vanishing cream is a suspension of stearic acid in water, to which a stabilizer has been added to prevent the ingredients from separating. The stabilizer may be a soap, such as potassium stearate. These creams do not actually vanish; they merely spread as a smooth, thin covering over the skin.

Creams of various sorts may be used as the base for other cosmetic preparations; other ingredients are added to give desired properties to the creams. As an example, hydrated aluminum chloride can be added to prepare a cream deodorant.

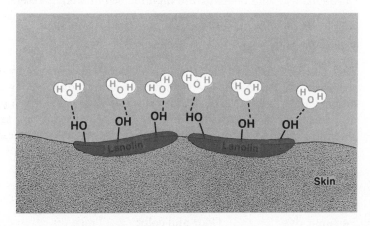

FIGURE 22–5 The hydroxyl groups of lanolin form hydrogen bonds with water and keep the skin moist. The fat parts of the molecule are "soluble" in the protein and fat layers of the skin.

Lipstick

The skin on our lips is covered by a thin corneal layer that is free of fat and consequently dries out easily. A normal moisture content is maintained from the mouth. In addition to being a beauty aid, lipstick can be helpful under harsh conditions that tend to dry lip tissue.

Lipstick consists of a solution or suspension of coloring agents in a mixture of high molecular weight hydrocarbons or their derivatives, or both. The material must be soft enough to produce an even application when pressed on the lips, yet the film must not be too easily removed, nor may the coloring matter run. Lipstick is perfumed to give it a pleasant odor. The color usually comes from a dye, or "lake," from the eosin group of dyes. A *lake* is a precipitate of a metal ion (Fe^{3+}, Ni^{2+}, Co^{3+}) with an organic dye. The metal ion enhances the color or changes the color of the dye and keeps the dye from dissolving.

Two suitable dyes, used in admixture and with their lakes, are dibromo-fluorescein (yellow-red) and tetrabromofluorescein (purple):

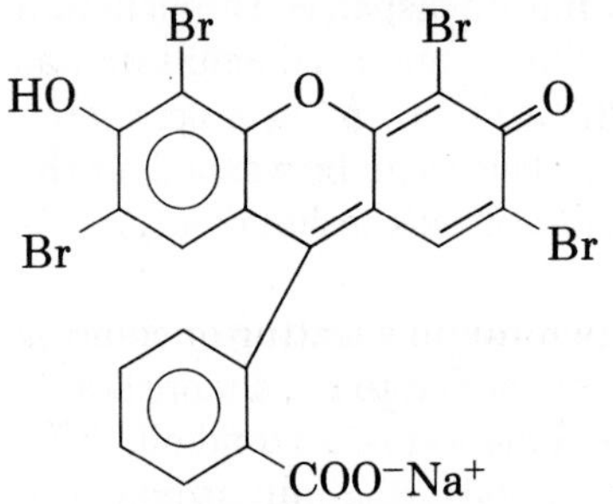

A *lake* is a coloring agent made up of an organic dye adhering to an inorganic substance called a mordant. Some lakes are also approved as food colors.

TETRABROMOFLUORESCEIN (EOSIN)
(SODIUM SALT)

The ingredients in a typical formulation of lipstick include:

Dye	Furnishes color	4–8%
Castor oil, paraffins or fats	Dissolves dye	50%
Lanolin	Emollient	25%
Carnauba wax } Beeswax }	{ Makes stick stiff by raising the melting point }	18%
Perfume	Imparts pleasant odor	Small amount

Carnauba wax and beeswax are high molecular weight esters.

Face Powder

Face powder is used to give the skin a smooth appearance by covering up any oil secretions, which would otherwise give it a shiny look. The powder must have some hiding ability, but if it is too opaque it will look too obvious. A powder that has the proper appearance, sticking properties, absorbence for oily skin secretions, and spreading ability usually requires several ingredients. A typical formula is:

Astringents shrink tissue, restricting fluid flow.

Talc	Absorbent	56%
Precipitated chalk	Absorbent	10%
Zinc oxide	Astringent	20%
Zinc stearate	Binder	6%
Perfume, dye	Odor and color	Trace

Eye Makeup

There are several types of eye makeup: eyebrow pencils, mascaras for eyelashes, and shading, among others. Eyebrow pencils are very much like lipstick, but they contain a different coloring matter. The coloring matter is a pigment such as lampblack; the other ingredients include fats, oils, petrolatum, and lanolin, blended to give the desired melting point, which may be raised by the addition of beeswax or paraffin. Petrolatum is a semisolid mixture of hydrocarbons (saturated, $C_{16}H_{34}$ to $C_{32}H_{66}$; and unsaturated, $C_{16}H_{32}$; etc.; melting point, 34–54°C). Brown pencils are made by using iron oxide pigments in combination with lampblack.

Mascara is used to darken eyelashes and give them a longer appearance. The same colors as in eyebrow pencils are used, as well as other mineral coloring matters such as chromic oxide (dark green) and ultramarine (blue pigment of variable composition; probably a double silicate of sodium and aluminum silicate with some sodium sulfide). The coloring matter is suspended in a foundation that is a mixture of a soap, oils, fats, and waxes. The mascara may be water-soluble or water-resistant, depending on the composition of the foundation. A typical formula consists of about 40% wax (beeswax, carnauba wax, and paraffin, adjusted for hardness), 50% soap, 5% lanolin, and 5% coloring matter.

Eye shadow is a formulation of a coloring matter suspended in an oily-fatty-waxy base. A formula that has been used for this purpose is 60% petroleum jelly, 6% lanolin, 10% fats and waxes (approximately equal amounts of cocoa butter, beeswax, and spermaceti), and the balance zinc oxide (white) plus tinting or coloring dyes. Cocoa butter (melting point, 30–35°C) is composed of glycerides of stearic, palmitic, and lauric acids. It is obtained from natural products by compression of cacao seed. Spermaceti (melting point, 45°C) is chiefly cetyl palmitate, $CH_3(CH_2)_{14}COO(CH_2)_{15}CH_3$. It is taken from the solid fat from the head of the sperm whale.

Titanium dioxide, a white powder, is also used as a base for many eye makeup preparations.

Perfume

A typical perfume has at least three components of somewhat different volatility and molecular weight. (Recall that lower molecular weight compounds are generally more volatile.) The first, called the **top note,** is the most volatile and is the most obvious odor when the perfume is first applied. The second, called the **middle note,** is less volatile and is generally a flower extract (violet, lilac, etc.). The last, or **end note,** is least volatile, and is usually a resin or waxy polymer.

Perfumes are complex mixtures of odorous compounds.

Most perfumes contain many components, and chemically they are often complex mixtures. As the analysis of natural perfume materials has progressed, the use of pure synthetic organic compounds to duplicate specific odors has become very common. An example is the isolation of civetone (see structure, below), a cyclic ketone from civet, a secretion of the civet cat of Ethiopia. It is highly valued for perfumes.

Civetone is now available in a synthetic form. It is prepared by forming 8-hexadecene-1,16-dicarboxylic acid into a ring. The thorium ion (Th^{4+}) catalyzes the closure of the ring.

$$\begin{array}{l} HC{-}(CH_2)_7COOH \\ \parallel \\ HC{-}(CH_2)_7COOH \end{array} \xrightarrow[\Delta]{Th^{4+}} \begin{array}{l} HC{-}(CH_2)_7 \\ \parallel \qquad\qquad\diagdown \\ HC{-}(CH_2)_7 \diagup \end{array}\!\!C{=}O + CO_2 + H_2O$$

CIVETONE

Musk is obtained from the
musk deer.

Civet is a collection of sex attractants, as is musk. The perfumes that "work" have these sex attractants cleverly masked by herbaceous and floral odors. The initial attraction comes from the pleasant odor, but the "basic effect" is from the civetone or musk.

Other compounds used in perfumes include high molecular weight alcohols and esters. An example is geraniol (bp 230°C), a principal component of Turkish geranium oil.

After-shave lotions and
colognes are diluted
perfumes, about one tenth
(or less) as strong.

$$H_3C$$
$$\diagdown$$
$$C{=}CH{-}CH_2{-}CH_2{-}C{-}CH_3$$
$$\diagup \qquad\qquad\qquad \|$$
$$H_3C \qquad\qquad\qquad HC{-}CH_2OH$$

GERANIOL

Esters of this alcohol are used to make synthetic rose aromas for perfumes. For example, the ester formed by reaction between geraniol and formic acid has a rose odor.

$$\underset{\substack{\text{FORMIC}\\\text{ACID}}}{H{-}\overset{\overset{\textstyle O}{\|}}{C}{-}OH} + \underset{\text{GERANIOL}}{HOCH_2{-}\overset{\overset{\textstyle H}{|}}{C}{=}R} \rightarrow \underset{\text{GERANYL FORMATE}}{H{-}\overset{\overset{\textstyle O}{\|}}{C}{-}O{-}CH_2CH{=}\overset{\overset{\textstyle CH_3}{|}}{C}(CH_2)_2CH{=}C(CH_3)_2}$$

Ethyl alcohol is a major
constituent of most
perfumes.

Typical perfumes are 10% to 25% perfume essence and 75% to 80% alcohol and a fixative to retain the essential oils. Perfumes are added to most cosmetics to give the products desirable odors; they also mask the natural odors of other constituents such as sex attractants. They are mildly bactericidal and antiseptic because of their alcohol content.

Suntan Lotions

Ultraviolet radiation tans
skin.

One of the agents most harmful to skin is the short wavelength (ultraviolet) light from the Sun. It is considered desirable to exclude the shorter, more harmful wavelengths, while transmitting enough less energetic, longer wavelength ultraviolet to permit gradual tanning.

The variety of suntan products ranges from lotions, which selectively filter out the higher energy ultraviolet rays of the sun, to preparations that essentially dye light-colored skin a tan color.

The lotions that filter out the ultraviolet rays are more accurately described as sunscreens, and their ingredients are often mixed with other materials, to give a lotion that both screens and tans. A common ingredient in preparations used to *prevent* sunburn is p-aminobenzoic acid.

Sunbathers refer to
p-aminobenzoic acid as
PABA.

$$H_2N{-}\!\left\langle\!\bigcirc\!\right\rangle\!{-}COOH$$

p-AMINOBENZOIC ACID
(PABA)

Like most aromatic compounds, it absorbs strongly in the ultraviolet region of the spectrum (Fig. 22–6).

In tanning, the skin is stimulated to increase its production of the pigment **melanin.** At the same time, the skin thickens and becomes more resistant to deep burning.

Increased melanin
protects sensitive lower
layers of skin.

Preparations for the relief of sunburn pain are solutions of local anesthetics such as benzocaine.

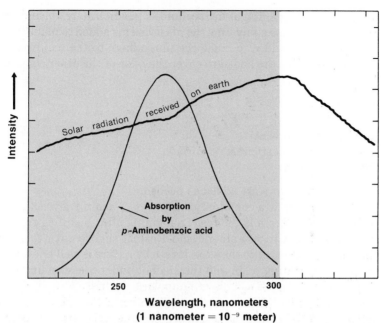

FIGURE 22–6 Absorption spectrum of *p*-aminobenzoic acid and its relationship to solar ultraviolet radiation received on Earth. Maximum absorption occurs at 265 nanometers, although it absorbs at other wavelengths as shown. The maximum of the deep-burning ultraviolet radiation received on Earth is at about 308 nanometers.

SELF-TEST 22–A

1. The surface of the skin epidermis is known as the _____ layer.
2. The oily secretion of skin is _____.
3. Keratin is a protein found in _____, _____, and _____.
4. Bridges between protein chains may be _____ linkages or _____ bonds.
5. The two calcium minerals in teeth are _____ and _____.
6. Lanolin is a skin softener or a(n) _____.
7. A skin cream is either an oil-in-water or a water-in-oil _____.
8. A cosmetic that is colored with "a lake" is _____.
9. The major mineral ingredient in face or body powder is _____.
10. Civetone is likely to be the attracting odor in a perfume. ( ) True or ( ) False
11. An example of an active chemical that will absorb ultraviolet light in a sunscreen lotion is _____.
12. Skin darkens because of an increased concentration of the skin pigment _____.

NAIL POLISH AND POLISH REMOVER

Nail polish is essentially a lacquer or varnish. It can be made of nitrocellulose, a plasticizer, a resin, a solvent, and perhaps a dye. The nitrocellulose can be replaced by another polymer molecule that possesses similar qualities. The evaporation of

Nail polish and hair sprays are formulated very much alike.

the solvent leaves a film of nitrocellulose, plasticizer, resin, and dye. The nitrocellulose furnishes the shiny film; the plasticizer is added to make the film less brittle; and the resin is added to make the film adhere to the nail better and to prevent flaking. Perfumes are added to cover the odor of the other constituents. A typical formulation is:

Nitrocellulose	15%
Acetone (solvent)	45%
Amyl acetate (solvent)	30%
Butyl stearate (plasticizer)	5%
Ester gum (resin)	5%

Perfumes and colors are added as needed.

Ester gum is a combination of esters — mainly glyceryl, methyl, and ethyl esters of rosin. Rosin is the resin remaining after distilling turpentine from pine exudate. It is 80% to 90% abietic acid. Rosin is slightly toxic to mucous membranes and slightly irritating to the skin. Its sticky nature is well known. The ester gum is prepared by heating rosin and the alcohol under pressure until the esterification occurs. The gums are soluble in nonpolar solvents.

Nail polish removers are simply solvents that dissolve the film left by the nail polish. They consist largely of acetone or ethyl acetate, or both, to which small amounts of butyl stearate and diethylene glycol monomethyl ether have been added to reduce the drying effect of the solvent. However, some formulations contain combinations of amyl acetate, butyl acetate, ethyl acetate, benzene, olive oil, lanolin, and alcohol. Both nail polish and nail polish removers are very flammable, and care should be taken never to use them in the presence of open flames or lighted cigarettes.

Cuticle softeners are primarily wetting agents and alkalies used to soften skin around the fingernails so it can be shaped as desired. The use of alkali to soften and swell protein is well known. A typical cuticle softener contains potassium hydroxide (3%), glycerol (12%), and water (85%). A cuticle softener may also contain sodium carbonate (an alkali), triethanol amine (a detergent, used as wetting agent), and trisodium phosphate (an alkali).

ABIETIC ACID
(MAIN COMPONENT OF ROSIN)

$CH_2-CH_2OCH_2-CH_2$
$|$ $|$
OH OCH_3
DIETHYLENE GLYCOL
MONOMETHYL ETHER

CURLING, COLORING, GROWING, AND REMOVING HAIR

The curl, color, and presence of human hair are matters of personal choice that vary considerably from person to person. Chemical means are available to alter these conditions except the growing of hair on skin where none is present or the prevention of natural baldness.

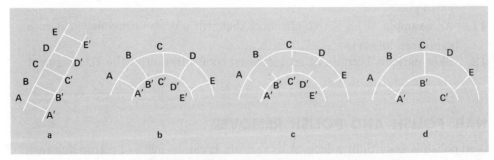

FIGURE 22–7   A schematic diagram of a permanent wave.

Changing the Shape of Hair

When hair is wet, it can be stretched to one and a half times its dry length because water (pH 7) weakens some of the ionic bonds and causes swelling of the keratin. Imagine the disulfide crosslinks remaining between two protein chains in hair as in Figure 22–7a. Winding the hair on rollers causes tension to develop at the crosslinks (b). In "cold" waving, these crosslinks are broken by a reducing agent (c), relaxing the tension. Then, an oxidizing agent regenerates the crosslinks (d) and the hair holds the shape of the roller. The chemical reactions in simplified form are shown in Figure 22–8.

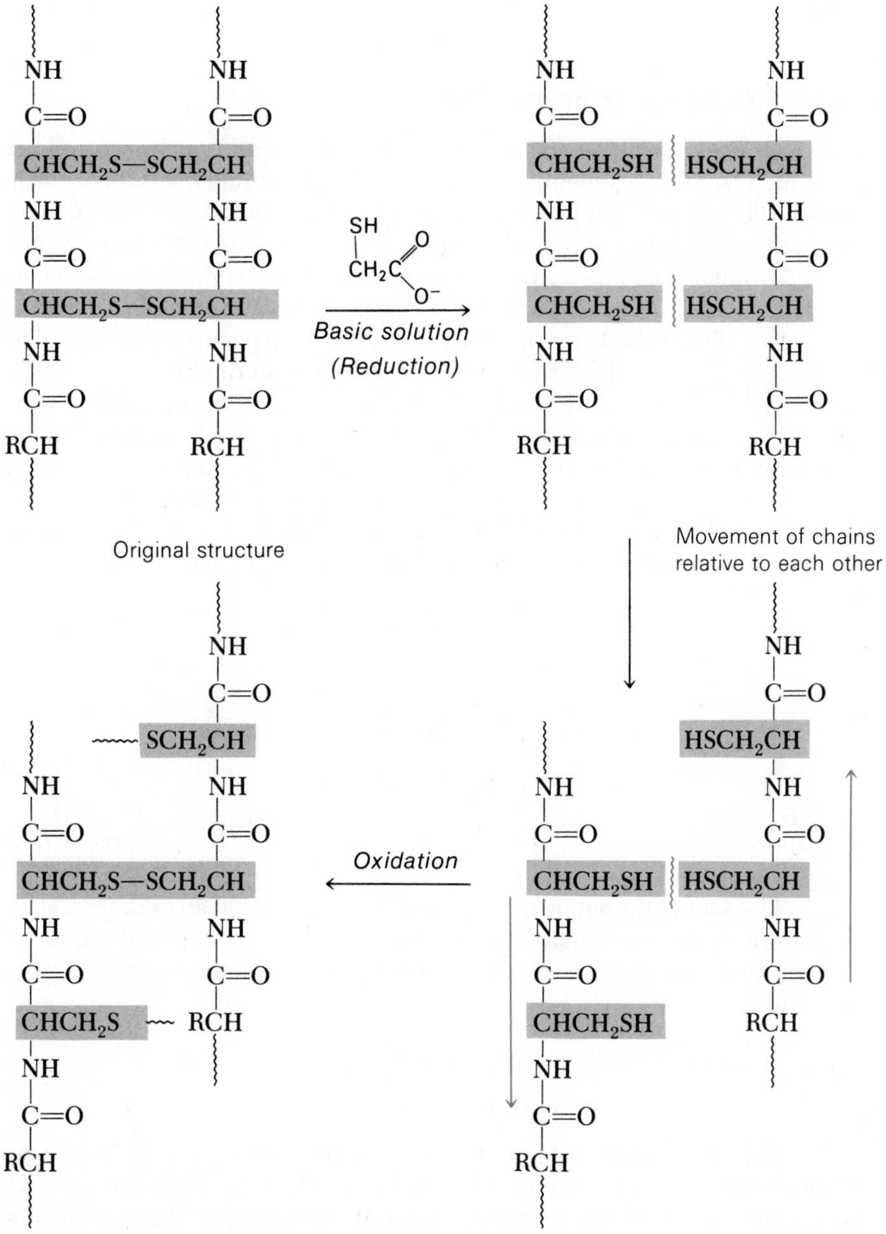

FIGURE 22–8 Structural changes that occur in hair during a permanent wave.

$$CH_2 - C \begin{array}{c} O \\ \diagdown \ OH \end{array}$$
$$| \\ SH$$

THIOGLYCOLIC
ACID

Hair can be straightened
by the same solutions. It is
simply "neutralized" (or
oxidized) while straight
(no rolling up).

The most commonly used reducing agent is thioglycolic acid. The common oxidizing agents used include hydrogen peroxide, perborates ($NaBO_2 \cdot H_2O_2 \cdot 3\,H_2O$), and sodium or potassium bromate ($KBrO_3$). A typical neutralizer solution contains one or more of the oxidizing agents dissolved in water. The presence of water and strong base in the oxidizing solution also helps to break and re-form hydrogen bonds between adjacent protein molecules. However, too-frequent use of strong base causes hair to become brittle and lifeless.

Various additives are present in both the oxidizing and the reducing solutions in order to control pH, odor, and color, and for general ease of application. A typical waving lotion contains 5.7% thioglycolic acid, 2.0% ammonia, and 92.3% water.

Coloring and Bleaching Hair

Melanin — black.

Iron pigment — red.

Hair contains two pigments: brown-black melanin and an iron-containing red pigment. The relative amounts of each actually determine the color of the hair. In deep black hair melanin predominates and in light-blond, the iron pigment predominates. The depth of the color depends upon the size of the pigment granules.

Some of the hair dyes are
suspected of being
carcinogenic.

Formulations for dyeing hair vary from temporary coloring (removable by shampoo), which is usually achieved by means of a water-soluble dye that acts on the surface of the hair, to semipermanent dyes, which penetrate the hair fibers to a great extent (Fig. 22–9). These often consist of cobalt or chromium complexes of dyes dissolved in an organic solvent. Permanent dyes are generally "oxidation" dyes. They penetrate the hair, and then are oxidized to give a colored product that is permanently attached to the hair by chemical bonds or that is much less soluble than the reactant molecule. Permanent hair dyes generally are derivatives of phenylenediamine. Phenylenediamine dyes hair black. A blond dye can be formulated with *p*-aminodiphenylaminesulfonic acid or *p*-phenylenediaminesulfonic acid.

NH_2

p-PHENYLENEDIAMINE

NH_2 — DIPHENYLAMINESULFONIC ACID — SO_3H

p-PHENYLENEDIAMINE-
SULFONIC ACID

Just about any shade of
hair color can be prepared
by varying the modifying
groups on certain basic
dye structures.

The active compounds are applied in an aqueous soap or detergent solution containing ammonia to make the solution basic. The dye material is then oxidized by hydrogen peroxide to develop the desired color. The amines are oxidized to nitro compounds.

$$-NH_2 + 3\,H_2O_2 \xrightarrow{\text{oxidation}} -NO_2 + 4\,H_2O$$
AMINE   NITRO
COMPOUND

Hair can be bleached by a more concentrated solution of hydrogen peroxide, which destroys the hair pigments by oxidation. The solutions are made basic with ammonia to enhance the oxidizing power of the peroxide. Parts of the chemical process are given in Figure 22–10. This drastic treatment of hair does more than

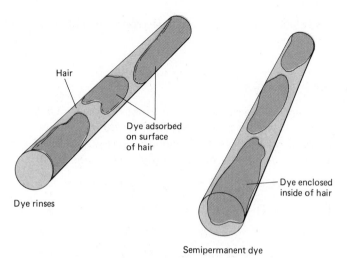

Hair

Dye adsorbed
on surface
of hair

Dye rinses

Dye enclosed
inside of hair

Semipermanent dye

FIGURE 22–9 Methods of dyeing hair.

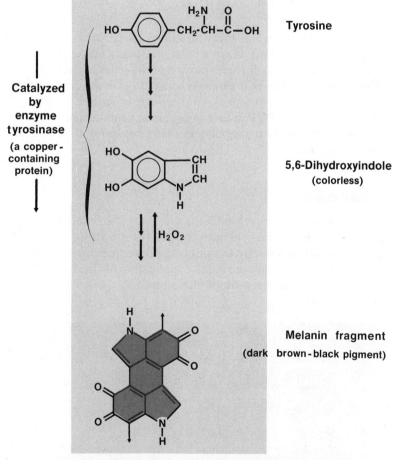

**Catalyzed
by
enzyme
tyrosinase**
(a copper-
containing
protein)

Tyrosine

5,6-Dihydroxyindole
(colorless)

H_2O_2

Melanin fragment
(dark brown-black pigment)

FIGURE 22–10 Bleaching of the hair by hydrogen peroxide. There are several chemical intermediates between the amino acid—tyrosine—and the hair pigment—melanin, which is partly protein. Hydrogen peroxide oxidizes melanin back to colorless compounds, which are stable in the absence of tyrosinase (found only in the hair roots). Melanin is a high-molecular-weight polymeric material of unknown structure. The structure shown here is only a segment of the total structure.

just change the color. It may destroy sufficient structure to render the hair brittle and coarse.

Hair Sprays

Hair spray coats the hair with a plastic film.

Hair sprays are essentially solutions of a resin in a volatile solvent whose purpose, when sprayed on hair, is to furnish a film with sufficient strength to hold the hair in place after the solvent has evaporated (Fig. 22–11). A common resin in hair sprays is the addition polymer, polyvinylpyrrolidone (PVP).

POLYVINYLPYRROLIDONE (PVP)

There are several dangers in breathing the vapor of hair sprays, such as the danger of possible carcinogens acting on delicate lung tissue and the danger of asphyxiation by the plastic coating lining the lungs.

The resin is blended in hair spray formulations with a plasticizer, a water repellent, and a solvent-propellant mixture. The plasticizer makes the plastic more pliable. The resin concentration of hair sprays is of the order of 4%. Other additives, such as silicone oils, are often put into hair sprays to give the hair a sheen.

Since PVP tends to pick up moisture, other less hygroscopic polymers are beginning to replace PVP in hair sprays. For example, significantly better moisture control is obtained with a copolymer made from a 60/40 ratio of vinylpyrrolidone and vinyl acetate.

Growing Hair

Minoxidil is a peripheral vasodilator that is used to reduce blood pressure. Applied under skin patches, this chemical causes the growth of fine, baby-like hair anywhere hair follicles exist on the skin. Oral doses produce the effect body-wide. The effect, though temporary during the application of the drug, offers some possibilities for the controlled growth of human hair.

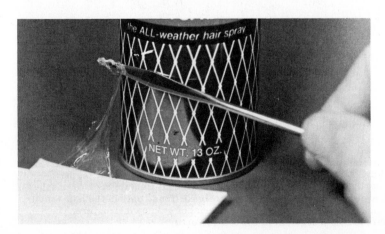

FIGURE 22–11 Film of hair spray. Hair spray was allowed to dry on white surface and was then pulled up to reveal film.

Depilatories

The purpose of a depilatory is to remove hair chemically. Since skin is sensitive to the same kind of chemical attack as hair, such preparations should be used with caution and, even then, some attack on the skin is almost unavoidable. Because of this, the interval between applications of a depilatory should be of the order of a week or so. It should never be used on skin that is infected or that has a rash and should not be followed by application of a deodorant with its *astringent* (contracting) action. If the sweat pores are closed by the deodorant, the caustic chemicals are retained and can do considerable harm. If the sweat pores are open, the body fluids will dilute and wash the caustic chemicals to the outside of the body.

The chemicals used as depilatories include sodium sulfide, calcium sulfide, strontium sulfide (water-soluble sulfides), and calcium thioglycolate [$Ca(HSCH_2COO)_2$], the calcium salt of the compound used to break S—S bonds between protein chains in permanent waving. A typical cream depilatory contains calcium thioglycolate (7.5%), calcium carbonate (filler, 20%), calcium hydroxide (provides basic solution, 1.5%), cetyl alcohol ([$CH_3(CH_2)_{15}OH$], skin conditioner, 6%), sodium lauryl sulfate (detergent, 0.5%), and water (64.5%).

The water-soluble sulfides are all strong bases in water, as indicated by the hydrolysis of the sulfide ion:

$$\underset{\text{SULFIDE}}{S^{2-}} + H_2O \rightarrow HS^- + \underset{\text{HYDROXIDE}}{OH^-}$$

For example, a 0.1 M solution of Na_2S has a pH of about 13, a strongly basic solution. The compounds act chemically on the hair to disrupt bonds in the protein chains and cause it to disintegrate by hydrolyzing to soluble amino acids and small peptides, which may be removed.

0.1 M means 0.1 mole of Na_2S (7.81 g) dissolved in sufficient water to make one liter of solution.

The area on which a depilatory has been used should be washed with soap and water, dried, and then treated with small amounts of talcum powder.

DEODORANTS

The 2 million sweat glands on the surface of the body are primarily used to regulate body temperature via the cooling effect produced by the evaporation of the water they secrete. This evaporation of water leaves solid constituents, mostly sodium chloride, as well as smaller amounts of proteins and other organic compounds. Body odor results largely from amines and hydrolysis products of fatty oils (fatty acids, acrolein, etc.) emitted from the body and from bacterial growth within the residue from sweat glands. Sweating is both normal and necessary for the proper functioning of the human body; sweat itself is quite odorless, but the bacterial decomposition products are not.

Body odor is promoted by bacterial action.

There are three kinds of deodorants: those that directly "dry up" perspiration or act as astringents, those that have an odor to mask the odor of sweat, and those that remove odorous compounds by chemical reaction. Among those that have astringent action are hydrated aluminum sulfate, hydrated aluminum chloride ($AlCl_3 \cdot 6\,H_2O$), aluminum chlorohydrate [actually aluminum hydroxychloride, $Al_2(OH)_5Cl \cdot 2\,H_2O$ or $Al(OH)_2Cl$ or $Al_6(OH)_{15}Cl_3$], and alcohols. Those compounds that act as deodorizing agents include zinc peroxide, essential oils and perfumes, and a variety of mild antiseptics. Zinc peroxide removes odorous compounds by oxidizing the amines and fatty acid compounds. The essential oils and

An astringent closes the pores, thus stopping the flow of perspiration.

perfumes absorb or otherwise mask the odors, and the antiseptics are generally oxidizing or reducing agents that kill the odor-causing bacteria.

DISINFECTANTS

Harmful microorganisms (germs) are always with us in large numbers. Fortunately, there are so many nonpathogenic microorganisms among them there is little room left for the pathogenic kinds. Depending on body location, healthy skin may host as many as a million microorganisms per square centimeter.

Pathogenic organisms cause disease.

Things we do to our skin, such as shaving, can upset the balance between harmful and harmless bacteria and promote infection by pathogenic microorganisms. *Disinfectants* kill these organisms before they can overcome the skin's defenses. Most commonly used are the alcohols. These are the only disinfectants used in many aftershave preparations. Maximum effectiveness of ethanol is reached at a concentration of 70%, while isopropyl (rubbing) alcohol is most effective at a concentration of 50%. These alcohols kill germs apparently by hydrogen bonding with water, which dehydrates the cellular structure of the germ.

Alcohols dehydrate microbes.

PHENOL
(CARBOLIC ACID)

To denature a protein is to break down its structure.

One widely used disinfectant is phenol; its aqueous solution is known as carbolic acid. An —OH group attached to the benzene ring is slightly acidic. The disinfectant action of phenol was discovered in 1867 by Sir Joseph Lister, who introduced it into surgery. It appears that phenol kills bacteria by denaturing cellular proteins. Today, about one third of all toilet soaps sold contain some derivative of phenol.

SELF-TEST 22–B

1. Nail polish is essentially a paint. The two types in common use are a(n) _____ and a(n) _____ .

2. Two chemicals widely used in nail polish removers are _____ and _____ .

3. What acid is used to reduce hair chemically (break the crosslinks between protein molecules)? _____

4. The last step in the production of a permanent wave in hair is ( ) oxidation or ( ) reduction. _____

5. What color is imparted to hair by phenylenediamine dyes? _____

6. Hair sprays leave a film of _____ on the hair.

7. Name three chemicals that are used as depilatories. _____ , _____ , and _____

8. What is the purpose of aluminum chlorohydrate in deodorants? _____

9. The most widely used disinfectant in cosmetics is _____ .

10. Another common name for phenol is _____ acid.

CLEANSING AGENTS

Dirt has been defined as matter in the wrong place. Tomato catsup is esteemed as a palatable food, but on your shirt it is dirt. There is a large number of cleansing, or **surface-active,** agents capable of removing the dirt without harm to the shirt (Fig. 22–12). Indeed, radio and television advertising might lead us to believe that the soaps and detergents we have today are unique and vastly superior to the products of a year or a century ago. This is not always so. Soap, for example, has always been made by a time-tested recipe that dates back at least to the second century of the Christian era. Galen, the great Greek physician, mentions that soap was made from fat, ash lye, and lime. Moreover, Galen stated that soap not only served as a medicament but also removed dirt from the body and clothes. What **is** new is the greater purity of soap, the improvement in its cleaning action by numerous additives, and the advent of the relatively new synthetic detergents.

Surface-active agents stabilize suspensions of nonpolar materials in polar solvents or vice versa. Examples include soaps, detergents, wetting agents, and foaming agents.

Soap

Fats and oils can be hydrolyzed in strongly basic solutions to form glycerol and salts of the fatty acids. Such hydrolysis reactions are called **saponification** reactions; the sodium or potassium salts of the fatty acids formed are **soaps.** Pioneers prepared their soap by boiling animal fat with an alkaline solution obtained from the ashes of hard wood. The resulting soap could be "salted out" by adding sodium chloride, making use of the fact that soap is less soluble in a salt solution than in water.

Review the molecular structure of fats and oils in Chapter 13.

Soaps can be made by treating fats or oils with sodium hydroxide.

$$CH_3(CH_2)_{16}COO—CH_2$$
$$CH_3(CH_2)_{16}COO—CH + 3\ NaOH \rightarrow 3\ CH_3(CH_2)_{16}COO^-Na^+ + HO—CH$$
$$CH_3(CH_2)_{16}COO—CH_2 \qquad\qquad HO—CH_2$$

TRISTEARIN
(GLYCERYL TRISTEARATE)   SODIUM STEARATE
(A SOAP)   GLYCEROL

The cleansing action of soap can be explained in terms of its molecular structure. Material that is water soluble can be readily removed from the skin or a surface by simply washing with an excess of water. To remove a sticky sugar syrup

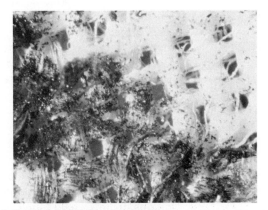

FIGURE 22–12 Photomicrograph of clean cotton cloth *(left)* and soiled cotton cloth *(right).* The proper application of surface-active agents should return the soiled cloth to its original state.

PRINCIPAL FATS AND OILS USED FOR MAKING SOAP

- Tallow or animal fat from beef or mutton is primarily an ester of stearic acid [$CH_3(CH_2)_{16}COOH$]. It is usually mixed with coconut oil in making soap to prevent the product from being too hard.
- Coconut oil is a low melting solid. It is primarily an ester of lauric acid [$CH_3(CH_2)_{10}COOH$]. A soap made from coconut oil alone is very soluble in water and will lather even in sea water.
- Palm oil contains a very high concentration of free fatty acids, about 45% to 50% of which is oleic acid [$CH_3(CH_2)_7CH{=}CH(CH_2)_7COOH$]. It is an important constituent in toilet soaps.
- Olive oil is used in making Castile soap. It has a larger percentage (70% to 85%) of esters of oleic acid than palm oil.
- Bone grease is an animal fat of somewhat lower melting point than tallow, and it comes from a variety of sources. It is a relatively cheap source of fat. The esters of oleic acid (41% to 51%) are most prominent.
- Cottonseed oil is also a cheap source of glycerides for making soap. Its esters are mostly of linoleic acid [$CH_3(CH_2)_4(CH{=}CHCH_2)_2(CH_2)_6COOH$].

---

from one's hands, the sugar is dissolved in water and rinsed away. Many times the material to be removed is oily, and water will merely run over the surface of the oil. Since the skin has natural oils, even substances such as ordinary dirt that are not oily themselves can cover the skin in a greasy layer. The cohesive forces (forces between like molecules tending to hold them together) within the water layer are too large to allow the oil and water to intermingle (Fig. 22–13). When present in an oil-water system, soap molecules such as sodium stearate

Soap, water, and oil together form an emulsion, with the soap acting as the emulsifying agent.

$$CH_3CH_2CH_2CH_2CH_2CH_2CH_2CH_2CH_2CH_2CH_2CH_2CH_2CH_2CH_2CH_2CH_2C\overset{\displaystyle O}{\underset{\displaystyle O^-Na^+}{\big<}}$$

will move to the interface between the two liquids. The carbon chain, which is a nonpolar organic structure, will mix readily with the nonpolar grease molecules whereas the highly polar $—COO^-Na^+$ group enters the water layer because it becomes hydrated (Fig. 22–13b). The soap molecules will then tend to lie across the oil-water interface. The grease is then broken up into small droplets by agitation, each droplet surrounded by hydrated soap molecules (Fig. 22–13c). The surrounded oil droplets cannot come together again since the exterior of each droplet is covered with $—COO^-Na^+$ groups that interact strongly with the surrounding water. If enough soap and water are available, the oil will be swept away, forming a clean and water-wet surface.

TOILET SOAPS Toilet soaps generally have little or no filler and a minimal amount of free base, if any. Often much of the glycerol released in the saponification process is left in the soap. Perfumes, dyes, and medicinal agents may be added prior to casting the soap into a solid form. Floating soaps have air beaten into them as they solidify. A hard soap is obtained if a high percentage of a sodium salt of a relatively long-chain fatty acid, such as stearic acid, is present. A soft or liquid soap is obtainable by saponification with potassium hydroxide, with the liquidity increasing as the chain length of the fatty acid decreases. Fatty acids with chains as short as C_{12} or shorter are not used because the resultant soaps irritate the skin. They are

Sodium — hard soap.

Potassium — soft soap.

Ammonium — liquid soap.

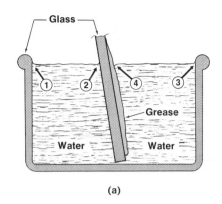

(a)

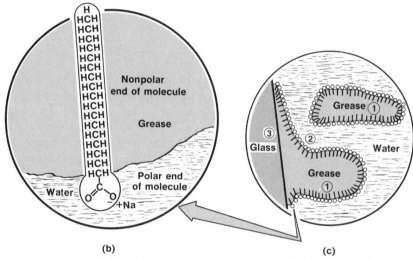

(b)   (c)

FIGURE 22–13 The cleansing action of soap. (*a*) A piece of glass coated with grease inserted in water gives evidence for the strong adhesion between water and glass at 1, 2, and 3. The water curves up against the pull of gravity to wet the glass. The relatively weak adhesion between oil and water is indicated at 4 by the curvature of the water away from the grease against the force tending to level the water. (*b*) A soap molecule, having oil-soluble and water-soluble ends, will become oriented at an oil-water interface such that the hydrocarbon chain is in the oil (with molecules that are electrically similar, nonpolar) and the COO$^-$Na$^+$ group is in the water (highly charged polar groups interacting electrically). (*c*) In an idealized molecular view, a grease particle, 1, is surrounded by soap molecules, which in turn are strongly attracted to the water. At 2 another droplet is about to break away. At 3 the grease and clean glass interact before the water moves between them.

more volatile and create an odor problem. Fatty acids with chains longer than stearic acid (18 carbon atoms) tend to give very insoluble soaps.

SHAMPOOS Shampoos are often mixtures of several ingredients designed to satisfy a number of requirements. In addition to soaps, condensation products from diethanolamine and lauric acid are used often. These are essentially a type of detergent obtained by the following reaction:

$$HN(CH_2CH_2OH)_2 + CH_3(CH_2)_{10}COOH \rightarrow CH_3(CH_2)_{10}\overset{\overset{O}{\|}}{C}-\overset{\overset{H}{|}}{N}(CH_2CH_2OH)_2$$

DIETHANOLAMINE   LAURIC ACID   AN AMIDE DETERGENT

Some shampoos contain anionic detergents, which are less damaging to the eyes than cationic detergents. Sodium lauryl sulfate is an example of an anionic detergent.

$$CH_3(CH_2)_{11}OSO_3^-Na^+$$
SODIUM LAURYL SULFATE

The hair is more manageable and has a better sheen if all the shampoo is removed. An anionic detergent can be removed by rinsing with a cationic detergent (about a 1% solution), which neutralizes the anions and facilitates their removal. Caution should be exercised with rinses, since cationic detergents are damaging to the eyes. Types of cationic detergents are described in the next section.

Shampoos also contain compounds to prevent the calcium or magnesium ions in hard water from forming a precipitate; EDTA is often used for this purpose. Lanolin and mineral oil are often added to keep the scalp from drying out and scaling. The presence of these oils is indicated by a cloudy appearance.

Synthetic Detergents

Synthetic detergents (**"syndets"**) are derived from organic molecules that have been designed to have the same cleansing action, but less reaction than soaps with the cations found in hard water, such as Ca^{2+}, Mg^{2+} and Fe^{3+}. As a consequence, synthetic detergents are more effective in hard water than soap, which gives a precipitate in the presence of Ca^{2+}, Mg^{2+}, or Fe^{3+} ions. Since such precipitates have no cleansing action and tend to stick to laundry, their presence is very undesirable.

There are many different synthetic detergents on the market. Their molecular structure consists of a long oil-soluble (hydrophobic) group and a water-soluble (hydrophilic) group. The hydrophilic groups include the sulfate ($-OSO_3-$), sulfonate ($-SO_3-$), hydroxyl ($-OH$), ammonium ($-NH_3^+$), and phosphate $[-OPO(OH)_2]$ groups.

$$R-O-\overset{\displaystyle O}{\underset{\displaystyle O}{\overset{\uparrow}{\underset{\downarrow}{S}}}}-O^-$$
SULFATE GROUP

$$R-\overset{\displaystyle O}{\underset{\displaystyle O}{\overset{\uparrow}{\underset{\downarrow}{S}}}}-O^-$$
SULFONATE GROUP

The early synthetic detergents were mostly sodium alkyl sulfate. The preparation of sodium lauryl sulfate is given here to illustrate the chemical processes involved. The principal starting material is a suitable vegetable oil, such as cottonseed oil or coconut oil. The first step is hydrogenation:

$$\begin{matrix} RCOOCH_2 \\ | \\ RCOOCH \\ | \\ RCOOCH_2 \end{matrix} + 6\,H_2 \xrightarrow{\text{Catalyst}} 3\,RCH_2OH + \begin{matrix} CH_2OH \\ | \\ CHOH \\ | \\ CH_2OH \end{matrix}$$

COCONUT OIL   HYDROGEN   [MAINLY LAURYL ALCOHOL, $CH_3(CH_2)_{11}OH$]   GLYCEROL

The second step involves esterification of the $-OH$ group on the end of the lauryl alcohol hydrocarbon chain. This is accomplished by treating the lauryl alcohol with sulfuric acid:

$$CH_3(CH_2)_{11}OH + H_2SO_4 \rightarrow CH_3(CH_2)_{11}OSO_3H + H_2O$$
LAURYL ALCOHOL   SULFURIC ACID   LAURYL HYDROGEN SULFATE

The final step involves neutralizing the acidic lauryl hydrogen sulfate with sodium hydroxide:

$$CH_3(CH_2)_{11}OSO_3H + NaOH \rightarrow CH_3(CH_2)_{11}OSO_3^-Na^+ + H_2O$$
SODIUM LAURYL SULFATE

Other synthetic detergents (also called **"surfactants,"** from surface-active agents) are the alkylbenzenesulfonates. They are prepared by putting large alkyl groups on a benzene ring and then sulfonating the benzene ring with sulfuric acid. Before use, they are transformed into their sodium salts. The reactions involved are:

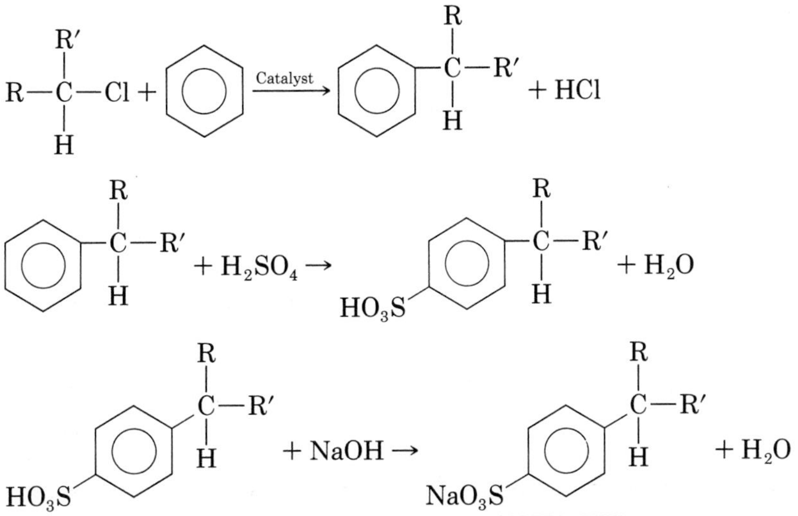

$$R-\underset{\underset{H}{|}}{\overset{\overset{R'}{|}}{C}}-Cl + \bigcirc \xrightarrow{\text{Catalyst}} \bigcirc\underset{\underset{H}{|}}{\overset{\overset{R}{|}}{C}}-R' + HCl$$

$$\bigcirc\underset{\underset{H}{|}}{\overset{\overset{R}{|}}{C}}-R' + H_2SO_4 \rightarrow HO_3S\bigcirc\underset{\underset{H}{|}}{\overset{\overset{R}{|}}{C}}-R' + H_2O$$

$$HO_3S\bigcirc\underset{\underset{H}{|}}{\overset{\overset{R}{|}}{C}}-R' + NaOH \rightarrow NaO_3S\bigcirc\underset{\underset{H}{|}}{\overset{\overset{R}{|}}{C}}-R' + H_2O$$

SODIUM ALKYL-
BENZENESULFONATE

All surfactants consist of a long hydrophobic chain and a highly polar group that interacts strongly with water.

In addition to the anionic (negatively charged) synthetic detergents already described, there are also detergents in which the polar group at the end of the hydrocarbon chain is positive or neutral.

Cationic (positively charged) detergents are almost all quaternary ammonium halides

$$R_1-\underset{\underset{R_4}{|}}{\overset{\overset{R_2}{|}}{N^+}}-R_3 \quad X^-$$

where one of the R groups is a long hydrocarbon chain and another frequently includes an —OH group. In these the water-soluble portion is positively charged; so they are sometimes called invert soaps (in soaps the water-soluble portion is negatively charged). They are prepared by treating the appropriate amine with an alkyl chloride:

$$R_1-\underset{\underset{R_4}{|}}{\overset{\overset{R_2}{|}}{N}}-R_3 + R_4Cl \rightarrow R_1-\underset{\underset{R_4}{|}}{\overset{\overset{R_2}{|}}{N^+}}-R_3 \quad Cl^-$$

Cationic detergents frequently exhibit pronounced bactericidal qualities. Cationic detergents are incompatible with anionic detergents. When they are brought together, a high molecular-weight insoluble salt precipitates out, and this has none of the desired detergent properties of either starting material:

Cationic detergents act as disinfectants.

$$R_1\!-\!\underset{\underset{R_4}{\mid}}{\overset{\overset{R_2}{\mid}}{N}}\!-\!R_3^+ \quad Cl^- + Na^+ \ {}^-O_3SOR_5 \rightarrow R_1\!-\!\underset{\underset{R_4}{\mid}}{\overset{\overset{R_2}{\mid}}{N}}\!-\!R_3^+ \ {}^-O_3SR_5 + Na^+ + Cl^-$$

CATIONIC DETERGENT   ANIONIC DETERGENT   PRECIPITATE

Nonionic detergents have a polar, but not an ionic, grouping attached to a large organic grouping of low polarity. A typical example is a material prepared by the reaction of an organic acid with ethylene oxide:

$$RCOOH + (x + 2)CH_2\!\!\underset{\underset{O}{\diagdown\diagup}}{-}\!\!CH_2 \rightarrow R\!-\!\overset{\overset{O}{\parallel}}{C}\!-\!O\!-\!(CH_2)_2O(CH_2CH_2O)_xCH_2CH_2OH$$

In a typical nonionic detergent, $R = C_{12}H_{25}$ and $x = 2$. The large number of weakly polar $C\!-\!O\!-\!C$ bonds has an effect similar to that of a single ionic group, and this end of the molecule provides the water solubility.

The nonionic detergents have several advantages over ionic detergents. Since they contain no ionic groups, they cannot form salts with calcium and magnesium ions and consequently are unaffected by hard water. For the same reason, nonionic detergents do not react with acids and may be used even in strong acid solutions.

In general, the nonionic detergents foam less than ionic surface active agents, a property which is desirable where nonfoaming detergents are required, as in dishwashing. Nonionics do suffer from one drawback. They cannot be dried to solid powders. They are heavy liquids with melting points below room temperature. Consequently, nonionic detergents are almost always available in liquid form.

FILLERS OR BUILDERS A number of materials are added to soap powders for laundry purposes. These materials are often quite basic, and their addition gives the soap a greater detergent action. Commonly added materials include sodium carbonate, sodium phosphates, sodium polyphosphates, and sodium silicate. Rosin neutralized with sodium hydroxide is also commonly added to laundry soaps in large amounts. The rosin is mostly abietic acid. The neutralized acid has the nonpolar (hydrocarbon) part and polar end required for a soap. Such soaps are not to be recommended for use on the human skin. Phosphates, carbonates, and silicates hydrolyze to give OH^- ions, which react with grease to make soaps.

$$PO_4^{3-} + H_2O \rightarrow HPO_4^{2-} + OH^-$$
$$CO_3^{2-} + H_2O \rightarrow HCO_3^- + OH^-$$
$$SiO_3^{2-} + H_2O \rightarrow HSiO_3^- + OH^-$$

ABIETIC ACID

Hard water contains metal ions that react with soaps and give precipitates.

Soft water—less than 65 mg of metal ion per gallon
Slightly hard—65–228 mg
Moderately hard—228–455 mg
Hard—455–682 mg
Very hard—above 682 mg

Builders also assist in negating the effect of the ions (Mg^{2+}, Ca^{2+}, Fe^{3+}) that cause hard water. Since the phosphate, carbonate, and hydroxide compounds of these ions are insoluble, these ions are precipitated. It is fortunate that these precipitates are powdery and easily rinsed away. In contrast, the soap precipitates form scum that sticks to the material being washed.

$$HCO_3^- + OH^- \rightarrow CO_3^{2-} + H_2O$$
$$Ca^{2+} + CO_3^{2-} \rightarrow CaCO_3\!\downarrow$$
$$Mg^{2+} + 2\,OH^- \rightarrow Mg(OH)_2\!\downarrow$$

OTHER CLEANSERS A very large number of special cleaners or cleansing agents is available. Simple abrasive cleansers contain a large percentage of an abrasive such as silica (SiO_2) or pumice (65% to 75% SiO_2, 10% to 20% Al_2O_3), a variable amount of soap, and generally some polyphosphates. They may also contain some synthetic detergent and a bleaching agent. All-purpose solid cleansers may contain one or more of a variety of salts which react with water to produce a basic solution: trisodium phosphate, sodium carbonate, sodium bicarbonate, sodium pyrophosphate, or sodium tripolyphosphate, plus a detergent and perhaps pine oil to give an attractive odor. Metal cleansers may contain strong acid or strong base to dissolve impurities. Many cleaning liquids contain organic solvents such as perchloroethylene, 1,1,1-trichloroethane, and the like. The vapors of these are quite toxic so the cleansers must be used in a ventilated area.

Soaps containing pumice (finely powdered volcanic ash) will wash out ground-in dirt.

WHITER WHITES Bleaching agents are compounds that are used to remove color from textiles. Most commercial bleaches are oxidizing agents such as sodium hypochlorite. Optical brighteners are quite different, since they act by converting a portion of the invisible ultraviolet light, which impinges on them, into visible blue or blue-green light, which is emitted. Together or separately, these two classes of compounds find their way into commercial laundry and cleaning preparations, since they seem to be making clothes cleaner.

In earlier times textiles were bleached by exposure to sunlight and air. In 1786, the French chemist Berthollet introduced bleaching with chlorine, and subsequently this process was carried out with sodium hypochlorite, an oxidizing agent prepared by passing chlorine into aqueous sodium hydroxide:

$$2\,Na^+ + 2\,OH^- + Cl_2 \rightarrow \underset{\text{SODIUM HYPOCHLORITE}}{Na^+ + OCl^-} + Na^+ + Cl^- + H_2O$$

Shortly after this, hydrogen peroxide was introduced as a textile bleach. Later, a number of other oxidizing agents based on chlorine were developed and introduced.

One way to decolorize materials is to remove or immobilize those electrons in the material which are activated by visible light. The hypochlorite ion is capable of removing electrons from many colored materials. In this process, the hypochlorite ion is reduced to chloride and hydroxide ions:

Chlorine produces hypochlorite when it reacts with water:
$H_2O + Cl_2 \rightarrow$
$HOCl + HCl$

$$ClO^- + H_2O + 2\,e^- \rightarrow Cl^- + 2\,OH^-$$

As stated previously, optical brighteners are compounds that transform incident ultraviolet light into emitted visible light; this is a type of fluorescence. When optical brighteners are incorporated into textiles or paper, they make the material appear brighter and whiter (Fig. 22–14).

An example of such a brightener has this structure, and its absorption and emission spectra are presented in outline form in Figures 22–14 and 22–15.

A fluorescent material absorbs shorter wavelength light and emits light of a longer wavelength.

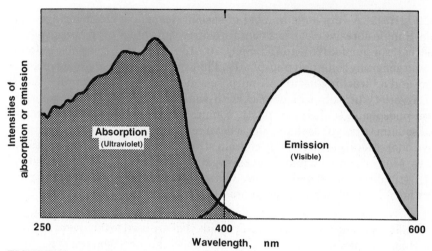

FIGURE 22–14   Absorption and emission spectra of a typical optical brightener.

Spot and Stain Removers

To a large extent, stain removal procedures are based on solubility patterns or chemical reactions. Many stains, such as those due to chocolate or other fatty foods, can be removed by treatment with the typical dry-cleaning solvents such as tetrachloroethylene, $Cl_2C\!\!=\!\!CCl_2$.

Stain removers for the more resistant stains are almost always based on a chemical reaction between the stain and the essential ingredients of the stain remover. A typical example is an iodine stain remover, which is simply a concentrated solution of sodium thiosulfate. The reaction here is

$$I_2 \;\; + 2\,Na_2S_2O_3 \rightarrow \underbrace{2\,NaI + Na_2S_4O_6}$$

IODINE   SOLUBLE IN WATER
                                       (COLORLESS)

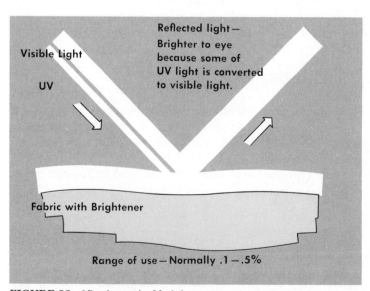

FIGURE 22–15   An optical brightener converts ultraviolet energy to visible light; hence, more light can be detected by the eye.

TABLE 22–1   Some Common Stains and Stain Removers*

STAIN	STAIN REMOVER
Coffee	Sodium hypochlorite
Lipstick	Isopropyl alcohol, isoamyl acetate, Cellosolve ($HOCH_2CH_2OCH_2CH_3$), chloroform
Rust and ink	Oxalic acid, methyl alcohol, water
Airplane cement	50/50 amyl acetate and toluene or acetone
Asphalt	Benzene or carbon disulfide
Blood	Cold water, hydrogen peroxide
Berry, fruit	Hydrogen peroxide
Grass	50/50 amyl acetate and benzene or sodium hypochlorite or alcohol
Nail polish	Acetone
Mustard	Sodium hypochlorite or alcohol
Antiperspirants	Ammonium hydroxide
Perspiration	Ammonium hydroxide, hydrogen peroxide
Scorch	Hydrogen peroxide
Soft drinks	Sodium hypochlorite
Tobacco	Sodium hypochlorite

* Before any of these stain removers are used on clothing, the possibility of damage should be checked on a portion of the cloth that ordinarily is hidden.

Some stain removers, such as benzene and chloroform, are suspected carcinogens, and some are toxic, for example, methanol and carbon disulfide.

Many stains can be removed by an appropriate solvent or chemical reagent.

Iron stains are removed by treatment with oxalic acid, which forms a soluble coordination compound with the iron:

$$Fe_2O_3 + 6\ H_2C_2O_4 \rightarrow \underbrace{3\ H_2O + 2\ Fe(C_2O_4)_3^{3-} + 6\ H^+}_{\text{SOLUBLE IN WATER}}$$

OXALIC ACID

Citric acid and tartaric acid will also remove iron stains and are less toxic than oxalic acid.

Mildew stains can be removed by hydrogen peroxide or laundry bleach (sodium hypochlorite), which oxidizes the fungus responsible for the mildew. Blood stains on cotton can be removed by hypochlorite solution. Bleach should not be used on wool because it reacts chemically with the nitrogen atoms present in the peptide chains. The chemicals used to remove a few common stains are listed in Table 22–1.

TOOTHPASTE

The structure of tooth enamel is essentially that of a stone, a stone composed of calcium carbonate and calcium hydroxy phosphate (apatite) (Fig. 22–4). Such structures are readily attacked by acid. Since the decay of some food particles produces acids and since bacteria will convert plaque, a deposit of dextrins, to acids, it is important to keep teeth clean and free from prolonged contact with acids if the hard, stonelike enamel is to be preserved.

The two essential ingredients in toothpaste are a detergent and an abrasive. The abrasive serves to cut into the surface deposits, and the detergent assists in suspending the particles in a water medium to be carried away in the rinse. Abrasives commonly used in toothpaste formulations include hydrated silica (a form of sand, $SiO_2 \cdot nH_2O$); hydrated alumina, $Al_2O_3 \cdot nH_2O$; and calcium carbonate, $CaCO_3$. It is a difficult choice to select an abrasive hard enough to cut the surface contamination and yet not so hard the abrasive will cut the tooth enamel. The choice of detergent is easier; any good detergent such as sodium lauryl sulfate will do quite well.

Since the necessary ingredients in toothpaste are not very palatable, it is not surprising to see the inclusion of flavors, sweeteners, thickeners, and colors to appeal to our senses.

One addition to the toothpaste mixture has made a significant difference in the amount of tooth decay in our population; it is the addition of stannous fluoride, SnF_2, to provide a low level of fluoride ion concentration in the brushing medium. The fluoride ion actually replaces the hydroxide ions in the hydroxyapatite structure, $Ca_{10}(PO_4)_6(OH)_2$ (Fig. 22–4), to form fluoroapatite, $Ca_{10}(PO_4)_6F_2$. The fluoride ion forms a stronger ionic bond in the crystalline structure, and as a result, the fluoroapatite is harder and less subject to acid attack than the hydroxyapatite. Hence, there is less tooth decay. The fluoride ion is also introduced into drinking water on a wide-scale basis for this same purpose.

Most teeth are lost as a result of gum disease, which results from the lack of proper massage, deposits below the gum line, and bacterial infection in these deposits. More attention is being given to toothpastes containing disinfectants such as peroxides in addition to the soap and abrasive.

SELF-TEST 22–C

1. A fat is a tri-ester of glycerol and _____ acids.
2. To make soap, a fat is treated with _____.
3. Is the acid or salt group in a soap molecule more soluble in water or in oil? _____
4. What oil is used in making Castile soap? _____
5. What makes floating soap float? _____
6. Which is more likely to precipitate the hard-water ions (Ca^{2+}, Mg^{2+}, Fe^{3+}) as a sticky precipitate, traditional soaps or synthetic detergents? _____
7. *Surfactant* is a short term for _____ _____ _____.
8. Why should a cationic detergent not be mixed with an anionic detergent in a laundry blend? _____
9. Which foams the least, cationic, anionic, or nonionic detergents? _____
10. Give two purposes for adding a detergent builder to a laundry product. _____ and _____
11. Optical brighteners transform ultraviolet light into _____ light.

12. What are the two fundamental ingredients in a toothpaste? _____ and _____

13. A compound of what element is added to toothpaste to replace some of the hydroxide ions in apatite? _____

MATCHING SET

_____ **1.** Keratin

_____ **2.** Melanin

_____ **3.** Sodium lauryl sulfate

_____ **4.** Polyvinylpyrrolidone

_____ **5.** Alcohol

_____ **6.** Hydrated aluminum chloride

_____ **7.** Soap

_____ **8.** Fat

_____ **9.** Cholesterol

_____ **10.** Thioglycolic acid

_____ **11.** *p*-aminobenzoic acid

_____ **12.** Glycerol

_____ **13.** Sodium tripolyphosphate

_____ **14.** Whiteners

_____ **15.** Sodium hypochlorite

_____ **16.** Calcium carbonate

_____ **17.** Tin (II) fluroide (stannous fluoride)

a. Salt of fatty acid
b. Reducing agent in wave lotion
c. Hair spray resin
d. Abrasive in toothpaste
e. Holds moisture in skin
f. Common alcohol in lanolin
g. Ultraviolet absorber in suntan lotion
h. Alcohol produced in saponification of fat or oil
i. Detergent builder
j. Radiates a different wavelength of light than that absorbed
k. A synthetic detergent
l. Laundry bleach
m. Dark pigment
n. Deodorant component
o. Furnishes fluoride for stronger teeth
p. Dehydrates skin microbes
q. Skin and hair protein

QUESTIONS

1. You read in the newspaper about a new compound that will break disulfide bonds in proteins. What potential use might it have?

2. a. What is the purpose of an emulsifier?
  b. In which of the following cosmetics is an emulsifier important: suntan lotion, hair spray, cold cream?

3. Which one of each of the following pairs of properties would be appropriate for a hair spray propellant? Why?
  a. High or low boiling point
  b. Soluble or insoluble in the active ingredients
  c. Capable or incapable of chemical reaction with the active ingredients
  d. Odorous or odorless
  e. Toxic or nontoxic

4. Describe an ionic bond that holds protein chains together.

5. What is the purpose of each of the following?
  a. Detergent in toothpastes
  b. Polyvinylpyrrolidone in hair sprays
  c. Aluminum chloride in deodorants
  d. *p*-aminobenzoic acid in suntan lotion

6. What specific substance is broken down during the bleaching of hair?

7. Use the structures of the constituents of lanolin to justify its ability to emulsify face creams.

8. Hydrogen bonding is a very handy theoretical tool. Name three applications of hydrogen bonding in cosmetics and cleansing agents.

9. If you were going to formulate a suntan lotion, what particular spectral property would you look for in choosing the active compound?

10. Why are detergents better cleansing agents than soaps in regions where the water supply contains calcium or magnesium salts?

11. Why is a soap from coconut oil more soluble in water than a soap made from palm oil?

12. Suggest ways of removing each of the following from clothing:
    a. Motor oil
    b. Iodine stain
    c. Lard
    d. Copper sulfate

13. Explain why vinegar is able to remove some stains that are soluble in weak acids.

14. Name and give the functions for four chemical ingredients in perfume.

15. Explain why Grandma's lye soap produced rough, red hands.

16. Explain how an optical brightener in a detergent works.

17. What do skin, hair, and nails have in common?

18. What is the major difference between keratin in the hair and other proteins?

19. Why do the lips dry so easily?

20. What is the structure of the monomer unit in polyvinylpyrrolidone?

21. Describe in chemical terms what happens when a person gets a permanent.

22. What three types of chemical bonds hold hair proteins together?

23. Is hair curl the result of primary, secondary, or tertiary protein structure?

24. What happens to the solar energy absorbed by *p*-aminobenzoic acid in a suntan lotion or oil?

25. If a substance is astringent, what is its action?

26. What is the purpose of talc in face powder?

27. Name an astringent widely used in deodorants.

28. What is the major ingredient in lipstick?

29. What is the chemical action of hydrogen peroxide on hair?

30. Commercial lanolin comes from what animal?

31. A fat is a(n): (a) acid, (b) alcohol, (c) ester, (d) alkane.

32. Vegetable oils can be used as well as animal oils to make soap. True or False

33. Is the hydrocarbon end of the soap molecule polar or nonpolar? Explain.

34. Which is more soluble in water, calcium stearate or sodium stearate?

35. The oxide stains of what metal can be removed with a solution of oxalic acid?

36. How important is advertising in the sale of cosmetics? What part does advertising play in consumer satisfaction?

Chapter 23

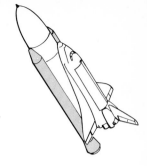

TRANSPORTATION CHEMISTRY — FROM HERE TO THERE AND QUICKLY ABOUT IT!

Transportation has long been an important factor in our lives. Our ancestors had to move to discover and settle our nation. We depend on transportation today to provide us with our food, clothing, and certain aspects of our social lives. We have become a nation of transporters and travelers. We use jet aircraft to go to jobs and vacations and to send packages to another city overnight. We use automobiles in every conceivable way. Over 200 million automobiles are on our nations' highways. We can go virtually anywhere we want to go any time we want using automobiles, buses, trains, planes, and boats. We are also beginning to venture away from our home planet using rockets. Whether travel into space will progress as rapidly as surface transportation has developed remains to be seen. If the rapid development of air travel during this century can be used as an indicator, many of us may live to see travel to the moon and beyond become quite popular. Don't forget that only 66 years passed between the very first flight (Wright brothers — 1903) and the touch-down on the moon (Armstrong — 1969).

Most developments in transportation have been connected to advances in science and technology during the past century. The discovery of new alloying methods has led to high-strength alloys. The discovery of petroleum has led to plastics and fuels. The discoveries in the chemistry of elements like silicon have led to microelectronic devices (chips), which function in numerous ways in transportation from controlling traffic lights and giving eyes to the air traffic controller to metering the flow of the fuel-air mixture into the automobile engine.

The advances have not been without a price, however. We are now psychologically dependent on readily available transportation. Our lives are dependent on it. We are "hooked." We pay the price in the showroom, at the gas pump, at the ticket counter, and at the tax assessor's office. There are so many aspects to transportation that whole books are written on specific areas of transportation. Classes are taught in college, and one can major in several specific aspects of transportation.

This chapter will discuss how chemistry plays a role in two types of transportation, the automobile — because it represents surface transportation that is primarily dependent on air oxidation of fuels — and the rocket (or spacecraft) — because it represents transportation of the future and because its chemistry offers an interesting extension of what we have learned about oxidizing agents and reducing agents (fuels). Aircraft really represent only a special case of surface transportation since it is the physics of the lift phenomenon that makes the aircraft different from the automobile, not the type of engine or construction details.

THE AUTOMOBILE

The automobile is an important part of most Americans' lives. Even city dwellers who may not own a personal automobile can relate to the ease of transportation and the flexibility the automobile offers over mass transit. Since the automobile consumes so much of our time and income, it is worth considering those factors that relate chemistry to this mode of transportation. Armed with a better understanding of how the automobile works, how it is made, and how it can be maintained, we can make better decisions about this important aspect of our lives — our personal transportation.

THE COSTS OF A CAR

Most of us eventually come to realize that an automobile costs money to operate as well as to purchase. The automobile is an investment that usually begins to decline in value as time passes. This is caused by factors beyond our control such as styling changes and buyer acceptance. Other factors such as rust, cleanliness, and wear on moving parts are more within our control. Operating costs are also within our control. The factors that affect operating costs (what we pay out for fuel to get us there and back) are as follows:

1. Weight of the automobile — Do you like large cars?
2. Engine characteristics — Fast engines burn more fuel.
3. Drive-train — Do you want to shift gears?

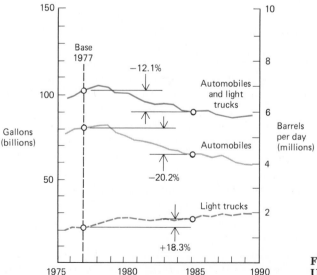

FIGURE 23–1 Estimated U.S. fuel economy for automobiles and light trucks.

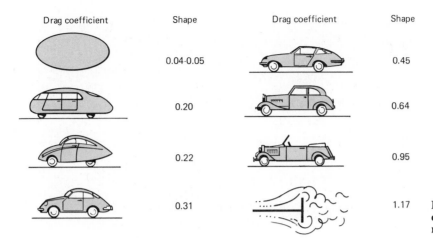

Drag coefficient	Shape	Drag coefficient	Shape
0.04-0.05		0.45	
0.20		0.64	
0.22		0.95	
0.31		1.17	

FIGURE 23–2 How drag coefficient is related to design.

4. Aerodynamics — Styling may be everything.
5. Rolling resistance — Are fancy tires worth it?
6. Driving cycle — All town or all city?
7. Driving habits — Quick start your bag?

Of these seven factors, the first five are directly controllable by automobile designs. Since 1977, a 20% decrease in fuel consumption by the American driving public has resulted from decreased weight by downsizing automobiles and by using lighter materials along with increased engine efficiency and improved aerodynamics. This is shown in Figure 23–1.

DRAG

Automobiles have undergone radical design changes over the years. Many of these have been for the purpose of reducing aerodynamic drag. In 1933, W. E. Lay at the University of Michigan showed that a rectangular box with wheels on it had a drag coefficient of 0.86. By rounding the edges of the box, he lowered the drag coefficient to 0.46. By comparison, a flat plate moving upright through the air has a drag of 1.17. Figure 23–2 shows how the drag coefficient is related to design.

Figure 23–3 shows the effects of drag coefficient on fuel economy for several speeds — the higher the speed, the greater the resistance to air flow past the automobile body and hence the more fuel consumed. Figure 23–4 shows an old automobile with a relatively high drag coefficient.

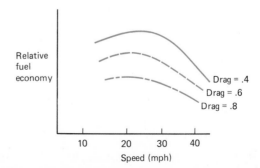

FIGURE 23–3 The effects of drag coefficient and speed on fuel economy.

FIGURE 23–4 A 1939 Chrysler with a drag coefficient of about 0.60. (The Bettmann Archive)

OUR DRIVING HABITS

A factor having perhaps the greatest influence on our car's fuel economy is our driving habits. The typical trip taken in an automobile is surprisingly short. Data taken by the U.S. Environmental Protection Agency (EPA) show this fact (Fig. 23–5).

When trip length is related to the fuel economy of cold-started automobile engines (Fig. 23–6), it becomes apparent that fuel is wasted on short trips. When the engine is cold, frictional effects are greatest and the thermal efficiency of the engine is lower than when the engine is at its designed operating temperature. Note also in Figure 23–6 that automobile engines reach their maximum economy more quickly when the weather is warm rather than cold. All these factors mean that winter fuel economy is going to be less than that for spring, summer, or fall.

CHANGES IN THE MATERIALS USED IN AUTOMOBILES

When automobiles were first made, they were actually motorized carriages, made of wood and steel. As automobiles became loaded with passengers, more attention was paid to performance, passenger comforts, appearance, and durability, and new

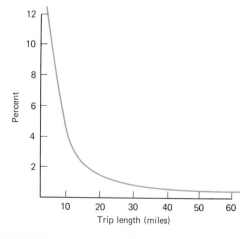

FIGURE 23–5 Trip frequency compared with trip length in the United States.

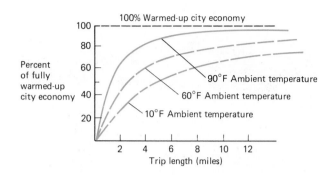

FIGURE 23–6 Fuel economy and operating temperature.

materials began to replace the old materials. In the 1920s and 1930s automobiles had powerful engines to move the massive machines around. Typical weights of limousines and touring cars were in the 4000- to 8000-pound range (2 to 4 tons). Engine efficiencies gave acceptable fuel economies, especially when gasoline prices were low. In the 1950s and 1960s automobiles became somewhat like rolling living rooms with air conditioning, power seats, power windows, and plush carpeting, all of which added weight, used engine power, and thus lowered fuel economy. Gasoline use continued to increase as more automobiles crowded America's highways. This increasing number of automobiles was accompanied by decreasing oil production in this country, and by 1972, U.S. oil imports totaled $5 billion.

In 1973 the member countries of the Oil Producing and Exporting Countries (OPEC) embargoed oil shipments to the United States and by 1980 the cost of imported oil had risen to $80 billion, or roughly $1000 for every American household. Something had to be done. The V-8 gasoline engine practically disappeared, diesel automobiles became popular, small imported automobiles captured a large portion of the U.S. market, and American automobiles began downsizing to reduce the weight.

Today, a typical automobile contains a different mix of metals, glass, rubber, and plastics than did a typical automobile of 20 years ago (Table 23–1).

Besides downsizing of an automobile, weight savings have also been achieved by using materials with greater strength characteristics per pound. High-strength alloy steels have been used rather than cold rolled steel in order to get more strength per pound. Additional weight savings have been achieved by using

TABLE 23–1 Composition of Typical Automobiles in 1966 and 1986

MATERIAL	1966 (%)	1986 (%)
Steel	47	15
High-strength steel	10	30
Cast iron	20	15
Aluminum	5	15
Thermosetting plastics	3	10
Plastics	2	2
Rubber	2	2
Glass	2	2
Zinc	5	5
Miscellaneous	4	4

FIGURE 23–7
Aluminum alloy auto-
mobile body developed
jointly by Audi and
Alcoa.

aluminum alloys. Figure 23–7 shows an automobile body that weighs 48% less than a comparable steel body. In addition, the aluminum body is 13.8% stiffer than the steel body. Although aluminum requires more energy to produce than steel and therefore costs more, automobile manufacturers have calculated that the additional cost of aluminum over steel would be recovered in about 15,000 miles of driving with the increased fuel economy obtained with the lower vehicle weight.

Plastics and other polymers are used in automobiles for both weight savings and cosmetic purposes. In addition, there are some places where no other material will function. For example, automobiles with front-wheel drive use a rubber boot to protect the constant velocity (CV) joint (Fig. 23–8). Most CV boots are made of polychloroprene (neoprene) rubber, and some are made of silicone rubber. These boots must be flexible at both high and low temperatures, resistant to grease, and resistant to weathering. In 1985, over 100 million CV joint boots were installed using about 13 million pounds of these polymers. This is just one part in a list of hundreds of polymeric parts used in a modern automobile. Figure 23–9 shows other locations where polymers are used in automobiles. Table 23–2 gives the names and uses of some common polymers used in automobiles.

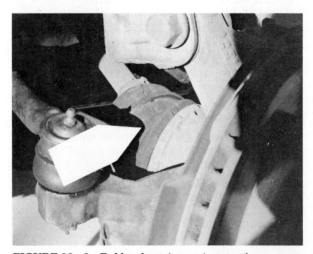

FIGURE 23–8   Rubber boot *(arrow)* protecting a constant-velocity joint of a front wheel drive automobile. This rubber boot is shown with the wheel and tire removed. These boots should remain flexible for the life of the car. The rubber is compounded to withstand extremes in temperature as well as chemical attack. Should the boot crack and allow entry of dirt and moisture, lubrication of the joint would diminish, causing the joint to fail.

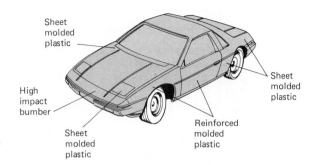

FIGURE 23–9 How polymers are used in automobiles.

SELF-TEST 23–A

1. Which would probably not improve fuel economy? (a) lighter vehicle weight, (b) decreased aerodynamic drag, (c) increased trip length, (d) high-speed driving _____

2. An automobile should get better city fuel economy in warm weather. ( ) True or ( ) False

3. Name two polymers that are used in automobiles to reduce weight. _____ and _____

4. Which would give better fuel economy, city driving or highway driving? _____

5. Which has a lower drag coefficient, a box or an egg? _____

AUTOMOTIVE ENGINES AND FUELS

Historically, automobiles have been powered by batteries, internal combustion engines (gasoline and diesel) or steam engines. Of these types, only steam power has fallen into disuse. Most automobiles are currently powered by gasoline-burning engines, but there are vast numbers of diesel and battery-powered vehicles world-wide. Each of these power plants has its good and bad features.

Gasoline engines burn mixtures of gasoline and air that are ignited by a high-voltage spark (Fig. 23–10). Ignition timing as well as valve timing is important for proper performance. Gasoline engines are more complex than diesel or steam engines, which means that they are more expensive to build and repair, but gasoline engines have good cold weather starting characteristics, rapid power output, and good throttle response. These factors endeared gasoline engines to the American driving public.

TABLE 23–2 Some Polymers Used in Automobiles

CHEMICAL NAME	USES
Epichlorohydrin polymer	MacPherson strut bushings
Polychloroprene	Boots and bellows
Terpolymer of ethylene, propylene, and butadiene	Window seals, bumpers
Cis-polyisoprene	Body mounts, tires
Silicone rubber	Fuel lines, ignition wires
Fluoropolymer	Valve stem seals, fuel lines

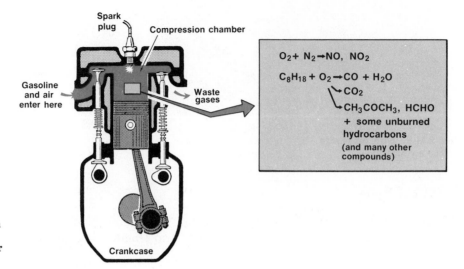

FIGURE 23–10
Diagram of combustion chamber and some of the reactions that occur in it.

Preignition occurs when the gasoline air mixture ignites on the compression stroke prior to the spark igniting the mixture. In some cases the explosion occurs early enough to retard severely the upward movement of the piston, with a knocking sound being produced in the engine.

Review the discussion of octane rating in Chapter 12.

Early gasoline engines suffered from poor performing fuel that was composed of straight cuts of petroleum crude without the benefits of catalytic reforming (see Chapter 12). These simple gasolines burned with severe preignition and detonation in the early engines. As a result, the early automobiles sputtered, lurched, and backfired a great deal.

In 1904 Professor Bertrum Hopkinson of Cambridge University showed that hot surfaces within the combustion chamber caused detonation, which in turn prematurely raised the temperature of the entire combustion chamber. This phenomenon was the cause of *preignition* or *knocking*. One of Hopkinson's students, H. R. Ricardo built a research engine in which he described the "antidetonant" characteristics of pure benzene. Benzene is used today in modern gasolines to improve combustion characteristics even though it is both toxic and carcinogenic. From these studies on preignition and knocks came the **octane rating** system for motor fuels.

In 1921 Thomas Midgley discovered that tetraethyl lead had strong antiknock characteristics in gasoline blends. When tetraethyl lead burns,

$$C_2H_5-\underset{\underset{C_2H_5}{|}}{\overset{\overset{C_2H_5}{|}}{Pb}}-C_2H_5$$

TETRAETHYL LEAD

lead monoxide, PbO, forms as an aerosol of finely divided particles. These lead oxide particles are effective in breaking the chain reaction involving hydrocarbon free radicals during the detonation process. Since only small quantities of tetraethyl lead proved effective in raising the octane rating of a fuel, tetraethyl lead became quite popular as a fuel additive.

However, beginning in 1975, new automobiles were required to use lead-free gasoline. This requirement was a result of the addition of catalytic converters (see Fig. 23–13) to the automobile's exhaust systems in order to reduce nitrogen oxides, unburned hydrocarbons, and carbon monoxide from the exhaust emissions (Table 23–3). Since lead compounds can **poison,** or deactivate, the platinum-based cata-

TABLE 23–3 Levels of Auto Emissions in Grams per Mile

	HYDROCARBONS, HC	CARBON MONOXIDE, CO	NITROGEN OXIDES, NO_x
Prior to control	11	80	4.0
EPA standards for 1977 model cars	1.5	15.0	2.0
California standards, 1978	0.41	9.0	1.5
EPA standards for 1982 model cars	0.41	7.0	1.0

lyst in the catalytic converters, automobiles equipped with catalytic converters require unleaded gasoline. Although leaded gasolines are still available for use in older cars, EPA regulations were initiated in the 1970s to phase out leaded gasoline gradually.

A discussion of leaded gasoline can be found in Chapter 12.

The Diesel Engine

There are several other fuels that can be used in internal combustion engines. Hydrocarbons containing 6 to 16 carbon atoms make a fuel that will smoothly autoignite when the fuel is mixed with air and injected into a combustion chamber during the compression stroke. The German Rudolph Diesel discovered this type of engine, which he patented in 1892 and 1893 (Fig. 23–11). Today, diesel engines are used in automobiles because of their high efficiency and fuel economy. Because of the high operating temperatures of diesel engines, more nitrogen oxides are formed than with gasoline engines burning the same amount of fuel. Also, diesel engines produce soot particles, which are discharged into the atmosphere. These particles, being quite small, can be breathed into the deepest parts of the lungs, possibly carrying with them harmful chemicals formed in the combustion process.

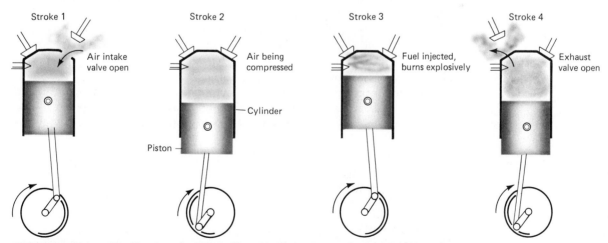

FIGURE 23–11 The diesel engine. This diagram shows how a single cylinder of a four-stroke engine operates. In stroke 1, the air intake valve opens, allowing air to enter the chamber. In stroke 2, the air is compressed and reaches a temperature of about 900°F. At the top of this stroke, oil (diesel fuel) is injected. The high temperature of the air makes the oil ignite, causing the downstroke (stroke 3). Stroke 4 removes the combustion gases when the exhaust valve is opened.

Gaseous Fuels for Automobiles

Another fuel that has been burned in internal combustion engines is natural gas. At present, natural gas is relatively abundant in this country compared with liquid petroleum. This means that natural gas enjoys an availability when petroleum supplies are interrupted. Using a nonimported fuel is also better for the U.S balance of payments. Compared with gasoline, natural gas has several disadvantages and a few advantages in addition to its availability.

Methane (CH_4), the principal component of natural gas, actually produces more energy per gram than octane (C_8H_{18}), a component of gasoline, produces.

$$CH_4 + 2\ O_2 \rightarrow CO_2 + 2\ H_2O + 13.3 \text{ kcal/g}$$
$$C_8H_{18} + 25/2\ O_2 \rightarrow 8\ CO_2 + 9\ H_2O + 11.8 \text{ kcal/g}$$

Thus it would appear that methane (natural gas) would be a better fuel. This is not the case, however, since liquefied natural gas has a lower density than gasoline and hence a lower energy content per gallon.

	Density		Energy Content
Natural gas at $-162°C$	0.42 g/mL or 1589 g/gal		21,133 kcal/gal
Gasoline at $20°C$	0.74 g/mL or 2800 g/gal	about	33,040 kcal/gal

This means that gasoline, even with its lower energy of combustion per gram compared with methane, still contains more energy per gallon since a gallon of gasoline contains more combustible material. As a result, a gallon of natural gas will move an automobile a shorter distance than a gallon of gasoline. However, the advantages of natural gas may outweigh this disadvantage. Natural gas has excellent cold-start characteristics because it is already a gas or can be so easily vaporized from the liquid state. In addition, the small hydrocarbon molecules in natural gas have good antiknock properties and burn cleanly, which results in low exhaust emissions. The hydrocarbon molecules in natural gas also show low chemical reactivity when exhausted, unburned, into the atmosphere. Natural gas may be stored and transported as an automotive fuel as either liquefied natural gas (LNG) or compressed natural gas (CNG).

LNG must be stored as a cryogenic fluid at $-162°C$. Like all cryogenic cylinders, LNG containers must be vented, which makes storage of LNG hazardous. In addition, the low temperature of LNG requires more training for its safe handling than for handling normal liquid fuels.

CNG is contained in high-pressure cylinders. These can be constructed of low-density metals like aluminum. While the high pressures associated with this form of storage may seem dangerous, these cylinders are actually stronger than the

All gases can be liquefied eventually to liquids by cooling and (or) applying pressure. Gases that must be cooled to less than $-150°F$ to bring about their liquefaction are called **cryogens,** from the Greek *kyros,* meaning "icy cold."

TABLE 23–4   Comparison of Gasoline, LNG, and CNG as Test Fuels in a 2900-Pound Automobile

GASOLINE	LNG	CNG
Volatile liquid at ambient temperature	Cryogenic liquid at $-162°C$	Compressed gas at ambient temperature
18-gallon tank has range of about 430 miles (23.9 mpg).	18-gallon tank has range of about 250 miles (13.9 mpg).	Three cylinders with total capacity of 32 gallons have range of about 200 miles.

automobile body itself and are more likely to withstand the impact of collision. There are currently very few compression stations in the United States that can dispense CNG. However, home compressors have been developed that would fill an automobile's storage tanks overnight by using natural gas piped to the home. Table 23–4 compares the fuel properties of gasoline, LNG, and CNG.

SELF-TEST 23 – B

1. What are the two main combustion products of gasoline burning in air? _____ and _____
2. Detonation in a combustion chamber during the compression stroke is called ringing or knocking? _____
3. What is the name of the lead compound that has been used for years to raise the octane number in gasolines? _____
4. Diesel engines are fuel efficient. (   ) True or (   ) False

AIR POLLUTION AND THE AUTOMOBILE

The automobile is a special case of air pollution simply because there are so many of them — more than 200 million now travel the U.S. roadways. Collectively, automobiles are *the* major source of air pollution (See Table 18–1).

Air pollutants may enter the atmosphere from three major locations in the automobile (Fig. 23–12): from the exhaust, from the crankcase blowby (gases that escape around the piston rings), and by evaporation from the fuel tank and the carburetor.

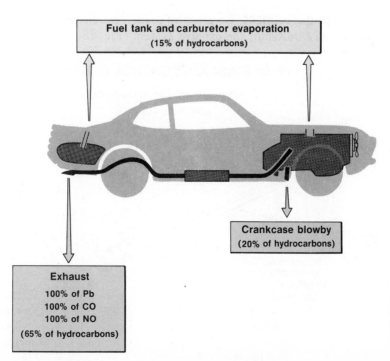

Fuel tank and carburetor evaporation
(15% of hydrocarbons)

Crankcase blowby
(20% of hydrocarbons)

Exhaust

100% of Pb
100% of CO
100% of NO
(65% of hydrocarbons)

FIGURE 23–12
Sources of automobile pollutant emissions. The percentages show the relative amounts of the various types of pollutants from the three major automobile sources for automobiles that have no pollution controls.

The mixture of pollutants emitted by a given automobile depends on the composition of the gasoline and the effectiveness (or lack) of pollution controls. Generally when gasoline is burned, the products are CO, CO_2, H_2O, a variety of nitrogen oxides (principally NO and a little NO_2), newly formed hydrocarbons, and some unburned, original hydrocarbons. If tetraethyllead is present in the gasoline, additional products are PbBrCl, $PbBr_2$, and $PbCl_2$, which are formed when lead reacts with the purposeful gasoline additive 1,2-dibromoethane and 1,2-dichloro-ethane. If pollution controls are absent or imperfect, gasoline evaporates from the carburetor and the gas tank.

Ways to Control Pollutant Emissions

As an impetus, both state and federal governments have passed legislation that regulates automobile emissions and establishes standards. Governmental standards for hydrocarbons, CO, and nitrogen oxides are summarized in Table 23–3. Standards are being met by mechanical changes and changes in fuel. Fuel modifications are more effective because they apply to all cars, old or new.

Mechanical changes in cars for the purpose of decreasing air pollutant emissions include the following devices and adjustments (Fig. 23–13).

POSITIVE CRANKCASE VENTILATION VALVE (PCV)   On all cars since 1963, fresh air (drawn by a vacuum system) sweeps the crankcase blowby gases through the PCV valve and connecting hose into the reaction cylinder of the engine for a second chance at combustion.

CONTROL VALVES ON GAS TANK CAPS   Air can flow into the gas tank, but gasoline vapors cannot flow out into the air.

CATALYTIC CONVERTERS   On most car models produced beginning in 1975, a platinum-based catalyst in the exhaust system (Fig. 23–14) converts hydrocarbons,

1985 EMISSION CONTROL SYSTEM

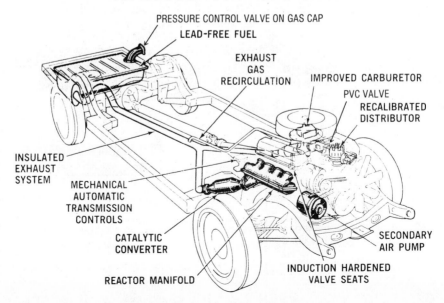

FIGURE 23–13
Automotive hardware
for pollution control.

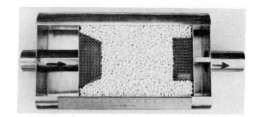

FIGURE 23–14 Cutaway view of catalytic exhaust muffler showing catalyst pellets.

CO, and NO to CO_2, H_2O, and N_2. Since the catalyst is inactivated by lead, lead-free gasoline must be used in these cars.

AFTERBURNER TECHNIQUE Air is injected into the hot gases as they exit through the exhaust valves. A portion of the gas completes the combustion process. The energy emitted contributes to the problem of dissipation of energy without contributing any power to the engine.

ADJUSTMENT OF AIR-FUEL RATIO This adjustment is an interesting trade-off type of chemical problem. For example, the *complete* combustion of one mole of octane or an isomer of octane requires 12.5 moles of oxygen.

$$C_8H_{18} + 12.5\ O_2 \rightarrow 8\ CO_2 + 9\ H_2O + \text{heat}$$

Experimentally the maximum power is obtained when the oxygen-fuel molar ratio is 12.5 : 1, but this ratio does not give the lowest emissions of CO and hydrocarbons. In the split instant of combustion, not all of the fuel is burned and only part of the carbon goes all the way to CO_2; some stops at CO. A higher ratio of oxygen to fuel (15.1 : 1) produces about eight times less CO and slightly fewer hydrocarbons. Therefore, the 15.1 : 1 oxygen-fuel ratio gives a more efficient use of the fuel — more complete burning. The extra heat from the more complete burning when put with the extra oxygen (and *nitrogen*) produces more nitrogen oxides. (Reactions of nitrogen and oxygen to form nitrogen oxides require heat; they are endothermic.) It's a case of trying to have your cake and eat it, too. To get less CO and hydrocarbons, you get more nitrogen oxides — and less power. The reduced power comes from energy being used to form the nitrogen oxides and to heat the extra air. The problem of nitrogen oxides is partially offset by recycling part of the exhaust. The combustion temperature is reduced and so is the amount of nitrogen oxides that form. Recirculation of about 15% of the exhaust reduces the nitrogen oxides by 80%, accompanied by a 16% cut in power output and a 15% decrease in fuel economy. Changes in timing can increase power and economy, but the amount of nitrogen oxides increases. Fuel injected specifically at the spark plug gap produces low hydrocarbon, CO, and nitrogen oxide emissions, but then particulate emissions become a problem. As you can see, the burning of hydrocarbon fuels to meet fuel economy and performance requirements while at the same time keeping emissions of pollutants at a minimum is a formidable task. Sometimes it seems easier to try some other means of propulsion.

Isomers of octane are the major components of gasoline.

Endothermic:
$N_2 + O_2 + 4.32\text{ kcal} \rightarrow 2\text{ NO}$

Exothermic:
$C + O_2 \rightarrow CO_2 + 94.1\text{ kcal}$

The formation of nitrogen oxides deducts heat energy that could be used for expansion. This reduces power.

ELECTRIC AUTOMOBILES

One of the very few ways we can eliminate combustion processes is to use electricity in some way to power our transportation devices. Early attempts to develop electric cars in the United States were made in Boston in 1888, 28 years after the develop-

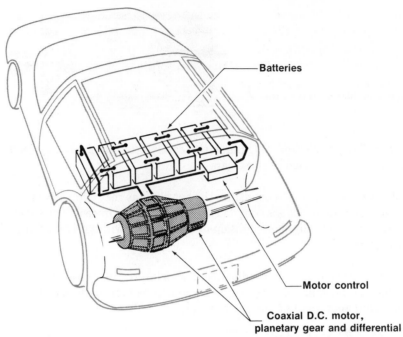

Batteries

Motor control

Coaxial D.C. motor,
planetary gear and differential

FIGURE 23–15   Experimental electric car. Powered by conventional lead-acid batteries, it has a range of 50 miles. Lighter, more energetic batteries would increase its range significantly. Present-day batteries account for 30% of the total weight of the car.

At the turn of the century, a Baker Electric Coupe cost $2600 and a Borland Electric Delux could be bought for $5500. A model T Ford cost $300 in 1913.

ment of the first storage battery in 1860. By 1912, about 6000 electric passenger and 4000 commercial vehicles were being manufactured annually in this country. The electric car lost out in competition with the internal combustion engine during the 1920s. The problems of short range (about 20 miles), low speeds (20 miles per hour maximum), 8 to 12 hours recharge time for the batteries, and relatively high price combined to eliminate the electric car from the race for leader in transportation.

At present, more than 100,000 rider-type, battery-powered, materials-handling vehicles are operating in U.S. plants and warehouses where it is vital to avoid air pollution from internal combustion engines. Despite this positive start, attempts to employ electric vehicles for street and highway use generally have been commercial failures. The energy storage capacities of conventional batteries (lead-acid, nickel-iron, and silver-zinc) are too limited or expensive to provide an acceptable energy source for electric passenger cars (Fig. 23–15). New battery designs may prove competitive for short-range urban travel. Nickel-zinc, nickel-iron, zinc-fluoride, and aluminum-air batteries are all being studied seriously as possible successors to the lead-acid batteries now in use.

SELF-TEST 23–C

1. What device is placed on an automobile to control carbon monoxide emissions? _____

2. Which pollutant would be produced in greater amount if the engine were run hotter? _____

3. What causes carbon monoxide in automotive exhaust? _____
4. What compound in automobile exhaust contains the hydrogen originally found in the gasoline hydrocarbons? _____
5. What causes hydrocarbons to be released in automotive exhaust?

_____

AUTOMOTIVE CONSUMER PRODUCTS

Proper fuels and lubricants must be used to keep an automobile in top running condition. The resale value of an automobile is also enhanced if cleaners and other products are used to keep the automobile surface cleaned and polished. All this adds up to choices for the consumer. Most products on the market designed for these purposes function adequately. Any differences noticed are probably due to differences in individual tastes and slight differences in the paints and plastics used for the automobile surface.

Gasoline Additives

The production of gasoline from petroleum and the antiknock properties of gasoline were discussed earlier. In addition to tetraethyl lead and/or aromatic compounds which reduce preignition or "knock" in an automobile engine, numerous other chemicals are added to gasoline to improve its properties.

Other chemicals are added to gasoline to prevent the ignition of new fuel by glowing particles from a previous ignition. These additives are called **deposit modifiers.** Phosphorus compounds such as tricresyl phosphate (one trade name is TCP) and, more recently, boron compounds, have been used for this purpose. These alter the composition of the deposits in the combustion chamber and make them less likely to glow. The phosphorus compounds also prevent spark plug deposits from becoming so electrically conductive that the charge leaks away instead of firing the plug.

$$\left(CH_3-\hexagon-O\right)_3 PO$$

TRICRESYL PHOSPHATE

Antioxidants such as phenylenediamine, aminophenols, dibutyl-p-cresol, and ortho-alkylated phenols, are added to prevent the formation of peroxides that lead to knock and gum formation. About 2 or 3 pounds of these additives are added to every 1000 barrels of gasoline. The mechanism of antioxidation of automotive oils is very similar to the mechanism described for BHA and BHT in food additives (Chapter 20). Because copper ions catalyze gum formation, metal ion scavengers such as ethylenediamine are added to chelate the trace amounts of copper ions and render them ineffective. The copper gets into the gasoline from the copper tubing used for fuel lines and from brass parts of the engine.

Brass is an alloy of copper and zinc.

To inhibit water from corroding and rusting storage tanks, pipelines, tankers, and fuel systems of engines, **antirust agents** are added. Four compounds used to prevent corrosion are trimethyl phosphate, sodium and calcium sulfonates, and N,N'-di-sec-butyl-p-phenylenediamine. All these compounds have a polar or ionic end and a nonpolar end in the molecule. These agents coat metal surfaces with a very thin protective film that keeps water from contacting the surfaces, as shown in Figure 23–16. This also helps prevent gummy deposits in the carburetor and combats carburetor icing during cold weather.

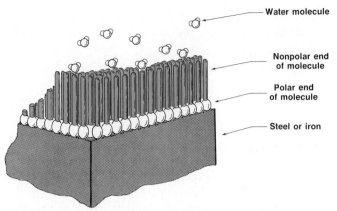

FIGURE 23–16   The action of surface active agents, such as rust inhibitors, mild antiwear agents, and some deicing agents.

OH OH
| |
H—C—C—H
| |
H H
ETHYLENE GLYCOL

Antiicing agents coat metal surfaces, as do antirust agents, and thus prevent ice particles from accumulating on surfaces, and/or depress the freezing point. The small ice particles pass harmlessly through the carburetor and into the engine where the heat converts them to water vapor that eventually exits with the exhaust gases. The freezing point depressants, which include alcohols and glycols, act in the same manner as the antifreeze in the engine's cooling system.

Detergents, which include alkylammonium dialkyl phosphates, are added to prevent the accumulation of high-boiling components on the walls of the carburetor. These deposits interfere with the air flow into the carburetor and cause rough idling, frequent stalls, poor performance, and increased fuel consumption. The effectiveness of these detergents stems from their surface active properties, as shown in Figure 23–17. The film of detergent provides a thin, nonpolar coating on the metal surfaces which prevents high molecular weight, nonpolar gums from forming thick deposits on the surfaces.

FIGURE 23–17
Possible defoaming action of substances like methylsilicone polymers in oil.

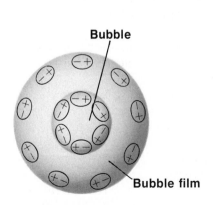

Bubble

Bubble film

Lubricants and Greases

Lubricants have been used to separate moving surfaces, and thus minimize friction and wear, for a long time. Even before 1400 B.C., animal tallow was used to lubricate chariot wheels. Petroleum lubricating oils and greases came into widespread use after the famous Drake well was drilled at Titusville, Pennsylvania, in 1859. The use of additives in lubricants has progressed rapidly since about 1930; synthetic lubricants have been developed largely since World War II.

Lubricating oils from petroleum consist essentially of complex mixtures of hydrocarbon molecules. These generally range from low viscosity oils, having molecular weights as low as 250 amu to very viscous lubricants with molecular weights as high as about 1000 amu. The viscosity of an oil can often determine its use. For example, if the oil is too viscous, it offers too much resistance to the metal parts moving against each other. On the other hand, if the oil is not viscous enough, it will be squeezed out from between the metal surfaces, and consequently offer insufficient lubricating power. For these and other reasons, motor oil is often a mixture of oils with varying viscosities. The common 10W-30 oil, for instance, combines the low-temperature viscosity of the Society of Automotive Engineers (SAE) 10W classification for easy low-temperature starting with SAE 30 high-temperature viscosity for better load capacity in bearings at the normal engine running temperature.

Viscous means resistant to flow, like molasses.

The SAE scale for rating motor oils is based on the viscosity of the oils. The viscosity criteria for the SAE scale are given in Table 23–5.

During distillation of petroleum crude oils, the lubricating crude oil fractions boil off after the lower boiling gasoline, kerosene, and fuel oils are removed. Most of the aromatic compounds are then removed from the lubrication oil fraction by solvent extraction to prevent the formation of sludge during high-temperature operation. Paraffin wax is then removed by low-temperature filtration to give the oil better flow characteristics. The final refining step is contact with an activated clay such as Fuller's earth, which absorbs many of the colored particles and pro-

TABLE 23–5 Viscosity Data at −18°C and 99°C for the SAE Method of Rating Motor Oils

| MOTOR OIL | VISCOSITY (SUS)* | | | |
| | −18°C | | 99°C | |
	Min.	Max.	Min.	Max.
5W		4 000	39	
10W	6000	12,000	39	
20W	12,000	48,000	39	
20			45	58
30			58	70
40			70	85
50			85	110

* SUS is the Saybolt Universal Second, which is the time in seconds required for 60 ml of oil to empty out of the cup in a Saybolt viscometer through a carefully specified capillary opening. Note the extremely shortened time for the outflowing of the hotter oil.

HO O O OH

C C

CH$_2$—CH

Polar groups

Nonpolar group (long alkyl chain)

ALKYLSUCCINIC ACID

vides an oil with a light color. Lubricating oils with the desired properties are then made by blending one or more refined stocks with the proper additives.

After the motor oil has been separated and refined, its usefulness is improved by the addition of substances such as antiwear agents, oxidation inhibitors, rust inhibitors, detergents, viscosity improvers, and foam inhibitors.

When two metal surfaces contact under heavy load and high temperature, as in the differentials of most cars, the friction produces intense heat which renders organic lubricant films ineffective. Under extreme conditions the metals can weld together. To combat this, *extreme pressure lubricants* were developed. These contain inorganic compounds as additives which react at the high contact temperatures to form high-melting inorganic lubricant films, such as lead sulfide and iron sulfide, on the metal surfaces; the presence of these films inhibits breakdown. These additives generally consist of sulfur, chlorine, phosphorus, and lead compounds which act either by providing layers that are hard and difficult to wear away or by serving as fluxing agents to contaminate the metal surface and prevent welding.

Under conditions of less severe friction, mildly polar organic acids, such as the alkylsuccinic type, and organic amines are often added as **antiwear agents.** These compounds provide an adherent, adsorbed film over metallic surfaces and reduce shearing of the metal. In somewhat more severe conditions about 1% tricresyl phosphate (TCP) or zinc dialkyldithiophosphate is widely used.

Rust inhibitors, such as the mild antiwear agents and antirust agents in gasoline, are preferentially adsorbed as a film on iron and steel surfaces to protect them from attack by moisture. If only a little water is present in a large amount of oil, mildly polar organic compounds such as alkylsuccinic acids and organic amines are often used. Where severe conditions are anticipated, more strongly adherent organic phosphates (TCP, for example), polyhydric alcohols, and sodium and calcium sulfonates are used.

Oil oxidation is thought to involve a chain reaction mechanism with hydroperoxide formation (—OOH) as the initiating process which eventually leads to the formation of organic acids and other products. **Oxidation inhibitors** appear to interrupt the chain reaction by tying up the hydroperoxide. This action delays the formation of sludge, varnish, and acids for extended operating periods and minimizes corrosion problems with the zinc-, cadmium-, and copper-containing alloys, which are corroded by organic acids in oxidized oils. Zinc, barium, and calcium thiophosphates are frequently used to prevent oxidation of the oils.

Detergents are widely used in a 2% to 20% concentration in motor oils to prevent or remove deposits of oil-insoluble sludge, varnish, carbon, and lead compounds. The detergents are adsorbed on the insoluble particles, keeping them suspended in the oil so as to minimize deposits on rings, valves, and cylinder walls. The action is similar to the action of soap (or detergent) in removing grease from clothes or hands, as described in Chapter 22. A basic difference exists, however, because the polar end of the detergent molecule generally is attached to the particle, and its hydrocarbon end extends into the medium (oil). (In soapy water, the hydrocarbon end of the soap is in the oil or grease particle and the polar end is in the medium, water.) Barium and calcium sulfonates and phenoxides are used extensively as detergents in automotive motor oils.

The viscosity of motor oils can be adjusted with additives. Polymethacrylate is added in small amounts (1% or less) to prevent wax, which condenses out at low temperatures, from forming a network of crystals that would immobilize the oil.

O

Na$^+$ + $^-$O—S

O

A SODIUM ALKYL SULFONATE

The prefix *thio* denotes sulfur replacing oxygen, as in sodium monothiophosphate, Na$_3$PSO$_3$, compared with sodium phosphate, Na$_3$PO$_4$.

Adsorb means to attach to the surface, while *absorb* means to penetrate into the interior.

Ba(O—⟨O⟩)$_2$

BARIUM PHENOXIDE

The additive appears to adsorb on crystal faces, which prevents the interlocking crystal growth. These additives are ineffective in normally high viscosity oils. The viscosity of oils can be increased by adding linear polymers in the molecular weight range of about 5000 to 20,000 amu. The three types most commonly used are polyisobutylenes, polymethacrylates, and polyalkylstyrenes (Chapter 13). The entanglement of the long chains prevents easy flow of the oil. With use, the chains are broken into smaller fragments and the oil assumes its base viscosity.

Severe churning and mixing of oil with air may cause foam and an oil overflow from the engine; failure of the machine may eventually ensue. Methyl silicone polymers (Chapter 13) in concentrations of only a few parts per million are effective for defoaming oil. Since the silicone additive is not completely soluble in the oil, it functions by forming minute droplets of low surface tension, which aid in breaking up foam bubbles to release the trapped air (Fig. 23–17).

Greases are essentially lubricating oils thickened with a gelling agent such as fatty acid soaps of lithium, calcium, sodium, aluminum, or barium. The fatty acids are usually oleic, palmitic, stearic, or other carboxylic acids derived from tallow, hydrogenated fish oil, castor oil, or, less often, wool grease and rosin. The soaps form a network of fibers that entrap the oil molecules within the interlacing fiber structure. Carbon black, silica gel, and clay are also used to thicken petroleum greases. Chemical additives similar to those used in lubricating oils and gasolines are added to greases to improve oxidation resistance, rust protection, and extreme pressure properties. Synthetic greases are being developed that deteriorate so slowly that longer intervals between grease jobs are now possible. Silicone greases have a useful life of up to 1000 hours at 450°F (232°C). Unfortunately, silicone greases provide relatively poor lubrication for gears and other sliding devices. Diester greases such as di(2-ethylhexyl) sebacate have found extensive use among synthetic greases. Lithium soaps dissolve well in the diester oil and form a grease with equal or better lubrication characteristics and a considerably longer useful life than petroleum greases. Blends of silicone oil and diester oil provide greases with good low-resistance lubricating power even at low temperatures (−73°C).

Antifreeze

An antifreeze is a substance that is added to a liquid, usually water, to lower its freezing point. Although various substances have been used as antifreezes in the past, nearly all of the current market is supplied by ethylene glycol and methyl alcohol.

ETHYLENE GLYCOL   METHYL ALCOHOL

More than 95% of the antifreeze on the market is "permanent" antifreeze, having ethylene glycol as the major constituent. The largest use of antifreeze is in protection of water-cooled automobile and truck engines. Water has been selected as the coolant for these engines because of its universal availability, low cost, and good heat transfer properties; however, it has two serious disadvantages. First, it has a relatively high freezing point and, second, under normal operating conditions, it is corrosive. Modern antifreeze mixtures effectively counteract these problems.

DI(2-ETHYLHEXYL) SEBACATE

Ethylene comes from cracking petroleum. Ethylene glycol is prepared from ethylene, $CH_2{=}CH_2$, by an oxidation reaction

$$2\ CH_2{=}CH_2 + O_2 \rightarrow 2\ CH_2{-}CH_2 \diagdown O$$

ETHYLENE OXIDE

followed by hydrolysis:

$$CH_2{-}CH_2 + H_2O \rightarrow CH_2{-}CH_2$$

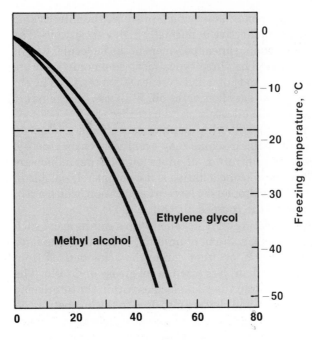

FIGURE 23–18
Freezing point depres-
sion of water by
ethylene glycol and
methyl alcohol.

The temperature in the United States, except for Alaska, seldom, if ever, falls below $-40°F$ ($-40°C$). Both ethylene glycol and methyl alcohol can prevent water from freezing at these temperatures, as shown in the graph of Figure 23–18. Methyl alcohol more effectively lowers the freezing point, but because of its volatility and combustibility it is seldom used.

For ethylene glycol, methyl alcohol, and other nonelectrolytes, the depression of the freezing point in dilute solutions is proportional to the concentration of the solute and nearly independent of the nature of the solute. That is,

$$\Delta T_f = K_f m$$

where ΔT_f is the freezing point depression, K_f is the proportionality constant peculiar to each solvent ($1.86°C$ for water), and m is the number of moles of solute per 1000 g of water (molal concentration). In concentrated solutions this relationship can be used to obtain only a rough estimate of the freezing point.

Antifreeze protection charts always show the temperatures at which the first ice crystals form. Below this temperature the antifreeze solution turns to slush. If the slush is unable to circulate through the radiator, overheating, boiling and engine damage can result. For this reason it is best to add sufficient antifreeze to prevent the formation of the first ice crystals even at the lowest anticipated temperature (Fig. 23–19).

The density of an antifreeze solution can be used to determine its freezing temperature.

A service station attendant measures the effectiveness of the antifreeze in your car by reading the position at which a hydrometer floats in a portion of a radiator solution. He is really measuring the solution's density, which varies with the amount of antifreeze present.

In 1960, automobile radiators were equipped with caps able to withstand pressures of 13 to 17 pounds per square inch; this change allowed a 20 to $28°C$

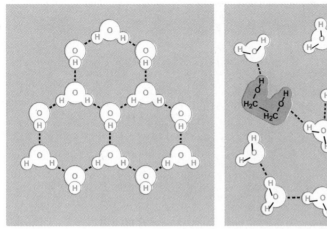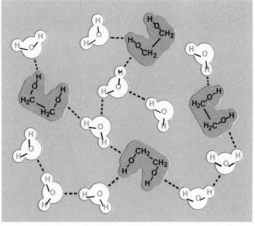

a   b

FIGURE 23–19   **How foreign molecules prevent water from forming its normal crystal structure. For example, by forming hydrogen bonds to water molecules, ethylene glycol (b) prevents water molecules from assuming their places in the ice structure (a).**

increase in maximum coolant temperatures. Thermostats now operate between 85 and 99°C, whereas in the past the range was 60 to 70°C. Ethylene glycol raises the boiling point of water as well as lowers its freezing point, as shown in Figure 23–20. Ethylene glycol reduces the vapor pressure of water, thus requiring a higher temperature for the solution to boil. Therefore, it is good policy to keep the antifreeze in the radiator all year. It prevents freezing in the winter and boiling over in the summer.

Commercial antifreeze contains various additives to prevent corrosion, leaks, damage to rubber, and foaming. A typical permanent antifreeze contains more than 95% ethylene glycol, several reducing agents to prevent corrosion, a substance to stop small leaks in the cooling system, and an antifoaming agent.

The prevention of corrosion is the second most important job of antifreeze. Metals in the cooling system which are subject to corrosion are copper, steel, cast iron, aluminum, solder (lead and tin), and brass (copper and zinc). The presence of oxygen, along with high temperatures, pressures, and flow rates, increases the possibility of general corrosion. Although corrosion can "eat" through the walls of the cooling system, the most general trouble is overheating caused by flakes of metal clogging the radiator. A large assortment of substances is used to inhibit corrosion. Some inhibit corrosion by acting as reducing agents (e.g., nitrites), some as ion scavengers (e.g., phosphates), and some as surface-active agents (detergents; e.g., triethanolamine). Most antifreezes contain two or more inhibitors for the different metals. All inhibitors are depleted with use.

Radiator sealants have been on the market for years. Modern sealants in antifreeze include asbestos fiber and polystyrene spheres. When a radiator springs a leak, the coolant penetrates the crack because the pressure inside the cooling system is greater than atmospheric pressure. Asbestos fibers are often too big to squeeze through and thus get caught in the crack, plugging the hole in the radiator. On the other hand, the larger polystyrene particles initially plug up most of the leak while the smaller ones build up behind them. The spheres fuse together under the pressure and temperature conditions that exist and form a solid plug in the crack

Good auto antifreeze formulations also contain rust inhibitors.

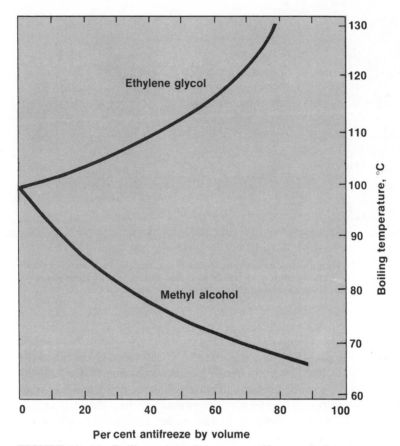

FIGURE 23–20   Boiling points of aqueous antifreeze solutions.

(Fig. 23–21). This is effective in stopping up holes or cracks up to 0.5 millimeter width, which includes about 90% of all radiator leaks.

Foaming is caused by one or more of several factors: air leaks in hoses, water pump, or radiator; exhaust gas leaking into the cooling system; failure to drain out cooling system cleansers; or extended use of antifreeze. Foaming can be corrected by tightening the system or by use of antifoam additives, such as silicones, polygly-

FIGURE 23–21   How polystyrene works to plug radiator leaks. Details are given in the text.

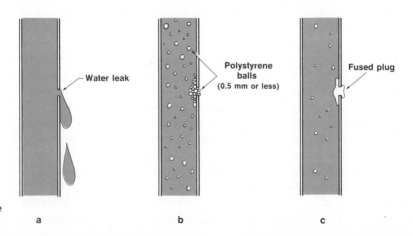

cols, mineral oils, high molecular weight alcohols, organic phosphates, alkyl lactates, castor oil soaps, and calcium acetate. These substances reduce the surface tension so bubbles cannot hold together (Fig. 23–17).

Since cooling system corrosion begins in earnest when the pH of the coolant drops below 7, most commercial antifreeze contains an extra amount of alkali. As time passes this alkali is exhausted and corrosion begins. A check of your antifreeze's corrosion-fighting capability can be made with a piece of litmus paper. Pink means acid and trouble through corrosion. Adding a can of rust inhibitor (which contains an alkali) will adjust the pH back to a value above 7. A retest with litmus paper shows blue (basic). Your car is now protected as long as the mixture composition doesn't change due to boil-away or leaks.

> Pink litmus — acidic.
> Blue litmus — basic.

Deicing fluids are a type of antifreeze, but they also will melt ice and frost. They are chiefly employed in removing ice and frost from parked aircraft and car windows. The glycol-alcohol type used on cars came on the market in 1959. The better formulations contain the following ingredients: ethylene or propylene glycol (for protection against refogging); one or more of the lower alcohols, such as 2-propanol, denatured alcohol, and so on (for low viscosity and good spray pattern); water (to minimize inside fogging by diminishing the evaporative-cooling effect of the formulation); nitrite or other inhibitor (to prevent corrosion of the container); and wetting agent, such as ethanolamine (to break surface tension so fluid will attack the ice readily).

> Deicers lower the freezing point of water.

A simple alcohol-water mixture works satisfactorily for quick frost removal and to defog the inside of glass, but it evaporates quickly and refogging occurs unless glycol is present. Before a deicer formulation is applied, the snow should be removed and the ice should be scored to permit faster penetration.

SELF-TEST 23–D

1. Antiicing agents are similar in function to what other automotive chemical? _____

2. What is necessary to form a grease? _____ and _____
3. Name a nonpermanent antifreeze that was once used in automobiles. _____

4. What is the main ingredient in permanent antifreeze? _____
5. Which oil weight represents the highest viscosity, 10W or 50W? _____

ROCKET TRANSPORTATION

Rocket travel into space is not new. Many college students have grown up in a world where the moon has already been explored (beginning in the summer of 1969) and earth-orbiting space stations have been built. The Russians started space exploration, the United States raced to catch up, and now both countries are busy establishing their continuing space programs for exploration and possible commercial and political exploitation.

The first mention of travel into space appeared in a work by a Greek author named Lukian. His *True History,* written in 180 A.D., described a voyage to the moon. In 1634 the astronomer Johann Kepler wrote *Sleep,* a guidebook for emigrants to the moon. Later, Jules Verne, Alexandre Dumas, and H. G. Wells wrote of space travel. These dramatic tales, interesting as they were, gave few details about how to do space travel. This all changed in 1903 — the same year Orville and Wilbur Wright first flew their airplane — with the publication of a scientific journal article entitled "Exploration of Space with Reactive Devices." The author of the article was a Russian named Konstantin Tsiolkovsky. His article can be summarized by five points:

1. Space travel is possible.
2. Rockets will be necessary since they are the only type of propulsion that will work in empty space.
3. Gunpowder rockets (the kind commonly known at the time) could not be used since they would not have sufficient energy.
4. Certain liquid fuels could be used in rockets of some new design.
5. Liquid hydrogen and liquid oxygen would make an ideal fuel-oxidizer combination.

Tsiolkovsky's statement regarding liquid hydrogen as a fuel and liquid oxygen as an oxidizer were especially interesting since oxygen had been first liquified only 20 years earlier by a Polish scientist, Z. F. Wroblewski, and hydrogen had been liquified just five years earlier by the English scientist James Dewar. Tsiolkovsky wrote other articles on space travel in which he accurately described the conditions of weightlessness in space and speculated about various fuels, but he never tried any of them in actual working models of rockets. It was the American, Robert H. Goddard, a professor at Clark University, who began experimenting with rockets in

Goddard patented his first rocket design in 1914.

FIGURE 23–22 Robert Goddard's first working rocket. (The Bettmann Archive)

1909. On March 16, 1926, Goddard flew his first test rocket a distance of 184 feet in 2.5 seconds (Fig. 23–22). The fuel was gasoline, and the oxidizer was liquid oxygen. By 1932 Goddard's rockets were reaching altitudes of 7500 feet.

ROCKET BASICS

What was learned about rockets during those early years? What are the best fuels? Why, after all the many experiments on many different fuels, have rocket engineers selected the hydrogen-oxygen propellant system for such rockets as the space shuttle? The answers to these questions lie in the basics of rocketry. In a rocket motor a chemical reaction occurs in which heat is given off (exothermic reaction). The heat of this reaction causes gaseous reaction products to expand. The hotter the gases are, the greater the expansive force. This is a basic property that all gases exhibit.

Exothermic reactions were introduced in Chapter 6.

Any chemical reaction capable of producing hot gaseous products could be a candidate for a rocket propulsion system. A nuclear reactor heat could also heat gases, but these types of processes will not be discussed here. In general, a rocket propulsion system requires an **oxidizing agent** and a **reducing agent.** For safety, if liquid propellants are used, the oxidizing agent is generally stored in one container and the reducing agent is stored in another container prior to mixing and ignition (Fig. 23–23). In 1932, an Italian, Luigi Crocco, experimented with compounds like nitroglycerine, which has both the oxidizing agent and the reducing agent in the same molecule. A compound like nitroglycerine is classed as a **monopropellant,** since there is only one ingredient in the propellant system. Of course, most people recognize nitroglycerine as an explosive and as such had only a short life as a rocket propellant. The nitroglycerine tended to explode and destroy the test rocket on ignition. Perhaps it was fortunate for Crocco that his money for experiments with nitroglycerine ran out before his luck ran out!

Oxidizing agents and reducing agents were discussed in Chapter 10.

$$H_2C-ONO_2$$
$$|$$
$$HC-ONO_2$$
$$|$$
$$H_2C-ONO_2$$
NITROGLYCERINE

The principle of having all that explosive power available, but in somewhat more controlled form, is what rocket propulsion is all about. So, what kinds of oxidizing agent–reducing agent combinations would work? All of the oxidizing agents and reducing agents shown in Table 23–6 have been tried in all of the possible combinations. More exotic mixtures have also been tried but will not be discussed here.

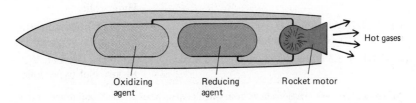

Oxidizing agent   Reducing agent   Rocket motor   Hot gases

FIGURE 23–23 Basic configuration of propellant storage tanks and rocket motor.

TABLE 23–6   Some Oxidizing and Reducing Agents That Have Been Used in Rocket Propulsion Systems*

OXIDIZING AGENTS	REDUCING AGENTS
Oxygen, O_2	Hydrogen, H_2
Fluorine, F_2	Hydrazine, N_2H_4
Nitrogen tetroxide, N_2O_4	Kerosene, $CH_{1.95}$ (approx)
Oxygen difluoride, OF_2	Diborane, B_2H_6
Chlorine trifluoride, ClF_3	Pentaborane, B_5H_9
Nitric acid, HNO_3	Ethanol, C_2H_5OH
Nitrogen trifluoride, NF_3	Acetylene, C_2H_2
	Methylhydrazine, CH_3NHNH_2
Aluminum perchlorate, $Al(ClO_4)_3$	Unsymmetrical dimethylhydrazine, $(CH_3)_2NHNH_2$
	Aluminum powder, Al

* All of the materials listed are handled as liquids except aluminum perchlorate and powdered aluminum, which are solids.

All of these reactants produce large amounts of heat and gaseous reaction products. An interesting property that some propellant liquids exhibit is the tendency to ignite immediately on contact. Liquids that behave this way are termed **hypergolic.** The rocket engineer who has chosen two hypergolic liquids does not need to design an ignition system into the rocket motor; the liquids simply ignite when they are pumped together. Some of the possible reactions are shown in Table 23–7.

Other factors that govern the choice of chemicals used in rocket propulsion systems include the physical properties and toxicity of both the reactants and the combustion products. In general, substances that are liquids at room temperature and that exhibit low toxicity would be preferred over other substances. The problem is that only a few such materials exist. Of the substances listed in Table 23–7, only ethanol is a liquid at room temperature and meets low-toxicity criteria. Table 23–8 shows some of the undesirable properties of the common rocket propellant ingredients.

TABLE 23–7   Some Rocket Propulsion Systems

$2\,H_2 + O_2 \rightarrow 2\,H_2O + 3.8$ kcal*	Used in Space Shuttle. Not hypergolic.
$2\,N_2H_4 + N_2O_4 \rightarrow 4\,H_2O + 3\,N_2 + 1.9$ kcal	Used in attitude-adjusting rockets in shuttle and as main rocket propellant in some military rockets. Hypergolic.
$CH_3NHNH_2 + 1.25\,N_2O_4 \rightarrow$ $CO_2 + 3\,H_2O + 2.25\,N_2 + 2.2$ kcal	Used in attitude-adjusting rockets. Hypergolic.
$C_2H_5OH + 3.5\,O_2 \rightarrow CO_2 + H_2O + 1.6$ kcal	Used in the German V-2 rocket. Not hypergolic.

* Energies for these reactions have been calculated on a per gram of total propellant. For example, 36 grams of hydrogen and oxygen were used in the first reaction.

TABLE 23-8   Summary of Properties of Some Rocket Propellants

INGREDIENT	BOILING POINT	COMMENTS
Fluorine	−188°C	Highly reactive, toxic. Hydrogen fluoride formed in combustion is also highly toxic. Liquid fluorine must be handled as a cryogenic liquid.
Oxygen	−183°C	Highly reactive, but not as much so as fluorine. Liquid oxygen is also a cryogenic liquid.
Nitrogen tetroxide	21.2°C	Highly reactive and toxic. Low boiling point makes handling the liquid difficult.
Hydrogen	−252.8°C	Very flammable. Must be handled as a cryogenic liquid. Causes embrittlement of some metals it is in contact with.
Hydrazine	113.5°C	Flammable and toxic. Can be unstable under conditions in which it is in contact with catalysts like metal salts.
Unsymmetrical dimethyl hydrazine	63°C	Flammable and toxic.

When the rocket motor is designed with a nozzle to restrict outward flow, the reaction gases flow through the nozzle at a rapid velocity and expand into a lower ambient pressure (Fig. 23-23). The **thrust** of a rocket motor depends on the difference in pressure inside the combustion chamber and the outside or ambient pressure, but this factor is minor compared to the forces produced by the mass flow (in pounds per second) and its velocity out of the nozzle. The thrust of the rocket motor therefore depends primarily on the nature of the pressure-producing reaction (heat of reaction and gaseous products) and the design of the nozzle. In short, the rocket engineer wants the highest attainable temperature and the highest possible exhaust velocity. The highest attainable temperature is determined by the nature of the reactants and practical design considerations such as the ability of the rocket motor to withstand the high temperatures produced during combustion. Higher temperatures can generally be attained when greater amounts of propellants are pumped into the rocket motor's combustion chamber, but the escaping gases also carry away heat energy. If the nozzle restricts the flow of gases too much, the rocket motor can be melted.

To measure the performance of rocket motors, one measures the **specific impulse.** The specific impulse measures the thrust delivered by the rocket per unit weight of propellant consumed per second. For example, if 200 pounds of thrust are produced by the consumption of 1 pound of propellant per second, the specific impulse is 200 seconds.

Often rocket designers hold back on the output of the engine for various reasons. For example, Robert Goddard's early rockets burned gasoline in pure oxygen. The specific impulses of these rockets were seldom above 170 seconds. He probably designed his rockets this way in order not to burn up the engines so they could be reused.

Table 23-9 gives the maximum attainable specific impulses for several liquid propellants using liquid oxygen and liquid fluorine as oxidizing agents. The fluorine values are shown only for comparison to illustrate the stronger oxidizing properties of fluorine.

TABLE 23–9   **Specific Impulse Values for Several Liquid Fuels with Oxygen and Fluorine as Oxidizers**

FUEL	OXYGEN		FLUORINE	
	T_c (K)*	Specific Impulse (seconds)	T_c (K)	Specific Impulse (seconds)
Hydrogen, H_2	2980	391	4117	410
Diborane, B_2H_6	3846	344	4934	371
Acetylene, C_2H_2	4172	327	3967	325
Hydrazine, N_2H_4	3410	313	4687	360
Methyl hydrazine, CH_3NHNH_2	3581	311	4419	345
Unsymmetrical dimethyl hydrazine, $(CH_3)_2NHNH_2$	3623	310	4183	338
Kerosene approx $CH_{1.95}$	3687	301	3917	317
Ammonia, NH_3	3104	295	4576	358
Ethanol, C_2H_5OH	3467	287	4184	330

* T_c is the combustion temperature inside the rocket motor (in degrees Kelvin).

Looking at Table 23–9 you can see that fluorine is definitely a better oxidizer than oxygen and produces a higher specific impulse for almost all of the fuels except acetylene. A problem with fluorine, however, is its reactivity. It is just too difficult to handle safely. Fluorine is highly toxic. It destroys human tissue on contact, it reacts with most common metals, some even explosively, and most important, hydrogen fluoride, which is formed from the oxidation of hydrogen containing reducing agents, is a highly corrosive and toxic compound. So the benefits gained by the more energetic fluorine-hydrogen propellant system must be unrealized in favor of safer systems that use oxygen instead.

Notice also that in Table 23–9 all of the fuels are low molecular weight compounds or elements. The heat energy of the combustion process increases the kinetic energy of the gaseous combustion product molecules. Since the kinetic energy of these molecules is expressed by the product $1/2\ mV^2$, the lower molecular weight combustion products will be forced out of the combustion chamber at a greater velocity. Then, according to Newton's third law of motion, for every action there is an opposite but equal reaction, the gases go out the nozzle, and the rocket engine is forced to move in the opposite direction.

In practice, the hydrogen-oxygen propellant system is maximized by using a **nonstoichiometric** ratio of hydrogen to oxygen. Instead of the ratio called for in the reaction

$$2\ H_2 \quad + \quad O_2 \quad \rightarrow 2\ H_2O$$
4 pounds   32 pounds

the actual ratio producing the highest specific impulse is 4 pounds hydrogen to 16 pounds oxygen. This means not enough oxygen is present to burn all of the hydrogen. Only 2 pounds of hydrogen can be burned by 16 pounds of oxygen. The effect of the unburned hydrogen is to lower the overall average molecular weight to about 10 instead of 18 if only water were produced. The combustion products of a large hydrogen-oxygen rocket engine are water and hydrogen molecules in a 1 to 1 ratio.

The lower average molecular weight means the velocity of the exhaust gases out of the rocket motor is greater and hence the rocket has more thrust.

OLD ROCKETS TO NEW ROCKETS

Practically all of the work with rockets during the 1930s and 1940s was directed toward developing weapons of war. The Germans developed their V-2 rockets, which were flown against England during World War II (Fig. 23–24). These rockets were propelled by liquid oxygen and ethyl alcohol. Later, some new designs used nitric acid to oxidize diesel fuel. Much of the German rocket development was directed by Wernher von Braun, who had earned his Ph.D. degree in 1937 with his experimental rocket work. Besides designing war rockets, von Braun and his co-workers began to design multistage rockets and describe accurately the requirements for working space stations and rocket planes that would be capable of returning to Earth after a prolonged Earth orbit mission. After the war, von Braun came to America to continue his rocket research. In 1952 he published *Across the Space Frontier,* which contained the forerunner design of the present day space shuttle. Rocket designs by Goddard, von Braun, and others led to the development of the Redstone, Jupiter-C, Atlas, Thor, Titan, and Saturn rockets, which were used in the early days of the U.S. space program. In early 1968 the National

FIGURE 23–24 German V-2 rocket used in World War II. These rockets were 46 feet long and 5 feet in diameter. They carried 1 ton of explosives over a range of 200 miles at speeds of 3000 miles per hour. The first V-2 rocket hit England on September 8, 1944. Before the war ended, 1100 had been launched against England, killing over 2800 persons. (The Bettmann Archive)

Aeronautics and Space Administration (NASA) began to design the space shuttle, and by 1972 the present design was fixed.

THE SPACE SHUTTLE

The first space shuttle, Columbia, was completed in 1977. This space craft was designed to be a reusable space ferry for a number of different missions. To lift its 150,000 pounds into orbit required building a massive fuel storage tank (Fig. 23–25) holding 224,000 pounds (378,378 gallons) of liquid hydrogen and 1,332,000 pounds (139,623 gallons) of liquid oxygen. The three main engines of the shuttle each develop 375,000 pounds of thrust, using 1122 pounds of propellant per second at liftoff. The weight of the entire spacecraft and its fuel tank is so great that two booster rockets are used. These booster rockets are attached to the external fuel tank. The propellant for these booster rockets is a mixture of aluminum perchlorate and aluminum metal, along with some iron oxide to act as a catalyst and a polymeric binder to hold the mixture together. Each booster has a thrust of 2.9 million pounds (Fig. 23–26). These solid fuel booster rockets were made to be reused and were designed in segments so each one could be loaded with propellant and the rocket assembled in a short time. About 2 minutes and 20 seconds into the launch, these solid rocket boosters separate from the external tank and parachute into the sea, where they are recovered. After about 8 minutes into the launch, the external tank is separated from the shuttle and falls back to Earth, where it breaks up on reentry into the atmosphere over the Indian Ocean.

In all, the fully loaded space shuttle with filled external fuel tanks and solid boosters weighs 4.4 million pounds. The combined thrust of all five rockets at launch time is 6,925,000 pounds so there is ample thrust to give the spacecraft the

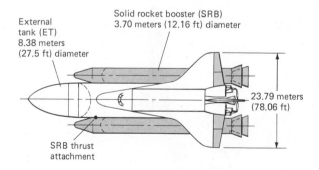

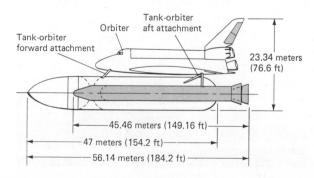

FIGURE 23–25 Diagram of the space shuttle showing the attached external tank holding the hydrogen and oxygen propellants for the shuttle main engines and the two solid rocket boosters.

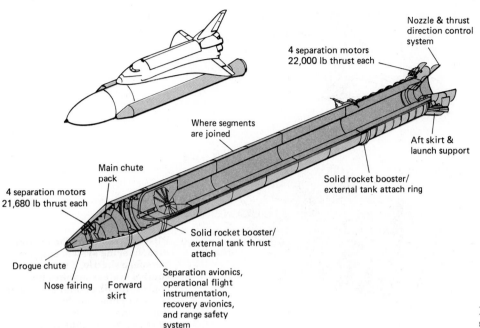

Nozzle & thrust direction control system

4 separation motors 22,000 lb thrust each

Where segments are joined

Main chute pack

4 separation motors 21,680 lb thrust each

Aft skirt & launch support

Solid rocket booster/ external tank attach ring

Solid rocket booster/ external tank thrust attach

Drogue chute

Nose fairing   Forward skirt

Separation avionics, operational flight instrumentation, recovery avionics, and range safety system

FIGURE 23–26   The solid rocket booster.

required velocity of 25,000 feet per second to achieve Earth orbit. The payload, or the weight of crew and materials that can be placed in orbit by the shuttle, varies from a high of 65,000 pounds when a due-east launch is performed from Cape Canaveral, Florida — in that direction the Earth's rotation boosts the orbit speed by 1000 miles per hour. At Vandenberg, California, the payload drops to 40,000 pounds when the shuttle is launched in a north-south orbit. A due-west orbit would drop the payload to only 32,000 pounds.

In practice, payloads this large have never been attempted. One of the largest was 46,615 pounds on April 4, 1983, on the spacecraft Challenger.

Once at orbital velocity, the space shuttle commander uses two orbital maneuvering engines (Fig. 23–27), which use the hypergolic mixture nitrogen tetroxide and methylhydrazine to control the exact velocity of the spacecraft to achieve the precise orbit called for in the mission plan. Most of the space shuttle orbits have been in the 150- to 180-mile range. When it is time for the space shuttle to return to Earth, the commander turns the spacecraft around and fires the orbital maneuvering engines for about 2.5 minutes. The spacecraft then glides to a landing on a long aircraft runway. Getting the spacecraft back to Earth safely presents other problems.

The early spacecraft used by the United States used organic epoxy resins as **heat shields** to help dissipate the frictional heat generated by the atmosphere on reentry. These heat shields burned up on reentry and therefore could not be used again. This was acceptable with the early spacecraft because they were not designed to be reused. When the shuttle was designed, its reusable nature required that some other type of heat shield material — one that was reusable — be employed. The temperatures on some surfaces of the spacecraft were calculated to reach 2700°C, while other surfaces were expected to reach lower temperatures of only 800 to 900°C. The upper parts of the spacecraft were expected to remain relatively cool. It was decided to cover most of the spacecraft surface with ceramic tiles consisting of fibers of silica (SiO_2) (Fig. 23–28) and stiffened with clay. The tiles would be contoured exactly to the surface of the spacecraft and held in place

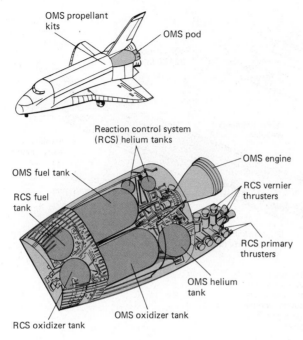

OMS propellant kits

OMS pod

Reaction control system (RCS) helium tanks

OMS engine

OMS fuel tank

RCS vernier thrusters

RCS fuel tank

RCS primary thrusters

OMS helium tank

OMS oxidizer tank

RCS oxidizer tank

FIGURE 23–27 The orbital maneuvering system engines on the space shuttle. The reaction control system supplies fine control for direction. Helium gas is used to push the hypergolic liquids into the combustion chamber where they ignite.

with an epoxy glue. These heat shield tiles almost became the undoing of the space shuttle program. Attaching the more than 31,000 tiles to the fuselage and wings took 18 months longer than planned. In addition, many of the tiles fell off the first space shuttle when it was flown from California to Florida in 1979. When the Columbia was first launched on April 2, 1981, two years behind schedule, some of the tiles fell off on the launch pad and others were recovered on the nearby beaches. It turned out that the lost tiles were in relatively unimportant places near the rear of the spacecraft, and the mission was not lost because of them (Fig. 23–29).

Travel into space is not as simple as surface transportation. You must literally take all of your physical needs with you when you venture off the surface of this planet. All of your food, water, and, most important, oxygen must be somehow

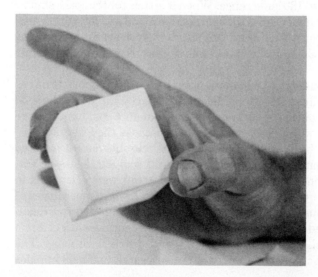

FIGURE 23–28 A silica tile like the ones used for thermal protection on the space shuttle. This tile has just been removed from a 1000°C oven.

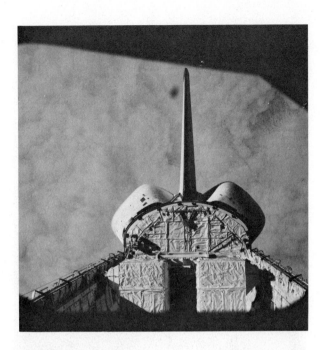

FIGURE 23–29   NASA photo showing thermal protection tiles lost on the port (sunlight) side orbital maneuvering engine pod.

stored on your spacecraft. The technical complexity of a launch requires thousands of ground support personnel. A space travel launch is hundreds of times more complex than an airplane takeoff and thousands of times more complex than leaving for a vacation with the family in the station wagon.

Like surface transportation, space travel is dangerous. Several Soviet astronauts have died in takeoffs and landings. Three American astronauts died in 1967 when the interior of their spacecraft cabin was destroyed by fire because faulty wiring caused an ignition in the pure oxygen atmosphere that was being used at the time. And in January 1986 seven astronauts died when the space shuttle Challenger was destroyed by an explosion caused by a faulty solid rocket booster. The best evidence indicates that a rubber O-ring seal used to secure the segmented solid booster rocket failed (Fig. 23–30). This seal failure was caused either by unusually

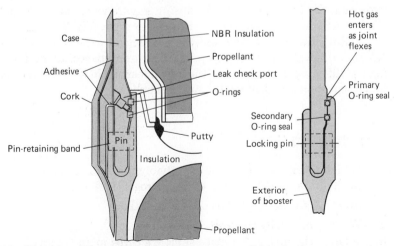

FIGURE 23–30   Diagram of the sealing mechanism where solid rocket booster segments were joined.

FIGURE 23–31   Photo of the Challenger space shuttle explosion. (UPI/Bettmann Newsphotos)

cold weather on the morning of the launch, which caused the rubber material to harden and possibly crack, or by vibrations in the booster rocket itself. When the seal broke, superhot rocket exhaust gases broke through and struck the external tank, causing it to rupture. Since that part of the external tank holds the liquid hydrogen, the hydrogen ignited, causing a massive explosion that destroyed the Challenger spacecraft and its seven-member crew (Fig. 23–31).

Both the fire in 1967 and the Challenger explosion in 1986 have caused NASA to reevaluate its safety procedures and spacecraft design. Future space flights will be safer as a result of the lesson learned in the past, but the possibilities for catastrophic accidents will never disappear.

SELF-TEST 23–E

1. Which of the following would probably not be a rocket fuel oxidizer?
(a) fluorine, F_2 (b) oxygen, O_2 (c) nitrogen trifluoride, NF_3
(d) ethyl alcohol, C_2H_5OH _____

2. What is meant by the term *hypergolic*? _____

3. What is the fuel of the space shuttle's three main engines?
(a) kerosene, (b) hydrogen, (c) ethyl alcohol, (d) hydrazine _____

4. What fuel was used by Robert Goddard in the first liquid-fueled rocket?

    (a) kerosene, (b) gasoline, (c) hydrazine, (d) hydrogen _____

5. Which oxidizing agent, in general, produces the higher specific impulse when
    used with a given fuel, fluorine or oxygen? _____

6. In the space shuttle main engines, not all of the hydrogen is consumed.
    (   ) True or (   ) False

MATCHING SET

_____ 1. Diesel engine

_____ 2. Gasoline engine

_____ 3. Soybean oil

_____ 4. Tetraethyl lead

_____ 5. Benzene

_____ 6. Shape of the automobile

_____ 7. Hypergolic

_____ 8. Specific impulse

_____ 9. Aluminum

_____ 10. Pinging

_____ 11. LNG

_____ 12. Liquid oxygen

_____ 13. Nitrogen tetroxide

_____ 14. Grease

_____ 15. Ethylene glycol

_____ 16. Fluorine

_____ 17. Air

a. Oil with a soap added
b. Preignition
c. Oxidant used to burn gasoline in an automobile engine
d. Permanent antifreeze
e. Ignite on contact
f. Can burn in a diesel engine
g. Has no spark plug
h. Low-density metal used to lower weight in autos, aircraft, and spacecraft
i. Used in the orbital maneuvering rockets on the space shuttle
j. Strongest oxidizer
k. Has a spark plug for ignition
l. Compound that the U.S. EPA has ruled must not be in gasolines
m. Used to raise the octane rating in gasolines
n. Aerodynamic drag
o. Cryogenic liquid that can be used as an automotive fuel
p. Cryogenic fluid used in the shuttle main engine
q. The thrust in pounds developed by a rocket motor per pound of propellant consumed per second

QUESTIONS

1. Explain what causes aerodynamic drag and how drag affects automobile fuel economy.

2. Name three factors that you can control that can increase the fuel economy of your automobile.

3. Explain why short trips affect fuel economy.

4. Look at your car or that of a friend and try to list all the different materials that you can find.

5. Look at the octane rating on the gasoline pumps at a nearby service station. What are the octane ratings of the different gasolines sold there? Which have the highest octane numbers, and which have the lowest? Suggest reasons for these differences.

6. What are the undesirable components of gasoline engine exhaust?

7. What are the undesirable components of diesel engine exhaust?

8. What is one health hazard associated with diesel engine exhaust not found in gasoline engine exhaust?

9. What problems do you see if vegetable oils were

mixed with diesel fuel or burned alone in diesel engines? List the technical problems separately from any social problems you might foresee.

10. List three advantages liquefied natural gas may have over gasoline when used as an automotive fuel.

11. List three disadvantages liquefied natural gas may have compared with gasoline when used as an automotive fuel.

12. What purpose does a lubricant like motor oil serve in an automobile engine?

13. How does a grease differ from a motor oil? How are they similar?

14. How does an antifreeze work?

15. What is the difference between a permanent antifreeze and a nonpermanent antifreeze?

16. Suggest a reason why nitroglycerine does not make a very good rocket propellant.

17. What is specific impulse as applied to rocketry?

18. If someone came to you and said that they had a new rocket propellant that could launch a rocket into Earth orbit and, while they were extoling the virtues, remarked that the average molecular weight of the exhaust gases was some number in excess of 100, would you be suspicious?

19. If a rocket produces 5000 pounds of thrust while consuming 25 pounds of propellant per second, what is its specific impulse?

20. The space shuttle's orbital maneuvering rocket produces 6000 pounds of thrust. Just prior to reentry, the rocket is fired for 155 seconds. If the rocket has a specific impulse of 290 seconds, how many pounds of propellant are consumed?

21. Suggest a reason why the solid rocket boosters produce so much thrust compared with the hydrogen-oxygen shuttle main engines. Compare the total propellant weight for the shuttle main engines with the propellant weight of the two solid rocket boosters.

22. Use Table 23–8 to arrive at the next answer. Hydrazine and dimethyl hydrazine are being considered as fuels to be oxidized by nitrogen tetroxide. Hydrazine is much more toxic and difficult to handle and store than the dimethyl derivative. Which one would you use?

23. Suggest a reason why the German V-2 rockets used alcohol as a fuel instead of hydrogen, which would have given the rockets greater payload and range.

24. Use information found in the text, and calculate the weight ratio of hydrogen to oxygen in the external storage tank used by the space shuttle.

Chapter 24

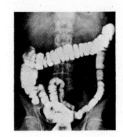

CHEMISTRY OF IMAGING — REPRESENTATIONS OF REALITY

An imaging device converts photons reflected from or transmitted by the object into a temporary or permanent image of the object. Lenses of the imaging device focus photons from the object onto a photosensitive material or surface. This may be the retina of the eye, the film in a camera, or the photodetector surface of a TV camera.

PHOTOSENSITIVE SUBSTANCES

Images can be observed by reflection from a surface (movie screen) or recorded on a photosensitive surface such as photographic film. A **photosensitive** (light-sensitive) substance is a substance that undergoes a chemical or physical change when exposed to photons. The three types of photosensitive substances used in imaging devices can be classified as **photochemicals, photoconductors,** and **phosphors.** Examples are given in Table 24–1.

Photochemistry deals with the chemical changes produced by absorbed radiant energy.

TABLE 24–1   Photosensitive Substances Used in Imaging Devices

SUBSTANCE	IMAGING DEVICE
Photochemicals	
11-cis-retinal goes to 11-trans-retinal	Eye retina,
Silver ion is reduced to silver atom	camera film
Photoconductors	Photodetectors,
Silicon	photomultipliers
Germanium	in TV cameras,
Cesium antimonide, Cs_3Sb	medical imaging
Cadmium sulfide	devices
Phosphors	
Zinc sulfide	Cathode-ray tube,
Calcium halophosphate	video displays
Strontium halophosphate	
Color Phosphors	
Copper-activated zinc sulfide	Green in color TV
Silver-activated zinc sulfide	Blue in color TV
Europium-activated yttrium vanadate	Red in color TV

Review the discussion of semiconductors in Chapter 8.

Photochemical substances undergo chemical change when exposed to photons. For example, most photographic films are based on the photochemical reduction of silver halides to produce a black-and-white effect.

Photoconductors are semiconductors such as silicon or germanium that conduct electric current when exposed to light. Photodetectors in imaging devices use photoconductors to convert a light signal into an electrical signal, which, in turn, can be processed into an image.

Phosphors are luminescent materials that emit a well-defined set of wavelengths of visible light when activated by a high-energy photon or electron beam. Luminescence includes both fluorescence and phosphorescence. In fluorescence, the absorbed light energy is reemitted almost immediately at a longer wavelength, whereas, in phosphorescence, there is a time delay between the absorption and emission of the electromagnetic energy.

Fluorescent lights are an example of the use of phosphors as a source of bright light.

Although we are most familiar with visible light because our eyes respond to photons of visible light, many imaging devices, particularly those used in medicine, use photons from other parts of the electromagnetic spectrum.

ELECTROMAGNETIC RADIATION

The speed of light in English units is 186,000 miles per second, about seven times around the circumference of the earth in 1 second.

All electromagnetic radiation travels at the speed of 3.00×10^8 meters per second in a vacuum. Max Planck (1858–1947) showed in 1900 that the electromagnetic radiation is quantized and the energy of this radiation can be represented by

$$E = h\nu$$

where h is Planck's constant (6.63×10^{-34} joule seconds) and ν is the frequency. Since the frequency of light is related to wavelength (λ) by

$$\lambda\nu = c$$

where c is the speed of light, the energy of photons can also be related to the wavelength.

$$E = hc/\lambda$$

The solar radiation reaching the Earth is made up mostly of photons from the ultraviolet, visible, and infrared regions.

The electromagnetic spectrum, shown in Figure 24–1, ranges from high-energy, short-wavelength gamma rays to low-energy, long-wavelength radio waves.

CHEMISTRY OF VISION

Visible light photons correspond to colors or wavelengths of light from violet, with $\lambda = 400$ nm, to deep-red, with $\lambda = 700$ nm.

The most complex imaging device is the eye-brain combination. Considerable progress has been made in understanding how the eye transforms photons of visible light into signals that are sent to the brain for organization and interpretation as an image.

The outer parts of the eye, particularly the lens, focus the photons on the retina, a light-sensitive material with the thickness of tissue paper. The retina has two types of light-sensitive substances or **photoreceptor cells, rods** and **cones.** The human eye contains about 1 billion rods and 3 million cones. Cones work in bright light and are sensitive to color, whereas rods function in dim light and are unable to distinguish colors. This is why only shades of gray and not colors are distinguished in moonlight. The sensitivity and adaptability of the eye is phenomenal. A person can sense as few as five photons of light in a darkened room and then

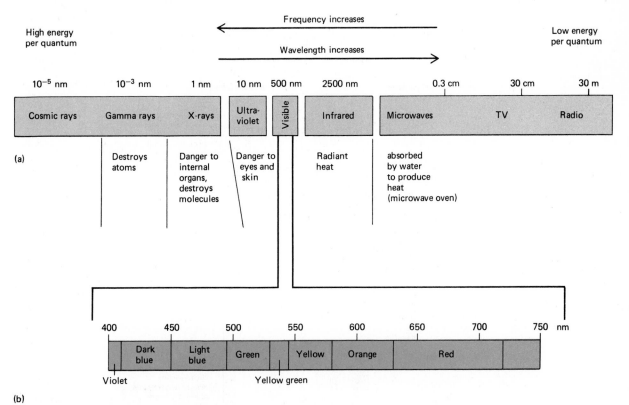

FIGURE 24–1 *(a)* **Electromagnetic radiation spectrum. One nanometer (nm) equals 10^{-9} meters (m).** *(b)* **Visible spectrum.**

adjust within a minute to the myriad of photons in bright light without being blinded.

The photosensitive material in the photoreceptor cells is **rhodopsin,** which includes the protein **opsin** and the compound **11-cis-retinal.** Photons focused on the retina isomerize the 11-cis-retinal to **all-trans-retinal** (Fig. 24–2). This isomerization causes a series of other molecular changes, which result in the dissociation of rhodopsin into opsin and all-trans-retinal. These structural changes trigger an electrical signal, which is carried by the optic nerve to the brain. Later in the visual cycle all-trans-retinal is converted back to 11-cis-retinal, which then combines with opsin to yield rhodopsin. A schematic of the complete visual cycle is shown in Figure 24–3.

Retinal is derived from vitamin A, which is why a shortage of vitamin A in the diet can lead to night blindness. Eating carrots is good for your eyes because β-carotene is converted to vitamin A in your body (Fig. 24–4).

SELF-TEST 24–A

1. The photosensitive substances that undergo chemical change are (   ) photochemicals, (   ) photoconductors, or (   ) phosphors.

2. _____ in carrots is converted to vitamin A in your body.

3. The equation for relating energy of photons to wavelength is _____.

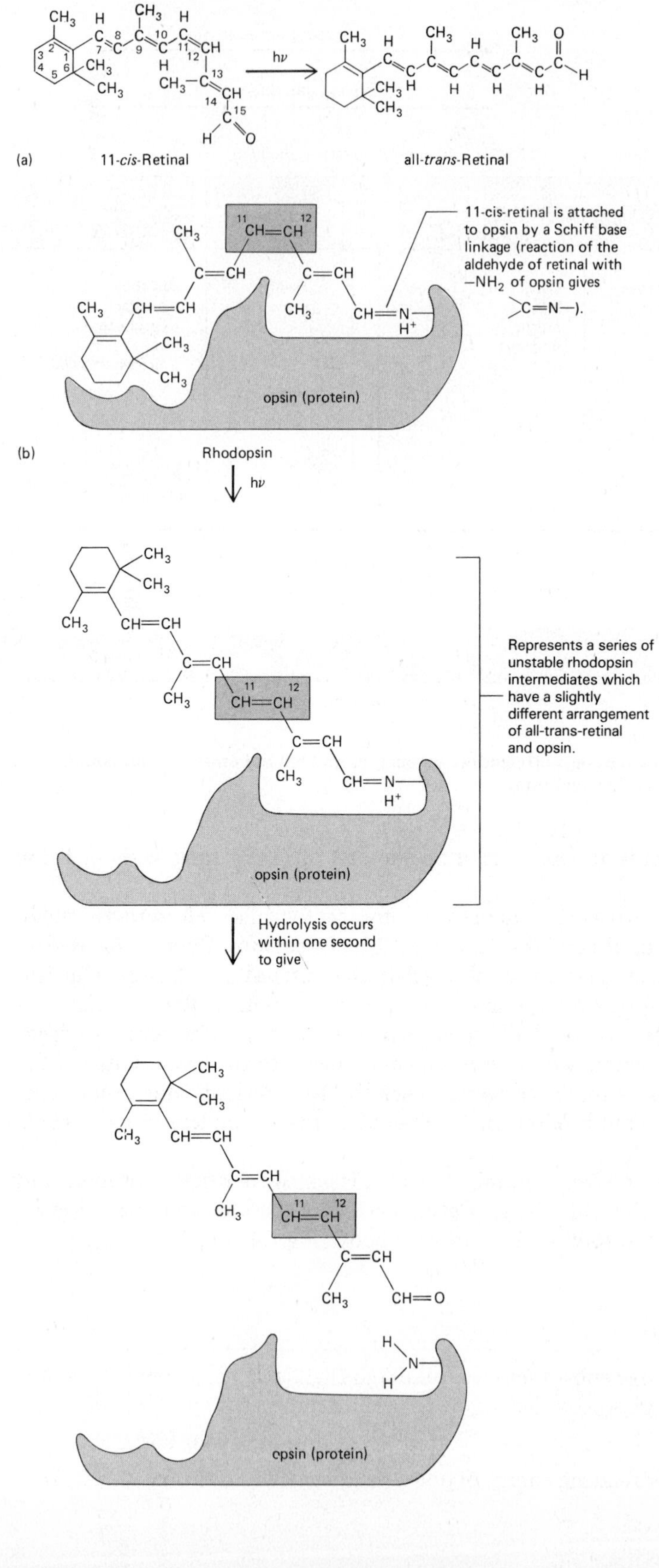

FIGURE 24–2 *(a)* Isomerization of 1′1-cis-retinal to all-trans-retinal by a photon of light. *(b)* Structural change in rhodopsin results in the separation of all-trans-retinal from opsin.

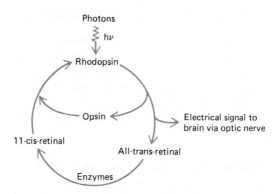

FIGURE 24-3 Schematic of the visual cycle of rhodopsin.

4. The photochemical change that occurs on the retina of the eye is the isomerization of _____ to _____.
5. Arrange the following forms of electromagnetic radiation in order of increasing wavelength: microwaves, radiowaves, infrared light, visible light.

_____

6. The shorter the wavelength, the ( ) lower or ( ) higher the energy of the
photons.
7. The photosensitive material in the photoreceptor cells of the eye retina is

_____.

8. Photoreceptor cells in the retina are referred to as _____ and

_____.

9. _____ are sensitive to different colors; _____ are not sensitive to colors.

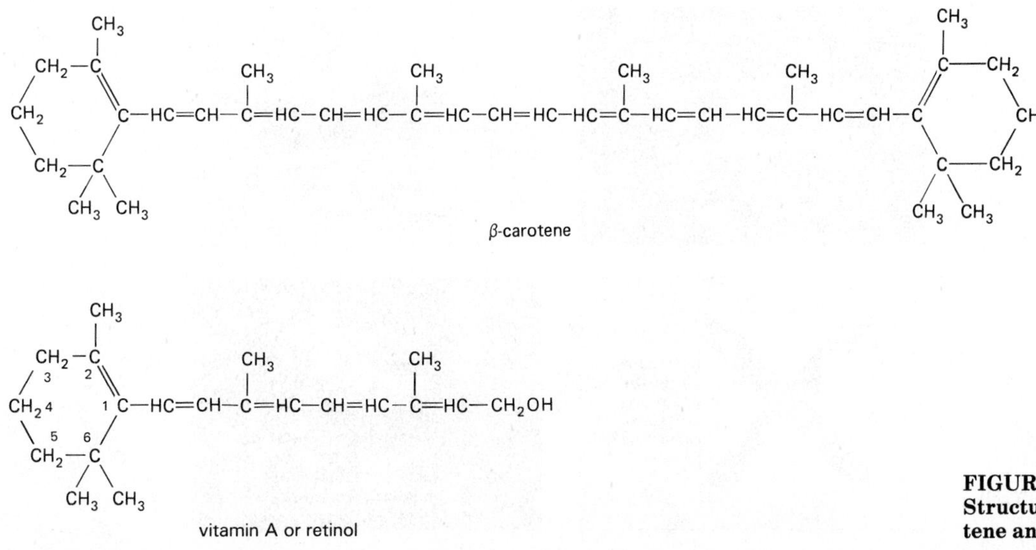

β-carotene

vitamin A or retinol

**FIGURE 24-4
Structures of β-caro-
tene and vitamin A.**

CHEMISTRY OF PHOTOGRAPHY

Comparisons are often made between the pupil of the eye and the f-stop of a camera or between the retina of the eye and photographic film. The eye is not only much more complex than a camera and its film, but the two imaging devices actually function by different mechanisms. The photographer regulates the f-stop opening and time of exposure to match the sensitivity of film, while the retina sensitivity of the eye adjusts to correspond to the light level of the scene.

Black and White Photography

J. H. Schulze observed in 1727 that a mixture of silver nitrate and chalk darkened on exposure to light. The first permanent images were obtained in 1824 by Nicéphore Niepce, a French physicist, by means of glass plates coated with a coal derivative (called bitumen) containing silver salts. In the early 1830s, Niepce's partner, Louis Daguerre, discovered by accident that mercury vapor was capable of developing an image from a silver-plated copper sheet that had been sensitized by iodine vapor. The **daguerreotype** image was rendered permanent by washing the

Pronunciation of daguerreotype: (dä-ger ´-ō-tīp).

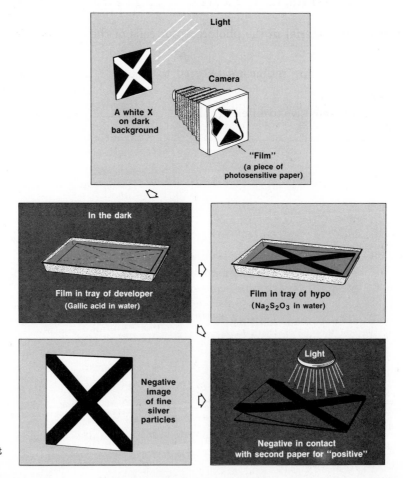

FIGURE 24–5 Talbot process.

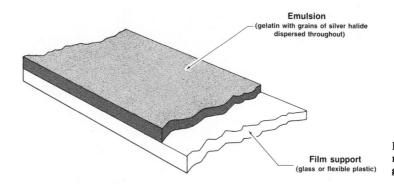

Emulsion
(gelatin with grains of silver halide dispersed throughout)

Film support
(glass or flexible plastic)

FIGURE 24–6   A modern photosensitive gelatin emulsion.

plate with hot concentrated salt solution. In 1839 Daguerre demonstrated his photographic process to the Academy of Sciences in Paris. The process was improved by using sodium thiosulfate to wash off the unexposed silver salts.

In 1841, an Englishman, William Henry Fox Talbot, announced the calotype process. The Talbot process (Fig. 24–5) involved a paper made sensitive to light by silver iodide. The light-sensitive paper could be developed into a negative image with gallic acid in a development process essentially the same as is used today. When made with semitransparent paper, Talbot's negatives could be laid over another piece of photographic paper which, when exposed and developed, yielded a "positive," or direct copy of the original. Although the Talbot process required less time than the Daguerre process, the Talbot images were not sharp. It was obvious that some way of holding the silver halides on a transparent material would have to be devised.

At first, the silver salts were held on glass with egg white as binder. This provided sharp, though easily damaged, pictures. By 1871, the problem had been solved by an amateur photographer and physician, Dr. R. L. Maddox. He discovered a way to make a gelatin emulsion of silver salts and apply it to glass. In 1887, George Eastman introduced the Kodak, a camera using film made by attaching a gelatin emulsion to a plastic (cellulose nitrate) base (Fig. 24–6). The camera could take 100 pictures and then camera and film had to be sent to Rochester, New York, for processing. The age of modern photography had arrived.

Cellulose acetate replaced easily combustible cellulose nitrate as the film support in 1951.

Photochemistry of Silver Salts

To understand the chemistry of photography, we must first look at the photochemistry of silver salts. A typical photographic film contains tiny crystallites called **grains** (Fig. 24–7), which are composed of a slightly soluble silver salt, such as silver bromide, AgBr. The grains are suspended in gelatin, and the resulting gelatin emulsion is melted and applied as a coating on glass plates or plastic film.

When light of an appropriate wavelength strikes one of the grains, a series of reactions begins that leaves a small amount of free silver in the grain. Initially, a free bromine atom is produced when the bromide ion absorbs the photon of light:

$$Ag^+Br^- \xrightarrow{\text{light absorption}} Ag^+ + Br^0 + e^-$$

The silver ion can combine with the electron to produce a silver atom.

$$Ag^+ + e^- \rightarrow Ag^0$$

In order for an exposed AgBr grain to be developable, it will need a minimum of four silver atoms as Ag_4^0.

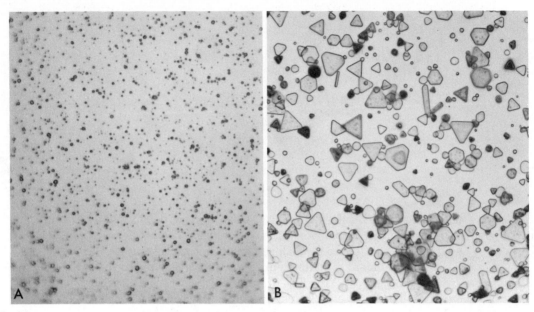

FIGURE 24–7 *(a)*
**Photomicrograph of the
grains of a slow
positive emulsion.** *(b)*
**The grains of a
high-speed negative
emulsion at the same
magnification.**

The latent image is the
"invisible developable
image" stored in the silver
halide grains.

Association within the grains produces species such as Ag_2^+, Ag_2^0, Ag_3^+, Ag_3^0, Ag_4^+, and Ag_4^0. The presence of this free silver in the exposed silver bromide grains provides the **latent image,** which is later brought out by the development process. The grains containing the free silver in the form of Ag_4^0 are readily reduced by the developer to form relatively massive amounts of free silver; hence a dark area appears at that point on the film (Fig. 24–8). The unexposed grains are not reduced by the developer under the same critical conditions.

Film sensitivity is related to grain size and to the halide composition. As the grain size in the emulsion increases, the effective light sensitivity of the film

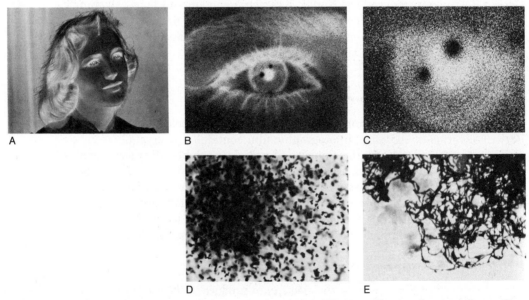

FIGURE 24–8   Grain in a photographic image is illustrated by five degrees of magnification *(a)* **original size;** *(b)* $\times 25$; *(c)* $\times 250$; *(d)* $\times 2500$; **and** *(e)* $\times 25,000$ **(by electron microscope).**

increases (up to a point; Fig. 24–7). The reason for this is that the same number of silver atoms are needed to initiate reduction of the entire grain by the developer despite the grain size.

Film sensitivity is rated on the American Standards Association (ASA) scale. The larger the number, the more sensitive the film is to light.

Amplification of the Latent Image — Development

Silver halides are not the most photosensitive materials known. Why, then, are they effective image producers? The answer lies in the fact that the impact of a single photon on a silver halide grain produces a nucleus of at least four silver atoms, and this effect is amplified as much as a billion times by the action of a proper reducing agent **(developer).**

When an exposed film is placed in developer, the grains that contain silver atom nuclei are reduced faster than those grains that do not. The more nuclei present in a given grain, the faster the reaction. The reduction reaction is

$$\underset{\substack{\text{IN GRAINS}\\\text{CONTAINING Ag}_4^0}}{Ag^+} + e^- \rightarrow Ag^0$$

Factors such as temperature, concentration of the developer, pH, and the total number of nuclei in each grain determine the extent of development and the intensity of free silver (blackness) deposited in the film emulsion in a given time.

The blackness on the negative is due to free silver atoms, Ag^0.

Not only must the developer be capable of reducing silver ions to free silver, but it must be selective enough not to reduce the unexposed grains, a process known as "fogging." Table 24–2 lists some substances that are used as developing agents.

TABLE 24–2 Some Compounds Used as Photographic Developers

NAME	FORMULA
Gallic acid	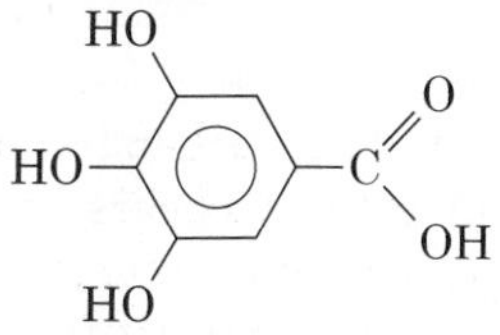
o-Aminophenol	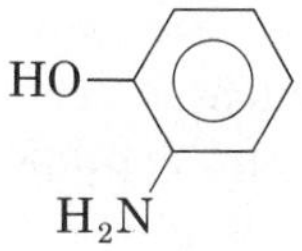
Hydroquinone	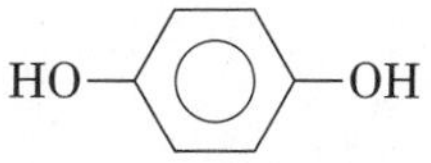
p-Methylaminophenol (Metol)	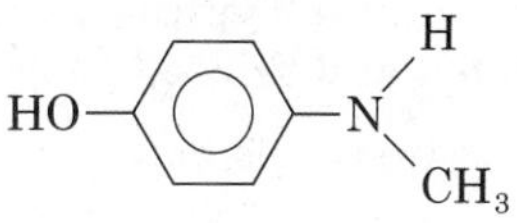
1-Phenyl-3-pyrazolidone (Phenidone)	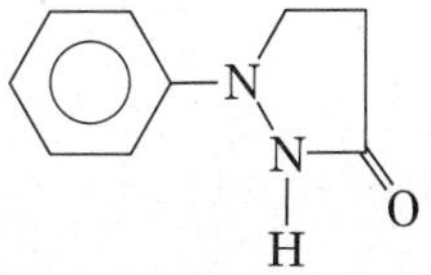

Most developers used for black and white photography are composed of hydroquinone and Metol or hydroquinone and Phenidone. A typical developer consists of a developing agent (or two), a preservative to prevent air oxidation, and an alkaline buffer to prevent the actual reduction reaction from being retarded (Table 24–3). Other chemicals might be added but they are not absolutely necessary.

When hydroquinone acts as a developer, quinone is formed. Two hydrogen ions are also produced for every two silver atoms.

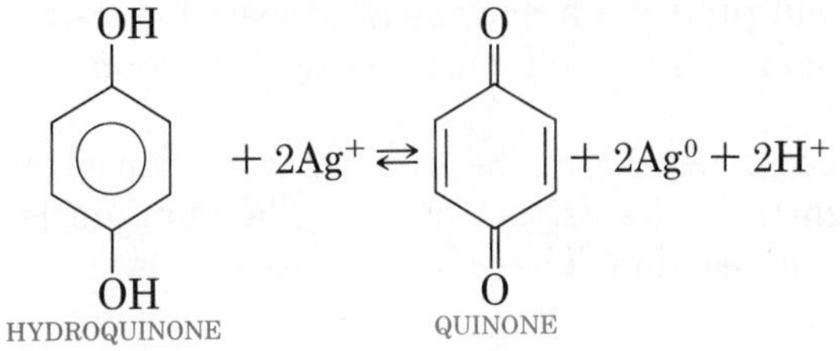

Since this reaction is reversible, a buildup of either hydrogen ions or quinone would impede the development process. The sodium sulfite reacts with quinone and destroys its ability to revert back to hydroquinone.

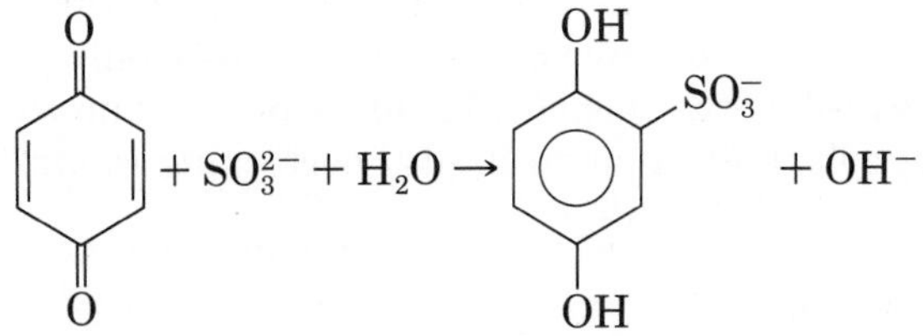

The hydrogen ions are neutralized effectively by the hydroxide ions.

$$H^+ + OH^- \rightarrow H_2O$$

If development proceeds either too long or at a higher temperature than recommended, sufficient fogging occurs to render a negative useless. Since the rates of the development reactions increase with increasing temperature, the photographer usually controls the temperature of the development bath very carefully.

The development process is terminated by a **stop bath.** The stop bath usually contains a weak acid such as acetic acid, which decreases the pH. The action of a stop bath is to build up the proportion of hydrogen ions quickly, which effectively stops the hydroquinone → quinone reaction.

TABLE 24–3   Formula for a Typical Developer for Black and White Films

750 ml water at 50°C; dissolve in this water:	
Metol	2.0 g
Hydroquinone	5.0
Sodium sulfite	100.0
Borax, $Na_2B_4O_7 \cdot 10\ H_2O$	2.0
Add cold water to make 1000 ml of solution.	

Fixing

One of the principal problems in the early days of photography was the lack of permanence of the image. If development only produces free silver where the light intensity was greatest and nothing further is done to the negative, the undeveloped silver halide will be exposed the instant it is taken into the light. After that, almost any reducing agent will completely fog the negative. In order to overcome this problem, a suitable substance had to be found to remove the unreduced silver halides. The most commonly used **fixing agent** in black and white photography is the thiosulfate ion ($S_2O_3^{2-}$). Thiosulfate ions form stable complexes with silver ions in aqueous solution:

$$\underset{\substack{\text{INSOLUBLE}\\\text{SALT}\\\text{(UNDEVELOPED)}}}{AgBr(s)} \;\; + \underset{\substack{\text{FROM}\\\text{"HYPO"}\\\text{SOLUTION}}}{2\,S_2O_3^{2-}} \rightarrow \underset{\substack{\text{WATER-SOLUBLE}\\\text{COMPLEX}}}{Ag(S_2O_3)_2^{3-}} + Br^-$$

Solutions of sodium thiosulfate, $Na_2S_2O_3$, known as "hypo," were first used by Sir J. W. F. Herschel to "fix" negatives.

Instant Black and White Pictures

In 1947, Dr. Edwin H. Land introduced his invention of a process that produced a finished picture in 1 minute. Since their introduction, the Polaroid and similar processes have been popular for just this unique feature.

After exposure, a Polaroid film is brought into contact with a piece of receiver paper; at the same time a pod containing both a developer and a silver solvent is broken and the pasty mixture is spread over the film. As the developer reduces the exposed silver halide grains in the film emulsion, the silver solvent picks up the unexposed silver ions, which then diffuse across the boundary onto the receiver paper. There, in contact with minute grains of silver already in the paper, the developer reduces the silver in the hypo complex to free silver and a **positive** image. This positive image forms since black areas in the original object photographed exposed no silver grains in the emulsion, and it was this silver, as Ag^+ ions, which was then carried onto the receiver paper and reduced (Fig. 24–9).

Spectral Sensitivity

Probably the most important ingredients in a black and white photographic emulsion, other than the silver halide salts themselves, are the spectral sensitizing dyes. Silver halides are most sensitive to blue light or higher energy electromagnetic

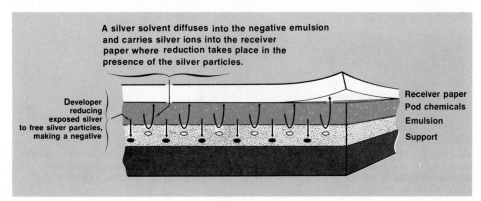

FIGURE 24–9 A schematic diagram of the chemistry of the Polaroid process.

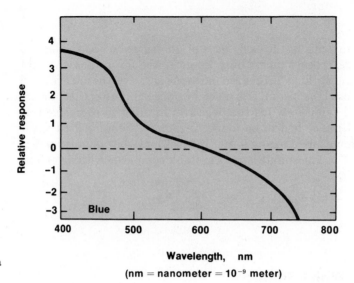

FIGURE 24–10
Spectral sensitivity of a typical AgBr emulsion.

Wavelength, nm

(nm = nanometer = 10^{-9} meter)

radiation such as ultraviolet light (Fig. 24–10). A film manufactured with only silver halides as the photosensitive agents will be only blue-sensitive and will not "see" reds, yellows, greens, and so on, as ordinary colors.

In 1873, while trying to eliminate light scattering problems in photographing the solar spectrum, W. H. Vogel, a German chemist, added a yellow dye to his emulsion. To his surprise he discovered that he could now record images in the green region of the visible spectrum. Later, in 1904, another German, B. Homolka, discovered a dye, pinacyanol (Fig. 24–11), which when added to a silver halide emulsion rendered it sensitive to the entire visible spectrum (Fig. 24–12). Films of this type are called **panchromatic** or "pan" films.

The mechanism by which a dye molecule can impart spectral sensitivity to silver halide grains seems to involve initially the absorption of a photon of light by the dye molecule. Next, the excited molecule ejects an electron into the silver halide grain, where a free silver atom is formed. The electron-deficient dye molecule then oxidizes the bromide ion, producing a bromine atom:

$$\text{Dye} \xrightarrow{\text{light}} \text{Dye}^+ + \text{e}^-$$
$$\text{Ag}^+ + \text{e}^- \rightarrow \text{Ag}^0 \rightarrow \text{Silver nuclei}$$
$$\text{Dye}^+ + \text{Br}^- \rightarrow \text{Dye} + \text{Br}^0$$

By adding spectral sensitizing dyes, photographic emulsions can be made that are sensitive to selected regions of the spectrum with wavelengths from 100 nanometers (ultraviolet) to 1300 nanometers (infrared).

A panchromatic film is sensitive to the entire range of visible wavelengths.

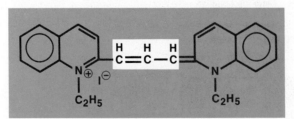

FIGURE 24–11   Pinacyanol, a cyanine dye. The conjugated group (double bond, single bond, etc.—one sequence in the white block) produces the color of the dye. Other cyanine dyes have more CH groups, absorb longer wavelengths of light, and shift the film sensitivity toward the red.

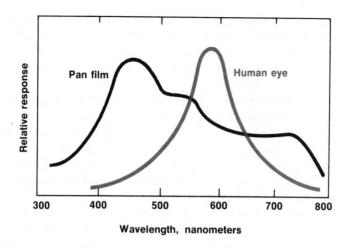

FIGURE 24–12
Spectral sensitivity of a panchromatic film compared to that of the human eye.

Thus, the process is effectively the same as a photon striking the bromide in the grain itself, the dye serving as a catalyst.

Color Photography

The chemistry of color photography, which dates back to 1861, is more complicated than that of black and white photography. James Clerk Maxwell, the famous English scientist, was the first to photograph an object in color. He used three exposures through three primary color filters and then resynthesized the color of the object by projecting the images through the same filters. This experiment actually predated panchromatic film, but because of a peculiarity of his particular emulsion, it was essentially panchromatic nonetheless. The important point is that the results were consistent with the then emerging theory of color vison and thus led to other more significant results.

Additive and Subtractive Primary Colors

As early as 1611, De Dominis showed that the visible spectrum could be approximated by three fundamental colors: red, green, and blue (known as **additive primaries**). This concept has since proved useful in the development of color vision theory and in color photography. After 1861 the idea slowly evolved that in order to reproduce color images, a film would have to be made with three different layers, each layer sensitive to one of the three primary colors. After the discovery of color-sensitizing dyes and panchromatic black and white film, several different techniques for color photography were developed; but not until 1935, when Kodachrome was placed on the market, did the products reach the consumer. The Kodachrome process produces transparencies (or slides) that are viewed with transmitted light.

If additive primary colors are used to form an image by superposition, such as we might expect in a color transparency, problems with light transmittance arise. Combinations of additive primary-color filters produce black. In order to overcome these problems and obtain a color slide (or color negative, for that matter), another system of primary-color filters was developed, known as **sub-**

$$\left.\begin{array}{l}\text{Blue light +}\\\text{Green light}\end{array}\right\}\text{Cyan}$$

$$\left.\begin{array}{l}\text{Green light +}\\\text{Red light}\end{array}\right\}\text{Yellow}$$

$$\left.\begin{array}{l}\text{Blue light +}\\\text{Red light}\end{array}\right\}\text{Magenta}$$

tractive primary colors. These colors are produced by dyes that absorb the additive primary colors. Thus a dye that absorbs red light transmits or reflects the remainder of the spectrum and appears greenish-blue *(cyan)*. Absorption of the blue light renders a dye yellow, and absorption of green light makes the dye appear bluish-red *(magenta)*.

When the proper mixture of subtractive primary dyes is formed in a photographic emulsion during the development process, an image is produced with the desired color. For example, a mixture of magenta and cyan dyes would appear blue, since the magenta dye absorbs green light and the cyan dye absorbs red light, leaving only blue to be transmitted out of the three components of white light.

The use of subtractive primaries in color photography was suggested as early as 1869, but it was much later before the chemistry was worked out in enough detail to yield good results. The problem is to get the right amount of the correct subtractive primary dye in the right place to reproduce the correct true-to-life color. White is produced by the absence of all three subtractive primaries, and black is produced by an equal balance of all three.

Color Film

Generally, a color film consists of a support and three color-sensitive emulsion layers. The blue-sensitive layer is usually on the top since silver halides are inherently blue-sensitive. Next, a yellow colored filter layer is added. This layer absorbs blue light and serves to protect the lower emulsion layers from blue light. A green-sensitive layer is added and followed by a red-sensitive layer and the support (Fig. 24–13). These layers are rendered color-sensitive by dyes similar to those in the cyanine class, which render black and white film panchromatic. It should be realized, however, that the color-sensitizing dyes are *not* generally involved in producing the final primary colors responsible for the color of the image. It is the final processing of the color film that yields the color image.

The thickness of the entire emulsion of color film is only about 0.0254 mm (0.001 inch).

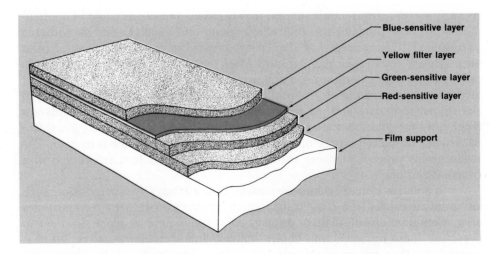

Blue-sensitive layer

Yellow filter layer

Green-sensitive layer

Red-sensitive layer

Film support

FIGURE 24–13   A typical arrangement of color-sensitive emulsion layers in a color film.

Color Development

Most color films are developed with the aid of a dye-forming color process first introduced by a German chemist, R. Fischer, in 1912. The basis for this process is the oxidation of the developer to a dye-forming substance, which is then allowed to react with a molecule called a **coupler** to form the dye.

Color developers are generally substituted amines and as such are reducing agents. An example is N,N-diethyl-p-phenylenediamine:

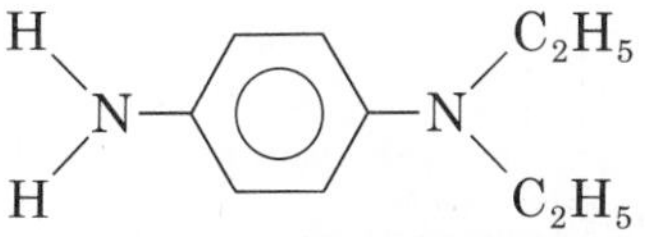

N,N-DIETHYL-p-PHENYLENEDIAMINE
(A COLOR DEVELOPER)

To form a cyan dye during the development process, a phenol compound such as α-naphthol acts as a coupler:

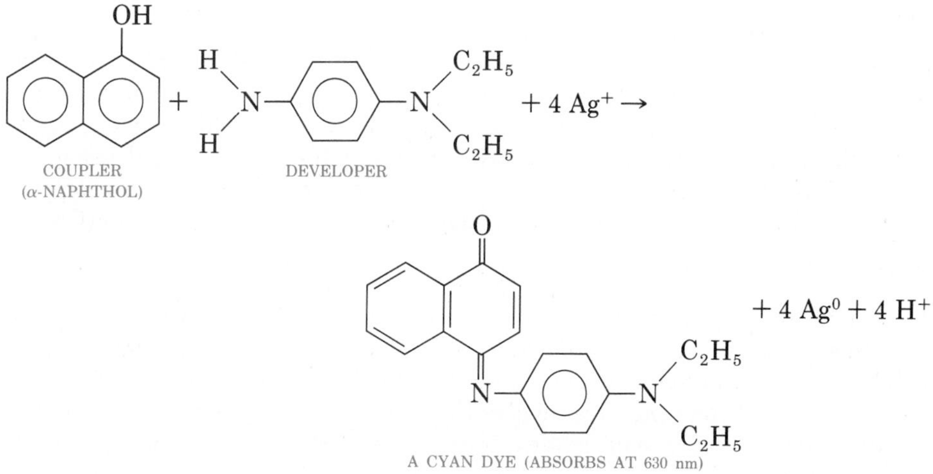

COUPLER
(α-NAPHTHOL)   DEVELOPER   + 4 Ag$^+$ →

+ 4 Ag^0 + 4 H$^+$

A CYAN DYE (ABSORBS AT 630 nm)

Thus, in the development of an exposed silver halide grain in the red-sensitive emulsion layer, a small amount of cyan dye is produced. The free silver must be bleached out prior to finishing.

Light with a wavelength of 630 nm is red.

The Kodachrome Process

An interesting example of a widely used color photography system is the Kodachrome process of the Eastman Kodak Company. The Kodachrome process is a **reversal** process; this means colors are reproduced in terms of their correct values and not their negative or complementary colors. The first developer in the Kodachrome process is a black and white developer. By careful temperature control, development of the exposed silver halide is made essentially complete.

The remaining unexposed silver halide in the three color-sensitive emulsions is a positive record of the original exposure. For example, red light striking the film would, upon black and white development, leave free silver in the red-sensitive layer (Fig. 24–14). Since no other color-sensitive layers were exposed by the origi-

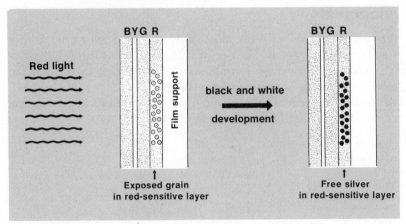

FIGURE 24–14 Simplified color image-forming process. B—blue-sensitive layer; Y—yellow filters; G—green-sensitive layer; R—red-sensitive layer.

All the silver is bleached out of color negative and reversal films during processing.

nal image, they contain no information. Now, selective reexposure and color development will produce free silver throughout the emulsion layers, along with the colored dyes, *except* where the red light originally struck the film. No dye forms there since the silver was previously reduced with a black and white developer.

Next, all the silver in the three emulsion layers, as well as the yellow-colored protective layer, is bleached with an oxidant such as cyanoferrate ion, $Fe(CN)_6^{3-}$:

$$Ag^0 + Fe(CN)_6^{3-} \rightarrow Ag^+ + Fe(CN)_6^{4-}$$

Once oxidized, the silver is treated with hypo and washed from the emulsion. The resulting emulsion is colored, but transparent. Considering that red light originally exposed the film, we see that the transmitted light will appear red (Fig. 24–15).

"Instant" color pictures involve the same principles described above, but require a delicate balance of light exposure, photochemical reagents, dyes, developers, and couplers.

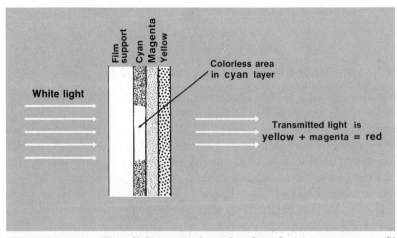

FIGURE 24–15 White light passes through a three-layer transparency. Since blue light and green light are absorbed, the transmitted light is red.

SELF-TEST 24–B

1. What chemical composes the dark regions on a negative? _____
2. Name a chemical that can be used as a black and white developer.

   _____
3. In the development process, silver is ( ) oxidized or ( ) reduced.
4. Complete the following equation for the fixing process.

   $Ag^+ + 2\,S_2O_3^{2-} \rightarrow$ _____
5. Which are more sensitive to light, very small or somewhat larger grains of AgBr? _____
6. Photographic developers are ( ) oxidizing or ( ) reducing agents.

   _____
7. Hypo is another name for _____ .
8. a. Cyan + magenta = _____

   b. Cyan + yellow = _____

   c. Magenta + yellow = _____
9. The primary additive colors are _____ , _____ , and

   _____ .
10. A Kodachrome color transparency has no silver in it. ( ) True or ( ) False
11. In the formation of a dye in color film, silver is ( ) oxidized or ( ) reduced.
12. Film sensitive to all light in the visible range is known as _____ film.

PHOTOCOPYING

Photocopying also uses photosensitive materials. Although there are different photocopying methods, one of the more common ones is xerography (Fig. 24–16). In this method, a plate coated with a photoconductor such as zinc oxide or selenium receives a positive electrical charge. The image of the original document is projected through a lens onto the charged plate. The electric charge is partially or completely drained in areas that are exposed to light because of the photoconducting properties of zinc oxide or selenium. The remaining charged area represents the latent image that is developed by applying a negatively charged graphite powder (toner, "dry ink"). The toner is picked up by the charged surface (electrostatic attraction) and is baked into the paper with a heater. Hence, the absence of light produces black. High-quality equipment is very discriminating in the production of grays and produces almost photographic quality in copies. Colored toners are also becoming quite popular.

VIDEO DISPLAY

The cathode-ray tube, described in Chapter 3, was the forerunner of video display devices such as the TV picture tube. Figure 3–6 illustrates the use of a phosphor material such as zinc sulfide to detect an electron beam. Magnetic fields can be used to focus the electron beam.

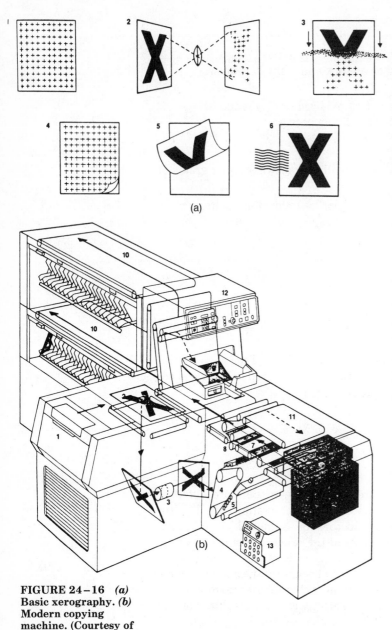

Basic Xerography

(1) A photoconductive surface is given a positive electrical charge (+).

(2) The image of a document is exposed on the surface. This causes the charge to drain away from the surface in all but the image area, which remains unexposed and charged.

(3) Negatively charged powder is cascaded over the surface. It electrostatically adheres to the positively charged image area making a visible image.

(4) A piece of plain paper is placed over the surface and given a positive charge.

(5) The negatively charged powder image on the surface is electrostatically attracted to the positively charged paper.

(6) The powder image is fused to the paper by heat.

After the photoconductive surface is cleaned, the process can be repeated.

The 9400

The original document automatically moves from the document handler (1) to the platen (2) where it is exposed by lamps and mirrors through a lens (3) focusing the image (in same or selected reduced sizes) onto the photoreceptor belt (4). Magnetic rollers (5) brush the belt with dry ink which clings to the image area.

A sheet of copy paper moves from either the main or auxiliary tray (6) to the belt, where the dry ink is transferred to it (7). The copy then goes between two rollers (8), where the dry ink image is fused to it by heat and pressure.

A copy of a single-page document emerges in the receiving tray (9). Copies of multi-page documents go to the sorter (10) for collating into as many as 999 sets. If the sheet is to be copied on both sides, it returns by conveyor (11) to the auxiliary tray to repeat the process.

The 9400 also features a control console (12) with lighted instructions guiding the operator on all jobs, and a maintenance module (13) allowing easy adjustment and testing of the unit's systems.

FIGURE 24–16 *(a)* **Basic xerography.** *(b)* **Modern copying machine. (Courtesy of Xerox Corporation)**

Use a magnifying glass to examine a picture in the newspaper to "see" a picture made of dots.

TV images are constructed on the screen at a rate of 30 per second, which is fast enough that our eyes sense only the full screen image.

Imagine that the image focused by the lens of the TV camera onto the photosensitive surface is an array of points. Each point in the image is referred to as a picture element or **pixel.** The normal television image is 572 by 572 pixels. The TV camera is a scanning device that uses photodetectors to convert the intensity of the photon signal at each pixel into an electrical signal. The camera rapidly scans each image, line by line, 30 images per second, and produces a succession of electrical signals that correspond to the different light intensity signals. At the television receiver, the signal is converted back to an image on the screen by imposing the electrical signal on an electron beam (similar to the beam in a cathode-ray tube), which then strikes a phosphor coating on the screen. The image is reassembled on the TV screen as a 572-by-572 array of points just as in the case of the camera scan.

The variation in the electrical signal affects the brightness of the phosphor coating at a given point.

In summary, photodetectors in imaging devices use photoconducting materials to convert photons into an electric signal that can be processed to obtain information about the image. The photodetectors in a scanning device such as a TV camera replace the eye or the photographic film by scanning the image plane point by point.

Pictures of the Earth received from satellites in outer space use the same principles as those described here for the TV camera and video display.

MEDICAL IMAGING

A variety of medical imaging devices is used in medical diagnosis. The only one available until the 1950s was X-ray imaging, which has been in use since 1896. In the 1950s, gamma-emitting radioisotopes were introduced to medical diagnosis. Since then, the succession of new techniques has included ultrasound, X-ray computed tomography (CAT or CT), positron emission tomography (PET), and nuclear magnetic resonance (NMR) or magnetic resonance imaging (MRI). All except the ultrasound imaging devices make use of photons from various regions of the electromagnetic spectrum shown in Figure 24–1. They include gamma-ray photons in nuclear medicine imaging, X-ray photons in X-ray imaging, and radiofrequency photons in NMR imaging. Ultrasound imaging uses high-frequency sound waves.

A major difference between medical imaging devices and the eye or camera is the source of photons. The eye and the camera receive photons reflected by the surface of opaque objects. Medical imaging devices include the energy source of photons as part of the device, and the energy is either transmitted or reflected by structures within the body.

Review gamma-emitting radioisotopes in Chapter 7.

NMR imaging is now being referred to as magnetic resonance imaging (MRI) to avoid confusion with nuclear medicine. As you will see, NMR has nothing to do with radioactive isotopes.

X-Ray Imaging

The part of the body being examined is exposed to an X-ray source. The X-ray photons that pass through the body strike a fluorescent screen sensitive to X-ray photons. The visible light photons given off by the fluorescent screen expose the light-sensitive film placed in close contact with the screen, thereby producing an image. Contrast between bone and soft tissue occurs because photons that hit atoms lose their energy by ionizing the hit atoms and therefore fail to reach the film. Bones have a greater density of atoms than does soft tissue. Only 1% of the incident X-ray photons pass through the body unabsorbed.

Although X-ray film techniques are effective for imaging parts of the body with high contrast, such as lungs and bones, organs composed of soft tissue cannot be distinguished unless some contrast agent is used. Since X rays scatter more from heavier atoms than from light atoms, introduction of heavy elements can provide contrast. For example, patients are given a suspension of barium sulfate ($BaSO_4$) to outline the digestive tract (Fig. 24–17). Another contrasting agent is iodine, which allows blood vessels to be contrasted when it is injected into the bloodstream.

X-ray computed tomography, or computed axial tomography, confines the beam of X rays to a thin "slice" of the body (Fig. 24–18), and images or tomograms of many individual slices are taken. The X-ray beam is narrowly focused, and this decreases X-ray scatter, which improves the image contrast. Figure 24–18 is a diagram of one arrangement used for CT in which the X-ray detectors travel in

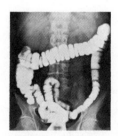

FIGURE 24–17
$BaSO_4$ is used in intestinal X-ray studies to provide contrast.

Although barium is a toxic metal, barium sulfate is safe to use because it is insoluble in water and passes through the body with no ill effect.

Tomo in tomography means "section."

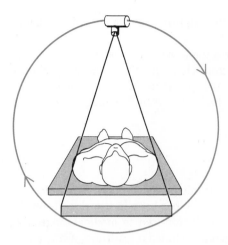

FIGURE 24–18 Drawing illustrates one possible arrangement for CT. The X-ray tube and the detectors rotate in unison around the body.

synchronization with the X-ray source. The signal from the X-ray detectors is fed into a computer for mathematical construction of the image. CT imaging is more time consuming than X-ray film imaging because irradiation and data acquisition usually take 2 to 10 seconds per slice and the calculations take 15 to 30 seconds. The resulting CT image may be viewed on a video screen or stored on magnetic tape for permanent record. The quality of CT images is superior to that obtained by X-ray film techniques. CT imaging is useful for imaging the brain, diagnosing hemorrhages and strokes, evaluating certain kinds of cancers, and determining the size and position of cerebral ventricles. Figure 24–25a shows a CT image of the brain of a patient who has multiple sclerosis. The dark spots of hardened tissue in the brain are barely visible above the background.

Nuclear Imaging

Review the discussion in Chapter 7 on nuclear medicine.

The **m** in technetium-99^{m} stands for *metastable*. ^{99m}Tc is a gamma-emitter with a half-life of 6 hours, which is ideal for medical purposes.

The use of nuclear medicine in diagnosis and therapy was described in Chapter 7. In nuclear imaging, a gamma-emitting isotope, either as the element or in a compound, is chosen on the basis of its tendency to accumulate at the site of the disorder. Table 7–10 lists some of the gamma-emitting isotopes that will be discussed here. For example, iodine-131 and technetium-99^{m} are used to study the thyroid. Technetium-99^{m} is also used to detect brain tumors and bone tumors. Thallium-201 is used for coronary heart disease studies.

The method is based on the emission of gamma rays from the target organ (Fig. 24–19). The gamma rays strike photosensitive sodium iodide in the imaging device. The photon signal emitted from the sodium iodide is converted to an electric signal and amplified with photomultiplier tubes. The resulting signal is processed by a computer and fed to a video display for construction of the image on the screen. Figure 24–20 illustrates the use of nuclear imaging to detect tumors in the thyroid.

A positron has the positive charge of a proton but only the mass of an electron.

Positron emission tomography is a form of nuclear imaging that uses positron emitters, such as carbon-11, fluorine-18, nitrogen-13, or oxygen-15, that are incorporated into compounds that enter metabolic reactions. For example, deoxyglucose labelled with positron-emitting fluorine-18 is injected intravenously. After 40 minutes, a tomographic image of the brain reveals the local metabolic activity. The method uses the scanning techniques described for CT imaging. A compound containing the positron-emitting isotope is inhaled or injected prior to the scan.

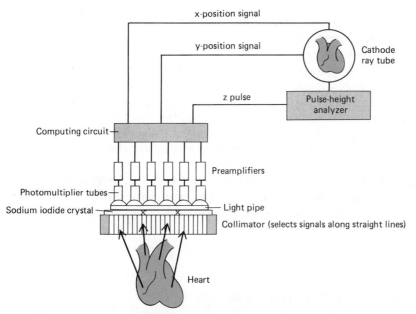

FIGURE 24–19   The gamma rays emitted from the organ of interest strike a sodium iodide crystal. The photon signal from the sodium iodide crystal is converted to an electrical signal, amplified, and then processed by a computer for construction of an image on the cathode-ray tube. (Redrawn from C. Carl Jaffe: Medical imaging. *American Scientist,* **Vol. 70, p. 579, 1982, with permission of Sigma Xi)**

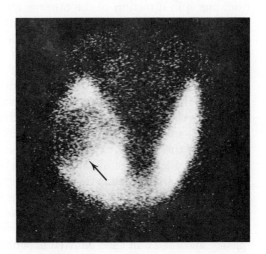

FIGURE 24–20   Nuclear image of the thyroid with technetium-99ᵐ. The darkened area on the left is a tumor. (Courtesy of the Department of Radiology, Vanderbilt University)

The emitted positron collides with a neighboring electron, and two gamma rays are produced.

$$_{+1}^{0}e + _{-1}^{0}e \rightarrow 2\gamma$$

These gamma photons travel in opposite directions and are detected by gamma-ray detectors 180 degrees apart. The imaging process is then the same as that described previously for gamma rays.

Although the spatial resolution of PET is not as good as CT, PET offers the advantage of being applicable to metabolic processes taking place in various organs. PET is particularly useful for brain scans. For example, PET has detected differences in glucose metabolism in patients suffering from Alzheimer's disease, epilepsy, schizophrenia, or stroke. Researchers have also been able to use this technique to determine what areas of the brain are active when subjects are engaged in verbal analysis or spatial analysis.

A major limitation to widespread use of PET is the requirement of a nearby cyclotron to produce positron-emitting isotopes since their half-lives are in the range of minutes. As a result, there are only a few locations where PET can be used as a diagnostic method.

Nuclear Magnetic Resonance Imaging

s = +½

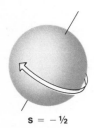

s = −½

Felix Bloch and Edward Purcell were awarded the Nobel prize in 1952 for their discovery of nuclear magnetic resonance.

In Chapter 5, the magnetic field created by the spinning electron was mentioned in connection with the quantized spin of electrons (clockwise or counterclockwise spin, represented by quantum numbers $+1/2$ and $-1/2$). Since the charge on the proton (hydrogen nucleus) is $+1$, a spinning proton also has a magnetic field associated with it. When an external magnetic field is applied, there is a slight preference for the hydrogen nuclei to spin in one direction, or orientation, with the magnetic field of the nucleus aligned with the external field, but the energy difference between the two orientations of spin is small. The energy for flipping nuclei from one spin direction to the other is quantized so Planck's equation, $E = h\nu$, applies. The energy per quantum is low enough that the frequencies in the radio-frequency region of the electromagnetic spectrum can change the direction of spin of hydrogen nuclei whose spins have been aligned by a strong magnetic field.

In 1945, physicists Felix Bloch of Stanford University and Edward M. Purcell of Harvard University, working independently, discovered the phenomenon of NMR. They did this by placing a sample of water in a strong magnetic field and then exposing the sample to a radiofrequency source. By scanning different radiofrequencies, they discovered a signal was emitted that depended on the strength of the magnetic field and the chemical environment (kind of hydrogen compound). Picture the aligned hydrogen nuclei absorbing a particular radiofrequency and being flipped to the less stable spin direction. When the nuclei flip back to the more stable spin direction, the radiofrequency is emitted. The emitted radiofrequency can be measured with a radio receiver.

The discovery of NMR led to extensive use by chemists since the radiofrequency absorbed and then emitted depends on the chemical environment of the hydrogen atom. Chemists could deduce what kind of atoms were bonded to hydrogen as well as the number of hydrogen atoms present. Hence, NMR has been a valuable structural and analytical tool.

Nuclei other than hydrogen can be examined by NMR. These include many nuclei of biological interest such as ^{1}H, ^{31}P, ^{13}C, and ^{23}Na. The radiofrequency required to change nuclear spin directions is quite different for different nuclei. However, the signal is much weaker for nuclei other than hydrogen, and this is more of a problem in magnetic resonance imaging than in the chemical use of NMR.

Since our bodies are 70% water and hydrogen gives a strong NMR signal, ^{1}H is the most logical candidate for NMR imaging. The first use of NMR for imaging was not reported until 1973. The reason for the delay is related to the technical difficulties associated with getting a uniform magnetic field with sufficient diameter to hold a patient. In addition, the advances in computer technology for analysis of data and construction of an image from this data were needed.

NMR imaging is based on the time it takes for the hydrogen nuclei in the unstable high-energy position to "relax" or flip back to the low-energy position after the nuclei have been flipped. These times are different for hydrogen nuclei in fat, muscle, blood, and bone, and these differences are the basis for the contrast in the NMR image.

In medical NMR imaging (or, as it's now known, MRI), the patient is placed in an opening of a large magnet (Fig. 24–21 and Fig. 24–22). The magnetic field aligns the magnetic spin of hydrogen nuclei (as well as other magnetic nuclei). A radiofrequency transmitter coil is placed in position near the region of the body to be examined. The radiofrequency energy absorbed by the spinning hydrogen nuclei causes the aligned nuclei to flip to the less stable, high-frequency spin direction. When the nuclei flip back to the more stable spin state, a radiofrequency signal is emitted. The intensity of the emitted signal is related to the density of hydrogen nuclei in the region being examined, and the time it takes for the signal to be emitted (relaxation time) is related to the type of tissue. The emitted radiofrequency signal is received by a radio receiver coil, which then sends it to a computer for mathematical construction of the image, similar to the mathematical treatment for data from CT images.

Contrast in MRI is obtained by making small changes in the magnetic field for different parts of the "slice" being examined. Since the signal is field dependent, a well-defined gradient in the magnetic field produces different relaxation times. Figure 24–23 shows the excellent anatomical detail that is possible with MRI.

FIGURE 24–21 Magnetic resonance imaging machine. The patient is placed on the platform and rolled forward into the magnet opening, which also contains the radiofrequency coils. (Courtesy of the Department of Radiology, Vanderbilt University)

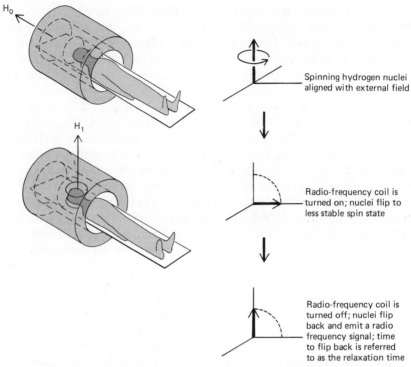

Spinning hydrogen nuclei aligned with external field

Radio-frequency coil is turned on; nuclei flip to less stable spin state

Radio-frequency coil is turned off; nuclei flip back and emit a radio frequency signal; time to flip back is referred to as the relaxation time

FIGURE 24–22   Schematic of magnetic resonance imaging method. The external magnetic field H_0, is used to align the spinning nuclei. The radiofrequency coil is turned on, and the resulting radiofrequency, H_1, flips the nuclei to the less stable spin state. A radiofrequency signal is generated when the nuclei flip back to the more stable spin state.

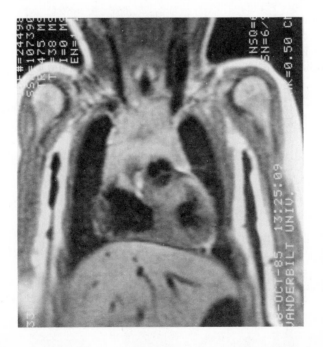

FIGURE 24–23   Magnetic resonance image of a normal two-year-old child. Front view shows the excellent anatomical detail that is possible with MRI. (Courtesy of the Department of Radiology, Vanderbilt University)

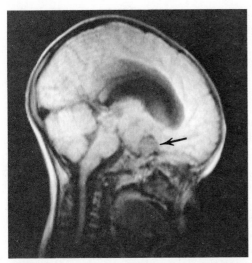
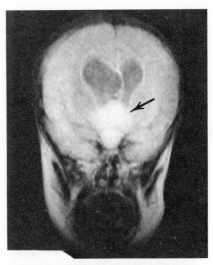

FIGURE 24–24   Magnetic resonance image of the skull of an 18-month-old child. *(a)* Side view — dark mass at the arrow is a brain tumor. *(b)* Top view — use of different instrument parameters shows brain tumor as a white mass. (Courtesy of the Department of Radiology, Vanderbilt University)

NMR imaging can readily distinguish brain tumors from normal brain tissue, not only on the basis of altered anatomy but also on the basis of high water content due to hypervascularity or reactive edema. Figure 24–24 shows two views of an 18-month-old child's skull. Note the tumor mass is dark gray in Figure 24–24a and white in Figure 24–24b. Differences in contrast are obtained by using different instrument parameters.

NMR images of the skull are clearly superior to those provided by CT imaging, and the process is safer because ionizing radiation isn't used. This is illustrated in Figure 24–25, which shows CT and NMR images of the head of an adult patient with multiple sclerosis. Patches of hardened tissue in the brain are

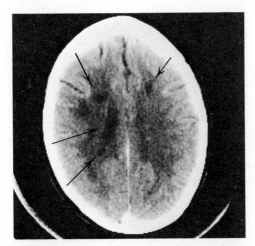
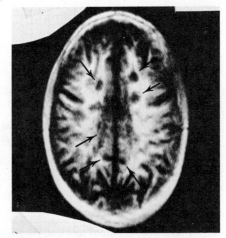

FIGURE 24–25   CT image and magnetic resonance image of the top of the skull of an adult with multiple sclerosis. *(a)* Black spots of hardened tissue in the CT image of the brain are barely visible in the gray background. *(b)* Black spots of hardened tissue are clearly visible against the white background in the magnetic resonance image. (Courtesy of the Department of Radiology, Vanderbilt University)

barely visible above the background in the CT image. The MRI image shows the black spots clearly against the white background of softer tissue.

ELECTRON MICROSCOPE

An atom of 3×10^{-8} cm in diameter could only be magnified to 6×10^{-3} cm on the viewing screen, too small to be seen, but giant molecules can be "seen" with these diffraction patterns.

Another important imaging device is the electron microscope. This imaging device is capable of magnification up to 200,000 times. The electron microscope uses an electron beam rather than a light beam. The electron beam is focused with electric and magnetic fields instead of optical lenses. The object is placed in a vacuum chamber, where the focused beam of electrons is directed at the surface of the object.

The wave properties of electrons were described in Chapter 3, where it was pointed out that electron beams with velocities close to the speed of light can be diffracted by the layers of atoms in a metal such as silver. This is the basis for the electron microscope. The difference between the electron beams in the cathode ray tube or television picture tube and the electron beam in the electron microscope is the velocity of the electron beam. Wave properties of electrons are observed experimentally only for electrons that have a velocity approaching the speed of light. For example, an electron beam moving at 3×10^6 meters per second, 0.01 of the speed of light, has a wavelength of 0.242 nanometers, which is similar to the interatomic distance in metals.

Image contrasts are formed by the scattering of electrons by the object. Denser portions scatter more electrons and hence appear darker in the image. When the electron beam hits the object, it knocks electrons off the object's atoms into the vacuum, where they are attracted to an electronic sensor or photodetector as described earlier. This electronic signal is processed by a computer and projected on a video display in the same way as that described for other video displays. Figure 1–7 is an example of electron microscope magnification.

SELF-TEST 24–C

1. Each point or picture element of an image is referred to as a _____ .
2. TV cameras use _____ to convert the photon signal into an electrical signal.
3. The screen of a TV receiver has a _____ coating to convert an electrical signal into a light signal.
4. Nuclear imaging is based on radioactive isotopes, which emit _____ .
5. In NMR imaging, the energy of the electromagnetic radiation needed to flip the direction of spinning hydrogen nuclei aligned with a magnetic field is in the _____ region.
6. The safest medical method is _____ because it does not use ionizing radiation.
7. The imaging device that uses the wave properties of the electron is the _____ .

MATCHING SET

_____ 1. Negative
_____ 2. Larger grain size in film
_____ 3. Daguerreotype
_____ 4. George Eastman
_____ 5. Hydroquinone
_____ 6. Acid solution
_____ 7. Sodium thiosulfate
_____ 8. Panchromatic
_____ 9. Red, green, and blue
_____ 10. Pinacyanol
_____ 11. Edwin H. Land
_____ 12. Zinc sulfide
_____ 13. Green phosphor
_____ 14. Germanium

a. Introduced Kodak
b. Developer
c. Stop bath
d. More light sensitive film
e. Additive primary colors
f. Early photographic form
g. A cyanine dye
h. Introduced instant photography
i. Exposed areas are dark
j. Film that is sensitive to broad band of wavelengths
k. Fixer
l. Subtractive primary colors
m. Photoconductor
n. Phosphor in cathode-ray tube
o. Copper-activated zinc sulfide

QUESTIONS

1. Explain the difference between photochemicals, photoconductors, and phosphors, and give one example of each.
2. Diagram the components of NMR imaging, and explain the function of each component.
3. What advantage does NMR imaging offer over CT?
4. A number of the medical imaging devices use a group of letters to represent the method. Identify the methods that are represented by NMR, CT, CAT, PET.
5. Why is MRI replacing NMR as the symbolism for the medical imaging method that uses the principles of nuclear magnetic resonance?
6. Give an example of how computer technology is essential to medical imaging.
7. What is the difference in principle of operation between a film camera and a TV camera for recording an image?
8. Visible light causes a reversible chemical change in the eye. However, ultraviolet light can cause permanent blindness. Explain on the basis of the size of the electromagnetic quantum.
9. What would be the effect of lower pH on a typical developer?
10. What is the chemical explanation for fogging on film?
11. Explain the term _fixing_. What is a fixer, and why is it important in photography?
12. Describe what happens chemically from the time the shutter is opened on a camera until the latent image is developed on the film.
13. Write a chemical equation for the development of silver ions by Metol.
14. Explain what the stop bath solution does in the development process.
15. What are the subtractive primary colors?
16. Explain how a red dot would be photographed with a color reversal film such as Kodachrome.
17. What color is produced by the superposition of the following subtractive primary colors?
    a. Magenta and yellow
    b. Cyan and yellow
    c. Magenta and cyan
    d. Magenta, cyan, and yellow
18. Is all of the silver reduced in the development of a black and white film? Explain.
19. What different support materials have been used for the light-sensitive emulsion in photography?
20. What is the purpose of sodium sulfite in black and white film development?
21. Where are the developer and fixer in the black and white Polaroid film?

22. What was done to increase the spectral range of black and white film?

23. Explain the action of a coupler in color photography.

24. Relate the concept of the half-life of a radioactive isotope to the limited use of positron emission tomography (PET).

25. Is silver essential to the photographic industry at present?

26. In a completely different system of black-and-white photography that does not use silver salts, what chemicals would you use?

EPILOGUE

What does the future hold, chemically speaking? No one knows! However, this will not keep thoughtful persons from extrapolating from past and current events to the future for a variety of reasons. Some enjoy the intellectual challenge, others are dedicated to the solution of current and developing human problems, while many simply want to position themselves in a strong financial capability in the ebb and flow of the appetite expressed by our society.

Barring catastrophic disasters such as a thermonuclear war or a contagious, uncontrolled disruption of "normal" cellular chemistry, molecular manipulations and chemical understandings will cause and (or) allow us to:

1. Produce vastly greater amounts of food through both the modification of the food-producing organism and the chemical environment in which it lives, and the chemical transformation of previously unacceptable food raw materials into usable foodstuffs.

2. Use to near extinction the petroleum reserves of the world for its burnable energy content, necessitating renewed and increasing interest in the clean use of coal. (Engineers are presently working on new designs for coal-fired trains.)

3. Generate a larger percentage of electricity with nuclear power, generating ever larger stockpiles of radioactive wastes that require centuries of protected storage.

4. Proceed slowly with alternative sources of energy, such as solar, geothermic, and wind, as long as fossil fuels can be burned to produce cheaper energy.

5. Move toward environmental controls that will include the cost of waste disposal and cleanup in the cost of the item or material produced.

6. Produce an almost endless array of new materials that will revolutionize structural and facade materials in construction and in transportation and that will offer many new choices in the materials we use in clothing, personal tools, and surroundings for pleasure and comfort.

7. Continue the explosion in our ability to store and process information by the controlled molecular changes on the surface of semiconductor materials. The transmittal of knowledge will be increasingly cheap, allowing more human energy to be devoted to understandings and value choices; this will expand human creativity.

8. Expand our knowledge and control of genetic engineering and genetic coding of life controlling information, producing microbes designed to control specific

chemical changes in and out of other life forms, and modifying the chemistry of complex organisms to affect gross functions. Considerable environmental risk will be taken in this area as we try to sort out "cause and effect" at the same time we are modifying organisms that have the ability to reproduce at an exponential rate.

9. Continue to improve health and conquer disease by using recently acquired chemical knowledge to design and synthesize drugs to alleviate cancer, athero-sclerosis, hypertension, and disorders of the central nervous and immune systems.

Human control of chemical change is neither good nor bad in and of itself relative to most of our value systems; rather, it is the use made of the controlled change that can be classified as good or bad. It is evident that it can go either way in mass (possibly from the lack of purposeful choice or the lack of the understandings of consequential results) or for the individual who can dramatically alter his or her life through chemical choices.

Perhaps the most important question of all is the choice of the chooser! Who should be entrusted with these fateful choices? (1) Should it be the person or company that stands to make a financial profit from the change? (2) Should it be a government agent or agency that is properly schooled to make such selections for those who are represented or controlled? (3) Should it be a scientist, or a group of them, who gain their position in history by advancing new ideas (hopefully correct ones) and by having them accepted? The answer to this question of who is to be the chooser of the chemical choices is not altogether obvious because of personal limitations in knowledge and understanding. However, we believe that the most important scientific attribute, after intelligence, of course, is skepticism. Scientific skepticism calls for relatively little regard for human authority in explaining nature and total acceptance of natural displays as the final authority in understanding what is and what is not. Theory from any source, even though it can many times predict fact, must always be subservient to observable phenomena. If the mass in society can understand this most fundamental working in this and the other sciences, the public will realize that it cannot depend on any vested interest group to make the societal chemical choices and will, through government, seek to force the common sense of consensus choices. It is because of this belief that our future depends on this high level of societal chemical responsibility. We have presented to you the story of chemistry, how it works, and what its potentialities are. We hope you are convinced that you do not have to be a chemist to participate, through good citizenship and personal choices, in the control of the unfolding chemical story.

Appendices

Appendix A

THE INTERNATIONAL SYSTEM OF UNITS (SI)

Since 1960, a coherent system of units known as the Système International (SI system), bearing the authority of the International Bureau of Weights and Measures, has been in effect and is gaining acceptance among scientists. It is an extension of the metric system that began in 1790, with each physical quantity assigned a unique SI unit. An essential feature of both the older metric system and now the newer SI is a series of prefixes that indicate a power of ten multiple or submultiple of the unit.

Units of Length

The standard unit of length is the *meter.* It was originally meant to be one ten-millionth of the distance along a meridian from the North Pole to the equator. However, the lack of precise geographical information necessitated a better definition. For a number of years the meter was defined as the distance between two etched lines

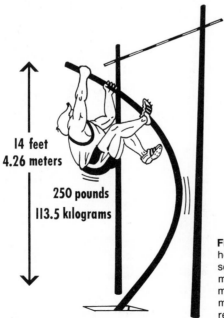

14 feet
4.26 meters

250 pounds
113.5 kilograms

Figure A-1   The pole-vaulter is easily recognized as hefty when described by 250 pounds and his jump something less than a record 14 feet. As Americans move closer to the use of the system of international measurements, the 113.5 kilograms and the 4.26 meters will produce similar conceptualizations related to previous experience.

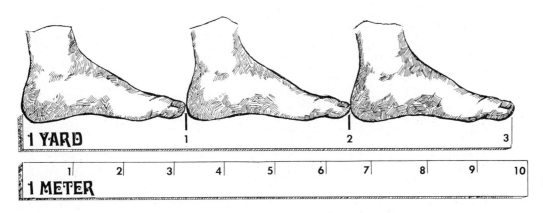

Figure A-2 A meter equals 1.094 yards.

on a platinum-iridium bar kept at 0°C (32°F) in the International Bureau of Weights and Measures at Sèvres, France. The inability to measure this distance as accurately as desired prompted a recent redefinition of the meter as being a length equal to 1,650,763.73 times the wavelength of the orange-red spectrographic line of $^{86}_{36}Kr$.

The meter (39.37 inches) is a convenient unit with which to measure the height of a basketball goal (3.05 meters), but it is unwieldy for measuring the parts of a watch or the distance between continents. For this reason, prefixes are defined in such a way that, when placed before the meter, they define distances convenient for our particular purposes. Some of the prefixes with their meanings are:

nano—1/1,000,000,000 or 0.000000001
micro—1/1,000,000 or 0.000001
milli—1/1,000 or 0.001
centi—1/100 or 0.01
deci—1/10 or 0.1
deka—10
hecto—100
kilo—1,000
mega—1,000,000

The corresponding units of length with their abbreviations are the following:

nanometer (nm)—0.000000001 meter
micrometer (μm)—0.000001 meter
millimeter (mm)—0.001 meter
centimeter (cm)—0.01 meter
decimeter (dm)—0.1 meter
meter (m)—1 meter
dekameter (dam)—10 meters
hectometer (hm)—100 meters
kilometer (km)—1,000 meters
megameter (Mm)—1,000,000 meters

Since the prefixes are defined in terms of the decimal system, the conversion from one metric length to another involves only shifting the decimal point. Mental calculations are quickly accomplished.

How many centimeters are in a meter? Think: Since a centimeter is the one-hundredth part of a meter, there would be 100 centimeters in a meter.

TABLE A–1 Conversion Factors*

Length:	1 inch (in.)	= 2.54 centimeters (cm)
	1 yard (yd.)	= 0.914 meter (m)
	1 mile (mi.)	= 1.609 kilometers (km)
Volume:	1 ounce (oz.)	= 29.57 milliliters (ml)
	1 quart (qt.)	= 0.946 liter (l)
	1.06 quart (qt.)	= 1 liter (l)
	1 gallon (gal.)	= 3.78 liters (l)
Mass (weight)†:	1 ounce (oz.)	= 28.35 grams (g)
	1 pound (lb.)	= 453.6 grams (g)
	1 ton (tn.)	= 907.2 kilograms (kg)

---

*Common English units are used.

†Mass is a measure of the amount of matter, whereas weight is a measure of the attraction of the earth for an object at the earth's surface. The mass of a sample of matter is constant, but its weight varies with position and velocity. For example, the space traveler, having lost no mass, becomes weightless in earth orbit. Although mass and weight are basically different in meaning, they are often used interchangeably in the environment of the earth's surface.

Conversion of measurements from one system to the other is a common problem. Some commonly used English-SI equivalents (conversion factors) are given in Table A–1.

Units of Mass

The primary unit of mass is the **kilogram** (1000 grams). This unit is the mass of a platinum-iridium alloy sample deposited at the International Bureau of Weights and Measures. One pound contains a mass of 453.6 grams (a five-cent nickel coin contains about 5 grams).

Conveniently enough, the same prefixes defined in the discussion of length are used in units of mass, as well as in other units of measure.

Units of Volume

The SI unit of volume is the **cubic meter** (m^3). However, the volume capacity used most frequently in chemistry is the liter, which is defined as 1 cubic decimeter (1 dm^3). Since a decimeter is equal to 10 centimeters (cm), the cubic decimeter is equal to $(10 \text{ cm})^3$ or 1,000 cubic centimeters (cc). One cc, then, is equal to one milliliter (the thousandth part of a liter). The ml (or cc) is a common unit that is often used in the measurement of medicinal and laboratory quantities. There are then 1,000 liters in a kiloliter or cubic meter.

Units of Energy

The SI unit for energy is the **joule** (J), which is defined as the work performed by a force of one newton acting through a distance of one meter. A newton is defined as that force which produces an acceleration of one meter per second per second when applied to a mass of one kilogram. Conversion units for energy are:

$$1 \text{ calorie} = 4.184 \text{ joules}$$
$$1 \text{ kilowatt-hour} = 3.5 \times 10^6 \text{ joules}$$

Other SI Units

Other SI units are listed below.

Time	second (s)
Temperature	Kelvin (K)
Electric current	ampere (A) $= 1$ coulomb per second
Amount of molecular substance	mole (mol) $= 6.023 \times 10^{23}$ molecules
Pressure	pascal (Pa) $= 1$ newton per square meter
Power	watt (W) $= 1$ joule per second
Electric charge	coulomb (C) $= 6.24196 \times 10^{18}$ electronic charges
	$= 1.036086 \times 10^5$ faradays

Further information on SI units can be obtained from "SI Metric Units—An Introduction," by H. F. R. Adams, McGraw-Hill Ryerson Ltd., Toronto, 1974.

Appendix B

TEMPERATURE SCALES

The system of measuring temperature that is used in scientific work is based on the *centigrade* or *Celsius* temperature scale. This temperature scale was defined by Anders Celsius, a Swedish astronomer, in 1742. The Celsius scale is based upon the expansion of a column of mercury that occurs when it is transferred from a cold standard temperature to a hot standard temperature. The cold standard temperature is the temperature of melting ice and is defined as 0°C. The hot standard temperature is the temperature at which water boils under standard conditions of pressure; it is defined as 100°C. The expansion of the mercury is assumed to be linear over this range, and the distance through which the mercury column expands between 0°C and 100°C is divided into 100 equal parts, each corresponding to one degree.

On the scale commonly used in the United States, the Fahrenheit scale (developed by and named after Gabriel Daniel Fahrenheit, a German-Dutch physicist of the early eighteenth century), the freezing mark is 32°F and the boiling mark is 212°F. Obviously, the ice is no hotter when a Fahrenheit thermometer is stuck into the system than when a Celsius thermometer is used. The marks on the tubes are simply different names for the same thing.

The ideal gas law predicts another temperature scale based on the temperature-volume behavior of a gas (Charles' law). Lord Kelvin reasoned that if an ideal gas lost $\frac{1}{273}$ of its volume during a temperature decrease from 0°C to −1°C, then at −273°C, theoretically at least, there should be no volume at all. All gases liquefy before this temperature is reached. However, this temperature, −273°C, appears to be a zero temperature defined by nature. Thus the Kelvin temperature scale was developed. Minus 273°C becomes 0 Kelvin and 0°C becomes 273 Kelvin. (More accurately, 273.15 K.) Note that current SI usage does not write °K, but K.

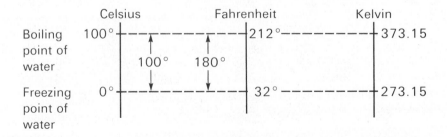

Because all these systems are now in common use, it is necessary at times to convert from one system to another.

Note that 100 Celsius degrees = 180 Fahrenheit degrees or
1 C degree = 1.8 (or ⁹⁄₅) F degrees.

But if we wish to convert temperature from one scale to another, we must remember that 0°C is the same as 32°F. Therefore, 32 must be added to the calculated number of degrees Fahrenheit in order to revert to the start of the counting in the Fahrenheit system.

Therefore,

$$°F = \tfrac{9}{5}\,(°C) + 32$$

which can be arranged to

$$°C = \tfrac{5}{9}\,(°F - 32)$$

Example 1
Convert 50°F to the corresponding Celsius temperature:

$$°C = \tfrac{5}{9}\,(°F - 32)$$
$$= \tfrac{5}{9}\,(50 - 32)$$
$$= \tfrac{5}{9}\,(18)$$
$$= 10°C$$

Example 2
Convert 25°C to the corresponding Fahrenheit temperature:

$$°F = \tfrac{9}{5}\,(°C) + 32$$
$$= \tfrac{9}{5}\,(25) + 32$$
$$= 45 + 32$$
$$= 77°F$$

To convert Celsius to Kelvin:

$$K = °C + 273$$

Example 3
Convert 23°C to Kelvin:

$$K = 23° + 273$$
$$= 296\ K$$

Rather than memorizing formulas, many students prefer to figure out temperature conversions using common sense. A common-sense approach is shown in Table B–1.

TABLE B–1 Common-Sense Method for Converting Temperatures

Starting point: $\begin{cases} \text{freezing point (fp) of water is } 0°C = 32°F \\ \text{boiling point (bp) of water is } 100°C = 212°F \end{cases}$

Other information needed: 100 C degrees = 180 F degrees

General Procedure	Example 1 (50°F = ?°C)	Example 2 (25°C = ?°F)	Example 3 (98.6°F = ?°C)
1. Select a reference point.	1. fp of water	1. fp of water	1. bp of water
2. Determine relationship of temperature of interest to reference point.	2. 50°F is 18 F degrees above fp of water	2. 25°C is 25 C degrees above fp of water	2. 98.6°F is 113.4 F degrees below bp of water
3. Convert the number of degrees.	3. 18 F degrees is $18 \text{ F}° \times \dfrac{100 \text{ C}°}{180 \text{ F}°} =$ 10 C degrees	3. 25 C degrees is $25 \text{ C}° \times \dfrac{180 \text{ F}°}{100 \text{ C}°} =$ 45 F degrees	3. 113.4 F degrees is $113.4 \text{ F}° \times \dfrac{100 \text{ C}°}{180 \text{ F}°} =$ 63.0 C degrees
4. Express temperature, taking into account the selected reference point.	4. The temperature is 10 C degrees above the fp of water (which is 0°C), so the temperature is *10°C*	4. The temperature is 45 F degrees above the fp of water (which is 32°F), so the temperature is *45 + 32 = 77°F*	4. The temperature is 63.0 C degrees below the bp of water (which is 100.0°C), so the temperature is *100.0 − 63.0 = 37.0°C*

Appendix C

FACTOR-LABEL APPROACH TO CONVERSION PROBLEMS

For converting a measurement from one system of units to another, the following factor-label method of solution is straightforward and does not require the decision of whether to divide or multiply to obtain the proper answer. This method makes the decision for you, a decision that is sometimes difficult when dealing with new units with which you have had little experience.

Example

How many liters are in 6 quarts?

1. Write down the unit to which you are converting. The question mark indicates the number of liters to be determined.

   1. ? liters =

2. On the right-hand side of the equal sign, write down the quantity given. Write both the number and the name or label.

   2. ? liters = 6 quarts

3. Now look at the two units. Recall a conversion between these two units. These conversions must be learned or looked up.

   3. 1 liter = 1.06 quarts, 1 liter per 1.06 quarts

   $$\text{or } \frac{1 \text{ liter}}{1.06 \text{ quarts}}$$

   $$\text{or } \frac{1.06 \text{ quarts}}{\text{liter}}$$

4. Write the conversion factor on the right-hand side so that the unwanted units will cancel; that is, a unit in the numerator will cancel the same unit in the denominator, and only the unit you want will remain. Do not, of course, cancel the numbers, just the units.

   4. ? liters =

   $$6 \text{ quarts} \times \frac{1 \text{ liter}}{1.06 \text{ quarts}}$$

5. Do the indicated multiplication and division. The line, of course, means divided by. Check the units on both sides of the equation. The units should be the same and in the same position (numerator or denominator).

   5. ? liters = $\dfrac{6}{1.06}$ liters

   = 5.66 liters

Now, suppose you needed to convert 3 pints to ml, and you do not know a conversion factor that will make the conversion in one step.

Proceed as before:

? ml = 3 pints

Note this is a volume conversion from the English system to the metric system as before. Recall the volume conversion factor that you know between these systems.

1 liter = 1.06 quarts

Convert in the system of the given quantity until you reach the unit in the conversion factor for this system. Then write down the conversion factor between systems so that like units in separate factors will cancel.

$$? \text{ ml} = 3 \text{ pints} \times \frac{1 \text{ quart}}{2 \text{ pints}} \times \frac{1 \text{ liter}}{1.06 \text{ quarts}}$$

Cancel units and recall a conversion factor that converts the units you have left to the unit you want. Do the indicated multiplication and division.

$$? \text{ ml} = \frac{3 \times 1 \times 1 \text{ liter}}{2 \times 1.06} \times \frac{1000 \text{ ml}}{1 \text{ liter}}$$

$$= \frac{3 \times 1 \times 1 \times 1000}{2 \times 1.06 \times 1} \text{ ml}$$

$$= 1420 \text{ ml}$$

In brief, the method is very simple:

1. Write the units you want, an equal sign, and the quantity you have given.
2. Write conversion factors so the unwanted factors will cancel.
3. Keep in one system until you come to units in a familiar intersystem conversion factor.
4. Do the indicated arithmetic and check the units.
5. Examine the size of the answer for reasonableness.

Note: This approach to problem solving has wide applicability to many other types of problems. It is especially useful in problems pertaining to weight relationships in chemical reactions. See Appendix D.

Appendix D

CALCULATIONS WITH CHEMICAL EQUATIONS

The bases for calculations with chemical equations were presented in Chapter 6. Problems of a more complex nature and a systematic approach to their solution are presented in the following examples. Finally, a list of exercise problems is given for further study.

Example 1

Balanced Equations Express Number Ratios for Particles

In the reaction of hydrogen with oxygen to form water, how many molecules of hydrogen are required to combine with 19 oxygen molecules?

Solution: A chemical equation can be written for a reaction only if the reactants and products are identified and the respective formulas determined. In this problem, the formulas are known and the unbalanced equation is:

$$H_2 + O_2 \longrightarrow H_2O$$

It is evident that one molecule of oxygen contains enough oxygen for two water molecules and the equation, as written, does not account for what happens to the second oxygen atom. As it is, the equation is in conflict with the conservation of atoms in chemical changes. This conflict is easily corrected by balancing the equation:

$$2H_2 + O_2 \longrightarrow 2H_2O$$

Now all atoms are accounted for in the equation and it is obvious that two hydrogen molecules are required for each oxygen molecule. In other words, two hydrogen molecules are equivalent to one oxygen molecule in their usage. This can be expressed as follows:

$$2 \text{ hydrogen molecules } \{\text{are equivalent to}\} \text{ 1 oxygen molecule;}$$

or,

$$2H_2 \text{ molecules} \sim O_2 \text{ molecule;}$$

or,

$$\frac{2H_2 \text{ molecules}}{O_2 \text{ molecule}},$$

which can be read as two hydrogen molecules per one oxygen molecule.

Using now the factor-label approach developed in Appendix C, the solution is readily achieved.

? H_2 molecules = $19O_2$ molecules

? H_2 molecules = $19\,\cancel{O_2\text{ molecules}} \times \dfrac{2H_2 \text{ molecules}}{\cancel{O_2 \text{ molecule}}}$

$\qquad\qquad = 38H_2$ molecules

Note: The reader is likely to say at this point that the method is cumbersome and that he can quickly see the answer to be $38H_2$ molecules without "the method." However, problems to follow are made much easier if a systematic method of approach is used.

Example 2

Laboratory Mole Ratios Identical with Particle Number Ratios

How many moles of hydrogen molecules must be burned in oxygen (the reaction of Example 1) to produce 15 moles of water molecules (about a glassful)?

Solution: The balanced equation

$$2H_2 + O_2 \longrightarrow 2H_2O$$

tells us that two molecules of hydrogen produce two molecules of water; or,

2 molecules hydrogen $\sim$ 2 molecules water,

and therefore,

1 molecule hydrogen $\sim$ 1 molecule water.

It is obvious then that the number of water molecules produced will be equal to the number of hydrogen molecules consumed regardless of the actual number involved. Therefore,

6.02×10^{23} molecules of hydrogen $\sim 6.02 \times 10^{23}$ molecules of water

Since 6.02×10^{23} is a number called the mole, it follows that one mole of hydrogen molecules will produce one mole of water molecules. The general conclusion, then, is the following: the ratio of particles in the balanced equation is the same as the ratio of moles in the laboratory. The solution to the problem logically follows:

$$\left.\begin{array}{c}? \text{ moles of hydrogen}\\ \text{molecules}\end{array}\right\} = \left\{\begin{array}{c}15\,\cancel{\text{moles}}\\ \cancel{\text{water molecules}}\end{array}\right\} \times \dfrac{1 \text{ mole hydrogen molecules}}{1 \;\cancel{\text{mole water molecules}}}$$

$$= 15 \text{ moles hydrogen molecules}$$

Note: Again the solution to the problem looks simple enough without resorting to the factor-label method. However, in Examples 3 and 4, the numbers become such that a quick mental solution is not readily achieved by most students.

Example 3

Mole Weights Yield Weight Relationships

How many grams of oxygen are necessary to react with an excess of hydrogen to produce 270 grams of water?

Solution: From the balanced equation

$$2H_2 + O_2 \longrightarrow 2H_2O$$

the mole ratio between oxygen and water is immediately evident and is one mole of oxygen molecules per two moles of water molecules, or

$$\frac{1 \text{ mole oxygen molecules}}{2 \text{ moles water molecules}}$$

This mole ratio can be changed into a weight ratio since the mole weight can be easily calculated from the atomic weights involved. One molecule of oxygen (O_2) weighs 32 amu (16 amu for each oxygen atom). Therefore, a mole of oxygen molecules weighs 32 grams. Similarly, two moles of water weigh 36 grams $[2(16 + 1 + 1)]$. Therefore, the weight ratio is:

$$\frac{1 \text{ mole oxygen molecules} \times \dfrac{32 \text{ grams oxygen}}{\text{mole oxygen molecules}}}{2 \text{ moles water molecules} \times \dfrac{18 \text{ grams water}}{\text{mole water molecules}}}$$

or,

$$\frac{32 \text{ grams oxygen}}{36 \text{ grams water}}$$

This weight relationship is exactly the conversion factor needed to answer the original question:

$$? \text{ grams oxygen} = 270 \text{ grams water} \times \frac{32 \text{ grams oxygen}}{36 \text{ grams water}}$$
$$= 240 \text{ grams oxygen}$$

Note: It should be observed that a weight relationship could be established between any two of the three pure substances involved in the reaction, regardless of whether they are reactants or products.

Example 4

How many molecules of water are produced in the decomposition of 8 molecules of table sugar? The unbalanced equation is as follows:

$$C_{12}H_{22}O_{11} \longrightarrow C + H_2O$$

Solution: Balance the equation

$$C_{12}H_{22}O_{11} \longrightarrow 12C + 11H_2O$$

$$? \text{ molecules of water} = 8 \text{ molecules sugar} \times \frac{11 \text{ molecules water}}{1 \text{ molecule sugar}}$$
$$= 88 \text{ molecules of water}$$

Example 5

How many grams of mercuric oxide are necessary to produce 50 grams of oxygen? Mercuric oxide decomposes as follows:

$$2HgO \longrightarrow 2Hg + O_2$$

Solution:

Weight of two moles of HgO = 2(201 + 16) = 2(217) = 434 g

Weight of 1 mole of O_2 = 2(16) = 32 g

? g HgO = 50 g~~oxygen~~ $\times \dfrac{434 \text{ g mercuric oxide}}{32 \text{ g } \text{oxygen}}$

= 678 g mercuric oxide

Example 6

How many pounds of mercuric oxide are necessary to produce 50 pounds of oxygen by the reaction:

$2HgO \longrightarrow 2Hg + O_2$

Solution: Note that the problem is the same as Example 5 except for the units of chemicals. Also note that the conversion factor of Example 5

$\dfrac{434 \text{ g mercuric oxide}}{32 \text{ g oxygen}}$

can be converted to any other units desired:

$$\dfrac{434 \text{ g mercuric oxide} \times \dfrac{1 \text{ pound}}{454 \text{ g}}}{32 \text{ g oxygen} \times \dfrac{1 \text{ pound}}{454 \text{ g}}} = \dfrac{434 \text{ pounds mercuric oxide}}{32 \text{ pounds oxygen}}$$

It is evident that the ratio, $\dfrac{434}{32}$, expresses the ratio between weights of mercuric oxide and oxygen in this reaction regardless of the units employed.

? pounds mercuric oxide = 50 ~~pounds oxygen~~ $\times \dfrac{434 \text{ pounds mercuric oxide}}{32 \text{ pounds oxygen}}$

= 678 pounds of mercuric oxide

Problems

1. What weight of oxygen is necessary to burn 28 g of methane, CH_4? The equation is:

$CH_4 + 2O_2 \longrightarrow CO_2 + 2H_2O$ *Ans.* 112 g oxygen

2. Potassium chlorate, $KClO_3$, releases oxygen when heated according to the equation:

$2KClO_3 \longrightarrow 2KCl + 3O_2$

What weight of potassium chlorate is necessary to produce 1.43 g of oxygen?   *Ans.* 3.65 g $KClO_3$

3. Fe_3O_4 is a magnetic oxide of iron. What weight of this oxide can be produced from 150 g of iron?

Ans. 207 g oxide

4. Steam reacts with hot carbon to produce a fuel called water gas; it is a mixture of carbon monoxide and hydrogen. The equation is:

$$H_2O + C \longrightarrow CO + H_2$$

What weight of carbon is necessary to produce 10 g of hydrogen by this reaction?

Ans. 60 g carbon

5. Iron oxide, Fe_2O_3, can be reduced to metallic iron by heating it with carbon.

$$2Fe_2O_3 + 3C \longrightarrow 4Fe + 3CO_2$$

How many tons of carbon would be necessary to reduce 5 tons of the iron oxide in this reaction?

Ans. 0.56 ton carbon

6. How many grams of hydrogen are necessary to reduce 1 pound (454 g) of lead oxide (PbO) by the reaction:

$$PbO + H_2 \longrightarrow Pb + H_2O$$

Ans. 3.91 g hydrogen

7. Hydrogen can be produced by the reaction of iron with steam.

$$4H_2O + 3Fe \longrightarrow 4H_2 + Fe_3O_4$$

What weight of iron would be needed to produce one-half pound of hydrogen?

Ans. 10.5 lb iron

8. Tin ore, containing SnO_2, can be reduced to tin by heating with carbon.

$$SnO_2 + C \longrightarrow Sn + CO_2$$

How many tons of tin can be produced from 100 tons of SnO_2?

Ans. 79 tons tin

ANSWERS TO SELF-TEST QUESTIONS AND MATCHING SETS

Chapter 1

SELF-TEST 1–A

1. observed experimental facts
2. the same
3. the integrated circuit
4. *E. Coli*
5. (a) theories, (b) laws, (c) facts

MATCHING SET

1. l	6. a	10. d
2. f	7. j	11. e
3. g	8. c	12. i
4. h	9. d	13. k
5. b		

Chapter 2

SELF-TEST 2–A

1. dirt, wood, dusty air, salt water, etc.
2. operational definition
3. rain water, gold, quartz, diamond, etc.
4. false
5. physical, chemical, nuclear
6. solution
7. true

MATCHING SET

1. c	3. d
2. a	4. b

SELF-TEST 2–B

1. (a) metals: iron, copper, gold, silver, chromium, magnesium
   (b) nonmetals: oxygen, silicon, carbon, nitrogen, chlorine, fluorine
2. 109
3. false
4. burning coal, dissolving iron ore in an acid, making steel from iron ore, making aspirin
5. cutting diamond, blowing glass, molding plastic, slicing bread
6. chemical substance
7. Submicroscopic
8. (a) macroscopic, (b) microscopic, (c) molecular
9. (a) theories, (b) laws, (c) facts
10. nuclear
11. applied research
12. (a) oxygen, (b) iron, (c) hydrogen
13. chromatography, distillation, recrystallization, filtration
14. Students should list a dozen or more elements.

SELF-TEST 2–C

1. (a) a sodium atom, or a mole of sodium
   (b) two sodium atoms, or two moles of sodium
   (c) a hydrogen chloride molecule, or a mole of hydrogen chloride
   (d) reacts to form
   (e) a hydrogen molecule, or a mole of hydrogen
   (f) two units of sodium chloride, or two moles of sodium chloride
2. the element, an atom of the element, a mole of the elemental atoms
3. the elements present, the relative number of each type of atom
4. coefficient
5. false
6. seven

MATCHING SET

1. b	7. i	12. o
2. e	8. k	13. l
3. d	9. g	14. m
4. a	10. h	15. n
5. c	11. j	16. p
6. f		

Chapter 3

SELF-TEST 3–A

1. Leucippus, Democritus
2. (b) philosophy

3. gained, chemical reaction
4. CO, CO_2
5. (a) new compound, (b) 2:4:1, (c) law of multiple proportions
6. (d) Atoms are recombined into different arrangements.
7. (a) the same, (b) atoms
8. repel, attract
9. alpha (α), beta (β), gamma (γ), gamma (γ)

SELF-TEST 3 – B

1. protons, neutrons
2. small
3. nucleus
4. electrons, protons, neutrons
5. electrons, protons
6. atomic
7. about 1836
8. 33, 33, 42
9. different, identical
10. electrons
11. electrons, protons
12. false

SELF-TEST 3 – C

1. particles, waves
2. spectrum
3. farther from, closer to
4. (d) Wave nature of the electron
5. wavelength
6. 18, 2
7. (a) No, not the paths
   (b) In a way, yes. They are representations of the space in which we can expect to find the electron with 90% certainty.
   (c) Of a given type, yes, with 90% certainty

MATCHING SET

1. e	6. n	11. h
2. a	7. b	12. c
3. j	8. f	13. k
4. i	9. m	14. l
5. d	10. g	

Chapter 4

SELF-TEST 4 – A

1. IIA, VIII, VIIA, IIIA, IB
2. 2, 5, 7, 4

3. R, R, T, N, I
4. Mendeleev
5. atomic numbers
6. periodic
7. metals

SELF-TEST 4 – B

1. 3, $GaCl_3$; 2, $BaCl_2$; 2, $SeCl_2$, 1, ICl
2. metalloid, metal, nonmetal, metal, metal, nonmetal, nonmetal, metalloid
3. 1, 2, 7, 7, 6, 3, 4
4. ionization
5. He, F, Br, S
6. K, Br, S, In
7. IA, IVA, IIA

MATCHING SET

1. i	5. d	8. h
2. c	6. g	9. f
3. e	7. b	10. k
4. a		

Chapter 5

SELF-TEST 5 – A

1. ions
2. ionic
3. one
4. CaI_2
5. Cl^-
6. valence
7. losing
8. gaining
9. Rb, one electron lost; S, two electrons gained; Ca, two electrons lost; Mg, two electrons lost; K, one electron lost; Br, one electron gained

SELF-TEST 5 – B

1. (a) H_2, (b) HF
2. six
3. three
4. fluorine
5. (a) eight, (b) octet, (c) most of the time
6. fluorine
7. N, O, F
8. H_2O

SELF-TEST 5 – C

1. Nonbonding, bonding
2. linear, trigonal planar, tetrahedral, octahedral
3. two, two
4. three, one

MATCHING SET I

1. l	6. g	10. c
2. d	7. b	11. h
3. j	8. e	12. a
4. f	9. k	13. m
5. i		

MATCHING SET II

1. a	4. e	6. g
2. c	5. b	7. f
3. d		

Chapter 6

SELF-TEST 6 – A

1. (a) 6, 6, 1, 6 (The one is understood); (b) 6; (c) 6; (d) 44, 180; (e) 264 g
2. 136 kcal
3. (a) 2, 1, 2; (b) 1, 2, 1; (c) 4, 3, 2
4. (a) 2, 2, 1; (b) 18 g; (c) 16 g; (d) 18 tons

SELF-TEST 6 – B

1. temperature, because the rate of bacterial growth slows with decreasing temperatures
2. hydrogen, oxygen
3. hemoglobin and oxygen, and calcium oxide and water
4. (a) Catalysts are used to increase reaction rates. (b) Catalysts remain in original amounts.
5. no
6. Equilibrium constants vary with temperature.
7. the same flour in dust form
8. hemoglobin uptake and release of oxygen
9. the formation of any very stable compound such as water or table salt
10. Ordinary chemical change occurs with a minimum amount of change in the structure of the atoms, molecules, or ions involved.

MATCHING SET

1. g	5. c	9. h
2. f	6. a	10. l
3. e	7. j	11. i
4. b	8. k	

Chapter 7

SELF-TEST 7 – A

1. cloud chamber
2. $^{87}_{36}Kr$
3. $^{212}_{82}Pb$
4. 0.125 g
5. $^{8}_{4}Be$
6. James Chadwick; alpha-ray bombardment of beryllium
7. ionize, emits light
8. scintillation counter, because of its greater sensitivity to low-energy particles

SELF-TEST 7 – B

1. one billion (10^9)
2. cyclotron
3. neptunium (Np)
4. 109
5. $^{206}_{82}Pb$
6. less than 0.1 rad per year
7. uranium/lead dating
8. plutonium-239; plutonium would be very long-lived in the environment, since the longer the half-life the longer the material will stay in the environment.
9. Energies in the billions or trillions of electron volts are required.

SELF-TEST 7 – C

1. sterilization, tracers, medical diagnosis
2. (a) ^{60}Co
3. (d) all of these
4. (b) imaging
5. (c) metastable
6. (a) one eighth of the original dose
7. (b) 6-hour half-life isotope

MATCHING SET

1. b	4. k	7. m
2. i	5. e	8. d
3. a	6. c	9. p

10. o 13. f 16. j
11. h 14. l
12. n 15. g

Chapter 8

SELF-TEST 8–A

1. petroleum
2. natural gas
3. 42
4. CO, H_2, N_2
5. 33
6. coal, petroleum, natural gas
7. oxygen, water
8. true
9. work
10. (a) quantitatively
11. entropy
12. quality
13. calorie, joule, Btu
14. watt, kilowatt
15. aluminum, aluminum

SELF-TEST 8–B

1. fission
2. critical mass
3. uranium-235
4. (b) fusion
5. (a) containment of reactants at high temperature
6. tritium
7. plutonium-239, uranium-233
8. $^{85}_{35}Br$, $^{242}_{94}Pu$

MATCHING SET

1. f 6. c 11. p
2. l 7. d 12. h
3. b 8. e 13. k
4. j 9. g 14. m
5. n 10. a 15. o

Chapter 9

SELF-TEST 9–A

1. electrolyte
2. base, acid
3. base, acid
4. amphiprotic
5. neutral
6. water

7. acid, base
8. 12

SELF-TEST 9–B

1. 10, (b) basic; 7, (c) neutral; 3, (a) acidic
2. 1400 g, or 1.4 kg
3. NH_4^+, OH^-
4. H_3O^+, Cl^-
5. pH of 2
6. yes
7. buffers
8. CO_2
9. low
10. high

MATCHING SET

1. k 6. l 10. e
2. k 7. j 11. g
3. b 8. h 12. i
4. a 9. c 13. d
5. f

Chapter 10

SELF-TEST 10–A

1. burning/combustion
2. carbon dioxide, water
3. oxide
4. rust
5. water, carbon dioxide
6. carbon monoxide

SELF-TEST 10–B

1. oxygen
2. acetaldehyde, (c) hydrogen loss
3. (b) reduced
4. (b) oxidized
5. (b) reducing agent
6. (b) reduced
7. oxidation, reduction

SELF-TEST 10–C

1. reduction
2. oxidation
3. iron, oxygen, water
4. no
5. yes
6. fluorine
7. hydrogen, oxygen

MATCHING SET

1. h	7. g	13. a
2. o	8. n	14. d
3. k	9. e	15. j
4. q	10. p	16. b
5. f	11. c	17. m
6. l	12. i	

Chapter 11

SELF-TEST 11–A

1. oxygen
2. aluminum
3. Minnesota
4. limestone
5. copper
6. copper, aluminum, magnesium
7. slag
8. reduced
9. cathode
10. magnesium
11. positive ions

SELF-TEST 11–B

1. nitrogen, oxygen
2. liquid oxygen
3. cold materials in the temperature range of liquid air
4. oxygen: used in making steel, as an oxidizing agent, in welding, and in controlled atmospheres
nitrogen: used in cryosurgery, in an inert atmosphere, as a welding gas blanket, and in freezing food

SELF-TEST 11–C

1. silicon, oxygen
2. lead
3. silicon
4. clay, sand, feldspar
5. cement
6. sulfur
7. chlorine
8. air
9. sodium carbonate, sodium hydrogen carbonate (sodium bicarbonate)

MATCHING SET

1. i	5. h	8. a
2. j	6. e	9. c
3. b	7. f	10. d
4. g		

Chapter 12

SELF-TEST 12–A

1. organic
2. tetrahedral
3. 13, no
4. false
5. two (a, b, or d; and e or f)
6. double bond
7. 2,4-dimethylhexane
8. $-C_2H_5$

MATCHING SET

1. b, e
2. c, d, e
3. a, d, e

SELF-TEST 12–B

1. four
2. rotation of polarized light
3. 32
4. delocalized
5. 12 (6C, 6H)
6. 1,2,4-trimethylbenzene
7. false
8. methanol, ethanol, tertiary-butyl alcohol, benzene
9. Small amounts of moisture destabilize the methanol-gasoline mixture and cause engine corrosion. This problem is reduced by adding other alcohols to stabilize the methanol-gasoline mixture.
10. c

MATCHING SET

1. d	4. c	7. b
2. e	5. f	8. h
3. a	6. g	

Chapter 13

SELF-TEST 13–A

1. ethylene glycol
2. (a) methanol, (b) ethanol, (c) methanoic acid or formic acid, (d) 2-butanol, (e) ethanoic acid or acetic acid, (f) ethylene glycol, (g) acetaldehyde, (h) 2-bromopropane
3. fermentation of starch, hydration of ethylene
4. stearic acid, palmitic acid, or oleic acid, among others

5. (a) alcohol, (b) carboxylic acid, (c) aldehyde,
    (d) ketone
6. (a) C_2H_6, C_2H_5OH, CH_3CHO, CH_3COOH,
        $(C_2H_5)_2O$, $C_2H_5NH_2$
    (b) ethane, ethyl alcohol, acetaldehyde, acetic acid,
        diethyl ether, ethyl amine
    (c) $-C_2H_5$ or ethyl group in ethanol, diethyl ether,
        and ethyl amine; $-CH_3$ or methyl in acetalde-
        hyde and acetic acid
7. High molecular-weight acids are less volatile, so
    their odor is less noticeable.

SELF-TEST 13 – B

1. $CH_3CH_2CH_2O\overset{\overset{\displaystyle O}{\|}}{C}CH_2CH_3$ (an ester) $+ H_2O$
2. (a) A fat is a solid at room temperature while an oil
        is a liquid. The fat molecule has fewer double
        bonds.
    (b) by hydrogenation (addition of hydrogen to the
        $C{=}C$ double bonds)
3. false
4. (c) can be made from chlorobenzene
5. carbolic

MATCHING SET

1. g	4. i	7. b
2. e	5. a	8. c
3. h	6. d	9. f

Chapter 14

SELF-TEST 14 – A

1. monomers
2. (a) $H_2C{=}\underset{\underset{\displaystyle CH_3}{|}}{CH}$

    (b)

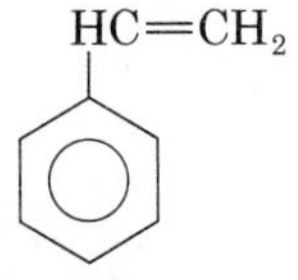

    (c) $F_2C{=}CF_2$
3. polyester
4. isoprene
5. copolymer
6. polyamide or condensation
7. water
8.

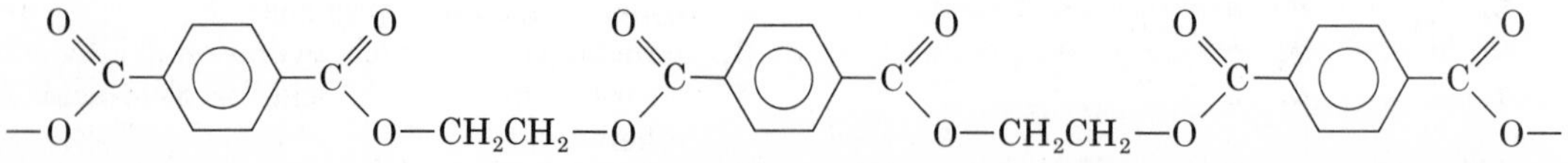

9. condensation
10. thinner, pigment
11. linseed
12. emulsions
13. alcohols, acids

SELF-TEST 14 – B

1. 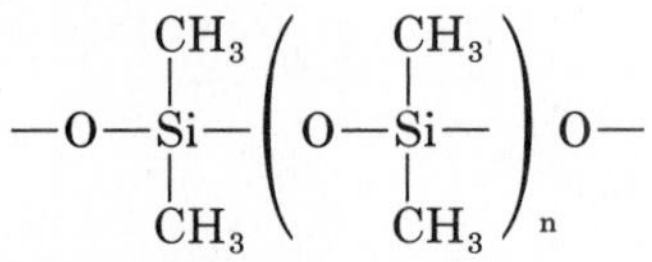 $-O-\underset{\underset{\displaystyle CH_3}{|}}{\overset{\overset{\displaystyle CH_3}{|}}{Si}}-\left(O-\underset{\underset{\displaystyle CH_3}{|}}{\overset{\overset{\displaystyle CH_3}{|}}{Si}}-\right)_n O-$
2. ultraviolet light
3. plasticizer
4. O (oxygen); it is a polymer held together by a net-
    work of Si—O bonds
5. HCl
6. Ultraviolet
7. polyethylene

MATCHING SET

1. l	6. g	11. p
2. d	7. b	12. m
3. j	8. c	13. e
4. k	9. f	14. n
5. a	10. i	

Chapter 15
SELF-TEST 15 – A

1. carbon, hydrogen, oxygen
2. monosaccharide
3. glucose, fructose
4. D-glucose
5. D-glucose
6. hydrogen

SELF-TEST 15 – B

1. amino acids
2. essential amino acids
3. $-\underset{\underset{\displaystyle O}{\|}}{\overset{\overset{\displaystyle H}{|}}{N-C}}-$
4. $R-\underset{\underset{\displaystyle NH_2}{|}}{CH}COOH$

5.

$$\text{H} - \overset{\overset{\displaystyle \text{H}}{|}}{\underset{\underset{\displaystyle \text{NH}_2}{|}}{\text{C}}} - \overset{\overset{\displaystyle \text{O}}{\|}}{\text{C}} - \overset{\overset{\displaystyle \text{H}}{|}}{\text{N}} - \text{CH}_2 - \text{C} \overset{\displaystyle \text{O}}{\underset{\displaystyle \text{OH}}{}}$$

6. (a) sequence of amino acids
   (b) helical structure in which the amino acid chains are coiled
   (c) the way in which the helical sections are themselves folded
   (d) the aggregation of subunits
7. (a) 27, (b) 6
8. Hydrogen atoms bond to nitrogen or oxygen atoms in an adjacent strand or at another location in the same strand of amino acids.

SELF-TEST 15–C

1. catalyst
2. key, lock
3. niacin
4. enzyme
5. enzyme
6. coenzyme
7. niacin
8. active site

SELF-TEST 15–D

1. the Sun
2. ATP
3. ADP, phosphate or phosphoric acid, energy
4. free
5. CO_2, H_2O, energy
6. NADPH
7. hydrolysis
8. lipids
9. emulsifying agents
10. ATP, lactic acid
11. CO_2, H_2O

SELF-TEST 15–E

1. (a) DNA, (b) transfer RNA
2. hydrogen
3. ATP
4. adenine: thymine or uracil; cytosine: guanine; guanine: cytosine; thymine: adenine; uracil: adenine
5. false
6. false
7. ribose, deoxyribose
8. phosphoric acid, a sugar (ribose or deoxyribose), a nitrogenous base
9. double helix
10. true, when it invades a host cell

MATCHING SET I

1. g	4. e	7. b
2. f	5. h	8. d
3. a	6. c	9. i

MATCHING SET II

1. c	5. a	9. k
2. d	6. b	10. h
3. g	7. f	11. e
4. i	8. j	

Chapter 16

SELF-TEST 16–A

1. dehydration, hydrolysis
2. oxidizing
3. hemoglobin, oxygen
4. false
5. CN^-, cytochrome oxidase, oxygen
6. fluoroacetic acid
7. heavy metal poisons, complex
8. true

SELF-TEST 16–B

1. neurotoxins
2. synapses
3. acetylcholine
4. genes, chromosomes
5. false
6. chimney sweeping
7. tetrahydrocannabinol (THC)
8. acetaldehyde
9. cancer, heart disease
10. teratogen

MATCHING SET

1. b	5. j	9. c
2. f	6. d	10. k
3. h	7. g	11. a
4. i	8. e	

Chapter 17

SELF-TEST 17–A

1. high heat capacity
2. chlorination
3. biochemical oxygen demand
4. Hydrogen bonding
5. 60
6. one-half gallon

7. 5% (based on average of 3 gallons for drinking and cooking)
8. sewage, fertilizers, pesticides, among others
9. distillation, sedimentation, aeration, filtration
10. aquifer
11. 70%

SELF-TEST 17 – B

1. Na^+, Mg^{2+}, Ca^{2+}, K^+
2. toxic
3. calcium, magnesium
4. chlorine
5. electric
6. two
7. more pure than
8. settling, filtration
9. metal, organics

MATCHING SET

1. e	5. c	8. j
2. h	6. b	9. d
3. i	7. g	10. f
4. a		

Chapter 18

SELF-TEST 18 – A

1. Parts per million, ten thousand, 920
2. carbon monoxide, hydrocarbons, nitrogen oxides, ozone, sulfur dioxide, among others
3. peroxy acyl nitrate, aldehydes, ketones, ozone; sulfuric acid, sulfurous acid, SO_3
4. aerosols
5. surface area
6. Cottrell precipitator
7. nitrogen dioxide
8. warm, cool
9. sulfur dioxide, nitrogen dioxide
10. sulfur dioxide
11. fuels
12. first
13. other elements
14. endothermic
15. oxides
16. false
17. false
18. SO_2, SO_3, NO_2

SELF-TEST 18 – B

1. limited
2. hemoglobin

3. false
4. benzo(α)pyrene (BaP)
5. ozone, O_3
6. polar
7. increases, increases

MATCHING SET

1. b	6. c	11. o
2. o, g	7. e	12. a
3. l	8. m	13. k
4. j	9. i	14. h
5. h	10. d	

Chapter 19

SELF-TEST 19 – A

1. soil
2. living matter, climate, parent rock, slope, time
3. clays
4. sour soils
5. (a) acidic, (b) basic
6. adsorbed
7. chemical composition, particle size
8. a trivalent ion
9. humus
10. silicate
11. calcium, magnesium, sulfur
12. oxidation
13. true

SELF-TEST 19 – B

1. false
2. lower
3. potassium nitrate
4. yes
5. gas
6. nitrous acid, nitric acid
7. denitrification
8. K_2CO_3
9. magnesium
10. 33%
11. DDT
12. chlordan
13. selective
14. 2,4-D
15. warfarin
16. an avacide
17. plant growth regulator

MATCHING SET

1. h	10. p	18. w
2. o	11. f	19. a
3. k	12. i	20. j
4. r	13. x or n	21. n or x
5. t	14. s	22. y
6. b	15. c	23. g
7. m	16. l	24. e
8. v	17. u	25. q
9. d		

Chapter 20

SELF-TEST 20-A

1. carbohydrates, fats, proteins, vitamins, minerals, water
2. oxygen, calcium
3. USRDA
4. only partially
5. fat
6. one, one
7. basal metabolic rate, weight

SELF-TEST 20-B

1. apoenzyme
2. protein
3. false
4. urea
5. lecithin, cholesterol
6. triglycerides
7. linoleic
8. prostaglandins, fatty acids
9. pork, chicken
10. fat, carbohydrates
11. fat
12. false
13. decreases
14. peroxides, free radicals, or polymers
15. false
16. sugar, flour

SELF-TEST 20-C

1. variety, whole, different places
2. iron, iron, hemoglobin
3. chromium, insulin
4. manganese
5. copper
6. zinc
7. zinc
8. zinc
9. potassium, sodium
10. greater than
11. magnesium
12. magnesium
13. calcium, bones or skeleton
14. kidney
15. iodine
16. mottled
17. three

SELF-TEST 20-D

1. B_{12}, cobalamin
2. false
3. oil, water
4. provitamin
5. carrots and green vegetables (provitamins), cod liver oil
6. A
7. antioxidant
8. vitamin E
9. K
10. antagonist
11. D
12. D
13. calcium
14. D
15. scurvy
16. coenzymes, growth
17. B_1, beriberi
18. niacin
19. diarrhea, dermatitis, dementia
20. B_5, pantothenic acid
21. B_6
22. B_6
23. vitamin B_6, or biotin
24. B_9, or folacin
25. B_{12}, or cobalamin
26. B_{12}

SELF-TEST 20-E

1. synergistic, potentiation
2. salting, drying
3. false
4. volatile
5. more
6. sequestrant
7. false
8. flavor enhancer
9. dehydrating
10. free radicals
11. saccharin
12. double bonds and aromatic rings
13. citric acids, lactic acids

14. carbon dioxide
15. breaks
16. generally recognized as safe
17. active

MATCHING SET I

1. e	**6.** m	**11.** o
2. c, j	**7.** p	**12.** g
3. l, n	**8.** i	**13.** a
4. f	**9.** b	**14.** h
5. d	**10.** k	

MATCHING SET II

1. b	**4.** e	**6.** a
2. c	**5.** f	**7.** h
3. d		

Chapter 21

SELF-TEST 21–A

1. generic
2. heart
3. (a) magnesium oxide, (d) sodium citrate
4. analgesic
5. poppy
6. true
7. antipyretic
8. liver
9. heroin

SELF-TEST 21–B

1. allergy
2. germacide
3. antibiotics
4. *p*-aminobenzoic acid
5. mold
6. fungus

SELF-TEST 21–C

1. cholesterol, triglycerides
2. faster
3. fatigue, lethargy, depression, confusion
4. relaxing
5. high-density lipoproteins
6. yes

SELF-TEST 21–D

1. testosterone
2. progesterone
3. muscle building

4. liver damage, liver cancer, acne, baldness, changes in sexual desire, enlargement of breasts
5. true
6. DNA
7. antimetabolites
8. leukemia

MATCHING SET

1. g	**7.** o	**13.** n
2. s	**8.** d	**14.** l
3. p	**9.** b	**15.** k
4. c	**10.** q	**16.** h
5. i	**11.** a	**17.** f
6. r	**12.** j	

Chapter 22

SELF-TEST 22–A

1. corneal
2. sebum
3. skin, hair, nails
4. disulfide, ionic
5. calcium carbonate, calcium hydroxyphosphate (apatite)
6. emollient
7. emulsion
8. lipstick
9. talc
10. false
11. *p*-aminobenzoic acid
12. melanin

SELF-TEST 22–B

1. lacquer, varnish
2. acetone, ethyl acetate
3. thioglycolic acid
4. oxidation
5. black
6. resin or plastic
7. sulfides of sodium, calcium, strontium
8. It acts as an astringent.
9. alcohol (ethyl and isopropyl)
10. carbolic

SELF-TEST 22–C

1. fatty
2. alkali or lye
3. water
4. olive oil
5. air
6. traditional soaps
7. surface active agents

8. It produces a precipitate.
9. nonionic detergents
10. increases soap action, precipitates hard-water ions
11. visible
12. detergent, abrasive
13. fluorine

MATCHING SET

1.	q	7.	a	13.	i
2.	m	8.	e	14.	j
3.	k	9.	f	15.	l
4.	c	10.	b	16.	d
5.	p	11.	g	17.	o
6.	n	12.	h		

Chapter 23

SELF-TEST 23–A

1. (d) high-speed driving
2. true
3. polychloroprene, silicone rubber
4. highway driving
5. egg

SELF-TEST 23–B

1. water, carbon dioxide
2. knocking
3. tetraethyllead
4. true

SELF-TEST 23–C

1. catalytic muffler
2. oxides of nitrogen
3. incomplete combustion of hydrocarbons
4. water
5. incomplete combustion

SELF-TEST 23–D

1. antifreeze
2. oil, soap
3. methyl alcohol (methanol)
4. ethylene glycol
5. 50W

SELF-TEST 23–E

1. (d) ethyl alcohol
2. ignite on contact
3. (b) hydrogen
4. (b) gasoline
5. fluorine
6. true

MATCHING SET

1.	g	7.	e	13.	i
2.	k	8.	q	14.	a
3.	f	9.	h	15.	d
4.	l	10.	b	16.	j
5.	m	11.	o	17.	c
6.	n	12.	p		

Chapter 24

SELF-TEST 24–A

1. photochemicals
2. β-carotene
3. $E = hc/\lambda$
4. 11-cis-retinal, all-trans-retinal
5. visible light, infrared light, microwaves, radiowaves
6. higher
7. rhodopsin
8. rods, cones
9. Rods, cones

SELF-TEST 24–B

1. silver
2. hydroquinone
3. reduced
4. $Ag(S_2O_3)_2^{3-}$
5. somewhat larger
6. reducing
7. sodium thiosulfate
8. (a) blue, (b) green, (c) red
9. red, blue, green
10. true
11. reduced
12. panchromatic

SELF-TEST 24–C

1. pixel
2. photodetectors
3. phosphor
4. gamma rays
5. radiowave
6. NMR
7. electron microscope

MATCHING SET

1.	i	6.	c	11.	h
2.	d	7.	k	12.	n
3.	f	8.	j	13.	o
4.	a	9.	e	14.	m
5.	b	10.	g		

INDEX

Note: d following a page number indicates a definition; i indicates an illustration or figure; s indicates a structure; and t indicates a table.

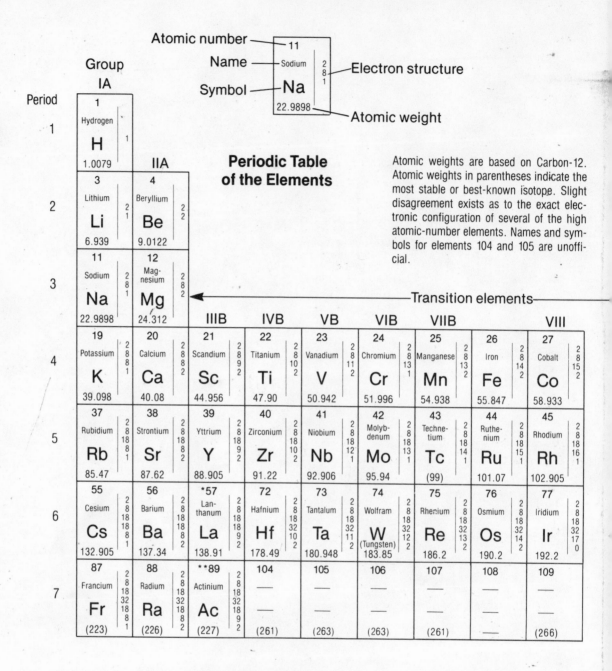

Atomic number — 11
Name — Sodium
Symbol — Na
Atomic weight — 22.9898
Electron structure — 2 8 1

Group
Period
IA

Periodic Table of the Elements

Atomic weights are based on Carbon-12. Atomic weights in parentheses indicate the most stable or best-known isotope. Slight disagreement exists as to the exact electronic configuration of several of the high atomic-number elements. Names and symbols for elements 104 and 105 are unofficial.

Transition elements

	IA	IIA	IIIB	IVB	VB	VIB	VIIB		VIII	
1	1 Hydrogen H 1.0079 (1)									
2	3 Lithium Li 6.939 (2,1)	4 Beryllium Be 9.0122 (2,2)								
3	11 Sodium Na 22.9898 (2,8,1)	12 Magnesium Mg 24.312 (2,8,2)								
4	19 Potassium K 39.098 (2,8,8,1)	20 Calcium Ca 40.08 (2,8,8,2)	21 Scandium Sc 44.956 (2,8,9,2)	22 Titanium Ti 47.90 (2,8,10,2)	23 Vanadium V 50.942 (2,8,11,2)	24 Chromium Cr 51.996 (2,8,13,1)	25 Manganese Mn 54.938 (2,8,13,2)	26 Iron Fe 55.847 (2,8,14,2)	27 Cobalt Co 58.933 (2,8,15,2)	
5	37 Rubidium Rb 85.47 (2,8,18,8,1)	38 Strontium Sr 87.62 (2,8,18,8,2)	39 Yttrium Y 88.905 (2,8,18,9,2)	40 Zirconium Zr 91.22 (2,8,18,10,2)	41 Niobium Nb 92.906 (2,8,18,12,1)	42 Molybdenum Mo 95.94 (2,8,18,13,1)	43 Technetium Tc (99) (2,8,18,14,1)	44 Ruthenium Ru 101.07 (2,8,18,15,1)	45 Rhodium Rh 102.905 (2,8,18,16,1)	
6	55 Cesium Cs 132.905 (2,8,18,18,8,1)	56 Barium Ba 137.34 (2,8,18,18,8,2)	*57 Lanthanum La 138.91 (2,8,18,18,9,2)	72 Hafnium Hf 178.49 (2,8,18,32,10,2)	73 Tantalum Ta 180.948 (2,8,18,32,11,2)	74 Wolfram W (Tungsten) 183.85 (2,8,18,32,12,2)	75 Rhenium Re 186.2 (2,8,18,32,13,2)	76 Osmium Os 190.2 (2,8,18,32,14,2)	77 Iridium Ir 192.2 (2,8,18,32,17,0)	
7	87 Francium Fr (223) (2,8,18,32,18,8,1)	88 Radium Ra (226) (2,8,18,32,18,8,2)	**89 Actinium Ac (227) (2,8,18,32,18,9,2)	104 — (261)	105 — (263)	106 — (263)	107 — (261)	108 —	109 — (266)	

Inner transition elements

* Lanthanide series 6

58 Cerium Ce 140.12 (2,8,18,20,8,2)	59 Praseodymium Pr 140.907 (2,8,18,21,8,2)	60 Neodymium Nd 144.24 (2,8,18,22,8,2)	61 Promethium Pm (147) (2,8,18,23,8,2)	62 Samarium Sm 150.35 (2,8,18,24,8,2)	63 Europium Eu 151.96 (2,8,18,25,8,2)	64 Gadolinium Gd 157.25 (2,8,18,25,9,2)

** Actinide series 7

90 Thorium Th 232.038 (2,8,18,32,18,10,2)	91 Protactinium Pa (231) (2,8,18,32,20,9,2)	92 Uranium U 238.03 (2,8,18,32,21,9,2)	93 Neptunium Np (237) (2,8,18,32,22,9,2)	94 Plutonium Pu (242) (2,8,18,32,23,9,2)	95 Americium Am (243) (2,8,18,32,25,8,2)	96 Curium Cm (247) (2,8,18,32,25,9,2)